Handbook of Signal Pro

Shuvra S. Bhattacharyya • Ed F. Deprettere
Rainer Leupers • Jarmo Takala
Editors

Handbook of Signal Processing Systems

Second Edition

Foreword by S.Y. Kung

 Springer

Editors
Shuvra S. Bhattacharyya
Department of Electrical
  and Computer Engineering
University of Maryland
College Park Maryland, USA

Ed F. Deprettere
Leiden Inst. Advanced Computer Science
Leiden Embedded Research Center
Leiden University
CA Leiden, Netherlands

Rainer Leupers
Software for Systems on Silicon
RWTH Aachen University
Aachen, Germany

Jarmo Takala
Department of Pervasive Computing
Tampere University of Technology
Tampere, Finland

ISBN 978-1-4939-4300-5   ISBN 978-1-4614-6859-2 (eBook)
DOI 10.1007/978-1-4614-6859-2
Springer New York Heidelberg Dordrecht London

To Milu

Shuvra Bhattacharyya

To Deirdre

Ed Deprettere

To Bettina

Rainer Leupers

To Auli

Jarmo Takala

Foreword

It gives me immense pleasure to introduce this timely handbook to the research/development communities in the field of signal processing systems (SPS). This is the first of its kind and represents the state-of-the-art coverage of research in this field. The driving force behind information technologies (IT) hinges critically upon the major advances in both *component integration* and *system integration*. The major breakthrough for the former is undoubtedly the invention of IC in the 1950s by Jack S. Kilby, the Nobel Prize Laureate in physics in 2000. In an integrated circuit, all components were made of the same semiconductor material. Beginning with the pocket calculator in 1964, there have been many increasingly complex applications followed. In fact, processing gates and memory storage on a chip have since then grown at an exponential rate, following Moore's Law. (Moore himself admitted that Moore's Law had turned out to be more accurate, longer lasting, and deeper in impact than he ever imagined.) With greater device integration, various signal processing systems have been realized for many killer IT applications. Further breakthroughs in computer sciences and Internet technologies have also catalyzed large-scale *system integration*. All these have led to today's IT revolution which has profound impacts on our lifestyle and overall prospect of humanity. (It is hard to imagine life today without mobiles or Internets!)

The success of SPS requires a well-concerted integrated approach from multiple disciplines, such as device, design, and application. It is important to recognize that *system integration* means much more than simply squeezing components onto a chip and, more specifically, there is a symbiotic relationship between applications and technologies. Emerging applications, e.g., 3D TV, will prompt modern system requirements on performance and power consumption, thus inspiring new intellectual challenges. The SPS architecture must consider overall system performance, flexibility, and scalability, power/thermal management, hardware–software partition, and algorithm developments. With greater integration, system designs become more complex and there exists a huge gap between what can be theoretically designed and what can be practically implemented. It is critical to consider, for instance, how to deploy in concert an ever increasing number of transistors with acceptable power consumption and how to make hardware effective for applications and yet friendly to the users (easy to program). In short, major advances in SPS must arise from close collaboration between application, hardware/architecture, algorithm, CAD, and system design.

The handbook has offered a comprehensive and up-to-date treatment of the driving forces behind SPS, current architectures, and new design trends. It has also helped seed the SPS field with innovations in applications; architectures; programming and simulation tools; and design methods. More specifically, it has provided a solid foundation for several imminent technical areas, for instance, scalable, reusable, and reliable system architectures, energy-efficient high-performance architectures, IP deployment and integration, on-chip interconnects, memory hierarchies, and future cloud computing. Advances in these areas will have greater impact on future SPS technologies.

It is only fitting for Springer to produce this timely handbook. Springer has long played a major role in academic publication on SPS, many of them have been in close cooperation with IEEE's signal processing, circuits and systems, and computer societies. For nearly 20 years, I have been the editor-in-chief of Springer's journal of signal processing systems, considered by many as a major forum for the SPS researchers. Nevertheless, the idea has been around for years that a single-volume reference book would very effectively complement the journal in serving this technical community. Then, during the 2008 IEEE Workshop on Signal Processing Systems, Washington D.C., Jennifer Evans from Springer and the editorial team led by Prof. Shuvra Bhattacharyya met to brainstorm implementation of such idea. The result is this handsome volume, containing contributions from renowned pioneers and active researchers.

I congratulate the authors and editors for putting together such an outstanding handbook. The fact that the second edition is released so soon after the first version is a clear attestation of its great demand from all the professional communities working in the broad area of signal processing systems (SPS). It is truly a timely contribution to the field of SPS for many decades to come.

Princeton, NJ   S.Y. Kung

Preface to the First Edition

This handbook provides a reference on key topics in the design, analysis, implementation, and optimization of hardware and software for signal processing systems. Plans for the handbook originated through valuable discussions with SY Kung (Princeton University) and Jennifer Evans (Springer). Through these discussions, we became aware of the need for an integrated reference focusing on the efficient implementation of signal processing applications. We were then fortunate to be able to enlist the participation of a diverse group of leading experts to contribute chapters in a broad range of complementary topics.

We hope that this handbook will serve as a useful reference to engineering practitioners, graduate students, and researchers working in the broad area of signal processing systems. It is also envisioned that selected chapters from the book can be used as core readings for seminar- or project-oriented graduate courses in signal processing systems. Given the wide range of topics covered in the book, instructors would have significant flexibility to orient such a course towards particular themes or levels of abstraction that they would like to emphasize.

The handbook is organized in four parts. Part I motivates representative applications that drive and apply state-of-the-art methods for design and implementation of signal processing systems; Part II discusses architectures for implementing these applications; Part focuses on compilers and simulation tools; and Part IV describes models of computation and their associated design tools and methodologies.

We are very grateful to all of the authors for their valuable contributions and for the time and effort they have devoted to preparing the chapters. We would also like to thank Jennifer Evans and SY Kung for their encouragement of this project, which was instrumental in getting the project off the ground, and Jennifer Maurer for her support and patience throughout the entire development process of the handbook.

College Park, MD   Shuvra S. Bhattacharyya
Leiden, The Netherlands   Ed Deprettere
Aachen, Germany   Rainer Leupers
Tampere, Finland   Jarmo Takala

Preface to the Second Edition

In this edition of the Handbook of Signal Processing Systems, many of the chapters from the first edition have been updated and several new chapters have been added. The new contributions include chapters on inertial sensors, real-time stream mining, stereoscopic and multiview 3D displays, wireless transceivers, radio astronomy, stereo vision, particle filtering architectures, and multidimensional dataflow graphs.

We hope that this updated edition of the handbook will continue to serve as a useful reference to engineering practitioners, graduate students, and researchers working in the broad area of signal processing systems. Selected chapters from the book can be used as core readings for seminar- or project-oriented graduate courses in signal processing systems. Given the wide range of topics covered in the book, instructors have significant flexibility to orient such a course towards particular themes or levels of abstraction that they would like to emphasize.

This new edition of the handbook is organized in three parts, where, due to their close relationship, the third and fourth parts from the first edition have been integrated into a single part (Part). Part I of the second edition motivates representative applications that drive and apply state-of-the-art methods for design and implementation of signal processing systems; Part II discusses architectures for implementing these applications; and Part focuses on compilers, as well as models of computation and their associated design tools and methodologies. In each part, chapters are ordered alphabetically based on the last name of the first author.

We are very grateful to all of the authors for their valuable contributions and for the time and effort they have devoted to preparing the chapters. We would also like to thank Courtney Clark for her support and patience throughout the entire development process of the handbook.

College Park, MD
Leiden, The Netherlands
Aachen, Germany
Tampere, Finland

Shuvra S. Bhattacharyya
Ed Deprettere
Rainer Leupers
Jarmo Takala

Contents

Contributors

Kiarash Amiri Rice University, Houston, TX, USA

Iuliana Bacivarov Computer Engineering and Networks Laboratory, ETH Zurich, Switzerland

Hyeon-min Bae KAIST, Daejeon, South Korea

Christian Banz Institute of Microelectronic Systems, Leibniz University of Hannover, Hannover, Germany

Twan Basten Eindhoven University of Technology and TNO-ESI, Eindhoven, The Netherlands

Lejla Batina KU Leuven, ESAT/COSIC and IBBT, Belgium
Radboud University Nijmegen, ICIS/Digital Security Group, The Netherlands

Shuvra S. Bhattacharyya University of Maryland, College Park, MD, USA

Holger Blume Institute of Microelectronic Systems, Leibniz University of Hannover, Hannover, Germany

Atanas Boev Department of Signal Processing, Tampere University of Technology, Tampere, Finland

Florian Brandner Applied Mathematics and Computer Science, Technical University of Denmark, Kongens Lyngby, Denmark

Robert Bregovic Department of Signal Processing, Tampere University of Technology, Tampere, Finland

Michael Brogioli Rice University, Houston, TX, USA

Luigi Carro Federal University of Rio Grande do Sul, Porto Alegre, RS, Brazil

Jeronimo Castrillon Institute for Software for Systems on Silicon, RWTH Aachen University, Aachen, Germany

Joseph R. Cavallaro Rice University, Houston, TX, USA

Liang-Gee Chen Graduate Institute of Electronics Engineering and Department of Electrical Engineering, National Taiwan University, Taipei, Taiwan, R.O.C.

Yanni Chen Marvell Semiconductor Inc., Santa Clara, CA, USA

Yu-Han Chen Graduate Institute of Electronics Engineering and Department of Electrical Engineering, National Taiwan University, Taipei, Taiwan, R.O.C.

Jussi Collin Department of Computer Systems, Tampere University of Technology, Tampere, Finland

Katherine Compton University of Wisconsin, Madison, WI, USA

Pavel Davidson Department of Computer Systems, Tampere University of Technology, Tampere, Finland

Elke De Mulder Cryptography Research, Inc., a division of Rambus, San Francisco, CA, USA

Bjorn De Sutter Ghent University, Gent, Belgium

Ed F. Deprettere Leiden University, Leiden, The Netherlands

Raphael Ducasse The Boston Consulting Group, Boston, MA, USA

Joachim Falk University of Erlangen-Nuremberg, Hardware-Software-Co-Design, Erlangen, Germany

Junfeng Fan KU Leuven, ESAT/COSIC and IBBT, Belgium

Björn Franke University of Edinburgh, School of Informatics, Informatics Forum, Scotland, United Kingdom

Marc Geilen Eindhoven University of Technology, Eindhoven, The Netherlands

Benedikt Gierlichs KU Leuven, ESAT/COSIC and IBBT, Belgium

Atanas Gotchev Department of Signal Processing, Tampere University of Technology, Tampere, Finland

Anthony Gregerson University of Wisconsin, Madison, WI, USA

Oscar Gustafsson Department of Electrical Engineering, Linköping University, Linköping, Sweden

Soonhoi Ha Seoul National University, Seoul, Republic of Korea

Wolfgang Haid Computer Engineering and Networks Laboratory, ETH Zurich, Switzerland

Timo D. Hämäläinen Tampere University of Technology, Tampere, Finland

Marko Hännikäinen Tampere University of Technology, Tampere, Finland

Christian Haubelt Applied Microelectronics and Computer Engineering, University of Rostock, Rostock-Warnemünde, Germany

Sangjin Hong Department of Electrical and Computer Engineering, Stony Brook University, Stony Brook, NY, USA

Nigel Horspool Department of Computer Science, University of Victoria, Victoria, BC, Canada

Yu Hen Hu Department of Electrical and Computer Engineering, University of Wisconsin, Madison, WI, USA

Kai Huang Computer Engineering and Networks Laboratory, ETH Zurich, Switzerland

Jörn W. Janneck Department of Computer Science, Lund University, Lund, Sweden

Markku Juntti Department of Communications Engineering and Centre for Wireless Communications, University of Oulu, Oulu, Finland

Ville Kaseva Tampere University of Technology, Tampere, Finland

Joachim Keinert Fraunhofer Institute for Integrated Circuits, Erlangen, Germany

Christoph W. Kessler Department of Computer Science (IDA), Linköping University, Linköping, Sweden

Martti Kirkko-Jaakkola Department of Computer Systems, Tampere University of Technology, Tampere, Finland

Miroslav Knežević NXP Semiconductors, Leuven, Belgium

Mikko Kohvakka Suntrica Ltd, Salo, Finland

Konstantinos Konstantinides Legal, Dolby Laboratories, Sunnyvale, CA

Andreas Krall Institut für Computersprachen, Technische Universität Wien, Vienna, Austria

Sun-Yuan Kung Department of Electrical Engineering, Princeton University, Princeton, NJ, USA

Andy Lambrechts IMEC, Heverlee, Belgium

Yong Ki Lee Samsung Electronics, Suwon, South Korea

Helena Leppäkoski Department of Computer Systems, Tampere University of Technology, Tampere, Finland

Rainer Leupers Institute for Software for Systems on Silicon, RWTH Aachen University, Aachen, Germany

William S. Levine Department of ECE, University of Maryland, College Park, MD, USA

Yiran Li Rensselaer Polytechnic Institute, Troy, NY, USA

Dake Liu Linköping University, Linköping, Sweden

Roel Maes KU Leuven, ESAT/COSIC and IBBT, Belgium

Marco Mattavelli SCI-STI-MM Lab, EPFL, Lausanne, Switzerland

John McAllister Institute of Electronics, Communications and Information Technology (ECIT), Queen's Universtiy Belfast, Belfast, UK

Hyunok Oh Hanyang University, Seoul, Korea

Seong-Jun Oh College of Information & Communication, Division of Computer and Communications Engineering, Korea University, Seoul, Korea

Bryan E. Olivier ACE Associated Compiler Experts bv., Amsterdam, The Netherlands

Yangyang Pan Rensselaer Polytechnic Institute, Troy, NY, USA

Keshab K. Parhi Department of Electrical and Computer Engineering, University of Minnesota, Minneapolis, MN, USA

Peter Pirsch Institute of Microelectronic Systems, Leibniz University of Hannover, Hannover, Germany

William Plishker University of Maryland, College Park, MD, USA

Praveen Raghavan IMEC, Kapeldreef Heverlee, Belgium

Mickaël Raulet IETR/INSA Rennes, Rennes, France

Markku Renfors Department of Communications Engineering, Tampere University of Technology, Tampere, Finland

Mateus Beck Rutzig Federal University of Santa Maria, Santa Maria, RS, Brazil

Jason Schlessman Department of Electrical Engineering, Princeton University, Princeton, NJ, USA

Michael J. Schulte Advanced Micro Devices, Austin, TX, USA

Naresh Shanbhag University of Illinois at Urbana-Champaign, Urbana, IL, USA

Raj Shekhar Children's National Medical Center, Washington, DC, USA

Weihua Sheng Institute for Software for Systems on Silicon, RWTH Aachen University, Aachen, Germany

Andrew Singer University of Illinois at Urbana-Champaign, Urbana, IL, USA

Jukka Suhonen Tampere University of Technology, Tampere, Finland

Yang Sun Rice University, Houston, TX, USA

Wonyong Sung Department of Electrical and Computer Engineering, Seoul National University, Gwanak-gu, Seoul, Republic of Korea

Jarmo Takala Tampere University of Technology, Tampere, Finland

Jürgen Teich Hardware-Software-Co-Design, University of Erlangen-Nuremberg, Erlangen, Germany

Bart D. Theelen Embedded Systems Innovation by TNO, Eindhoven, The Netherlands

Lothar Thiele Computer Engineering and Networks Laboratory, ETH Zurich, Switzerland

Olli Vainio Tampere University of Technology, Tampere, Finland

Mikko Valkama Department of Communications Engineering, Tampere University of Technology, Tampere, Finland

Mihaela van der Schaar University of California, Los Angeles, CA, USA

Alle-Jan van der Veen TU Delft, Fac. EEMCS, Mekelweg, CD Delft, The Netherlands

Ingrid Verbauwhede KU Leuven, ESAT/COSIC and IBBT, Belgium Electrical Engineering, University of California, Los Angeles, CA, USA

Sven Verdoolaege Department of Computer Science, Katholieke Universiteit Leuven, Leuven, Belgium

Vivek Walimbe GE Healthcare, Waukesha, WI, USA

Jian Wang Linköping University, Linköping, Sweden

Lars Wanhammar Department of Electrical Engineering, Linköping University, Linköping, Sweden

Stefan J. Wijnholds Netherlands Institute for Radio Astronomy (ASTRON), Oude Hoogeveensedijk, PD Dwingeloo, The Netherlands

Marilyn Wolf School of Electrical and Computer Engineering, Georgia Institute of Technology, Atlanta, GA, USA

Roger Woods ECIT Institute, Queen's University of Belfast, Queen's Island, Belfast, UK

Christian Zebelein Applied Microelectronics and Computer Engineering, University of Rostock, Rostock-Warnemünde, Germany

Tong Zhang Rensselaer Polytechnic Institute, Troy, NY, USA

Part II
Architectures

Architectures for Stereo Vision

Christian Banz, Holger Blume, and Peter Pirsch

Abstract Stereo vision is an elementary problem for many computer vision tasks. It has been widely studied under the two aspects of increasing the quality of the results and accelerating the computational processes. This chapter provides theoretic background on stereo vision systems and discusses architectures and implementations for real-time applications. In particular, the computationally most intensive part, the stereo matching, is discussed on the example of one of the leading algorithms, the semi-global matching (SGM). For this algorithm two implementations are presented in detail on two of the most relevant platforms for real-time image processing today: Field Programmable Gate Arrays (FPGAs) and Graphics Processing Units (GPUs). Thus, the major differences in designing parallelization techniques for extremely different image processing platforms are being illustrated.

1 Introduction

The field of stereo vision is highly inspired by the capabilities of the human imaging system. It encompasses all aspects of computer vision processing data from stereo image pairs in one way or another. The goal is to estimate 3D information about the observed scene, which can be used for a number of applications such as e.g. distance measurement, 3D reconstruction, and arbitrary view interpolation. Crucial for stereo vision is the task of stereo matching which identifies the projection points of the same 3D real world point in both images of the stereo pair. The location

C. Banz (✉) • H. Blume • P. Pirsch
Institute of Microelectronic Systems, Leibniz University of Hannover Appelstr. 4,
30167 Hannover, Germany
e-mail: christian.banz@alumni.uni-hannover.de; blume@ims.uni-hannover.de;
pirsch@ims.uni-hannover.de

S.S. Bhattacharyya et al. (eds.), *Handbook of Signal Processing Systems*,
DOI 10.1007/978-1-4614-6859-2_16, © Springer Science+Business Media, LLC 2013

Fig. 1 Results for the stereo correspondence problem: (**a**) left rectified input image (raw input images taken from [20]), (**b**) disparity map after left/right check where *white* denotes disparities marked as invalid, (**c**) false color representation of the disparity map, (**d**) untextured 3D view generated from the disparity map of (**b**)

difference (*the disparity*) in conjunction with a known stereo camera calibration allows to infer the depth information. Figure 1 gives an example.

The importance of stereo matching has been underlined by Szeliski and Scharstein stating that it is "one of the most widely studied and fundamental problems of computer vision" [83]. Active research in this field has resulted in a wide range of disparity estimation algorithms using radically different approaches. Recently, a general taxonomy has been introduced [81] including a comprehensive survey, what resulted in the on-going online *Middlebury benchmark* [80]. Further surveys evaluated different algorithms and variations thereof [39, 86]. Major focus was the quality of the stereo matching in terms of accuracy, density of the disparity map, and robustness.

However, advances in robustness and accuracy were accompanied with significant increases in complexity and computational requirements making the use of specialized implementations for many of today's real-time applications an absolute necessity. Surveys on efficient implementations for selected types of algorithms have been conducted [29, 59, 66, 86] and many more specialized implementations and architectures for individual algorithms and applications have been proposed.

Considering all aspects (algorithmic performance, implementation performance, architectures) a huge design space is unfolded. For embedded systems the choice is invariably on low-power solutions, e.g. based on application specific architectures implemented on FPGAs or ASICs. However, with the recent rise of GPUs for high performance computing, GPUs offer a cost-efficient alternative for stationary systems where power consumption is not an issue.

This chapter addresses high performance disparity estimation considering both, algorithmic and implementation performance. The chapter is structured into an algorithmic and an architectural section; these being Sects. 2 and 3. An introduction to the fundamental principles of the stereo image matching (*epipolar geometry*) and a minimal practical stereo vision system is given in Sect. 2.1. The algorithmic and architecture sections both give a comprehensive overview of recent works. It is followed by a detailed discussion of the semi-global matching algorithm (SGM) [37] (Sect. 2.3) and two exemplary implementations on FPGA (Sect. 3.7) and GPU (Sect. 3.6), respectively.

2 Algorithms

A minimal system for disparity estimation from a real camera setup consists of two processing steps: The first step is camera lens undistortion and *rectification* of non-ideal stereo camera setup (Sect. 2.1) while the second step is the actual stereo matching (Sect. 2.2). All other image preprocessing steps (e.g. noise reduction, equalization) and disparity map post-processing steps (e.g. whole filling, interpolation of pixel with missing stereo information) are optional.

2.1 Epipolar Geometry and Rectification

The objective is to find corresponding pixels in the two images of a stereo camera setup. Due to the underlying epipolar geometry [34, 83] of a stereo camera setup, the search space for corresponding pixels is one-dimensional. As shown in Fig. 2a, for a given pixel in the base image all potential correspondences project onto the *epipolar line* (e_{bm}) in the match image and vice versa. Strictly speaking the possible projections are bound by the *epipole* and the viewing rays for a real-world point at infinity.

For efficient correspondence search implementations a preprocessing step, the *rectification*, is employed. Both images are warped such that epipolar lines in both images are parallel to the scanlines and are row-aligned, i.e. corresponding pixels are in the same horizontal line [34, 103]. Thus, efficient memory access patterns and parallelism over independent scanlines can be obtained. After rectification, the focal axis are parallel to each other and perpendicular to the line joining the two camera centers (*baseline*) and the disparity for points at infinity is 0.

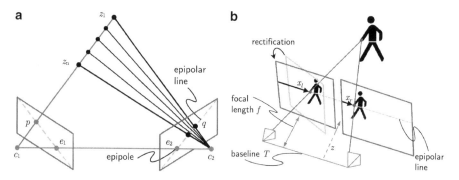

Fig. 2 Epipolar geometry: in (**a**) an unrectified setup and in (**b**) a rectified setup is shown. The rectification process in (**b**) to achieve row-aligned search space is illustrated only for the left projection plane

The rectified stereo setup is shown in Fig. 2b and the disparity is purely horizontal offset $d = (x_l - x_r)$ [pixel]. With the rectified focal length f [pixel], the baseline T [m] of the camera pair, the distance z [m] between the baseline and the 3D point can be calculated as

$$z = \frac{fT}{(x_l - x_r)} = \frac{fT}{d}. \tag{1}$$

This is also referred to as *standard rectified geometry* [83]. Thus, extracting depth information from a stereo camera setup becomes estimating the *disparity map* $d(x,y)$.

In addition to a non-ideal camera setup, stereo vision systems have to handle camera-inflicted image distortions, of which the most common are radial lens distortion, sensor tilting and offset from focal axis [11]. These must be compensated *before* rectification. However, when applying undistortion and rectification to a sequence of input images both steps can be combined. Reverse mapping assigns every pixel in the undistorted and rectified image a sub-pixel accurate origin in the input image. The rectified pixels are obtained using any desired pixel interpolation method. The bilinear interpolation for example, exhibits a reasonable trade-off between image quality and hardware implementation costs. Alternative interpolation methods are spline interpolation, which has higher silicon area requirements, and nearest-neighbor, which does not provide the required resolution for disparity estimation. Intermediate results from the processing steps are shown in Fig. 3.

The displacement vectors for undistortion and rectification are calculated using the intrinsic and extrinsic matrices, the tangential and the radial distortion parameters. These can be obtained by separate camera calibration steps (e.g. [104]) using a calibration pattern, such as a chessboard pattern employed in OpenCV [11]. Alternatively, or additionally, camera self-calibration from scene structure can be employed for particular camera parameters. For latter use in e.g. cars, camera self-calibration or at least updating of the intrinsic parameters from scene structure is mandatory.

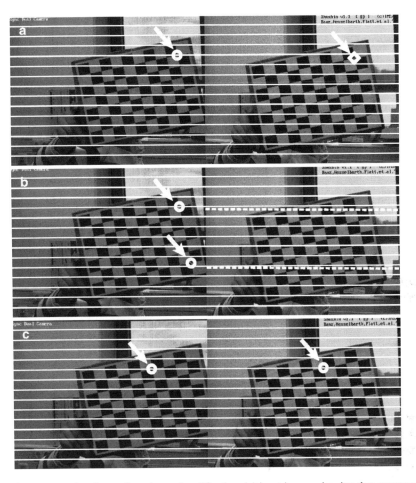

Fig. 3 Image results after undistortion and rectification: (**a**) input images showing that correspondences are not aligned (*circle* and *square*). (**b**) Undistorted images showing the epipolar lines (*dashed lines*) for two exemplary points (*circles*). Here, the effect is minor but the epipolar lines are clearly not aligned to the scanlines (i.e. horizontal pixel rows, *white*). (**c**) Final, undistorted, rectified images with row aligned epipolar lines

2.2 Stereo Correspondence

The origins of stereo correspondence were *sparse*, feature-based methods processing only a set of potentially highly discriminative image points. Today, most algorithms are *dense* methods, trying to infer a complete disparity map even for texture-less regions. Dense methods are typically classified into *local* and *global* approaches. However, for both classes of dense methods a common taxonomy and categorization has been introduced in [81]. Generally, a disparity estimation algorithm consists of the four processing steps:

1. Matching cost computation
2. Cost (support) aggregation
3. Disparity computation and optimization
4. Disparity refinement (or post processing)

It is of crucial importance to distinguish between the matching costs, which is the initial similarity measure between two pixels in the base and match image (or left and right image, respectively), and the aggregation method that uses these costs. The results of the matching cost computation are stored in the *disparity space image* $C(x, y, d)$. Cost aggregation of local (or area based) methods is performed on the information based in a local aggregation region (*support region*) from the matching costs $C(x, y, d)$. Global methods on the other hand perform one or more optimization steps on the matching costs often enforcing some kind of smoothness criterion. Depending on the algorithm, steps have varying importance and some might even be omitted. An example will be given in Sect. 2.3.

For matching cost computation a number of different window-based similarity measures can be employed. With rectified input images similarity of potentially corresponding pixels must be computed at location $\mathbf{p} = [x, y]^T$ in the left image and $\mathbf{q} = [x - d, y]^T$ in the right image. Initially often used and inspired by other areas of video processing are sum of squared intensity differences (SSD), sum of absolute differences (SAD), normalized cross correlation (NCC) and their respective zero mean variations. More recently, measures specifically for stereo matching have been proposed. For example, rank and census transform [99] are non-parametric transforms, and are thus robust to a certain amount of intensity differences. A vast number of other measures based on gradients, phase correlations, ordinal measures and dense feature descriptors exist. Entropy based measures (e.g. mutual information [37, 51]) have also been proposed. For those measures that compare absolute difference values, the approach of Birchfeld and Tomasi (BT) [9] can be used to include sampling insensitivity. For a more complete list of similarity measures refer to [83]. Detailed studies on the performance of the similarity measures in conjunction with different aggregation methods have been conducted [39, 81, 90].

The emphasis of local methods is on the cost aggregation step (step 2). A recent, comprehensive comparison of aggregation methods can be found in [86] and of selected methods for GPU implementation in [29]. Support regions can be two- or three-dimensional windows from the disparity space with fixed or adaptive window sizes, shapes, anchor points or weights. Adaptation may e.g. be performed by a full search through multiple windows or from a number of cues, e.g. constant disparity constraints and color-based segmentation. A more complete list may also be found in [83]. After cost aggregation, follows the disparity computation (step 3) of which the most basic form is selecting the disparity with minimal aggregated cost value for each pixel. Local methods are often well suited for hardware implementation due to the implicit parallelism and local data dependencies.

With global methods, the cost aggregation (step 2) is often omitted because the global smoothness constraints, which are enforced by the optimization process during the disparity computation (step 3), perform similar functions [83]. Global methods are often formulated within an energy-minimization framework:

$$E(D) = E_d(D) + \lambda E_s(D). \tag{2}$$

The objective is to find a solution d that minimizes the total energy E for a disparity map D, where E_d is the data term representing how well the solution fits to the input image and E_s represents the smoothness constraints made by the algorithm. These *regularization* or variational formulations are also employed in many others areas of image processing. In stereo processing it is important to formulate $E_s(d)$ to allow for *discontinuity preservation* in the disparity map. Algorithms to find the solution to (2) include belief propagation, graph cuts, and total variation among others [83]. Unfortunately, the problem is NP-hard for many discontinuity preserving if E_s is formulated two-dimensionally [10]. Reducing E_s to one-dimension along the scanlines, allows for independent, parallel *scanline optimization* but suffers from streaking (inconsistency between scanlines). Other global methods are based on *dynamic programming*, which performs global optimization for independent scanlines. Dynamic programming also suffers from streaking, but several works have addressed this problem (e.g. [78]).

For each approach several algorithms have been proposed and minute details influence the performance. As mentioned in Sect. 1, comparative studies have been performed (e.g. [29, 39, 81]) and a widely used benchmark exists [80]. For a stereo vision system with high performance in terms of robustness, accuracy and processing speed, several aspects have to be weighted against each other. While some local methods are more efficiently implementable, they can be challenged by areas with low or repetitive textures due to a high level of ambiguity [39]. Iterative, global minimization methods are often computationally intensive. However, Tombari et al. [86] express, that with sophisticated cost aggregation some local methods yield performance comparable with many global methods. The semi-global optimization strategy [37] is a solution resident in between by accumulating optimization results from multiple independent one-dimensional directions for each pixel. It produces very high quality results, even though not *the* best in the Middlebury benchmark [80]. Further, it is robust and it can be implemented efficiently for a global method. For robustness of the entire disparity estimation a suitable similarity measure must be chosen. Among the more robust measures in [38, 39] were census, rank, ordinal measures, and hierarchical mutual information.

Disparity refinement (step 4) often includes sub-pixel refinement, confidence or integrity checks, and interpolation measures. Since most stereo methods compute disparities at integer level, a *sub-pixel* refinement is necessary for many applications. An easy and computationally efficient way is to fit a curve through the discrete disparity space around the selected disparity. Interpolation functions are investigated in [32]. Several arising issues are discussed in [84]. An often computationally prohibitively expensive alternative is to start the computation with a disparity space already discretized to sub-pixel accuracy.

Foreground objects in the scene occlude different parts of the background when seen from the two camera perspectives. Consequently disparities cannot be computed for these occluded areas of the image due to missing stereo correspondences. This is visible in Fig. 1 by the halos around the foreground objects. It is often desirable to exclude these areas and areas with low confidence from the disparity map and optionally process them with sophisticated hole filling algorithms. Identification of these areas is performed with a *left/right check*, where the disparity maps for the left and right perspective are computed and only matching depth information from both perspectives to a 3D world point is allowed. With respect to the camera-to-camera projection in a rectified stereo pair the constraint for a valid disparity in the base image can be formulated as

$$
D_{b.\text{check}}(x,y) = \begin{cases} D_b & \text{if } |D_b(x,y) - D_m(e_{mb}(x,D_b(x,y)),y)| \leq \delta \\ invalid & \text{otherwise} \end{cases} \tag{3}
$$

with D_b and D_m are the disparity maps from the base and match perspective, respectively.

Further post-processing of the disparity map can be performed using basic median filtering to remove single outliers, peak removal and sophisticated whole filling algorithms, such as surface fitting. However, without a dense, highly accurate initial disparity map, post-processing will not provide reliable disparities.

2.3 Algorithm Example: Semi-global Matching

As a specific example disparity estimation based on the highly relevant and top-performing combination of rank transform [99] and semi-global matching algorithm (SGM) [37] will be used to illustrate the matter of the previous sections. Simultaneously, SGM will be used as a case study for implementations on FPGA and GPU.

The matching costs $C(x,y,d)$ (step 1) are calculated from the rank transform of the base and match image R_b and R_m with absolute difference comparison:

$$
C(\mathbf{p},d) = |R_b(p_x,p_y) - R_m(p_x - d,p_y)|. \tag{4}
$$

It is $\mathbf{p} = [p_x,p_y]^T$ the pixel location in the left image. The rank transform is defined as the number of pixels $\mathbf{p}'$ in a square $M \times M$ neighborhood $A(\mathbf{p})$ of the center pixel $\mathbf{p}$ with a luminous intensity I less than $I(\mathbf{p})$

$$
R(\mathbf{p}) = \|\{\mathbf{p}' \in A(\mathbf{p}) \mid I(\mathbf{p}') < I(\mathbf{p})\}\|. \tag{5}
$$

Fig. 4 The path cost
aggregation is performed
from eight cardinal directions
to every pixel

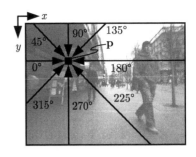

These initial pixel-wise calculated matching costs (i.e. locally calculated) yield non-unique or wrong correspondences due to low texture and ambiguity. Therefore, semi-global matching introduces global consistency constraints by aggregating matching costs along several independent, one-dimensional *paths* from different cardinal directions as shown in Fig. 4. A path $\mathbf{r}$ is formulated recursively by the definition of the path costs $L_r(\mathbf{p}, d)$.

$$L_r(\mathbf{p},d) = C(\mathbf{p},d) + \min\left[L_r(\mathbf{p}-\mathbf{r},d)\,,\right.$$
$$L_r(\mathbf{p}-\mathbf{r},d-1)+P_1,$$
$$L_r(\mathbf{p}-\mathbf{r},d+1)+P_1,$$
$$\min_i L_r(\mathbf{p}-\mathbf{r},i)+P_2\Big] -$$
$$\min_l L_r(\mathbf{p}-\mathbf{r},l) \tag{6}$$

The first term, $C(\mathbf{p},d)$, describes the initial matching costs. The second term adds the minimal path costs of the previous pixel $\mathbf{p} - \mathbf{r}$ including a penalty P_1 for disparity changes and P_2 for disparity discontinuities, respectively. Discrimination of small changes $|\Delta d| = 1$ pixel (*px*) and discontinuities $|\Delta d| > 1$ px allows for slanted and curved surfaces on the one hand and preserves disparity discontinuities on the other hand. The last term prevents constantly increasing path costs. For a detailed discussion refer to [37]. P_1 is an empirically determined constant. P_2 can also be an empirically determined constant or can be adapted to the image content. The selection of these penalty functions is investigated in [5] in detail.

Path costs are calculated from several cardinal directions to each pixel, as shown in Fig. 4 and are summed. The aggregated sum costs S are the sum of the path costs

$$S(\mathbf{p},d) = \sum_{\mathbf{r}} L_r(\mathbf{p},d). \tag{7}$$

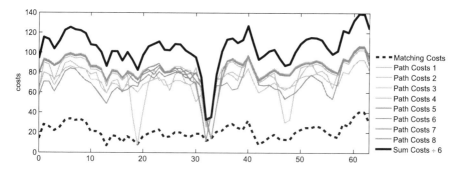

Fig. 5 Effect of path cost aggregation: matching costs, aggregated path costs, and sum costs (scaled by factor 1/6 for better presentation) for the pixel $\mathbf{p} = [183; 278]$ of the Teddy test image [82] calculated with SGM

By (6) and (7) SGM aims to approximate the following global energy minimization problem:

$$E(D) = \underbrace{\sum_{\mathbf{p}} C(\mathbf{p}, d)}_{E_d(D)}$$

$$+ \underbrace{\sum_{\mathbf{p}} \left(\sum_{\mathbf{p}' \in A} P_1 \mathrm{T} \left[|D_{\mathbf{p}} - D_{\mathbf{p}'}| = 1 \right] + \sum_{\mathbf{p}' \in A} P_2 \mathrm{T} \left[|D_{\mathbf{p}} - D_{\mathbf{p}'}| > 1 \right] \right)}_{E_s(D)} \quad (8)$$

where E_s contains the 2D smoothness constraints on the disparity map. For a derivation of (8) see [37]. The resulting method of the approximation resembles a scanline optimization approach but with excellent regard to interscanline consistencies.

Final disparity selection (step 3) is performed by a *winner-take-all* (WTA) approach. The disparity map $D_b(p_x, p_y)$ from the perspective of the base camera is calculated by selecting the disparity with the minimal aggregated costs

$$\min_d S(p_x, p_y, d) \quad (9)$$

for each pixel. For calculating the disparity map from the perspective of the match camera $D_m(q_x, q_y)$, the minimal aggregated costs along the corresponding epipolar lines are selected:

$$\min_d S(q_x + d, q_y, d). \quad (10)$$

Alternatively, SGM can be applied again, but with the other image as base image.

The effect of the path costs aggregation and the disparity selection is illustrated in Fig. 5. The initial matching costs $C(\mathbf{p}, d)$ (dashed line) exhibit a high level of ambiguity. Seven of the eight aggregated paths costs $L_r(\mathbf{p}, d)$ already show distinct

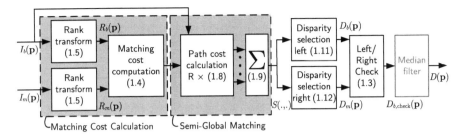

Fig. 6 Processing steps for disparity estimation using rank transform, semi-global matching, and optional median filter. Numbers in parenthesis refer to the respective equations and R denotes the number of paths

minima. The summed path costs $S(\mathbf{p},d)$ (thick black line) clearly identify the minimum at a disparity level of 32 resolving all ambiguities. However, the cost difference for the positions 32 and 33 is minimal indicating that the correct position is located a sub-pel precision.

Finally, left/right check according to (3) and post processing can be applied, e.g. a median filter in its most basic form. An overview of the processing steps is given in Fig. 6.

3 Architectures

The variety of architectures and implementations to compute the stereo correspondence easily rivals the variety of the underlying stereo matching algorithms. Today very efficient implementations for local and global stereo methods are available on FPGAs, ASICs, GPUs, and DSPs. For real-time image throughput, local methods have been and continue to be favored by many researches because of their efficient implementation possibilities. However, with advances in computational power, many global methods are also implementable in real-time.

Early work includes a complete stereo vision system from 1996 featuring rectification and stereo matching with an SSD variant on a custom hardware board consisting of off-the-shelf components and a DSP array [49]. At 30 fps 200×200 images with 5 bit depth resolution could be computed. Other noteworthy early implementations have been presented in 1993 using a DSP array [21] and in 1997 using a single DSP [52]. In [52] also an early overview of implementations is provided. An FPGA array was used in 1997 to implement a census transform stereo matching method [95]. All of these implementations directly compute the disparity information from the matching costs without cost aggregation. Early work includes a complete stereo vision system from 1996 featuring rectification and stereo matching with an SSD variant on a custom hardware board consisting of off-the-shelf components and an DSP array [49]. At 30 fps 200×200 images with

5 bit depth resolution could be computed. Other noteworthy early implementations have been presented in 1993 using a DSP array [21] and in 1997 using a single DSP [52]. In [52] also an early overview of implementations is provided. An FPGA array was used in 1997 to implement a census transform stereo matching method [95]. All of these implementations directly compute the disparity information from the matching costs without cost aggregation.

The following three subsections aim to give an overview of the conducted research on architectures and implementations for disparity estimation. Each subsection focuses on one specific hardware platform. Some key throughput values will be highlighted but without indication of algorithmic performance. A fair comparison must take into account architecture specific features, scalability, algorithmic performance under challenging imaging conditions (which are not present in the standard Middlebury data set), termination criterions on data dependent algorithms (e.g. belief propagation), and varying post processing steps. Thus, a comparison is an extremely complex task and beyond the scope of this chapter. The interested reader may consult the references themselves or one of the comparison studies.

3.1 GPU-Based Implementations

Commodity graphics processing hardware, nowadays superseded by *general purpose graphics processing units* (GPUs), have been used quite early to outsource, at first, part of the computation and then the entire stereo matching. For the calculation of disparity maps with an image size of 200×200 pixels and 50 disparity levels (abbreviated 200×200×50), 106 ms were achieved using a variable window SSD method on a NVIDIA GeForce4 in 2003 [76]. Early implementations for scene reconstruction are [77] on a Nvidia GeForce4 and [100] on a ATI Radeon 9700Pro from 2002 and 2003, respectively.

For belief propagation (BP) an efficient technique has been proposed in [22] and implemented for CPUs. It has been extended with occlusion handling and adapted for GPU implementation [12]. The same technique is used in recent implementations [43, 96] reaching 2.75 s for a 640×480×33 image on Nvidia GeForce GTX 280 and 93.98 s on an Intel Core 2 Duo (2.13 GHz). A fast converging hierarchical belief propagation is proposed and implemented in [97] reaching 16 fps for 320×240×16 images. New message passing schemes for BP have been applied in [55] for a GPU and VLSI implementation.

A dynamic programming solution with extensive use of MMX instructions on the CPU using color based cost aggregation has been presented in [23]. In 2005 the dynamic programming optimization step was still slower on the GPU than on the CPU [30] due to limitations of general purpose computation capabilities of the GPU (e.g. branching). Consequently, mixed CPU/GPU implementations performing cost aggregation on the GPU and dynamic programming optimization on the CPU have been presented [30, 56]. Scanline optimization in [101] also shows mixed performance results when comparing GPU versus CPU implementations. Recently,

[48] presented a multi-resolution symmetric dynamic programming variant on a GTX 295 reaching 14 fps for 2048×2048×256 images. A total variation algorithm with GPU implementation has been presented requiring between 15 and 60 s per image [73].

Variants of local methods examining the different techniques of adaptive weights or adaptive support regions have received much attention. Recent local approaches are census based with basic box filter cost aggregation [92] and a local truncated laplacian kernel approximation with adaptive cost aggregation [44]. Locally adaptive support regions have been used and speeded up with bitwise voting in [50]. Further work on local variants with adaptive cost aggregation methods includes [45, 63] and [40]. Instead of adaptive support regions on the input images [61] use edge-preserving filtering on the matching costs. A comparison of six local methods in terms of algorithmic and computational performance on GPUs has been conducted [29]. A plane sweep algorithm with local depth connectivity in order to retain depth discontinuities has been examined in [16].

For SGM various implementations have been presented on a GeForce 8800 Ultra [19] (0.0057 fps at 640×480×128), a Quadro FX5600 [27], a GTX 280 without [31] and with increased depth accuracy [67], and on a Tesla C2050 [4], which is the highest performing implementation with 63 fps for 640×480×128 images. This allows a very interesting retrospective on the evolution of GPUs. Especially some of the new features of Nvidia's compute capability 2.0 graphics cards allow radically different parallelization schemes, which was exploited in [4]. We will have a detailed look at this implementation in Sect. 3.6. Furthermore, a combination of adaptive support regions with a reduced version of SGM is proposed in [62] reaching 10 fps for 450×375×64 images.

3.2   Dedicated Architectures (FPGA and VLSI)

For dedicated architectures targeting FPGAs or ASICs, local methods are often favored because of potentially very small designs. This goes as far as to omit the cost aggregation altogether despite the drawbacks in accuracy and robustness. Nevertheless, new cost aggregation concepts have also been investigated and incorporated in hardware. In the following implementations without cost aggregation are indicated with "w/o CA".

Some examples of early architectures using SAD based matching w/o CA are [2, 54, 64]. An SAD based stereo vision system with three cameras has been presented in [98]. Depending on the emphasis of the referenced work, the results vary in throughput and resolution up to 640×480×64 and 31 fps. The so-called Tyzx ASIC for color-image census-based stereo-matching (w/o CA) achieves 200 fps for 512×480 images and 52 disparity levels [93]. It forms the basis of an extended stereo vision system in [94].

Also for recent implementations local methods with and without cost aggregation are still popular. This includes [46] where a census transform (w/o CA) is employed as basis of an entire stereo vision system on an FPGA. Another complete system based on SAD (w/o CA) is presented in [91]. Census with aggregation cues from the original and gradient images is investigated in [1]. Color SAD with a fuzzy logic disparity selection has been proposed and implemented on an FPGA [26]. Methods and architectures using adaptive support weights have been proposed in [14] employing a census variant and in [89] employing an absolute differences variant. In order to reduce the amount of data to be processed [88] works on sobel filtered images, which goes into the direction of sparse matching.

In [60], the architecture of [17], which is based on a local, phase-based method, is extended to large disparity ranges without significant additional hardware cost by adapting an offset of the smaller disparity search window across multiple frames. After large disparity changes, a latency of several frames occurs before correct disparity information can be regained. A bio-inspired method based on gabor filters is introduced in [18].

Among the implementations of dynamic programming approaches a trellis-based implementation, using a single interline consistency constraint has been investigated [68]. A dynamic programming approach based on a maximum-likelihood method is implemented in [78] achieving 64 fps at 640×480 px with 128 disparity levels. And a symmetric dynamic programming variant, similar to the GPU implementation of [48], has been implemented on an FPGA [65].

An FPGA architecture for memory efficient belief propagation for stereo matching has been proposed in [71]. New concepts and architectures for the message passing in BP are proposed [87].

For semi-global matching two architectures have been proposed. The implementation of [25] utilizes a SGM variant with depth adaptive sub-sampling. It achieves 27 fps at 320×200 px and 64 px disparity range. A parameterizable parallelization scheme for SGM and a corresponding FPGA architecture have been proposed in [7, 8]. It achieves, depending on the degree of parallelism, up to 176 fps for VGA images with 128 disparity levels and 4 SGM paths. This architecture will be studied in more detail in Sect. 3.7.

3.3 Other Architectures

The use of programmable architectures besides GPUs has also been investigated in some depth. Mühlmann et al. [66] investigated memory layout schemes for the disparity space and implementations schemes including MMX optimizations for SAD-based matching without cost aggregation (w/o CA).

A number of publications specifically target programmable embedded solutions: An SSD with multiple window selection has been implemented on the ClearSpeed CSX700 architecture (250 MHz, 9 W) which provides massively parallel SIMD in multiple parallel processing elements [41]. The same algorithm has been implemented [79] on the Tilera TILEPro64, which is a MIMD architecture with 64 integer processing cores organized in a two dimensional mesh network running at MHz.

A SAD w/o CA is also investigated on the Tilera TilePro 64 and on many-core CPUs [75]. SAD (w/o CA) for a VLIW processor (Texas Instruments TMS320C6414T, 1.0 GHz) has been shown in [13].

Application specific processors (ASIP) have been investigated in two cases: For semi-global matching an instruction set extension for the Tensilica LX2 DSP template has been proposed [6] reporting 20 fps for 640×480×64 images with reduced number of paths when run at 373 MHz, which is possible with the targeted TSMC 90 nm process. Similarly for SGM, architecture optimizations for a VLIW processor template, the MOAI, have been investigated in [69] reaching 30 fps when running at 400 MHz.

Apart from the original CPU implementation of SGM running at 1.3 s for 450× 375×64 images [36], a variant with depth adaptive sub-sampling has been proposed running at 14 fps for 320×160 images [24].

The cell broadband engine has been utilized for belief propagation and dynamic programming, both taking few seconds to process an image pair [58]. An SAD (w/o CA) implementation on the cell achieves 30 fps for VGA images with 48 disparity levels.

3.4   Comparison Studies

In addition to the algorithmic studies mentioned earlier, studies also taking into account the computational performance have been conducted. An evaluation of cost aggregation for local methods with focus on algorithmic performance and run-time on CPU can be found in [86]. Selected algorithms (various SAD variants, belief propagation, and dynamic programing) have been compared on a CPU in [59]. An evaluation of local algorithms on the GPU has been conducted in [29, 57].

An implementation of belief propagation on GPU and for VLSI has been compared in [55]. And symmetric dynamic programing on GPU and FPGA has been compared in [47]. Comparison of a census based approach (w/o CA) on a DSP (TI C6416), a GPU (GeForce 9800 GT), and a CPU (Intel Core2Quad) has been conducted in [42]. And in [75] SAD (w/o CA) has been studied on a GPU, two multi-core CPUs and the MIMD Tilea architecture. Further, in many of the references in the previous sections the GPU or FPGA implementation is compared to a regular CPU implementation. However, these are too numerous to list them here.

3.5   Current Trends

When targeting real-world applications, an everlasting question is to improve algorithmic performance while reducing computational requirements. This has already been addressed in many of the above references. A recent research direction is to integrate the computation of various information retrieval image processing tasks (e.g. disparity estimation with optical flow). In [28] an algorithm for joint computation of disparity estimation and optical flow is proposed and implemented on the GPU. A holistic architecture for phase based disparity estimation, optical flow, and more is presented [85] and implemented on an FPGA. An holistic architecture for disparity estimation and motion estimation based on SAD is presented in [102].

3.6   Implementation Example: Semi-global Matching on the GPU

An example implementation of the semi-global matching algorithm for GPUs will be given based on the works in [4]. Since GPUs are becoming more and more common, an introduction of the architecture and the terminology will be skipped. Please refer to the Nvidia manuals and [35] for a detailed background on GPU architecture or directly to [4] for a short sketch. The evaluation platform in the following is a Nvidia Tesla C2050 with compute capability 2.0 providing 3 GB DDRRAM global memory with a maximum theoretical bandwidth of 144 GB/s.

3.6.1   Parallelization Principles

Banz et al. [4] formulate the following performance limiting factors for a kernel:

- **Effective memory bandwidth usage** for the payload data which is e.g. reduced by nonaligned, overhead-producing memory access
- **Instruction throughput** defined as the number of instructions performing arithmetics for the core computation and other non-ancillary instructions per unit of time
- **Latency of the memory interface** occurring e.g. when accessing scattered memory locations even if aligned and coalesced, warp-wise access is performed
- **Latency of the arithmetic pipeline** of the ALUs inside the GPU cores if arithmetic instructions depend on each other and can only be executed with the result from the previous instruction

Accordingly, kernels can be memory bound, compute bound or latency bound. Kernels that are not limited by any of the three bounds are ill-adapted for GPU implementation and can be classified as bound by their parallelization scheme.

An efficient parallelization scheme guarantees inherently *aligned* and *coalesced* data access schemes without instruction overhead. Coalesced memory access is the simultaneous memory access to consecutive memory locations of all threads of a warp. It further includes a combination of parallel and sequential processing with independent arithmetic computation steps. An inner (sequential) loop in the otherwise parallel threads working on a set of data that is kept in shared memory or

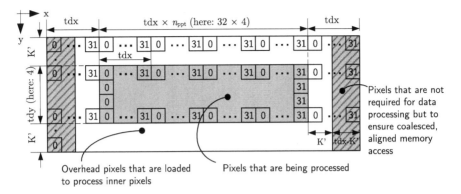

Fig. 7 Data fetching and accessing scheme for the 2D filter kernels processing $tdx \cdot n_{ppt} \times tdy$ kernel windows with a radius of K' where n_{ppt} is the number of pixels processed per thread and the launch configuration, which determines how threads are grouped and executed on the streaming multiprocessors, is $tdx \times tdy$. *Each square* represents a pixel and the number inside is the x-dimension thread ID which fetches the pixel from global memory

register facilitates data reuse, increases the instruction ratio, and keeps the pipeline filled. Further, coherent access schemes are ensured for the memory interface if results are written out with each loop iteration. Apart from an inner loop, executing several warps per streaming multiprocessor increases pipeline utilization.

3.6.2 Rank Transform and Median Filter Kernel

The rank transform and median filter are both non-linear, non-separable 2D image transforms. To generate the result of one output pixel, the data of a local $N \times N$-neighborhood from the input image is required.

The kernel for rank transform and median filter are based on the same principle which is based on the implementation of a separable convolution in [74]. It prefetches data of a two-dimensional spatial locality from global memory into shared memory. Thus, data reuse is maximized because all filter kernels that fully reside in this spatial locality can be processed by a block of threads without additional global memory access. An aligned group of pixels is processed by a two-dimensional block of threads first loading the neighboring center pixels of all kernels. Left and right pixels outside the center area are always loaded with the warp width. Even though this causes minimal data to be loaded which is not used by the current block, it ensures inherent coalesced memory access without instruction overhead or warp divergence. An inner loop allows to process several pixels per thread (n_{ppt}) with a stride of the warp width. Adjusting n_{ppt} and the *launch configuration*, i.e. the number of threads per block in x-dimension (*tdx*) and y-dimension (*tdy*), allows to navigate between the different optimization principles. Figure 7 shows the data layout and thread access scheme.

Fig. 8 Performance of the 3×3 median filter: on 1280×960 images as the parallelization configuration changes. Block width is fixed to *tdx* = 32. Best performance is achieved with *tdx*×*tdy* = 32×4 and n_{ppt} = 4

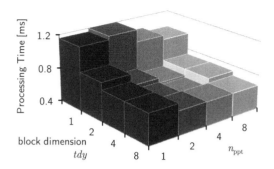

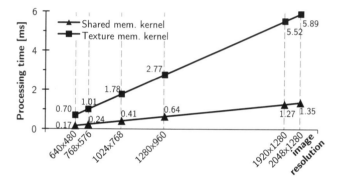

Fig. 9 Performance of the 3×3 median filter: comparison of the texture memory kernel and the proposed shared memory kernel on a Tesla C2050 GPU for the best-performing parallelization configuration

The median filter is always compute bound and performs best with $tdx \times tdy = 32 \times 4$ threads and $n_{\mathrm{ppt}} = 4$. The results of the parameter study for $tdx = 32$ are shown in Fig. 8. Configurations with $n_{\mathrm{ppt}} = 8$ perform slightly worse although redundant memory access is further reduced because of inefficient pipeline utilization. Processing times for a 3×3 median filter (i.e. kernel radius $K' = 1$) are given in Fig. 9 resulting in 0.64 ms for the new shared memory based kernel. For a texture-memory based kernel, which is the most often suggested way of implementing a 2D non-separable filter, processing time is which is 2.77 ms. In comparison, this yields a speed-up of 4.3 when processing a 1280×960 image.

For a 9×9 rank transform (i.e. $K' = 4$) experiments showed that a block size of $tdx \times tdy = 32 \times 4$ with $n_{\mathrm{ppt}} = 4$ yields best performance. A speed up of 4.0 is obtained switching from the texture-based kernel (3.13 ms) to the shared memory kernel (0.78 ms) for 1280×960 images.

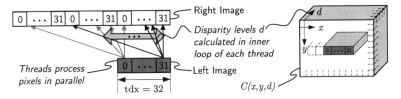

Fig. 10 Memory access scheme for calculating the matching costs for *tdx* pixels in parallel in a *tdx*-thread wide warp and *tdy* = 1. The location of the results in the 3D matching cost space is shown. Again the numbers in the *squares* represent the thread ID that fetches the according pixels from global memory

3.6.3 SGM Kernel

For every pixel location $\mathbf{p}$, calculation of the matching cost $C(\mathbf{p}, d)$ according to Eq. (4) results in a vector with one entry for each disparity level d. Thus, the spatial directions (x and y) and the disparity range span the three-dimensional disparity space. The matching cost (*MC*) calculation for every point in this space can be performed independently allowing for parallelization in all three dimensions.

A straightforward parallelization is to assign each thread with the calculation of one entry in the 3D cost space of $C(\mathbf{p}, d)$. This kernel (mc_unaligned) reaches 16.3 ms and 48.6 GB/s which is far from the bandwidth limit due to inefficient, often misaligned memory access, lack of data reuse, and little latency hiding possibilities. This kernel is latency bound which can only be eliminated by a new parallelization scheme.

The new kernel (mc_proposed) processes all disparity levels of a group of *tdx* neighboring pixels synchronously in *tdx* threads. The disparity dimension itself is further separated into *tdy* groups each processing d_{range}/tdy disparity levels with an inner loop in the kernel. By adjusting *tdy* thread parallelism is substituted with inner loop complexity. Pixels from the base image are read aligned and coalesced over the *tdx* threads. The required pixels from the right image are loaded in groups of *tdx* aligned, coalesced pixels into the shared memory where they can be accessed and reused by all threads. The parallelization scheme is shown in Fig. 10. Furthermore, only 8 bit precision is required. Since performing arithmetic in non-native GPU data types (i.e. other than 32-bit integer and float) is slow, input images and computation are based on 32-bit integer and type conversion to uchar is performed just before writing out the result. Consequently, type conversion to uchar is performed just before writing out the result. Choosing *tdx* as a multiple of the warp size (i.e. 32) results in always aligned memory access. This kernel adheres the optimization approach of Sect. 3.6.1 by providing inherently aligned memory access, high data reuse, and efficient use of the arithmetic pipeline. With an obtained performance of 1.8 ms and 111.2 GB/s this is a speed-up of factor 9.2. A performance summary is given in Table 1.

The path costs (*PC*) calculation according to Eq. (6) is performed by individually traversing along each of the eight path directions updating the matching cost values

Table 1 Performance results of the optimized kernels (*MC*: matching costs, *PC*: path costs, *WTA*: winner-takes-all) with optimal launch configuration for computing the semi-global matching algorithm for images with 1280×960 pixels and 128 disparity levels

Kernel	Time (ms)	Bandwidth (GB/s)	Bound by
MC unaligned	16.32	48.6	Parallelization scheme
MC proposed (uchar4)	1.80	107.3	Pipeline latency
MC+PC 8 path dir. (sequential)	75.68	20.9	Pipeline latency
MC+PC 8 path dir. (concurrent)	39.81	39.7	Pipeline latency
Sum, WTA left disp. map	15.09	117.4	Memory bandwidth

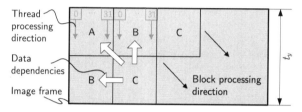

Fig. 11 Image tiling for the 45° path and $t_y = 2$ allowing divergent processing direction and path direction while tiles with the same letter can be processed in parallel. The processing direction ensures coalesced and aligned memory access

and resulting in a new 3D cost space for each path direction. PC calculation must be done sequentially along the respective path direction (e.g. from left to right) because the previous pixel's path costs must be known. The parallel minimum search over the disparity levels has been implemented similar to the parallel reduction scheme from [33]. Although the MCs are common to all PC directions and it seems obvious to separate MC and PC calculation, it is faster to integrate MC and PC calculation and recalculate the MCs on-the-fly for each PC direction. This drastically reduces pressure on the performance-limiting memory bandwidth since the MC data is never transferred via the external memory but can be kept locally in shared memory. All eight path directions are executed concurrently using the CUDA concurrent kernel execution on Fermi architectures.

Due to the coalesced memory access necessity only a group of horizontally neighboring pixels can be efficiently accessed in memory. The path costs kernels must be modified according to their path direction in order to maintain efficient memory access. For each diagonal path direction, processing is separated into rectangular tiles. Within each tile the processing direction is along the image columns, i.e. misaligned to the path direction, but ensuring aligned memory access. Tiles not sharing data dependencies can be processed in parallel as independent thread blocks. This is similar to the intrablock encoding scheme for video streams proposed in [53]. An example of the parallel processing order is shown in Fig. 11. Since block synchronization does not exist on GPUs, correct execution order is established by sequentially launching a kernel for each diagonal tile front (identical letters in Fig. 11) causing some minor time overhead. The seeming alternative

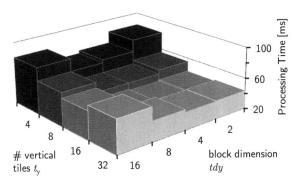

Fig. 12 Impact of the parallelization configuration on the performance of the concurrent path cost calculation for eight paths of the SGM for 1280×960 images and 128 disparity levels. Block width and tile width are both fixed to $tdx = 32$. Best performance is achieved with $tdx \times tdy = 32 \times 4$ (i.e. each inner loop processes 32 disparity levels) and $t_y = 16$

of keeping the processing implementation unchanged but rearranging the data in the memory creates an inherently contradictory situation: if the GPU is used to rearrange the data, the re-sorting causes additional memory access with is not even coalesced.

Again, parameters adjustment allows to navigate between the performance optimization principles. The first parameter (tdy) trades thread parallelism against sequential computation in the inner loop for all kernels. The second parameter (t_y) trades the number of parallelly processable blocks versus launch overhead and memory overhead for the four diagonal paths. Figure 12 shows the result of the parameter study. Choosing $tdy = 4$ and $t_y = 16$ results in best performance (39.8 ms and 39.7 GB/s) for a 1280×960 image. If the concurrent kernel execution is not used, performance is approximately halved (75.7 ms and 20.9 GB/s). Both kernel sets, concurrent and sequential, are latency bound.

Summation of the eight path cost spaces (7) and winner-takes-all disparity selection (9) can be performed independently for each pixel allowing for the same parallelization scheme as for the MC calculation. This kernel (sum_wta) requires 15.1 ms and is memory bound with 117.4 GB/s.

3.6.4   Performance

The processing time for the complete disparity estimation including rank transform, semi-global matching for eight paths, disparity map generation (without left/right check (3)) and median filtering on a Tesla C2050 Fermi architecture GPU is summarized in Table 2. Overall, a 1280×960 image with 128 disparity levels requires 56.2 ms. The processing times do not include data transfer between host and GPU because it can be effectively hidden using concurrent data transfer when processing image streams. When processing 1280×960 image sets ca. 5 ms additional transfer time is required.

Table 2 Performance results for the entire disparity estimation algorithm using rank transform, semi-global matching and median filtering on a Nvidia Tesla C2050 GPU

Image size	$d_{range} = 64$	128	256
640×480	9.7 ms	16.0 ms	29.0 ms
1024×768	21.5 ms	35.9 ms	67.1 ms
1280×960	32.9 ms	56.2 ms	105.7 ms

Results are k-mean values over multiple runs and images

3.7 Implementation Example: VLSI Architecture for Semi-global Matching

In this section a parallelization scheme and corresponding VLSI architecture for semi-global matching will be discussed. It is based on the works of [7, 8].

3.7.1 Parallelization

A crucial point for VLSI-implementation is the mapping of the algorithm into a *parallel-processable* and *stream-based* flow that only requires a single-pass across the input images. Further important aspects are *regularity* and *locality* of the architecture that implements this flow [72]. Challenges are imposed by the semi-global matching due to the recursively defined paths and their orientations within the images (see Sect. 2.3), which are not aligned to a stream-based flow.

First, the two-dimensional parallelization concept that enables stream-based processing will be introduced. Afterwards, an extension of the concept into the third dimension is presented, what significantly increases processing speed and throughput. The two-dimensional parallelization concept is shown in Fig. 13 and will be presented for the path directions of $0°$, $45°$, $90°$, $135°$. Pixels are processed from left to right along the image row ($0°$ path). After processing pixel $\mathbf{p}_{-1} = [x-1, y]$ of the upper row, all path costs over d of all directions are available in the path costs buffers (z^{-1}). Path costs are delayed, according to their path directions of $90°$ and $45°$ by one and two *additional* processing steps, respectively. Afterwards, path costs of $L_{45°}(x-3, y, d)$, $L_{90°}(x-2, y, d)$, $L_{135°}(x-1, y, d)$ are available at the output of the path cost buffers. These are exactly those path costs needed for parallel and synchronous calculation of all path costs of all orientations for pixel $\mathbf{p}_2 = [x-2, y+1]$. Synchronous calculation allows direct summation of path costs in a pipeline that returns the aggregated costs S.

Therefore, all paths to the pixels $\mathbf{p}_1 = [x, y]$ and $\mathbf{p}_2 = [x-2, y+1]$ are calculated in parallel in a single processing step. This concept is extendable to an arbitrary number of rows. An additional delay by two pixels is introduced for each new row, as illustrated in Fig. 13. Images are separated into image slices of N parallel rows in order to process whole images. Path costs of the last row of an image slice need to be stored and made available to the first row of the next slice.

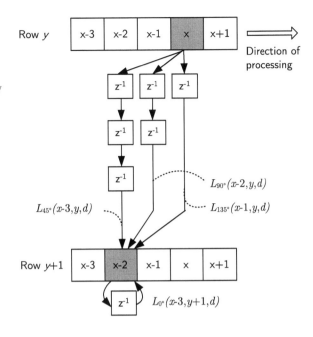

Fig. 13 Synchronized and parallel calculation of the path costs of the four paths $L_{0°}$, $L_{45°}$, $L_{90°}$ and $L_{135°}$ for the two pixels $\mathbf{p}_1 = [x, y]$ and $\mathbf{p}_2 = [x - 2, y + 1]$. Each delay element stores the respective path costs over all disparity levels for the duration of one processing step

Generalization of this concept is only limited by the fact that the maximum angle range must be within the half-closed interval $[0, 180°)$. This means that no paths in opposite directions can be directly supported without additional hardware. The two-dimensional parallelization allows regular data accesses of the input images and all intermediate values and will further be referred to as row parallelism. Moreover, this concept is independent of the processing method of the disparity levels, which can be either serial or parallel. Processing the disparity levels in parallel establishes a third dimension of parallelism, which will be referred to as disparity level parallelism. An approach of particular interest for dedicated hardware implementations is not to choose either extreme (none or all disparity levels in parallel) but to process the disparity levels in small groups (e.g. 2, 4, or 8). In this case the size of the path cost buffers, as specified above, remains constant while the throughput increases linearly with the number of parallelized disparity levels. However, some additional logic for the arithmetic computation of n paths in parallel will be required. The increase of logic requirements vs. performance of disparity level parallelism and row level parallelism will be investigated in Sect. 3.7.3.

3.7.2 Architecture

The hardware architecture for the entire stereo matching algorithm is given in Fig. 14. Computation of the rank transform of both images and calculation of the data dependent penalty term P_2 is done in parallel and synchronously utilizing the same data path.

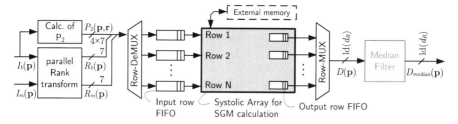

Fig. 14 Hardware architecture for calculation of disparity maps using rank-transform and semi-global matching. The median filter is optional

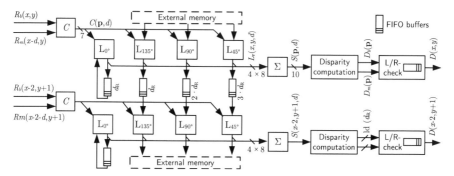

Fig. 15 Hardware architecture of the systolic array for parallel path cost calculation of the semi-global matching for two parallel rows

An N-row buffer provides this data to the systolic array, which calculates the disparities of all N rows in parallel according to the parallelization concept introduced above. As a basic post processing step, a median filter is employed for outlier suppression.

A heterogeneous, completely synchronized systolic array realizes the parallelization concept for the semi-global matching utilizing path directions from $0°$, $45°$, $90°$, $135°$. Figure 15 shows the corresponding block diagram without utilization of disparity level parallelism. In this case processing of a pixel $\mathbf{p}$ is carried out sequentially over all disparities of this pixel. The first processing elements (C-PEs) calculate the matching costs $C(\mathbf{p}, d)$. Each of the following PEs (L-PEs) calculates the path costs L_r along a path $\mathbf{r}$ according to Eq. (6). The results are buffered in the appropriate path cost buffers. All L-PEs are completely identical and the path orientations are solely defined by the delays introduced by the path cost buffers. Path costs are summed to S and then processed by disparity computation PEs (D-PEs). D-PEs locate the minimum, i.e. the correct disparity, for the disparity maps D_b and D_m of the base and match camera, respectively. A final L/R-Check-PE projects the disparity map D_m to the perspective of the base camera, executes the left/right check including occlusion detection, and marks pixels accordingly. A local single row buffer is needed for the projection. It functions simultaneously as an output buffer.

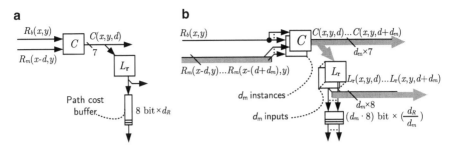

Fig. 16 Architectural extension (**b**) of the 2D-systolic array (**a**) for introducing disparity level parallelism where d_m specifies the number of disparity levels processed in parallel

In order to introduce disparity level parallelism in addition to the row level parallelism, the C-PEs and L-PEs are extended to process several consecutive disparity levels in parallel. These groups of parallel disparity levels are processed serially. This leads to an approximately linear increase in throughput. Further, it is area efficient for two reasons. First, additional logic is only required for parts of the processing units. And second, the absolute size of local buffers does not change—only the depth-to-width ratio. This is a major advantage of disparity level parallelism. The architectural extension for disparity level parallelism is shown in Fig. 16.

Boundary treatment for pixels with missing stereo overlap (i.e. $x < d_{max}$) significantly reduces the number of entries of the cost spaces $C(\mathbf{p},d)$, $L_r(\mathbf{p},d)$, $S(\mathbf{p},d)$, and, consequently, leads to a computing time reduction. For VGA images and a disparity range of 128 px the reduction is 9.9% (without disparity level parallelism).

An external interim memory is required for storing the path costs of the three non-horizontal paths of the last row of an image slice and providing them to the first row of the consecutive image slice. Due to the extremely regular data transfer, obeying the FIFO-principle, and the low transfer rates, external SSRAM and SDRAM-memories can be used. Alternatively, on-chip memory can be considered due to the quite low absolute memory requirements.

3.7.3 Performance

Performance of the complete system and scalability of the SGM unit are analyzed with the minimum clock frequency required to fulfill a fixed throughput constraint. This metric, i.e. the clock frequency normalized for a fixed throughput, allows direct and accurate comparison, and reflects the importance of performance while being independent from varying operating clock frequencies [70]. This also models a typical design constraint of real-world applications, where the required throughput is usually specified by external circumstances (e.g. by the cameras, required depth resolution, etc.). In this case, throughput-normalized metrics for clock frequency,

Table 3 Minimum required clock frequencies of the SGM unit (including rank transform and median filter) for a fixed resolution of 640×480 px with 128 disparity levels at 30 fps and resource usage on a Xilinx Virtex-5 FPGA

$p_r \setminus d_m$	Min. clock frequency (MHz)				LUTs			
	1	2	4	8	1	2	4	8
5	219.6	112.3	58.5	31.6	5,652	6,621	10,110	17,214
10	111.9	57.4	30.0	16.8	11,595	13,398	20,565	34,589
20	58.3	30.2	17.2	13.4	23,379	26,986	41,292	69,578
30	40.7	21.4	14.9	12.4	35,119	40,700	61,930	103,504

The number of parallel rows and parallel disparity levels is denoted p_r and d_m, respectively

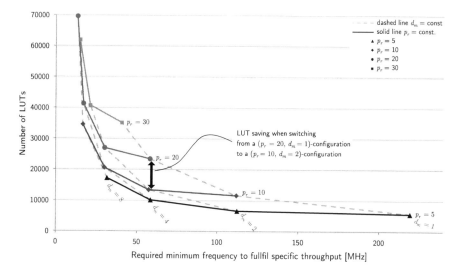

Fig. 17 Required number of LUTs of the systolic array of the SGM unit over the minimum required clock frequency to process 640×480 px with 128 disparity levels at 30 fps. The number of parallel rows and parallel disparity levels is denoted p_r and d_m, respectively. The diagrams effectively show the impact in area and performance when varying row parallelism and/or disparity level parallelism. The *lower left border* in the diagram reflects the Pareto-optimum configuration points

resource usage, power, and latency enable straightforward identification of the Pareto-optimal point of operation. Table 3 provides the results for the SGM unit for a typical parameter set of 640×480 px at 30 fps. As metric for required silicon area, only Virtex5 LUTs are used. For more information (e.g. BRAMs) please refer to [8].

Interesting insights can be gained by studying the row parallelism vs. disparity parallelism trade-off. With increasing degree of parallelism, the SGM unit can be clocked with lower frequencies at the price of higher area requirements. However, there are significant differences between row level parallelism and disparity level parallelism. Each point in Fig. 17 is a specific configuration of the design represent-ing the LUT requirements over the normalized clock frequency. This representation

Fig. 18 Hardware setup of the stereo vision system with the system board and the stereo camera rig. On the *right* is the input image of lab scene and the computed raw disparity map before false color visualization and sending to display is conducted

is considerably different from a typical AT-diagram, which would be inadequate for this comparison as it would not reflect the throughput constraint.

For a small number of parallel disparity levels, increasing disparity level parallelism is very efficient since it has a significantly smaller influence on the total resource usage than increasing row level parallelism. However, row parallelism is the key concept for stream-based processing and crucial for a high base performance but increases linearly with the number of rows. The full potential of the parallelism approaches is exploited when using a combination of both, i.e. by using a small number of parallel rows and additionally introducing disparity level parallelism up to the configuration that does not yet require additional memory resources. For example, starting from the $(p_r = 10, d_m = 1)$-configuration a performance increase of approximately factor two can be achieved by doubling the number of either parallel rows or disparity levels. Increasing disparity level parallelism does not increase BRAM requirements (not shown, see [8]) and results in a LUT saving of factor 1.8. The major benefit of increasing disparity level parallelism is that local memory requirements remain constant for both, path costs buffers and input/output buffers.

A stereo vision system covering the entire stereo vision process including image acquisition, noise reduction, rectification, disparity estimation, post processing, and visualization has been integrated into a single FPGA. The system has been integrated on a custom build hardware platform show in Fig. 18. This work shows that it is possible to implement an algorithmically extremely high performing disparity matching algorithm in an FPGA with true real-time performance. More details on the implementation can be found in [8].

4 Summary

There has been and continues to be tremendous research in the field of computer vision, both on the algorithmic side and on the hardware side. Nowadays, many implementations for GPUs, FPGAs, ASICs, DSPs, and ASIPs are available. These cover a huge variety of algorithms and design aspects (e.g. algorithmic performance vs. silicon area). The two example implementations on the GPU and the FPGA for semi-global matching based disparity estimation show, that it is possible to realize high quality stereo correspondance search in real-time. The GPU implementation enables SGM processing with eight paths but without left/right check with more than 62 fps of images with a resolution of 640×480 and 128 disparity levels on Nvidia Fermi architecture GPUs. The VLSI architecture is scalable and allows exact adaptation to the particular application. For the same image resolution frame rates of 1.7 fps to 319 fps are achieved at a operating frequency of 133 MHz. Which of both architectures is the more suitable solution depends on the external parameters.

5 Further Reading

A detailed algorithmic overview is provided in the textbook [83] and the surveys [29,81,85]. Epipolar geometry and rectification is covered in [34,103]. The OpenCV library provides many functions for stereo processing [11]. For multi-view stereo and 3D reconstruction [83] is a good starting point.

Dedicated image processing architectures including rectification and many more are covered in [3] and RTL hardware design in [15]. Various kinds of computer architectures including GPUs are found in the newest edition of [35].

References

1. Ambrosch, K., Kubinger, W.: Accurate hardware-based stereo vision. Computer Vision and Image Understanding, Elsevier **114**, 1303–1316 (2010)
2. Arias-Estrada, M., Xicotencatl, J., Brebner, G., Woods, R.: Multiple stereo matching using an extended architecture. Proc. Field-Programmable Logic and Applications **2147**, 203–212 (2001)
3. Bailey, D.G.: Design for embedded image processing on FPGAs. John Wiley & Sons, Singapore (2011)
4. Banz, C., Blume, H., Pirsch, P.: Real-time semi-global matching disparity estimation on the GPU. Proc. IEEE Intl. Conf. Computer Vision Workshops pp. 514–521 (2011)
5. Banz, C., Blume, H., Pirsch, P.: Evaluation of penalty functions for SGM cost aggregation. Intl. Archives of Photogrammetry and Remote Sensing (2012)
6. Banz, C., Dolar, C., Cholewa, F., Blume, H.: Instruction set extension for high throughput disparity estimation in stereo image processing. Proc. IEEE Intl. Conf. Architectures and Processors Application Specific Systems pp. 169–175 (2011)

7. Banz, C., Hesselbarth, S., Flatt, H., Blume, H., Pirsch, P.: Real-time stereo vision system using semi-global matching disparity estimation: Architecture and FPGA-implementation. Proc. IEEE Intl. Conf. Embedded Computer Systems: Architectures, Modeling, and Simulation pp. 93–101 (2010)

8. Banz, C., Hesselbarth, S., Flatt, H., Blume, H., Pirsch, P.: Real-time stereo vision system using semi-global matching disparity estimation: Architecture and FPGA-implementation. Trans. High-Performance Embedded Architectures and Compilers, Springer (2012)

9. Birchfield, S., Tomasi, C.: A pixel dissimilarity measure that is insensitive to image sampling. IEEE Trans. Pattern Analysis and Machine Intelligence **20**(4), 401–406 (1998)

10. Boykov, Y., Veksler, O., Zabih, R.: Fast approximate energy minimization via graph cuts. Proc. IEEE Intl. Conf. Computer Vision **1**, 377–384 (1999)

11. Bradski, G., Kaehler, A.: Learning OpenCV, 1 edn. O'Reilly, Sebastopol (2008)

12. Brunton, A., Chang, S., Roth, G.: Belief propagation on the GPU for stereo vision. Proc. Canadian Conf. Computer and Robot Vision p. 76 (2006)

13. Chang, N., Lin, T.M., Tasi, T.H., Tseng, Y.C., Chang, T.S.: Real-time DSP implementation on local stereo matching. Proc. IEEE Intl. Conf. Multimedia and Expo pp. 2090–2093 (2007)

14. Chang, N., Tasi, T.H., Hsu, B., Chen, Y., Chang, T.S.: Algorithm and architecture of disparity estimation with mini-census adaptive support weight. IEEE Trans. Circuits and Systems for Video Technology **20**(6), 792–805 (2010)

15. Chu, P.P.: RTL hardware design using VHDL: Coding for efficiency, portability, and scalability. Wiley-Interscience, Hoboken and N.J (2006)

16. Cornells, N., van Gool, L.: Real-time connectivity constrained depth map computation using programmable graphics hardware. Proc. IEEE Conf. Computer Vision and Pattern Recognition **1**, 1099–1104 (2005)

17. Darabiha, A., MacLean, W., Rose, J.: Reconfigurable hardware implementation of a phase-correlation stereo algorithm. Machine Vision and Applications, Springer **17**, 116–132 (2006)

18. Diaz, J., Ros, E., Carrillo, R., Prieto, A.: Real-time system for high-image resolution disparity estimation. IEEE Trans. Image Processing **16**(1), 280–285 (2007)

19. Ernst, I., Hirschmüller, H.: Mutual information based semi-global stereo matching on the GPU. Proc. Intl. Symp. Visual Computing **5358**, 228–239 (2008)

20. Ess, A., Leibe, B., Schindler, K., van Gool, L.: A mobile vision system for robust multi-person tracking. Proc. IEEE Conf. Computer Vision and Pattern Recognition pp. 1–8 (2008)

21. Faugeras, O., Viéville, T., Theron, E., Vuillemin, J., Hotz, B., Zhang, Z., Moll, L., Bertin, P., Mathieu, H., Fua, P., Berry, G., Proy, C.: Real-time correlation-based stereo: Algorithm, implementations and applications (1993)

22. Felzenszwalb, P.F., Huttenlocher, D.P.: Efficient belief propagation for early vision. Intl. Journal of Computer Vision, Springer **70**, 41–54 (2006)

23. Forstmann, S., Kanou, Y., Jun, O., Thuering, S., Schmitt, A.: Real-time stereo by using dynamic programming. Proc. IEEE Conf. Computer Vision and Pattern Recognition Workshop p. 29 (2004)

24. Gehrig, S., Rabe, C.: Real-time semi-global matching on the CPU. Proc. IEEE Conf. Computer Vision and Pattern Recognition Workshop pp. 85–92 (2010)

25. Gehrig, S.K., Eberli, F., Meyer, T.: A real-time low-power stereo vision engine using semi-global matching. Proc. Intl. Conf. Computer Vision Systems **5815**, 134–143 (2009)

26. Georgoulas, C., Andreadis, I.: A real-time fuzzy hardware structure for disparity map computation. Journal of Real-Time Image Processing, Springer **6**(4), 257–273 (2011)

27. Gibson, J., Marques, O.: Stereo depth with a unified architecture GPU. Proc. IEEE Conf. Computer Vision and Pattern Recognition Workshop pp. 1–6 (2008)

28. Gong, M.: Real-time joint disparity and disparity flow estimation on programmable graphics hardware. Computer Vision and Image Understanding, Elsevier **113**(1), 90–100 (2009)

29. Gong, M., Yang, R., Wang, L., Gong Mingwei: A performance study on different cost aggregation approaches used in real-time stereo matching. Intl. Journal of Computer Vision, Springer **75**, 283–296 (2007)

30. Gong, M., Yang, Y.H.: Near real-time reliable stereo matching using programmable graphics hardware. Proc. IEEE Conf. Computer Vision and Pattern Recognition 1, 924–931 (2005)
31. Haller, I., Nedevschi, S.: GPU optimization of the SGM stereo algorithm. Proc. IEEE Intl. Conf. Intelligent Computer Communication and Processing pp. 197–202 (2010)
32. Haller, I., Nedevschi, S.: Design of interpolation functions for subpixel-accuracy stereo-vision systems. IEEE Trans. Image Processing 21(2), 889–898 (2012)
33. Harris: Optimizing parallel reduction in CUDA (2007). Whitepaper included in Nvidia Cuda SDK 4.0
34. Hartley, R.I., Zisserman, A.: Multiple view geometry in computer vision, 2. ed., 7. print. edn. Cambridge Univ. Press, Cambridge (2010)
35. Hennessy, J.L., Patterson, D.A.: Computer architecture: A quantitative approach, 5 edn. Morgan Kaufmann, San Francisco and Calif and Oxford (2011)
36. Hirschmüller, H.: Accurate and efficient stereo processing by semi-global matching and mutual information. Proc. IEEE Conf. Computer Vision and Pattern Recognition 2, 807–814 (2005)
37. Hirschmüller, H.: Stereo processing by semiglobal matching and mutual information. IEEE Trans. Pattern Analysis and Machine Intelligence 30(2), 328–341 (2008)
38. Hirschmüller, H., Scharstein, D.: Evaluation of cost functions for stereo matching. Proc. IEEE Conf. Computer Vision and Pattern Recognition pp. 1–8 (2007)
39. Hirschmüller, H., Scharstein, D.: Evaluation of stereo matching costs on images with radiometric differences. IEEE Trans. Pattern Analysis and Machine Intelligence 31(9), 1582–1599 (2009)
40. Hosni, A., Bleyer, M., Rhemann, C., Gelautz, M., Rother, C.: Real-time local stereo matching using guided image filtering. Proc. IEEE Intl. Conf. Multimedia and Expo pp. 1–6 (2011)
41. Hosseini, F., Fijany, A., Safari, S., Fontaine, J.: Fast implementation of dense stereo vision algorithms on a highly parallel SIMD architecture. Journal of Real-Time Image Processing, Springer pp. 1–15 (2011)
42. Humenberger, M., Zinner, C., Kubinger, W.: Performance evaluation of a census-based stereo matching algorithm on embedded and multi-core hardware. Proc. Intl. Symp. Image and Signal Processing and Analysis pp. 388–393 (2009)
43. Ivanchenko, V., Shen, H., Coughlan, J.: Elevation-based MRF stereo implemented in real-time on a GPU. Workshop Applications of Computer Vision pp. 1–8 (2009)
44. Jiangbo, L., Rogmans, S., Lafruit, G., Catthoor, F.: Real-time stereo correspondence using a truncated separable laplacian kernel approximation on graphics hardware. Proc. IEEE Intl. Conf. Multimedia and Expo pp. 1946–1949 (2007)
45. Jiangbo, L., Zhang, K., Lafruit, G., Catthoor, F.: Real-time stereo matching: a cross-based local approach. Proc. IEEE Intl. Conf. Acoustics, Speech and Signal Processing pp. 733–736 (2009)
46. Jin, S., Cho, J., Pham, X.D., Lee, K.M., Park, S.K., Kim, M., Jeon, J.W.: FPGA design and implementation of a real-time stereo vision system. IEEE Trans. Circuits and Systems for Video Technology 20(1), 15–26 (2010)
47. Kalarot, R., Morris, J.: Comparison of FPGA and GPU implementations of real-time stereo vision. Proc. IEEE Conf. Computer Vision and Pattern Recognition Workshop pp. 9–15 (2010)
48. Kalarot, R., Morris, J., Gimel'farb, G.: Performance analysis of multi-resolution symmetric dynamic programming stereo on GPU. Proc. Intl. Conf. Image and Vision Computing New Zealand pp. 1–7 (2010)
49. Kanade, T., Yoshida, A., Oda, K., Kano, H., Tanaka, M.: A stereo machine for video-rate dense depth mapping and its new applications. Proc. IEEE Conf. Computer Vision and Pattern Recognition pp. 196–202 (1996)
50. Ke, Z., Jiangbo, L., Qiong, Y., Lafruit, G., Lauwereins, R., van Gool, L.: Real-Time and Accurate Stereo: A Scalable Approach With Bitwise Fast Voting on CUDA. IEEE Trans. Circuits and Systems for Video Technology 21(7), 867–878 (2011)

51. Kim, J., Kolmogorov, V., Zabih, R.: Visual correspondence using energy minimization and mutual information. Proc. IEEE Intl. Conf. Computer Vision pp. 1033–1040 (2003)
52. Konolige, K.: Small vision systems: Hardware and implementation. Proc. Intl. Symp. Robotic Research (1997)
53. Kung, M., Au, O., Wong, P., Chun, H.L.: Block based parallel motion estimation using programmable graphics hardware. Proc. Intl. Conf. Audio , Language and Image Processing pp. 599–603 (2008)
54. Lee, S.H., Yi, J., Kim, J.S.: Real-time stereo vision on a reconfigurable system. Proc. Intl. Conf. Embedded Computer Systems: Architectures, Modeling, and Simulation Workshops **3553**, 299–307 (2005)
55. Liang, C., Cheng, C., Lai, Y., Chen, L., Chen, H.: Hardware-efficient belief propagation. IEEE Trans. Circuits and Systems for Video Technology **21**(5), 525–537 (2011)
56. Liang, W., Miao, L., Minglun, G., Ruigang, Y., Nister, D.: High-quality real-time stereo using adaptive cost aggregation and dynamic programming. Proc. Intl. Symp. 3D Data Processing, Visualization, and Transmission pp. 798–805 (2006)
57. Liang, W., Mingwei, G., Minglun, G., Ruigang, Y.: How far can we go with local optimization in real-time stereo matching. Proc. Intl. Symp. 3D Data Processing, Visualization, and Transmission pp. 129–136 (2006)
58. Liu, J., Xu, Y., Klette, R., Chen, H., Vaudrey, T.: Disparity Map Computation on a Cell Processor (2009)
59. van der Mark, W., Gavrila, D.: Real-time dense stereo for intelligent vehicles. IEEE Trans. Intelligent Transportation Systems **7**(1), 38–50 (2006)
60. Masrani, D., MacLean, W.: A real-time large disparity range stereo-system using FPGAs. Proc. Intl. Conf. Computer Vision Systems p. 13 (2006)
61. Mattoccia, S., Viti, M., Ries, F.: Near real-time Fast Bilateral Stereo on the GPU. Proc. IEEE Conf. Computer Vision and Pattern Recognition Workshop pp. 136–143 (2011)
62. Mei, X., Sun, X., Zhou, M., Jiao, S., Wang, H., Zhang, X.: On building an accurate stereo matching system on graphics hardware. Proc. IEEE Intl. Conf. Computer Vision Workshops pp. 467–474 (2011)
63. Minglun, G., Ruigang, Y.: Image-gradient-guided real-time stereo on graphics hardware. Proc. Intl Conf. 3D Digital Imaging and Modeling pp. 548–555 (2005)
64. Miyajima, Y., Maruyama, T.: A real-time stereo vision system with FPGA. Proc. Intl. Conf. Field Programmable Logic And Application **2778**, 448–457 (2003)
65. Morris, J., Jawed, K., Gimel'farb, G., Khan, T.: Breaking the 'Ton': Achieving 1% depth accuracy from stereo in real time. Proc. Intl. Conf. Image and Vision Computing New Zealand pp. 142–147 (2009)
66. Mühlmann, K., Maier, D., Hesser, J., Manner, R.: Calculating dense disparity maps from color stereo images, an efficient implementation. Intl. Journal of Computer Vision, Springer **47**(1–3), 79–88 (2002)
67. Pantilie, C., Nedevschi, S.: SORT-SGM: Subpixel optimized real-time semiglobal matching for intelligent vehicles. IEEE Trans. Vehicular Technology **61**(3), 1032–1042 (2012)
68. Park, S., Jeong, H.: Real-time stereo vision FPGA chip with low error rate. Proc. Intl. Conf. Multimedia and Ubiquitous Engineering pp. 751–756 (2007)
69. Paya Vaya, G., Martin Langerwerf, J., Banz, C., Giesemann, F., Pirsch, P., Blume, H.: VLIW architecture optimization for an efficient computation of stereoscopic video applications. Proc. Intl. Conf. Green Circuits and Systems pp. 457–462 (2010)
70. Paya-Vaya, G., Martin-Langerwerf, J., Pirsch, P.: A multi-shared register file structure for VLIW processors. Journal of Signal Processing Systems, Springer **58**(2), 215–231 (2010)
71. Perez, J., Sanchez, P., Martinez, M.: High memory throughput FPGA architecture for high-definition Belief-Propagation stereo matching. Proc. Intl. Conf. Signals, Circuits and Systems pp. 1–6 (2009)
72. Pirsch, P.: Architectures for digital signal processing. John Wiley & Sons, Inc., Chichester (2008)

73. Pock, T., Schoenemann, T., Graber, G., Bischof, H., Cremers, D.: A convex formulation of continuous multi-label problems. Proc. European Conference on Computer Vision **5304**, 792–805 (2008)
74. Podlozhnyuk, V.: Image Convolution with CUDA (2007). Whitepaper included in Nvidia Cuda SDK 4.0
75. Ranft, B., Schoenwald, T., Kitt, B.: Parallel matching-based estimation - a case study on three different hardware architectures. Proc. IEEE Intelligent Vehicles Symposium pp. 1060–1067 (2011)
76. Ruigang, Y., Pollefeys, M.: Multi-resolution real-time stereo on commodity graphics hardware. Proc. IEEE Conf. Computer Vision and Pattern Recognition **1**, I–211–I–217 (2003)
77. Ruigang, Y., Welch, G., Bishop, G.: Real-time consensus-based scene reconstruction using commodity graphics hardware. Proc. Pacific Conf. Computer Graphics and Applications pp. 225–234 (2002)
78. Sabihuddin, S., Islam, J., MacLean, W.: Dynamic programming approach to high frame-rate stereo correspondence: A pipelined architecture implemented on a field programmable gate array. Proc. Canadian Conf. Electrical and Computer Engineering pp. 001,461–001,466 (2008)
79. Safari, S., Fijany, A., Diotalevi, F., Hosseini, F.: Highly parallel and fast implementation of stereo vision algorithms on MIMD many-core Tilera architecture. Proc. IEEE Aerospace Conference pp. 1–11 (2012)
80. Scharstein, D., Szeliski, R.: The Middlebury Stereo Pages. http://vision.middlebury.edu/stereo/
81. Scharstein, D., Szeliski, R.: A taxonomy and evaluation of dense two-frame stereo correspondence algorithms. Intl. Journal of Computer Vision, Springer **47**(1), 7–42 (2002)
82. Scharstein, D., Szeliski, R.: High-accuracy stereo depth maps using structured light. Proc. IEEE Conf. Computer Vision and Pattern Recognition **1**, I–195–I–202 (2003)
83. Szeliski, R.: Computer vision: Algorithms and applications. Springer, London and New York (2011)
84. Szeliski, R., Scharstein, D.: Sampling the disparity space image. IEEE Trans. Pattern Analysis and Machine Intelligence **26**(3), 419–425 (2004)
85. Tomasi, M., Vanegas, M., Barranco, F., Daz, J., Ros, E.: Massive parallel-hardware architecture for multiscale stereo, optical flow and image-structure computation. IEEE Trans. Circuits and Systems for Video Technology **22**(2), 282–294 (2012)
86. Tombari, F., Mattoccia, S., Di Stefano, L., Addimanda, E.: Classification and evaluation of cost aggregation methods for stereo correspondence. Proc. IEEE Conf. Computer Vision and Pattern Recognition pp. 1–8 (2008)
87. Tseng, Y.C., Chang, T.S.: Architecture design of belief propagation for real-time disparity estimation. IEEE Trans. Circuits and Systems for Video Technology **20**(11), 1555–1564 (2010)
88. Ttofis, C., Hadjitheophanous, S., Georghiades, A., Theocharides, T.: Edge-directed hardware architecture for real-time disparity map computation. IEEE Trans. Computers **PP**(99), 1 (2012)
89. Ttofis, C., Theocharides, T.: Towards accurate hardware stereo correspondence: A real-time FPGA implementation of a segmentation-based adaptive support weight algorithm. Proc. Conf. Design, Automation & Test in Europe pp. 703–708 (2012)
90. Vaish, V., Levoy, M., Szeliski, R., Zitnick, C., Sing, B.K.: Reconstructing occluded surfaces using synthetic apertures: Stereo, focus and robust measures. Proc. IEEE Conf. Computer Vision and Pattern Recognition **2**, 2331–2338 (2006)
91. Villalpando, C., Morfopolous, A., Matthies, L., Goldberg, S.: FPGA implementation of stereo disparity with high throughput for mobility applications. Proc. IEEE Aerospace Conference pp. 1–10 (2011)
92. Weber, M., Humenberger, M., Kubinger, W.: A very fast census-based stereo matching implementation on a graphics processing unit. Proc. IEEE Intl. Conf. Computer Vision Workshops pp. 786–793 (2009)

93. Woodfill, J., Gordon, G., Buck, R.: Tyzx DeepSea high speed stereo vision system. Proc. IEEE Conf. Computer Vision and Pattern Recognition Workshop p. 41 (2004)
94. Woodfill, J., Gordon, G., Jurasek, D., Brown, T., Buck, R.: The Tyzx DeepSea G2 vision system, a taskable, embedded stereo camera. Proc. IEEE Conf. Computer Vision and Pattern Recognition Workshop p. 126 (2006)
95. Woodfill, J., Herzen, B.v.: Real-time stereo vision on the PARTS reconfigurable computer. Proc. IEEE Symp. FPGAs for Custom Computing Machines pp. 201–210 (1997)
96. Xu, Y., Chen, H., Klette, R., Liu, J., Vaudrey, T.: Belief propagation implementation using CUDA on an Nvidia GTX 280. Proc. Advances in Artificial Intelligence **5866**, 180–189 (2009)
97. Yang, Q., Wang, L., Yang, R., Wang, S., Liao, M., Nister, D.: Real-time global stereo matching using hierarchical belief propagation. Proc. The British Machine Vision Conference pp. 989–998 (2006)
98. Yunde, J., Xiaoxun, Z., Mingxiang, L., Luping: A miniature stereo vision machine (MSVM-III) for dense disparity mapping. Proc. IEEE Intl. Conf. Pattern Recognition **1**, 728–731 (2004)
99. Zabih, R., Woodfill, J.: Non-parametric local transforms for computing visual correspondence. Proc. European Conference on Computer Vision pp. 151–158 (1994)
100. Zach, C., Klaus, A., Hadwiger, M., Karner, K.: Accurate dense stereo reconstruction using graphics hardware. Proc. EUROGRAPHICS pp. 227–234 (2003)
101. Zach, C., Sormann, M., Karner, K.: Scanline optimization for stereo on graphics hardware. Proc. Intl. Symp. 3D Data Processing, Visualization, and Transmission pp. 512–518 (2006)
102. Zatt, B., Shafique, M., Bampi, S., Henkel, J.: Multi-level pipelined parallel hardware architecture for high throughput motion and disparity estimation in Multiview Video Coding. Proc. Conf. Design, Automation & Test in Europe pp. 1–6 (2011)
103. Zhang, Z.: Determining the epipolar geometry and its uncertainty: A review. Intl. Journal of Computer Vision, Springer **27**(2), 161–195 (1998)
104. Zhang, Z.: A flexible new technique for camera calibration. IEEE Trans. Pattern Analysis and Machine Intelligence **22**(11), 1330–1334 (2000)

Multicore Systems on Chip

Luigi Carro and Mateus Beck Rutzig

Abstract This chapter discusses multicore architectures for DSP applications. We explain briefly the main challenges involved in future processor designs, justifying the need for thread level parallelism exploration, since instruction-level parallelism is becoming increasingly difficult and unfeasible to explore given a limited power budget. We discuss, based on an analytical model, the tradeoffs on using multiprocessor architectures over high-end single-processor design regarding performance and energy. Hence, the analytical model is applied to a traditional DSP application, illustrating the need of both instruction and thread exploration on such application domain. Some successful MPSoC designs are presented and discussed, indicating the different trend of embedded and general-purpose processor market designs. Finally, we produce a thorough analysis on hardware and software open problems like interconnection mechanism and programming models.

1 Introduction

Industry competition in the current wide electronics market makes the design of a device increasingly complex. New marketing strategies have been focusing on increasing the product functionalities to attract consumer's interest. However, the convergence of different functions in a single device produces new design challenges, making the device implementation even more difficult. To worsen this scenario, designers should handle well known design constraints as energy consumption and process costs, all mixed in the difficult task to increase the

L. Carro (✉)
Federal University of Rio Grande do Sul, Bento Gonçalves 9500, Porto Alegre, Brazil
e-mail: carro@inf.ufrgs.br

M.B. Rutzig
Federal University of Santa Maria, Av. Roraima 1000, Santa Maria, Brazil
e-mail: mateus@inf.ufsm.br

S.S. Bhattacharyya et al. (eds.), *Handbook of Signal Processing Systems*,
DOI 10.1007/978-1-4614-6859-2_17, © Springer Science+Business Media, LLC 2013

processing capability, since users desire nowadays the equivalent of a portable supercomputer. Therefore, the fine tuning of these requirements is fundamental to produce the best possible product, which consequently will obtain a wider market.

The convergence of functionalities to a single product enlarges the application software layer over the hardware platform, increasing the range of heterogeneous code that the processing element should handle. The clear examples of such convergence with mixed behavior are iPhone and Android phones. Aiming to balance manufacturing costs and to avoid overloading of the original hardware platform with extra processing capabilities, way beyond the original hardware design scope, there is a natural lifecycle that guides the functionality execution scheme in the platform. Here, the functionality lifecycle is divided into three phases: introduction, growth and maturity. Commonly, during the introduction phase, which reflects the time when a novel functionality is launched in the market, due to doubts about the consumer acceptance, the logical behavior of the product is described using well known high level software languages like C++, Java and .NET. This execution strategy avoids costs, since the target processing element, usually a general-purpose processor (GPP), is the very same of (or very close to) the previous product. After market consolidation the growth and maturity phase start. At this time, the functionality is used in a wide range of products, and its execution tends to be closer to the hardware to achieve better energy efficiency, speedup, and to avoid overloading in the general purpose processor.

Generally, two methods are used to approach the required functionality of a product to the underlying hardware, aiming to explore the benefits of a joint development. The first technique evaluates small applications parts, which possibly have huge impact in the whole application execution. For example, this used to be the scenario of embedded systems some time ago, or of some very specialized applications in general-purpose processors. After the profiling and evaluation phase, chosen code parts are moved to specialized hardwired instructions that will extend the processor instruction set architecture (ISA) and assist a delimited range of applications. MMX, SSE and 3DNow! are successful ISA extensions that have been created aiming to support certain application domains, in those cases, multimedia processing.

A second technique uses a more radical approach to close the gap between the hardware and the required functionality. Its entire logic behavior is implemented in hardware, aiming to build an application specific integrated circuit (ASIC). ASIC development can be considered a better design solution than ISA extensions, since it provides better energy efficiency and performance.

Due to several reasons the current scenario of the electronics market is changing, since over 1.5 billion of embedded devices were shipped in 2011 [20] showing an increasing by 11% in comparison with 2010. Besides the already explained drawbacks regarding the traditional product manufacturing, such designs need to worry about battery life, size and, for critical applications, reliability. Aiming to achieve their hard requirements, system-on-a-chip (SoC) is largely used in the embedded domain. The main proposal of a SoC design is to integrate in a single die the processing element (in most cases a GPP is employed), memory,

Fig. 1 Different MPSoC approaches

communication, peripheral and external interfaces. This eases the deployment of a product, by freeing designers from the hard system validation task. The mobileGT SoC platform [6] exemplifies a successful employment of a SoC for the embedded domain. BMW, Ford, General Motors, Hyundai and Mercedes Benz use the cited SoC in their cars to manage the information and entertainment applications.

However, SoC designs do not sustain the required performance for the growing market of heterogeneous software behaviors, which has been running on top of the available GPP processor. Moreover, the instruction level parallelism (ILP) exploration is no longer an efficient technique, in terms of energy consumption, to improve performance of those processors, due the limited ILP available in the application code [22]. Despite the great advantages shown on ISA extensions employment, like the MMX strategy, this approach relies on a high design and validation time, which goes against the need of a fast time-to-market of embedded systems. ASIC employment usually depends on a complex and costly design process that could also affect the time-to-market required by the company. In addition, this architecture attacks only a specific application domain, and can fail to deliver the required performance to software behaviors that are out of this domain. However, both ASIC and ISA extension are supplied by their high-performance under a limited application domain.

This critical scenario dictates the need for changes on the hardware platform development paradigm. A new organization that appears as a possible solution for the current design requirements is the Multiprocessor System-on-chip (MPSoC). Many advantages can be obtained by combining different processing elements into a single die. The computation time can clearly benefit since, at the same time, several different programs could be executed in the available processing elements. In addition, the flexibility to combine different processing elements appears as a solution to the heterogeneous software execution problem, since designers can select the set of processing elements that best fit in their design requirements. Figure 1 illustrates three examples of MPSoC with different set of processing elements. MPSoC #1 is composed of only general-purpose processors (GPP) with the same processing capability. In contrast, the MPSoC #2 aggregates GPP with distinct computation capability. As another example, MPSoC #3 is assembled to another computation purpose, since it contains three different processing paradigms. More details about each MPSoC approach are explained in the next sections.

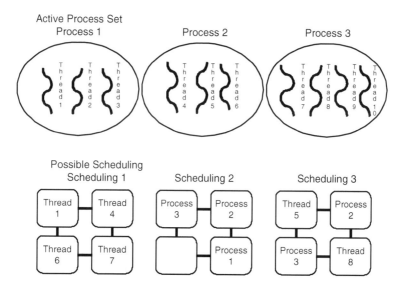

Fig. 2 Scheduling methodologies for a given active process set

MPSoC provides several advantages, and three highlight among all: performance, energy consumption and validation time. MPSoC design introduces a new parallel execution paradigm aiming to overcome the performance barrier created for the instruction level parallelism limitation. In contrast to the ILP exploration paradigm that can be supported both for hardware or software, current coarser parallelism detection strategies are made only for the software team. The hardware team is only responsible for encapsulating and making the communication infrastructure to build a single die with a suitable number of processing elements. The software team splits and distributes the paralyzed code among the MPSoC processing elements. Although both teams might work together, the reality is that there is no strong methodology nowadays that can support software development for multiprocessing chips for any kind of application.

We define parallel code as composed of threads or processes. A thread is contained inside a process, and differs from this by its lightweight context information. Moreover, a process has a private address space, while a set of threads shares the same address space used to make communication with each other in an easier fashion. Figure 2 illustrates an active process set environment composed of three threaded processes. Scheduling 1 reflects the traditional Thread-Level Parallelism (TLP), exploiting the parallel execution of active processes parts. Scheduling 2 methodology supports only single-threaded parallelism exploitation, in Process-Level Parallelism (PLP) only processes are scheduled, regardless of their internal thread number. Finally, there is mixed parallelism exploitation, illustrated by Scheduling 3, working both in thread and process scheduling levels. The last approach is widely used in the commercial MPSoC due to its greater flexibility on scheduling methodology. The scheduling policies rely on the MPSoC processing elements set (Fig. 1). Therefore, the above scheduling approaches can be associated with the previous MPSoC processing element sets. Supposing that all GPP of MPSoC #1 and MPSoC #2 have the same ISA, Scheduling 1 methodology only fits in this hardware modeling due the binary compatibility of threads code. In contrast, the three examples of MPSoC illustrated in Fig. 1 support the Scheduling 2 and 3.

Commonly, parallel execution in multiprocessing platforms is not only used to achieve better speedup, but energy consumption is a big reason for its usage as well. Several factors, both in hardware and software level, contribute to the lower energy consumption of MPSoC architectures. Considering the architecture design, there is no dedicated hardware to split and distribute the application code over the processing elements, since the software team does such work during application development time. A responsible for software distribution is needed, and this is the main problem.

Embedded platforms employ several techniques to save energy in multiprocessing designs. Dynamic voltage and frequency scaling (DVFS) is applied at the entire chip, changing, at runtime, the global MPSoC voltage and frequency [26]. The management policy is based on some system tradeoff like processing elements load, battery charge or chip temperature. Commonly, DVFS is controlled by software, which provides unsuitable time overhead to change the processing element status. Nevertheless, DVFS can affect the performance and even cause drawbacks on real-time based systems. DVFS can also be applied independently on each processing element. Software team during profiling can insert frequency scaling instructions into code, to avoid hardware monitoring for DVFS. Hence, the computational load of each processor is individually monitored, making chip power management more accurate [2, 18]. More recently, aiming to decrease the delay of power management based on software, Core i7, a multiprocessing platform from Intel, employs a dedicated built-in microcontroller to handle DVFS [9, 31]. The built-in chip power manager frequently looks at the temperature and power consumption aiming to turn off independent cores when they are not being used. The DVFS microcontroller can be seen as an ASIC that supplies only the power management routine, which makes clear the convergence of dedicated processing elements to a single die.

Consumers are always waiting for products equipped with exciting features. Time-to-market appears as an important consumer electronics constraint that should be carefully handled. Researches explain that 70% of the design time is spent in the platform validation [1] being an attractive point for optimization. Regarding this subject, MPSoC employment softens the hard task to shrink time-to-market. Commonly, an MPSoC is built by the combination of validated processing elements that are aggregated into a single die as a puzzle game. Since each puzzle block reflects a validated processing element, the remaining design challenge is to assemble the different blocks. Actually, the designer should only select a communication mechanism to connect the entire system, easing the design process by the use of standard communication mechanisms.

Up to now, only explanations about hardware characteristics of MPSoC were mentioned. However, software partitioning is a key feature in an MPSoC system. Considering software behavior, a computational powerful MPSoC becomes useless

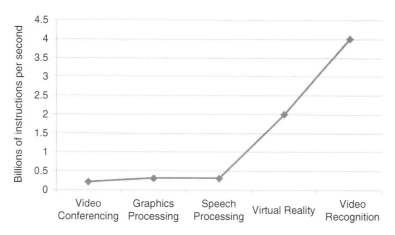

Fig. 3 Throughput requirements of traditional DSP applications, adapted from [7]

if the application partitioning is poorly performed, or if application does not provide a minimum thread or process level parallelism to be explored by the computational resources. Amdahl's law argues that an application speedup is limited by its sequential part. Therefore, if an application needs 1 h to execute, being 5 min sequential (almost 9% of entire application code), the maximum speedup provided for a multiprocessing system is 12 times, no matter how many processing elements are available. Therefore, the designer ability to perform software partitioning and the limited parallel code are the main challenges on an MPSoC design development.

Today, we are usually in touch with functionalities like video conferencing, graphics and speech processing. These functionalities are widespread in digital cameras, MP3, DVD, cell phones and generic communication devices. To support their executions several DSP algorithms are running inside of these devices. Fast Fourier Transform (FFT), Finite Impulse Response (FIR), Discrete Cosine Transform (DCT) and IEEE 802.11 communication standard are examples of these algorithms. The growing convergence of different functionalities to a single device has been increasing, and it is difficult for a single hardware platform to reach the performance requirements of these applications.

Many DSP functionalities and their throughput requirements are shown in Fig. 3. Let us suppose a mobile device composed by all the functionalities shown in Fig. 3, and also let us suppose that they are running on a 1 GHz 8-issue superscalar processor. The processor throughput needed to support the execution is almost seven billion instructions per second. The application execution scenario becomes hard for a general-purpose processor, since the maximum performance of the baseline processor reaches eight billion of instructions per second (with perfect ILP exploitation and perfect branch prediction). In this way, DSP processors are always present in real life products, aiming to achieve, with their specialized DSP instructions, the necessary performance for the DSP applications. However, even DSP processors are suffering to reach the high performance required when many

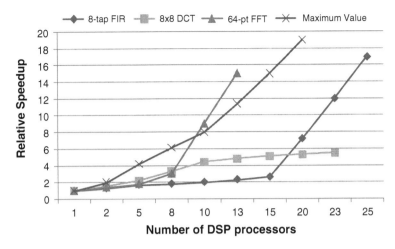

Fig. 4 Speedup of traditional DSP application on MPSoC system [8]

applications are deployed. Hence, the multiprocessing system appears, also in the DSP domain, as a solution to supply the high performance requirements. The major advantage of multiprocessing elements on the DSP domain is the inherent Thread-Level Parallelism, not present in most traditional applications [15].

Figure 4 illustrates the speedup provided by a multiprocessing system for DSP applications. The speedup for the Maximum Value application increases almost linearly with the number of processor. The eight-tap FIR and 64-pt FFT also show optimistic speedups when the number of processing elements increases. On the other hand, the 8 × 8 Discrete Cosine Transform provides the smallest speedup among the application workload. Clearly, this application is a good example for Amdahl's law, demonstrating that multiprocessing systems can fail to accelerate applications that have a meaningful sequential part.

The rest of this chapter is divided into five major sections. First, a more detailed analytical modeling is provided to show the actual design space of multiprocessing devices in the DSP field. Next, an MPSoC hardware taxonomy is presented, aiming to explain the different architectures and applicability domain. MPSoC systems are classified taking into account two points of view: architecture and employed organization. A dedicated section demonstrates some successful MPSoC designs. A section discusses open problems in MPSoC design, covering from software development, communication infrastructure. Finally, future challenges like how can one overcome Amdahl's law limitations are covered.

2 Analytical Model

In this sub-section, we try to figure out the potential of single parallelism exploitation by modeling a multiprocessing architecture (*MP-MultiProcessor*) composed of

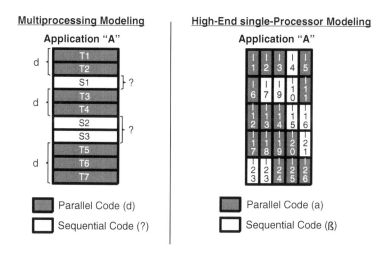

Fig. 5 Modeling of the (**a**) multiprocessor system and the (**b**) high-end single-processor

many simple and homogeneous cores without any capability to explore instruction level parallelism (ILP) to elucidate the advantages of Thread Level Parallelism (TLP) exploitation. We also compare its Execution Time (ET) to a high-end single processor *SHE* (*Single High-End*) model, which is able to exploit only ILP available in applications.

We have considered different amounts of fine- (instruction) and coarse- (thread) level parallelism available in the application code to investigate the performance potentials of the both aforementioned architectures.

Considering a part of a given application code, we classify it in four different ways:

- α – the instructions that can be executed in parallel in a single processor;
- β – the instructions that cannot be executed in parallel in a single processor;
- δ – the amount of instructions that can be distributed among the processors of the multiprocessor environment.
- γ – the amount of instructions that cannot be split, and therefore must be executed in one of the processors among those in the multiprocessor environment.

Figure 5 exemplify how the previously stated classification, considering a certain application "A", would be applied. In the example shown in this figure, when the application is executed in the multiprocessor system (Fig. 5a), 70% of the application code can be parallelized at some degree (i.e. divided in threads) and executed on different cores at the same time, so $\delta = 0.7$ and $\gamma = 0.3$. On the other hand, when the very same application "A" is executed on the high-end single-processor (Fig. 5b), in 64% of the application code instructions can be executed in parallel at some degree, so $\alpha = 0.64$ and $\beta = 0.36$.

2.1   Performance Comparison

Let us start with the basic equation relating Execution Time (ET) with instructions,

$$ET = Instructions * CPI * CycleTime, \tag{1}$$

where *CPI* is the mean number of cycles necessary to execute an instruction, and *Cycletime* is the clock period of the processor.

In this model, information about cache accesses, performance of the disk or any other I/O operation is not considered. However, although simple, this model can provide interesting performance clues on the potential of multiprocessing architectures and aggressive instruction level parallelism exploitation for a wide range of different applications classes.

2.1.1   Low End Single Processor

Based on Eq. (1), for a Low-End Single *SLE* (*Single Low End*) processor, the execution time can be written as:

$$ET_{SLE} = Instructions \left(\propto CPI_{SLE} + \beta CPI_{SLE} \right) CycleTime_{SLE}. \tag{2}$$

Since the low-end processor is a single-issue processor, it is not able to exploit ILP. Therefore, classifying instructions in α and β as previously stated does not make much sense. In this case, α is always equal to zero and β equal to one, but we will keep the notation and their meaning for comparison purposes.

2.1.2   High End Single Processor

In the case of a high-end ILP exploitation architecture, based on Eqs. (1) and (2), one can state that ET_{SHE}(*Execution Time of the High End Single Processor*) is given by the following equation:

$$ET_{SHE} = Instructions \left(\propto CPI_{SHE} + \beta \ CPI_{SLE} \right) CycleTime_{SHE}. \tag{3}$$

As already explained, coefficients α and β refer to the percentage of instructions that can be executed in parallel or not (this way, $\alpha + \beta = 1$), respectively. *CycleTime_{SHE}* represents the clock cycle time of the high-end single processor.

The CPI_{SHE} is usually smaller than 1, because a single high-end processor can exploit high levels of ILP, thanks to the replication of functional units, branch prediction, speculative execution, mechanisms to handle false data dependencies and so on. A typical value of CPI_{SHE} for a current high-end single processor is 0.62

[8], which shows that more than one instruction can be issued and executed per cycle. The CPI$_{SHE}$, could also be written as $\frac{\propto CPI_{SLE}}{issue}$, where *Issue* is the number of instructions that can be issued in parallel to the functional units, when considering the average situation (i.e. a High-End Single processor would have the same CPI as the CPI of a Low-End Processor divided by the mean number of instructions issued per cycle). Thus, based on Eq. (3), one gets:

$$ET_{SHE} = Instructions \left(\frac{\propto CPI_{SLE}}{issue} + \beta \ CPI_{SLE} \right) CycleTime_{SHE}. \qquad (4)$$

Having stated the equation to calculate the performance of both high-end and low-end single processor models, now the potential of using a homogeneous multiprocessing architecture to exploit TLP is studied. Because we are considering that such architecture is built by the replication of low-end processors (so that a large number of them can be integrated within the same die), a single low-end processor does not have any capability to exploit the available ILP of each thread.

If one considers that each application has a certain number of sequences of instructions that can be split (transformed to threads) to be executed on several processors, one could write the following equation, based on Eqs. (1) and (2):

$$ET_{MP} = Instructions \left(\frac{\delta}{P} + \gamma \right) (\alpha CPI_{SLE} + \beta CPI_{SLE}) \ CycleTime_{MP}, \qquad (5)$$

where δ is the amount of sequential code that can be parallelized (i.e. transformed into multithreaded code), while γ is the part of the code that must be executed sequentially (so no TLP is exploited). P is the number of low-end processors that is available in the chip. As can be observed in the second term of the Eq. (5), because the single low end processor is considered, the multiprocessor architecture does not exploit ILP ($\alpha = 0$ and $\beta = 1$). Therefore, when one increases the number of processors P, only the part of code that presents TLP (δ) will benefit from the extra processors.

2.2 High-End Single Processor Versus Homogeneous Multiprocessor Chip

Based on the above reasoning, now we compare the performance of the high-end single processor to the multiprocessor architecture. Since power is crucial in an embedded system design, we have chosen a certain total power budget as a fair performance factor to compare both designs. Thus, based on Eqs. (3) and (5), one can consider the following equation:

$$\frac{ET_{SHE}}{ET_{MP}} = \frac{\left[Instructions \left(\propto \frac{CPI_{SLE}}{issue} + \beta CPI_{SLE}\right) CycleTime_{SHE}\right]}{\left[Instructions \left(\frac{\delta}{P} + \gamma\right) \left(\propto CPI_{SLE} + \beta CPI_{SLE}\right) CycleTime_{MP}\right]}; \quad (6)$$

If one considers that in the model of the multiprocessor environment, a single low end processor is not capable of exploiting instruction level parallelism, and then $\propto = 0$, one can reduce the Eq. (6) to:

$$\frac{ET_{SHE}}{ET_{MP}} = \frac{\left[Instructions \left(\propto \frac{CPI_{SLE}}{issue} + \beta CPI_{SLE}\right) CycleTime_{SHE}\right]}{\left[Instructions \left(\frac{\delta}{P} + \gamma\right) \left(0 * CPI_{SLE} + 1 * CPI_{SLE}\right) CycleTime_{MP}\right]}, \quad (7)$$

and, by simplifying (7), one gets

$$\frac{ET_{SHE}}{ET_{MP}} = \frac{\left[Instructions \left(\propto \frac{CPI_{SLE}}{issue} + \beta CPI_{SLE}\right) CycleTime_{SHE}\right]}{\left[Instructions \left(\frac{\delta}{P} + \gamma\right) \left(CPI_{SLE}\right) CycleTime_{MP}\right]}. \quad (8)$$

We are also considering that, as a homogeneous multiprocessor design is composed of several low-end processors with a very simple organization, those processors could run at much higher frequencies than a single and complex high-end processor. Therefore, we will assume that

$$\left(\frac{1}{CycleTime_{MP}}\right) = K * \left(\frac{1}{CycleTime_{SHE}}\right), \quad (9)$$

where K is the frequency adjustment factor to equal the power consumption of the homogeneous multiprocessor with the high-end single processor.

By merging and simplifying Eqs. (8) and (9), one gets:

$$\frac{ET_{SHE}}{ET_{MP}} = \left[\frac{1}{\frac{\delta}{P} + \gamma}\right] \left[\frac{\propto \frac{CPI_{SLE}}{issue} + \beta CPI_{SLE}}{CPI_{SLE}}\right] K. \quad (10)$$

According to Eq. (10), a machine based on a high-end single core will be faster than a multiprocessor-based machine if $\left(\frac{ET_{SHE}}{ET_{MP}}\right) < 1$. This equation also shows that, although the multiprocessor architecture with low-end simple processors could have a faster cycle time (by a factor of K), that factor alone is not enough to define performance, as demonstrated in the second term between brackets in Eq. (10). Because the high-end processor can execute many instructions in parallel, better performance improvements can be obtained, as long as ILP is the dominant factor, instead of TLP.

To better illustrate this point, let us imagine the extreme case: $P = \infty$, meaning that infinite processors are available. In addition, if one considers that the multiprocessor design is composed of single low end processors that do not exploit ILP and, therefore, $\propto CPI_{SLE}$ is always equal to zero, it can be removed from the equation. Therefore, Eq. (10) reduces to:

$$\frac{ET_{SHE}}{ET_{MP}} = \left[\frac{\propto \frac{CPI_{SLE}}{issue} + \beta CPI_{SLE}}{\gamma CPI_{SLE}} \right] K. \qquad (11)$$

Let us consider that the execution of the very same application on both multiprocessor and single high-end architectures presents exactly the same amount of sequential code, so $\beta = \gamma$. In this case, the operating frequency (given by the K factor) will determine which architecture runs faster if the issue width of the high-end superscalar processor also tends to infinite.

In another example, if one applies the same Eq. (11) in a scenario where an application presents 10% of sequential code ($\beta = \gamma = 0.1$) and is executing on a four issue high-end single processor, the operating frequency of the four issue high-end single processor should be only 20% ($K = 0.8$) greater than the multiprocessor to achieve the very same execution time. On the other hand, if that the application now presents 90% of sequential code ($\beta = \gamma = 0.9$), the high-end single processor should run 3.2 times ($K = 0.31$) faster than the multiprocessor design. With these corner cases, one can conclude that when applications present small parts of its code that cannot be parallelized, both architectures running at the same operating frequency will present almost the same performance regardless the number of processors in a multiprocessor system. For applications with huge amount of sequential code, complex single processors must run at higher frequencies than multiprocessors systems to achieve the same performance.

2.3 Applying the Analytical Model in Real Processors

Given the analytical model, one can briefly experiment it with numbers based on real data. Let us consider a high-end single core: a 4-issue SPARC64 superscalar processor with CPI equal to 0.62 [8]; and a multiprocessor design composed of low-end single-issue TurboSPARC processors with CPI equal to 1.3 [8]. A comparison between both architectures is done using the equations of the aforementioned analytical model. In addition, we consider that the TurboSPARC has 5,200,000 transistors [16], and that the SPARC64 V design [24] requires 180,000,000 transistors to be implemented. For the multiprocessing design we add 37% of area overhead due to the intercommunication mechanism [25]. Therefore, aiming to make a fair performance comparison among the high-end single core and the multiprocessor system, we have devised an 18-Core design composed of low-end processors that has the same area of the 4-issue superscalar processor and consumes the same amount of power.

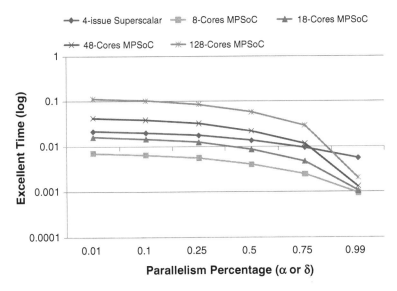

Fig. 6 Multiprocessor system and Superscalar performance regarding a power budget using different ILP and TLP; $\alpha = \delta$ is assumed

Figure 6 shows, in a logarithmic scale, the performance of the superscalar processor, when parameters α and β change, and the performance of the many in-order TurboSPARC cores, when the δ and γ and the number of processors (from 8 to 128) varies. The x-axis of Fig. 6 represents the amount of the instruction- and thread-level parallelism in the application, considering that the α factor is only valid for the superscalar processor, while δ is valid for all the multiprocessing systems' setups.

The goal of this comparison is to demonstrate which technique better explores its particular parallelism type at different levels, considering six values for both ILP and TLP. For instance, $\delta = 0.01$ means that a hypothetic application only shows 1% of TLP available within its code (in the case of the multiprocessing systems). In the same way, when $\alpha = 0.01$, it is assumed that only 1% of the total number of instructions can be executed in parallel on the superscalar processor. In these experiments, we considered the same power budget for the high-end single core and the multiprocessor approaches. In order to normalize the power budget of both approaches we have tuned the adjustment factor K of Eq. (9). For that, we fixed the power consumption of the 4-issue superscalar to use it as the reference, changing the operating frequency (K factor) of the remaining approaches to achieve the same power consumption.

Thus, the operating frequency of the 8-Core multiprocessing system must be 3 times higher than the one of the four-issue superscalar processor. For the 18-Core setup, the operating frequency must be a 25% higher than the reference value. Since a considerable number of cores is employed in the 48-Core setup, it must execute 2 times slower than the superscalar processor to operate under the same power budget.

Finally, the operating frequency of the 128-Core design must be 5.3 times lower than the superscalar.

In the leftmost side of Fig. 6, one considers any application that has a minimum amount of instruction ($\alpha = 0.01$) and thread ($\delta = 0.01$) level parallelism. In this case, the superscalar processor is slower than the 8- and 18-Core designs since the parallelism is insignificant, the higher operating frequency of both multiprocessing system is responsible for faster execution. Moreover, when the application shows higher parallelism levels ($\alpha > 0.25$ and $\delta > 0.25$), the 18- and 8-Core better handles the extra TLP available than the superscalar does with the ILP, presenting more performance. So, considering only the 18-Core design, the multiprocessing system achieve better performance with the same area and power budget in the whole spectrum of parallelism available.

However, as more cores are added in a multiprocessor design, the overall clock frequency tends to decrease, since the adjustment factor K must be decreased to respect the power budget. Therefore, the performance of applications that present low TLP (small δ) worsens when the number of cores increases. Applications with $\delta = 0.01$ in Fig. 6 are good examples of this case: performance is significantly affected as the number of cores increases. As another representative example, even when almost the whole application presents high TLP ($\delta > 0.99$), the 128-Core design takes longer than the other multiprocessor designs. Figure 6 concludes that the increasing on the number of cores not always produces a satisfactory tradeoff among energy, performance and area.

2.4   Energy Comparison

If the superscalar processors usually does not perform well in most cases, when one measures energy consumption, several factors combine to further affect its employment on embedded systems: power in CMOS circuits is proportional to the switching capacitance, to the operating frequency and to the square of the power supply. In this simple model, we will assume that the power is dissipated only in the data path. This is clearly overly optimistic for what regards the power dissipated by a superscalar, but this can also give an idea of the lower bound of energy dissipation in the high-end single processor.

The power dissipated by a high-end single processor can be written as

$$P_{SHE} \approx issue * C * \left(\frac{1}{CPI_{SHE}} \right) * V_{SHE}^2, \tag{12}$$

where C is the capacitance switching of a single issue processor, and V_{ss} is the voltage the processor is operating on. The term $\left(\frac{1}{CPI_{SHE}} \right)$ is included to consider the extra power needed during the speculation process to sustain performance with a CPI smaller than 1. The energy of the high-end single processor is given by:

$$E_{SHE} = P_{SHE} * T_{SHE}. \tag{13}$$

Power consumed by a homogeneous multiprocessing system is given by

$$P_{MP} \approx P * C * \left(\frac{1}{CPI_{SLE}} \right) * V_{MP}^2. \tag{14}$$

Again, as in the case of superscalar processor, the term considering the CPI of the single low-end processor $\left(\frac{1}{CPI_{SLE}} \right)$ has been also included, while its energy is given by

$$E_{MP} = P_{MP} * T_{MP}. \tag{15}$$

When one compares both, it is possible to write

$$\frac{E_{SHE}}{E_{MP}} = issue * \left[\left(\propto \frac{CPI_{SLE}}{issue} + \beta CPI_{SLE} \right) * \frac{1}{CPI_{SHE}} * \frac{1}{K} \right]$$

$$= P * \left[\left(\frac{\delta}{P} + \gamma \right) * (\alpha CPI_{SLE} + \beta CPI_{SLE}) \right] * \frac{1}{CPI_{MP}}, \tag{16}$$

and, by simplifying (16), one gets

$$\frac{E_{SHE}}{E_{MP}} = \frac{V_{SHE}^2 \left[(\propto + issue * \beta) * \frac{1}{K} \right]}{V_{MP}^2 \left[(\delta + P\gamma) \right]}. \tag{17}$$

Equation (17) demonstrates that both approaches unnecessarily spend power when there is no ILP or TLP available since there is no power management technique modeled to reduce power supply (V_{SHE}^2 and V_{MP}^2). Figure 7 shows the energy results considering the same power budget, as it was already done in the performance model. For this first experiment, we do not consider the communication overhead for the multiprocessing environment that will be modeled later. In addition, we only show the energy of 8- and 18-Core Designs, since the conclusions of these setups are also valid for the rest of the setups.

The high-end single processor organization spends higher energy than the 18-Core multiprocessor the same amount of energy when considering all levels of available parallelism since the latter is faster than the former in all cases (Fig. 6).

To obey the given power budget, the 8-Core multiprocessor runs 3 times faster than four-issue superscalar and the 18-Core multiprocessor. Thus, as the 8-Core Design present 3 times lower execution time than the 4-issue superscalar, the former spends 3 times less energy. When the parallelism is more exposed the superscalar approaches to the 8-Core Design, since its execution time decreases. Multiprocessors composed of a significant number of cores present worst performance in applications with low/medium TLP (Fig. 6). Consequently, in those cases and if no power management techniques are considered (e.g., cores are turned off when not used), energy consumption of such multiprocessor designs tend to be higher than those with fewer cores. As can be seen in Fig. 7, the 8-Core multiprocessor

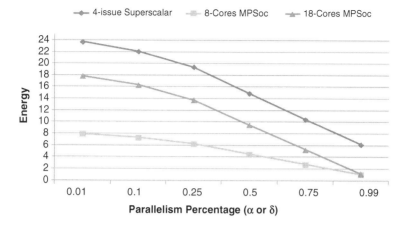

Fig. 7 Multiprocessing Systems and High-end single processor energy consumption; $\alpha = \delta$ is assumed

consumes less energy than the 18-Core for low/medium TLP values ($\delta < 0.75$). However, when applications present greater thread level parallelism ($\delta > 0.9$), the energy consumed by the 18-Core multiprocessor reaches the same values as the 8-Core design, thanks to the better usage of the available processors.

2.5 DSP Applications on a MPSoC Organization

We evaluate the superscalar and MPSoC performance considering an actual DSP application execution. An 18-tap FIR filter is used to measure the performance of both approaches handling a traditional DSP application. The C-like description of the FIR filter employed in this experiment is illustrated in Fig. 8. Superscalar machines explore the FIR filter instruction level parallelism in a transparent way, working on the original binary code. Unlike the superscalar approach, to explore the potential of the MPSoC architecture there is a need to make manual source code annotations in order to split the application code among many processing elements. In this way, some code highlights are shown in Fig. 8 to simulate annotations, indicating the necessary number of cores to explore the ideal thread level parallelism of each part of the FIR filter code. For instance, the first annotation considers a loop controlled for IMP_SIZE value, which depends on the number of FIR taps. In this case, 54 loop iterations are done since the experiment regards an 18-tap FIR filter.

The OpenMP [4] programming language provides specific code directives to easily split loop iterations among processing elements. Using OpenMP directives, the ideal exploration of this loop is done through 54-core MPSoC, each one being responsible for running single loop iteration. However, when the amount of processing elements is lower than the number of loop iterations, OpenMP combines

```
#define NTAPS 18
#define IMP_SIZE (3 * NTAPS)
static const double h[NTAPS] = {1.0, 2.0, 3.0, 4.0, 5.0,
6.0 };
static double h2[2 * NTAPS], z[2 * NTAPS], imp[IMP_SIZE];
double output;
int ii, state;

    /* make impulse input signal */                    54 Cores
    for (ii = 0; ii < IMP_SIZE; ii++) {
        imp[ii] = 0;

    }
    imp[5] = 1.0;
    /* create a SAMPLEd h */                            18 Cores
    for (ii = 0; ii < NTAPS; ii++) {
        h2[ii] = h2[ii + NTAPS] = h[ii];

    }

    /* clear z */                                       18 Cores
    for (ii = 0; ii < NTAPS; ii++) {
        z[ii] = 0;

    }

    for (ii = 0; ii < IMP_SIZE; ii++) {                 54 Cores

        z[0] = imp[ii];
    output = 0;

    /* calc FIR */                                      54 Cores
    for (ii = 0; ii < IMP_SIZE; ii++) {
        output += h[ii] * z[ii];

    }
    /* shift delay line */
    for (ii = IMP_SIZE - 2; ii >= 0; ii--) {
        z[ii + 1] = z[ii];
    }
```

Fig. 8 C-like FIR filter

them in groups to distribute tasks among the available resources. Hence, regarding the execution of 54 loop iterations into 18-core MPSoC, OpenMP creates 18 groups, each one composed of three iterations. Figure 8 demonstrates that almost the entire FIR code can be parallelized, since its code is made up of several loops. In general, traditional DSP applications (ex: FFT and DCT) have a loop-based code behavior suitable for OpenMP loop parallelization. However, some loops cannot be parallelized due data dependency among iterations. For instance, the last loop of the FIR filter description shown in Fig. 8 demonstrates this behavior, since shifting array values presents dependencies among all loop iterations.

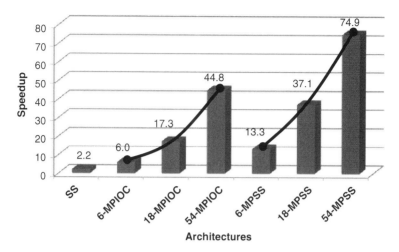

Fig. 9 Speedup provided in 18-tap FIR filter execution for superscalar, MPSoC and a mix of both approaches

Aiming to illustrate the impact on performance of TLP and ILP exploration on DSP applications, we evaluated the 18-tap FIR execution over three different architectures: a four-issue Superscalar (SS); 6- 18- and 54-core MPSoCs based on pipelined cores, with no ILP exploration capabilities (MP_{IOC}). Finally, in order to have a glimpse on the future, we imagined a 6- 18- and 54-Cores MPSoCs based on a four-issue superscalar processor, able to explore both ILP and TLP (MP_{SS}). We have extracted the speedup with a tool [16] that makes all data dependence graphs of the application. After, considering the characteristics of the evaluated architectures, the execution time of each graph is measured in order to obtain their speedup over the baseline processor. It is important to point out that instruction and thread communication overhead has not been taken into account in this experiment.

The results shown in Fig. 9 reflect the speedup provided over a single pipelined core performance running the C-like description of the 18-tap FIR filter presented in Fig. 8. The leftmost bar shows the speedup provided for the ILP exploration of a four-issue superscalar processor. In this case, the execution time of the Superscalar processor is 2.2 times lower than that of a pipelined core, showing that the FIR filter has neither high nor low ILP, since a four-issue superscalar processor could potentially achieve up to 4 times the performance of a pipelined core.

Considering the MPSoC composed of pipelined cores, the 6-core machine provides almost a linear speedup, decreasing by 5.96 times the single pipelined core execution time. This behavior is maintained when more pipelined cores are inserted. However, when 18-tap FIR filter is explored for the maximum TLP (54-MP_{IOC}), a speed up of only 44.8 times is achieved, showing that even applications which are potentially suitable for TLP exploration could present non-linear speedups. This can be explained by the sequential code present inside of each loop iteration.

Amdahl's Law shows that it is not sufficient to build architectures with a large number of processors, since most parallel applications contain a certain amount of sequential code [23]. Hence, there is a need to balance the number of processors with a suitable ILP exploration approach to achieve greater performance. The MP_{SS} approach combines TLP with ILP exploration of four-issue superscalar aiming to show that simple TLP extraction is not enough to achieve linear speedups even for DSP applications with high TLP. Figure 9 illustrates the speedup of the MP_{SS} approach. As it can be noticed, 6-MP_{SS} accelerates the 18-tap FIR filter more than twice 6-MP_{IOC}, since higher ILP is explored for the superscalar processor. In this case, when the first loop of Fig. 8 is split among the MPSoC, each core executes nine loop iterations providing a large room for ILP exploration. However, when the number of cores increases, the sequential code decreases, making less room for the ILP optimization. Nevertheless, the 18-tap FIR filter execution in 54-MP_{SS} is 66% faster than 54-MP_{IOC} execution showing the need for a mixed parallelism exploration.

Summarizing, most DSP applications benefit of thread level parallelism exploration thanks to their loop-based behavior. However, even applications with high TLP could still obtain some performance improvement by also exploiting ILP. Hence, in a MPSoC design ILP techniques also should be investigated to conclude what is the best fit considering the design requirements. Finally, this subsection also shows that replications of simple processing elements leaves a significant optimization possibility unexplored, indicating that heterogeneous MPSoC could be a possible solution to balance the performance of the system.

3 MPSoC Classification

Due to the large amount of application domains that the MPSoC design principle can be applied to, some methods to define the advantages/disadvantages of each one should be defined. Commonly, a MPSoC design is classified only considering its architecture type, as homogeneous or heterogeneous. In this section the classification is done from two points of view: architecture and organization modeling.

3.1 Architecture View Point

When processors are classified under the architecture view point, the instruction set architecture (ISA) is the parameter used to distinguish them. The first step of a processor design is the definition of its basic operation capabilities, aiming to create an ISA to interface those operations to the high level programmer. Generally, an ISA is typically classified as reduced instruction set computer (RISC) or complex instruction set computer (CISC). The strategy of RISC architectures is making the processor data path control as simple as possible. Their instructions set are non-specialized, covering only the basic operations needed to make simple computations. Besides its reduced instruction set, the genuine RISC architectures provide additional features like instructions issued at every clock cycle execution, special instructions to perform memory accesses, register-based operations and a small clock cycle, thanks to their pipelined micro-operations. On the other hand, the CISC architecture strategy aims at supporting all required operations on hardware, building specialized instructions to better support high level software abstractions. In fact, a single CISC instruction reflects the behavior of a several RISC instructions, with different amount of execution cycles and memory accesses. Hence, the data path control becomes more complex and harder to manage than RISC architectures. Both strategies are widely employed in processor architectures, always guided by the requirements and constraints of the design. DSP processors better fit the CISC characteristics, since there are many specialized instructions to provide efficient signal processing execution. Multiply-accumulate, round arithmetic, specialized instructions for modulo addressing in circular buffers and bit-reversed addressing mode for FFT are some examples of specialized CISC instructions focused on accelerating specific and critical parts of the DSP application domain.

In the architecture point of view, there are several ways to combine processing elements inside a die. Figure 1 presents three examples of MPSoC with different processing element arrangement. As can be noticed in this figure, MPSoC #1 is composed of four processing elements with the very same ISA, and this MPSoC arrangement is classified as a homogeneous MPSoC architecture. The same classification is given for the MPSoC #2 that only differs from the first for the diversity on computing time of its processing elements. Details about this organization fashion are given in the next section.

Homogeneous MPSoC architectures have been widespread in general-purpose processing domain. As already explained, the current difficulties found on increasing the applications performance by using ILP techniques, aggregated to the large density provided by the high transistor level integration, encouraged the coupling of several GPP into a single die. Since, in most cases, the baseline processor is already validated, the inter-processor communication design is the unique task to be performed on the development of a homogeneous MPSoC architecture (although this turned out to be not such an easy task). In the software side, programming models are used to split the code among the homogeneous cores, making the hard task of coding parallel software easier. OpenMP [4] and MPI [14] are the most used programming models providing a large number of directives that facilitate such a hard task. More details about programming languages are provided in the last section.

For almost 5 years now, Intel and AMD have been using this approach to speed up their high-end processors. In 2006, Intel has shipped its first MPSoC based on homogeneous architecture strategy. Intel Core Duo is composed of two processing elements that make communication among themselves through an on-chip cache memory. In this project, Intel has thought beyond the benefits of MPSoC employment and created an approach to increase the process yield. A new processor market line, called Intel Core Solo, was created aiming to increase the process yield by selling even Core Duo dies with manufacturing defects. In this way, Intel Core Solo has the very same two-core die as the Core Duo, but only one core is defect free.

Recently, embedded general-purpose RISC processors are following the trend of high-end processors coupling many processing elements, with the same architecture, on a single die. Early, due to the hard constraints of these designs and the few parallel applications that would benefit from several GPP, homogeneous MPSoC were not suitable for this domain. However, the embedded software scenario is getting similar to a personal computer one due the large amount of simultaneous applications running over its GPP. ARM Cortex-A9 MPSoC processor is the pioneer to employ homogeneous multiprocessing approach into embedded domain, coupling four Cortex-A9 cores into a single die. Each processing element uses powerful techniques for ILP exploration, as superscalar execution and SIMD instruction set extensions, which closes the gap between the embedded processor design and high-end general purpose processors.

Heterogeneous MPSoC architectures support different market requirements than homogeneous ones. Generally, the main goal of the homogeneous approach is to provide efficiency on a large range of applications behaviors, being widespread used in GPP processor. On the other hand, heterogeneous approach aims to supply efficient execution on a defined range of application behavior, including some particular ISA extensions to handle specific software domains. MPSoC #3 (Fig. 1) better exemplifies a heterogeneous architecture composed of three different ISAs. This architecture style is widely used in the mobile embedded domain, due to the need for high performance with energy efficiency. The heterogeneous MPSoC assembly depends a lot on the performance and power requirements of the application. Supposing a current cell phone scenario where there are several applications running over the MPSoC. Commonly, designers explore the design space by trying to find critical parts of code that could be efficiently executed in hardware. After, these parts are moved to hardware, achieving better performance and energy efficiency. In an MPSoC used by a current cell phone there are many heterogeneous processing elements providing efficient execution for specific tasks, like video decompression and audio processing.

Samsung S3C6410 better illustrates the embedded domain trend to use MPSoCs. This heterogeneous architecture handles the most widely used applications on embedded devices like multimedia and digital signal processing. Samsung's MP-SoC is based on an ARM 1176 processor, which includes several ISA extensions, such as DSP and SIMD processing, aiming to increase the performance of those applications. Its architecture is composed of many other dedicated processing elements, called hard-wired multimedia accelerators. Video Codec (MPEG4, H.264 and VC1), Image Codec (JPEG), 2D and 3D accelerators have become hard-wired due to the performance and energy constraints of the embedded systems. This drastic design change illustrates the growth and maturity phase of the functionality lifecycle discussed in the beginning of this chapter. In this phase, the electronic consumer market already has absorbed these functionalities, and their hard-wired execution is mandatory for energy and performance efficiency.

Fig. 10 Examples of
homogeneous and
heterogeneous organizations

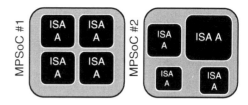

3.2   Organization View Point

In contrast with the previous section, this part of the chapter explores the heterogeneous versus homogeneous arrangement considering the MPSoC organization
as a classification parameter. This means that all MPSoC designs discussed here
have the same ISA, and they are classified by the use of similar/dissimilar processor
organization.

The homogeneous MPSoC organization, as already explained, has been
widespread in general-purpose processing. Intel and AMD replicate their cores into
a single die to explore the process/thread parallelism, moving the responsibility
of process/thread scheduling to the operating system or software team. Also, the
ARM Cortex design has introduced MPSoC in the embedded domain with a
homogeneous MPSoC organization employment. When a simple replication of the
very same processing elements occurs, the design is classified as homogeneous,
both in architecture and organization. Their processing elements have the same
processing capability, area and energy consumption, which delimits the design space
exploration, since no special task scheduling could be employed.

The process/thread scheduling, in an MPSoC design, is an important feature
to be explored, in order to create a tradeoff between the requirements and design
constraints, aiming to find the right balance between performance and power
consumption. Supposing the two approaches illustrated in Fig. 10, MPSoC #1
is composed by replications of the same processing element, being classified as
homogeneous both in architecture and organization. On the other hand, MPSoC
#2, from the organization point of view, demonstrates different processing elements
despite having the same ISA. The last MPSoC fashion is classified as a heterogeneous MPSoC in the organization point of view, and homogeneous MPSoC in
the architecture point of view. MPSoC #1 supports only naive task scheduling
approaches, since all processing elements have the same characteristics. In this
organization fashion, the task is always allocated in the same processing element
organization, meaning that there is no choosing for a smart task scheduling to
achieve certain design requirements, such as energy efficiency. In contrast, MPSoC
#2 provides flexibility on task scheduling. An efficient algorithm could be developed
to support design tradeoffs regarding requirements and constraints of the device.
Supposing a scenario with three threads (thread#1, thread#2, and thread #3) that
require the following performance level: low, very low and very high. The MPSoC
#1 scheduling approach simply ignores the computation requirements, executing
the three threads into their powerful processing elements, causing a waste of

energy resources. An efficient scheduling algorithm can be employed in MPSoC #2 to avoid the previous drawback, exploiting its heterogeneous organization. Supposing a dynamic task scheduling algorithm that saves information about threads execution, one can fit the thread computation need regarding the different processing capabilities available in MPSoC #2 which allows an energy-efficient design. For the given example, thread#1 is executed in the middle-size processing element, thread#2 in the smallest one and thread#3 in the largest processing element.

4 Successful MPSoC Designs

Many successful MPSoC solutions are in the market, for both high-end performance personal computers and highly-constrained mobile embedded devices. Commonly, as already explained, high-end performance personal computers provide robust GPP use with architecture and organization shaped on a homogeneous way due to the large variation of software behavior that runs over these platforms. On the other hand, MPSoCs for mobile embedded devices supply many ISAs, through application specific instruction set processor to provide an efficient execution on a restricted application domain. This subsection delimits MPSoC exploration to embedded domain due their widely usage of DSP processors. Some successful embedded MPSoC platforms are discussed, demonstrating the trends for mobile multiprocessors.

4.1 OMAP Platform

In 2002, Texas Instruments has launched in the market an Innovator Development kit (IDK) targeting high performance and low power consumption for multimedia applications. IDK provides an easy design development, with open software, based on a customized hardware platform called open multimedia applications processor (OMAP). Since its launching, OMAP is a successful platform being used by the embedded market leaders like Nokia with its N90 cell phones series, Samsung OMNIA HD and Sony Ericsson IDOU. Currently, due to the large diversity found on the embedded consumer market, Texas Instruments has divided the OMAP family in two different lines, covering different aspects. The high-end OMAP line supports the current sophisticated smart phones and powerful cell phone models, providing pre-integrated connectivity solutions for the latest technologies (3G, 4G, WLAN, Bluetooth and GPS), audio and video applications (WUXGA), including also high definition television. The low-end OMAP platforms cover down-market products providing older connectivity technologies (GSM/GPRS/EDGE) and low definition display (QVGA).

Launched in 2009, OMAP4440 covers the connectivity besides high-quality video, image and audio support. This mobile platform came to supply the need for

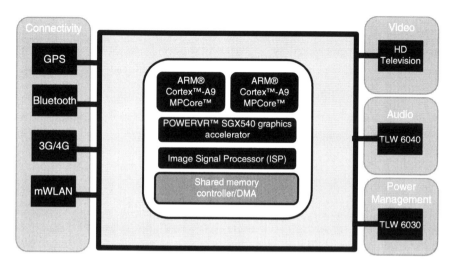

Fig. 11 OMAP4440 block diagram

the increasingly multimedia applications convergence in a single embedded device. As can be seen in Fig. 11, this platform incorporates a dual-core ARM Cortex A9 MPCore providing higher mobile general-purpose computing performance. A novel power management hardware technique available in ARM Cortex A9 MPCore balances the power consumption with the performance requirements, activating only the cores that are needed for a particular execution. Also, due to the high performance requirement of today smart phones, up to eight threads can be concurrently fired in the MPCore, since each core is composed of four single-core Cortex A9. The single-core ARM Cortex A9 implements speculative issue out-of-order superscalar execution, SIMD instruction set and DSP extensions, showing almost the same processing power as a personal computer into an embedded mobile platform. Excluding the ARM Cortex MPCore, the remainder processing elements are dedicated to multimedia execution.

The PowerVR™ SGX540 multi-threaded graphics accelerator employs state-of-art support for 2D and 3D graphics, including OpenGL and EGL APIs. Several image enhancements like digital anti-aliasing, digital zoom and auto-focus are provided by the Image Signal Processor (ISP), which focus on increasing the quality of the digital camera. The processing capabilities of ISP enables picture resolutions of up to 20-megapixel, with an interval of 1 s between shots. The high memory bandwidth needed for ISP is supplied by a DMA controller that achieves a memory-to-memory rate of 200 megapixel/s at 200 MHz. To conclude the robust multimedia accelerators scenario, OMAP4440 allows a true 1080p multi-standard high definition record and playback at 30 frames per second rate, by its IVA3 hardware accelerator. IVA DSP processor supports a broad of multimedia codec standards, including MPEG4 ASP, MPEG-2 MP, H.264 HP, ON2 VP7 and VC-1 AP.

The large amount of OMAP4440 processing elements makes its power dissipation rates unfeasible for mobile embedded devices. Texas Instruments has handled this drawback by integrating on the OMAP 440 design a power management technology. Smartreflex2 supports, both at the hardware and software level, the following power management techniques: Dynamic Voltage and Frequency Scaling (DVFS), Adaptive Voltage Scaling (AVS), Dynamic Power Switching (DPS), Static Leak Management (SLM) and Adaptive Body Bias (ABB). DVFS supports multiple voltage values, aiming to balance the power supply and frequency regarding the performance needed in a given execution time. AVS is coupled to DVFS to provide a greater balance on power consumption, considering hardware features like temperature. The coupling of both techniques delivers the maximum performance of OMAP4440 processing elements with the minimum possible voltage at every computation. Regarding leakage power, the 45 nm OMAP4440 MPSoC die already provides a significant reduction in the consumption. DPS and SLM techniques allow lowest stand-by power mode avoiding leakage power. Finally, ABB works at transistor level enabling dynamic changes on transistor threshold voltages aiming to reduce leakage power. All techniques are supported by the integrated TLW6030 power management chip illustrated in Fig. 11. In order to support energy efficient audio processing, Texas has included the TWL6040 integrated chip that acts as an audio back-end chip, providing up to 140 h of music playback in airplane mode.

Recently, Texas Instrument released its latest high-end product, OMAP543x family platform [27]. Several technological improvements over the OMAP4440 were coupled to OMAP5430. The ARM Cortex-A9 was replaced to ARM Cortex-A15 running at 2 GHz. ARM reports that the former performs 40% faster the latter. Two ARM Cortex-M4 were inserted in the SoC to provide low power consumption when offload computation mode is activated. The multicore PowerVR SGX544-MPx replaced the single core SGX540 graphics accelerator, being capable of encapsulate up to 16 processing elements, now supporting OpenGL and DirectX application programming interfaces. Its 28 nm fabrication process produces lower power consumption than the 45 nm of OMAP4440, which improves the battery life. The OMAP543x family is divided in two target markets: area-sensitive and cost-sensitive. The former includes the OMAP5430 platform and it is targeted to smartphones and tablets encapsulating several imaging interfaces. The latter includes the OMAP5432 platform being focused on mobile computing which requires higher bandwidth and lower interface to handle images. In terms of organization and architecture, Texas Instrument keeps both heterogeneous encapsulating more specialized processing elements to achieve low-power computation.

4.2 Samsung SoCs

As OMAP, Samsung MPSoC designs are focused on multimedia-based development. Their projects are very similar due to the increasing market demand for powerful multimedia platforms, which stimulates the designer to take the same

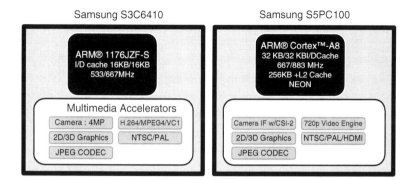

Fig. 12 Samsung S3C6410 and S5PC100 MPSoC block diagrams

decision to achieve efficient multimedia execution. Commonly, the integration of specific accelerators is used, since this reduces the design time avoiding validation and testing time.

In 2008, Samsung has launched the most powerful of the Mobile MPSoC family. At first, S3C6410 was a multimedia MPSoC like OMAP4440. However, after its deployment in the Apple iPhone 3G employment, it has become one of the most popular MPSoCs, shipping three million units during the first month after the launching. Recently, Apple has developed iPhone 3GS, which assures better performance with lower power consumption. These benefits are supplied by the replacement of the S3C6410 architectures with the high-end S5PC100 version. In this subsection, we will explore both architecture highlighting the platform improvements between S3C6410 and S5PC100. The S5PC100 was the last Samsung SoC employed by Apple in the iPhone. After that, Apple encapsulates his own chips in iPhone 4 and 4S, the Apple A4 and Apple A5 chips, respectively. However, these chips still being fabricated by Samsung.

Following the multimedia-based MPSoC trend, Samsung platforms are composed of several application specific accelerators building heterogeneous MPSoC architectures. S3C6410 and S5PC100 have a central general-purpose processing element, in both cases ARM-based, surrounded by several multimedia accelerators tightly targeted to DSP processing. As can be noticed in Fig. 12, both platforms skeleton follows the same execution strategy changing only the processing capability of their IP cores. This is an interesting design modeling approach, since while traditional architecture designs suffer for long design/validate/test phases to build a new architecture, MPSoC ones shorter these design times only by replacing their low-end cores for high-end ones, creating a new MPSoC architecture. This can be verified in the Fig. 12, small platform changes are done from S3C6410 to S5PC100 aiming to increase the MPSoC performance. More specifically, a nine-stage pipelined ARM 1176JZF-S core with SIMD extensions is replaced for a 13-stage superscalar pipelined ARM Cortex A8 providing greater computation capability for general-purpose applications. Besides its double-sized L1 cache compared with ARM1176JZF-S, ARM Cortex A8 also includes a 256KB L2 cache avoiding external memory accesses due L1 cache misses. NEON ARM technology is included in ARM Cortex A8 to provide flexible and powerful acceleration for intensive multimedia applications. Its SIMD based-execution accelerates multimedia and signal processing algorithms such as video encode/decode, 2D/3D graphics, speech processing, image processing at least 2× of previous SIMD technology.

Regarding multimedia accelerators, both MPSoCs are able to provide suitable performance for any high-end mobile devices. However, S5PC100 includes the latest codec multimedia support using powerful accelerators. A 720p multi format codec (MFC) video is included aiming to provide high quality capture and playback at 30 frames per second rate. Also, previous video codec as MPEG4, VCI and H.264 are available in this platform. Besides the S3C6410 TV out interfacing PAL and NTSC, the new MPSoC version includes high definition multimedia interface (HDMI) standard allowing high definition video formats. Summarizing, all technologies included in S5PC100 platform accelerate multimedia applications employing DSP approaches.

In 2010, Samsung launched the first SoC of its *Exynos* platforms, named as *Exydos* 3110 [28], which was employed in the high-end cell phones and tablets developed in such year (i.e. Samsung Galaxy S, Samsung Galaxy Tab). The second SoC of this family, named as *Exynos* 4210 [29], keeps being the most employed SoC in the high-end cell phones of Samsung. The best seller, Galaxy S II (i9100), contain an *Exynos* 4210 composed of dual-core Cortex A9 running at 1.2 GHz coupled to an ARM 4-Core Mali-400MP graphics processing unit. However, the newer version of this cell phone, Galaxy S II (i9100G), changes the *Exynos* 4210 to an already discussed SoC, a 1.2 GHz dual core TI OMAP 4430 with a PowerVR SGX540 GPU.

4.3   Apple SoCs

In March of 2010, Apple has announced its first SoC named as Apple A4, an ARM Cortex-A8 running at 1 GHz and a PowerVR SGX535 GPP were coupled in a single die to push up the performance of the iPAD, iPhone 4 and iTouch. In 2011, Apple introduced the multicore concept in their devices launching Apple A5 SoC. This SoC improves the performance over the Apple A4 by integrating a dual-core ARM Cortex-A9 together with a dual-core PowerVR SGX543MP, iPAD 2 and iPhone 4S explore such powerful performance. Recently, Apple has announced its latest SoC, named as Apple A5X. The changes over the previous version happen only in the GPU. Apple A5X encapsulates a quad-core PowerVR SGX544MP4 which drastically increases its video processing capabilities. The new iPAD, equipped with such engine, achieves a display resolution of 2,048 × 1,536, meaning a million more pixels than the HDTV standard.

4.4   Other MPSoCs

Other MPSoCs have already been released in the market, with different goal from the architectures discussed before. Sony, IBM and Toshiba have worked together to design the Cell Broadband Engine Architecture [3]. The Cell architecture combines a powerful central processor with eight SIMD-based processing elements. Aiming to accelerate a large range of application behaviors, the IBM PowerPC architecture is used as GPP processor. Also, this processor has the responsibility to manage the processing elements surrounding it. These processing elements, named as synergistic processing elements (SPE), are built to support streaming applications with SIMD execution. Each SPE has a local memory, but no hardware is employed to manage it, which avoid that these memories directly access the system memory. These facts make the software development for the Cell processor even more difficult, since the software team should be aware of this local memory, and manage it at the software level to better explore the SPE execution. Despite its high processing capability, the Cell processor does not yet have a large market acceptance due to the intrinsic difficulty to produce software. Sony has lost parts of the gaming entertainment market after the Cell processor was deployed in the Playstation console family, since game developers had not enough knowledge to efficiently explore the complex Cell architecture.

Homogeneous MPSoC organization is also explored in the market, mainly for personal computers with general-purpose processors, because of the huge amount of different applications these processors have to face, and hence it is more difficult to define specialized hardware accelerators. In 2005, Sun Microsystems announced its first homogeneous MPSoC, composed of up to eight processing elements executing the full RISC SparcV9 instruction set. UltraSparc T1, also called Niagara [11], is the first multithreaded homogeneous MPSoC, and each processing element is able to execute four threads concurrently. In this way, Niagara can handle, at the same time, up to 32 threads. Recently, with the deployment of UltraSparc T2 [10], this number has grown to 64 concurrent threads. Niagara MPSoC family targets massive data computation with distributed tasks, like the market for web servers, database servers and network file systems.

Intel has announced its first MPSoC homogeneous prototype with 80 cores, which is capable of executing 1 trillion floating-point operations per second, while consuming 62 W [21]. Hence, the x86 instruction set architecture era could be broken, since their processing elements is based on the very long instruction word (VLIW) approach, letting to the compiler the responsibility for the parallelism exploration. The interconnection mechanism used on the 80-core MPSoC uses a mesh network to communicate among its processing elements [5]. However, even employing the mesh communication turns out to be difficult, due to the great amount of processing elements. In this way, this ambitious project uses a 20 MB stacked on-chip SRAM memory to improve the processing elements communication bandwidth.

Graphic processing unit (GPU) is another MPSoC approach aiming to accelerate graphic-based software. However, this approach has been arising as promise architecture also to improve general-purpose software. Intel Larrabee [17] attacks both applications domain thanks to its CPU- and GPU-like architecture. In this project Intel has employed the assumption of energy efficiency by simple cores replication. Larrabee uses several P54C-based cores to explore general-purpose applications. In 1994, P54C was shipped in CMOS 0.6 μm technology reaching up to 100 MHz and does not include out-of-order superscalar execution. However, some modifications have been done in the P54C architecture, like supporting of SIMD execution aiming to provide more powerful graphic-based software execution. The SIMD Larrabee execution is similar to but powerful than the SSE technology available in the modern x86 processors. Each P54C is coupled to a 512-bit vector pipeline unit (VPU), capable of executing, in one processor cycle, 16 single precision floating point operations. Also, Larrabee employs a fixed-function graphics hardware that performs texture sampling tasks like anisotropic filtering and texture decompression. However, in 2009, Intel discontinued Larrabee project.

NVIDIA Tesla [12] is another example of MPSoC based on the concept of a general-purpose graphic processor unit. Its massive-parallel computing architecture provides support to Compute Unified Device Architecture (CUDA) technology. CUDA, the NVIDIA's computing engine, supports the software developer by easing the application workload distribution and by providing software extensions. Also, CUDA provides permission to access the native instruction set and memory of the processing elements, turning the NVIDIA Tesla to a CPU-like architecture. Tesla architecture incorporates up to four multithreaded cores communicating through a GDDR3 bus, which provides a huge data bandwidth.

In 2011, NVIDIA introduced the project named Tegra 3 [30], also called Kal-el mobile processor. This project is the first to encapsulate four processors in a single die for mobile computation. The main novelty introduced by this project is the Variable Symmetric Multiprocessing (vSMP) technology. vSMP introduces a fifth processor named "Companion Core" that executes tasks a low frequency for active standby mode, as mobile systems tend to keep in this mode for most time. All five processors are ARM Cortex-A9, but the companion core is built in a special low power silicon process. In addition, all cores can be enabled/disabled individually and when the active standby mode is on, only the "Companion Core" works, so battery life can significant improved. NVIDIA reports that the switching from the companion core to the regular cores are supported only by hardware and take less than 2 ms being not perceptible to the end users. In comparison with Tegra 2, NVIDIA previous platform, vSMP achieves up to 61% of energy savings on running HD video playback. Tegra 3 is inside of several tablets developed by Asus and Acer, such as Asus Eee Pad Trasformer Prime, Trasformer Pad Infinity and Acer Iconia.

Summarizing all MPSoC discussed before, Table 1 compares their main characteristics showing their differences depending on the target market domain. Heterogeneous architectures, such as OMAP, Samsung and Cell, incorporate several specialized processing elements to attack specific applications for highly-

Table 1 Summarized MPSoCs

	Architecture	Organization	Cores	Multithreadedcores	Intercon-nection
OMAP 4440	Heterogeneous	Heterogeneous	2 ARM Cortex A91 PowerVR540 Image Signal Processors	Yes (GPU)	Integrated Bus
OMAP543x	Heterogeneous	Heterogeneous	2 ARM Cortex A152 ARM Cortex M41 PowerVR SGX544 Image Signal Processors	Yes (GPU)	Integrated Bus
Samsung S3C6410/S5PC100	Heterogeneous	Heterogeneous	ARM1176JZF-S5 multimedia accelerators	No	Integrated Bus
Samsung Exynos 3110	Heterogeneous	Heterogeneous	1 ARM Cortex A81 PowerVR540	No	Integrated Bus
Samsung Exynos 4210	Heterogeneous	Heterogeneous	2 ARM Cortex A91 ARM Mali 400 MP4	No	Integrated Bus
NVIDIA Tegra 3	Heterogeneous	Heterogeneous	1Quad ARM Cortex-A15 MPCore + companion core1 GeForce GPU	No	Integrated Bus
Cell	Heterogeneous	Heterogeneous	1 PowerPC 8 SPE	No	Integrated Bus
Niagara	Homogeneous	Homogeneous	8 Sparc V9 ISA	Yes (4 threads)	Crossbar
Intel 80-Cores	Homogeneous	Homogeneous	80 VLIW	No	Mesh
Intel Larrabee	Homogeneous	Homogeneous	n p54C x86 cores SIMD execution	No	Integrated Bus
NVIDIA Tesla (GeForce8800)	Homogeneous	Homogeneous	128 Stream Processors	Yes (up to 768 threads)	Network

constrained mobile or portable devices. These architectures have multimedia-based processing elements, following the trend of embedded systems. Homogeneous architectures still use only homogeneous organizations, coupling several processing elements with the same ISA and the processing capabilities. Heterogeneous organizations have not been used on homogeneous architectures, since the variable processing capability is supported dynamically by power management like DVFS. Unlike heterogeneous architectures, homogeneous ones aims at the general-purpose processing market, handling a wide range of applications behavior by replicating general purpose processors.

5   Open Problems

Due to the short history of MPSoCs, several design points are still open. In this section, we discuss two open problems of MPSoC design: interconnection mechanism and MPSoC programming models.

5.1   Interconnection Mechanism

The interconnection mechanism has a very important role in an MPSoC design, since it is responsible for supporting the exchange of information between all MPSoC components, typically between processing elements or processing elements and some storage component. The development of the interconnection mechanism in a MPSoC should take into account the following aspects: parallelism, scalability, testability, fault tolerance, reusability, energy consumption and communication bandwidth [13]. However, there are several communication interconnection approaches that provide different qualitative levels regarding the cited aspects. As can be noticed in Table 1, there is no agreement on the interconnection mechanism, since each design has specifics constraints and requirements that guide the choices of the communication infrastructure, always considering its particular aspects.

Commonly, buses are the most used mechanism on current MPSoC designs. Current buses can achieve high speeds, and buses have additional advantages like low cost, easy testability and high communication bandwidth, encouraging the use of buses in MPSoC designs. The weak points of this approach are poor scalability, no fault tolerance and no parallelism exploitation. However, modifications on its original concept can soften these disadvantages, but could also affect some good characteristics. Segmented bus is an original bus derivative to increase the performance, communication parallelism exploration and energy savings [7]. This technique divides the original bus in several parts, enabling concurrent communication inside of each part. However, bus segmentation impacts on the scalability, and makes the communication management between isolated parts harder. Besides their disadvantages, for obvious historical reasons buses are still widely used in MPSoC designs. Intel and AMD are still using integrated buses to make the communication infrastructure on their high-end MPSoC, due to the easier implementation and high bandwidth provided.

Crossbars are widely used on network hardware like switches and hubs. Some MPSoC designers have been employing this mechanism to connect processing elements [10]. This interconnection approach provides huge performance, allowing communication between any processing elements and the smallest possible time. However, high area cost, energy consumption and poor scalability discourage its employment. AMD Opteron family and Sun Niagara use crossbars to support high communication bandwidth within their general-purpose processors.

Network-on-chip has been emerging as a solution to couple several processing elements [21]. This approach provides high communication parallelism, since several connecting paths are available for each node. In addition, as the technology scales, wire delays increase (because of the increased resistance derived form the smaller cross-section of the wire), and hence shorter wires, as used in NoCs could soften this scaling drawback. Also, its explicit modular shape positively affects the processing elements scalability, and can be explored by a power management technique to support the simple turning off of idle components of the network on chip. NoC disadvantages include the excessive area overhead and high latency of the routers. Intel 80-core prototype employs a mesh style network-on-chip interconnection to supply the communication of its 80 processing elements.

5.2 MPSoC Programming Models

For decades, many ILP exploration approaches were proposed to improve the processor performance. Most of those works employed dynamic ILP exploration at hardware level, becoming an efficient and adaptive process used in superscalar architectures, for instance. Also, traditional ILP exploration free software developers from the hard task to explicit, in the source code, those parts that can be executed in parallel. Some works [19] can even translate code with enough ILP into TLP, so that more than one core can execute the code, exploiting ILP also at the single issue core level, when embedded in a multiprocessing device. However, MPSoC employment relies on manual source code changes to split the parallel parts among the processing elements. Currently, software developers must be aware of the MPSoC characteristic like the number of processing elements to allocate the parallel code. Thus, the sequential programming knowledge should migrate to the parallel programming paradigm. Due to these facts, parallel programming approaches have been gaining importance on computing area, since easy and efficient code production is fundamental to explore the MPSoC processing capability.

Communication between processing elements is needed whenever information exchange among the threads is performed. Commonly, this communication is based either on message passing or on shared memory techniques. Message passing leaves the complex task of execution management to the software description level.

The software code should contain detailed description of the parallelization process since the application developer has the complete control to manage the process. The meticulous message passing management by the software becomes slower than shared memory. However, this approach enables robust communication employment providing the complete control to the software developer to make this job. Message Passing Interface (MPI) [14] is a widely used standard protocol that employs message passing communication mechanism. MPI provides an application programming interface (API) that specifies a set of routines to manage inter processes communication. The advantage of MPI employment over other mechanisms is that both data and task parallelism can be explored, with the cost of more code changes to achieve parallel software. In addition, MPI relies on network communication hardware, which guides the performance of the entire system.

Shared memory communication uses a storage mechanism to hold the necessaries information for threads communication. This approach provides simpler software development, since thanks to a global addressing system most of the communication drawbacks are transparent to the software team. The main drawback of shared memory employment is the bottleneck between processing element and memory, since several threads could try to concurrently access the same storage element at a certain time. Memory coherency also can be a bottleneck for shared memory employment, since some mechanism should be applied to guarantee data coherency. OpenMP [4] employs shared memory communication mechanism to manage the parallelization process. This approach is based on a master and slave mechanism, where the master thread forks a specified number of slave threads to execute in parallel. Thus, each thread executes a parallelized section of the application into a processing element independently. OpenMP provides easier programming than MPI, with greater scalability, since a smaller number of code changes should be done to increase the number of spawned threads. However, in most cases OpenMP code coverage is limited to highly parallel parts and loops. This approach is strongly based on a compiler, which must translate the OpenMP directives to recognize what section should be parallelized.

6   Future Research Challenges

As discussed throughout this chapter, although MPSoCs can be considered a consolidated strategy for the development of high performance and low energy products, there are still many open problems that required extra research effort. Software partitioning is one of the most important open issues. That is, taking a program developed in the way programs have been developed for years now, and making it work in an MPSoC environment by the use of automatic partitioning is very difficult already for the homogeneous case, not to mention the heterogeneous one. Matching a program to an ISA and to a certain MPSoC organization is an open and challenging problem.

Another research direction should contemplate hardware development. From the results shown in Sect. 2 it is clear that heterogeneous SoCs, capable of exploring parallelism at the thread and also at the instruction level, are the most adequate to obtain real performance and energy gains. Unfortunately, in an era where fabrication costs demand huge volumes, the big question to be answered regards the right amount of heterogeneity to be embedded in the MPSoC fabric.

References

1. Anantaraman, A., Seth, K., Patil, K., Rotenberg, E., Mueller, F.: Virtual simple architecture (VISA): Exceeding the complexity limit in safe real-time systems. In: Proc. 30th Annual Int. Symposium Computer Architecture, pp. 350–361. San Diego, CA (2003)
2. Bergamaschi, R., Han, G., Buyuktosunoglu, A., Patel, H., Nair, I., Dittmann, G., Janssen, G., Dhanwada, N., Hu, Z., Bose, P., Darringer, J.: Exploring power management in multi-core systems. In: Proc. Asia and South Pacific Design Automation Conf. Seoul, Korea (2008)
3. Chen, T., Raghavan, R., Dale, J.N., Iwata, E.: Cell broadband engine architecture and its first implementation: A performance view. IBM J. Res. Dev. **51**(5), 559–572 (2007)
4. Dagum, L., Menon, R.: OpenMP: An industry-standard API for shared-memory programming. IEEE Comput. Sci. Eng. **5**(1), 46–55 (1998)
5. Du, J.: Inside an 80-corechip: The on-chip communication and memory bandwidth solutions (2007). URL http://blogs.intel.com/research/2007/07/inside_the_terascale_many_core.php
6. Freescale, Inc. URL http://www.freescale.com/webapp/sps/site/ homepage.jsp
7. Guo, J., Papanikolaou, A., Marchal, P., Catthoor, F.: Physical design implementation of segmented buses to reduce communication energy. In: Proc. Asia and South Pacific Design Automation Conference, pp. 42–47. Yokohama, Japan (2006)
8. Guthaus, M.R., Ringenberg, J.S., Ernst, D., Austin, T.M., Mudge, T., Brown, R.B.: MiBench: A free, commercially representative embedded benchmark suite. In: Proc. IEEE Int. Workshop Workload Characterization, pp. 3–14. Austin, TX (2001)
9. Intel core i7 processor SDK webinar: Q and A transcript from the 8:00 a.m. PST. URL http://software.intel.com/sites/webinar/corei7-sdk/intel-core-i7-8am.doc
10. Johnson, T., Nawathe, U.: An 8-core, 64-thread, 64-bit power efficient Sparc SoC (Niagara2). In: Proc. Int. Symposium Physical Design, p. 2. Austin, TX (2007)
11. Kongetira, P., Aingaran, K., Olukotun, K.: Niagara: A 32-way multithreaded Sparc processor. IEEE Micro 25(2), 21–29 (2005)
12. Lindholm, E., Nickolls, J., Oberman, S., Montrym, J.: NVIDIA Tesla: A unified graphics and computing architecture. IEEE Micro 28(2), 39–55 (2008)
13. Marcon, C., Borin, A., Susin, A., Carro, L., Wagner, F.: Time and energy efficient mapping of embedded applications onto NoCs. In: Proc. Asia and South Pacific-Design Automation Conference (2005)
14. Pacheco, P.S.: Parallel Programming with MPI. Morgan Kaufmann Publishers, Inc. (1996)
15. Reifel, M., Chen, D.: Parallel digital signal processing: An emerging market. Application Note (1994). URL http://focus.ti.com/lit/an/spra104/spra104.pdf
16. Rutzig, M.B., Beck, A.C., Carro, L.: Dynamically adapted low power ASIPs. In: Reconfigurable Computing: Architectures, Tools and Applications, Lecture Notes In Computer Science, vol. 5453, pp. 110–122. Springer-Verlag, Berlin, Germany (2009)
17. Seiler, L., Carmean, D., Sprangle, E., Forsyth, T., Abrash, M., Dubey, P., Junkins, S., Lake, A., Sugerman, J., Cavin, R., Espasa, R., Grochowski, E., Juan, T., Hanrahan, P.: Larrabee: A many-core x86 architecture for visual computing. In: ACM SIGGRAPH 2008 Papers, pp. 1–15. Los Angeles, CA (2008)
18. Shi, K., Howard, D.: Challenges in sleep transistor design and implementation in low-power designs. In: Proc. 43rd Annual Conf. Design Automation, pp. 113–116 (2006)

19. Song, Y., Kalogeropulos, S., Tirumalai, P.: Design and implementation of a compiler framework for helper threading on multi-core processors. In: Proceedings of the 14th international Conference on Parallel Architectures and Compilation Techniques, pp. 99–109 (2005)

20. IDC Analyze the Future. Available at http://www.idc.com/getdoc.jsp?containerId= prUS23297412

21. Vangal, S., Howard, J., Ruhl, G., Dighe, S., Wilson, H., Tschanz, J., Finan, D., Iyer, P., Singh, A., Jacob, T., Jain, S., Venkataraman, S., Hoskote, Y., Borkar, N.: An 80-tile 1.28TFLOPS network-on-chip in 65nm CMOS. In: Digest of Technical Papers IEEE Int. Solid-State Circuits Conf., pp. 98–589 (2007)

22. Wall, D.W.: Limits of instruction-level parallelism. SIGPLAN Not. **26**(4), 176–188 (1991)

23. Woo, D.H., Lee, H.H.: Extending Amdahl's law for energy-efficient computing in the many-core era. Computer 41(12), 24–31 (2008)

24. Diefendorff, K. Hal Makes Sparcs Fly. [S.l.]. 1999

25. Intel. Inside an 80-core chip: The on-chip communication and memory bandwidth solution, 2007. Available at: http://blogs.intel.com/research/2007/07/inside_the_terascale_many_core.php.

26. Zhao, G., Kwan, H.K., Lei, C.U., Wong, N.: Processor frequency assignment in three-dimensional MPSoCs under thermal constraints by polynomial programming. In: Proc. IEEE Asia Pacific Conf. Circuits Syst., pp. 1668–1671 (2008)

27. OMAP 5430 white paper. Available at http://www.ti.com/lit/an/swpt048/swpt048.pdf

28. Exynos 3 Single platform. Available at http://www.samsung.com/global/business/semiconductor/minisite/Exynos/products3110.html

29. Exynos 4 Dual platform. Available at http://www.samsung.com/global/business/semiconductor/minisite/Exynos/products4210.html

30. NVIDIA white paper. Available at http://www.nvidia.com/content/PDF/tegra_white_papers/tegra-whitepaper-0911b.pdf

31. Intel Corp. White paper. Available at http://www.intel.com/content/dam/doc/white-paper/intel-microarchitecture-white-paper.pdf

Coarse-Grained Reconfigurable Array Architectures

Bjorn De Sutter, Praveen Raghavan, and Andy Lambrechts

Abstract Coarse-Grained Reconfigurable Array (CGRA) architectures accelerate the same inner loops that benefit from the high ILP support in VLIW architectures. Unlike VLIWs, CGRAs are designed to execute only the loops, which they can hence do more efficiently. This chapter discusses the basic principles of CGRAs and the wide range of design options available to a CGRA designer, covering a large number of existing CGRA designs. The impact of different options on flexibility, performance, and power-efficiency is discussed, as well as the need for compiler support. The ADRES CGRA design template is studied in more detail as a use case to illustrate the need for design space exploration, for compiler support and for the manual fine-tuning of source code.

1 Application Domain

Many embedded applications require high throughput, meaning that a large number of computations needs to be performed every second. At the same time, the power consumption of battery-operated devices needs to be minimized to increase their autonomy. In general, the performance obtained on a programmable processor for a certain application can be defined as the reciprocal of the application execution time. Considering that most programs consist of a series P of consecutive phases with different characteristics, performance can be defined in terms of the operating frequencies f_p, the instructions executed per cycle IPC_p and the instruction counts

B. De Sutter (✉)
Ghent University, Sint-Pietersnieuwstraat 41, 9000 Gent, Belgium
e-mail: bjorn.desutter@elis.ugent.be

P. Raghavan • A. Lambrechts
IMEC, Kapeldreef 75, 3001 Heverlee, Belgium
e-mail: ragha@imec.be; lambreca@imec.be

S.S. Bhattacharyya et al. (eds.), *Handbook of Signal Processing Systems*,
DOI 10.1007/978-1-4614-6859-2_18, © Springer Science+Business Media, LLC 2013

IC_p of each phase, and in terms of the time overhead involved in switching between the phases $t_{p \to p+1}$ as follows:

$$\frac{1}{\text{performance}} = \text{execution time} = \sum_{p \in P} \frac{IC_p}{IPC_p \cdot f_p} + t_{p \to p+1}. \tag{1}$$

The operating frequencies f_p cannot be increased infinitely because of power-efficiency reasons. Alternatively, a designer can increase the performance by designing or selecting a system that can execute code at higher IPCs. In a power-efficient architecture, a high IPC is reached for the most important phases $l \in L \subset P$, with L typically consisting of the compute-intensive inner loops, while limiting their instruction count IC_l and reaching a sufficiently high, but still power-efficient frequency f_l. Furthermore, the time overhead $t_{p \to p+1}$ as well as the corresponding energy overhead of switching between the execution modes of consecutive phases should be minimized if such switching happens frequently. Note that such switching only happens on hardware that supports multiple execution modes in support of phases with different characteristics.

Course-Grained Reconfigurable Array (CGRA) accelerators aim for these goals for the inner loops found in many digital signal processing (DSP) domains, including multimedia and Software-Defined Radio (SDR) applications. Such applications have traditionally employed Very Long Instruction Word (VLIW) architectures such as the TriMedia 3270 [74] and the TI C64 [70], Application-Specific Integrated Circuits (ASICs), and Application-Specific Instruction Processors (ASIPs). To a large degree, the reasons for running these applications on VLIW processors also apply for CGRAs. First of all, a large fraction of the computation time is spent in manifest nested loops that perform computations on arrays of data and that can, possibly through compiler transformations, provide a lot of Instruction-Level Parallelism (ILP). Secondly, most of those inner loops are relatively simple. When the loops include conditional statements, this can be implement by means of predication [45] instead of with complex control flow. Furthermore, none or very few loops contain multiple exits or continuation points in the form of, e.g., break or continue statements as in the C-language. Moreover, after inlining the loops are free of function calls. Finally, the loops are not regular or homogeneous enough to benefit from vector computing, like on the EVP [6] or on Ardbeg [75]. When there is enough regularity and Data-Level Parallelism (DLP) in the loops of an application, vector computing can typically exploit it more efficiently than what can be achieved by converting the DLP into ILP and exploiting that on a CGRA. So in short, CGRAs (with limited DLP support) are ideally suited for applications of which time-consuming parts have manifest behavior, large amounts of ILP and limited amounts of DLP.

In the remainder of this chapter, Sect. 2 presents the fundamental properties of CGRAs. Section 3 gives an overview of the design options for CGRAs. This overview help designers in evaluating whether or not CGRAs are suited for their applications and their design requirements, and if so, which CGRA designs are most suited. After the overview, Sect. 4 presents a case study on the ADRES CGRA architecture. This study serves two purposes. First, it illustrates the extent to which source code needs to be tuned to map well onto CGRA architectures. As we will show, this is an important aspect of using CGRAs, even when good compiler support is available and when a very flexible CGRA is targeted, i.e., one that puts very few restrictions on the loop bodies that it can accelerate. Secondly, our use case illustrates how Design Space Exploration is necessary to instantiate optimized designs from parameterizable and customizable architecture templates such as the ADRES architecture template. Some conclusions are drawn in Sect. 5.

2 CGRA Basics

CGRAs focus on the efficient execution of the type of loops discussed in the previous section. By neglecting non-loop code or outer-loop code that is assumed to be executed on other cores, CGRAs can take the VLIW principles for exploiting ILP in loops a step further to consume less energy and deliver higher performance, without compromising on available compiler support. Figures 1 and 2 illustrate this.

Higher performance for high-ILP loops is obtained through two main features that separate CGRA architectures from VLIW architectures. First, CGRA architectures typically provide more Issue Slots (ISs) than typical VLIWs do. In the CGRA literature some other commonly used terms to denote CGRA ISs are Arithmetic-Logic Units (ALUs), Functional Units (FUs), or Processing Elements (PEs). Conceptually, these terms all denote the same: logic on which an instruction can be executed, typically one per cycle. For example, a typical ADRES [9–11, 20, 46, 48–50] CGRA consists of 16 issue slots, whereas the TI C64 features 8 slots, and the NXP TriMedia features only 5 slots. The higher number of ISs directly allows to reach higher IPCs, and hence higher performance, as indicated by Eq. (1). To support these higher IPCs, the bandwidth to memory is increased by having more load/store ISs than on a typical VLIW, and special memory hierarchies as found on ASIPs, ASICs, and other DSPs. These include FIFOs, stream buffers, scratch-pad memories, etc. Secondly, CGRA architectures typically provide a number of direct connections between the ISs that allow data to "flow" from one IS to another without needing to pass data through a Register File

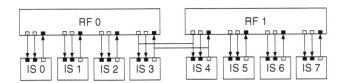

Fig. 1 An example clustered VLIW architecture with two RFs and eight ISs. *Solid directed edges* denote physical connections. *Black* and *white small boxes* denote input and output ports, respectively. There is a one-to-one mapping between input and output ports and physical connections

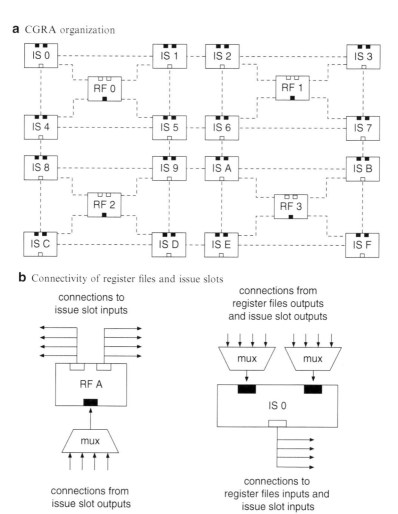

Fig. 2 Part (**a**) shows an example CGRA with 16 ISs and 4 RFs, in which *dotted edges* denote conceptual connections that are implemented by physical connections and muxes as in part (**b**)

(RF). As a result, less register copy operations need to be executed in the ISs, which reduces the IC term in Eq. (1) and frees ISs for more useful computations.

Higher energy efficiency is obtained through several features. Because of the direct connections between ISs, less data needs to be transferred into and out of RFs. This saves considerable energy. Also, because the ISs are arranged into a 2D matrix, small RFs with few ports can be distributed in between the ISs as depicted in Fig. 2. This contrasts with the many-ported RFs in (clustered) VLIW architectures, which basically feature a one-dimensional design as depicted in Fig. 1. The distributed CGRA RFs consume considerably less energy. Finally, by not supporting control flow, the instruction memory organization can be simplified. In statically reconfigurable CGRAs, this memory is nothing more than a set of configuration bits that remain fixed for the whole execution of a loop. Clearly this is very energy-efficient. Other CGRAs, called dynamically reconfigurable CGRAs, feature a form of distributed level-0 loop buffers [43] or other small controllers that fetch new configurations every cycle from simple configuration buffers. To support loops that include control flow and conditional operations, the compiler then replaces that control flow by data flow by means of predication [45] or other mechanisms. In this way CGRAs differ from VLIW processors that typically feature a power-hungry combination of an instruction cache, instruction decompression and decoding pipeline stages and a non-trivial update mechanism of the program counter.

There are two main drawbacks to CGRA architectures. Firstly, because they can only execute loops, they need to be coupled to other cores on which all other parts of the program are executed. In some designs, this coupling introduces run-time and design-time overhead. Secondly, as clearly visible in the example CGRA of Fig. 2, the interconnect structure of a CGRA is vastly more complex than that of a VLIW. On a VLIW, scheduling an instruction in some IS automatically implies the reservation of connections between the RF and the IS and of the corresponding ports. On CGRAs, this is not the case. Because there is no one-to-one mapping between connections and input/output ports of ISs and RFs, connections need to be reserved explicitly by the compiler or programmer together with ISs, and the data flow needs to be routed explicitly over the available connections. This can be done, for example, by programming switches and multiplexors (a.k.a. muxes) explicitly, like the ones depicted in Fig. 2b. Consequently more complex compiler technology than that of VLIW compilers is needed to automate the mapping of code onto a CGRA. Moreover, writing assembly code for CGRAs ranges from being very difficult to virtually impossible, depending on the type of reconfigurability and on the form of processor control.

Having explained these fundamental concepts that differentiate CGRAs from VLIWs, we can now also differentiate them from Field-Programmable Gate Arrays (FPGAs), where the name CGRA actually comes from. Whereas FPGAs feature bitwise logic in the form of Look-Up Tables (LUTs) and switches, CGRAs feature more energy-efficient and area-conscious word-wide ISs, RFs and interconnections. Hence the name *coarse-grained array* architecture. As there are much fewer ISs on a CGRA than there are LUTs on an FPGA, the number of bits required to configure the CGRA ISs, muxes, and RF ports is typically orders of magnitude smaller than on FPGAs. If this number becomes small enough, dynamic reconfiguration can be possible every cycle. So in short, CGRAs can be seen as statically or dynamically reconfigurable coarse-grained FPGAs, or as 2D, highly-clustered loop-only VLIWs with direct interconnections between ISs that need to be programmed explicitly.

3   CGRA Design Space

The large design space of CGRA architectures features many design options. These
include the way in which the CGRA is coupled to a main processor, the type of
interconnections and computation resources used, the reconfigurability of the array,
the way in which the execution of the array is controlled, support for different forms
of parallelism, etc. This section discusses the most important design options and the
influence of the different options on important aspects such as performance, power
efficiency, compiler friendliness, and flexibility. In this context, higher flexibility
equals placing fewer restrictions on loop bodies that can be mapped onto a CGRA.

Our overview of design options is not exhaustive. Its scope is limited to the most
important features of CGRA architectures that feature a 2D array of ISs. However,
the distinction between 1D VLIWs and 2D CGRAs is anything but well-defined.
The reason is that this distinction is not simply a layout issue, but one that also
concerns the topology of the interconnects. Interestingly, this topology is precisely
one of the CGRA design options with a large design freedom.

3.1   Tight Versus Loose Coupling

Some CGRA designs are coupled loosely to main processors. For example, Fig. 3
depicts how the MorphoSys CGRA [44] is connected as an external accelerator to a
TinyRISC Central Processing Unit (CPU) [1]. The CPU is responsible for executing
non-loop code, for initiating DMA data transfers to and from the CGRA and the
buffers, and for initiating the operation of the CGRA itself by means of special
instructions added to the TinyRISC ISA.

This type of design offers the advantage that the CGRA and the main CPU can be
designed independently, and that both can execute code concurrently, thus delivering
higher parallelism and higher performance. For example, using the double frame
buffers [44] depicted in Fig. 3, the MorphoSys CGRA can be operating on data in
one buffer while the main CPU initiates the necessary DMA transfers to the other
buffer for the next loop or for the next set of loop iterations. One drawback is that
any data that needs to be transferred from non-loop code to loop code needs to
be transferred by means of DMA transfers. This can result in a large overhead,
e.g., when frequent switching between non-loop code and loops with few iterations
occurs and when the loops consume scalar values computed by non-loop code.

By contrast, an ADRES CGRA is coupled tightly to its main CPU. A simplified
ADRES is depicted in Fig. 4. Its main CPU is a VLIW consisting of the shared
RF and the top row of CGRA ISs. In the main CPU mode, this VLIW executes
instructions that are fetched from a VLIW instruction cache and that operate on
data in the shared RF. The idle parts of the CGRA are then disabled by clock-gating
to save energy. By executing a start_CGRA instruction, the processor switches to
CGRA mode in which the whole array, including the shared RF and the top row

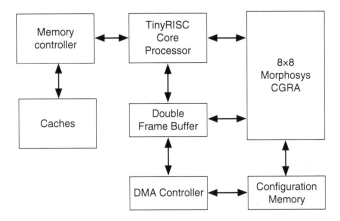

Fig. 3 A TinyRISC main processor loosely coupled to a MorphoSys CGRA array. Note that the main data memory (cache) is not shared and that no IS hardware or registers is are shared between the main processor and the CGRA. Thus, both can run concurrent threads

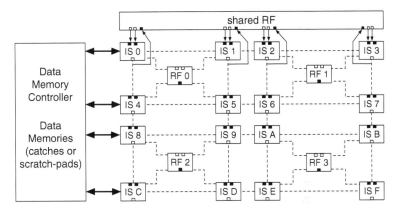

Fig. 4 A simplified picture of an ADRES architecture. In the main processor mode, the *top row* of ISs operates like a VLIW on the data in the shared RF and in the data memories, fetching instructions from an instruction cache. When the CGRA mode is initiated with a special instruction in the main VLIW ISA, the whole array starts operating on data in the distributed RFs, in the shared RF and in the data memories. The memory port in IS 0 is also shared between the two operating modes. Because of the resource sharing, only one mode can be active at any point in time

of ISs, executes a loop for which it gets its configuration bits from a configuration memory. This memory is omitted from the figure for the sake of simplicity.

The drawback of this tight coupling is that because the CGRA and the main processor mode share resources, they cannot execute code concurrently. However, this tight coupling also has advantages. Scalar values that have been computed in non-loop code, can be passed from the main CPU to the CGRA without any overhead because those values are already present in the shared RFs or in the shared memory banks. Furthermore, using shared memories and an execution model of exclusive execution in either main CPU or CGRA mode significantly eases the automated co-generation of main CPU code and of CGRA code in a compiler, and it avoids the run-time overhead of transferring data. Finally, on the ADRES CGRA, switching between the two modes takes only two cycles. Thus, the run-time overhead is minimal.

Silicon Hive CGRAs [12,13] do not feature a clear separation between the CGRA accelerator and the main processor. Instead there is just a single processor that can be programmed at different levels of ILP, i.e., at different instruction word widths. This allows for a very simple programming model, with all the programming and performance advantages of the tight coupling of ADRES. Compared to ADRES, however, the lack of two distinctive modes makes it more difficult to implement coarse-grained clock-gating or power-gating, i.e., gating of whole sets of ISs combined instead of separate gating of individual ISs.

Somewhere in between loose and tight coupling is the PACT XPP design [53], in which the array consist of simpler ISs that can operate like a true CGRA, as well as of more complex ISs that are in fact full-featured small RISC processors that can run independent threads in parallel with the CGRA.

As a general rule, looser coupling potentially enables more Thread-Level Parallelism (TLP) and it allows for a larger design freedom. Tighter coupling can minimize the per-thread run-time overhead as well as the compile-time overhead. This is in fact no different from other multi-core or accelerator-based platforms.

3.2 CGRA Control

There exist many different mechanisms to control how code gets executed on CGRAs, i.e., to control which operation is issued on which IS at which time and how data values are transferred from producing operations to consuming ones. Two important aspects of CGRAs that drive different methods for control are reconfigurability and scheduling. Both can be static or dynamic.

3.2.1 Reconfigurability

Some CGRAs, like ADRES, Silicon Hive, and MorphoSys are fully dynamically reconfigurable: exactly one full reconfiguration takes place for every execution cycle. Of course no reconfiguration takes places in cycles in which the whole array is stalled. Such stalls can happen, e.g., because memory accesses take longer than expected in the schedule as a result of a cache miss or a memory bank access conflict. This cycle-by-cycle reconfiguration is similar to the fetching of one VLIW instruction per cycle, but on these CGRAs the fetching is simpler as it only iterates through a loop body existing of straight-line CGRA configurations without control flow. Other CGRAs like the KressArray [29–31] are fully statically reconfigurable, meaning that the CGRA is configured before a loop is entered, and

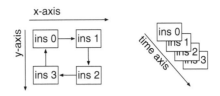

Fig. 5 The *left* part shows a spatial mapping of a sequence of four instructions on a statically reconfigurable 2×2 CGRA. *Edges* denote dependencies, with the edge from instruction 3 to instruction 0 denoting that instruction 0 from iteration i depends on instruction 3 from iteration $i - 1$. So only one out of four ISs is utilized per cycle. The *right* part shows a temporal mapping of the same code on a dynamically reconfigurable CGRA with only one IS. The utilization is higher here, at 100 %

no reconfiguration takes place during the loop at all. Still other architectures feature a hybrid reconfigurability. The RaPiD [18, 23] architecture features partial dynamic reconfigurability, in which part of the bits are statically reconfigurable and another part is dynamically reconfigurable and controlled by a small sequencer. Yet another example is the PACT architecture, in which the CGRA itself can initiate events that invoke (partial) reconfiguration. This reconfiguration consumes a significant amount of time, however, so it is advised to avoid it if possible, and to use the CGRA as a statically reconfigurable CGRA.

In statically reconfigured CGRAs, each resource performs a single task for the whole duration of the loop. In that case, the mapping of software onto hardware becomes purely spatial, as illustrated in Fig. 5a. In other words, the mapping problem becomes one of placement and routing, in which instructions and data dependencies between instructions have to mapped on a 2D array of resources. For these CGRAs, compiler techniques similar to hardware synthesis techniques can be used, as those used in FPGA placement and routing [7].

By contrast, dynamic reconfigurability enables the programmer to use hardware resources for multiple different tasks during the execution of a loop or even during the execution of a single loop iteration. In that case, the software mapping problem becomes a spatial and temporal mapping problem, in which the operations and data transfers not only need to be placed and routed on and over the hardware resources, but in which they also need to be scheduled. A contrived example of a temporal mapping is depicted in Fig. 5b. Most compiler techniques [20, 22, 25, 48, 52, 54, 55] for these architectures also originate from the FPGA placement and routing world. For CGRAs, the array of resources is not treated as a 2D spatial array, but as a 3D spatial-temporal array, in which the third dimension models time in the form of execution cycles. Scheduling in this dimension is often based on techniques that combine VLIW scheduling techniques such as modulo scheduling [39, 61], with FPGA synthesis-based techniques [7]. Still other compiler techniques exist that are based on constraint solving [67], or on integer linear programming [2, 41, 77].

The most important advantage of static reconfigurability is the lack of reconfiguration overhead, in particular in terms of power consumption. For that reason, large arrays can be used that are still power-efficient. The disadvantage is that even in the large arrays the amount of resources constrains which loops can be mapped.

Dynamically reconfigurable CGRAs can overcome this problem by spreading the computations of a loop iteration over multiple configurations. Thus a small dynamically reconfigurable array can execute larger loops. The loop size is then not limited by the array size, but by the array size times the depth of the reconfiguration memories. For reasons of power efficiency, this depth is also limited, typically to tens or hundreds of configurations, which suffices for most if not all inner loops.

A potential disadvantage of dynamically reconfigurable CGRAs is the power consumption of the configuration memories, even for small arrays, and of the configuration fetching mechanism. The disadvantage can be tackled in different ways. ADRES and MorphoSys tackle it by not allowing control flow in the loop bodies, thus enabling the use of very simple, power-efficient configuration fetching techniques similar to level-0 loop buffering [43]. Whenever control flow is found in loop bodies, such as for conditional statements, this control flow then first needs to be converted into data flow, for example by means of predication and hyperblock formation [45]. While these techniques can introduce some initial overhead in the code, this overhead typically will be more than compensated by the fact that a more efficient CGRA design can be used.

The MorphoSys design takes this reduction of the reconfiguration fetching logic even further by limiting the supported code to Single Instruction Multiple Data (SIMD) code. In the two supported SIMD modes, all ISs in a row or all ISs in a column perform identical operations. As such only one IS configuration needs to be fetched per row or column. As already mentioned, the RaPiD architecture limits the number of configuration bits to be fetched by making only a small part of the configuration dynamically reconfigurable. Kim et al. provide yet another solution in which the configuration bits of one column in one cycle are reused for the next column in the next cycle [37]. Furthermore, they also propose to reduce the power consumption in the configuration memories by compressing the configurations [38].

Still, dynamically reconfigurable designs exist that put no restrictions on the code to be executed, and that even allow control flow in the inner loops. The Silicon Hive design is one such design. Unfortunately, no numbers on the power consumption overhead of this design choice are publicly available.

A general rule is that a limited reconfigurability puts more constraints on the types and sizes of loops that can be mapped. Which design provides the highest performance or the highest energy efficiency depends, amongst others, on the variation in loop complexity and loop size present in the applications to be mapped onto the CGRA. With large statically reconfigurable CGRAs, it is only possible to achieve high utilization for all loops in an application if all those loops have similar complexity and size, or if they can be made so with loop transformations, and if the iterations are not dependent on each other through long-latency dependency cycles (as was the case in Fig. 5). Dynamically reconfigurable CGRAs, by contrast, can also achieve high average utilization over loops of varying sizes and complexities, and with inter-iteration dependencies. That way dynamically reconfigurable CGRAs can achieve higher energy efficiency in the data path, at the expense of higher energy consumption in the control path. Which design option is the best depends also on the process technology used, and in particular on the ability to perform clock or power gating and on the ratio between active and passive power (a.k.a. leakage).

3.2.2 Scheduling and Issuing

Both with dynamic and with static reconfigurability, the execution of operations and of data transfers needs to be controlled. This can be done statically in a compiler, similar to the way in which operations from static code schedules are scheduled and issued on VLIW processors [24], or dynamically, similar to the way in which out-of-order processors issue instructions when their operands become available [66]. Many possible combinations of static and dynamic reconfiguration and of static and dynamic scheduling exist.

A first class consists of dynamically scheduled, dynamically reconfigurable CGRAs like the TRIPS architecture [28, 63]. For this architecture, the compiler determines on which IS each operation is to be executed and over which connections data is to be transferred from one IS to another. So the compiler performs placement and routing. All scheduling (including the reconfiguration) is dynamic, however, as in regular out-of-order superscalar processors [66]. TRIPS mainly targets general-purpose applications, in which unpredictable control flow makes the generation of high-quality static schedules difficult if not impossible. Such applications most often provide relatively limited ILP, for which large arrays of computational resources are not efficient. So instead a small, dynamically reconfigurable array is used, for which the run-time cost of dynamic reconfiguration and scheduling is acceptable.

A second class of dynamically reconfigurable architectures avoids the overhead of dynamic scheduling by supporting VLIW-like static scheduling [24]. Instead of doing the scheduling in hardware where the scheduling logic then burns power, the scheduling for ADRES, MorphoSys and Silicon Hive architectures is done by a compiler. Compilers can do this efficiently for loops with regular, predictable behavior and high ILP, as found in many DSP applications. As for VLIW architectures, software pipelining [39,61] is a very important to expose the ILP in software kernels, so most compiler techniques [20, 22, 25, 48, 52, 54, 55] for statically scheduled CGRAs implement some form of software pipelining.

A final class of CGRAs are the statically reconfigurable, dynamically scheduled architectures, such as KressArray or PACT (neglecting the time-consuming partial reconfigurability of the PACT). The compiler performs placement and routing, and the code execution progress is guided by tokens or event signals that are passed along with data. Thus the control is dynamic, and it is distributed over the token or event path, similar to the way in which transport-triggered architectures [17] operate. These statically reconfigurable CGRAs do not require software pipelining techniques because there is no temporal mapping. Instead the spatial mapping and the control implemented in the tokens or event signals implement a hardware pipeline.

We can conclude by noting that, as in other architecture paradigms such as VLIW processing or superscalar out-of-order execution, dynamically scheduled CGRAs can deliver higher performance than statically scheduled ones for control-intensive code with unpredictable behavior. On dynamically scheduled CGRAs the code path that gets executed in an iteration determines the execution time of that iteration, whereas on statically scheduled CGRAs, the combination of all possible execution paths (including the slowest path which might be executed infrequently) determines the execution time. Thus, dynamically scheduled CGRAs can provide higher performance for some applications. However, the power efficiency will then typically also be poor because more power will be consumed in the control path. Again, the application domain determines which design option is the most appropriate.

3.2.3   Thread-Level and Data-Level Parallelism

Another important aspect of control is the possibility to support different forms of parallelism. Obviously, loosely-coupled CGRAs can operate in parallel with the main CPU, but one can also try to use the CGRA resources to implement SIMD or to run multiple threads concurrently within the CGRA.

When dynamic scheduling is implemented via distributed event-based control, as in KressArray or PACT, implementing TLP is relatively simple and cheap. For small enough loops of which the combined resource use fits on the CGRA, it suffices to map independent thread controllers on different parts of the distributed control.

For architectures with centralized control, the only option to run threads in parallel is to provide additional controllers or to extend the central controller, for example to support parallel execution modes. While such extensions will increase the power consumption of the controller, the newly supported modes might suit certain code fragments better, thus saving in data path energy and configuration fetch energy.

The TRIPS controller supports four operation modes [63]. In the first mode, all ISs cooperate for executing one thread. In the second mode, the four rows execute four independent threads. In the third mode, fine-grained multi-threading [66] is supported by time-multiplexing all ISs over multiple threads. Finally, in the fourth mode each row executes the same operation on each of its ISs, thus implementing SIMD in a similar, fetch-power-efficient manner as is done in the two modes of the MorphoSys design. Thus, for each loop or combination of loops in an application, the TRIPS compiler can exploit the most suited form of parallelism.

The Raw architecture [69] is a hybrid between a many-core architecture and a CGRA architecture in the sense that it does not feature a 2D array of ISs, but rather a 2D array of tiles that each consist of a simple RISC processor. The tiles are connected to each other via a mesh interconnect, and transporting data over this interconnect to neighboring tiles does not consume more time than retrieving data from the RF in the tile. Moreover, the control of the tiles is such that they can operate independently or synchronized in a lock-step mode. Thus, multiple tiles can cooperate to form a dynamically reconfigurable CGRA. A programmer can hence partition the 2D array of tiles into several, potentially differently sized, CGRAs that each run an independent thread. This provides very high flexibility to balance the available ILP inside threads with the TLP of the combined threads.

The Polymorphic Pipeline Array (PPA) [56] integrates multiple tightly-coupled ADRES-like CGRA cores into a larger array. Independent threads with limited amounts of ILP can run on the individual cores, but the resources of those individual cores can also be configured to form larger cores, on which threads with more ILP can then be executed. The utilization of the combined resources can be optimized dynamically by configuring the cores according to the available TLP and ILP at any point during the execution of a program.

Other architectures do not support (hardware) multi-threading within one CGRA core at all, like the current ADRES and Silicon Hive. The first solution to run multiple threads with these designs is to incorporate multiple CGRA accelerator cores in a System-on-Chip (SoC). The advantage is then that each accelerator can be customized for a certain class of loop kernels. Also, ADRES and Silicon Hive are architecture templates, which enables CGRA designers to customize their CGRA cores for the appropriate amount of DLP for each class of loop kernels, in the form of SIMD or subword parallelism.

Alternatively, TLP can be converted into ILP and DLP by combining, at compile-time, kernels of multiple threads and by scheduling them together as one kernel, and by selecting the appropriate combination of scheduled kernels at run time [64].

3.3   Interconnects and Register Files

3.3.1   Connections

A wide range of connections can connect the ISs of a CGRA with each other, with the RFs, with other memories and with IO ports. Buses, point-to-point connections, and crossbars are all used in various combinations and in different topologies.

For example, some designs like MorphoSys and the most common ADRES and Silicon Hive designs feature a densely connected mesh-network of point-to-point interconnects in combination with sparser buses that connect ISs further apart. Thus the number of long power-hungry connections is limited. Multiple studies of point-to-point mesh-like interconnects as in Fig. 6 have been published in the past [11,36,40,47]. Other designs like RaPiD feature a dense network of segmented buses. Typically the use of crossbars is limited to very small instances because large ones are too power-hungry. Fortunately, large crossbars are most often not needed, because many application kernels can be implemented as systolic algorithms, which map well onto mesh-like interconnects as found in systolic arrays [59].

Unlike crossbars and even busses, mesh-like networks of point-to-point connections scale better to large arrays without introducing too much delay or power consumption. For statically reconfigurable CGRAs, this is beneficial. Buses and

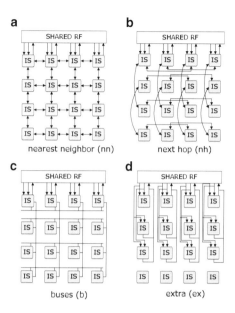

Fig. 6 Basic interconnects that can be combined. All bidirectional edges between two ISs denote that all outputs of one IS are connected to the inputs of the other IS and vice versa. Buses that connect all connected IS outputs to all connected IS inputs are shown as *edges without arrows*

other long interconnects connect whole rows or columns to complement short-distance mesh-like interconnects. The negative effects that such long interconnects can have on power consumption or on obtainable clock frequency can be avoided by segmentation or by pipelining. In the latter case, pipelining latches are added along the connections or in between muxes and ISs. Our experience, as presented in Sect. 4.2.2 is that this pipelining will not necessarily lead to lower IPCs in CGRAs. This is different from out-of-order or VLIW architectures, where deeper pipelining increases the branch misprediction latency [66]. Instead at least some CGRA compilers succeed in exploiting the pipelining latches as temporary storage, rather than being hampered by them. This is the case in compiler techniques like [20,48] that are based on FPGA synthesis methods in which RFs and pipelining latches are treated as interconnection resources that span multiple cycles instead of as explicit storage resources. This treatment naturally fits the 3D array modeling of resources along two spatial dimensions and one temporal dimension. Consequently, those compiler techniques can use pipelining latches for temporary storage as easily as they can exploit distributed RFs. This ability to use latches for temporary storage has been extended even beyond pipeline latches, for example to introduce retiming chains and shift registers in CGRA architectures [71].

3.3.2   Register Files

Compilers for CGRA architectures place operations in ISs, thus also scheduling them, and route the data flow over the connections between the ISs. Those connections may be direct connections, or latched connections, or even connections that go through RFs. Therefore most CGRA compilers treat RFs not as temporary storage, but as interconnects that can span multiple cycles. Thus the RFs can be treated uniformly with the connections during routing. A direct consequence of this compiler approach is that the design space freedom of interconnects extends to the placement of RFs in between ISs. During the Design Space Exploration (DSE) for a specific CGRA instance in a CGRA design template such as the ADRES or Silicon Hive templates, both the real connections and the RFs have to be explored, and that has to be done together. Just like the number of real interconnect wires and their topology, the size of RFs, their location and their number of ports then contribute to the interconnectivity of the ISs. We refer to [11, 47] for DSEs that study both RFs and interconnects.

Besides their size and ports, another important aspect is that RFs can be rotating [62]. The power and delay overhead of rotation is very small in distributed RFs, simply because these RFs are small themselves. Still they can provide an important functionality. Consider a dynamically reconfigurable CGRA on which a loop is executed that iterates over x configurations, i.e., each iteration takes x cycles. That means that for a write port of an RF, every x cycles the same address bits get fetched from the configuration memory to configure the address set at that port. In other words, every x cycles a new value is being written into the register specified by that same address. This implies that values can stay in the same register for at most x cycles; then they are overwritten by a new value from the next iteration. In many loops, however, some values have a life time that spans more than x cycles, because it spans multiple loop iterations. To avoid having to insert additional data transfers in the loop schedules, rotating registers can be used. At the end of every iteration of the loop, all values in rotating registers rotate into another register to make sure that old values are copied to where they are not overwritten by newer values.

3.3.3   Predicates, Events and Tokens

To complete this overview on CGRA interconnects, we want to point out that it can be very useful to have interconnects of different widths. The data path width can be as small as 8 bits or as wide as 64 or 128 bits. The latter widths are typically used to pass SIMD data. However, as not all data is SIMD data, not all paths need to have the full width. Moreover, most CGRA designs and the code mapped onto them feature signals that are only one or a few bits wide, such as predicates or events or tokens. Using the full-width datapath for these narrow signals wastes resources. Hence it is often useful to add a second, narrow datapath for control signals like tokens or events and for predicates. How dense that narrow datapath has to be, depends on the type of loops one wants to run on the CGRA. For example, multimedia coding and decoding typically includes more conditional code than SDR baseband processing. Hence the design of, e.g., different ADRES architectures for multimedia and for SDR resulted in different predicate data paths being used, as illustrated in Sect. 4.2.1.

At this point, it should be noted that the use of predicates is fundamentally not that different from the use of events or tokens. In KressArray or PACT, events and tokens are used, amongst others, to determine at run time which data is selected to be used later in the loop. For example, for a C expression like x + (a>b) ? y + z : y - z one IS will first compute the addition y+z, one IS will compute the subtraction y-z, and one IS will compute the greater-than condition a>b. The result of the latter computation generates an event that will be fed to a multiplexor to select which of the two other computer values y+z and y-z is transferred to yet another IS on which the addition to x will be performed. Unlike the muxes in Fig. 2b that are controlled by bits fetched from the configuration memory, those event-controlled multiplexors are controlled by the data path.

In the ADRES architecture, the predicates guard the operations in ISs, and they serve as enable signals for RF write ports. Furthermore, they are also used to control special `select` operations that pass one of two input operands to the output port of an IS. Fundamentally, an event-controlled multiplexor performs exactly the same function as the `select` operation. So the difference between events or tokens and predicates is really only that the former term and implementation are used in dynamically scheduled designs, while the latter term is used in static schedules.

3.4   Computational Resources

Issue slots are the computational resources of CGRAs. Over the last decade, numerous designs of such issue slots have been proposed, under different names, that include PEs, FUs, ALUs, and flexible computation components. Figure 7 depicts some of them. For all of the possible designs, it is important to know the context in which these ISs have to operate, such as the interconnects connecting them, the control type of the CGRA, etc.

Figure 7a depicts the IS of a MorphoSys CGRA. All 64 ISs in this homogeneous CGRA are identical and include their own local RF. This is no surprise, as the two MorphoSys SIMD modes (see Sect. 3.2.1) require that all ISs of a row or of a column execute the same instruction, which clearly implies homogeneous ISs.

In contrast, almost all features of an ADRES IS, as depicted in Fig. 7b, can be chosen at design time, and can be different for each IS in a CGRA that then becomes heterogeneous: the number of ports, whether or not there are latches between the multiplexors and the combinatorial logic that implements the operations, the set of operations supported by each IS, how the local registers file are connected to ISs and possibly shared between ISs, etc. As long as the design instantiates the ADRES template, the ADRES tool flow will be able to synthesize the architecture and to generate code for it. A similar design philosophy is followed by the Silicon Hive tools. Of course this requires more generic compiler techniques than

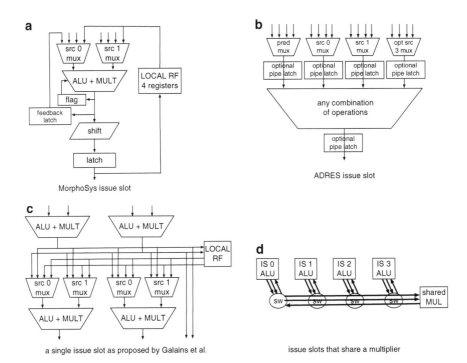

Fig. 7 Four different structures of ISs proposed in the literature. Part (**a**) displays a fixed MorphoSys IS, including its local RF. Part (**b**) displays the fully customizable ADRES IS, that can connect to shared or non-shared local RFs. Part (**c**) depicts the IS structure proposed by Galanis et al. [27], and (**d**) depicts a row of four RSPA ISs that share a multiplier [33]

those that generate code for the predetermined homogeneous ISs of, e.g., the MorphoSys CGRA. Given the state of the art in compiler technology for this type of architecture, the advantages of this freedom are (1) the possibility to design different instances optimized for certain application domains, (2) the knowledge that the features of those designs will be exploited, and (3) the ability to compile loops that feature other forms of ILP than DLP. DLP can still be supported, of course, by simply incorporating SIMD-capable (a.k.a. subword parallel) ISs of, e.g., 4×16 bits wide. The drawback of this design freedom is that, at least with the current compiler techniques, these techniques are so generic that they miss optimization opportunities because they do not exploit regularity in the designed architectures. They do not exploit it for speeding up the compilation, nor do they for producing better schedules. An other potential problem of specializing the ISs for an application domain is overspecialization. While extensive specialization will typically benefit performance, it can also have negative effects, in particular on energy consumption [72].

Figure 7c depicts the IS proposed by Galanis et al. [27]. Again, all ISs are identical. In contrast to the MorphoSys design, however, these ISs consist of several

ALUs and multipliers with direct connections between them and their local RFs. These direct connections within each IS can take care of a lot of data transfers, thus freeing time on the shared bus-based interconnect that connects all ISs. Thus, the local interconnect within each IS compensates for the lack of a scaling global interconnect. One advantage of this clustering approach is that the compiler can be tuned specifically for this combination of local and global connections and for the fact that it does not need to support heterogeneous ISs. Whether or not this type of design is more power-efficient than that of CGRAs with more design freedom and potentially more heterogeneity is unclear at this point in time. At least, we know of no studies from which, e.g., utilization numbers can be derived that allow us to compare the two approaches.

Some architectures combine the flexibility of heterogeneous ADRES ISs with clustering. For example, the CGRA Express [57] and the expression-grained reconfigurable array (EGRA) [3] architectures feature heterogeneous clusters of relatively simple, fast ALUs. Within the clusters, those ALUs are chained by means of a limited number of latchless connections. Through careful design, the delay of those chains is comparable to the delay of other, more complex ISs on the CGRA that bound the clock frequency. So the chaining does not effect the clock frequency. It does allow, however, to execute multiple dependent operations within one clock cycle. It can therefore improve performance significantly. As the chains and clusters are composed of existing components such as ISs, buses, multiplexers and connections, these clustered designs do not really extend the design space of non-clustered CGRAs like ADRES. Still it can be useful to treat clusters as a separate design level in between the IS component level and the whole array architecture level, for example because it allows code generation algorithms in compilers to be tuned for there existence [57].

A specific type of clustering was proposed to handle floating-point arithmetic. While most research on CGRAs is limited to integer and fixed-point arithmetic, Lee et al. proposed to cluster two ISs to handle floating-point data [41]. In their design, both ISs in the cluster can operate independently on integer or fixed-point data, but they can also cooperate by means of a special direct interconnect between them. When they cooperate, one IS in the cluster consumes and handles the mantissas, while the other IS consumes and produces the exponents. As a single ISs can thus be used for both floating-point and integer computations, Lee et al. are able to achieve high utilization for integer applications, floating-point applications, as well as mixed applications.

With respect to utilization, it is clear that the designs of Fig. 7a, b will only be utilized well if a lot of multiplications need to be performed. Otherwise, the area-consuming multipliers remain unused. To work around this problem, the sharing of large resources such as multipliers between ISs has been proposed in the RSPA CGRA design [33]. Figure 7d depicts one row of ISs that do not contain multipliers internally, but that are connected to a shared multiplier through switches and a shared bus. The advantage of this design, compared to an ADRES design in which each row features three pure ALU ISs and one ALU+MULT IS, is that this design allows the compiler to schedule multiplications in all ISs (albeit only one per cycle), whereas this scheduling freedom would be limited to one IS slot in the ADRES design. To allow this schedule freedom, however, a significant amount of resources in the form of switches and a special-purpose bus need to be added to the row. While we lack experimental data to back up this claim, we firmly believe that a similar increase in schedule freedom can be obtained in the aforementioned 3+1 ADRES design by simply extending an existing ADRES interconnect with a similar amount of additional resources. In the ADRES design, that extension would then also be beneficial to operations other than multiplications.

The optimal number of ISs for a CGRA depends on the application domain, on the reconfigurability, as well as on the IS functionality and on the DLP available in the form of subword parallelism. As illustrated in Sect. 4.2.2, a typical ADRES would consist of 4×4 ISs [10, 46]. TRIPS also features 4×4 ISs. MorphoSys provides 8×8 ISs, but that is because the DLP is implemented as SIMD over multiple ISs, rather than as subword parallelism within ISs. In our experience, scaling dynamically reconfigurable CGRA architectures such as ADRES to very large arrays (8×8 or larger) does not make sense even with scalable interconnects like mesh or mesh-plus interconnects. Even in loops with high ILP, utilization drops significantly on such large arrays [51]. It is not yet clear what is causing this lower utilization, and there might be several reasons. These include a lack of memory bandwidth, the possibility that the compiler techniques [20, 48] simply do not scale to such large arrays, or the fact that the relative connectivity in such large arrays is lower. Simply stated, when a mesh interconnects all ISs to their neighbors, each IS not on the side of the array is connected to 4 other ISs out of 16 in a 4×4 array, i.e., to 25 % of all ISs, while it is connected to 4 out of 64 ISs on an 8×8 array, i.e., to 6.25 % of all ISs.

To finalize this section, we want to mention that, just like in any other type of processor, it makes sense to pipeline complex combinatorial logic, e.g., as found in multipliers. There are no fundamental problems to do this, and it can lead to significant increases in utilization and clock frequency.

3.5   Memory Hierarchies

CGRAs have a large number of ISs that need to be fed with data from the memory. Therefore the data memory sub-system is a crucial part of the CGRA design. Many reconfigurable architectures feature multiple independent memory banks or blocks to achieve high data bandwidth.

The RAW architecture features an independent memory block in each tile for which Barua developed a method called modulo unrolling to disambiguate and assign data to different banks [4]. However, this technique can only handle array references through affine index expression on loop induction variables.

MorphoSys has a 256-bit wide frame buffer between the main memory and a reconfigurable array to feed data to the ISs operating in SIMD mode [44]. The efficient use of such a wide memory depends by and large on manual data placement and operation scheduling. Similar techniques for wide loads and stores have also been proposed in regular VLIW architectures for reducing power [60]. Exploiting that hardware requires manual data layout optimizations as well.

Both Silicon Hive and PACT feature distributed memory blocks without a crossbar. A Silicon Hive programmer has to specify the allocation of data to the memory for the compiler to bind the appropriate load/store operations to the corresponding memories. Silicon Hive also supports the possibility of interfacing the memory or system bus using FIFO interfaces. This is efficient for streaming processing but is difficult to interface when the data needs to be buffered on in case of data reuse.

The ADRES architecture template provides a parameterizable Data Memory Queue (DMQ) interface to each of the different single-ported, interleaved level-1 scratch-pad memory banks [19]. The DMQ interface is responsible for resolving bank access conflicts, i.e., when multiple load/store ISs would want to access the same bank at the same time. Connecting all load/store ISs to all banks through a conflict resolution mechanism allows maximal freedom for data access patterns and also maximal freedom on the data layout in memory. The potential disadvantage of such conflict resolution is that it increases the latency of load operations. In software pipelined code, however, increasing the individual latency of instructions most often does not have a negative effect on the schedule quality, because the compiler can hide those latencies in the software pipeline. In the main processor VLIW mode of an ADRES, the same memories are accessed in code that is not software-pipelined. So in that mode, the conflict resolution is disabled to obtain shorter access latencies.

Alternatively, a data cache can be added to the memory hierarchy to complement the scratch-pad memories. By letting the compiler partition the data over the scratch-pad memories and the data cache in an appropriate manner, high throughput can be obtained in the CGRA mode, as well as low latency in the VLIW mode [32].

3.6   Compiler Support

Apart from the specific algorithms used to compile code, the major distinctions between the different existing compiler techniques relate to whether or not they support static scheduling, whether or not they support dynamic reconfiguration, whether or not they rely on special programming languages, and whether or not they are limited to specific hardware properties. Because most compiler research has been done to generate static schedules for CGRAs, we focus on those in this section. As already indicated in Sects. 3.2.1 and 3.2.2, many algorithms are based on FPGA placement and routing techniques [7] in combination with VLIW code generation techniques like modulo scheduling [39, 61] and hyperblock formation [45].

Whether or not compiler techniques rely on specific hardware properties is not always obvious in the literature, as not enough details are available in the descriptions of the techniques, and few techniques have been tried on a wide range of CGRA architectures. For that reason, it is very difficult to compare the efficiency (compilation time) and effectiveness (quality of generated code) of the different techniques.

The most widely applicable static scheduling techniques use different forms of Modulo Resource Routing Graphs (MRRGs). RRGs are time-space graphs, in which all resources (space dimension) are modeled with vertices. There is one such vertex per resource per cycle (time dimension) in the schedule being generated. Directed edges model the connections over which data values can flow from resource to resource. The schedule, placement, and routing problem then becomes a problem of mapping the Data Dependence Graph (DDG) of some loop body on the RRG. Scheduling refers to finding the right cycle to perform an operation in the schedule, placement refers to finding the right IS in that cycle, and routing refers to finding connections to transfer data from producing operations to consuming operations. In the case of a modulo scheduler, the modulo constraint is enforced by modeling all resource usage in the modulo time domain. This is done by modeling the appropriate modulo reservation tables [61] on top of the RRG, hence the name MRRG.

The granularity of its vertices depends on the precise compiler algorithm. One modulo graph embedding algorithm [54] models whole ISs or whole RFs with single vertices, whereas the simulated-annealing technique in the DRESC [20, 48] compiler that targets ADRES instances models individual ports to ISs and RFs as separate vertices. Typically, fewer nodes that model larger components lead to faster compilation because the graph mapping problem operates on a smaller graph, but also to lower code quality because some combinations of resource usage cannot be modeled precisely.

Several types of code schedulers exist. In DRESC, simulated annealing is used to explore different placement and routing options until a valid placement and routing of all operations and data dependencies is found. The cost function used during the simulated annealing is based on the total routing cost, i.e., the combined resource consumption of all placed operations and of all routed data dependencies. In this technique, a huge number of possible routes is evaluated, as a result of which the technique is very slow. Other CGRA scheduling techniques [25, 52, 54, 55] operate much more like (modulo) list schedulers [24]. In the best list-based CGRA schedulers, the used heuristics also rely heavily on routing costs. However, as these techniques try to optimize routes of individual data dependencies in a greedy manner, with little backtracking, they explore orders of magnitude fewer routes. Those schedulers are therefore one to two orders of magnitude faster than the simulated-annealing approach. They are also less effective, but for at least some architectures [52], the difference in generated code quality is very small.

As an alternative to simulated-annealing, Lee et al. use quantum-inspired evolutionary algorithms [41]. In their compiler, they use this algorithm to improve a suboptimal schedule already found by a list scheduler. Their evaluation is limited to rather small loops and a CGRA with limited dynamic reconfigurability, so at this point it is impossible to compare their algorithm's effectiveness and efficiency to those of the other algorithms.

MRRG-based compiler techniques are easily retargetable to a wide range of architectures, such as those of the ADRES template, and they can support many programming languages. Different architectures can simply be modeled with different MRRGs. It has even been demonstrated that by using the appropriate modulo constraints during the mapping of a DDG on a MRRG, compilers can generate a single code version that can be executed on CGRAs of different sizes [58]. This is particularly interesting for the PPA architecture that can switch dynamically between different array sizes [56]. To support different programming languages like C and Fortran, the techniques only require a compiler front-end that is able to generate DDGs for the loop bodies. Obviously, the appropriate loop transformations need to be applied before generating the DDG in order to generate one that maps well onto the MRRG of the architecture. Such loop transformations are discussed in detail in Sect. 4.1.

The aforementioned algorithms have been extended to not only consider the costs of utilized resources inside the CGRA during scheduling, but also consider bank conflicts that may occur because of multiple memory accesses being scheduled in the same cycle [34, 35].

Many other CGRA compiler techniques have been proposed, most of which are restricted to specific architectures. Static reconfigurable architectures like RaPiD and PACT have been targeted by compiler algorithms [14, 22, 76] based on placement and routing techniques that also map DDGs on RRGs. These techniques support subsets of the C programming language (no pointers, no structs, ...) and require the use of special C functions to program the IO in the loop bodies to be mapped onto the CGRA. The latter requirement follows from the specific IO support in the architectures and the modeling thereof in the RRGs.

For the MorphoSys architecture, with its emphasis on SIMD across ISs, compiler techniques have been developed for the SA-C language [73]. In this language the supported types of available parallelism are specified by means of loop language constructs. These constructs are translated into control code for the CGRA, which are mapped onto the ISs together with the DDGs of the loop bodies.

CGRA code generation techniques based on integer-linear programming have been proposed for the several architectures, both for spatial [2] and for temporal mapping [41, 77]. Basically, the ILP formulation consists of all the requirements or constraints that must be met by a valid schedule. This formulation is built from a DDG and a hardware description, and can hence be used to compile many source languages. It is unclear, however, to what extent the ILP formulation and its solution rely on specific architecture features, and hence to which extent it would be possible to retarget the ILP-formulation to different CGRA designs. A similar situation occurs for the constraint-based compilation method developed for the Silicon Hive architecture template [67], of which no detailed information is public. Furthermore, ILP-based compilation is known to be unreasonably slow. So in practice it can only be used for small loop kernels.

Code generation techniques for CGRAs based on instruction-selection pattern matching and list-scheduling techniques have also been proposed [26, 27]. It is unclear to what extent these techniques rely on a specific architecture because we know of no trial to use them for different CGRAs, but these techniques seem to rely heavily on the existence of a single shared-bus that connects ISs as depicted in Fig. 7c. Similarly, the static reconfiguration code generation technique by Lee et al. relies on CGRA rows consisting of identical ISs [42]. Because of this assumption, a two-step code generation approach can be used in which individual placements within rows are neglected in the first step, and only taken care of in the second step. The first step then instead focuses on optimizing the memory traffic.

Finally, compilation techniques have been developed that are really specialized for the TRIPS array layout and for its out-of-order execution [16].

One rule-of-thumb covers all the mentioned techniques: more generic techniques, i.e., techniques that are more flexible in targeting different architectures or different instances of an architecture template, are less efficient and often less effective in exploiting special architecture features. In other words, techniques that rely on specific hardware features, such as interconnect regularities or specific forms of IS clustering, while being less flexible, will generally be able to target those hardware features more efficiently and often also more effectively. Vice versa, architectures with such features usually need specialized compiler techniques. This is similar to the situation of more traditional DSP or VLIW architectures.

4   Case Study: ADRES

This section presents a case study on one specific CGRA design template. The purpose of this study is to illustrate that it is non-trivial to compile and optimize code for CGRA targets, and to illustrate that within a design template, there is a need for hardware design exploration. This illustrates how both hardware and software designers targeting CGRAs need a deep understanding of the interaction between the architecture features and the used compiler techniques.

ADRES [5,9–11,19,20,46,48–50] is an architecture design template from which dynamically reconfigurable, statically scheduled CGRAs can be instantiated. In each instance, an ADRES CGRA is coupled tightly to a VLIW processor. This processor shares data and predicate RFs with the CGRA, as well as memory ports to a multi-banked scratch-pad memory as described in Sect. 3.1. The compiler-supported ISA of the design template provides instructions that are typically found in a load/store VLIW or RISC architecture, including arithmetic operations, logic operations, load/store operations, and predicate computing instructions. Additional domain-specific instructions, such as SIMD operations, are supported in the programming tools by means of intrinsics [68]. Local rotating and non-rotating, shared and private local RFs can be added to the CGRA as described in the previous sections, and connected through an interconnect consisting of muxes, buses and point-to-point connections that are specified completely by the designer. Thus, the ADRES architecture template is very flexible: it offers a high degree of design freedom, and it can be used to accelerate a wide range of loops.

4.1 Mapping Loops on ADRES CGRAs

The first part of this case study concerns the mapping of loops onto ADRES CGRAs, which are one of the most flexible CGRAs supporting a wide range of loops. This study illustrates that many loop transformations need to be applied carefully before mapping code onto ADRES CGRAs. We discuss the most important compiler transformations and, lacking a full-fledged loop-optimizing compiler, manual loop transformations that need to be applied to source code in order to obtain high performance and high efficiency. For other, less flexible CGRAs, the need for such transformations will even be higher because there will be more constraints on the loops to be mapped in the first place. Hence many of the discussed issues not only apply to ADRES CGRAs, but also to other CGRA architectures. We will conclude from this study that programming CGRAs with the existing compiler technology is not compatible with high programmer productivity.

4.1.1 Modulo Scheduling Algorithms for CGRAs

To exploit ILP in inner loops on VLIW architectures, compilers typically apply software pipelining by means of modulo scheduling [39,61]. This is no different for ADRES CGRAs. In this section, we will not discuss the inner working of modulo scheduling algorithms. What we do discuss, are the consequences of using that technique for programming ADRES CGRAs.

After a loop has been modulo-scheduled, it consists of three phases: the prologue, the kernel and the epilogue. During the prologue, stages of the software-pipelined loop gradually become active. Then the loop executes the kernel in a steady-state mode in which all software pipeline stages are active, and afterwards the stages are gradually disabled during the epilogue. In the steady-state mode, a new iteration is started after every II cycles, which stands for Initiation Interval. Fundamentally, every software pipeline stage is II cycles long. The total cycle count of a loop with $iter$ iterations that is scheduled over ps software pipeline stages is then given by

$$cycles_{prologue} + II \cdot (iter - (ps - 1)) + cycles_{epilogue}. \tag{2}$$

In this formula, we neglect processor stalls because of, e.g., memory access conflicts or cache misses.

For loops with a high number of iterations, the term $II \cdot iter$ dominates this cycle count, and that is why modulo scheduling algorithms try to minimize II, thus increasing the IPC terms in Eq. (1).

The minimal II that modulo scheduling algorithms can reach is bound by $minII = \max(RecMII, ResMII)$. The first term, called resource-minimal II ($ResMII$) is determined by the resources required by a loop and by the resources provided by the architecture. For example, if a loop body contains nine multiplications, and there are only two ISs that can execute multiplications, then at least $\lceil 9/2 \rceil = 5$ cycles will be needed per iteration. The second term, called recurrence-minimal

II (*RecMII*) depends on recurrent data dependencies in a loop and on instruction latencies. Fundamentally, if an iteration of a loop depends on the previous iteration through a dependency chain with accumulated latency *RecMII*, it is impossible to start that iteration before at least *RecMII* cycles of the previous iteration have been executed.

The next section uses this knowledge to apply transformations that optimize performance according to Eq. (1). To do so successfully, it is important to know that ADRES CGRAs support only one thread, for which the processor has to switch from a non-CGRA operating mode to CGRA mode and back for each inner loop. So besides minimizing the cycle count of Eq. (2) to obtain higher IPCs in Eq. (1), it is also important to consider the terms $t_{p \to p+1}$ in Eq. (1).

4.1.2 Loop Transformations

Loop Unrolling

Loop unrolling and the induction variable optimizations that it enables can be used to minimize the number of iterations of a loop. When a loop body is unrolled x times, *iter* decreases with a factor x, and *ResMII* typically grows with a factor slightly less than x because of the induction variable optimizations and because of the ceiling operation in the computation of *ResMII*. By contrast, *RecMII* typically remains unchanged or increases only a little bit as a result of the induction variable optimizations that are enabled after loop unrolling.

In resource-bound loops, *ResMII* > *RecMII*. Unrolling will then typically have little impact on the dominating term $II \cdot iter$ in Eq. (2). However, the prologue and the epilogue will typically become longer because of loop unrolling. Moreover, an unrolled loop will consume more space in the instruction memory, which might also have a negative impact on the total execution time of the whole application. So in general, unrolling resource-bound loops is unlikely to be very effective.

In recurrence-bound loops, $RecMII \cdot iter > ResMII \cdot iter$. The right hand side of this inequality will not increase by unrolling, while the left hand side will be divided by the unrolling factor x. As this improvement typically compensates for the longer prologue and epilogue, we can conclude that unrolling can be an effective optimization technique for recurrence-bound loops if the recurrences can be optimized with induction variable optimizations. This is no different for CGRAs than it is for VLIWs. However, for CGRAs with their larger number of ISs, it is more important because more loops are recurrence-bound.

Loop Fusion, Interchange, Combination and Data Context Switching

Fusing adjacent loops with the same number of iterations into one loop can also be useful, because fusing multiple recurrence-bound loops can result in one resource-bound loop, which will result in a lower overall execution time. Furthermore, less switching between operating modes takes place with fused loops, and hence the terms $t_{p \to p+1}$ are minimized. Furthermore, less prologues and epilogues need to be executed, which might also improve performance. This improvement will usually be limited, however, because the fused prologues and epilogues will rarely be much shorter than the sum of the original ones. Moreover, loop fusion does result in a loop that is bigger than any of the original loops, so it can only be applied if the configuration memory is big enough to fit the fused loop. If this is the case, less loop configurations need to be stored and possibly reloaded into the memory.

Interchanging an inner and outer loop serves largely the same purpose as loop fusion. As loop interchange does not necessarily result in larger prologues and epilogues, it can be even more useful, as can be the combining of nested loops into a single loop. Data-context switching [8] is a very similar technique that serves the same purpose. That technique has been used by Lee et al. for statically reconfigurable CGRAs as well [42], and in fact most of the loop transformations mentioned in this section can be used to target such CGRAs, as well as any other type of CGRA.

Live-in Variables

In our experience, there is only one caveat with the above transformations. The reason to be careful when applying them is that they can increase the number of live-in variables. A live-in variable is a variable that gets assigned a value before the loop, which is consequently used in the loop. Live-in variables can be manifest in the original source code, but they can also result from compiler optimizations that are enabled by the above loop transformations, such as induction variable optimizations and loop-invariant code motion. When the number of live-in variables increases, more data needs to be passed from the non-loop code to the loop code, which might have a negative effect on $t_{p \to p+1}$. The existence and the scale of this effect will usually depend on the hardware mechanism that couples the CGRA accelerator to the main core. Possible such mechanisms are discussed in Sect. 3.1. In tightly-coupled designs like that of ADRES or Silicon Hive, passing a limited amount of values from the main CPU mode to the CGRA mode does not involve any overhead: the values are already present in the shared RF. However, if their number grows too big, there will not be enough room in the shared RF, which will result in much less efficient passing of data through memory. We have experienced this several times with loops in multimedia and SDR applications that were mapped onto our ADRES designs. So, even for tightly-coupled CGRA designs, the above loop transformations and the enabled optimizations need to be applied with great care.

Predication

Modulo scheduling techniques for CGRAs [20, 22, 25, 48, 54, 55] only schedule loops that are free of control flow transfers. Hence any loop body that contains conditional statements first needs to be if-converted into hyperblocks by means of predication [45]. For this reason, many CGRAs, including ADRES CGRAs, support predication.

Hyperblock formation can result in very inefficient code if a loop body contains code paths that are executed rarely. All those paths contribute to *ResMII* and potentially to *RecMII*. Hence even paths that get executed very infrequently can slow down a whole modulo-scheduled loop. Such loops can be detected with profiling, and if the data dependencies allow this, it can be useful to split these loops into multiple loops. For example, a first loop can contain the code of the frequently executed paths only, with a lower *II* than the original loop. If it turns out during the execution of this loop that in some iteration the infrequently executed code needs to be executed, the first loop is exited, and for the remaining iterations a second loop is entered that includes both the frequently and the infrequently executed code paths.

Alternatively, for some loops it is beneficial to have a so-called inspector loop with very small *II* to perform only the checks for all iterations. If none of the checks are positive, a second so-called executor loop is executed that includes all the computations except the checks and the infrequently executed paths. If some checks were positive, the original loop is executed.

One caveat with this loop splitting is that it causes code size expansion in the CGRA instruction memories. For power consumption reasons, these memories are kept as small as possible. This means that the local improvements obtained with the loop splitting need to be balanced with the total code size of all loops that need to share these memories.

Kernel-Only Loops

Predication can also be used to generate so-called kernel-only loop code. This is loop code that does not have separate prologue and epilogue code fragments. Instead the prologues and epilogues are included in the kernel itself, where predication is now used to guard whole software pipeline stages and to ensure that only the appropriate software pipeline stages are activated at each point in time. A traditional loop with a separate prologue and epilogue is compared to a kernel-only loop in Fig. 8. Three observations need to be made here.

The first observation is that kernel-only code is usually faster because the pipeline stages of the prologue and epilogue now get executed on the CGRA accelerator, which typically can do so at much higher IPCs than the main core. This is a major difference between (ADRES) CGRAs and VLIWs. On the latter, kernel-only loops are much less useful because all code runs on the same number of ISs anyway.

Secondly, while kernel-only code will be faster on CGRAs, more time is spent in the CGRA mode, as can be seen in Fig. 8. During the epilogue and prologue, the

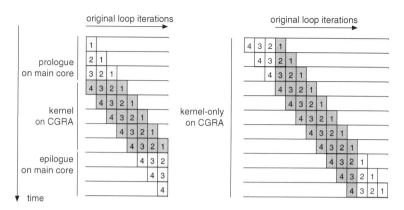

Fig. 8 On the *left* a traditional modulo-scheduled loop, on the *right* a kernel-only one. *Each numbered box* denotes one of four software pipeline stages, and *each row* denotes the concurrent execution of different stages of different iterations. *Grayed boxes* denote stages that actually get executed. On the *left*, the *dark grayed boxes* get executed on the CGRA accelerator, in which exactly the same code is executed every *II* cycles. The *light grayed boxes* are pipeline stages that get executed outside of the loop, in separate code that runs on the main processor. On the *right*, kernel-only code is shown. Again, the *dark grey boxes* are executed on the CGRA accelerator. So are the *white boxes*, but these get deactivated during the prologue and epilogue by means of predication

whole CGRA is active and thus consuming energy, but many ISs are not performing useful computations because they execute operations from inactive pipeline stages. Thus, kernel-only is not necessarily optimal in terms of energy consumption.

The third observation is that for loops where predication is used heavily to create hyperblocks, the use of predicates to support kernel-only code might over-stress the predication support of the CGRA. In domains such as SDR, where the loops typically have no or very little conditional statements, this poses no problems. For applications that feature more complex loops, such as in many multimedia applications, this might create a bottleneck even when predicate speculation [65] is used. This is where the ADRES template proves to be very useful, as it allowed us to instantiate specialized CGRAs with varying predicate data paths, as can be seen in Table 1.

4.1.3 Data Flow Manipulations

The need for fine-tuning source code is well known in the embedded world. In practice, each compiler can handle some loop forms better than other forms. So when one is using a specific compiler for some specific VLIW architecture, it can be very beneficial to bring loops in the appropriate shape or form. This is no different when one is programming for CGRAs, including ADRES CGRAs.

Table 1 Main differences between two studied ADRES CGRAs

	multimedia CGRA	SDR CGRA
# issue slots (FUs)	4x4	4x4
# load/store units	4	4
ld/st/mul latency	6/6/2 cycles	7/7/3 cycles
# local data RFs	12 (8 single-ported) of size 8	12 (8 single-ported) of size 4
data width	32	64
config. word width	896 bits	736 bits
ISA extensions	2-way SIMD, clipping, min/max	4-way SIMD, saturating arithm.
interconnect	Nearest Neighbor (NN) + 8 predicate buses	NN + next-hop + 8 data buses
pipelining	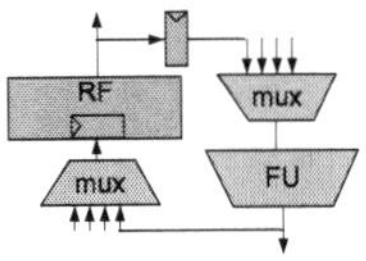	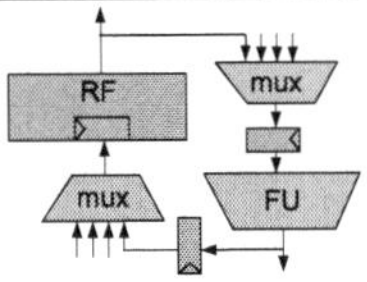
power, clock, and area	91 mW at 300 MHz for 4mm^2	310 mW at 400 MHz for 5.79 mm^2

Power, clock and area include the CGRA and its configuration memory, the VLIW processor for non-loop code, including its 32K L1 I-cache, and the 32K 4-bank L1 data memory. These numbers are gate-level estimates

(a) original 15-tap FIR filter

```
const short c[15] = {-32, ..., 1216};
for (i = 0; i < nr; i++) {
  for(value = 0, j = 0; j < 15; j++)
    value += x[i+j]*c[j];
  r[i] = value;
}
```

(b) filter after loop unrolling, with hard-coded constants

```
const short c00 = -32, ..., c14 = 1216;
for (i = 0; i < nr; i++)
  r[i] = x[i+0]*c00 + x[i+1]*c01 + ... + x[i+14]*c14;
```

(c) after redundant memory accesses are eliminated

```
int i, value, d0, ..., d14;
const short c00 = -32, ..., c14 = 1216;
for (i = 0; i < nr+15; i++) {
  d0 = d1;  d1 = d2;  ... ; d13 = d14; d14 = x[i];
  value = c00 * d0 +  c01 * d1 + ... + c14 * d14;
  if (i >= 14) r[i - 14 ] = value;
}
```

Fig. 9 Three C versions of a FIR filter

Apart from the above transformations that relate to the modulo scheduling of loops, there are important transformations that can increase the "data flow" character of a loop, and thus contribute to the efficiency of a loop. Three C implementations of a Finite Impulse Response (FIR) filter in Fig. 9 provide an excellent example.

Table 2 Number of execution cycles and memory accesses (obtained through simulation) for the FIR-filter versions compiled for the multimedia CGRA, and for the TI C64+ DSP

Program	Cycle count		Memory accesses	
	CGRA	TI C64+	CGRA	TI C64+
FIR (a)	11,828	1,054	6,221	1,618
FIR (b)	1,247	1,638	3,203	2,799
FIR (c)	664	10,062	422	416

Figure 9a depicts a FIR implementation that is efficient for architectures with few registers. For architectures with more registers, the implementation depicted in Fig. 9b will usually be more efficient, as many memory accesses have been eliminated. Finally, the equivalent code in Fig. 9c contains only one load per outer loop iteration. To remove the redundant memory accesses, a lot of temporary variables had to be inserted, together with a lot of copy operations that implement a delay line. On regular VLIW architectures, this version would result in high register pressure and many copy operations to implement the data flow of those copy operations. Table 2 presents the compilation results for a 16-issue CGRA and for an 8-issue clustered TI C64+ VLIW. From the results, it is clear that the TI compiler could not handle the latter code version: its software-pipelining fails completely due to the high register pressure. When comparing the minimal cycle times obtained for the TI C64+ with those obtained for the CGRA, please note that the TI compiler applied SIMDization as much as it could, which is fairly orthogonal to scheduling and register allocation, but which the experimental CGRA compiler used for this experiment did not yet perform. By contrast, the CGRA compiler could optimize the code of Fig. 9c by routing the data of the copy operations over direct connections between the CGRA ISs. As a result, the CGRA implementation becomes both fast and power-efficient at the same time.

This is a clear illustration of the fact that, lacking fully automated compiler optimizations, heavy performance-tuning of the source code can be necessary. The fact that writing efficient source code requires a deep understanding of the compiler internals and of the underlying architecture, and the fact that it frequently includes experimentation with various loop shapes, severely limits the programming productivity. This has to be considered a severe drawback of CGRAs architectures.

Moreover, as the FIR filter shows, the optimal source code for a CGRA target can be radically different than that for, e.g., a VLIW target. Consequently, the cost of porting code from other targets to CGRAs or vice versa, or of maintaining code versions for different targets (such as the main processor and the CGRA accelerator), can be high. This puts an additional limitation on programmer productivity.

4.2 ADRES Design Space Exploration

In this part of our case study, we discuss the importance and the opportunities for DSE within the ADRES template. First, we discuss some concrete ADRES instances that have been used for extensive experimentation, including the fabrication of working silicon samples. These examples demonstrate that very power-efficient CGRAs can be designed for specific application domains.

Afterwards, we will show some examples of DSE results with respect to some of the specific design options that were discussed in Sect. 3.

4.2.1 Example ADRES Instances

During the development of the ADRES tool chain and design, two main ADRES instances have been worked out. One was designed for multimedia applications [5, 46] and one for SDR baseband processing [9, 10]. Their main differences are presented in Table 1. Both architectures have a 64-entry data RF (half rotating, half non-rotating) that is shared with a unified three-issue VLIW processor that executes non-loop code. Thus this shared RF has six read ports and three write ports. Both CGRAs feature 16 FUs, of which four can access the memory (that consists of four single-ported banks) through a queue mechanism that can resolve bank conflicts. Most operations have latency one, with the exception of loads, stores, and multiplications. One important difference between the two CGRAs relates to their pipeline schemes, as depicted for a single IS (local RF and FU) in Table 1. As the local RFs are only buffered at their input, pipelining registers need to be inserted in the paths to and from the FUs in order to obtain the desired frequency targets as indicated in the table. The pipeline latches shown in Table 1 hence directly contribute in the maximization of the factor f_p in Eq. (1). Because the instruction sets and the target frequencies are different in both application domains, the SDR CGRA has one more pipeline register than the multimedia CGRA, and they are located at different places in the design.

Traditionally, in VLIWs or in out-of-order superscalar processors, deeper pipelining results in higher frequencies but also in lower IPCs because of larger branch misprediction penalties. Following Eq. (1), this can result in lower performance. In CGRAs, however, this is not necessarily the case, as explained in Sect. 3.3.1. To illustrate this, Table 3 includes IPCs obtained when generating code for both CGRAs with and without the pipelining latches.

The benchmarks mapped onto the multimedia ADRES CGRA are a H.264AVC video decoder, a wavelet-based video decoder, an MPEG4 video coder, a black-and-white TIFF image filter, and a SHA-2 encryption algorithm. For each application at most the ten hottest inner loops are included in the table. For the SDR ADRES CGRA, we selected two baseband modem benchmarks: one WLAN MIMO Channel Estimation and one that implements the remainder of a WLAN SISO receiver. All

Table 3 Results for the benchmark loops

Benchmark	CGRA	Loop	#ops	ResMII	Pipelined			Non-pipelined		
					RecMII	II	IPC	RecMII	II	IPC
AVC decoder	Multimedia	MBFilter1	70	5	2	6	11.7	1	6	11.7
		MBFilter2	89	6	7	9	9.9	6	8	11.1
		MBFilter3	40	3	3	4	10.0	2	3	13.3
		MBFilter4	105	7	2	9	11.7	1	9	11.7
		MotionComp	109	7	3	10	10.9	2	10	10.9
		FindFrameEnd	27	4	7	7	3.9	6	6	4.5
		IDCT1	60	4	2	5	12.0	1	5	12.0
		MBFilter5	87	6	3	7	12.4	2	7	12.4
		Memset	10	2	2	2	5.0	1	2	5.0
		IDCT2	38	3	2	3	12.7	1	3	12.7
		Average					**10.0**			**10.5**
Wavelet	Multimedia	Forward1	67	5	5	6	11.2	5	5	13.4
		Forward2	77	5	5	6	12.8	5	6	12.8
		Reverse1	73	5	2	6	12.2	1	6	12.2
		Reverse2	37	3	2	3	12.3	1	3	12.3
		Average					**12.1**			**12.7**
MPEG-4 encoder	Multimedia	MotionEst1	75	5	2	6	12.5	1	6	12.5
		MotionEst2	72	5	3	6	12.0	2	6	12.0
		TextureCod1	73	5	7	7	10.4	6	6	12.2
		CalcMBSAD	60	4	2	5	12.0	1	5	12.0
		TextureCod2	9	1	2	2	4.5	1	2	4.5
		TextureCod3	91	6	2	7	13.0	1	7	13.0
		TextureCod4	91	6	2	7	13.0	1	7	13.0
		TextureCod5	82	6	2	6	13.7	1	6	13.7
		TextureCod6	91	6	2	7	13.0	1	7	13.0
		MotionEst3	52	4	3	4	13.0	2	5	10.4
		Average					**11.7**			**11.6**
Tiff2BW	Multimedia	Main loop	35	3	2	3	11.7	1	3	11.7
SHA-2	Multimedia	Main loop	111	7	8	9	12.3	8	9	12.3
MIMO	SDR	Channel2	166	11	3	14	11.9	1	14	10.4
		Channel1	83	6	3	8	10.4	1	8	10.7
		SNR	75	5	4	6	12.5	2	6	12.5
		Average					**11.6**			**11.2**
WLAN	SDR	DemapQAM64	55	4	3	6	9.2	1	6	9.2
		64-point FFT	123	8	4	10	12.3	2	12	10.3
		Radix8 FFT	122	8	3	10	12.2	1	12	10.2
		Compensate	54	4	4	5	10.8	2	5	10.8
		DataShuffle	153	14	3	14	10.9	1	16	9.6
		Average					**11.1**			**10.0**

First, the target-version-independent number of operations (#ops) and the ResMII. Then for each target version the RecMII, the actually achieved II and IPC (counting SIMD operations as only one operation), and the compile time

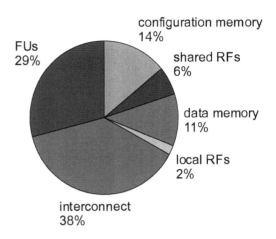

Fig. 10 Average power consumption distribution of the ADRES SDR CGRA in CGRA mode

applications are implemented in standard ANSI C using all language features such as pointers, structures, different loop constructs (while, for, do-while), but not using dynamic memory management functions like `malloc` or `free`.

The general conclusions to be taken from the mapping results in Table 3 are as follows. (1) Very high IPCs are obtained at low power consumption levels of 91 and 310 mW and at relatively high frequencies of 300 and 400 MHz, given the standard cell 90 nm design. (2) Pipelining seems to be bad for performance only where the initiation interval is bound by *RecMII*, which changes with pipelining. (3) In some cases pipelining even improves the IPC.

Synthesizable VHDL is generated for both processors by a VHDL generator that generates VHDL code starting from the same XML architecture specification used to retarget the ANSI C compiler to different CGRA instances. A TSMC 90 nm standard cell GP CMOS (i.e. the General-Purpose technology version that is optimized for performance and active power, not for leakage power) technology was used to obtain the gate-level post-layout estimates for frequency, power and area in Table 1. More detailed results of these experiments are available in the literature for this SDR ADRES instance [9,10], as well as for the multimedia instance [5,46]. The SDR ADRES instance has also been produced in silicon in samples of a full SoC SDR chip [21]. The two ADRES cores on this SoC proved to be fully functional at 400 MHz, and the power consumption estimates have been validated.

One of the most interesting results is depicted in Fig. 10, which displays the average power consumption distribution over the ADRES SDR CGRA when the CGRA mode is active in the above SDR applications. Compared to VLIW processor designs, a much larger fraction of the power is consumed in the interconnects and in the FUs, while the configuration memory (which corresponds to an L1 VLIW instruction cache), the RFs and the data memory consume relatively little energy. This is particularly the case for the local RFs. This clearly illustrates that by focusing on regular loops and their specific properties, CGRAs can achieve higher performance and a higher power-efficiency than VLIWs. On the CGRA, most of the

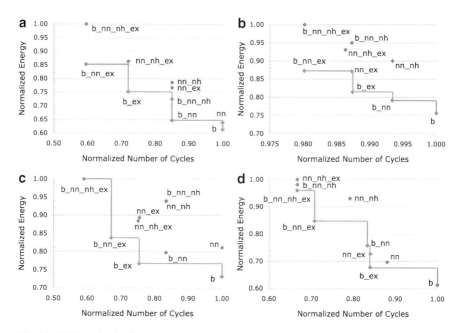

Fig. 11 DSE results for four microbenchmarks on 4×4 CGRAs with fixed ISs and fixed RFs, but with varying interconnects (**a**) MIMO; (**b**) AVC interpolation; (**c**) Viterbi; (**d**) AVC motion estimation

power is spent in the FUs and in the interconnects, i.e., on the actual computations and on the transfers of values from computation to computation. The latter two aspects are really the fundamental parts of the computation to be performed, unlike the fetching of data or the fetching of code, which are merely side-effects of the fact that processors consist of control paths, data paths, and memories.

4.2.2 Design Space Exploration Example

Many DSEs have been performed within the ADRES template [5, 11, 15, 40, 46, 51]. We present one experimental result [40] here, not to present absolute numbers but to demonstrate the large impact on performance and on energy consumption that some design choices can have. In this experiment, a number of different interconnects have been explored for four microbenchmarks (each consisting of several inner loops): a MIMO SDR channel estimation, a Viterbi decoder, an Advanced Video Codec (AVC) motion estimation, and an AVC half-pixel interpolation filter. All of them have been compiled with the DRESC compiler for different architectures of which the interconnects are combinations of the four basic interconnects of Fig. 6, in which distributed RFs have been omitted for the sake of clarity.

Figure 11 depicts the relative performance and (estimated) energy consumption for different combinations of these basic interconnects. The names of the different architectures indicate which basic interconnects are included in its interconnect. For example, the architecture b_nn_ex includes the buses, nearest neighbor interconnects and extra connections to the shared RF. The lines connecting architectures in the charts of Fig. 11 connect the architectures on the Pareto fronts: these are the architectures that have an optimal combination of cycle count and energy consumption. Depending on the trade-off made by a designer between performance and energy consumption, he will select one architecture on that Pareto front.

The lesson to learn from these Pareto fronts is that relatively small architectural changes, in this case involving only the interconnect but not the ISs or the distributed RFs, can span a wide range of architectures in terms of performance and energy-efficiency. When designing a new CGRA or choosing for an existing one, it is hence absolutely necessary to perform a good DSE that covers ISA, ISs, interconnect and RFs. Because of the large design space, this is far from trivial.

5 Conclusions

This chapter on CGRA architectures presented a discussion of the CGRA processor design space as an accelerator for inner loops of DSP-like applications such as software-defined radios and multimedia processing. A range of options for many design features and design parameters has been related to power consumption, performance, and flexibility. In a use case, the need for design space exploration and for advanced compiler support and manual high-level code tuning have been demonstrated. The above discussions and demonstration support the following main conclusions. Firstly, CGRAs can provide an excellent alternative for VLIWs, providing better performance and better energy efficiency. Secondly, design space exploration is needed to achieve those goals. Finally, existing compiler support needs to be improved, and until that happens, programmers need to have a deep understanding of the targeted CGRA architectures and their compilers in order to manually tune their source code. This can significantly limit programmer productivity.

References

1. Abnous, A., Christensen, C., Gray, J., Lenell, J., Naylor, A., Bagherzadeh, N.: Design and implementation of the "Tiny RISC" microprocessor. Microprocessors & Microsystems 16(4), 187–193 (1992)
2. Ahn, M., Yoon, J.W., Paek, Y., Kim, Y., Kiemb, M., Choi, K.: A spatial mapping algorithm for heterogeneous coarse-grained reconfigurable architectures. In: DATE '06: Proceedings of the Conference on Design, Automation and Test in Europe, pp. 363–368 (2006)
3. Ansaloni, G., Bonzini, P., Pozzi, L.: EGRA: A coarse grained reconfigurable architectural template. IEEE Transactions on Very Large Scale Integration (VLSI) Systems 19(6), 1062–1074 (2011)

4. Barua, R.: Maps: A compiler-managed memory system for software-exposed architectures. Ph.D. thesis, Massachusetss Institute of Technology (2000)
5. Berekovic, M., Kanstein, A., Mei, B., De Sutter, B.: Mapping of nomadic multimedia applications on the ADRES reconfigurable array processor. Microprocessors & Microsystems **33**(4), 290–294 (2009)
6. van Berkel, k., Heinle F. amd Meuwissen, P., Moerman, K., Weiss, M.: Vector processing as an enabler for software-defined radio in handheld devices. EURASIP Journal on Applied Signal Processing **2005**(16), 2613–2625 (2005). DOI 10.1155/ASP.2005.2613
7. Betz, V., Rose, J., Marguardt, A.: Architecture and CAD for Deep-Submicron FPGAs. Kluwer Academic Publishers (1999)
8. Bondalapati, K.: Parallelizing DSP nested loops on reconfigurable architectures using data context switching. In: DAC '01: Proceedings of the 38th annual Design Automation Conference, pp. 273–276 (2001)
9. Bougard, B., De Sutter, B., Rabou, S., Novo, D., Allam, O., Dupont, S., Van der Perre, L.: A coarse-grained array based baseband processor for 100Mbps+ software defined radio. In: DATE '08: Proceedings of the Conference on Design, Automation and Test in Europe, pp. 716–721 (2008)
10. Bougard, B., De Sutter, B., Verkest, D., Van der Perre, L., Lauwereins, R.: A coarse-grained array accelerator for software-defined radio baseband processing. IEEE Micro **28**(4), 41–50 (2008). DOI http://doi.ieeecomputersociety.org/10.1109/MM.2008.49
11. Bouwens, F., Berekovic, M., Gaydadjiev, G., De Sutter, B.: Architecture enhancements for the ADRES coarse-grained reconfigurable array. In: HiPEAC '08: Proceedings of the International Conference on High-Performance Embedded Architectures and Compilers, pp. 66–81 (2008)
12. Burns, G., Gruijters, P.: Flexibility tradeoffs in SoC design for low-cost SDR. Proceedings of SDR Forum Technical Conference (2003)
13. Burns, G., Gruijters, P., Huiskens, J., van Wel, A.: Reconfigurable accelerators enabling efficient SDR for low cost consumer devices. Proceedings of SDR Forum Technical Conference (2003)
14. Cardoso, J.M.P., Weinhardt, M.: XPP-VC: A C compiler with temporal partitioning for the PACT-XPP architecture. In: FPL '02: Proceedings of the 12th International Conference on Field-Programmable Logic and Applications, pp. 864–874 (2002)
15. Cervero, T., Kanstein, A., López, S., De Sutter, B., Sarmiento, R., Mignolet, J.Y.: Architectural exploration of the H.264/AVC decoder onto a coarse-grain reconfigurable architecture. In: Proceedings of the International Conference on Design of Circuits and Integrated Systems (2008)
16. Coons, K.E., Chen, X., Burger, D., McKinley, K.S., Kushwaha, S.K.: A spatial path scheduling algorithm for EDGE architectures. In: ASPLOS '06: Proceedings of the 12th International Conference on Architectural Support for Programming Languages and Operating Systems, pp. 129–148 (2006)
17. Corporaal, H.: Microprocessor Architectures from VLIW to TTA. John Wiley (1998)
18. Cronquist, D., Franklin, P., Fisher, C., Figueroa, M., Ebeling, C.: Architecture design of reconfigurable pipelined datapaths. In: Proceedings of the Twentieth Anniversary Conference on Advanced Research in VLSI (1999)
19. De Sutter, B., Allam, O., Raghavan, P., Vandebriel, R., Cappelle, H., Vander Aa, T., Mei, B.: An efficient memory organization for high-ILP inner modem baseband SDR processors. Journal of Signal Processing Systems **61**(2), 157–179 (2010)
20. De Sutter, B., Coene, P., Vander Aa, T., Mei, B.: Placement-and-routing-based register allocation for coarse-grained reconfigurable arrays. In: LCTES '08: Proceedings of the 2008 ACM SIGPLAN-SIGBED Conference on Languages, Compilers, and Tools for Embedded Systems, pp. 151–160 (2008)
21. Derudder, V., Bougard, B., Couvreur, A., Dewilde, A., Dupont, S., Folens, L., Hollevoet, L., Naessens, F., Novo, D., Raghavan, P., Schuster, T., Stinkens, K., Weijers, J.W., Van der Perre, L.: A 200Mbps+ 2.14nJ/b digital baseband multi processor system-on-chip for SDRs. In: Proceedings of the Symposium on VLSI Systems, pp. 292–293 (2009)

22. Ebeling, C.: Compiling for coarse-grained adaptable architectures. Tech. Rep. UW-CSE-02-06-01, University of Washington (2002)
23. Ebeling, C.: The general RaPiD architecture description. Tech. Rep. UW-CSE-02-06-02, University of Washington (2002)
24. Fisher, J., Faraboschi, P., Young, C.: Embedded Computing, A VLIW Approach to Architecture, Compilers and Tools. Morgan Kaufmann (2005)
25. Friedman, S., Carroll, A., Van Essen, B., Ylvisaker, B., Ebeling, C., Hauck, S.: SPR: An architecture-adaptive CGRA mapping tool. In: FPGA '09: Proceeding of the ACM/SIGDA International symposium on Field Programmable Gate Arrays, pp. 191–200. ACM, New York, NY, USA (2009). DOI http://doi.acm.org/10.1145/1508128.1508158
26. Galanis, M.D., Milidonis, A., Theodoridis, G., Soudris, D., Goutis, C.E.: A method for partitioning applications in hybrid reconfigurable architectures. Design Automation for Embedded Systems 10(1), 27–47 (2006)
27. Galanis, M.D., Theodoridis, G., Tragoudas, S., Goutis, C.E.: A reconfigurable coarse-grain data-path for accelerating computational intensive kernels. Journal of Circuits, Systems and Computers pp. 877–893 (2005)
28. Gebhart, M., Maher, B.A., Coons, K.E., Diamond, J., Gratz, P., Marino, M., Ranganathan, N., Robatmili, B., Smith, A., Burrill, J., Keckler, S.W., Burger, D., McKinley, K.S.: An evaluation of the TRIPS computer system. In: ASPLOS '09: Proceeding of the 14th International Conference on Architectural Support for Programming Languages and Operating Systems, pp. 1–12 (2009)
29. Hartenstein, R., Herz, M., Hoffmann, T., Nageldinger, U.: Mapping applications onto reconfigurable KressArrays. In: Proceedings of the 9th International Workshop on Field Programmable Logic and Applications (1999)
30. Hartenstein, R., Herz, M., Hoffmann, T., Nageldinger, U.: Generation of design suggestions for coarse-grain reconfigurable architectures. In: FPL '00: Proceedings of the 10th International Workshop on Field Programmable Logic and Applications (2000)
31. Hartenstein, R., Hoffmann, T., Nageldinger, U.: Design-space exploration of low power coarse grained reconfigurable datapath array architectures. In: Proceedings of the International Workshop - Power and Timing Modeling, Optimization and Simulation (2000)
32. Kim, H.s., Yoo, D.h., Kim, J., Kim, S., Kim, H.s.: An instruction-scheduling-aware data partitioning technique for coarse-grained reconfigurable architectures. In: LCTES '11: Proceedings of the 2011 ACM SIGPLAN-SIGBED Conference on Languages, Compilers, Tools and Theorie for Embedded Systems, pp. 151–160 (2011)
33. Kim, Y., Kiemb, M., Park, C., Jung, J., Choi, K.: Resource sharing and pipelining in coarse-grained reconfigurable architecture for domain-specific optimization. In: DATE '05: Proceedings of the Conference on Design, Automation and Test in Europe, pp. 12–17 (2005)
34. Kim, Y., Lee, J., Shrivastava, A., Paek, Y.: Operation and data mapping for CGRAs with multi-bank memory. In: LCTES '10: Proceedings of the 2010 ACM SIGPLAN-SIGBED Conference on Languages, Compilers, and Tools for Embedded Systems, pp. 17–25 (2010)
35. Kim, Y., Lee, J., Shrivastava, A., Yoon, J., Paek, Y.: Memory-aware application mapping on coarse-grained reconfigurable arrays. In: HiPEAC '10: Proceedings of the 2010 International Conference on High Performance Embedded Architectures and Compilers, pp. 171–185 (2010)
36. Kim, Y., Mahapatra, R.: A new array fabric for coarse-grained reconfigurable architecture. In: Proceedings of the IEEE EuroMicro Conference on Digital System Design, pp. 584–591 (2008)
37. Kim, Y., Mahapatra, R., Park, I., Choi, K.: Low power reconfiguration technique for coarse-grained reconfigurable architecture. IEEE Transactions on Very Large Scale Integration (VLSI) Systems 17(5), 593–603 (2009)
38. Kim, Y., Mahapatra, R.N.: Dynamic Context Compression for Low-Power Coarse-Grained Reconfigurable Architecture. IEEE Transactions on Very Large Scale Integration (VLSI) Systems 18(1), 15–28 (2010)
39. Lam, M.S.: Software pipelining: An effecive scheduling technique for VLIW machines. In: Proc. PLDI, pp. 318–327 (1988)

40. Lambrechts, A., Raghavan, P., Jayapala, M., Catthoor, F., Verkest, D.: Energy-aware interconnect optimization for a coarse grained reconfigurable processor. In: Proceedings of the International Conference on VLSI Design, pp. 201–207 (2008)
41. Lee, G., Choi, K., Dutt, N.: Mapping multi-domain applications onto coarse-grained reconfigurable architectures. IEEE Trans. on Computer-Aided Design of Integrated Circuits and Systems **30**(5), 637–650 (2011)
42. Lee, J.e., Choi, K., Dutt, N.D.: An algorithm for mapping loops onto coarse-grained reconfigurable architectures. In: LCTES '03: Proceedings of the 2003 ACM SIGPLAN Conference on Languages, Compilers, and Tools for Embedded Systems, pp. 183–188 (2003)
43. Lee, L.H., Moyer, B., Arends, J.: Instruction fetch energy reduction using loop caches for embedded applications with small tight loops. In: ISLPED '99: Proceedings of the 1999 International symposium on Low power electronics and design, pp. 267–269. ACM, New York, NY, USA (1999). DOI http://doi.acm.org/10.1145/313817.313944
44. Lee, M.H., Singh, H., Lu, G., Bagherzadeh, N., Kurdahi, F.J., Filho, E.M.C., Alves, V.C.: Design and implementation of the MorphoSys reconfigurable computing processor. J. VLSI Signal Process. Syst. **24**(2/3), 147–164 (2000)
45. Mahlke, S.A., Lin, D.C., Chen, W.Y., Hank, R.E., Bringmann, R.A.: Effective compiler support for predicated execution using the hyperblock. In: MICRO 25: Proceedings of the 25th annual International symposium on Microarchitecture, pp. 45–54. IEEE Computer Society Press, Los Alamitos, CA, USA (1992). DOI http://doi.acm.org/10.1145/144953.144998
46. Mei, B., De Sutter, B., Vander Aa, T., Wouters, M., Kanstein, A., Dupont, S.: Implementation of a coarse-grained reconfigurable media processor for AVC decoder. Journal of Signal Processing Systems **51**(3), 225–243 (2008)
47. Mei, B., Lambrechts, A., Verkest, D., Mignolet, J.Y., Lauwereins, R.: Architecture exploration for a reconfigurable architecture template. IEEE Design and Test of Computers **22**(2), 90–101 (2005)
48. Mei, B., Vernalde, S., Verkest, D., Lauwereins, R.: Design methodology for a tightly coupled VLIW/reconfigurable matrix architecture: A case study. In: DATE '04: Proceedings of the Conference on Design, Automation and Test in Europe, pp. 1224–1229 (2004)
49. Mei, B., Vernalde, S., Verkest, D., Man, H.D., Lauwereins, R.: ADRES: An architecture with tightly coupled VLIW processor and coarse-grained reconfigurable matrix. In: Proc. of Field-Programmable Logic and Applications, pp. 61–70 (2003)
50. Mei, B., Vernalde, S., Verkest, D., Man, H.D., Lauwereins, R.: Exploiting loop-level parallelism for coarse-grained reconfigurable architecture using modulo scheduling. IEE Proceedings: Computer and Digital Techniques **150**(5) (2003)
51. Novo, D., Schuster, T., Bougard, B., Lambrechts, A., Van der Perre, L., Catthoor, F.: Energy-performance exploration of a CGA-based SDR processor. Journal of Signal Processing Systems (2008)
52. Oh, T., Egger, B., Park, H., Mahlke, S.: Recurrence cycle aware modulo scheduling for coarse-grained reconfigurable architectures. In: LCTES '09: Proceedings of the 2009 ACM SIGPLAN/SIGBED Conference on Languages, Compilers, and Tools for Embedded Systems, pp. 21–30 (2009)
53. PACT XPP Technologies: XPP-III Processor Overview White Paper (2006)
54. Park, H., Fan, K., Kudlur, M., Mahlke, S.: Modulo graph embedding: Mapping applications onto coarse-grained reconfigurable architectures. In: CASES '06: Proceedings of the 2006 International Conference on Compilers, architecture and synthesis for embedded systems, pp. 136–146 (2006)
55. Park, H., Fan, K., Mahlke, S.A., Oh, T., Kim, H., Kim, H.S.: Edge-centric modulo scheduling for coarse-grained reconfigurable architectures. In: PACT '08: Proceedings of the 17th International Conference on Parallel Architectures and Compilation Techniques, pp. 166–176 (2008)
56. Park, H., Park, Y., Mahlke, S.: Polymorphic pipeline array: A flexible multicore accelerator with virtualized execution for mobile multimedia applications. In: MICRO '09: Proceedings of the 42nd Annual IEEE/ACM International Symposium on Microarchitecture, pp. 370–380 (2009)

57. Park, Y., Park, H., Mahlke, S.: CGRA express: accelerating execution using dynamic operation fusion. In: CASES '09: Proceedings of the 2009 International Conference on Compilers, Architecture, and Synthesis for Embedded Systems, pp. 271–280 (2009)
58. Park, Y., Park, H., Mahlke, S., Kim, S.: Resource recycling: putting idle resources to work on a composable accelerator. In: CASES '10: Proceedings of the 2010 International Conference on Compilers, Architectures and Synthesis for Embedded Systems, pp. 21–30 (2010)
59. Petkov, N.: Systolic Parallel Processing. North Holland Publishing (1992)
60. P.Raghavan, A.Lambrechts, M.Jayapala, F.Catthoor, D.Verkest, Corporaal, H.: Very wide register: An asymmetric register file organization for low power embedded processors. In: DATE '07: Proceedings of the Conference on Design, Automation and Test in Europe (2007)
61. Rau, B.R.: Iterative modulo scheduling. Tech. rep., Hewlett-Packard Lab: HPL-94-115 (1995)
62. Rau, B.R., Lee, M., Tirumalai, P.P., Schlansker, M.S.: Register allocation for software pipelined loops. In: PLDI '92: Proceedings of the ACM SIGPLAN 1992 Conference on Programming Language Design and Implementation, pp. 283–299 (1992)
63. Sankaralingam, K., Nagarajan, R., Liu, H., Kim, C., Huh, J., Burger, D., Keckler, S.W., Moore, C.R.: Exploiting ILP, TLP, and DLP with the polymorphous TRIPS architecture. SIGARCH Comput. Archit. News **31**(2), 422–433 (2003). DOI http://doi.acm.org/10.1145/871656.859667
64. Scarpazza, D.P., Raghavan, P., Novo, D., Catthoor, F., Verkest, D.: Software simultaneous multi-threading, a technique to exploit task-level parallelism to improve instruction- and data-level parallelism. In: PATMOS '06: Proceedings of the 16th International Workshop on Integrated Circuit and System Design. Power and Timing Modeling, Optimization and Simulation, pp. 107–116 (2006)
65. Schlansker, M., Mahlke, S., Johnson, R.: Control CPR: A branch height reduction optimization for EPIC architectures. SIGPLAN Notices **34**(5), 155–168 (1999). DOI http://doi.acm.org/10.1145/301631.301659
66. Shen, J., Lipasti, M.: Modern Processor Design: Fundamentals of Superscalar Processors. McGraw-Hill (2005)
67. Silicon Hive: HiveCC Databrief (2006)
68. Sudarsanam, A.: Code optimization libraries for retargetable compilation for embedded digital signal processors. Ph.D. thesis, Princeton University (1998)
69. Taylor, M., Kim, J., Miller, J., Wentzla, D., Ghodrat, F., Greenwald, B., Ho, H., Lee, M., Johnson, P., Lee, W., Ma, A., Saraf, A., Seneski, M., Shnidman, N., Frank, V., Amarasinghe, S., Agarwal, A.: The Raw microprocessor: A computational fabric for software circuits and general purpose programs. IEEE Micro **22**(2), 25–35 (2002)
70. Texas Instruments: TMS320C64x Technical Overview (2001)
71. Van Essen, B., Panda, R., Wood, A., Ebeling, C., Hauck, S.: Managing short-lived and long-lived values in coarse-grained reconfigurable arrays. In: FPL '10: Proceedings of the 2010 International Conference on Field Programmable Logic and Applications, pp. 380–387 (2010)
72. Van Essen, B., Panda, R., Wood, A., Ebeling, C., Hauck, S.: Energy-efficient specialization of functional units in a coarse-grained reconfigurable array. In: FPGA '11: Proceedings of the 19th ACM/SIGDA International Symposium on Field Programmable Gate Arrays, pp. 107–110 (2011)
73. Venkataramani, G., Najjar, W., Kurdahi, F., Bagherzadeh, N., Bohm, W., Hammes, J.: Automatic compilation to a coarse-grained reconfigurable system-on-chip. ACM Trans. Embed. Comput. Syst. **2**(4), 560–589 (2003). DOI http://doi.acm.org/10.1145/950162.950167
74. van de Waerdt, J.W., Vassiliadis, S., Das, S., Mirolo, S., Yen, C., Zhong, B., Basto, C., van Itegem, J.P., Amirtharaj, D., Kalra, K., Rodriguez, P., van Antwerpen, H.: The TM3270 media-processor. In: MICRO 38: Proceedings of the 38th annual IEEE/ACM International Symposium on Microarchitecture, pp. 331–342. IEEE Computer Society, Washington, DC, USA (2005). DOI http://dx.doi.org/10.1109/MICRO.2005.35

75. Woh, M., Lin, Y., Seo, S., Mahlke, S., Mudge, T., Chakrabarti, C., Bruce, R., Kershaw, D., Reid, A., Wilder, M., Flautner, K.: From SODA to scotch: The evolution of a wireless baseband processor. In: MICRO '08: Proceedings of the 2008 41st IEEE/ACM International Symposium on Microarchitecture, pp. 152–163. IEEE Computer Society, Washington, DC, USA (2008). DOI http://dx.doi.org/10.1109/MICRO.2008.4771787
76. Programming XPP-III Processors White Paper (2006)
77. Yoon, J., Ahn, M., Paek, Y., Kim, Y., K., C.: Temporal mapping for loop pipelining on a MIMD-style coarse-grained reconfigurable architecture. In: Proceedings of the International SoC Design Conference (2006)

Arithmetic

Oscar Gustafsson and Lars Wanhammar

Abstract In this chapter fundamentals of arithmetic operations and number representations used in DSP systems are discussed. Different relevant number systems are outlined with a focus on fixed-point representations. Structures for accelerating the carry-propagation of addition are discussed, as well as multi-operand addition. For multiplication, different schemes for generating and accumulating partial products are presented. In addition to that, optimization for constant coefficient multiplication is discussed. Division and square-rooting are also briefly outlined. Furthermore, floating-point arithmetic and the IEEE 754 floating-point arithmetic standard are presented. Finally, some methods for computing elementary functions, e.g., trigonometric functions, are presented.

1 Number Representation

The way we select to represent our numbers have a profound impact on the corresponding computational units. Here we consider number representations based on a positional weight (radix) of two. It is worth noting that as the main application area considered here is DSP, we will where required choose to consider numbers that are fractional rather than integer. This will mainly affects the numbering of the indices.

O. Gustafsson (✉) • L. Wanhammar
Department of Electrical Engineering, Linköping University, SE-581 83 Linköping, Sweden
e-mail: oscarg@isy.liu.se; larsw@isy.liu.se

S.S. Bhattacharyya et al. (eds.), *Handbook of Signal Processing Systems*,
DOI 10.1007/978-1-4614-6859-2_19, © Springer Science+Business Media, LLC 2013

1.1 Binary Representation

An unsigned binary number, X, with W_f fractional bits can be written

$$X = \sum_{i=1}^{W_f} x_i 2^{-i} \tag{1}$$

where $x_i \in \{0,1\}$. Denoting the weight of the least significant position Q, in this case $Q = 2^{-W_f}$, one can see that the range of X is $0 \le X \le 1 - Q$. Q is sometimes referred to as unit of least significant position, *ulp*. As an example, the number 0.25_{10} is written using $W_f = 3$ as $.010_2$ and $Q = 2^{-3} = 0.125_{10}$.

1.2 Two's Complement Representation

To represent negative numbers, there are several different number representations proposed. The most common one is the two's complement (2C) representation. Here, a binary number, X, with W_f fractional bits is written

$$X = -x_0 + \sum_{i=1}^{W_f} x_i 2^{-i}. \tag{2}$$

This gives a numerical range as $-1 \le X \le 1 - Q$. It is worth noting that the range is not symmetric. This will cause problems when implementing certain arithmetic operations as discussed later. x_0, the sign bit, is one if $X < 0$ and zero otherwise. A number -0.25_{10} is represented as 1.11_{2C} with $W_f = 2$, while 0.25_{10} is 0.01_{2C} .

A beneficial property of two's complement arithmetic is the fact that an arbitrary long sequence of numbers can be added in arbitrary order as long as the result is known to be in the range of the representation. Any overflows/underflows in the intermediate computations will cancel. This is related to the fact that computations in two's complement number representation are performed modulo $2W_S$, where W_S is the weight of the sign bit. For the representation in (2) $W_S = 1$ so all computations are performed modulo 2.

1.3 Redundant Representations

A redundant representation is a representation where a number may have more than one representation. As we will see later the fact that we can select the representation will provide a number of advantages, most importantly the ability to perform addition in constant time.

Fig. 1 (**a**) Full adder: truth table. (**b**) Full adder: symbol

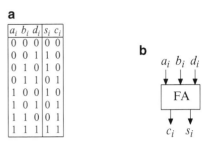

a

a_i	b_i	d_i	s_i	c_i
0	0	0	0	0
0	0	1	1	0
0	1	0	1	0
0	1	1	0	1
1	0	0	1	0
1	0	1	0	1
1	1	0	0	1
1	1	1	1	1

1.3.1 Signed-Digit Representation

In a signed-digit (SD) number representation, the digits may have either positive or negative sign. For a radix-2 representation we have $x_i \in \{-1, 0, 1\}$. Using W_f fractional bits as in (1) the range of X is $-1 + Q \leq X \leq 1 - Q$. A number may now have more than one representation. Consider the number 0.25_{10} which can be written as $.01_{SD}$ or $.1\bar{1}_{SD}$, where $\bar{1}$ is used to denote -1.

In some applications as, for instance, digital filters [30], FFTs [4] and DCTs [28] as well as general DSP algorithms [44], it is of interest to find a signed-digit representation with a minimum number of non-zero positions to simplify the corresponding multiplication (see Sect. 3.5). This is referred to as a minimum signed-digit (MSD) representation. However, in general it is non-trivial to determine if a representation is minimum. A specific minimum signed-digit representation is obtained if the constraint $x_i x_{i+1} = 0, \forall i$ is imposed. The resulting representation is called canonic signed-digit (CSD) representation and is apart from being minimum also unique (as the name indicates).

1.3.2 Carry-Save Representation

The carry-save representation stems from the ripple-carry adder, which will be further discussed in Sect. 2.1. Instead of propagating the carries in the addition, these bits are stored and the data is represented using two vectors. This also leaves an additional input of the full adder cells unused, so it is possible to add three vectors. A full adder cell is illustrated in Fig. 1b and the truth table is given in Fig. 1a. From this we can see that

$$s_i = \text{parity}(a_i \oplus b_i \oplus d_i) = a_i \oplus b_i \oplus d_i \qquad (3)$$

where $\oplus$ is the exclusive-OR operation, and

$$c_i = \text{majority}(a_i, b_i, d_i) = a_i b_i + a_i d_i + b_i d_i. \qquad (4)$$

Fig. 2 Redundant binary
adder with two positive and
one negative inputs: truth
table

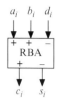

Fig. 3 Redundant binary
adder with two positive and
one negative inputs: symbol

a_i	b_i	d_i	s_i	c_i	Result
0	0	0	0	0	0
0	0	1	1	0	-1
0	1	0	1	1	1
0	1	1	0	0	0
1	0	0	1	1	1
1	0	1	0	0	0
1	1	0	0	1	2
1	1	1	1	1	1

Assuming two's complement representation of the vectors, a number is represented as

$$X = S + C = -s_0 + \sum_{i=1}^{W_f} s_i 2^{-i} - c_0 + \sum_{i=1}^{W_f} c_i 2^{-i}. \tag{5}$$

It is possible to represent numbers in the range $-2 \leq X \leq 2 - 2Q$. However, it is common to let X span the same numerical range as the two's complement representation since it usually will be converted into a non-redundant representation. The carry-save representation can also be seen as representation with radix-2 and digits $x_i \in \{0, 1, 2, 3\}$ where $x_i = 2c_i + s_i$.

The concept of carry-save representation can be generalized to binary redundant representations by replacing the relation $x_i = 2c_i + s_i$ with either $x_i = 2c_i - s_i$ where $x_i \in \{-1, 0, 1, 2\}$ or $x_i = -2c_i + s_i$ where $x_i \in \{-2, -1, 0, 1\}$. It is then possible to use cells corresponding to full adders, but with designated signs of the inputs, such that the bits with negative weights should be connected to $'-'$-inputs. This is illustrated in Fig. 2 while the computational rules for an adder with two positive and one negative input are shown in Fig. 3. Note that both types of representations need to be used as otherwise there would be an imbalance as each adder produces one positive and one negative vector but inputs either one positive and two negative or two positive and one negative.

1.4   Shifting and Increasing the Word Length

When adding signed numbers it is for most number representations important that the vectors have the same lengths. For two's complement representation we must extend the sign by copying the sign-bit at the MSB side of the vector. On the LSB side it is enough to extend with zeros.[1]

The operation of shifting, i.e., multiplying or dividing by a power of two, is in fact the same as increasing the word length. Hence, shifting a two's complement number two positions to the right (dividing by 4) requires copying of the sign-bit to the two introduced positions.

For carry-save representation the extension of the word length on the MSB side require that the sign-bits of the carry and sum vectors are corrected to avoid what is referred to as carry overflow. The origin of this effect is that once the carry and sum vectors are added there is an overflow in the addition. As the resulting carry bit from the MSB position is neglected the result still remains valid, in fact this is a required property for two's complement representation to work. However, if the vectors are shifted once before the final addition, the effect of the overflow will occur in an earlier position.

1.5 Negation

Negation is a useful operation in itself, but also in a subtraction, $Z = X - Y$, which can be seen as an addition of the negated value, $Z = X + (-Y)$.

To negate a two's complement number one inverts all the bits and add a one to the LSB position. As will be seen later on, this does not necessarily have to be performed explicitly, but can rather be integrated with an addition.

1.6 Finite Word Length Effects

In recursive algorithms it is necessary to use non-linear operations to maintain a finite word length. This introduces small re-quantization errors, so called granularity errors. In addition, very large overflow errors will occur if the finite number range is exceeded. These errors will not only cause distortion, but may also be the cause of parasitic oscillations in recursive algorithms [5, 13, 46].

The effect of these errors depends on many factors, for example, type of quantization, algorithm, type of arithmetic, representation of negative numbers, and properties of the input signal [31]. The analysis of the influence of round-off errors in floating-point arithmetic is very complicated [32, 45, 57], because quantization of products as well as addition in floating-point arithmetic and subtraction causes errors that depend on the signal values. Quantization errors in fixed-point arithmetic are with few exceptions independent of the signal values and can therefore be analyzed independently. Lastly, the high dynamic range provided by floating-point

---

[1]It is worth noticing that for one's complement the sign-bits are inserted at the LSB side.

Fig. 4 Overflow
characteristic of two's
complement arithmetic

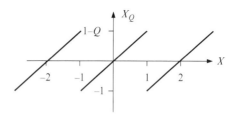

Fig. 5 Overflow
characteristic of saturation
arithmetic

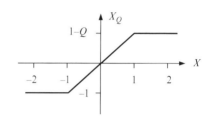

arithmetic is not really needed in good filter algorithms, since filter structures with
low coefficient sensitivity also utilize the available number range efficiently.

Fixed-point arithmetic is predominately used in application-specific ICs since the
required hardware is much simpler, faster, and consumes less power compared with
floating-point arithmetic. We will therefore focus on fixed-point arithmetic.

1.6.1 Overflow Characteristics

A two's complement representation of negative numbers is usually employed in digital hardware. The overflow characteristic of the two's complement representation
is shown in Fig. 4.

As discussed earlier, the largest and smallest numbers in two's complement
representation are $1 - Q$ and -1, respectively. A two's complement number, X,
larger than $1 - Q$ will be interpreted as $X - 2$, while a number slightly smaller than
-1 will be interpreted as $X + 2$. Hence, very large overflows errors are incurred.
A common scheme to reduce the size of overflow errors and mitigate their harmful
effect is to limit numbers outside the normal range to either the largest or smallest
representable number. This scheme is referred to as *saturation arithmetic* and is
shown in Fig. 5.

Many standard signal processors provide addition and subtraction instructions
with inherent saturation. Another saturation scheme, which may be simpler to
implement in hardware, is to invert all bits when overflow occurs. Using fixed-point
arithmetic the signal levels need to be adjusted at the input of multiplications with
non-integer coefficients. Note that, as discussed earlier, a sum of several numbers,
using two's complement representation, may have intermediate overflows as long as
the final value (sum) is within the valid range.

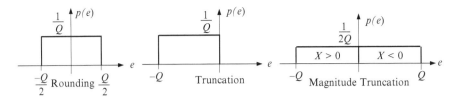

Fig. 6 Error distributions for fixed-point arithmetic

1.6.2 Truncation

Quantizing a binary number, X, with infinite word length to a number, X_Q, with finite word length yields an error

$$e = X_Q - X. \tag{6}$$

Truncation of the binary number is performed by removing the bits with index $i > W_f$. The resulting error density distribution is shown in the center of Fig. 6. The variance is $\sigma^2 = \frac{Q^2}{12}$ and the mean value is $-Q/2$ where Q refer to the weight of the last bit position.

1.6.3 Rounding

Rounding is, in practice, performed by adding $2^{-(W_f+1)}$ to the non-quantized number before truncation. Hence, the quantized number is the nearest approximation to the original number. However, if the word length of X is $W_f + 1$, the quantized number should, in principle, be rounded upwards if the last bit is 1 and downwards if it is 0, in order to make the mean error zero. This special case is often neglected in practice. The resulting error density distribution is shown to the left in Fig. 6. The variance is $\sigma^2 = \frac{Q^2}{12}$ and the mean value is zero.

1.6.4 Magnitude Truncation

Magnitude truncation quantizes the number so that

$$|X_Q| \leq |X|. \tag{7}$$

Hence, e is ≤ 0 if X is positive and ≥ 0 if X is negative. This operation can be performed by adding $2^{-(W_f+1)}$ before truncation if X is negative and 0 otherwise. That is, in two's complement representation adding the sign bit to the last position. The resulting error density distribution is shown to the right in Fig. 6. The error analysis of magnitude truncation becomes very complicated since the error and sign of the signal are correlated [31].

Magnitude truncation is needed to suppress parasitic oscillation in wave digital filters [13].

1.6.5   Quantization of Products

The effect of a quantization operation, except for magnitude truncation, can be modeled with a white additive noise source that is independent of the signal and with the error density functions as shown in Fig. 6. This model can be used if the signal varies from sample to sample over several quantization levels in an irregular way. However, the error density function is a discrete function if both the signal and the coefficient have finite word lengths. The difference is significant only if a few bits are discarded by the quantization. The mean value and variance for the errors are

$$m = \begin{cases} \frac{Q_c}{2}Q, & \text{rounding} \\ \frac{Q_c-1}{2}Q, & \text{truncation} \end{cases} \tag{8}$$

and

$$\sigma_e 2 = k_e (1 - Q_c^2) \frac{Q^2}{12} \tag{9}$$

where

$$k_e = \begin{cases} 1, & \text{rounding or truncation} \\ 4 - \frac{6}{\pi}, & \text{magnitude truncation} \end{cases} \tag{10}$$

where Q and Q_c refer to the signal and coefficient, respectively. For long coefficient word lengths the average value is close to zero for rounding and $-Q/2$ for truncation. Correction of the average value and variance is only necessary for short coefficient word lengths, for example, for the scaling coefficients.

2   Addition

The operation of adding two or more numbers is in many ways the most fundamental arithmetic operation since most other operations in one or another way are based on addition.

The methods discussed here concerns either two's complement or unsigned representation. Then, the major problem is to efficiently speed up the carry-propagation in the adder. There are other schemes than those presented here, for more details we refer to e.g., [58]. It is also possible to perform addition in constant time using redundant number systems such as the previously discussed signed-digit or carry-save representations. An alternative is to use residue number systems (RNS), that split the carry-chain into several shorter ones [39].

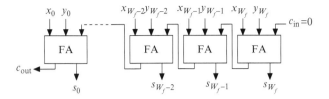

Fig. 7 Ripple-carry adder

Fig. 8 Example of addition in two's complement arithmetic using a ripple-carry adder

Value	0 1 2 3 4 5 6 7	Signal
$-53/256$	1 1 0 0 1 0 1 1	x_i
$94/256$	0 1 0 1 1 1 1 0	y_i
	1 1 0 1 1 1 1 0	c_i
$41/256$	0 0 1 0 1 0 0 1	s_i

2.1 Ripple-Carry Addition

The probably most straightforward way to perform addition of two numbers is to perform bit-by-bit addition using a full adder (FA) cell and propagate the carry bit to the next stage. This is called ripple-carry addition and is illustrated in Fig. 7. This type of adder can add both unsigned and two's complement numbers. However, for two's complement numbers the result must have the same number of bits as the inputs, while for unsigned numbers the carry bits acts as a possible additional bit to increase the word length. The operation of adding two two's complement numbers is outlined in Fig. 8 for the example of numbers $-53/128$ and $94/128$.

The major drawback with the ripple-carry adder is that the worst case delay is proportional to the word length. Also, typically the ripple-carry adder will produce many glitches due to the full adder cells having to wait for the correct carry. This situation is improved if the delay for the carry bit is smaller than that of the sum bit [22]. However, due to the simple design the energy per computation is still reasonable small [58].

2.2 Carry-Lookahead Addition

To speed up the addition several different methods have been proposed, see, for instance, [58]. Methods typically referred to as carry-lookahead methods are based on the following observation. The carry output of a full adder sometimes depends on the carry input and sometimes it is determined without the need of the carry input. This is illustrated in Table 1. Based on this we can define the propagate signal, p_i, and the generate signal, g_i, as

$$p_i = a_i \oplus b_i \text{ and } g_i = a_i b_i. \tag{11}$$

Table 1 Cases for
carry-propagation in a full
adder cell

a_i	b_i	c_{i-1}	Case
0	0	0	No carry-propagation (kill)
0	1	c_i	Carry-propagation (propagate)
1	0	c_i	Carry-propagation (propagate)
1	1	1	Carry-generation (generate)

Fig. 9 Illustration of N-stage
carry-lookahead carry
propagation

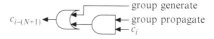

Now, the carry output can be expressed as

$$c_{i-1} = g_i + p_i c_i. \tag{12}$$

For the next stage the expression becomes

$$c_{i-2} = g_{i-1} + p_{i-1} c_{i-1} = g_{i-1} + p_{i-1}(g_i + p_i c_i) = g_{i-1} + p_{i-1} g_i + p_{i-1} p_i c_i. \tag{13}$$

For $N + 1$:th stage we have

$$c_{i-(N+1)} = g_{i-N} + p_{i-N} g_{i-(N-1)} + p_{i-N} p_{i-N-1} g_{i-(N-2)} + \cdots + p_{i-N} \cdots p_{i-1} p_i c_i. \tag{14}$$

The terms containing g_k and possibly p_k terms are called group generate, as they together acts as a merged generate signal for all the bits i to $i - N$. The subterm $p_{i-N} \cdots p_{i-1} p_i$ is similarly called group propagate. Both the group generate and group propagate signals are independent of any carry signal. Hence, (14) shows that it is possible to have the carry propagate N stages with a maximum delay of one AND-gate and one OR-gate as illustrated in Fig. 9. However, the complexity and delay of the precomputation network grows with N, and, hence, a careful design is required to not make the precomputation the new critical path.

The carry-lookahead approach can be generalized using dot-operators. Adders using dot-operators are often referred to as parallel prefix adders. The dot-operator operates on a pair of generate and propagate signals and is defined as

$$\begin{bmatrix} g_k \\ p_k \end{bmatrix} = \begin{bmatrix} g_i \\ p_i \end{bmatrix} \bullet \begin{bmatrix} g_j \\ p_j \end{bmatrix} \triangleq \begin{bmatrix} g_i + p_i g_j \\ p_i p_j \end{bmatrix}. \tag{15}$$

The group generate from position k to position l, $k < l$, can be denoted by $G_{k:l}$ and similarly the group propagate as $P_{k:l}$. These are then defined as

$$\begin{bmatrix} G_{k:l} \\ P_{k:l} \end{bmatrix} \triangleq \begin{bmatrix} g_k \\ p_k \end{bmatrix} \bullet \begin{bmatrix} g_{k+1} \\ p_{k+1} \end{bmatrix} \bullet \cdots \bullet \begin{bmatrix} g_l \\ p_l \end{bmatrix}. \tag{16}$$

Fig. 10 Illustration of the idempotency property for group generate and propagate signals

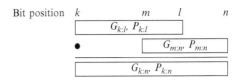

The dot operator is associative but not commutative. Furthermore, the dot-operation is idempotent. This means that

$$\begin{bmatrix} g_k \\ p_k \end{bmatrix} = \begin{bmatrix} g_k \\ p_k \end{bmatrix} \bullet \begin{bmatrix} g_k \\ p_k \end{bmatrix}. \tag{17}$$

For the group generate and propagate signals this leads to that

$$\begin{bmatrix} G_{k:n} \\ P_{k:n} \end{bmatrix} = \begin{bmatrix} G_{k:l} \\ P_{k:l} \end{bmatrix} \bullet \begin{bmatrix} G_{m:n} \\ P_{m:n} \end{bmatrix}, \ k \le l, m \le n, m \le l - 1. \tag{18}$$

This is illustrated in Fig. 10. Hence, we can form the group generate and group propagate signals by combining smaller, possibly overlapping, subgroup generate and propagate signals.

The carry signal in position k can be written as

$$c_k = G_{(k+1):l} + P_{(k+1):l}c_l \tag{19}$$

Similarly, the sum signal in position k for an adder using W_f fractional bits is then expressed according to (3) as

$$s_k = a_k \oplus b_k \oplus d_k = p_k \oplus (G_{(k+1):W_f} + P_{(k+1):W_f}c_{in}) = p_k \oplus c_k \tag{20}$$

From this, one can see that it is of interest to compute all group generate and group propagate originating from the LSB position, i.e., $G_{k:W_f}$ and $P_{k:W_f}$ for $1 \le k \le W_f$.

A straightforward way of obtaining this is to do a sequential operation as shown in Fig. 11. However, again the delay will be linear in the word length, as for the ripple-carry adder. Indeed, the adder in Fig. 11 is a ripple-carry adder where the full adder cells explicitly compute p_i and g_i.

Based on the properties of the dot operator, we can possibly find ways to interconnect the adders such that the depth is reduced by computing different group generate and propagate signals in parallel. This is illustrated for an 8-bits adder in Fig. 12. This particular structure of interconnecting the dot-operators are referred to as a Ladner–Fischer parallel prefix adder [27]. Often one uses a simplified structure to represent the parallel prefix computation, as shown in Fig. 13a. Comparing Figs. 12 and 13a it is clear that dots in Fig. 13a correspond to dot-operators in Fig. 12. In fact, the parallel prefix graph in Fig. 13a works for any associative operation.

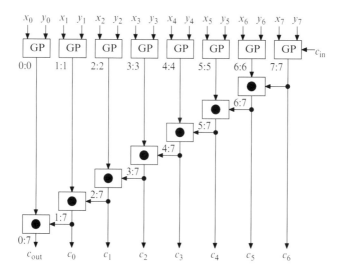

Fig. 11 Sequential computation of group generate and propagate signals

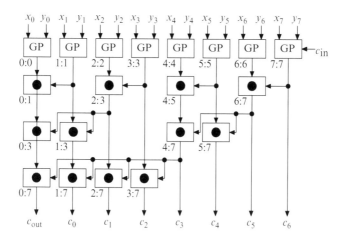

Fig. 12 Parallel computation of the group generate and propagate signals

Over the years there has been a multitude of different schemes for parallel prefix addition trading the depth, number of dot-product operations, and fan-out of the dot-product cells. In Fig. 13b–d, three of schemes for 16-bits parallel prefix computations are illustrated, namely Ladner–Fischer [27], Kogge–Stone [25], and Brent–Kung [3], respectively. Unified views of all possible parallel prefix schemes have been proposed in [19, 23].

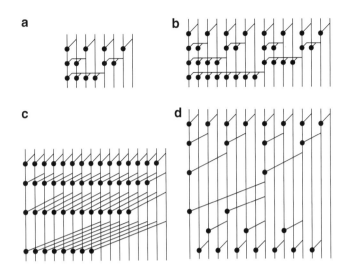

Fig. 13 Different parallel prefix schemes for, (**a**) An 8-bit Ladner–Fischer adder [27] as shown in Fig. 12 and for 16-bits adders: (**b**) Ladner–Fischer [27], (**c**) Kogge–Stone [25], and (**d**) Brent–Kung [3]

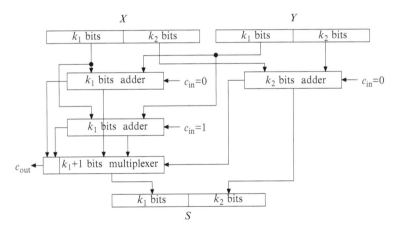

Fig. 14 Two-stage carry-select adder

2.3 Carry-Select and Conditional Sum Addition

The fundamental idea of carry-select addition is to split the adder into two or more stages. For all stages except the stage with the least significant bits one uses two adders. It is assumed that the incoming carry bit is zero one of the adders and one for the other. Then, a multiplexer is used to select the correct result and carry to the next stage once the incoming carry is known. A two-stage carry-select adder is shown in Fig. 14. The length of the stages should be designed such that the delay of

Fig. 15 Principle of a
multi-operand adder

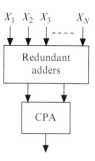

the stage is equivalent to the delay of the first stage plus the number of multiplexers
that the carry signal passes through. Hence, the actual values are determined by
the relative adder and multiplexer delays, as well as the fan-out of the multiplexer
control signals.

For each smaller adder in the carry-select adder, it is possible to apply the same
idea of splitting each smaller adder into even smaller adders. For example, each of
the two k_1 bit adders can be split into two k_3 bits and one k_4 bits adders, where $k_1 =
k_3 + k_4$, in a similar way. Note, however, that only four smaller adders are required
instead of six as the same two k_3 bits adders can be used. If this is applied until
only 1-bit adders remain, we obtain a conditional sum adder. There are naturally, a
wide range of intermediate adder structures based on the ideas of carry-select and
conditional sum adders.

2.4 Multi-operand Addition

When several operands are to be added, it is beneficial to avoid several carry-
propagations. Especially, when there are delay constraints it is inefficient to use
several high-speed adders. Instead it is common to use a redundant intermediate
representation and a fast final carry-propagation adder (CPA). The basic concept is
illustrated in Fig. 15.

For performing multi-operand addition, either counters or compressors or a
combination of counters and compressors can be used. A counter is a logic gate that
takes a number of inputs, add them together and produce a binary representation of
the output. The simplest counter is the full adder cell shown in Fig. 1b. In terms of
counters, it is a 3:2 counter, e.g., it has three inputs and produce a 2-bits output word.
This can be generalized to $n : k$ counters, having n inputs of the same weight and
producing a k bit output corresponding to the number of ones in the input. Clearly,
n and k must satisfy $n \leq 2^k - 1$ or equivalently $k \geq \lceil \log_2(n + 1) \rceil$.

A compressor on the other hand does not produce a valid binary count of the
number of input bits. However, it does reduce the number of partial products, but at
the same time has several incoming and outgoing carries. The output carries should
be generated without any dependence on the input carries. The most frequently

Fig. 16 4:2 compressor
composed of full adders
(3:2 counters)

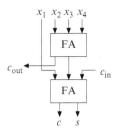

used compressor is the 4:2 compressor shown in Fig. 16, which is realized using full adders. Clearly, there is no major advantage using 4:2 compressors that are implemented as in Fig. 16 compared to using 3:2 counters (full adders). However, other possible realizations are available. These should satisfy

$$x_1 + x_2 + x_3 + x_4 + c_{in} = s + 2c + 2c_{out} \tag{21}$$

and c_{out} should be independent of c_{in}. There exist realizations with lower logic depth compared to full adders, and, hence, the total delay of the multi-operand addition may be reduced using 4:2 compressors.

It is important to note that an $n : k$ counter or compressor reduces the number of bits in the computation with exactly $n - k$. Hence, it is easy to estimate the required number of counters and compressors to perform any addition if the original number of bits to be added and the number of bits for the result are known. It should also be noted, that depending on the actual structure it is typically impossible to use only one type of compressors and adders. Specifically, half adders (or 2:2 counters) may sometimes be needed, despite not reducing the number of bits, to move bits to the correct weights for further additions.

3 Multiplication

The process of multiplication can be divided into three different steps: partial product generation that determines a number of bits to be added, summation of the generated partial products, and, for some of the summation structures, carry-propagation addition, usually called vector merging addition (VMA), as many summation structures produce redundant results.

3.1 Partial Product Generation

For unsigned binary representation, the partial product generation can be readily realized using AND-gates computing bit-wise multiplications as

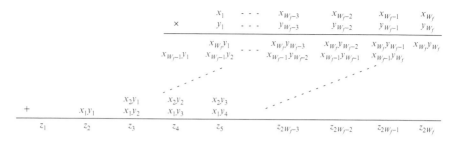

Fig. 17 Partial product array for unsigned multiplication

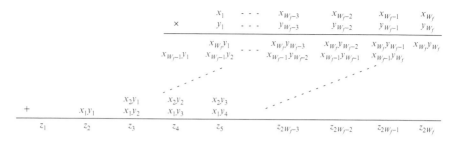

Fig. 18 Partial product array for two's complement multiplication

$$Z = XY = \sum_{i=1}^{W_{fX}} x_i 2^{-i} \sum_{j=1}^{W_{fY}} y_j 2^{-j} = \sum_{i=1}^{W_{fX}} \sum_{j=1}^{W_{fY}} x_i y_j 2^{-i-j} \tag{22}$$

This leads to a partial product array as shown in Fig. 17.

For two's complement data the result is very similar, except that the sign-bit causes some of the bits to have the negative sign. This can be seen from

$$
\begin{aligned}
Z &= XY \\
&= \left(-x_0 + \sum_{i=1}^{W_{fX}} x_i 2^{-i}\right)\left(-y_0 + \sum_{j=1}^{W_{fY}} y_j 2^{-j}\right) \\
&= x_0 y_0 - x_0 \sum_{j=1}^{W_{fY}} y_j 2^{-j} - y_0 \sum_{i=1}^{W_{fX}} x_i 2^{-i} + \sum_{i=1}^{W_{fX}} \sum_{j=1}^{W_{fY}} x_i y_j 2^{-i-j}
\end{aligned}
\tag{23}
$$

The corresponding partial product matrix is shown in Fig. 18.

3.1.1 Avoiding Sign-Extension

As previously stated, the word lengths of two two's complement numbers should be equal when performing the addition or subtraction. Hence, the straightforward way

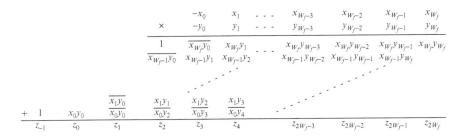

Fig. 19 Partial product array without sign-extension

of dealing with the varying word lengths in two's complement multiplication is to sign-extend the partial results to obtain the same word length for all rows.

To avoid this excessive sign-extension it is possible to either perform the summation from top to bottom and perform sign-extension of the partial results to match the next row to be added. This is further elaborated in Sects. 3.2.1 and 3.2.2. However, if we want to be able to add the partial products in an arbitrary order using a multi-operand adder as discussed in Sect. 2.4, the following technique, proposed initially by Baugh and Wooley [1], can be used. Note that for a negative partial product we have $-p = \bar{p} - 1$. Hence, we can replace all negative partial products with an inverted version. Then, we need to subtract a constant value from the result, but as there will be several constants, one from each negative partial product, we can sum these up and form a single compensation vector to be added. When this is applied we get the partial product array as shown in Fig. 19.

3.1.2 Reducing the Number of Rows

As discussed in Sect. 1.3.1, it is possible to reduce the number of non-zero positions by using a signed-digit representation. It would be possible to use e.g. a CSD representation to obtain a minimum number of non-zeros. However, the drawback is that the conversion from two's complement to CSD requires carry-propagation. Furthermore, the worst case is that half of the positions are non-zero, and, hence, one would still need to design the multiplier to deal with this case.

Instead, it is possible to derive a signed-digit representation that is not necessarily minimum but has at most half of the positions being non-zero. This is referred to modified Booth encoding [33] and is often described as being a radix-4 signed-digit representation where the recoded digits $r_i \in \{-2, -1, 0, 1, 2\}$. An alternative interpretation is a radix-2 signed-digit representation where $d_i d_{i-1}, i \in \{W_f, W_{f-2}, W_{f-4}, \dots\}$. The logic rules for performing the modified Booth encoding are based on the idea of finding strings of ones and replace them as $011\dots11 = 100\dots0\bar{1}$ and are illustrated in Table 2. From this, one can see that there is at most one non-zero digit in each pair of digits $(d_{2k} d_{2k+1})$.

Table 2 Rules for the radix-4 modified Booth encoding

x_{2k}	x_{2k+1}	x_{2k+2}	r_k	$d_{2k}d_{2k+1}$	Description
0	0	0	0	00	String of zeros
0	0	1	1	01	End of ones
0	1	0	1	01	Single one
0	1	1	2	10	End of ones
1	0	0	-2	$\bar{1}0$	Start of ones
1	0	1	-1	$0\bar{1}$	Start and end of ones ($\bar{1}0+01$)
1	1	0	-1	$0\bar{1}$	Start of ones
1	1	1	0	00	String of ones

Fig. 20 Generation of partial products for radix-4 modified Booth encoding

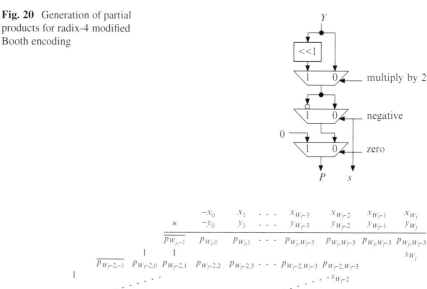

Fig. 21 Partial product array for radix-4 modified Booth encoded multiplier

Now, to perform the multiplication, we must be able to possibly negate and multiply the operand with 0, 1, or 2. This can conceptually be performed as in Fig. 20. As discussed earlier, the negation is typically performed by inverting the bits and add a one in the column corresponding to the LSB position. The partial product array for a multiplier using the modified Booth encoding is shown in Fig. 21.

It is possible to use the modified Booth encoding with higher radices than two. However, that requires the computations of non-trivial multiples such as 3 for radix-8 and 3, 5, and 7 for radix-16. The number of rows is reduced roughly by a factor of k for the radix-2^k modified Booth encoding.

3.1.3   Reducing the Number of Columns

It is common that the results after the multiplication are quantized to be represented with fewer bits than the original result. To reduce the complexity of the multiplication in these cases it has been proposed to perform the quantization at the partial product stage [29]. This is commonly referred to as fixed-width multiplication referring to the fact that (most of) the partial products rows have the same width.

Simply truncating the partial products will result in a rather large error. Several methods have therefore been proposed to compensate for the introduced error [42].

3.2   Summation Structures

The problem of summing up the partial products can be solved in three general ways; sequential accumulation where a subset of the partial products are accumulated in each cycle, array accumulation which gives a regular structure, and tree accumulation which gives the smallest logic depth but in general an irregular structure.

3.2.1   Sequential Accumulation

In so-called add-and-shift multipliers, the partial bit-products are generated sequentially and successively accumulated as generated. This type of multipliers are therefore slow, but the required chip area is small. The accumulation can be done using any of the bit-parallel adders discussed above or using digit-serial or bit-serial accumulators. An advantage of bit-serial over bit- parallel arithmetic is that it significantly reduces chip area. This is done in two ways. First, it eliminates wide buses and simplifies the wire routing. Second, by using small processing elements, the chip itself will be smaller and will require shorter wiring. A small chip can support higher clock frequencies and is therefore faster. Two's complement representation is suitable for DSP algorithms implemented with bit-serial arithmetic, since the bit-serial operations then can be done without knowing the sign of the numbers involved. Figure 22 shows a 5-bit serial/parallel multiplier where the bit-products are generated row-wise. In a serial/parallel multiplier, the multiplicand X arrive bit-serially while the multiplier a is applied in a bit-parallel format. Many different schemes for bit-serial multipliers have been proposed. They differ mainly in which order bit-products are generated and accumulated and in the way subtraction is handled.

Addition of the first set of partial bit-products starts with the products corresponding to the LSB of X. Thus, in the first time slot, at bit x_{W_f}, we simply add, $a \cdot x_{W_f}$ to the initially cleared accumulator. Next, the D flip-flops are clocked and the sum-bits from the FAs are shifted 1 bit to the right, each carry-bit is saved and added to the full-adder in the same stage, the sign-bit is copied, and 1 bit of the

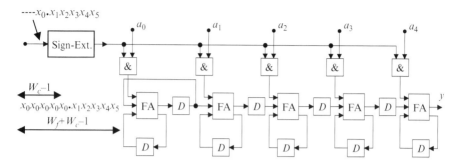

Fig. 22 Serial/parallel multiplier based on carry-save adders

product is produced at the output of the accumulator. These operations correspond to multiplying the accumulator contents by 2^{-1}. In the following clock cycle, the next bit of X is used to form the next set of bit-products, which are added to the value in the accumulator, and the value in the accumulator is again divided by 2. This process continues for W_f clock cycles, until the sign bit x_0 is reached, whereupon a subtraction must be done instead of an addition. During the first W_f clock cycles, the least significant part of the product is computed and the most significant is stored in the D flip-flops. In the next W_f clock cycles, zeros are therefore applied to the input so that the most significant part of the product is shifted out of the multiplier. Note that the accumulation of the bit-products is performed using a redundant representation, which is converted to a non-redundant representation in the last stage of the multiplier. A digit-serial multiplier, which accumulate several bits in each stage, can be obtained either via unfolding of a bit-serial multiplier or via folding of a bit-parallel multiplier.

3.2.2 Array Accumulation

Array multipliers use an array of almost identical cells for generation and accumulation of the bit-products. Figure 23 shows a realization of the Baugh–Wooley multiplier [1] with the multiplication time proportional to $2W_f$.

3.2.3 Tree Accumulation

The array structure provides a regular structure, but at the same time the delay grows linear with the word length. Considering Figs. 19 and 21, they both provide a number of partial products that should be accumulated. As mentioned earlier, it is common to accumulate the partial products such that there are at most two partial products of each weight and then use a fast carry-propagation adder to perform the final step.

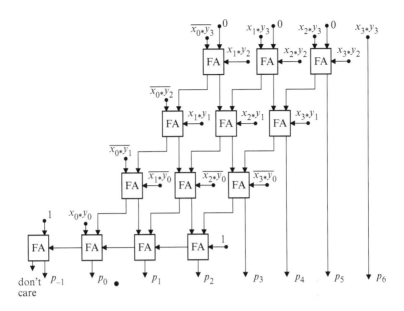

Fig. 23 A Baugh–Wooley multiplier

In Sect. 2.4 the problem of adding a number of bits was considered. Here, we will focus on structures using full and half adders (or 3:2 and 2:2 counters), although there are other structures proposed using different types of counters and compressors.

The first approach is to add as many full adders as possible to reduce as many partial products as possible. Then, we add as many half adders as possible to minimize the number of levels and try to shorten the word length for the vector merging adder. This approach is roughly the Wallace tree proposed in [53]. The main drawback of this approach is an excessive use of half adders. Dadda [7] instead proposed that full and half adders should only be used if required to obtain a number of partial products equal to a value in the Dadda series. The value of position n in the Dadda series is the maximum number of partial products that can be reduced using n levels of full adders. The Dadda series starts $\{3, 4, 6, 9, 13, 19, \ldots\}$. The benefit of this is that the number of half adders is significantly reduced while still obtaining a minimum number of levels. However, the length of the vector merging adder increases. A compromise between these two approaches is the Reduced Area heuristic [2], where similarly to the Wallace tree, as many full adders as possible are introduced in each level. Half adders are on the other hand only introduced if required to reach a number of partial products corresponding to a value in the Dadda series or if the least significant weight with more than one partial products is represented with exactly two partial products. In this way a minimum number of stages is obtained, while at the same time both the length of the vector merging adder and the number of half adders are kept small.

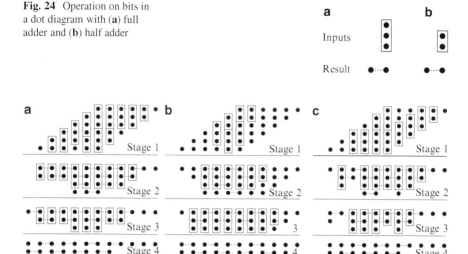

Fig. 24 Operation on bits in a dot diagram with (**a**) full adder and (**b**) half adder

Fig. 25 Reduction trees for a 6×6-bits multiplier: (**a**) Wallace [53], (**b**) Dadda [7], and (**c**) Reduced area [2]

Table 3 Complexity of the three reduction trees in Fig. 25

Tree structure	Full adders	Half adders	VMA length
Wallace [53]	16	13	8
Dadda [7]	15	5	10
Reduced area [2]	18	5	7

To illustrate the operation of the reduction tree approaches we use dot diagrams, where each dot corresponds to a bit (partial product) to be added. Bits with the same weight are in the same column and bits in adjacent columns have one position higher or lower weight, with higher weights to the left. The bits are manipulated by either full or half adders. The operation of these are illustrated in Fig. 24.

The reduction schemes are exemplified based on an unsigned 6×6-bits multiplication in Fig. 25. The complexity results are summarized in Table 3. It should be noted that the positioning of the results in the next level is done based on ease of illustration. From a functional point of view this step is arbitrary, but it is possible to optimize the timing by carefully utilizing different delays of the sum and carry outputs of the adder cells [38]. Furthermore, it is possible to reduce the power consumption by optimizing the interconnect ordering [40].

The reduction trees in Fig. 25 does not provide any regularity. This means that the routing is complicated and may become the limiting factor in an implementation. Reduction structures that provide a more regular routing, but still a small number of stages, include the Overturned stairs reduction tree [35] and the HPM tree [12].

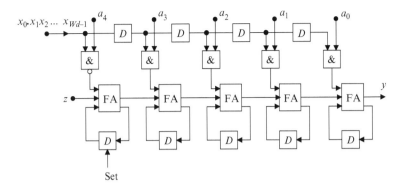

Fig. 26 Serial/parallel multiplier–accumulator

3.3 Vector Merging Adder

The role of the vector merging adder is to add the outputs of the reduction tree. In general, any carry-propagation adder can be used, e.g., those presented in Sect. 2. However, the different input signals to the adders will typically be available at different delays from the multiplier input values. Therefore is it possible to derive carry-propagation adders that utilize the different signal arrival times to optimize the adder delay [48].

3.4 Multiply–Accumulate

In many DSP algorithms computations of the form $Z = XY + A$ are common. These can be effectively implemented by simply adding another row corresponding to A in the partial product array. In many cases this will not increase the number of levels required.

For sequential operation, the modification of the first stage of the serial/parallel multiplier as shown in Fig. 26, makes it possible to add an input Z to be added to the product at the same level of significance as X.

3.5 Multiplication by a Constant

When the multiplier coefficient is known, it is possible to reduce the complexity of the corresponding circuit. First of all, no circuitry is required to generate the partial products. Second, there will in general be fewer partial products to add. This can easily be realized considering the partial product array in Fig. 17. For the coefficient

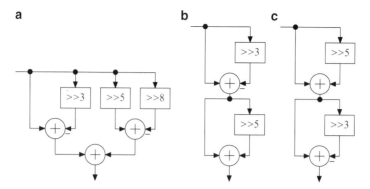

Fig. 27 Constant multiplication with $231/256 = 1.00\bar{1}0100\bar{1}$ using (**a**) no sharing, (**b**) sharing of the subexpression $100\bar{1}$, and (**c**) sharing of the subexpression 10001

bits that are zero, all the corresponding partial product bits will also be zero, and, hence, these are not required to be added. To obtain more zero positions, the use of a minimum signed digit representation such as CSD is useful.

It is also possible to utilize potential redundancy in the computations to further reduce the complexity. How this is done in detail depends on which type of addition is assumed to the basic operation. As both addition and subtraction have the same complexity, we will refer to both as addition. In the following we will assume carry-propagation addition, i.e., two input and one output, realized in any way discussed in Sect. 2. Furthermore, for ease of exposition, we will assume that the standard sign-extension is used. For carry-save addition we refer to [17].

Consider a signed-digit representation of a multiplier coefficient X such as shown in (1) with $x_i \in \{-1, 0, 1\}$. Each non-zero position will produce a partial result row and these partial result rows can be added in an arbitrary order. Now, if the same pattern of non-zero positions, called subexpression, occurs in more than one position of the representation, we only need to compute the corresponding partial result once and use it for all the instances where it is required. Figure 27a, b show examples of multiplication with the constant $231/256 = 1.00\bar{1}0100\bar{1}$ with and without utilizing redundancy, respectively. In this case we extract the subexpression $100\bar{1}$, but we might just as well have chosen 10001 and subtracted one of the subexpressions as shown in Fig. 27c.

This can be performed is a systematic way as described below. However, first we note that if we multiply the same data with several constant coefficients, as in a transposed direct form FIR filter [56], the different coefficients can share subexpressions. Hence, the systematic way is as follows [20, 43]:

1. Represent the coefficients in a given representation.
2. For each coefficient find and count possible subexpressions. A subexpression is characterized by the difference in non-zero position and if the non-zeros have the same or opposite signs.

3. If there are common subexpressions, select one to replace and replace instances of it by introducing a new symbol in place of the subexpression. The most common approach is to select the most frequent subexpression, thus, applying a greedy optimization approach, and replace all instances of it. However, it should be noted that from a global optimality point of view this is not always the best.
4. If there were subexpressions replaced, go to Step 2 otherwise the algorithm is done.

There are a number of sources of possible sub-optimality in the procedure described above. The first is that the results are representation dependent, and, hence, it will in general reduce the complexity trying several different representations. It may seem to make sense to use an MSD representation as it originally have few non-zero positions. However, the more non-zeros the more likely is it that common subexpressions will exist. For the single constant multiplication case this has been utilized for a systematic algorithm that searches all representations with the minimum number of non-zero digits plus k additional non-zero digits [18]. The second source is the selection of subexpression to replace. It is common to select the most frequent one, applying a greedy strategy, and replace all instances. However, it has been shown that from a global optimization point of view, it is not always beneficial to replace all subexpressions. Another issue is which subexpression to choose if there are more than one that are as frequent.

For single constant coefficients there is an optimal approach based on searching all possible interconnections of a given number of adders presented in [18]. It is shown that multiplication with all coefficients up to 19 bits can be realized using at most five additions, compared to up to nine additions using a straightforward CSD realization without sharing. The optimal approach avoids the issue of representation dependence by only considering the decimal value at each addition, independent of underlying representation. For the multiple constant multiplication case several effective algorithms have been proposed over the years that avoids the problem of representation dependence [14, 52]. Theoretical lower bounds for related problems have been presented in [15].

3.6 Distributed Arithmetic

Distributed arithmetic is an efficient scheme for computing inner products of a fixed and a variable data vector

$$Y = a^T X = \sum_{i=1}^{N} a_i X_i. \tag{24}$$

The basic principle is owed to Croisier et al. [6]. The inner product can be rewritten using two's complement representation

$$Y = \sum_{i=1}^{N} a_i \left[-x_{i0} + \sum_{k=1}^{Wf} x_{ik} 2^{-k} \right] \tag{25}$$

Table 4 Distributed arithmetic look-up table for $a_1 = (0.0100001)_{2C}$, $a_2 = (0.1010101)_{2C}$, and $a_3 = (1.1110101)_{2C}$

x_1	x_2	x_3	F_k	F_k
0	0	0	0	$(0.0000000)_{2C}$
0	0	1	a_3	$(1.1110101)_{2C}$
0	1	0	a_2	$(0.1010101)_{2C}$
0	1	1	$a_2 + a_3$	$(0.1001010)_{2C}$
1	0	0	a_1	$(0.0100001)_{2C}$
1	0	1	$a_1 + a_3$	$(0.0010110)_{2C}$
1	1	0	$a_1 + a_2$	$(0.1110110)_{2C}$
1	1	1	$a_1 + a_2 + a_3$	$(0.1101011)_{2C}$

Fig. 28 Block diagram for distributed arithmetic

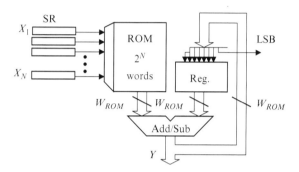

where x_{ik} is the kth bit in x_i. By interchanging the order of the two summations we get

$$Y = -\sum_{i=1}^{N} a_i x_{i0} + \sum_{k=1}^{Wf} \left[\sum_{i=1}^{N} a_i x_{ik} \right] 2^{-k} \tag{26}$$

$$= -F_0(x_{10}, x_{20}, \ldots, x_{N0}) + \sum_{k=1}^{Wf} F_k(x_{1k}, x_{2k}, \ldots, x_{Nk}) 2^{-k} \tag{27}$$

where

$$F_k(x_{1k}, x_{2k}, \ldots, x_{Nk}) = \sum_{i=1}^{N} a_i x_{ik}. \tag{28}$$

F_k is a function of N binary variables, ith variable being the kth bit in x_i. Since F_k can take on only 2^N values, it can be precomputed and stored in a look-up table. For example, consider the inner product $Y = a_1 X_1 + a_2 X_2 + a_3 X_3$ where $a_1 = (0.0100001)_{2C}$, $a_2 = (0.1010101)_{2C}$, and $a_3 = (1.1110101)_{2C}$. Table 4 shows the function F_k and the corresponding addresses.

Figure 28 shows a realization of (27) by Horner's method

$$y = ((\ldots((0 + F_{Wf})2^{-1} + \ldots + F_2)2^{-1} + F_1)2^{-1} - F_0) \tag{29}$$

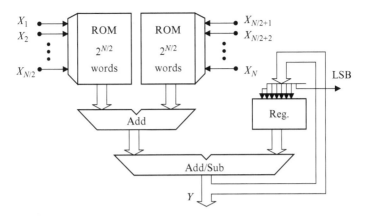

Fig. 29 Reducing the memory by partitioning

The inputs, $X_1, X_2, \ldots\ldots X_N$, are shifted bit-serially out from the shift registers with the least-significant bit first. Bits x_{ik} are used to address the look-up table. Since, the output is divided by 2, by the inherent shift, the circuit is called a shift-accumulator [56]. Computation of the inner product requires $W_f + 1$ clock cycles. In the last cycle, F_0 is subtracted from the accumulator register. Notice the resemblance with a shift-and-add implementation of a real multiplication.

A more parallel form of distributed arithmetic can also be realized by allocating several tables. The tables, which are identical, may be addressed in parallel and their appropriately shifted values.

3.6.1 Reducing the Memory Size

The memory requirement becomes very large for long inner products. There are mainly two ways to reduce the memory requirements. One of several possible ways to reduce the overall memory requirement is to partition the memory into smaller pieces that are added before the shift-accumulator, as shown in Fig. 29. The amount of memory is in this case reduced from 2^N words to $2 \cdot 2^{N/2}$ words. For example, for $N = 10$ we get $2^{10} = 1,024$ words, which is reduced to only $2 \cdot 2^5 = 64$ words at the expense of an additional adders.

Memory size can be halved by using the ingenious scheme[6] based on the identity $X = \frac{1}{2}[X - (-X)]$, which can be rewritten

$$X = -(x_0 - \bar{x}_0)2^{-1} + \sum_{k=1}^{W_f} (x_k - \bar{x}_k)2^{-k-1} - 2^{-(W_f+1)}. \tag{30}$$

Table 5 Look-up table contents using half-sized memory

x_1	x_2	x_3	F_k	u_1	u_2	A/S
0	0	0	$-a_1 - a_2 - a_3$	0	0	A
0	0	1	$-a_1 - a_2 + a_3$	0	1	A
0	1	0	$-a_1 + a_2 - a_3$	1	0	A
0	1	1	$-a_1 + a_2 + a_3$	1	1	A
1	0	0	$+a_1 - a_2 - a_3$	1	1	S
1	0	1	$+a_1 - a_2 + a_3$	1	0	S
1	1	0	$+a_1 + a_2 - a_3$	0	1	S
1	1	1	$+a_1 + a_2 + a_3$	0	0	S

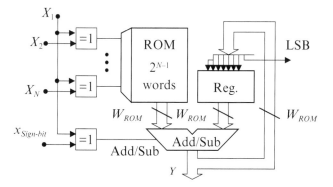

Fig. 30 Distributed arithmetic with half-sized memory

Notice that $(x_k - \bar{x}_k)$ can only take on the values -1 or $+1$. Inserting this expression into (24) yields

$$Y = \sum_{k=1}^{W_f} F_k(x_{1k}, \ldots, x_{Nk}) 2^{-k-1} - F_0(x_{10}, \ldots, x_N) 2^{-1} + F(0, \ldots, 0) 2^{-(W_f+1)} \quad (31)$$

where

$$F_k(x_{1k}, x_{2k} \ldots \ldots, x_{Nk}) = \sum_{i=1}^{N} a_i(x_{ik} - \bar{x_{ik}}). \quad (32)$$

The function F_k is shown in Table 5 for $N = 3$. Notice that only half the values are needed, since the other half can be obtained by changing the signs. To explore this redundancy we make the following address modification $u_1 = x_1 \oplus x_2$, $u_2 = x_1 \oplus x_3$, and $A/S = x_1 \oplus s_{\text{sign-bit}}$ where X_1 has been selected as control variable [56]. The control signal $x_{\text{sign-bit}}$ is zero at all times except when the sign bit of the inputs arrives. Figure 30 shows the resulting realization with halved look-up table. The XOR gates used for halving the memory can be merged with the XOR-gates that are needed for inverting F_k.

Table 6 ROM contents for a complex multiplier based on distributed arithmetic

a_i	b_i	F_1	F_2
0	0	$-(c-d)$	$-(c+d)$
0	1	$-(c+d)$	$(c-d)$
1	0	$(c+d)$	$-(c-d)$
1	1	$(c-d)$	$(c+d)$

3.6.2 Complex Multipliers

A complex multiplication requires three or four real multiplications and some additions but only two distributed arithmetic units, which from area, speed, and power consumption points of view are comparable to the real multiplier. Let $X = A + jB$ and $K = c + jd$ where K is the fixed coefficient and X is a variable. Now, the product of the two complex numbers can be written

$$
\begin{aligned}
K.X &= (cA - dB) + j(dA + cB) \\
&= \left\{ -c(a_0 - \bar{a}_0)2^{-1} + \sum_{k=1}^{W_f} c(a_k - \bar{a}_k)2^{-k-1} - c2^{-(W_f+1)} \right\} \\
&\quad - \left\{ -d(b_0 - \bar{b}_0)2^{-1} + \sum_{k=1}^{W_f} d(b_k - \bar{b}_k)2^{-k-1} - d2^{-(W_f+1)} \right\} \\
&\quad + j\left\{ -d(a_0 - \bar{a}_0)2^{-1} \sum_{k=1}^{W_f} d(a_k - \bar{a}_k)2^{-k-1} - d2^{-(W_f+1)} \right\} \\
&\quad + j\left\{ -c(b_k - \bar{b}_k)2^{-1} + \sum_{k=1}^{W_f} c(b_k - \bar{b}_k)2^{-k-1} - c2^{-(W_f+1)} \right\} \\
&= F_1(a_0, b_0)2^{-1} + \sum_{k=1}^{W_f} F_1(a_k, b_k)2^{-k-1} + F_1(0,0)2^{-(W_f+1)} \\
&\quad + j\left\{ F_2(a_0, b_0)2^{-1} + \sum_{k=1}^{W_f} F_2(a_k, b_k)2^{-k-1} + F_2(0,0)2^{-(W_f+1)} \right\}.
\end{aligned}
$$

Hence, the real and imaginary parts of the product can be computed using just two distributed arithmetic units. The content of the look-up table that stores F_1 and F_2 is shown in Table 6.

Obviously only two coefficients are needed, $(c+d)$ and $(c-d)$. If $a_j \oplus b_j = 1$, the F coefficients values are applied directly to the accumulators, and if $a_j \oplus b_j = 0$, the F coefficients values are interchanged and added or subtracted depending on the data bits a_k and b_k. The realization is shown in Fig. 31.

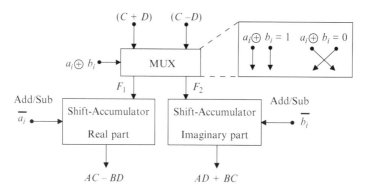

Fig. 31 Block diagram for a complex multiplier based on distributed arithmetic

4 Division

Of the four basic arithmetic operations, the division is the most complex to compute. Furthermore, the result of a division consists of two components, the quotient, Z, and the remainder, R, such that

$$X = ZD + R \tag{33}$$

where X is the dividend, $D \neq 0$ is the divisor, and $|R| < D$. By definition the sign of the remainder should be the same as that of the dividend. For the result to be a fractional number we must have $\left|\frac{X}{D}\right| \leq 1$. This can always be obtained by shifting the dividend and/or divisor. For ease of exposition we will initially start with unsigned data, but eventually introduce signed data. For further information on the methods presented here and others, we refer to [10].

4.1 Restoring and Nonrestoring Division

The simplest way to perform a division is to sequentially shift the dividend one position (multiply by two) and then check if the divisor has larger magnitude than the dividend. If so, the corresponding magnitude bit of the quotient is one and we subtract the divisor from the dividend. Conceptually, the comparison can be made by first subtracting the divisor from the dividend and then check if the result is positive (quotient bit is one) or negative (quotient bit is zero). If the result is negative we need to add the divisor again, which gives the name *restoring division*.

The computation in step i can be written as

$$r_i = 2r_{i-1} - z_i D \tag{34}$$

where $r_0 = X$. Therefore, if $2r_{i-1} - D$ is positive, we set $z_i = 1$, otherwise $z_i = 0$ and $r_i = 2r_{i-1}$. r_i is the remainder after iteration i and considering (33) we have $R = r_i 2^{-1}$. To compute a quotient with W_f fractional bits obviously W_f iterations of (34) are required.

Instead of restoring the remainder by adding the divisor, we can assign a negative quotient digit. This gives the *nonrestoring division* selection rule of the quotient digits, z_i, in (34) as

$$z_i = \begin{cases} 1, & r_{i-1}D \geq 0 \text{ i.e. same sign} \\ -1, & r_{i-1}D < 0 \text{ i.e. different signs.} \end{cases} \tag{35}$$

Note that with this definition of the selection rules the remainder will sometimes be positive, sometimes negative. Hence, division with a signed dividend and/or divisor is well covered within this approach. This also gives that the final remainder does not always have the same sign as the dividend. Hence, in that case we must compensate by adding or subtracting D to R and consequently subtracting or adding one LSB to Z.

The result from the nonrestoring division will be represented using a representation with $q_i \in \{-1, 1\}$. This representation is sometimes called *nega-binary* and is in fact not a redundant representation. The result should in most cases be converted into a two's complement representation. Naturally, one can convert this by forming a word with positive bits and one with negative bits and subtract the negative bits from the positive bits. However, for this all bits must be computed before the conversion can be done. Instead, it is possible to use the on-the-fly conversion technique in [9] to convert the digits into bits once they are computed.

Another consequence of the nega-binary representation is that if a zero remainder is obtained, this will not remain zero in the succeeding stages. Hence, a zero remainder should be detected and either the iterations stopped or corrected for at the end.

4.2 SRT Division

The *SRT* division scheme extends the nonrestoring scheme by allowing 0 as a quotient digit. Furthermore, by restricting the dividend to be in the range $1/2 \leq D < 1$, which can be obtained by shifting, the selection rule for the quotient digit in (34) can be defined as

$$z_i = \begin{cases} -1, & 2r_{i-1} < -1/2 \\ 0, & -1/2 \leq 2r_{i-1} < 1/2 \\ 1, & 1/2 \leq 2r_{i-1} \end{cases} \tag{36}$$

for the binary case. This has two main advantages: firstly, when $z_i = 0$ there is no need to add or subtract; secondly, comparing with $1/2$ or $-1/2$ only requires 3 bits of $2r_{i-1}$. There exists slightly improved selection rules that further reduce the number of additions/subtractions. However, the number of iterations are still W_f for a quotient with W_f fractional bits.

4.3 Speeding Up Division

While the number of additions/subtractions are reduced in the SRT scheme it would require an asynchronous circuit to improve the speed. In many situations this is not wanted. Instead, to reduce the number of cycles one can use division with a higher radix. Using radix $b = 2^m$ reduces the number of iterations to $\lceil \frac{W_f}{m} \rceil$. The iteration is now

$$r_i = br_{i-1} - z_i D \tag{37}$$

where $z_i \in \{0, 1, \ldots, b-1\}$ for restoring division. For SRT division $z_i \in \{-a, -a+1, \ldots, -1, 0, 1, \ldots, a\}$, where $\lceil (b-1)/2 \rceil \leq a \leq (b-1)$. The selection rules can be defined in several different ways similar to radix-2 discussed earlier. We can guarantee convergence by selecting the quotient digit such that $|r_i| < D$, which typically implies maximizing the magnitude of the quotient digit.

For SRT division it is possible to select the redundancy of the representation based on a. Higher redundancy leads to a larger overlap in the regions where one can select any of two different quotient digits. Having an overlap means that one can select the breakpoint such that few bits of r_i and D need to be compared. However, a higher redundancy means that there are more multiples of D that needs to be computed for the comparison. Hence, there is a trade-off between the number of bits that needs to be compared and the precomputations for the comparisons.

Even though higher-radix division reduces the number of iteration, each iteration still needs to be performed sequentially. In each step the current remainder must be known and the quotient digit selected before it is possible to start a new step. There are two different ways to overcome this limitation. First, it is possible to overlap the complete computation of the partial remainder in step i and the selection of the quotient digit in step $i + 1$. This is possible since not all bits of the remainder must be known to select the next quotient digit. Second, the remainder can be computed in a redundant number system.

4.4 Square Root Extraction

Computing the square root is in some ways a similar operation to division as one can sequentially iterate the remainder, initialized to the radicand, $r_0 = X$, with the partially computed square root $Z_i = \sum_{k=1}^{i} z_k 2^{-k}$ in a similar way as for division. More precisely, the iteration for square root extraction is

$$r_i = 2_r i - 1 - z_i(2z_i + z_i 2^{-i}) \tag{38}$$

where $Z_0 = 0$.

Schemes similar restoring, nonrestoring, and SRT division can be defined. For the quotient digit selection scheme similar to SRT division the square root is restricted to $1/2 \leq Z < 1$, which corresponds to $1/4 \leq X < 1$. The selection rule is then

$$z_i = \begin{cases} 1 & 1/2 \leq r_{i-1} < 2 \\ 0 & -1/2 < r_{i-1} < 1/2 \\ -1 & -2 \leq r_{i-1} \leq -2 \end{cases} \tag{39}$$

5   Floating-Point Representation

Floating-point numbers consists of two parts, the mantissa (or significand), M, and the exponent (or characteristic), E, with a number, X, represented as

$$X = Mb^E \tag{40}$$

where b is the base of the exponent. For ease of exposition we assume $b = 2$. With floating-point numbers we obtain a larger dynamic range, but at the same time a lower precision compared to a fixed-point number system using the same number of bits.

Both the exponent and the mantissa are typically signed integer or fractional numbers. However, their representation are often not two's complement. For the mantissa it is common to use a sign-magnitude representation, i.e., use a separate sign-bit, S, and represent an unsigned mantissa magnitude with the remaining bits. For the exponent it is common to use excess-k, i.e., add k to the exponent to obtain an unsigned number.

5.1   Normalized Representations

A general floating-point representation is redundant since

$$M2^E = \frac{M}{2} 2^{E+1} \tag{41}$$

Table 7 Special cases for the exponent in binary32

	$F = 0$	$F \neq 0$
$E = 0$	0	Denormalized
$E = 255$	$\pm\infty$	NaN

However, to use as much as possible of the dynamic range provided by the mantissa, we would like to use the representation without any leading zeros. This representation is called the *normalized* form.

Another benefit of normalized representations is that comparisons are simpler. It is possible to just compare the exponents, and only if the exponents are the same one will have to compare the mantissas. Also, as we know that there are no leading zero we make the first one in the representation explicit, and, hence, effectively add a bit to the representation.

5.2 IEEE 754

Before the emergence of the IEEE 754 floating-point standard, typically different computer systems had different floating-point standards making the transportation of binary data between different systems difficult. Nowadays, while some computer systems have their own floating-point representations, most have converged to the IEEE 754 standard. The most recent installment was released in August 2008 [21].

The IEEE 754-2008 standard defines three binary and two decimal formats, where we will focus on the 32-bit binary format, called binary32, as the other two binary formats follow along the same lines.

The binary32 format has a sign bit, eight exponent bits using excess-127 representation, and 23 bits for the mantissa plus a hidden leading one. The representation can be visualized as

$$\underbrace{s}_{\text{Sign}}\ \underbrace{e_{-7}e_{-6}e_{-5}e_{-4}e_{-3}e_{-2}e_{-1}e_0}_{E\ \text{8 bits biased exponent}}\ \underbrace{f_1 f_2 f_3 f_4 f_5 f_6 \cdots f_{22} f_{23}}_{F\ \text{23 bits unsigned fraction}}$$

The value of the floating-point number is given by

$$X = (-1)^s\, 1.F\, 2^{E-127}. \tag{42}$$

Note the hidden one due to the normalized number system, so $M = 1.F$. This means that the actual mantissa value will be in the range $1 \leq M < 2 - 2^{-23}$. Out of the 256 possible values for the exponent, two have special meanings to deal with zero value, $\pm\infty$, and undefined results (NaN). This is outlined in Table 7.

Table 8 The three binary formats in IEEE754-2008

Property	binary32	binary64	binary128
Total bits	32	64	128
Mantissa bits	$23 + 1$	$52 + 1$	$112 + 1$
Exponent bits	8	11	15
Bias	127	1,023	16,383
Ext. mantissa bits	≥ 32	≥ 64	≥ 128
Ext. exponent bits	11	15	20

The denormalized numbers are used to extend the dynamic range as the hidden one otherwise limits the smallest positive number to $2^{1-127} = 2^{-126}$. A denormalized number has a value of

$$X = (-1)^s 0.F \, 2^{-126}. \tag{43}$$

Using denormalized numbers it is possible to represent $2^{-23} 2^{-126} = 2^{-149}$. However, the implementation cost of denormalized numbers are high, and, hence, are not always included.

There is also an extended format defined that is used for intermediate results in certain complex functions. The extended binary32 format uses 11 bits for the exponent and at least 32 bits for the mantissa (now without a hidden bit).

The two other formats are defined similarly to binary32 and are outlined in Table 8.

5.3 Addition and Subtraction

Adding and subtracting floating-point values require that both operands have the same exponent. Hence, we have to shift the mantissa of the smaller operand as in (41) such that both exponents are the same. Then, assuming binary32 and $E_X \geq E_Y$, it is possible to factor out the exponent term as

$$Z = (-1)^{s_Z} M_Z \, 2^{E_Z - 127} = X \pm Y = \left((-1)^{s_X} M_X \pm (-1)^{s_Y} M_Y \, 2^{-(E_X - E_Y)} \right) 2^{E_X - 127} \tag{44}$$

where we can identify

$$(-1)^{s_Z} \hat{M}_Z = (-1)^{s_X} M_X \pm (-1)^{s_Y} M_Y \, 2^{-(E_X - E_Y)} \tag{45}$$

and

$$\hat{E}_Z = E_X. \tag{46}$$

Depending on the operation required and the sign of the two operand, either a subtraction or an addition of the mantissas are performed. If the effective operation is an addition, we have $1 \leq \hat{M}_Z < 4$, which means that we may need to right-shift once to obtain the normalized mantissa, M_Z, and at the same time increase $\hat{E}_Z$ by one to obtain E_Z. If the effective operation is a subtraction, the result is $0 \leq |\hat{M}_Z| < 2$. For this case we might have to right-shift to obtain the normalized number, M_Z, and correspondingly decrease the exponent to obtain E_Z.

It should be noted that adding or subtracting sign-magnitude numbers is more complex compared to adding or subtracting two's complement numbers as one will have to make decisions based on the sign and the magnitude of the operators to determine which the effective operation to be performed is. Also, in the case of subtraction one needs to either determine which the largest magnitude is and subtract the smaller from the larger or negate the result in the case it is negative.

5.4 Multiplication

The multiplication of two floating-point numbers (assumed to be in IEEE 754 binary32 format) is computed as

$$Z = (-1)^{s_Z} M_Z \, 2^{E_Z - 127} = XY = (-1)^{s_X} M_X \, 2^{E_X - 127} (-1)^{s_Y} M_Y \, 2^{E_Y - 127} \quad (47)$$

where we see that

$$s_Z = s_X \oplus s_Y \quad (48)$$

$$\hat{M}_Z = M_X M_Y \quad (49)$$

$$\hat{E}_Z = E_X + E_Y - 127. \quad (50)$$

As we have $1 \leq M_X, M_Y < 2$ for normalized numbers, we get $1 \leq \hat{M}_Z < 4$. Hence, it may be required to shift $\hat{M}_Z$ one position to the right to obtain the normalized value M_Z, which can be seen by comparing with (41). If this happens one will also need to add 1 to $\hat{E}_Z$ to obtain E_Z.

This gives that the multiplication of two floating-point numbers corresponds to one fixed-point multiplication, one fixed-point addition, and a simple normalizing step after the operations.

For multiply–accumulate it is possible to use a *fused* architecture with the benefit that the alignment of the operand to be added can be done concurrently with the multiplication. In this way it is possible to reduce the delay for the total MAC operation compared to using separate multiplication and addition. Furthermore, rounding is only performed for the final output.

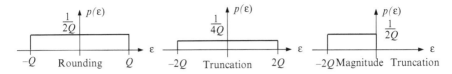

Fig. 32 Error distributions for floating-point arithmetic

5.5 *Quantization Error*

The quantization error in the mantissa of a floating-point number is

$$X_Q = (1 + \varepsilon)X \tag{51}$$

Hence, the error is signal dependent and the analysis becomes very complicated [32, 45, 57]. Figure 32 shows the error distributions of floating-point arithmetic. Also, the quantization procedure needed to suppress parasitic oscillation in wave digital filters is more complicated for floating-point arithmetic.

6 Computation of Elementary Functions

The need of computing non-linear functions arises in many different algorithms. The straightforward method of approximating an elementary function is of course to just store the function values in a look-up table. However, this will typically lead to large tables, even though the resulting area from standard cell synthesis grows slower than the number of memory bits [16]. Instead it is of interest to find ways to approximate elementary functions using a trade-off between arithmetic operations and look-up tables. In this section we briefly look at three different classes of algorithms. For a more thorough explanation of these and other methods we refer to [36].

6.1 *CORDIC*

The coordinate rotation digital computer (CORDIC) algorithm is a recursive algorithm to calculate elementary functions such as the trigonometric and hyperbolic (and their inverses) functions as well as magnitude and phase of complex vectors and was introduced by Volder [50] and generalized by Walther [54]. A summary of the development of CORDIC can be found in [34, 51, 55]. It revolves around the idea of rotating the phase of a complex number by multiplying it by a succession of constant values. However, these multiplications can all be made as powers of 2 and hence, in binary arithmetic they can be done using just shifts and adds. Hence, CORDIC is in general a very attractive approach when a hardware multiplier is not available.

Fig. 33 Similarity (rotation)
in the CORDIC algorithm

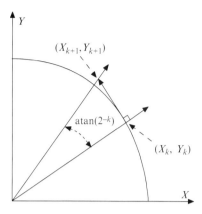

A rotation of a complex number $X + jY$ by an angle θ can be written as

$$\begin{bmatrix} X_r \\ Y_r \end{bmatrix} = \begin{bmatrix} \cos(\theta) & -\sin(\theta) \\ \sin(\theta) & \cos(\theta) \end{bmatrix} \begin{bmatrix} X \\ Y \end{bmatrix}. \tag{52}$$

The idea of the CORDIC is to decompose the rotation by θ in several steps such that each rotation is a simple operation. In the straightforward CORDIC algorithm we have

$$\theta = \sum_{k=0}^{\infty} d_k w_k, \; d_k = \pm 1, \; w_k = \arctan(2^{-k}) \tag{53}$$

Considering rotation k we get

$$\begin{bmatrix} X_{k+1} \\ Y_{k+1} \end{bmatrix} = \begin{bmatrix} \cos(d_k w_k) & -\sin(d_k w_k) \\ \sin(d_k w_k) & \cos(d_k w_k) \end{bmatrix} \begin{bmatrix} X_k \\ Y_k \end{bmatrix} = \cos(w_k) \begin{bmatrix} 1 & -d_k 2^{-k} \\ d_k 2^{-k} & 1 \end{bmatrix} \begin{bmatrix} X_k \\ Y_k \end{bmatrix}$$

$$\tag{54}$$

Now neglecting the $\cos(w_k)$ term we get a basic iteration which is a multiplication with 2^{-k} and an addition or subtraction. The sign of the rotation (d_n) is determined by comparing the required rotation angle θ with the currently rotated angle. This is typically done by using a third variable, Z_k, where $Z_0 = \theta$ and $Z_{k+1} = Z_k + d_k w_k$. Then

$$d_k = \begin{cases} 1 & Z_k \geq 0 \\ -1 & Z_k < 0 \end{cases} \tag{55}$$

The effect of neglecting the $\cos(w_k)$ in (54) is that the rotation is in fact not a proper rotation but instead a similarity [36]. Furthermore, as illustrated in Fig. 33, the magnitude of the vector is increased. The gain of the rotations depends on the number of iterations and can be written as

$$G(k) = \prod_{i=0}^{k} \sqrt{1 + 2^{-2i}} \tag{56}$$

Table 9 Variable selection for generalized CORDIC

Type	Parameters	Rot. mode, $d_k = \text{sign}(z_k)$	Vec. mode, $d_k = -\text{sign}(y_k)$
Circular	$m = 1$	$X_k \to G_k(X_0 \cos(Z_0) - Y_0 \sin(Z_0))$	$X_n \to G_k\sqrt{X_0^2 + Y_0^2}$
	$w_k = \arctan(2^{-k})$	$Y_k \to G_k(Y_0 \cos(Z_0) + X_0 \sin(Z_0))$	$Y_n \to 0$
	$\sigma(k) = k$	$Z_k \to 0$	$Z_n \to Z_0 + \arctan(\frac{Y_0}{X_0})$
Linear	$m = 0$	$X_k \to X_0$	$X_n \to X_0$
	$w_k = 2^{-k}$	$Y_k \to Y_0 + X_0 Z_0$	$Y_n \to 0$
	$\sigma(k) = k$	$Z_k \to 0$	$Z_n \to Z_0 + \frac{Y_0}{X_0}$
Hyperbolic	$m = -1$	$X_k \to \hat{G}_k(X_1 \cosh(Z_1) - Y_1 \sin(Z_1))$	$X_n \to \hat{G}_k\sqrt{X_1^2 - Y_1^2}$
	$w_k = \tanh^{-1}(2^{-k})$	$Y_k \to \hat{G}_k(Y_1 \cosh(Z_1) + X_1 \sinh(Z_1))$	$Y_n \to 0$
	$\sigma(k) = k - h_k$	$Z_k \to 0$	$Z_n \to Z_1 + \tanh^{-1}(\frac{Y_1}{X_1})$

For $k \to \infty$, $G \approx 1.6468$. Several schemes to compensate for the gain has been proposed and a survey can be found in [49].

The above application of the CORDIC algorithm is usually referred to as rotation mode and can be used to compute $\sin(\theta)$ and $\cos(\theta)$ or perform rotations of complex vectors. There are also a vectoring mode, where the rotation is performed such that the imaginary part, Y_k, becomes zero.

The generalized CORDIC iterations can be written as

$$X_{k+1} = X_k - m d_k Y_k 2^{-\sigma(k)}$$

$$Y_{k+1} = Y_k + d_k X_k 2^{-\sigma(k)}$$

$$Z_{k+1} = Z_k - d_k w_{\sigma(k)}. \qquad (57)$$

With an appropriate choice of m, d_k, w_k, and $\sigma(k)$ the CORDIC algorithm can perform a wide number of functions. These are summarized in Table 9, where three different types of CORDIC algorithms are introduced; Circular for computing trigonometric expressions, Linear for linear relationships, and Hyperbolic for hyperbolic computation. The scaling factor $\hat{G}_k$ for hyperbolic computations is

$$\hat{G}_k = \prod_{i=1}^{k} \sqrt{1 - 2^{-2(i-h_i)}} \qquad (58)$$

and the factor h_k is defined as the largest integer such that $3h_k + 1 + 2h_k - 1 \le 2n$. In practice this leads to that certain iteration angles, such that $k = (3^{i+1} - 1)/2$, are used twice to obtain convergence in the hyperbolic case.

The CORDIC computations can be performed in an iterative manner as in (57), but naturally also be unfolded. There has also been proposed Radix-4 CORDIC algorithms, performing two iterations in each step, as well as different approaches using redundant arithmetic to speed up each iteration.

Fig. 34 Block diagram for
Horner's method used for
polynomial evaluation

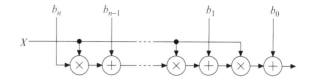

6.2 Polynomial and Piecewise Polynomial Approximations

It is possible to derive a polynomial $p(X)$ that approximates a function $f(X)$ by performing a Taylor expansion for a given point a such as

$$p(X) = \sum_{i=0}^{\infty} \frac{f^{(i)}(a)}{i!} (X - a)^i. \tag{59}$$

When the polynomial is restricted to a certain number of terms it is often better to optimize the polynomial coefficients as there is some accuracy to be gained. To determine the best coefficients is an approximation problems where typically there are more constraints (number of points for the approximation) than variables (polynomial order). This problem can be solved for a minimax solution using e.g. Remez' exchange algorithm or linear programming. For a least square solution, the standard methods to solve overdetermined systems can be applied. If fixed-point coefficients are required, the problem becomes much harder.

The polynomial approximations can be efficiently and accurately evaluated using Horner's method. This says that a polynomial

$$p(X) = b_0 + b_1 X + b_2 X^2 + \cdots + b_{n-1} X^{n-1} + b_n X^n \tag{60}$$

is to be evaluated as

$$p(X) = ((\ldots((b_n X + b_{n-1})X + b_{n-1})X + \cdots + b_2)X + b_1)X + b_0 \tag{61}$$

Hence, there no need to compute any powers of x explicitly and a minimum number of arithmetic operations is used. Polynomial evaluation using Horner's method maps nicely to MAC-operations. The resulting scheme is illustrated in Fig. 34.

The drawback of Horner's method is that the computation is inherently sequential. An alternative is to use Estrin's method [36], which by explicitly computing the terms X^{2^i} rearranges the computation in a tree structure, increasing the parallelism and reducing the longest computational path. Estrin's method for polynomial evaluation can be written as

$$p(X) = (b_3 X + b_2)X^2 + (b_1 X + b_0) \tag{62}$$

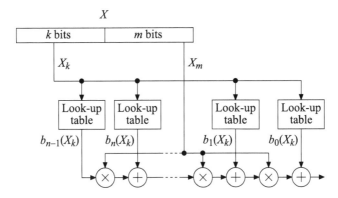

Fig. 35 Piecewise polynomial approximation using uniform segmentation based on the most significant bits

for a third-order polynomial. For a seventh-order polynomial it becomes

$$p(X) = ((b_7X + b_6)X^2 + (b_5X + b_4))X^4 + ((b_3X + b_2)X^2 + (b_1X + b_0)). \quad (63)$$

As can be seen, Estrin's method also maps well to MAC-operations.

The required polynomial order depends very much on the actual function that is approximated [36]. An approach to obtain a higher resolution despite using a lower polynomial order is to use different polynomials for different ranges. This is referred to as piecewise polynomials. A j segment n:th-order piecewise polynomial with segment breakpoints $x_k, k = 1, 2, \ldots, j + 1$ can be written as

$$p(X) = \sum_{i=0}^{n} b_{i,j}(X - x_j)^i, \; x_j \leq X < x_{j+1} \quad (64)$$

From an implementation point of view it is often practical to have 2^k uniform segments and let the k most significant bits determine the segmentation. However, it can be shown that in general the total complexity is reduced for non-uniform segments. An illustration of a piecewise polynomial approximation is shown in Fig. 35.

6.3 Table-Based Methods

The bipartite table method is based on splitting the input word, X, in three different subwords, X_0, X_1, and X_2. For ease of exposition we will assume that the length of these are identical, W_s and $W_f = 3W_s$, but in general it is possible to find a lower complexity realization by selecting non-uniform word lengths. Hence, we have

$$X = X_0 + 2^{-W_s}X_1 + 2^{-2W_s}X_2 \quad (65)$$

Fig. 36 Bipartite table
approximation structure

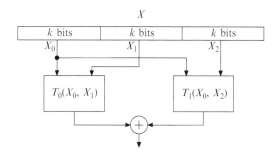

Now taking the first-order Taylor expansion of f at $X_0 + 2^{-W_s}X_1$ we get

$$f(X) \approx f(X_0 + 2^{-W_s}X_1) + 2^{-2W_s}X_2 f'(X_0 + 2^{-W_s}X_1) \qquad (66)$$

Again, we take the Taylor expansion, this time a zeroth-order expansion of $f'(X_0 + 2^{-W_s}X_1)$ at X_0 as

$$f'(X_0 + 2^{-W_s}X_1) \approx f'(X_0) \qquad (67)$$

This gives the bipartite approximation as

$$f(x) \approx T_1(X_0, X_1) + T_2(X_0, X_2) \qquad (68)$$

where

$$T_1(X_0, X_1) = f(X_0 + 2^{-W_s}X_1)$$
$$T_2(X_0, X_2) = 2^{-2W_s}X_2 f'(X_0) \qquad (69)$$

The functions T_1 and T_2 are tabulated and the results are added. The resulting structure is shown in Fig. 36.

The bipartite approximation can be seen as a piecewise linear approximation where the same slope tables are used in several intervals. Here, T_1 contains the offset values, and T_2 contains tabulated lines with slope $f'(X_0)$.

The accuracy of the bipartite approximation can be improved by instead performing the first Taylor expansion at $X_0 + 2^{-W_s}X_1 + 2^{-2W_s-1}$ and the second at $X_0 + 2^{-W_s-1}$ [47]. It is also possible to split the input word into more subwords yielding a multipartite table approximation [8].

7 Further Reading

Several books have been published on related subjects. For general digital arithmetic we refer to [11, 24, 26, 41]. For the specific cases of approximation of elementary functions and floating-point arithmetic, [36, 37] provide broad overviews, respectively.

References

1. C. R. Baugh and B. A. Wooley, "A two's complement parallel array multiplication algorithm," *IEEE Trans. Comput.*, vol. 22, pp. 1045–1047, Dec. 1973.
2. K. Bickerstaff, M. J. Schulte, and E. E. Swartzlander, Jr., "Parallel reduced area multipliers," *J. VLSI Signal Processing*, vol. 9, no. 3, pp. 181–191, Nov. 1995.
3. R. P. Brent and H. T. Kung, "A regular layout for parallel adders," *IEEE Trans. Comput.*, vol. 31, pp. 260–264, Mar. 1982.
4. S. C. Chan and P. M. Yiu, "An efficient multiplierless approximation of the fast Fourier transform using sum-of-powers-of-two (SOPOT) coefficients," *IEEE Signal Processing Lett.*, vol. 9, no. 10, pp. 322–325, Oct. 2002.
5. T. A. C. M. Claasen, W. F. G. Mecklenbräuker, and J. B. H. Peek, "Effects of quantization and overflow in recursive digital filters," *IEEE Trans. Acoust. Speech Signal Processing*, vol. 24, no. 6, Dec. 1976.
6. A. Croisier, D. J. Esteban, M. E. Levilion, and V. Rizo, "Digital filter for PCM encoded signals," U.S. Patent 3 777 130, Dec. 4, 1973.
7. L. Dadda, "Some schemes for parallel multipliers," *Alta Frequenza*, vol. 34, pp. 349–356, May 1965.
8. F. de Dinechin and A. Tisserand, "Multipartite table methods," *IEEE Trans. Comput.*, vol. 54, no. 3, pp. 319–330, Mar. 2005.
9. M. D. Ercegovac and T. Lang, "On-the-fly conversion of redundant into conventional representation," *IEEE Trans. Comput.*, vol. 36, pp. 895–897, July 1987.
10. _____, *Division and Square Root: Digit-Recurrence Algorithms and Implementations*, Kluwer Academic Publishers, 1994.
11. _____, *Digital Arithmetic*, Morgan Kaufmann Publishers, 2004.
12. H. Eriksson, P. Larsson-Edefors, M. Sheeran, M. Själander, D. Johansson, and M. Schölin, "Multiplier reduction tree with logarithmic logic depth and regular connectivity," in *Proc. IEEE Int. Symp. Circuits Syst.*, 2006.
13. A. Fettweis, and K. Meerkötter, "On parasitic oscillations in digital filters under looped conditions," *IEEE Trans. Circuits Syst.*, vol. 24, no. 9, pp. 475–481, Sept. 1977.
14. O. Gustafsson, "A difference based adder graph heuristic for multiple constant multiplication problems," in *Proc. IEEE Int. Symp. Circuits Syst.*, 2007.
15. _____, "Lower bounds for constant multiplication problems," *IEEE Trans. Circuits Syst. II*, vol. 54, no. 11, pp. 974–978, Nov. 2007.
16. O. Gustafsson and K. Johansson, "An empirical study on standard cell synthesis of elementary function look-up tables," in *Proc. Asilomar Conf. Signals Syst. Comput.*, 2008.
17. O. Gustafsson and L. Wanhammar, "Low-complexity and high-speed constant multiplications for digital filters using carry-save arithmetic" in *Digital Filters*, Intech, 2011.
18. O. Gustafsson, A. G. Dempster, K. Johansson, M. D. Macleod, and L. Wanhammar, "Simplified design of constant coefficient multipliers," *Circuits, Syst. Signal Processing*, vol. 25, no. 2, pp. 225–251, Apr. 2006.
19. D. Harris, "A taxonomy of parallel prefix networks", in *Proc. Asilomar Conf. Signals Syst. Comput.*, 2003.
20. R. I. Hartley, "Subexpression sharing in filters using canonic signed digit multipliers," *IEEE Trans. Circuits Syst. II*, vol. 43, no. 10, pp. 677–688, Oct. 1996.
21. IEEE 754-2008 Standard for Floating-Point Arithmetic
22. K. Johansson, O. Gustafsson, and L. Wanhammar, "Power estimation for ripple-carry adders with correlated input data," in *Proc. Int. Workshop Power Timing Modeling Optimization Simulation*, 2004.
23. S. Knowles, "A family of adders," in *Proc. Symp. Comput. Arithmetic*, 1990, pp. 30–34.
24. P. Kornerup and D. W. Matula, *Finite Precision Number Systems and Arithmetic*, Cambridge University Press, 2010.

25. P. M. Kogge and H. S. Stone, "A parallel algorithm for the efficient solution of a general class of recurrence equations," *IEEE Trans. Comput.*, vol. 22, pp. 786–793, 1973.

26. I. Koren, *Computer Arithmetic Algorithms*, 2nd edition, A. K. Peters, Natick, MA, 2002.

27. R. E. Ladner and M. J. Fischer, "Parallel prefix computation," *J. ACM*, vol 27, pp. 831–838. Oct. 1980.

28. J. Liang and T. D. Tran, "Fast multiplierless approximations of the DCT with the lifting scheme," *IEEE Trans. Signal Processing*, vol. 49, no. 12, pp. 3032–3044, Dec. 2001.

29. Y.-C. Lim, "Single-precision multiplier with reduced circuit complexity for signal processing applications," *IEEE Trans. Comput.*, vol. 41, no. 10, pp. 1333–1336, Oct. 1992.

30. Y.-C. Lim, R. Yang, D. Li, and J. Song, "Signed power-of-two term allocation scheme for the design of digital filters," *IEEE Trans. Circuits Syst. II*, vol. 46, no. 5, pp. 577–584, May 1999.

31. B. Liu, "Effect of finite word length on the accuracy of digital filters – a review," *IEEE Trans. Circuit Theory*, vol. 18, pp. 670–677, Nov. 1971.

32. B. Liu and T. Kaneko, "Error analysis of digital filters realized with floating-point arithmetic," *Proc. IEEE*, vol. 57, pp. 1735–1747, Oct. 1969.

33. O. L. MacSorley, "High-speed arithmetic in binary computers," *Proc. IRE*, vol. 49, no. 1, pp. 67–91, Jan. 1961.

34. P. K. Meher, J. Valls, T.-B. Juang, K. Sridharan, and K. Maharatna, "50 years of CORDIC: Algorithms, architectures and applications," *IEEE Trans. Circuits Syst. I*, vol. 56, no. 9, pp. 1893–1907, Sept. 2009.

35. Z.-J. Mou and F. Jutand, "'Overturned-Stairs' adder trees and multiplier design," *IEEE Trans. Comput.*, vol. 41, no. 8, pp. 940–948, Aug. 1992.

36. J.-M. Muller, *Elementary Functions: Algorithms and Implementation*, 2nd Edition, Birkhäuser Boston, 2006.

37. J.-M. Muller, N. Brisebarre, F. de Dinechin, C.-P. Jeannerod, V. Lefévre, G. Melquiond, N. Revol, D. Stehlé, and S. Torres, *Handbook of Floating-Point Arithmetic*, Birkhäuser Boston, 2010.

38. V. G. Oklobdzija, D. Villeger, and S. S. Liu, "A method for speed optimized partial product reduction and generation of fast parallel multipliers using an algorithmic approach," *IEEE Trans. Comput.*, vol. 45, no. 3, Mar. 1996.

39. A. Omondi and B. Premkumar, *Residue Number Systems: Theory and Implementation*, Imperial College Press, 2007.

40. S. T. Oskuii, P. G. Kjeldsberg, and O. Gustafsson, "A method for power optimized partial product reduction in parallel multipliers," in *Proc. IEEE Norchip Conf.*, 2007.

41. B. Parhami, *Computer Arithmetic: Algorithms and Hardware Designs*, 2nd edition, Oxford University Press, New York, 2010.

42. N. Petra, D. De Caro, V. Garofalo, E. Napoli, and A. G. M. Strollo, "Truncated binary multipliers with variable correction and minimum mean square error," *IEEE Trans. Circuits Syst. I*, 2010.

43. M. Potkonjak, M. B. Srivastava, and A. P. Chandrakasan, "Multiple constant multiplications: efficient and versatile framework and algorithms for exploring common subexpression elimination," *IEEE Trans. Computer-Aided Design*, vol. 15, no. 2, pp. 151–165, Feb. 1996.

44. M. Püschel, J. M. F. Moura, J. Johnson, D. Padua, M. Veloso, B. Singer, J. Xiong, F. Franchetti, A. Gacic, Y. Voronenko, K. Chen, R. W. Johnson, and N. Rizzolo, "SPIRAL: Code generation for DSP transforms," *Proc. IEEE*, vol. 93, no. 2, pp. 232–275, Feb. 2005.

45. B. D. Rao, "Floating point arithmetic and digital filters," *IEEE Trans. Signal Processing*, vol. 40, no. 1, pp. 85–95, Jan. 1992.

46. H. Samueli and A. N. Willson Jr., "Nonperiodic forced overflow oscillations in digital filters," *IEEE Trans. Circuits Syst.*, vol. 30, no. 10, pp. 709–722, Oct. 1983.

47. M. J. Schulte and J. E. Stine, "Approximating elementary functions with symmetric bipartite tables," *IEEE Trans. Comput.*, no. 8, vol. 48, pp. 842–847, Aug. 1999.

48. P. F. Stelling and V. G. Oklobdzija, "Design strategies for optimal hybrid final adders in a parallel multiplier," *J. VLSI Signal Processing*, vol. 14, no. 3, pp. 321–331, Dec. 1996.

49. D. Timmerman, H. Hahn, B. J. Hosticka, and B. Rix, "A new addition scheme and fast scaling factor compensation methods for CORDIC algorithms," *Integration, the VLSI Journal*, vol. 11, pp. 85–100, Nov. 1991.

50. J. E. Volder, "The CORDIC trigonometric computing technique," *IRE Trans. Elec. Comput.*, vol. 8, pp. 330–334, 1959.

51. _____, "The birth of CORDIC," *J. VLSI Signal Processing,* vol. 25, 2000.

52. Y. Voronenko and M. Püschel, "Multiplierless multiple constant multiplication," *ACM Trans. Algorithms* vol. 3, no. 2, May 2007.

53. C. Wallace, "A suggestion for a fast multiplier," *IEEE Trans. Electron. Comput.*, vol. 13, no. 1, pp. 14–17, Feb. 1964.

54. J. S. Walther, "A unified algorithm for elementary functions," *Spring Joint Computer Conf. Proc.,* vol. 38, pp. 379–385, 1971.

55. _____, "The story of unified CORDIC," *J. VLSI Signal Processing,* vol. 25, 2000.

56. L. Wanhammar, *DSP Integrated Circuits*, Academic Press, 1999.

57. B. Zeng and Y. Neuvo, "Analysis of floating point roundoff errors using dummy multiplier coefficient sensitivities," *IEEE Trans. Circuits Syst.*, vol. 38, no. 6, pp. 590–601, June 1991.

58. R. Zimmermann, *Binary adder architectures for cell-based VLSI and their synthesis*, Ph.D. Thesis, Swiss Federal Institute of Technology (ETH), Zurich, Hartung-Gorre Verlag, 1998.

Architectures for Particle Filtering

Sangjin Hong and Seong-Jun Oh

Abstract There are many applications in which particle filters outperform traditional signal processing algorithms. Some of these applications include tracking, joint detection and estimation in wireless communication, and computer vision. However, particle filters are not used in practice for these applications mainly because they cannot satisfy real-time requirements. This chapter discusses several important issues in designing an efficient resampling architecture for high throughput parallel particle filtering. The resampling algorithm is developed in order to compensate for possible error caused by finite precision quantization in the *resampling* step. Communication between the processing elements after resampling is identified as an implementation bottleneck, and therefore, concurrent buffering is incorporated in order to speed up communication of particles among processing elements. The mechanism utilizes a particle-tagging scheme during quantization to compensate possible loss of replicated particles due to the finite precision effect. Particle tagging divides replicated particles into two groups for systematic redistribution of particles to eliminate particle localization in parallel processing. The mechanism utilizes an efficient interconnect topology for guaranteeing complete redistribution of particles even in case of potential weight unbalance among processing elements. The architecture supports high throughput and ensures that the overall parallel particle filtering execution time scales with the number of processing elements employed.

S. Hong (✉)
Department of Electrical and Computer Engineering, Stony Brook University,
Stony Brook, NY 11794, USA
e-mail: sangjin.hong@stonybrook.edu

S.-J. Oh
College of Information & Communication, Division of Computer and Communications
Engineering, Korea University, Seoul 136-701, Korea
e-mail: seongjun@korea.ac.kr

S.S. Bhattacharyya et al. (eds.), *Handbook of Signal Processing Systems*,
DOI 10.1007/978-1-4614-6859-2_20, © Springer Science+Business Media, LLC 2013

1   Introduction

Particle filters are used in non-linear signal processing where the interest is in tracking and/or detection of random signals. Particle filters recursively generate random measures that approximate the distributions of the unknowns [1–4]. The random measures are composed of particles (samples) drawn from relevant distributions and of importance weights assigned to the particles [1, 5–8]. These random measures allow for computation of all sorts of estimates of the unknowns. In this chapter, we consider sample importance *resampling* (SIR) type of particle filters as a typical representative of particle filtering methods. The resampling operation is very important for accurate tracking [9–12]. The idea of resampling is to remove the trajectories that have small weights and to focus on trajectories that are dominating. Standard algorithms used for resampling are different variation of stratified resampling [6, 13]. The two most common methods for resampling are systematic and residual resampling [1, 14, 15]. While these algorithms are very effective in particle filtering, physical implementation of such algorithms pose challenges especially when parallel processing is sought for large number of particles. When the number of particles, M, becomes large, parallel processing of particle filtering is often considered to reduce its execution time of an iteration.

Commonly, the weights of the particles degenerate after several time instants. In that case, it is necessary to replicate the particles with large weights and to discard the ones with small weights. This is known as *resampling* [16, 17]. Since by *resampling*, we must process a large number of particles in a short period of time, an efficient mechanism is necessary for real-time applications. To improve throughput, multiple processing elements are necessary. While local computation can be improved with multiple processing elements executing in parallel, the resampling still becomes sequential because of non-deterministic (i.e., the number of particles transferred from a PE to the resampling unit is not determined and varies in every iteration) nature of data transfer between the processing units [18–20]. Thus, the architecture must support various data exchange patterns so that the performance can scale with the number of processing elements.

In this chapter, we consider a parallel particle filter architecture consisting of multiple processing elements (PEs) connected with a single central unit (CU) responsible for *resampling* [17]. While processing of particles in the sample and importance steps exploits spatial parallelism where feasible, the CU forces the filters to work sequentially. Thus, an efficient resampling mechanism is necessary in order to increase the speed and minimize the communication overhead among the PEs. We present a flexible resampling mechanism for high speed parallel particle filters. First, a low complexity resampling mechanism is discussed and then its implementation is presented.

While realization of particle filters on digital signal processors is feasible when the throughput requirement is not imposed, the entire process is sequential which limits the overall performance significantly [21]. It is also possible to realize the central unit on a digital signal processor (DSP) but due to the limited number of input and output pins that can be accessed in parallel, this implementation is not suitable for supporting multiple PEs. In addition, the resampling process is inherently memory-centric. Hence, typical DSP suffers from extensive memory accesses which seriously degrade the throughput of the particle filtering. Moreover, standard addressing schemes on standard buses are not suitable for handling non-deterministic data exchanges among the processing elements. On the other hand, commercial FPGAs are viable since they provide enough I/O pins for supporting concurrent data exchanges with the processing elements [16,22]. Moreover, FPGAs have fast logic elements, flexible interconnects, and memory. However, for high-throughput designs with the low-complexity that supports non-deterministic data exchanges among the processing elements, we consider VLSI implementations.

Here, we present a VLSI design and implementation of a flexible resampling mechanism. The architecture supports configurations with 2 or 4 PEs. With 4 PEs, three different subconfigurations are supported where the difference is in the performance and throughput tradeoff. The architecture is designed for tracking applications [23] but can be modified to support different particle filtering because the *resampling* process is identical. The main difference will be in the number of input and output pins, and the size of buffers. Static dual-ported SRAM is incorporated to maintain high throughput.

In this chapter, we also consider the fixed-point processing issue for multiple PEs. An efficient mechanism for single PE in fixed-point processing of a particle filter has been previously discussed [24]. It has also been shown that the execution time of a fully pipelined particle filtering including resampling is $2MT_{PE}$, where M is the total number of particles dedicated for the resampling, and T_{PE} is the execution clock period. Operational concurrency in particle filters, other than the resampling, can be exploited in the algorithm, which can be parallelized. However, the resampling requires a sequential processing, which negates the benefit of parallel processing. This is because the resampling has to consider all the M particles for their correct replication. For simple parallel processing with P PEs, the execution time for M particles can be represented as $[M/P + M]T_{PE}$ where $[M/P]T_{PE}$ is the time for concurrent parallel processing of filtering operations other than resampling, and MT_{PE} is the time required by resampling [20]. Thus, the overall execution throughput is lower bounded by MT_{PE}, even with infinite number of PEs. On the other hand, resampling can be done locally within each PE in parallel, where the PEs resample their own M/P particles. In this case, the execution time can be reduced to $[2M/P]T_{PE}$. However, such parallel processing has a serious limitation. Particles will be highly localized within each PE (i.e., bad particles will stay in the same PE if not enough replicated particles or some of the good particles will be discarded if there are more replicated particles in the PE). Thus, serious weight degeneracy may occur. For example, two particles in two different PEs may have the same weights, but their replication factors, which indicate the number of times that one particle should be replicated based on the decimal equivalent values of the weights, may differ significantly.

2 Resampling in Particle Filters

2.1 Theory of Operation

Particle filters belong to the class of Bayesian signal processing algorithms. They are applied to problems that can be described using dynamic state space (DSS) models. DSS models are used for modeling gradual changes of the system state in time and the observations as a function of the state. In comparison to classical signal processing algorithms, particle filters do not impose any restrictions to the DSS model, i.e., they can deal with any non-linear and/or non-Gaussian models, where ensuring distributions are computable.

In DSS models, signals are described using the state and the observation equations:

$$\begin{aligned}
\mathbf{x}_n &= \mathbf{f}_n(\mathbf{x}_{n-1}, \mathbf{u}_n) \\
\mathbf{z}_n &= \mathbf{g}_n(\mathbf{x}_n, \mathbf{v}_n)
\end{aligned} \tag{1}$$

where $n \in \mathbb{N}$ is a discrete-time index, $\mathbf{x}_n$ is a signal vector of interest, and $\mathbf{z}_n$ is a vector of observations. The symbols $\mathbf{u}_n$ and $\mathbf{v}_n$ are noise vectors, $\mathbf{f}_n$ is a signal transition function, and $\mathbf{g}_n$ is a measurement function. The analytical forms of $\mathbf{f}_n(\cdot)$ and $\mathbf{g}_n(\cdot)$ are assumed known. The densities of $\mathbf{u}_n$ and $\mathbf{v}_n$ are parametric and known, but their parameters may be unknown, and $\mathbf{u}_n$ and $\mathbf{v}_n$ are independent from each other. The objectives are to estimate *recursively* the signal $\mathbf{x}_n$, $\forall n$, from the observations $\mathbf{z}_{1:n}$, where $\mathbf{z}_{1:n} = \{\mathbf{z}_1, \mathbf{z}_2, \cdots, \mathbf{z}_n\}$.

The estimation of $\mathbf{x}_n$ is handled in Bayesian signal processing by computing the posterior distribution. In this chapter, we are actually interested in estimating the filtering density $p(\mathbf{x}_n|\mathbf{z}_{1:n})$ which allows us to compute the estimates of the states $\mathbf{x}_n$ given the whole history of observations $\mathbf{z}_{1:n}$. Commonly, the expressions for computing and updating filtering densities are not tractable and involve multidimensional integration. The computation of the estimates also involves integration. The estimate of the posterior expectation $E(\mathbf{f})$ of the function $\mathbf{f}(\mathbf{x}_n)$ is defined by

$$E(\mathbf{f}(\mathbf{x}_n)) = \int \mathbf{f}(\mathbf{x}_n) p(\mathbf{x}_n \mid \mathbf{z}_{1:n}) d\mathbf{x}_n. \tag{2}$$

Monte Carlo techniques allow for computing multidimensional integrals by converting the integration into summation. However, it is necessary to draw samples from the distribution $p(\mathbf{x}_n|\mathbf{z}_{1:n})$. If it is difficult to draw samples from the original distribution, importance sampling techniques can be applied. Then the samples are drawn from the importance sampling function $\pi(\mathbf{x}_n)$ from which it is easy to sample, and each sample has an associated weight. Each sample is obtained from $\pi(\mathbf{x}_n|\mathbf{z}_{1:n})$ and is weighted with respect to $p(\mathbf{x}_n|\mathbf{z}_{1:n})$. The basic operations of particle filters are the generation of new particles (*particle generate*) $\{\mathbf{x}_n^{(m)}\}_{m=1}^M$ and computation of particle weights (*particle update*) $\{w_n^{(m)}\}_{m=1}^M$.

Besides these two operations, the SIR filters perform the *resampling* step. This step is critical in every implementation of particle filtering because without it, the weights of many particles quickly become negligible that inference is made by using only a very small number of particles. The idea of *resampling* is to replicate particles with significant weights and to discard those with negligible weights. Typically, particles are replicated proportionally to their weights $r^{(m)} = w_n^{(m)} \cdot M$ times. The total number of replicated particles should remain M which means that the following condition should be satisfied

$$\Sigma_{m=1}^{M} r^{(m)} = M. \tag{3}$$

Since $r^{(m)}$ has to be an integer, $w_n^{(m)}$ has to be either rounded or truncated. In case of rounding or truncation, the total number of replicated particles will not be equal to M in general. There are several types of resampling algorithms which assure that the total number of replicated particles is M and that the replication is proportional and unbiased. Standard algorithms used for *resampling* are different variants of stratified sampling such as residual *resampling* (RR) [2], branching corrections [3], systematic *resampling* (SR) [1, 23] as well as resampling methods with rejection control [4]. As a result of resampling, a new set of resampled particles $\{\tilde{\mathbf{x}}_n^{(m)}\}_{m=1}^{M}$ with equal weights $1/M$ is formed.

The basic operations of the RR algorithm are shown in Fig. 1. The output r, is an array of replication factors which shows how many times each particle is replicated. We should note that RR is composed of two steps. In the first step, the number of replications of particles is calculated and the particles are stored in memory (blocks 1 and 2). Since this method does not guarantee that the number of resampled particles is M, the weights of the residual particles are computed together with the sum of residual particles, and the number of replicated particles M_r (blocks 3 and 4) is computed. The weights of residual particles represent the portion that is not used in the first step due to truncation.

The residual particles have to be resampled again in order to produce the remaining $M - M_r$ particles. This can be done by applying the SR algorithm for the remaining particles. A necessary step before applying SR is to normalize the weights which is shown in block 5. The normalized weights are used as input for the SR procedure. Finally the residual particles are formed and written to the memory.

2.2 Particle Sharing Scheme

A particle filter with 4 PEs is used to speed up the process by parallel operations. In order to support correct operation in multiple PEs configuration, an additional unit, which is the focus of this investigation, called the central unit (CU) is needed. The algorithms for parallel particle filtering called distributed resampling algorithms

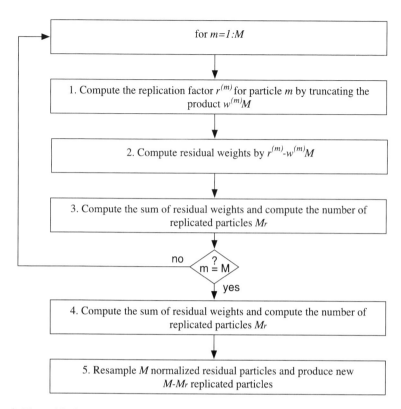

Fig. 1 The residual resampling algorithm

with proportional allocation (RPA) and non-proportional allocation (RNA) of particles are described in [25]. In this chapter, the CU is designed to support both of these algorithms with different modes.

Each PE performs in parallel the particle update and particle generate steps. Together with the weight computation (particle update), PE_i computes the local sum of weights sum_i for $i = 1, 2, \ldots, P$. Then the PEs send the local sum to the CU and receive the total sum sum_{total}. The total sum is used for weight normalization which is done locally within each PE. Then the PEs perform *resampling* in order to determine the replication factors $r^{(m)}$ for $m = 1, \ldots, M$. The CU is mainly responsible for two basic operations (Fig. 2). The first operation is done before *resampling* in which it gathers the sum of the weights generated from each PE and computes the summand (sum_{total}). In the second operation, the CU assists during the exchange of particles after resampling. This process is called *state update*.

We can distinguish between different algorithms and different modes based on how the normalization and particle exchange are performed. Different modes are shown in Table 1. There are a total of seven different configurations. The coding of each configuration (*mode*) is also shown in the table. These modes are illustrated in Fig. 3.

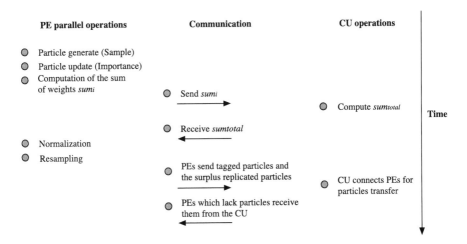

Fig. 2 The operations that are performed by the PEs and by the CU. The data exchanged between the PEs and the CU is also shown

Table 1 Illustration of particle sharing modes

Mode	Configuration	Resampling algorithm	*Mode*
0	Single PE	Resampling algorithm with tagged particles	000
1	2-Fixed	Only 2 PEs are used	001
2	2-2-Fixed	2 Separate particle filters each using 2 PEs	010 (2)
3	2-2-Fixed	RNA without regrouping	011
4	2-2-Mixed	RNA with regrouping with defined rules	100
5	2-2-Adaptive	RNA with adaptive regrouping	101
6	4-Mixed	RPA	111

The *state update* requires extensive data communication among the PEs and the CU. In the example shown in Fig. 4, a particle filter with $P = 4$ PEs and $M = 16$ particles is used. The particles and their weights are computed during the particle generate and particle update steps. In this example, it happened that most of the probability mass is contained in PE2. The normalized sums sum_i for the PEs 1 to 4 are: 0, 13/16, 0, 3/16. In this example, the algorithm with proportional allocation is used (RPA), so that each PE generates the number of particles as a result of resampling proportional to its local sum. So, PE1 generates 0 replicated particles, PE2 13 replicated particles and so on. It was shown in [25] that RPA produces the same resampling result as the sequential resampling algorithms. In order to start the next particle filtering iteration, the number of particles in each PE has to be the same, namely four. So, PE2 has to send its particle surplus to the PEs 1, 3, and 4. The way that it is done in RPA is shown in Fig. 5a. It can be noticed from this example that the communication pattern during the particle exchange is random. Namely, after each particle filtering operation, it is not known which PE will have surplus and which will lack particles. So, the direction and the amount of particles exchanged

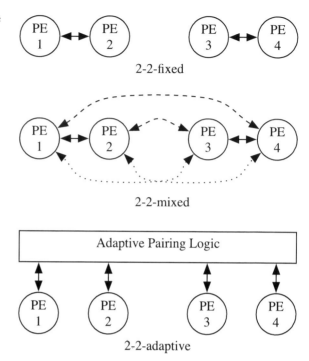

Fig. 3 Illustration of particle sharing for fixed, mixed and adaptive configurations

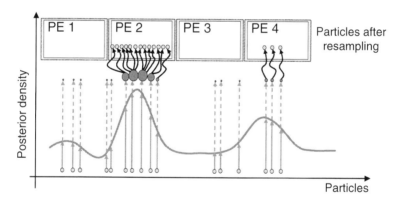

Fig. 4 In this example, a particle filter with 4 PEs and 16 particles is used. The posterior distribution of the unknown is presented. After resampling, the most of the probability mass is contained in PE2

among the PEs is unknown. We refer to the implementation of the RPA algorithm as 4-mixed configuration.

In order to reduce the amount of particles exchanged through the network as well as to make the communication deterministic, the parallel particle filtering algorithm with non-proportional allocation (RNA) is used. The main difference between RNA

Fig. 5 The example of
particle exchange for (**a**) RPA
and (**b**) RNA algorithm in
which the groups are formed
from two PEs

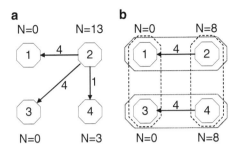

and RPA is that in RNA the PEs are grouped in groups of two and they act almost as separate particle filters. Let us assume that the sums sum_i for the PEs 1 to 4 are: $0, 13/16, 0, 3/16$ as in the previous example. Also, let us assume that PE1 and PE2 form one group and PE3 and PE4 form another group as it is shown in Fig. 5b using solid lines. The resampling is done inside these groups as if they are separate particle filters. Since all the probability mass is contained in PE2, it has to send four particles to PE1. The same happens in the second group, i.e., PE3 sends four particles to PE4. After resampling, the weights of the particles are not equal as in the RPA algorithm. They are set to the sum of the weights of the group. In the first group composed of PE1 and PE2, the sum of the weights is $13/16$ so that the weights of the replicated particles are $13/16$. The implementation of the RPA algorithm with fixed groups is referred to as 2-2 fixed (Mode 3) configuration.

If the groups are fixed, the weights in one group can become dominant in comparison to the weights from another group. In the case of unequally distributed weights, the particle filter performance will deteriorate because only one half (one group) of particle filter will contribute to the final estimate. This problem is solved by regrouping the PEs. In Fig. 5b, new groups are formed from PE1 and PE3, and PE2 and PE4 shown with dashed lines. Regrouping can be done using some defined rules or adaptively. The rule that is used in Fig. 5b is that groups are formed alternatively as PE1-PE2, PE3-PE4 and PE1-PE3, PE2-PE4. In adaptive configurations, the groups are formed in a way that the PE with the largest weight is grouped with the PE with the smallest weight. In this way, weights of particles are more evenly distributed among the PEs. The implementation of the RPA algorithm with regrouping with fixed rules is referred to as 2-2 mixed (Mode 4) and with adaptive regrouping as 2-2 adaptive (Mode 5).

Several additional modes are considered. For a single PE operation (Mode 0), the *resampling* is integrated into each PE and no particle sharing scheme is presented. Four independent particle filterings with different parameters can be processed concurrently. Although a single PE mode is supported by the resampling units, a PE with integrated resampling unit is more efficient in terms of data access and speed [26].

In Mode 1, only two PEs are active. In Mode 2, two fixed groups made of two PEs each are formed. There is no interaction among these two groups, and therefore two separate particle filters.

3 Parallel Resampling for Fast Particle Filtering

3.1 Weight Quantization Scheme

A weight of each particle, denoted as $w(m)$ for $m = [0 \cdots M - 1]$, is computed and these weights are normalized so that the sum of all weights is equal to one. This computation is done in the PEs before resampling. During the resampling, normalized particles are replicated according to the values of their weights in decimal representation. The most critical issue in replication of particles is that replicating exactly M particles with fixed-point processing is not guaranteed with traditional quantization and rounding [24].

As discussed in [24], these weights are quantized with K bits (excluding the sign bit), where $K = \log_2(M) + 2$. Thus, each weight is represented by

$$w = b_{K-1} b_{K-2} \cdots b_1 b_0 b_1' b_2'. \tag{4}$$

Two additional bits, $b_1' b_2'$ are used to simplify the rounding and truncation operations called *tagging*. The tagging is performed according to Table 2. Thus, when tagging is employed, rounding by any hardware is not necessary. When a particle is tagged, the particle is additionally replicated. This simplifies hardware complexity and speeds up the resampling process. Moreover, the tagging ensures that the total number of replicated particles be always larger than or equal to M. In the table, the entries of the first column are the last three bits of the binary representation of a weight, the entries of the second column represent the used rounding scheme, and the entries of the third column are the results due to rounding. After rounding, the last two bits are no longer used. The entries of the fourth column represent the tag status. Notice that the bit pattern 111 is not rounded but tagged. Without tagging, an adder is needed to incorporate carry propagation to the most significant bit. The tagging maintains the final replication factor without such hardware. However, the bit pattern 011 is rounded where a simple bit reversal is sufficient. The particles with 001 and 101 are also tagged.

Table 2 Rounding/truncation scheme and tagging method

Last Three Bits ($b_0 b_1' b_2'$)	Rounding Scheme	Final Bit (b_0)	Tag Status
000	Truncate	0	none
001	Truncate	0	tag
010	Truncate	0	tag
011	Round	1	none
100	Truncate	1	none
101	Truncate	1	tag
110	Truncate	1	tag
111	Truncate	1	tag

The scheme divides resampled particles into two sets of particle, R and T

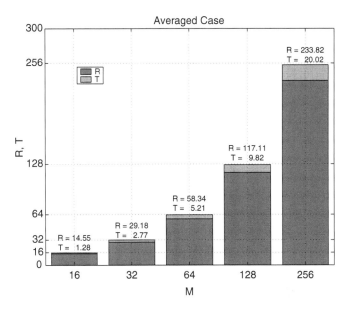

Fig. 6 Illustration of the average number of replicated particles classified as R and T. The results are based on 100 independent resamplings. In this plot we ignore 101 and 001 tags

Figure 6 illustrates the average of the total number of replicated particles classified as R and T without 101 and 001 tags (the particles with these ending will be included conditionally if more particles are necessary). R is the sum of the weights of particles according to the rounding and truncation schemes, and T is the total number of tagged particles. We can see that the T is resulted from proper rounding (round-up) of each weight. From the empirical study, the number of particles classified as R and T are 91.5% and 7.7% of M, respectively. As it can be seen in the figure, the total number of replicated particles is slightly less than M. Thus, this is the reason that particles with 101 and 001 in their last three bits are tagged to ensure that $R + T \geq M$. The number of particles due to this additional tagging is about 2.2% of M. This problem is introduced due to the finite precision processing and the problem will not appear when the resampling is performed with infinite precision processing. In the resampling, the tagged particles will have priority over the particles classified as R. Hence, about 1% of the particles classified as R will be eliminated in the actual processing. Therefore, the additional tagging will introduce some bias on the resampling performance. However, if the total number of replicated particles is less than M, the tracking performance will be worse than the performance of the resampling when tagging is used. We will show, with simulation, that the tagging has very little effects on resampling performance later in this chapter.

The above quantization scheme also supports a situation where one particle has a weight equal to 1.0 and the rest are all zero. Without any special modification, the scheme will get all the weights to zero since it considers only the K least significant

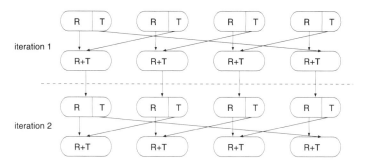

Fig. 7 Illustration of R and T particle sets redistribution

bits. The problem is avoided without having an additional bit. When the weight of 1.0 in decimal representation is $1 - 2^{-(K+1)}$, then the tagging method will guarantee that the total number of replicated particles is greater than or equal to M.

3.2 Particle Set Classification and Distribution

The tagging not only ensures enough particles to be replicated but also naturally classifies replicated particles into two groups of particles. In parallel resampling, particles classified as R and T are treated differently in particle redistribution among the PEs. Basically, all replicated particles in T will be used by other PEs in order to eliminate potential localization of particles (i.e., particles are not shared among the PEs).

Particles are distributed as shown in Fig. 7. Particle classified as R will be used by the same PE while particles classified as T will be used by an adjacent PE. While all the particles classified as T will always be used by an adjacent PE, any additional particles classified as R may go to the other PEs if enough particles are created at the PE. Thus, as iterations (i.e, successive resampling) of particle filtering continue, computations at all PEs are affected by all of the particles. Under this distribution scheme, localization of particles within any single PE is eliminated. We will provide an efficient mechanism that guarantees such particle redistribution in the next section. We will also show that classifying particles into two groups allows simple interconnection in the hardware implementation for high-speed redistribution among parallel PEs.

3.3 Architecture Support for Special-Cases

The effectiveness of particle redistribution in parallel resampling is measured by how the architecture supports the special-cases. In particle redistribution, the number of replicated particles is proportional to the sum of weights in each PE. Then, there are three special cases: (1) A single PE has the sum of weights equal to one, and the rest $P - 1$ PEs have the sum of weights equal to zero. (2) All PEs have the sum of weights larger than $1/P$, and one PE has the sum of weights equal to zero. (3) Each PE has the sum of weights equal to $1/P$, and the number of particles classified as T is zero.

The first two cases are directly related to the architecture efficiency in particle redistribution since collision occurs (i.e., many PEs send particles to one PE, or one PE sends particles to the other PEs). We fully utilize the concept of particle classification and efficient routing structure for supporting the case with no degradation in execution speed. In fact, we can achieve the execution speed of parallel resampling equal to $\lceil 2M/P \rceil T_{PE}$, which is the perfectly scalable architecture. The third case will never happen since we systematically create a finite amount of particles classified as T.

3.4   Particle Distribution Mechanism

A fast and efficient Central Unit (CU) mechanism is extremely crucial for real-time particle filtering with large M. The CU must guarantee that, (1) each PE will have M/P particles after the resampling, (2) the resampling completes with deterministic execution time, and (3) there is no deadlock in particle access operations. Moreover, the CU architecture should be scalable so that the mechanism works with a larger number of PEs.

The CU architecture, which performs particle redistribution for parallel particle filtering with P PEs, is shown in Fig. 8. For illustration, $P = 4$. Each PE executes independently but synchronously with its own M/P ($P = 4$) particles where M is the total number of particles. During the resampling, each PE interacts with the CU through a PE_CU interface. Thus, there are P such interfaces in the CU.

Each PE_CU interface (we will denote it as PE_CU$_i$ where i designates the i-th PE) consists of a set of buffers for storing particles in the CU. Each set of buffers contains two levels of buffer operations. The first level employs RB$_i$ and TB$_i$, that directly interface with the PE$_i$ using a bi-directional data bus for sending and receiving particles. A same bi-directional bus is used for transferring both the sum and particles. The second level buffers consist of RTB$_{iu}$ and RTB$_{id}$. The particles stored in RB$_i$ can be moved to RTB$_{id}$ where the subscript index id represents the index of the PE_CU interface and the direction of data movement is down (i.e., out of the PE_CU interface). Similarly, RB$_i$ can obtain particles from RTB$_{iu}$ where the direction of particle is up towards the PE_CU interface.

PE$_i$ sends all the particles classified as R to RB$_i$ and the particles classified as T to TB$_i$. The weights of the particles from each PE are also transferred to the CU through a bus so that particle replication is performed in CU. Each weight $(w = (r,t))$ consists of two parameters r and t, where r is the replication factor and t is the tagged factor. The value of t is either 0 or 1 whereas the value of r varies from 0 to

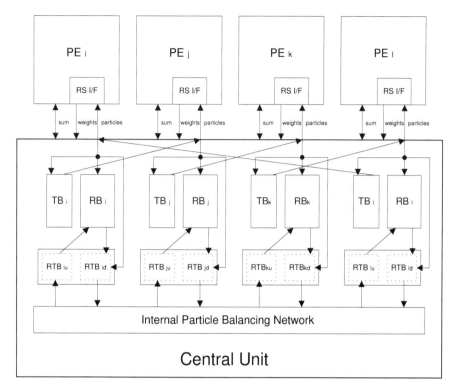

Fig. 8 The block diagram of parallel particle filter resampling. The central unit is responsible for correct particle exchanges

$M/P - 1$. When the particles classified in T are sent back to the PE, the particles in the adjacent TB buffer are transferred. This arrangement is to make sure that each PE receives particles from other PEs to eliminate particle localization.

In order to facilitate effective particle distribution, an efficient internal balancing network is necessary as shown in Fig. 9. In this architecture, we define physical connection types, *inner-connection* and *outer-connection*. The inner-connection uses buses shown on the top of the RTB buffers in the figure whereas the outer-connection uses buses shown at the bottom of the RTB buffers. In addition, there are logical connection types, *inter-connection* and *intra-connection*. The inter-connection represents connectivity between the RTBs in different PE_CU interfaces whereas intra-connection represents connectivity between the RTBs within the same PE_CU interface. With this reconfigurable balancing network, particle transfers between different PE_CU are carried out through this reconfigurable balancing network. Each connectivity can be classified uniquely based on the type of connection discussed above. The number of buses for inner-connection is fixed to 1 and the number of buses for outer-connection is equal to $P/2$ so that the particles can

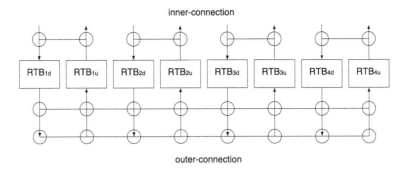

Fig. 9 The internal particle balancing network. $P = 4$ is assumed. Each *circle* indicates interconnect switch

be transferred in pairs of RTBs simultaneously. The number of these bi-directional busses, B, is a function of P where $B = \lceil P/2 \rceil$. This provides necessary one-to-one connection so that particles can be exchanged concurrently.

For handling of the clock speed mismatch between the PE and the CU, there are two FIFOs at the PE and CU interface. We assume that $T_{CU} \geq T_{PE}$ where T_{CU} and T_{PE} are execution clock speeds of the CU and PE, respectively. The particles from each PE are put into the FIFO before transferring them to the RB, TB, or RTB. The size of the input FIFO depends on the speed of the CU operation. Similarly, the output FIFO is to buffer the output particles from the RB and TB. In each PE_CU$_i$, there are counters that keep track of the number of particles in the CU. A detailed description of these counters is provided in the following sections.

3.5 Buffer Structure with Condition Generation

Buffers used in the CU need a special set of functions for the dual-port memory to maintain accurate state of the buffer activity. The structure of the buffer used in the CU is shown in Fig. 10. The buffer consists of a dual-port memory with read and write address generators. In addition, the buffer contains a track counter that keeps tracking of the number of particles in the memory. The track counter has two incrementers. One is for tracking the number of read particles and the other for tracking the number of written particles. These two separate incrementers are needed since one cannot track simultaneous by the read and write accesses of the memory. The difference between these two incrementers is then the number of particles in the buffer. The actual value that is incremented depends on the weights. Both incrementers are initialized to zero.

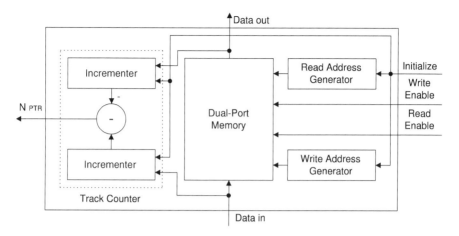

Fig. 10 Block diagram of the buffer used in the architecture

3.6 PE-CU-PE Interface Operation

Before resampling, the PE_i computes and accumulates the weights of the particles and sends its local sum of weights of M/P particles to the CU. Then the CU adds P sums of weights from all the PEs and returns the sum of weights of M particles, sum_{total}, back to each PE. Each PE normalizes the weights using sum_{total}. The total number of particles generated by the PE is approximately equal to the normalized sum of weights.

The resampling process is started as each PE starts to send a particle and its normalized weight to the CU. The interface between each PE and CU is shown in Fig. 11 where $P = 4$ is assumed. When the CU receives a particle from PE_i, it first checks whether its weight (r, t) is zero. The particles are stored in the input FIFO only if the weight is non-zero. At the output of the input FIFO, a particle is sent to the split operator is also stored in TB_i if it is tagged.

Before a particle is written to the memory, a particle split operation is performed as shown in Fig. 11. The main reason for using a split operation is that it is possible to have a particle with its replication factor larger than 1. Such a split operation always reduces the replication factor of a particle whenever a particle is written to a buffer except RTB_{id}. Since it is very critical to redistribute particle among P PEs, it is necessary that the largest replication factor of a particle must be less than or equal to M/P. After a particle is split, a FIFO is used to buffer the output of the split operator before it is written in a memory. Thus, when a particle is read from any PE, the CU has to make sure that the maximum replication factor of a particle is satisfied. This is accomplished by splitting a particle S times where $S = \log_2 P$.

The split operation can be described as follows. When a replication factor is larger than 1, it is divided by 2 (a simple shift operation is necessary). Then if the least significant bit of a replication factor before the split operation is 0, then no

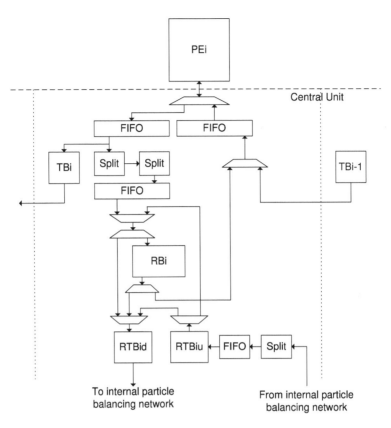

Fig. 11 Block diagram of the interface between each PE and CU

additional operation is done except writing the particle twice to the memory. If the least significant bit of the replication factor before split is 1, then 1 is added back to one of the two split particles. Consider a particle in which the replication factor is 4. In this case, the least significant bit is 0 and a simple division generates two identical particles with a replication factor of 2. On the other hand, if the replication factor is 3, this number is divided via a simple shift, which results in a replication factor of 1 each. Therefore, 1 is added back to one of the two particles. The splitting function at the RB is required to satisfy the granularity condition. But in order to further reduce the granularity, the split mechanism is also incorporated in the RTB. However, no mechanism is necessary in the TB since all the particles in the TB have a replication factor of 1.

Figure 12 illustrates execution time schedule of the PE-CU-PE activity. During particle reception, before M/P cycles of resampling, the particles from the split operator are written to the RB_i and RTB_{id}. If the total number of particles (i.e., sum of replication particles) in the RB_i is larger than M/P, the rest of the particles are transmitted to the RTB_{id} from the PE_i. The number of unique particles may be

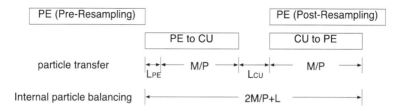

Fig. 12 Execution time schedule of the PE-CU-PE interface activity

less than M/P because a replication factor of a split particle may be greater than 1. At the same time, the CU has the information of the exact numbers of particles in each RB_i, TB_i and RTB_{id} (N_{RB_i}, N_{TB_i} and $N_{RTB_{id}}$, respectively) provided by their corresponding track counters, TC_{RB_i}, TC_{TB_i}, and $TC_{RTB_{id}}$, respectively. The CU handles the internal particle balancing such that $N_{RB_i} + N_{TB_l} \geq N_{PTR_i}$, where N_{RB_i} is the number of particles stored in the RB_i, N_{TB_l} is the number of particles stored in the adjacent TB_l, and N_{PTR_i} is the number of particles needed by the PE_i. Initially, each PE needs M/P particles so that $N_{PTR_i} = M/P$ but the number is decreased every time a particle is transferred to the PE_i.

The PE_CU_i sends particles back to the PE_i from the RB_i and TB_{i-1} after $M/P + L_{PE}$ cycles of resampling. L is the sum of latency incurred due to particle reception through by the CU via FIFO, denoted as L_{PE}, and the worst case latency due to the internal particle balancing operation, denoted as L_{CU}. These latencies are analyzed in Sect. 4.3. In order to guarantee that all tagged particles are exchanged, all the particles stored in TB_l are first transferred to its adjacent PE. Once all the particles in TB_l are read out (i.e., put into FIFO), the particles in RB_i are accessed by the PE_i through the output FIFO until $N_{PTR_i} = 0$. When a particle is sent to the output FIFO from the RB_i, each particle is written r consecutive cycles so that every particle will have its replication factor of 1.

3.7 Internal Particle Balancing Operation

The internal particle balancing operation begins as soon as the following conditions are satisfied for at least one PE_CU_i:

$$N_{RB_i} + N_{TB_l} + N_{RTB_{iu}} \geq N_{PTR_i}, \tag{5}$$

$$N_{RTB_{id}} > 0. \tag{6}$$

The expressions (5) and (6) indicate that at least one PE will have enough particles for itself and the additional particles are in RTB_{id}. Even if the above condition is

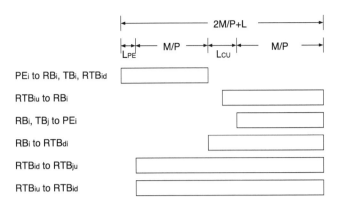

Fig. 13 Illustration of a valid range of all the buffer (including PE) operations

not met, the internal particle balancing begins after $M/P + L_{PE}$ cycles of resampling. The goal of the internal particle balancing is to guarantee that for PE_CU$_i$,

$$N_{RB_i} + N_{RTB_{iu}} + N_{TB_l} \geq N_{PTR_i}. \tag{7}$$

Thus, the internal particle balancing operation is ended when all the PE_CU$_i$ satisfy $N_{RB_i} + N_{TB_l} + N_{RTB_{iu}} \geq N_{PTR_i}$.

Note that (7) cannot be always satisfied if the particles in the RTBs are not allowed to be transferred to the RBs during the internal particle balancing. However, the particles in the RTB$_{iu}$ cannot be read out by the RB$_i$ before the first $M/P + L_{PE}$ cycles of the resampling operation because particles from the PE$_i$ may still be written into the RB$_i$. In order to avoid reading particles from an empty RTB$_{iu}$, the RB$_i$ can only read out particles from the RTB$_{iu}$ after $M/P + L_{PE} + L_{CU_1}$ cycles. The symbol L_{CU_1} denotes a component of L_{CU} and is discussed in Sect. 4.3.

Moreover, the particles from the RB$_i$ should be allowed to be transferred to RTB$_{id}$. However, during the first $M/P + L_{PE}$ cycles of the resampling operation, the RB$_i$ cannot send particles to RTB$_{id}$ since PE$_i$ may also write data to RTB$_{id}$. On the other hand, after $M/P + L_{PE}$ cycles of resampling, the RB$_i$ cannot write data to RTB$_{id}$ if RB$_i$ is sending particles to PE$_i$. Therefore, the only available time for the RB$_i$ to write data to RTB$_{id}$ is after $M/P + L_{PE}$ cycles of resampling (i.e., M/P particles have been written in PE$_i$ or TB$_l$ still has particles).

Valid particle transfers among the buffers and a PE are enumerated and also shown in Fig. 13. Internal particle balancing mechanism depends on the following conditions:

1. A particle is transferred from PE$_i$ to RB$_i$, RTB$_{id}$, or TB$_i$ whenever the particle weight is larger than 0 (i.e., $r > 0$ or $t \neq 0$) during the start of resampling until $M/P + L_{PE}$ cycles of resampling.
2. A particle is transferred from RTB$_{iu}$ to RB$_i$ after $M/P + L_{PE} + L_{CU_1}$ cycles of resampling, when $N_{RTB_{iu}} > 0$ and $N_{RB_i} + N_{TB_l} < N_{PTR_i}$ are satisfied. This does not interfere with condition 1.

Table 3 Connection strategy during internal particle exchange

flag$_{SD_i} = (n_1 n_2 n_3)$	Source/destination
000	Not available
001	Not available
010	Destination from j or source to i
011	Not available or source to i
100	Source to i or source to j
101	Source to j
110	Source to j
111	Source to j or source to i

i and j designate different PE_CU interfaces

3. A particle is transferred from RB$_i$ to PE$_i$ after $M/P + L$ cycles of resampling, when $N_{TB_l} = 0$ and $N_{PTR_i} > 0$ and $N_{RB_i} \neq 0$ are satisfied.
4. A particle is transferred from RB$_i$ to RTB$_{id}$ after $M/P + L_{PE}$ cycles of resampling, when $N_{RB_i} + N_{RTB_{iu}} + N_{TB_l} > N_{PTR_i}$ is satisfied.
5. A particle is transferred from RTB$_{id}$ to RTB$_{ju}$ during internal particle balancing operation. The connection is established based on a connection policy described later. The controller determines the source and the destination. Note that $i \neq j$.
6. A particle is transferred from RTB$_{iu}$ to RTB$_{id}$ when $N_{RB_i} + N_{TB_l} \geq N_{PTR_i}$, and neither PE$_i$ nor RB$_i$ is writing data into RTB$_{id}$. This is to move particles stored in RTB$_{iu}$ to RTB$_{id}$ directly. Note that whenever a particle is written to RTB$_{iu}$, the particle is splitted to reduce the value of the replication factor. This is to ensure that the particles have small values of replication factors. Without this, it is not possible to balance the particles perfectly.

During the internal particle balancing operation, each PE_CU$_i$ must be in one of two states, source or destination. Basically, a PE_CU$_i$ becomes a source if $N_{RTB_{id}} > 0$. On the other hand, a PE_CU$_i$ becomes a destination if $N_{RB_i} + N_{RTB_{iu}} + N_{TB_l} < N_{PTR_i}$. But we allow that PE_CU$_i$ can be either a source or a destination depending on the condition flag for PE_CU$_i$ interface, flag$_{SD_i}$.

Based on the status of each group of buffers, the CU controller sets up the connections between the RTBs. The flag$_{SD_i}$ is encoded with three bits of information n_1, n_2, and n_3, where $n_1 = 1$ if $N_{RTB_{di}} > 0$, $n_2 = 1$ if $N_{RTB_{ui}} > 0$, and $n_3 = 1$ if $N_{RB_i} + N_{TB_{i-1}} + N_{RTB_{ui}} > N_{PTR_i}$. A possible state for a given set of condition bits is summarized in Table 3.

When the buffers are accessed concurrently, it is possible that a deadlock may occur. For each set, it is possible to have multiple states. The reason for multiple states for each set of conditions is to avoid a deadlock. This is the case when all PE_CU$_i$s are determined as sources. If enough particles are in the RB$_i$, TB$_i$, and RTB$_{iu}$, no further particle balancing is needed. But if at least one of the PE_CU$_i$ needs more particles, a deadlock happens. On the other hand, a situation where all the PE_CU$_i$ are destinations, it will not be a problem since we have more than M particles in the CU by the quantization scheme discussed previously and some of the PE_CU interface will be changed to the source.

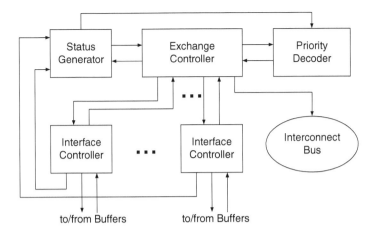

Fig. 14 PE-CU-PE controls

In order to avoid such deadlock, a priority decoder is needed to modify the state, so that internal particle balancing network can be established. A priority decoder is used to set and reset the connections between sources and destinations based on all flag$_{SD_i}$ so that data conflict does not occur. As shown in Table 3, each set of flag$_{SD_i}$ may have multiple possibilities.

The connection termination condition is when the source does not have any particles in the RTB$_{id}$ or the destination received enough particles. Once any one of these conditions is satisfied, the connection is reset and a new connection is established. In order not to get back to the original condition, the controller must ensure that at least one particle is transferred before terminating the connection. Otherwise it will get back to the original condition and cause an infinite loop without any particle transfer. If there are no particles in RTB$_{id}$ and $N_{RB_i} + N_{TB_{i-1}} + N_{RTB_{iu}} = N_{PTR_i} = 0$, the corresponding PE_CU$_i$ interface will not participate in the particle exchange.

3.8 Concurrent Execution Control

There are four major controller components for proper execution of the CU as shown in Fig. 14. A status generator generates all the necessary signals to be used by various components of the controller. The signals are mainly generated using the buffer counters as shown in Fig. 15. Since all the buffers are distributed, the status generator is physically distributed. These signals are used by the exchange controller, priority decoder, and independent PE_CU interface controller. The exchange controller decides the pairing of buffers for the particle transfers. The priority decoder generates correct pairing of each PE_CU interface based on the current state of the internal particle balancing.

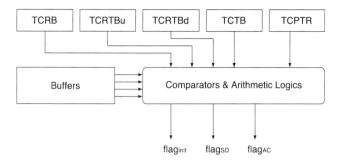

Fig. 15 Condition flags generator

There are two approaches in designing the overall exchange controller. One approach is to connect "ALL" the PE_CU interfaces at any given cycle. While this can connect as many PE_CU interfaces, the complexity of the controller is prohibitive because of the enormous number of possible states. The second approach is to connect "TWO" out of all the available PE_CU interfaces at any given cycle. The second approach, which is adopted in the mechanism, has a key benefit that it has very low complexity comparing to that of the first approach. Moreover, the controller is scalable where the only modification is on the priority decoder which suggest the best pairing based on the conditions from the status generator and the current activity of the PE_CU interfaces.

The priority Decoder is extremely important to maximize the concurrency of all the buffers so that the time it takes to transfer particles is minimized. Moreover, the Priority Decoder is responsible to avoid a possible deadlock due to sharing buffers. The Priority Decoder is ROM where the address of the ROM is set to flag$_{SD_i}$ and the activity condition of the PE_CU$_i$ interface. The activity condition indicates whether a particular PE_CU$_i$ interface is currently participating in the particle transfer. Thus based on these flags, final outputs suggesting the particle transfer topology are generated. When considering a possible connection, the inter-connection has a higher priority over intra-connection (i.e., particle transfers between different PE_CU interfaces have higher priorities). The output of the Priority Decoder is used by the Exchange Controller for the actual particle transfer. The size of the ROM depends on the number of PE employed.

At any given cycle, the Priority Decoder generates two sets of signals based on the information which it received as inputs. They are S_1 and S_2. These sets of signals signify two PE_CU interfaces in which the internal particle transfer is to be performed. One restriction is that both interfaces indicated by these signals cannot be sources and destinations. There are two cases for pairing. The source and destination are in different PE_CU interfaces and, the source and destination are within the same PE_CU interface. In the second case, we can establish two simultaneous connections. To support these capabilities, S_1 and S_2 consist of (s_1, s_2, s_3, s_4) where s_1 is an index of interface, s_2 indicates whether the connection is intra (1) or inter (0), s_3 indicates whether the corresponding interface is a source (1)

Table 4 Buffer activity status conditions of the PE_CU interface

$flag_{AC_i} = (a_1 a_2)$	Activity condition
00	Inner- and outer-connection available
01	Outer-connection available
10	Inner-connection available
11	No connection available

or a destination (0), and s_4 denotes whether the interface is valid or not. For example, if $S_1 = (s_1 = 2, s_2 = 1, s_3 = 1, s_4)$ and $S_2 = (s_1 = 1, s_2 = 1, s_3 = 0, s_4)$, it signifies that interface 2 is a source and interface 1 is a destination for inter-connection (i.e., RTB_{2d} to RTB_{1u}). In case of an intra-connection, two independent connections will be initiated at any given cycle. Since it is possible that only one intra-connection is available, we have to provide additional information from the priority decoder, which is s_4. If s_4 is 11, both S_1 and S_2 are valid. If it is 10, only S_1 is valid, if it is 01, only S_2 is valid. If it is 00, no connection should be made. In case of intra-connection, $s_3 = 1$ for connection from RTB_{iu} to RTB_{id}, and $s_3 = 0$ for connection from RTB_{id} to RTB_{iu}.

The Exchange Controller establishes connection between a pair of buffers. This is accomplished by providing signals to the switches that configure the routing. Once routing is established, the connection is connected at least one particle is transferred. Otherwise, undesired infinite loop may occur as mentioned previously. As discussed in the previous section, pairing is only established between RTBs. RTB_{id} to RTB_{ju} signifies particle transfers between different PE_CU interfaces, and RTB_{iu} to RTB_{id} signifies particle transfers within a PE_CU interface. We also indicated that particles in RB_i can be transferred to RTB_{id}. This process conflicts when this RTB_{id} is used as its own destination (i.e., RTB_{iu} is the source). So the Exchange Controller must generate appropriate signals to this PE_CU interface. On the other hand, when RTB_{iu} is transferring particles to RB_i, there will be no conflict since RTB_{iu} will never be chosen as a source to its own RTB_{id}. (i.e., short of particles). Since we have separate data ports, this RTB_{iu} can be a destination for RTB_{jd}. We have to make sure when both are active and that correct counting is accomplished. This is explained in Fig. 10.

The Exchange Controller provides controls and monitors execution of particle transfer. The Exchange Controller monitors the activity according to $flag_{AC_i}$, (a_1, a_2), described in Table 4. These flags are set and reset by both the Exchange Controller and PE_CU interface controller. a_1 is set if the inner-connection is not available and a_2 is set if the outer-connection is not available. When the Exchange Controller sets up a connection, the flags are modified. When the interface controller completes the buffer access, the flags are again modified. One important note is that since there is connection from RTB_{iu} to RB_i, when there is any particle transfer, the interface controller signals so that the Exchange Controller makes sure that the inner-connection is not available. This is only applicable for eliminating conflict with particle transfers from RTB_{iu} to RTB_{id}.

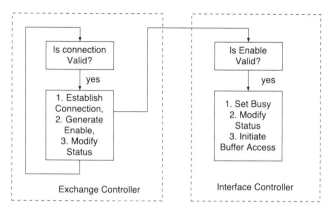

Fig. 16 Interaction between the exchange controller and interface controller

The PE_CU Interface Controller interfaces with the exchange controller as shown in Fig. 16. There is one Interface Controller for each PE_CU interface. Their functions are identical and operational under tight control by the Exchange Controller. Structurally, each Interface Controller contains address generation logic for each buffer in the PE_CU interface.

Interface controllers are mainly composed of counters and incrementors. The Interface Controller then reads and writes particles between a buffer pair. The initiation of buffer accesses by the Interface Controller is controlled by the Exchange Controller through the write enable and read enable signals. The PE_CU interface makes sure that local particle transfers, from RTB_{iu} to RB_i and from RB_i to RTB_{id}, are not active when connection between RTB_{iu} and RTB_{id} is established by the Exchange Controller. While buffers are being accessed, this interface controller sets busy flag to the Exchange Controller indicating that the corresponding PE_CU interface is not available. In turn, the Exchange Controller uses this information in establishing appropriate connection using the Priority Decoder. Once the Exchange Controller generates write enable and read enable for a particular buffer, each Interface Controller acknowledges back to the Exchange Controller. The Interface Controller reads or writes particles until the condition changes or p particles are transferred where p is small and constant. A reason for transferring finite number of particles at any connection is to minimize the non-deterministic latency (L_{CU_1}). The effects of p on the latency is explained in Sect. 4. When the Interface Controller reads particles from FIFO, an additional condition is that it reads if FIFO is not empty. Formally, possible conditions for terminating the entire particle transfer are: (1) there are no more particles to read from RTB, or (2) the source PE_CU interface has enough particles. If one of these conditions is met, the Interface Controller indicates that the operation is completed and returns to wait state.

Table 5 Multiple ROM usage table

PE_CU Index (1234)	ROM Information
0000	$PE\_CU_3PE\_CU_4$
0001	$PE\_CU_3PE\_CU_4$
0010	$PE\_CU_3PE\_CU_4$
0100	$PE\_CU_1PE\_CU_2$
1000	$PE\_CU_1PE\_CU_4$
0011	$PE\_CU_3PE\_CU_4$
0101	$PE\_CU_2PE\_CU_4$
0110	$PE\_CU_2PE\_CU_3$
1100	$PE\_CU_1PE\_CU_2$
1001	$PE\_CU_1PE\_CU_4$
1010	$PE\_CU_1PE\_CU_3$
0111	$PE\_CU_2PE\_CU_3$, $PE\_CU_2PE\_CU_4$, $PE\_CU_3PE\_CU_4$
1011	$PE\_CU_1PE\_CU_3$, $PE\_CU_1PE\_CU_4$, $PE\_CU_3PE\_CU_4$
1101	$PE\_CU_1PE\_CU_2$, $PE\_CU_1PE\_CU_4$, $PE\_CU_2PE\_CU_4$
1110	$PE\_CU_1PE\_CU_2$, $PE\_CU_1PE\_CU_3$, $PE\_CU_2PE\_CU_3$
1111	$PE\_CU_1PE\_CU_2$, $PE\_CU_1PE\_CU_4$, $PE\_CU_2PE\_CU_3$, $PE\_CU_3PE\_CU_4$

ROM index ij indicates that the entries contain S_1 and S_2 when PE_CU_i and PE_CU_j are used for particle transfer. The connection between the nearest PE_CU interfaces have high priority

3.9 ROM Reduction in the Priority Decoder

To generate S_1 and S_2, the input to the ROM requires P sets of flag$_{SD_i}$ and flag$_{AC_i}$. This implies that the size of ROM increases significantly as P becomes large. In order to reduce the ROM requirement, we employ a set of ROMs where only two sets of flag$_{SD_i}$ and flag$_{AC_i}$ are used. We select the appropriate ROM based on the availability of PE_CU interfaces. Table 5 illustrates the ROM section strategy. In the table the PE_CU_i-PE_CU_j pair implies that the corresponding ROM has priority information when the PE_CU_i and PE_CU_j are selected for particle transfers. If flag$_{AC_i} = 11$, PE_CU_i is not available (set to 0 in the table). Otherwise, PE_CU_i is available (set to 1). For illustration, $P = 4$ is chosen. Availability indicates if the PE_CU interface is available (i.e., 0 is not available, 1 is available). In case of multiple possible connections, a ROM is selected in Round Robin way. Also, note that the connection between the nearest PE_CU interfaces has a higher priority. As shown in the table, there are 6 unique ROMs. The number of ROM required by P PEs is given by

$$N_{ROM} = (P-1)!. \tag{8}$$

Thus, for $P = 4$, $(4 - 1)! = 6$ ROMs are needed. The number of ROMs increases rapidly as a function of P, and the total size of ROMs grows slowly rather than the case where 1 ROM is used.

For a case where the PU_CU index is 0000, the default ROM that provides PE_CU$_3$ and PE_CU$_4$ is used even though no connection will be made by the Exchange Controller. A case for 1111, three ROMs are used but not at the same time. When one ROM is used, the size of ROM becomes $2^{4(3+2)}$ (1M entries). On the other hand, when 6 ROMs are used, the total size of each ROM becomes $2^{2(3+2)}$ (10K entries).

4 Architecture Evaluation

4.1 Guaranteeing Complete Redistribution

As we have discussed in the previous section, as long as we can maintain $N_{RB_i} + N_{TB_{i-1}} + N_{RTB_{ui}} \geq N_{TPR_i}$ for all PE_CU$_i$, we can guarantee that each PE will get enough particles for the next iteration. During internal particle balancing, a replication factor of any particle may have a large value. This replication factor is modified whenever a particle is transferred (i.e., reduced by the splitting operator) so that the replication factor becomes small enough to balance the particles among the PE_CU$_i$s.

In addition to the number of particles requirement, the deadlock of internal particle balancing must be avoided. This occurs when all the PE_CU$_i$s become sources (i.e., they have particles to send out) or destinations (i.e., they all need particles). This deadlock is completely eliminated by a priority decoder, which decides the appropriate source and destination pair during the internal particle balancing. Thus, as long as the resampling provides a total number of replicated particles $(R + T)$ larger than M, which is guaranteed by the quantization scheme, a perfect redistribution is guaranteed.

4.2 Scalability of the Architecture

Thus far, we implicitly assume that the number of PEs is a power of two. However, the architecture is perfectly scalable in a sense that the execution time is M/P without overhead for any number of PEs as long as there are enough busses available between the PEs and the CU. The only condition, which is a function of the number of PEs (i.e., P), is the number of split operations required at the PE and the CU interface. The number of split operators required is given by

$$S = \log_2 P'$$ (9)

where P' is the smallest number in power of two that is larger than P. For example, if $P = 5$, then $P' = 8$. Thus the number of split operators before writing to the RB is 3.

When P increases, the number of busses in the balancing network also increases. To maintain full concurrency during the internal particle balancing operation, the number of busses that is contained in the internal particle balancing network is equal to $P/2$ so that the particles can be transferred in pairs of RTBs simultaneously. The number of bi-directional busses, B, is a function of P where $B = \lceil P/2 \rceil$.

4.3   Execution Time

We have shown in Fig. 12 that the completion time of resampling and redistribution varies depending on the initial particle distribution before resampling (i.e., the difference in weights among the different PEs). However, the mechanism guarantees that the new set of replicated particles is available after $M/P + L$ cycles to be used by each PE, even though the internal particle balancing may continue for the entire $2M/P + L$ cycles. This is because the architecture guarantees to generate a new set of particles for the PE to process in $M/P + L$.

The value of $L = L_{PE} + L_{CU}$ is finite and is not a function of M. L_{PE} is a finite latency due to the FIFO write and read at the PE_CU interface during the particle transmission to the CU. Thus, this latency is a constant and consists of one write and one read (i.e. two cycles). The latency $L_{CU} = L_{CU_1} + L_{CU_2}$ is also finite and is not a function of M, where L_{CU_1} and L_{CU_2} are the latencies that are required to support the worst-case situation. L_{CU_2} is finite due to the FIFO write and read at the PE_CU interface during the particle transmission back to the PE. Thus, this latency, again, consists of one write and one read (i.e. two cycles).

L_{CU_1} is a latency required to delay the operation to avoid particles access from an empty RB_i. These latencies may arise when TB_j and RB_i are empty after $M/P + L_{PE}$ cycles. Before sending particles to the PE, at least one particle must be present in the RB_i. The worst case is when a particle is coming from another RB through read and write of particle via RTB_{out} and RTB_{in}. When a connection is established, the latency involves read and write access of the source PE_CU interface, read and write between RTBs, and read and write at the destination interface. In order to maintain a fixed latency, a parameter of p is employed (i.e., this parameter has already been discussed). By restricting the maximum number of particle transfers at any given connection to p, the value of L_{CU_1} is $p + 4$, where p cycles are needed for particle transfers not involving the PE_CU interface in consideration. After the connection is established, four additional cycles are needed where one cycle is for reading a particle from the source RTB and one cycle is for writing to destination RTB, one cycle is for reading a particle from the destination RTB, and finally one cycle is for writing to the destination RTB. This is the reason that the Priority Decoder alternates selection of ROM accesses so that any one of the connections is connected for a long time. The value of p is usually chosen to be less than 10. Thus, the PE can

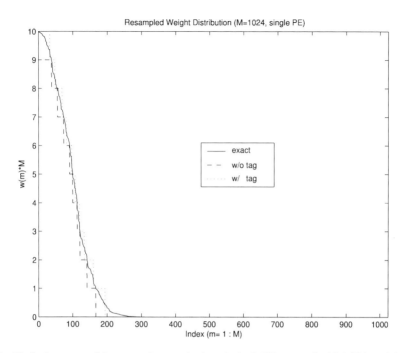

Fig. 17 Performance of the resampling mechanism. A single PE operated with 1,024 particles is considered

assume that a new set of particles (resampled particles) after $M/P + L$ cycles and the deterministic execution time is always $M/P + L$. The value of L is negligible when M is large.

4.4 Resampling Performance

We can measure a degree of resampling performance by comparing the replication of particles on single PE versus multiple PEs. In the evaluation, M weights are randomly generated and their weight distributions are compared with those of the full precision resampling. Figure 17 illustrates the performance of the resampling for the cases with and without using tagging method for $M = 1,024$. As shown in the figure, the number of replicated particles for the scheme with tagging method is closer to M (full precision resampling). On the other hand, the replication without tagging always produces less than M particles. In the figure, the stair-wise curves are due to integer values of the replication factors.

Figure 18 illustrates the performance of the resampling for the cases where four PEs are operated with and without the tagging method for $M = 1,024$. We assume that the weights are balanced among the PEs. Each PE operates with 256 particles

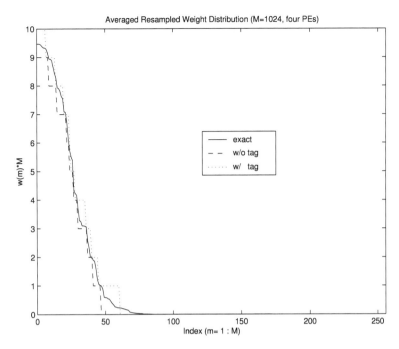

Fig. 18 Performance of the resampling mechanism. Four PEs operated with 1,024 particles is considered. A balanced weight distribution is assumed

and gets tagged particles from its adjacent PE. As shown in the figure, the number of replicated particles for the scheme with tagging method is also closer to M. Similarly, Fig. 19 illustrates the performance of the resampling for the worst-case where one PE generates all the weights, M, and the rest of the PEs have zero weight. Therefore, the PE that has the value of weights equal to M shares its particles to others. As shown in the figure, the number of replicated particles for the scheme with tagging method is still closer to M comparing to the operation without using the tagging method.

4.5 Memory Requirement and Bus Complexity

The memory complexity is a function of M, P, W and D where W is the wordlength used to represent particles and weights and D is the number of data that one particle contains. In addition, $V = \log_2 M + 1$ bits are necessary to represent the replication factor r and the tag factor t. For $P = 1$, the resampling operation can be incorporated into the particle filter [24]. When $P = 2$, only one RTB (instead of RTB_u and RTB_u) is necessary since the two PE_CU are directly connected.

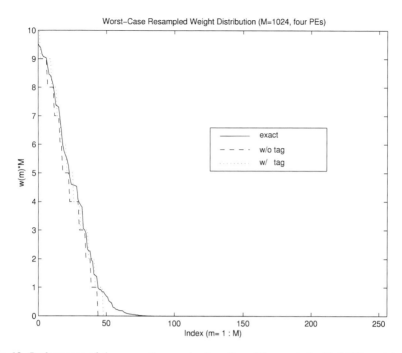

Fig. 19 Performance of the resampling mechanism. Four PEs operated with 1,024 particles is considered. The worst-case weight distribution is assumed

For example, for the particle filter used in the bearings-only tracking application, one particle contains four data where two are used to represent the position and the other two are used to represent the velocity of an object being tracked. The size of the memory should be large enough so that the particles can be buffered before the designated PE reads them out. Thus the total amount of memory for $P = 2$ and $P > 2$, respectively, is given by

$$MEM_{total} = 3M[DW + V], \tag{10}$$

$$MEM_{total} = 4M[DW + V]. \tag{11}$$

Bus complexity between the PE and the CU heavily depends on D, which is the dimension of the particle filter. In the bearings-only tracking problem, $D = 4$. The number of total bus wires between PEs and CU is given by

$$N_{wire} = P[DW + V]. \tag{12}$$

Thus when the value D increases, the wires required can be significant. The choice of P is then depending on wires availability and feasibility for integration. It is possible that particle transfers are time-multiplexed using a smaller ($D' < D$) set of wires. Then the speed of particle resampling is reduced by a factor of D/D'. As long as $P > D$, we can achieve speed up through parallelism. Otherwise, it is better to reduce the parallelism so that the required number of bus wires is decreased.

5 Conclusions

In parallel particle filters, *resampling* is the most critical unit that directly influences the performance and overall hardware complexity. In this chapter, we have presented a flexible resampling architecture that can be employed for high throughput particle filtering. A very efficient resampling mechanism is incorporated which reduces the overall resampling time by a factor of the number of PEs. The design supports up to four parallel PEs executing bearings-only tracking application with various modes of operations. The number of PEs is limited only by the number of input/output pins. The architecture presented in this chapter can be extended to other particle filters.

We have also presented a novel parallel resampling mechanism for perfect redistribution of particles. The mechanism utilizes a particle-tagging scheme during the quantization to compensate for a possible loss of replicated particles due to finite precision effect in weight computation. The architecture incorporates a very efficient interconnect topology for efficient particle redistribution. We have shown that the performance of multiple PE resampling is very close to that of a single PE. We have also shown that the mechanism ensures that the resampled particles are always available for further processing in deterministic time. Moreover, the overall parallel particle filtering execution time perfectly scales with the number of processing elements. The mechanism allows for realizing particle filtering with large M in parallel fashion, which has not been realized in the literature.

References

1. A. Doucet, N. de Freitas, and N. Gordon, Eds., *Sequential Monte Carlo Methods in Practice*, New York: Springer Verlag, 2001.
2. E. R. Beadle and P. M. Djurić, "A fast weighted Bayesian bootstrap filter for nonlinear model state estimation," *IEEE Transactions on Aerospace and Electronic Systems*, vol. 33, pp. 338–343, 1997.
3. D. Crisan, P. Del Moral, and T. J. Lyons, "Non-linear filtering using branching and interacting particle systems," *Markov processes and Related Fields*, vol. 5, no. 3, pp. 293–319, 1999.
4. J. S. Liu, R. Chen, and W. H. Wong, "Rejection control and sequential importance sampling," *Journal of American Statistical Association*, vol 93, no. 443, pp. 1022–1031, 1998.
5. M. S. Arulampalam, S. Maskell, N. Gordon, T. Clapp, "A Tutorial on particle filters for online nonlinear/non-Gaussian Bayesian tracking," *IEEE Journal of Signal Processing*, 2002.
6. J. Carpenter, P. Clifford, and P. Fearnhead, "An improved particle filter for non-linear problems," *IEE Proceedings-F: Radara, Sonar and Navigation*, vol. 146, pp. 2–7,1999.

7. N. J. Gordon, D. J. Salmond, and A. F. M. Smith, "A novel approach to nonlinear and non-Gaussian Bayesian state estimation," *IEE Proceedings-F: Radar, Sonar and Navigation*, vol. 140, pp. 107–113, 1993.

8. P. M. Djurić, J. H. Kotecha, J. Zhang, Y. Huang, T. Ghirmai, M. F. Bugallo, and J. Miguez, "Particle filtering," *IEEE Signal Processing Magazine*, vol. 20, no. 5, pp. 19–38, September 2003.

9. C. Berzuini, N. G. Best, W. R. Gilks, and C. Larizza, "Dynamic conditional independence models and Markov chain Monte Carlo methods," *Journal of the American Statistical Association*, vol. 92, pp. 1403–1412, 1997.

10. A. Kong, J. S. Liu, and W. H. Wong, "Sequential imputations and Beyesian missing data problems," *Journal of Americal Statistical Association*, vol. 89, no. 425, pp. 278–288, 1994.

11. J. S. Liu and R. Chen, "Blind convolution via sequential imputations," *Journal of American Statistical Association*, vol. 90, no. 430, pp. 567–576, 1995.

12. D. Crisan, "Particle filters - A theoretical perspective," *Sequential Monte Carlo Methods in Practice*, A. Doucet, J. F. G. de Freitas, and N. J. Gordon, Eds. New York: Springer-Verlag, 2001.

13. G. Kitagawa, "Monte Carlo Iteration and smoother for non-Gaussian nonlinear state space models," *Journal of Computational and Graphical Statistics*, 5:1–25, 1996.

14. E. Baker, "Reducing Bias and Inefficiency in The Selection Algorithms," *Proceedings of Second International Conference on Genetic Algorithms*, pp. 14–2, 1987.

15. J. S. Liu and R. Chen, "Sequential Monte Carlo methods for dynamic systems," *Journal of the American Statistical Association*, vol. 93, pp. 1032–1044, 1998.

16. S. Hong, S.-S. Shin, P. M. Djurić, and M. Bolić, "An efficient fixed-point implementation of residual systematic resampling scheme for high-speed particle filters," *VLSI Signal Processing*, vol. 44, no. 1, pp. 47–62, 2006.

17. S. Hong and P. M. Djurić, "High-throughput scalable parallel resampling mechanism for effective redistribution of particles," *IEEE Transactions on Signal Processing*, vol. 54, no. 3, pp. 1144–1155, 2006.

18. M. Bolić, P. M. Djurić, and S. Hong, "New resampling algorithms for particle filters", *IEEE ICASSP*, 2003.

19. M. Bolić, P. M. Djurić, and S. Hong, "Resampling algorithms for particle filters suitable for parallel VLSI implementation," *IEEE CISS*, 2003.

20. M. Bolić, P. M. Djurić, and S. Hong, "Resampling algorithms for particle filters: A computational complexity perspective," *EURASIP Journal of Applied Signal Processing*, no. 15, pp. 2267–2277, November 2004.

21. R. Tessier and W. Burleson, "Reconfigurable computing and digital signal processing: A survey," *Journal of VLSI Signal Processing*, May/June 2001.

22. Xilinx, "Virtex-II Platform FPGA Handbook," 2000.

23. N. J. Gordon, D. J. Salmond, and A. F. M. Smith, "A novel approach to nonlinear and non-Gaussian Bayesian state estimation," *IEE Proceedings F*, vol. 140, pp. 107–113, 1993.

24. S. Hong, M. Bolić, and P. M. Djurić, "An efficient fixed-point implementation of residual systematic resampling scheme for high-speed particle filters," *IEEE Signal Processing Letters*, vol. 11, no. 5, May 2004.

25. M. Bolić, P. M. Djurić, and S. Hong, "Resampling algorithms and architectures for distributed particle filters," *IEEE Transactions on Signal Processing*, vol. 53, no. 7, pp. 2442–2450, 2005.

26. S.-S. Chin and S. Hong, "VLSI design and implementation of high-throughput processing elements for parallel particle filters," *IEEE SCS*, 2003.

Application Specific Instruction Set DSP Processors

Dake Liu and Jian Wang

Abstract In this chapter, application specific instruction set processors (ASIP) for DSP applications will be introduced and discussed for readers who want general information about ASIP technology. The introduction includes ASIP design flow, source code profiling, architecture exploration, assembly instruction set design, design of assembly language programming toolchain, firmware design, benchmarking, and microarchitecture design. Special challenges from designing multicore ASIP are discussed. Two examples, design for instruction set level acceleration of radio baseband, and design for instruction set level acceleration of image and video signal processing, are introduced.

1 Introduction

1.1 ASIP Definition

An ASIP is an Application Specific Instruction-Set Processor designed for an application domain. ASIP instruction set is specifically designed to accelerate computationally heavy and most used functions. ASIP architecture is designed to implement the assembly instruction set with minimum hardware cost. An ASIP DSP is an application specific digital signal processor for computing extensive applications such as iterative data manipulation, transformation, and matrix computing.

It is important to understand when the use of an ASIP is appropriate. If the design parameters of an ASIP are well chosen, the ASIP will contribute benefits to a group of applications. An ASIP is suitable when: (1) the assembly instruction set will not be used as an instruction set for general purpose computing and control;

D. Liu (✉) • J. Wang
Linköping University, Linköping, Sweden
e-mail: dake@isy.liu.se; jianw@isy.liu.se

S.S. Bhattacharyya et al. (eds.), *Handbook of Signal Processing Systems*,
DOI 10.1007/978-1-4614-6859-2_21, © Springer Science+Business Media, LLC 2013

(2) the system performance or power consumption requirements cannot be achieved by general purpose processors, (3) the volume is high or the design cost is not a sensitive parameter, and (4) it is possible to use the ASIP for multiple products.

1.2   The Difference Between ASIP and General CPU

Designers of general-purpose processors think of both the maximum performance and maximum flexibility. The instruction set must be general enough to support general applications. The compiler should offer compilation for all programs and to adapt all programmers' coding behaviors.

ASIP designers have to think about applications and cost first. Usually the primary challenges for ASIP designers are the silicon cost and power consumption. Based on the carefully specified function coverage, the goal of an ASIP design is to reach the highest performance over silicon cost, the highest performance over power consumption, and the highest performance over the design cost. The requirement on flexibility should be sufficient instead of ultimate. The performance is application specific instead of the highest one. To minimize the silicon cost, a design of an ASIP aims usually to a custom performance requirement instead of an ultimate possible high performance.

Programs running in an ASIP might be relatively short, simple, with ultra high coding efficiency, requirements on tool qualities such as the quality of code compiler could be application specific. For example, for radio baseband, the requirement on compiler may not really be mandatory.

The main difference between a general-purpose processor and an ASIP DSP is the application domain. A general-purpose processor is not designed for a specific application class so that it should be optimized based on the performance of the "Application Mix". The application mix has been formalized and specified using general benchmarks such as SPEC (standard performance evaluation corporation).

The application domain of an ASIP is usually limited to a class of specific applications, for example video/audio decoding, or digital radio baseband. The design focus of an ASIP is on specific performance and specific flexibility with low cost for solving problems in a specific domain. A general-purpose microprocessor aims for the maximum average performance instead of specific performance.

Except for memories, there were two major families of integrated digital semiconductor components, microprocessors and ASIC. Microprocessors take the role of flexible computing of all programs running in desktops or laptops. An ASIC supplies application specific functionality, and its functional coverage and performance are fixed during the hardware design time, for higher performance, lower silicon area, and lower power consumption. In 1980s to 1990s, ASIC played dominant roles in communications and consumer electronics products, for example, the radio baseband in mobile phones was based on ASIC.

To get high performance and low power consumption at the same time, embedded system designers usually use ARM processors attached with non-programmable

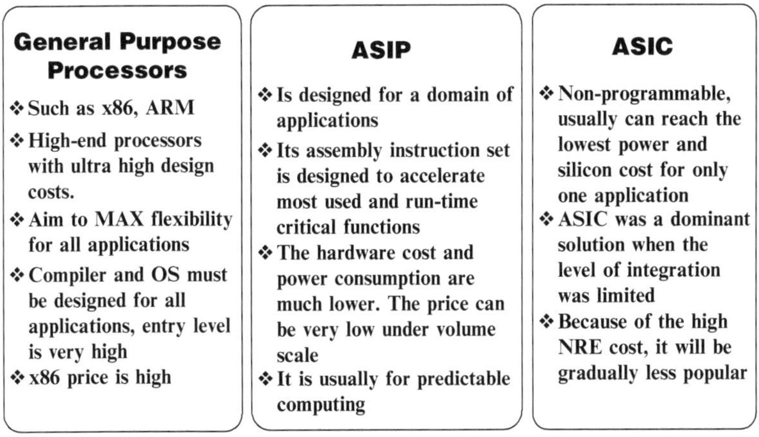

General Purpose Processors	ASIP	ASIC
❖ Such as x86, ARM ❖ High-end processors with ultra high design costs. ❖ Aim to MAX flexibility for all applications ❖ Compiler and OS must be designed for all applications, entry level is very high ❖ x86 price is high	❖ Is designed for a domain of applications ❖ Its assembly instruction set is designed to accelerate most used and run-time critical functions ❖ The hardware cost and power consumption are much lower. The price can be very low under volume scale ❖ It is usually for predictable computing	❖ Non-programmable, usually can reach the lowest power and silicon cost for only one application ❖ ASIC was a dominant solution when the level of integration was limited ❖ Because of the high NRE cost, it will be gradually less popular

Fig. 1 Comparing ASIP with general processors and ASIC

accelerators, such as a radio baseband modem, a video motion estimator, a DCT module etc. Following trends in industry, there are emerging two problems, not sufficient flexibility and too much NRE (no-return-engineering) cost.

A high-end Smartphone needs multi radio connections such as FDD-LTE, TDD-LTE, Wi-Fi, HSPA, WCDMA, EDGE, and GSM. More links could be needed such as DVB-T and GPS. Multiple baseband modems shall be integrated to support all connections listed. There will be many non-programmable modem accelerators consuming much silicon area, and the verification time will be much longer delaying the TTM (time to market). Moreover, whatever standard modifications and market changes will end up a new silicon design and tapeout, its full mask cost of advanced silicon technologies (less than 45 nm) will be more than 3 million US dollars.

It is obviously, the ASIP, or ASIP controlled by ARM, will be the dominant technology to achieve the high performance, low silicon cost, low power consumption, and low NRE cost. ASIC solutions will therefore be gradually replaced by ASIP. Advantages of ASIP include:

- Sufficient flexibility in an application domain.
- Higher performance from the accelerated instructions for most appearing functions.
- Low power consumption and low silicon cost from domain specific architecture.
- Design time can be very short based on ASIP design tools, such as NoGAP [4].
- Redesign and modification time can be even shorter because most modifications are on software.

To summarize, there are three kinds of essential functional components in embedded systems: the general purpose processor (CPU, MCU, and general purpose DSP), the non-programmable ASIC, and ASIP. Compares were summarized in Fig. 1.

2   ASIP Design Flow for DSP Applications

ASIP DSP design can be divided into three parts, as illustrated in Fig. 2. The first part is the "ASIP architecture design" including architecture exploration, design of the assembly instruction set and the instruction set architecture. The second part is the "design of programming tools for ASIP FW (firmware) coding", including compiler, assembler, linker, and instruction set simulator. The third part is the "ASIP firmware design" including the design of benchmarking codes and application codes. The divided three parts can be identified in the ASIP design flow depicted in Fig. 3. This ASIP design flow gives a guidance of ASIP design from requirement specification down to the microarchitecture specification and design.

Following the ASIP hardware design flow in Fig. 3, the starting point of an ASIP design is to specify the requirements including application coverage, performance, and costs. The inputs of the design are product specification and the collection of underlying standards. By analyzing markets, competitors, technologies, and standards, the output of the step is of course the product requirements including the function coverage, performance, and costs.

Fig. 2 Knowledge required in ASIP design

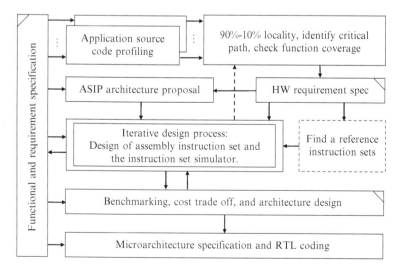

Fig. 3 DSP ASIP hardware design flow

After determining the functional coverage requirement of the ASIP, source code of applications will be collected and source code profiling will be conducted in order to further understand applications. The inputs of the source code profiling are the collected source code and application scenarios. The outputs of the source code profiling are the results of the code analysis including the estimated complexity, computing cost, data access cost, and memory cost. During source code profiling, the early partitioning of hardware and software can be proposed.

Hardware/software partitioning for an ASIP is the trade-off of function allocation to hardware or to software tasks. Implementing a task in software means to write a subroutine of the task in assembly language or in C. Implementing a task in hardware means to implement functions on an accelerated instruction or several accelerated instructions. ASIP instruction set design can be guided by the "10–90% code locality rule", which has been used as the rule of thumb in ASIP design in companies such as Coresonic [3].

The locality rule means that 10% of the instructions run at 90% time and 90% of the instructions appear only 10% of the execution time. In other words, the essential of ASIP design is to find the best instruction set architecture optimized for accelerating the 10% most frequently used instructions and to specify the rest of 90% instructions to fulfil functional coverage.

Based on the exposed 90–10% code locality, hardware requirements can be specified and a specific architecture of the ASIP can be proposed accordingly. ASIP architecture proposal could be based on a selected available architecture with modifications. If flexibility is an essential requirement, selecting an available architecture, such as VLIW architecture might be preferred. A typical case is to select VLIW or master-slave SIMD architecture for a multi-mode image and video signal processing.

An ASIP architecture could also be a dataflow architecture generated from the dataflow analysis during source code profiling. If the requirements on performance and computing latency are exceptionally tough, normal available architectures may not offer sufficient performance. A dataflow architecture might thus be a good choice. For example a dataflow architecture is suitable for a radio baseband processor in an advanced radio base station.

Following a proposed architecture, an assembly instruction set is proposed and its instruction set simulator is implemented for benchmarking. Cost trade-off analysis of the instruction set and the hardware architecture is performed. The inputs of the assembly instruction set design include the architecture proposal, profiling results of the source code, and the 90–10% code locality analysis. If a good reference instruction set can be found from a previous product or from a third party, the design of ASIP will be faster. The output of the assembly instruction set design is the assembly instruction set manual.

Instruction set design consists of the functional specification of instructions, allocation of functions to hardware, and coding of the instruction set. The coding of the instruction set includes the design of the assembly syntax and the design of the binary machine codes. Finally binary codes are executed by the instruction set simulator for benchmarking. The performance of the instruction set and the usage of each instruction is exposed for further optimization.

Before releasing an assembly instruction set, it should be evaluated via instruction set benchmarking. In processor design, benchmark is a type of program designed to measure the performance of a processor in a particular application. Benchmarking is the execution of such a program to allow processor users to evaluate the processor performance and the cost.

Benchmarking methods and benchmark scores of DSP processors can be found at BDTI [6]. More general benchmarking knowledge is available at EEMBC [7]. To support benchmarking, programming tools including compilers, assemblers, and instruction set simulators should be developed. Toolchain design and programming technique for benchmarking will be discussed later in this chapter.

After benchmarking and the instruction set release, the architecture can be designed. The inputs for architecture design are the assembly instruction set manual, the early architecture proposal, and the requirements on performance and costs. The output of the architecture level design is the architecture specification documents. The architecture design of a processor is a specification of the high level hardware modules (the datapath including RF, ALU, and MAC, the control path, bus system and memory subsystem), and interconnections between the top level hardware modules. During architecture design, function allocation to modules and scheduling of functions between modules is performed. The bandwidth and latency of busses is analyzed during the architecture design.

The inputs of microarchitecture design are the architecture specification, the assembly instruction manual, and knowledge of the selected silicon technology. The output of the processor microarchitecture design is the description of the bit accurate and cycle accurate implementation of hardware modules (such as ALU, RF, MAC, and control path) and inter-module connections for further RTL coding of the processor. As a document used in ASIP design flow, the output of microarchitecture design is the input for RTL coding of the processor.

During architecture design, functional modules are specified and the implementation of each functional module was not mentioned. It means that in the architecture level design, functional modules are usually treated as black boxes, in microarchitecture design, missing details in each module will be specified.

Microarchitecture design in general is to specify the implementation of functional modules in detail. Microarchitecture design includes function partition, allocation, connection, scheduling, and integration. Optimization of timing critical paths and minimization of power consumption will be specified during microarchitecture design.

Processor microarchitecture design, in particular, is to implement functions of each assembly instruction into involved hardware modules. It includes:

- The partitioning of each instruction into micro operations.
- The allocation of each micro operation to hardware modules.
- Scheduling of each micro operation into different pipeline stages.
- Performance optimization.
- Cost minimization.

The ASIP hardware design flow starts from the requirement specification and finishes after the microarchitecture design. The design of an ASIP is mostly based on experience. The essential goal of ASIP design flow is to minimize the cost of design iteration. Design of an ASIP is a risky and expensive business with much rewarding when the design is successfully implemented.

After specifying the microarchitecture of an ASIP, the VLSI implementation will be the same as all other silicon backend designs, and can be found from any backend VLSI book. RTL coding and silicon implementation are out of the scope of the chapter.

3   Architecture and Instruction Set Design

In this chapter, the design methodology of an ASIP will be refined. Architecture and assembly instruction set design methods will be introduced.

3.1   Code Profiling

The development of a large application may involve the experiences from hundreds of people over many years. Application engineers take years to learn and accumulate system design experience. Unlike application engineers, ASIP designers are hardware engineers who specialize on circuit specification and implementation as well the design of assembly coding tools. It is impossible that ASIP engineers can really understand all the design details of an application in a short time.

Source code profiling fills the gap between system design and hardware design. The purpose of profiling is not to understand the system design; it is rather to understand the cost of the design including dynamic behavior of code execution and the static behavior of the code structure, the hardware cost, and the runtime cost. Source code profiling is a technique to isolate the ASIP design from the design of applications.

Code profiling can be dynamic or static. Static profiling analyzes the code instead of running it, so it is data (or stimuli) independent. Static code profiling exposes the code structure and the critical (maximum, or worst case) execution time.

A static code profiler is based on a compiler frontend, the code analyzer. Code structure can be exposed in a control flow graph (CFG), the output of the code analyzer. All run time cost including computing and data access can be assigned to each branch of the graph and the total costs can be collected and estimated by accumulating the cost of each path. The critical path with the maximum execution time should be identified via source code analyzers for ASIP design.

However, a critical path may seldom or never be executed and dynamic profiling is thus needed. Dynamic profiling is to run the code by feeding selected data stimuli. Dynamic profiling can expose the statistics of the execution behavior if the stimuli

Fig. 4 Static and dynamic
profiling

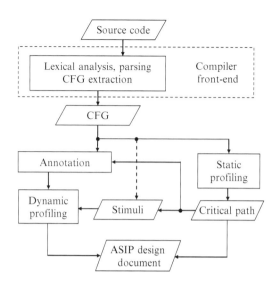

are correctly selected. It is usually used to identify the 90–10% code locality. By
placing probes in interesting paths, statistics on critical path appearances can be
achieved. Functions that are most timing critical are thus identified as inputs for
designing accelerated architectures and instructions. However, dynamic profiling
takes much time, so carefully selecting stimuli is an essential ASIP design skill.
More detailed information on profiling techniques can be found in Chap. 6 of [1].
An illustrative view of code profiling for ASIP design is given in Fig. 4.

3.2 Using an Available Architecture

Requirements on performance and function coverage were specified during code
profiling. If the required performance can be handled by an available architecture,
selecting and simplifying of an available architecture will be a good idea to reduce
the design complexity and costs.

Principally, there are so many different architectures to select that it is not an
easy task to decide a suitable architecture for a class of applications. That is why
architecture selection is mostly based on experiences.

Architecture selection can be based on architectural templates given in Fig. 5.
Four dimensions: performance, complexity handling, power efficiency and silicon
efficiency are taken into account in Fig. 5. Here the performance measurements
include arithmetic computing, addressing computing, load store, and control. The
control includes data quality control in software running on fixed point datapath
hardware, hardware resource control, and program flow control. Control complexity
may not be exposed during the source code profiling and it should be predicted
according to design experiences.

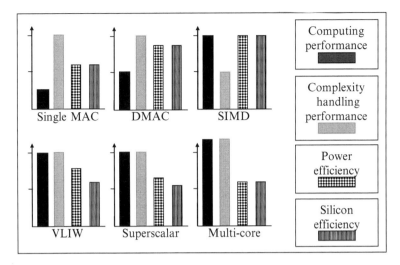

Fig. 5 Available architecture selection

Complexities can be further divided into data complexity, and control complexity. Data complexity stands for the irregularity of data, data storage, and data access. For general computing, the impact of the data irregularity can be hidden by using high-precision data types. For embedded computing a variety of data types can be used to minimize the silicon costs and power consumption.

The data complexity includes also the complexity of memory addressing and data access. Design of data storage and data access is based on the knowledge of memory subsystem (HW) and data structure (SW). Complexities of data storage and data access of DSP applications are different from the complexity of memory subsystem and the data structure of general computing. For general computing, the complexity comes from the requirements of flexible addressing, and large addressing space. The complexity is hidden by hardware cache.

For embedded DSP, the complexity of data storage and data access is induced by parallel data access for advanced algorithms. Successful handling of data complexity can give a competitive edge. Data access complexities cannot be exposed during source code profiling because the complexity of physical addressing is hidden in high level language.

Control complexity is exposed when handling program flow control and hardware resource management. On subroutine level, the control complexity is mixed up with data complexity, such as managing data dependencies. On high level in an application program, the complexity can be algorithm selection, data precision control, working mode (data rate, levels, profiles, etc.) selection, and asynchronous event handling (such as interrupt and exception handling). On top or system level, the complexity can be the hardware resource management through threading control and control for interprocessor communications. The management of control can be efficient if instructions for conditional execution, conditional branches, I/O control, and bit manipulations are carefully designed.

Power efficiency of a system is the average performance over the average power consumption. Low power ASIP design is the process of minimizing redundant operations (to minimize the dynamic power) and minimizing the circuit size (to minimize the static power). When the size of transistors shrinks, power efficiency and silicon efficiency are more related because the static power is no longer negligible and even becomes dominant.

In a modern high-end ASIP design, since the memory consumes most of the power, the design for power efficiency is almost the same as the design of efficient memory architecture. It is about optimizing the memory partition, minimizing the memory size, and minimizing memory transactions.

Silicon efficiency of a system is defined as the average performance over the ASIP silicon area. In general, as discussed previously, low power design is equivalent to designing for silicon efficiency. The on chip memory size of a silicon efficient solution is relatively small. If the on chip memory size is too small, however, the design may induce extra on-chip and off-chip data swapping and introduce extra power consumption.

Figure 5 gives architecture selection among six popular DSP processor architectures. The single MAC architecture is a typical low cost DSP processor architecture with single multiplication-and-accumulation unit in datapath. Only one iterative computing can be executed at a time in the processor. The performance may not be enough for advanced applications. The complexity handling capability can be relatively high if the instruction set is advanced. This architecture consumes the least silicon cost and power consumption is limited.

The dual MAC architecture is a kind of high performance and efficient architecture. There are two multiplication-and-accumulation units in datapath. Two iterative computations or two steps of an iterative computation can be executed at a time by Dual-MAC in the processor. By keeping the same complexity handling capability, the iterative computing performance of a Dual MAC machine can be double. Compared to the single MAC machine, the silicon cost and power consumption may be only few percentage higher, so the silicon efficiency and the power efficiency of a dual MAC machine is much better.

A SIMD (single instruction multiple data) machine can perform multiple operations while executing one instruction. The computing performance is high. However, a SIMD machine can handle data level parallel computing only when computation is regular. While handling irregular operations for complex control functions, a SIMD machine exposes its weaknesses. Power and silicon efficiency can be very high if a SIMD machine handles only regular vector data signal processing. In most systems, SIMD is a slave processor to accelerate signal processing with regular data structure. A master is used to handle complex control functions. Because control and parallel data processing are allocated to two processors, extra computing latencies induced by inter-processor communications should be taken into account.

When requirements on both the computing performance and the complexity handling are high at the same time, a VLIW (very long instruction word) machine may be a solution. A VLIW machine can execute multiple operations in one clock cycle. However, it does not have a data dependency checker in hardware. Advanced compiler technology must be available together with VLIW technology.

Because a VLIW machine handles both performance and complexity the same time. A VLIW machine is therefore popular for general purpose DSP with high throughput. However, when designing an ASIP, the application is known and a design based on master-slave SIMD or accelerators might be more efficient.

When applications are unknown and there is no sufficient compiler competence, VLIW machine architecture cannot be used. The superscalar architecture might thus be a choice. A superscalar processor can check data dependencies dynamically in hardware. At the same time, the extra hardware cost is needed for data dependency check. The hardware cost of a superscalar is therefore relatively higher. The power efficiency and silicon efficiency is relatively lower than that of VLIW. However superscalar processors offer higher performance while handling complex control. Superscalar ASIP can be found for searching based applications.

When control complexity cannot be separated from vector computing, a VLIW or a superscalar processor should be used. If control complexity can be separated from vector computing, and if there is a master processor available for handling the top-level DSP application program, a SIMD architecture is preferred to enhance the performance of vector/iterative processing. SIMD, VLIW, and superscalar all have only one control path issuing instructions, so these machines are ILP (instruction level parallel) machines.

Finally, a multicore solution is needed if the computation cannot be handled by one processor. A multicore solution introduces extra control complexities of inter-task dependence, control, and communications between processors. A multicore solution also introduces the partition induced overheads. Multicore ASIP is beyond the scope of the chapter.

A selected architecture can be a template of an ASIP. Architectural level modification and instruction level adaptation based on the selected template will be followed. The final architecture of the ASIP might be none of the architecture listed in Fig. 5.

3.3   Generating a Custom Architecture

A normal processor architecture is a control flow architecture (von Neumann architecture). A control flow architecture machine can handle only limited number of operations in one clock cycle. A data flow architecture might be introduced as an ASIP architecture to handle many operations, such as hundreds of operations, per clock cycle. In this case, optimized algorithm codes can be directly mapped or partly mapped to hardware to form a dataflow machine.

Task processing in a classical dataflow machine is controlled by a new event; it could be the system input "probe" or the data output "tag" of previous execution. The execution control of classical dataflow machine is not by program counter based FSM. However, dataflow architectures have been successfully merged into normal control flow architectures as slave machines of the main processor. In Fig. 6, a slave machine could be a function solver, a FSM driven datapath, or a programmable slave processor.

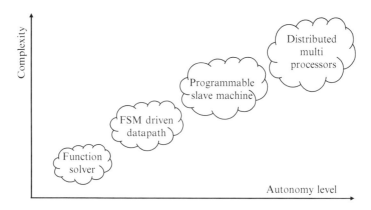

Fig. 6 Possible custom architectures

Except for data flow machines, there are other kinds of custom architectures. A function solver can be a part of datapath in an ASIP. It is a simple hardware module directly controlled by the control path of the processor. A typical function solver carries a single function such as 1/x, square root, functional module for complex data processing etc. It is usually driven by an I/O instruction or an accelerated instruction. The autonomy level of a function solver is low.

A FSM driven datapath usually offers autonomous execution of a single algorithm. The execution might be triggered by arriving data or by the master machine. This architecture is typically used to process functions requiring ultra high performance, or functions relatively independent of the master machine.

A programmable slave machine is controlled by a local programmable control path, and the autonomy level is rather high. However, task execution in a programmable slave machine is controlled by the master processor. A programmable slave machine is usually an ASIP. SIMD is a typical programmable slave machine.

Finally, a distributed system could be based on distributed ASIP processors. In this system, no one is assigned as master and no one is principally a slave. Each machine runs its own task and communications between tasks are based on message passing.

A FSM driven datapath in Fig. 7 could be a task flow machine. The task flow architecture is a typical custom architecture. It is also called function mapping hardware or algorithm mapping architecture. The architecture is the direct implementation of a DSP control flow graph. In Fig. 7, the method of hardware implementation of a control flow graph is depicted.

The left part in Fig. 7 is the behavioral task control flow graph of the DSP application. If the control flow graph is relatively simple and will not be changed through the life time of the product, the control flow graph on the left part of Fig. 7 can be implemented using HW on the right part of Fig. 7b. In Fig. 7b, the datapath is the mapping of the control flow graph in Fig. 7a. The FSM controls the execution of the DSP task in each hardware block step by step. Memories are used as the computing data buffer passing data between the hardware function blocks.

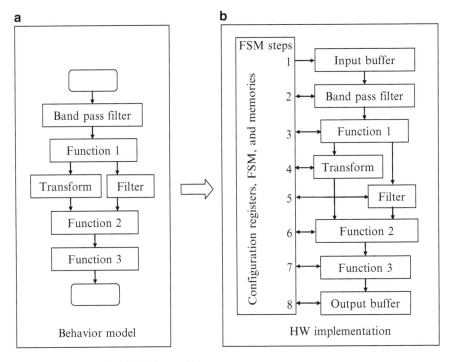

Fig. 7 Implementation of task flow architecture

The control flow graph architecture (a kind of dataflow architecture) is preferred when the complexity is not high and the data rate is very high. In a control flow graph architecture, hardware multiplexing may not be necessary, configuration of the hardware is easy, and programmability is not mandatory. The data flow architecture is mostly used for function accelerations at the algorithmic level.

Here configurability refers to the ability that the system functionality can be changed by an external device. Programmability stands for the ability of the hardware to run programs. The difference between configurability and programmability is the way the hardware is controlled; configuration control is relatively stable and it is not changed in each clock cycle and the running program changes the hardware function in each clock cycle.

4 Instruction Set Design

Different from a general purpose processor, an ASIP instruction set consists of two parts, the first part for function coverage and the second part for function acceleration. To simplify the ASIP design, the first part of the instruction set should

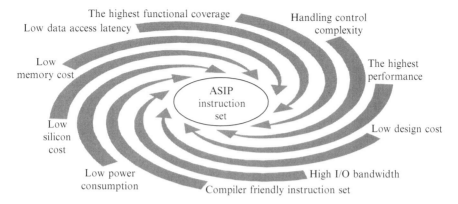

Fig. 8 Design space of an ASIP assembly instruction set

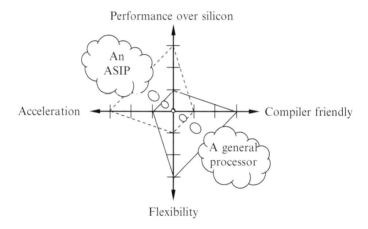

Fig. 9 Trade-off between performance and flexibility

be relatively simple. The second part of the instruction set offers both accelerations for computing extensive functions and for most frequently executed functions.

It is always expected that an assembly instruction set should be optimized for all possible applications. However, no instruction set can perfectly reach that goal. An optimization promotes some features and restrains other features. Figure 8 exposes different requirements to an ASIP instruction set. Obviously, not all can be reachable at the same time. When designing a processor, we specify the main goal and the cost constraints. Under limited power consumption and silicon cost, one can only focus on and optimize some features or parameters. At the same time, other features will be restrained. The trade-off among the four most important dimensions is illustrated in Fig. 9.

Here "Compiler friendly" means that the distance between IR (intermediate representation) and assembly language is relatively small. It also means that the

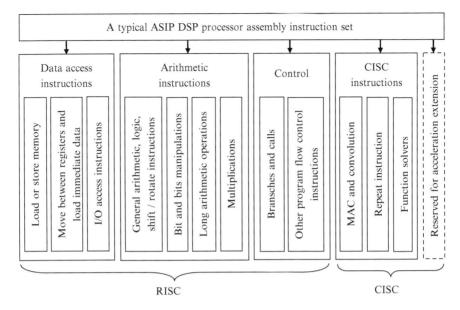

Fig. 10 Classification of an instruction set

compiler (usually a C compiler) can translate high level source code to assembly code and use the assembly instruction set in an efficient way. A "compiler friendly" instruction set is usually characterized by generality and orthogonality (for example, any instruction can use any register in the same way). Unfortunately in ASIP instruction sets may not be general enough if high performance and low power are required at the same time.

4.1 Instruction Set Specification

A DSP instruction set is usually divided into two groups, the RISC subset and the CISC subset, as illustrated in Fig. 10. All DSP processors, both ASIP DSP processors and general-purpose DSP processors, need RISC subset instructions for handling general arithmetic and control functions. The function coverage is offered by the RISC instruction subset. CISC subset instructions are used for function accelerations.

If a processor is required to supply a rather wide set of complex control functions, the RISC instruction subset could be similar to an instruction set of a general RISC processor. The RISC instruction subset is usually divided into three groups: instructions for basic load and store, instructions for basic arithmetic and logic operation and instructions for program flow control. Design of RISC instruction subset can be found from most computer architecture books.

4.2   Instructions for Functional Acceleration

The decision to introduce CISC instructions is based on 10–90% code locality analysis during source code profiling. CISC subset instructions can be divided into two categories: normal CISC instructions and instructions for accelerated extensions. The normal CISC instructions are specified in the assembly instruction set manual during the design of the ASIP core. The normal CISC instructions are completely decoded by the control path of the ASIP core.

ASIP might be designed as an SIP (Silicon Intellectual Property) for multi-products. Different products may have different requirements on function accelerations. Further instruction level function extension should be possible even after the design release of the ASIP RTL code and the programming toolchain. The method to add CISC instructions without touching the released ASIP design was discussed in detail in Chap. 17 of [1].

One group of CISC instructions are specified by merging a chain of available RISC arithmetic and load store instructions into one CISC instruction. In this way, acceleration is achieved without adding extra arithmetic hardware. The execution time is reduced to the time to run one instruction. The pipeline depth of the accelerated instruction is increased.

If the available hardware used by RISC instructions cannot support the acceleration, extra hardware should be added to support accelerated instructions. Another group of CISC instructions thus need extra hardware. By adding the first group of CISC instructions, we only increase the control complexity of the ASIP control path. By adding the second group of CISC instructions, we add complexity both to the control path and to the datapath.

A typical CISC instruction example is the MAC (multiplication and accumulation) instruction. It merges the multiplication instruction and accumulation instruction to one instruction with double pipeline depth. The execution of the instruction usually needs two clock cycles. The behavior description of the instruction is:

```
Buffer <= (Operand A) x (Operand B) //in the first clock cycle
ACR <= ACR + Buffer // in the second clock cycle
```

This is the most used arithmetic computation in DSP applications. There are opportunities to dramatically improve DSP performance by fusing several instructions into a "convolution instruction", an iterative loop, including memory address computing, data and coefficient loading, multiplication, accumulation, and loop control. By adding the convolution instruction, both data access overheads and loop control overhead are eliminated. A convolution instruction CONV N M1(AM) M2(A) is therefore introduced. The behavior description of the convolution instruction is:

```
01    // CONV instruction: iteration from N to 1
02    OPB <= TM (A); // load filter coefficient
03    OPA <= DM (AM);    // load sample
04    if AM == BUFFER_TOP then AM <= BOTTOM // Check FIFO bound
05       else INC (AM); // update sample pointer
06    INC (A);       // update coefficient pointer
07    BFR <= OPA * OPB; // Save compute product in buffer
```

```
08    ACR <= ACR + BFR;  // accumulate product
09    DEC (LOOP_COUNTER);         // DEC loop counter
10    if LOOP_COUNTER != 0 then jump to 01 // one more iteration
11      else Y <= Truncate (round (ACR));    // final result
12    end.
```

The loop counter LOOP_COUNTER, the data memory address pointer AM, and the coefficient memory address pointer A must be configured by running prolog instructions. In each operation step, two memory accesses, the multiplication, and the accumulation consumes one clock cycle, N taps of convolution will consume only N+2 clock cycles. Without the special CISC instruction CONV an iterative computing of a N-tap FIR could consume up to 12N clock cycles, 6N cycles for arithmetic computing, addressing, and data access, 3N clock cycles for checking the FIFO bound, and 3N clock cycles for checking the loop counter (consuming extra two clock cycles when a jump is taken).

4.3   Instruction Coding

Coding includes assembly coding and binary machine coding. Assembly coding is to name assembly instructions and make sure that the assembly language is human-readable. Hardware machines can only read binary machine code and the assembly code must be translated to binary machine code for execution.

The purpose of assembly coding is to let the assembly code be easy to use, easy to remember, and without ambiguity. Assembly coding is usually based on suggestions from the IEEE std. 694-1985. As the lowest level hardware dependent machine language, assembly code gives function descriptions based on micro-operations. Assembly code also specifies the usage of hardware such as the arithmetic computing units, physical registers, the physical data memory addresses, and the physical target address for jump instructions. The assembly language exposes micro-operations in the datapath, addressing path and memory subsystem, some micro-operations in the control path, such as test of jump conditions and target addressing.

It is not necessary to expose all micro-operations. Some micro-operations are not exposed in the assembly manual because they are not directly used by assembly programmers, such as bus transactions and the details of instruction decoding. In Fig. 11, explicit and implicit micro operations are classified.

All micro-operations that the assembly users (programmers) need to know will be exposed in the assembly language manual. However, to efficiently use the binary machine code, some micro-operations that are specified by the assembly language manual, will not be coded into assembly and binary machine code. Examples of such operations are flag operations for ALU computing and PC=PC+1 to fetch the next instruction. When writing instruction set simulator and the RTL code, all hidden micro-operations must be exposed and implemented.

All micro-operations related to an assembly instruction

Explicit micro-operations specified in assembly manual:

Explicit micro-operations specified in
assembly code and binary machine code:

| Implicit micro-operations not specified in assembly instruction set manual: bus usage, instruction decoding | Data memory addressing | Operands | Operations | Instruction configuration | Results | Target addressing | Implicit micro-operations, not necessary to specify them in assembly code: For example flag ops and PC<=PC+1 |

Fig. 11 Assembly and binary coding convention

Fig. 12 Traditional
embedded system design flow

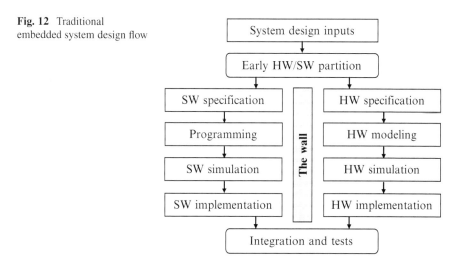

4.4 Instruction and Architecture Optimization

An ASIP instruction set shall be optimized for a group of applications with cost in mind. The optimization process is so called software–hardware co-design of the instruction set architecture.

Traditional embedded system design flow is given in Fig. 12. To simplify a design, system design inputs are partitioned and assigned to the HW design team and the SW design team during the functional design. The SW design team and HW design team design their SW and HW independently without much interwork. In the SW design team, functions mapped to SW will be implemented in program codes to be executed in hardware. The implemented functions will be verified during simulations and finally the simulated programs are translated to machine language during the SW implementation.

Fig. 13 Hardware/software
co-design flow

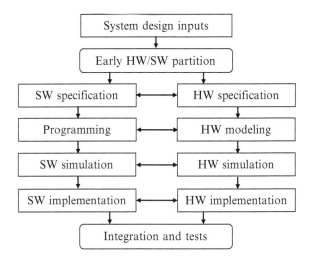

From HW design perspective, functions assigned to HW team are allocated to HW modules. Modules could be either processors or functional circuits. The behaviors of programmable HW modules can be described by assembly language simulator. The behaviors of non-programmable HW modules can be described by HW description languages. A function implemented in programmable hardware will be an assembly language instruction. Functions implemented in non-programmable hardware will be hardware modules or parts of a hardware module.

Finally, the implemented SW and HW is integrated. The implemented binary code of the assembly programs will be executed on the implemented HW. The design flow in Fig. 12 is principally correct because it follows the golden rule of design: divide-and-conquer. However, when designing a high quality embedded system, we actually do not know how to make a correct or optimized early partition. In other words, the early partition may not be good enough without iterative optimizations through the design.

To optimize a design, we need to interactively move functions between SW and HW design teams during the embedded system design. Under such a challenge, "HW/SW co-design" appeared in the early 1990s.

The HW/SW co-design flow is depicted in Fig. 13. Following the figure, the idea of the new design flow is to optimize the partition of HW and SW functions cooperatively at each design step during the embedded system design. HW/SW co-design is to trade-off function partition and implementation between SW and HW through embedded system design. Eventually, the results will be optimized following certain goals. Re-partition is not difficult. The difficulty is the fast and quantitative modeling for functional verification as well as performance and cost estimation of the new design after each re-partition. Fast processor prototyping is therefore a challenging research topic [4].

The implementation of an algorithm level function in ASIP HW is to design accelerated instructions for the specific function. Implementing a function in SW

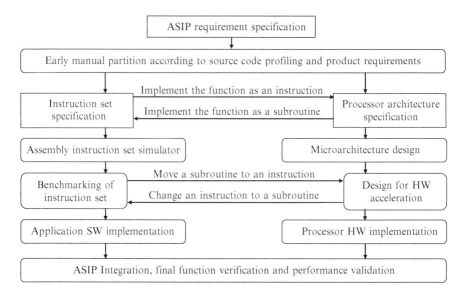

Fig. 14 Hardware/software co-design for an ASIP

means to design software subroutine of the function. After source code profiling, an
instruction set architecture can be proposed. Further HW/SW co-design of ASIP in
Fig. 14 is to modify the proposal by trading off HW/SW partition through all design
steps of an ASIP.

An ASIP assembly instruction set should be efficient enough to offer performance for the most frequently appearing functions and be general enough to cover
all functions in the specified application domains. When the performance is not
enough, the most used (kernel) subroutines or algorithms should be hardware-
accelerated. Those subroutines or algorithms that are seldom used should be
implemented as SW subroutines. Performance and coverage can be fulfilled by
carefully trading off the HW/SW partitioning.

However, functional coverage evaluation is not easy. So far, there is no better way
than running real applications. The cost is high and it takes a long time. HW/SW
co-design flow in Fig. 14 is part of the design flow in Fig. 3.

4.5 Design Automation of an Instruction Set Architecture

Researchers are working towards multi-objective-oriented evaluation tools to compare and select among different solutions. Many researchers also work on identification of functions to be accelerated and propose instructions for functional
acceleration. Also many researchers are working on ASIP synthesis.

There are two main approaches in research of ASIP synthesis. One is to synthesis an ASIP based on fixed template, and the other one is to synthesis an ASIP without template, only based on architecture description language (ADL) [4]. ASIP will have a bright future when it can be synthesized with or without a template architecture. In such a case, quantitative requirements including performance, flexibility, silicon costs, power consumption, and reliability can be reached. However, there are marketing, and project dependent problems that can not yet be modeled by EDA tools, such as:

- How to model market requirements
- How to model project requirements
- How to model human competence and availability
- How to balance SW design costs and HW design costs
- How to differentiate from competitors
- How to convince customers
- How to distribute weighting factors among requirements
- How to define the product life time
- How to trade-off between costs and flexibility.

At present, design of an ASIP DSP is based on experience. There are other non-technical decision factors. For example, the life time of ASIP may not be "the longer the better", as a longer product life time may block sales of new products.

5   An ASIP Instruction Set for Radio Baseband Processors

Baseband signal in a radio system is the signal which carries the information to be transmitted and the signal is not modulated onto the carrier wave of radio frequency. A baseband signal can be analog or digital. The sub system processing on a digital baseband signal is called digital baseband or DBB. The sub system between the DBB and the RF (radio frequency) sub system is called mixed analog baseband or ABB. In this section we only discuss DBB. A DBB in radio system is presented in Fig. 15.

A DBB module includes both the transmitter and the receiver. In a transmitter, the DBB codes binary streams from application payload to symbols on the radio channel. The coding includes efficient coding (the digital modulation, to code as many bits as possible in unit bandwidth) and redundant coding (to carry information for error correction on receiver side). The digital modulation is a way to carry multiple bits to a symbol.

The receiver recovers a radio signal from the analog to digital converter (ADC) in ABB to payload bits. The received waveforms carry both transmitted signals with distortions and interferences. Interferences are from external interference sources, such as glitches and white noise, and from the received signal itself. The received signal is a combination of the wave directly transmitted and waves reflected by different objects coming to the receiver at different times, giving interference

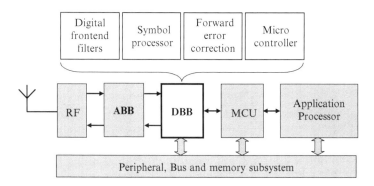

Fig. 15 A digital baseband DBB in a radio system

with each other. Signals arriving to the antenna at different times induce time domain distortion. Signals carried by different frequencies induce frequency domain distortion. Distortions and interferences must be cancelled before signal detection, which requires a pair of channel estimator and equalizer to remove the noise and distortions. Relative movement of a radio receiver and its transmitter changes the relative positions between the transmitter and the receiver. An estimated channel will not be valid as soon as the relative position is changed. If the mobility is high (e.g. a moving vehicle), the life time of a channel model will be on millisecond level, requiring recalculation of the channel in a millisecond.

To guard against glitches and white noise, forward error correction (FEC) algorithms must be executed. Computing features of advanced forward error correction algorithms include high computing load, computing based on short data precision, advanced addressing algorithms for parallel data access, and feasibilities for data and task level parallelization.

All these computing tasks, filtering signals, signal synchronizer, channel estimator, channel equalizer, signal detector, forward error correction, signal coding, and symbol shaping, need heavy computing. Moreover, radio baseband signal processing must conducted in real-time. A mobile channel should be estimated and updated within milliseconds, and all other signal processing for a receiver and transmitter must be done in each symbol period. For example, the symbol period of IEEE 802.11a/g/n is 4 μs, so several billions of arithmetic operations per second are required [2].

Some tasks have heavy requirements on computing performance and low requirements on flexibility, such as simple filtering, packet detection, and sampling rate adaptation. These tasks can be allocated to the DFE (digital frontend) module. Functions of these filters do not change in each clock cycle, thus configurability is sufficient for complexity handling of DFE. In the same way, error correction computing under acceptable low data precision can be allocated to FEC (forward error correction) modules without requiring programmability.

Table 1 Profiling and function allocation for IEEE 802.11a/g receiver

Tasks requiring high computing performance	Algorithms	Number of arithmetic operations per task	Appearance (μs)	Million arithmetic operations per second
Packet synchronization	Cross correlation	~ 3,000	~ 100	30
Sampling rate offset estimation	FFT/phase decision	2,500	100	25
F domain channel estimation	FFT, interpolation, 1/x	6,640	24	277
Payload Rotation	Vector product	~ 384	4	96
IFFT or FFT	64 points FFT/IFFT	1,984	4	496
F-domain channel equalization	Vector product	~ 384	4	96
Total cost				1,010

Table 2 Five most used task level (CISC) instructions accelerated on vector processing level

Instructions	Functional Specification
Conjugate complex convolution (auto correlation)	For I = 1 to N do {Complex REG = Complex REG + V1[i] * Conjugate of V1(or V2)[i]}
Complex convolution	For I = 1 to N do { Complex REG = Complex REG + V[i] * V2[i]}
Product of conjugate complex vectors	For I = 1 to N do {V3 [i] = V[i] * Conjugate of V2[i]}
Product of complex vectors	For I = 1 to N do {V3 [i] = V[i] * V2[i]}
Product of complex data vector and complex data scalar	For I = 1 to N do {V2 [i] = Scalar * V2[i]}

Except for DFE, FEC, and hardware control functions (to be allocated to the micro controller), the rest of the tasks are allocated to the symbol processor. Sufficient flexibility is required for signal processing on complex data. The main tasks allocated in the symbol processor include synchronization, channel estimation, time and frequency domain channel equalization, and transformation. The cost of 802.11a/g symbol signal processing in symbol processor is listed in Table 1. The required computing performances are calculated based on single precision integer data. Associated performance requirements for data access are not included in the table.

By offloading the majority of computing to DFE and FEC, the symbol processor only handles the small part of computing requiring programmability. However, the small part of the programmable computing requires more than 1 Giga arithmetic operations in a second, not yet including the data access cost. By analyzing tasks listed in Table 1, the most used ASIP task level instructions for signal processing of radio symbols are listed in Table 2.

In Table 2, V1[i], V2[i], and V3[i] are complex data vectors. If a datapath of complex data is used, one step of signal processing on complex data equals about 4–6 times of signal processing on integer data, and the performance will be enhanced 4–6 times. By further accelerating ASIP instructions on vector level, loop overheads can be eliminated by hardware loop controller. All functions listed in Table 1 can be processed on 200 MHz in one datapath with complex data type [2].

The digital baseband of IEEE802.11a/g is relatively easy. The computing cost for LTE and WiMAX will be at least 5–10 times higher. At least 5 more datapath modules for complex data computing are needed in an ASIP for LTE and WiMAX baseband signal processing.

6   An ASIP Instruction Set for Video and Image Compression

An image or video CODEC carries-out signal processing, eliminates redundancies of image and video data, and is a kind of applications of data compression [5]. Compressed images and video can be stored and transferred with low cost. The most frequently used image and video compression methods are lossy compression to get a high compression ratio. Generally, lossy compression techniques for image and video are based on 2D (two dimensional) signal processing, as illustrated in Fig. 16.

Figure 16 shows a simplified video compression flow, and the shaded part in Fig. 16 is a simplified image compression flow. The first step in image and video compression is color transform. The three original color planes R, G, B (R=red, G=green, B=blue) are translated to the Y, U, V (Y=luminance, U and V=chrominance) color planes. The human eye has high sensitivity to luminance ("brightness") and low sensitivity to chrominance ("color"). Normal people will not notice the difference of the resolution if the color planes U and V are down-sampled to $\frac{1}{4}$ of their original size. Therefore, a frame with a size factor of 3 (3 RGB

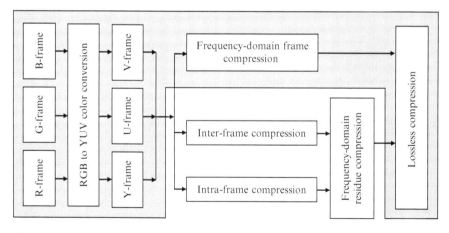

Fig. 16 Image and video compression

color planes) is down-sampled to a frame of YUV planes with the size factor of $1 + \frac{1}{4} + \frac{1}{4} = 1.5$, yielding a compression ratio of 2.

Frequency-domain compression is executed after the RGB to YUV transform. The discrete cosine transform (DCT) is used to translate the image from time-domain to frequency-domain. Because the human sensitivity to spatially high-frequency details is low, the information located in the higher frequency parts can be quantized with lower resolution. People will not notice that the resolution of the higher frequency part is low. After quantization, the resolution of the higher frequency part will only be 1 or 2 bits instead of 8 bits. The image or video signal in frequency domain is further compressed.

After the frequency-domain compression, continuous strings of 1's and 0's appear because of quantization. A lossless compression such as Huffman coding is finally used to compress the most frequent patterns, such as a chain of zeroes. Huffman coding assigns shorter bit strings for more frequent symbols and longer bit string for less probable symbols, and ensures that each code can be uniquely decoded.

Video compression is an extension of image compression, and it further utilizes two types of redundancies: spatial and temporal. Compression of the first video frame, the so called reference frame, is similar to the compression of a still image frame. The reference frame is used for the further compressing of later frames, the inter-frames. The difference between a reference frame and a neighbor frame is usually small. If only the identified inter-frame difference is compressed and transferred, the neighboring frame can be recovered by applying the differences to the reference frame. The size of inter-frame differences is very small compared to the size of the original frame. This coding method is temporal coding.

If one part of a video frame is the same or similar to another part in the same frame, only one part of the compressed code will be stored or transferred. It can be used to replace another part of the frame. Transferring or storing another part of the video frame will therefore not be necessary. Only the difference of the two parts will be transferred. This coding method is spatial coding.

In a video decoder, to comfort visual feeling of human, interpolation and de-blocking algorithms are introduced. To further improve video resolution, the corresponding sample is obtained using interpolation to generate non-integer positions. Non-integer position interpolation gives the best matching region comparing to integer motion estimation. Interpolation algorithms are needed on both the encoder and decoder sides. A de-blocking filter is applied to improve the decoded visual quality by smoothing the sharp edges between coding units.

The classical JPEG (joint picture expert group) compression for still images can reach 20:1 compression. Including the memory access cost, JPEG encoding consumes roughly 200 operations and decoding roughly 150 operations, for one RGB pixel (including three color components). If multiple pictures with high resolution should be compressed in a high resolution (e.g. $3,648 \times 2,736$) camera, accelerated instructions will be essential. Because the processing of Huffman coding and decoding can vary a lot between different images, the JPEG computing cost of a complete image cannot be accurately estimated.

Table 3 Accelerated algorithms for image and video signal processing

Algorithms	Kernel operations
SAD	Result=$\sum |Ai - Bi|$
Interpolation	Result=round $((a_1x_1 \pm a_2x_2 \pm a_3x_3 \pm a_4x_4 \pm a_5x_5 \pm a_6x_6)/32)$
De-blocking filter	Result=round$((a_1x_1 + a_2x_2 + a_3x_3 + a_4x_4 + a_5x_5 + a_6x_6)/8)$
8×8 Discrete cosine transform	Result=Integer butterfly computing and data access
Color transform	Result=$a_1x_1 \pm a_2x_2 \pm a_3x_3$

The classical MPEG2 (Moving Picture Expert Group) video compression standard can reach a 50:1 compression ratio on average. The advanced video codec H.264/AVC standard can increase this ratio to more than 100:1. The computing cost of a video encoder is very dependent on the complexity of the video stream and the motion estimation algorithm. Including the cost of memory accesses, a H.264 encoder may consume as much as 4,000 operations for a pixel and the decoder consumes about 500 operations for a pixel on average. As an example, for encoding a video stream with QCIF size (176×144) and 30 frames per second, the encoder requires about 3 Giga operations per second. The corresponding decoder requires about 400 Mega operations per second. A general DSP processor may not offer both the performance and low power consumption for such applications. An ASIP will be needed especially for handheld video encoding.

Obviously, function acceleration is needed both for encoding and decoding of images and video frames. The heaviest computing to accelerate is motion estimation. Many motion estimation algorithms have been proposed, most of them based on SAD (Sum of absolute difference). Some accelerated algorithms are listed in Table 3.

If data access and computing of the listed algorithms in Table 3 can be implemented with accelerated instructions, more than 80% of image/video signal processing can be accelerated.

7 Programming Toolchain and Benchmarking

As soon as an assembly instruction set is proposed, its toolchain should be provided for benchmarking the designed instruction set. An assembly programming toolchain includes the C-compiler, the assembler, the linker, the ISS (instruction set simulator), and the debugger. A simplified firmware design flow is given in the following Fig. 17a. The toolchain to support programming is given in Fig. 17b. Seven main design steps for translating the behavior C code to the qualified executable binary code are shown in Fig. 17a, and the six tools in the programmer's toolchain in Fig. 17b are used for the code translation and simulation. Each tool in the toolchain in Fig. 17b is marked with numbers. The ways that tools are used in each design step in the flow are annotated by numbers on the design steps in Fig. 17a.

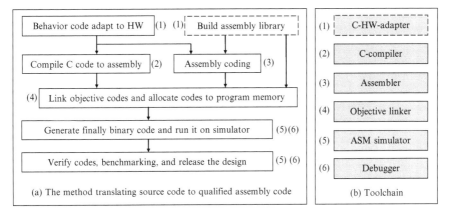

Fig. 17 Firmware design flow using programmer's toolchain

A C-code adapter might be a tool or a step of programming methodology. To get quality assembly code from C-compiler, the source C-code should adapt to hardware including:

- Change or modify algorithms to adapt to hardware.
- Adapt data types to hardware.
- Add annotations to guide instruction selections during compiling.

After adapting the source C-code to hardware, the C-code is compiled to assembly code by the compiler. A compiler consists of its frontend with source code analyzer and optimizer, and its target code synthesizer, or the backend. Most ASIP compilers are based on gcc (GNU Compiler Collection). Compiler design for ASIP DSP can be found in other chapters in this handbook.

The assembler further translates assembly code into binary code. The translation process has three major steps. The first step is to parse the code and set up symbol tables. The second step is to set up the relationship between symbolic names and addresses. The third step is to translate each assembly statement; opcode, registers, specifiers, and labels into binary code. The input of an assembler is the source assembly code. The output of an assembler is the relocatable binary code called object file. The output file from an assembler is the input file to a linker, which contains machine (binary) instructions, data, and book-keeping information (symbol table for the linker).

The assembly instruction set simulator executes assembly instructions (in binary machine code format) and behaves the same way as a processor. It is therefore called the behavioral model of the processor. A simulator can be a processor core simulator or a simulator of the whole ASIP DSP chip. The simulator of the core covers the functions of the basic assembly instructions, which should be "bit and cycle accurate" to the hardware of the core. The simulator of the ASIP chip covers the functions of both the core and the peripherals, which should be "bit, cycle, and pin accurate" meaning the result from the simulator should match the result from the whole processor hardware.

Once the firmware is developed, the programmer must debug it and make sure it works as expected. Firmware debugging can be based on a software simulator, or a hardware development board. In any cases, the programmer will need a way to load the data, run the software, and observe the outputs.

8 ASIP Firmware Design and Benchmarking

Firmware by definition is the fixed software in embedded products, for example an audio decoder (MP3 player) in a mobile phone, where it is visible to end users. On the other hand, firmware can also implicitly exist in a system not exposed to the user, e.g. the firmware for radio baseband signal processing. Firmware can be application firmware or system firmware. The latter is used to manage tasks and system hardware. A typical system firmware is the real-time operating system, RTOS. Qualified DSP Firmware design is based on deep understanding of applications, a DSP processor, and its assembly instruction set.

8.1 Firmware Design

The firmware design flow was briefly introduced on the left part in Fig. 17. It is further refined in Fig. 18. A complete firmware design process consists of behavior

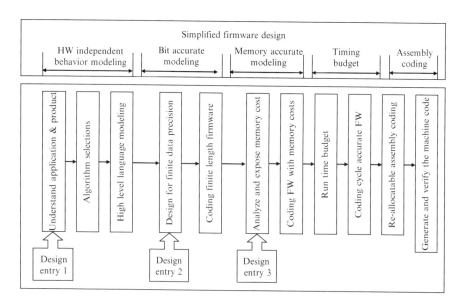

Fig. 18 Detailed firmware design flow

modeling, bit accurate modeling, memory accurate modeling, timing budget and real-time modeling, and finally assembly coding. Firmware design has three starting points according to the freedom of the design.

If designers have all freedoms including selection of algorithms, the design will start from design entry 1, the step of algorithm design and selection. For example, standards of radio baseband modems specify only the transmitter and do not regulate the implementation of receivers. During the implementation of radio baseband receivers, we have the freedom to select either a frequency-domain channel estimator or a time-domain channel estimator according to the cost-performance trade off, the computing latency, and the hardware architecture. Here the cost means mostly the computing cost, data access cost, and the memory cost. For example, some architecture will be suitable for transform and frequency domain algorithms. High level behavior of a DSP system is modeled based on high precision data types and a behavior modeling language, such as C.

However, if there is no freedom to select algorithms, a design starts from design entry 2. A typical example is video decoding for H.264 or audio decoding for MP3. Algorithms and C code are available from the standard. The C code is based on floating-point with excessive high data precision. For embedded applications, hardware with fixed point and limited precision will be used. The design following entry 2 starts from bit-accurate modeling. In this case, the freedom to decide the data precision is available. Data or intermediate results can be scaled and truncated during processing. During the bit accurate modeling phase, data quality control algorithms, such as data masking to emulate the finite hardware precision, signal level measurements, and gain control algorithms, will be added to the original source code.

Freedom of data precision control may not be available in some designs. When the freedoms of algorithm selection and data precision are not available, the firmware design entry is 3 in Fig. 18. In this case, the bit accurate model is available when the firmware design starts, for example the bit accurate source code could be from a standard committee. A typical case is the implementation of some voice decoder, which starts from available bit-accurate C code. Voice is usually compressed before transfer or storage to minimize the transfer bandwidth or storage cost. A voice decoder will decompresses the compressed voice and to recover the voice waveform at the voice user side.

In most ASIP, scratchpad memories instead of cache are used to minimize the data access latency. The time cost of data access and the memory cost is roughly exposed during source code profiling. The early estimate of memory cost might not be correct. Memory cost can be further exposed when the data types are decided after bit accurate modeling. A memory accurate model can therefore be achieved. Cache is not much used for embedded DSP computing because cache is expensive and data access in embedded computing is relatively explicit and predictable. A memory accurate model is an essential step to expose the data access cycle cost of scratchpad memories and finally to minimize it. The first step of memory accurate modeling is to re-model the data access in real hardware based on the constraint of the on chip memory sizes. After the first step, the extra memory transaction cost will be exposed. In the next step, moving data between memories and other storage devices will be modeled using program code or DMA transactions.

The firmware has runtime constraints when processing real-time data, especially streaming data. From profiling of the source code, runtime cost can be estimated. Extra run time cost is used to execute added subroutines for handling finite data precision, extra memory transactions, and task management. If the total runtime is less than the new data arrival time, the system implementation is feasible. Otherwise, enhancing hardware performance and selecting low cost algorithms are required.

Finally, the assembly code is generated and the functionality is verified. The assembly code as the firmware can be released when the final run time of the assembly code follows the specification and the precisions of the results are acceptable.

8.2 Benchmarking and Instruction Usage Profiling

Benchmarking by definition is some kind of measure on performance. In this context, benchmarking is used to measure the performance of the instruction set of an ASIP DSP. In general, benchmarking is to measure the run time (the number of clock cycles) for a defined task. Other measurement can also be important, for example memory usage and power consumption. Memory usage can be further divided into the cost of program memory and data memory required by a benchmark. DSP benchmarking is usually conducted by running the assembly benchmark code on the instruction set simulator of the ASIP.

DSP Benchmarking can be further divided into the benchmarking of DSP algorithm kernels and the benchmarking of DSP applications. By benchmarking DSP kernels, the 10% codes taking 90% runtime, essential performance and cost can be estimated. When benchmarking the kernel algorithms, kernel assembly code such as filters FIR, IIR, LMS (least mean square) adaptive filter, transforms (FFT and DCT), matrix computing, and function (1/x, for example) solvers are executed on the instruction set simulator of the processor. Benchmarking of an application runs the application codes on assembly instruction set simulator of the processor. The best way to evaluate a processor is to run applications because all overheads can be taken into account. However, the coding cost for an entire application can be very high. The benchmarking of an application should only be conducted on part of the application, the cost extensive part.

To benchmark a DSP instruction set, two kinds of cycle cost measurement are frequently used. One is the cycle cost per algorithm per sample data. For example "30 clock cycles are used to process a data sample by a 16-tap FIR filter". Another kind of measurement is the data throughput of an application firmware per mega Hertz. For example, "In one million clock cycles (1 MHz), up to 500 voice samples can be processed in a 2048-tap acoustic echo canceller". If the voice sampling rate is 8 kHz (it usually is), the computing cost of 16 MHz will be required in this case.

Table 4 Examples of kernel DSP algorithms running on a single MAC DSP

Algorithm kernel	Descriptions or specifications	Typical cycle cost
Block transfer	Move N data words from one memory to another	$\sim 3N + 3$
256p FFT	256 point FFT including computing and data access	$\sim 11,000$
Single FIR	A N-tap FIR filter running one sample	$\sim N + 12$
Frame FIR	A N-tap FIR filter running K samples	$\sim K(N+6)+8$
Complex data FIR	A N-tap complex data FIR filter running one sample	$\sim 8N + 15$
LMS Adaptive FIR	A N-tap least significant square adaptive filter	$\sim 3N + 10$
16/16 bits division	A positive 16bits divided by a 16bits positive data	~ 50
Vector add	C[i] $\leq$ A[i] + B[i] Here i is from 0 to N-1.	$\sim 3 + 3N$
Vector window	C[i] $\leq$ A[i] * B[i] Here i is from 0 to N-1.	$\sim 3 + 3N$
Vector Max	R $\leq$ MAX A[i] Here i is from 0 to N-1.	$\sim 2 + 2N$

The benchmarks of basic DSP algorithms are usually written in assembly language. However, if the firmware design time (time to market) is very short, benchmarks written in high-level language will be necessary. In this case, the benchmarking checks the mixed qualities of the instruction set and compiler.

BDTI (Berkeley Design Technology Incorporation [6]) supplies benchmarks based on handwritten assembly code. EEMBC (EDN embedded microprocessor benchmark consortium [7]) allows two scoring methods: Out-of-the-box benchmarking and Full-Fury benchmarking. Out-of-the-box (do not requiring any extra effort) benchmarking is based on the assembly code directly generated by the compiler. Full-Fury (also called optimized) benchmarking is based on assembly code generated and fine tuned by experienced programmers.

It is not easy to make an ideally fair comparison by benchmarking low level algorithm kernels on target processors. Each processor has dedicated features and is optimized for some algorithms, while not optimized for some other algorithms. A processor holding a poor benchmarking record of an application might have very good benchmarking record of another application. A typical case is that a radio baseband processor will never be used as a video decoder processor. For fair comparison, processors from different categories should not be compared.

DSP kernel algorithms consist of 10% of an application code that takes 90% of the runtime. Benchmarking on kernels is relevant because DSP kernel algorithms will take the majority of the execution time in most DSP applications. Well accepted DSP kernels are listed in Table 4. In the table, the typical cycle cost is measured on a DSP processor with single MAC unit and two separated memory blocks, a simple and typical DSP processor. It exposes the average performance among single MAC commercial DSP processors. If the benchmarking result of an ASIP designed by you is much behind scores in the table, you may need to understand why and try to improve your design.

Cycle cost and code cost consist of three parts while coding and running a kernel benchmarking subroutine, the prolog, the kernel, and the epilog. The prolog is the code to prepare and start running a kernel algorithm, it includes loading and configuring parameters of the algorithm. The kernel is the code of the algorithm, including loading data, algorithm computing, and storing intermediate results. The epilog is the code to terminate running of an algorithm, including storing the final result and restoring the changed parameters. Cycle cost for running the prolog, the kernel, and the epilog should be taken into account during benchmarking.

Instruction usage profiling is another measure on using an instruction set. Instead of measuring the performance, it measures the appearance of each instruction in application codes and how often it is executed in each million cycles. The usage profiling tells which instruction is used the most in an application. It also tells which instruction is the least used by an application. If several instructions are not used by a class of applications and the processor is only designed for this class of application, these instructions can be removed to simplify the hardware design and reduce the silicon costs. However, sometimes the saving of silicon cost and design cost is almost negligible while removing an instruction. The cost reduction could be significant only when many instructions are removed. On the other hand, since the most used instructions can be identified, optimizing them can significantly improve the performance or reduce the power consumption.

9 Microarchitecture Design for ASIP

The microarchitecture design of an ASIP specifies the design of hardware modules in an ASIP. In particular, it is the hardware implementation of the ASIP assembly instruction set to each function module of the core and peripheral modules of the ASIP. The input of the microarchitecture design is the ASIP architecture specification, the assembly instruction set manual, and the requirements on performance, power, and silicon cost. The output of the microarchitecture design is the microarchitecture specification for RTL coding, including module specifications of the core and interconnections between modules. The microarchitecture design of an ASIP can be divided into three steps.

(1) To partition each assembly instruction into micro-operations and to allocate each micro-operation to corresponding hardware modules and corresponding pipeline steps. A micro-operation is the lowest level hardware operation, for example, operand sign bit extension, operand inversion, addition, etc. If there is no such a module for some micro-operations, adding a module or adding functions to a module is needed.
(2) For each hardware module, collect all micro operations allocated in the module, allocate them to hardware components, and specify control for hardware multiplexing for RTL coding.
(3) Fine tune and implement inter-module specifications of the ASIP architecture and finalize the top-level connections and pipelines of the core.

Similar to architecture design, there are also two ways to design microarchitecture, one way is to use reference microarchitecture, which is an available design of a hardware module from publications, other design teams, and available IP.

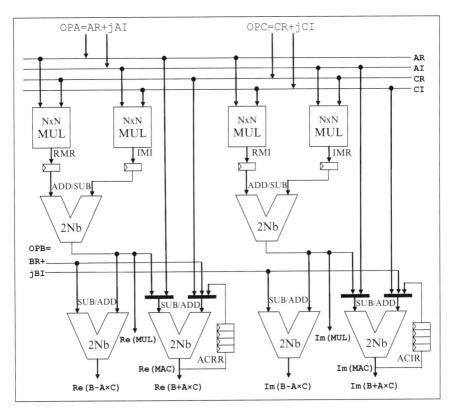

Fig. 19 A radix-2 data path for complex data computing

The reference microarchitecture is close to the requirements of the module to be designed and can execute most micro-operations allocated to the module.

Another method is to design a custom microarchitecture, which is to generate a custom architecture dedicated for a task flow of an application. This method is used when there is no reference microarchitecture or the reference microarchitecture is not good enough. The method is to map one or several CFG (control flow graph, the behavior flow chart of an application) to hardware. A typical case is to design accelerator modules for special algorithms, such as lossless compression algorithms, forward error correction algorithms, or packet processing protocols.

A typical example in Fig. 19 shows an example of a simplified datapath cluster for complex-data computing in a radio baseband signal processor. N in this figure is the word length of real or imaginary data. The cluster can execute the following functions one step in a clock cycle. These functions are computing a FFT butterfly, iterative computing for complex data multiplication and accumulation, complex data multiplication, and complex data addition/subtraction.

A Radix-2 decimation in time (DIT) butterfly can be executed in each clock cycle for FFT. The input data of the butterfly is data OPA = AR + jAI and OPB = BR

+ jBI, the input coefficient is OPC = CR + jCI. The outputs of a butterfly are Re(B+A*C) + jIm(B+A*C) and Re(B − A ∗ C) + jIm(B − A ∗ C). When executing the complex-data MAC (Multiplication and accumulation) or convolution, the input will be OPA and OPC. The output is thus Re(MAC) + jIm(MAC). When executing the complex multiplication, the input is also the OPA and OPC. The output is Re(MUL) + jIm(MUL). Finally, complex data addition and subtraction can be executed when the inputs are OPA, and OPC. Results are OPA + OPC = REA(MAC) + j Im(MAC) or OPA − OPC = REA(MAC) − j Im(MAC).

10 Multicore Parallel Processing

With the rapid development of radio standards, multimedia application, and graphics computing, the signal processing workload of modern embedded system has increased dramatically. For example, a high-end mobile phone that supports wideband radio communication and HD video codec demands a computation throughput of 100 GOPS [9]. The high computing load leads to multicore architecture with multiple ASIP processors to do parallel signal processing. The design of such a multiprocessor includes the design/selection of ASIP processors, memory hierarchy design, and interconnection architecture design.

10.1 Heterogeneous Chip Multiprocessing

Modern DSP system integrates multiple DSP cores to do signal processing in parallel. The wide range of applications and algorithms has led to the use of heterogeneous multiprocessor with specialized cores. Various ASIPs could be selected for different applications. For example, traditional single scalar DSP can be used for scalar algorithms like audio codec. SIMD style data parallel processors are used for computing tasks with rich vector operations such as radio baseband and multimedia. Since those ASIPs are optimized for data processing, a group of ASIPs are usually controlled by a microcontroller to form a DSP subsystem. The microcontroller handles the computing flow and data movements. Figure 20 shows a multiprocessor DSP system with one microcontroller and multiple application specific DSPs.

10.2 Memory Hierarchy

Multiprocessor DSPs usually utilize multi-level parallel memory hierarchy to improve performance by hiding memory access latency and avoiding access conflict. The memory system uses both distributed memory and shared memory, as shown in

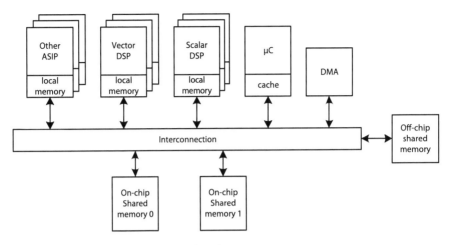

Fig. 20 Heterogeneous multiprocessor DSP subsystem

Fig. 20. The distributed memories are small and private on-chip memories of each ASIP core. The larger shared memory is mapped to the global address space of the system. The shared memory can have multiple levels; include both on-chip memory and off-chip memory. Unlike the general purpose processor uses cache to hide the complexity of data accesses, embedded DSP processors use scratch pad memory (SPM) as the local storage and rely on the software programmer to explicitly control the data movements.

The complex memory hierarchy and the use of SPM make a big challenge for software developments of multiprocessors. In a multiprocessor DSP system, it is common to have tens of memory modules in a single chip. Most of these memories are under explicit program control by software. A programmer needs to not only read and write memory data for computing but also use DMA functions to move data between these memories. The software programmer needs to handle multicore synchronization and data dependency. Multiprocessor platform should provide software development tools to trace memory access activities to help software programmers. Such tool can be found in the IBM Cell SDK [10] as an example.

10.3 Interconnection Architecture

Data communication in the multicore memory hierarchy is across the interconnection network. The design of interconnection architecture depends on requirements on the communication throughput and latency. If the communication throughput is not very high relative to the computing load on the processors, simple and low cost shared bus with narrow bus width can be used. If inter-processor communication is critical in a multiprocessor system, the interconnection design can use multi-connection and multi-layer busses (for example crossbar). The shared bus allows only one bus master to transfer data at one time. While a multi-connection and multi-layer bus allows concurrent data transfer between different master-slave pairs.

11 Further Reading

The ASIP DSP design flow was briefly introduced in this chapter. Architecture exploration and design, assembly instruction design, design for toolchain, firmware design, benchmarking, and microarchitecture design were introduced in the chapter. Two examples, radio baseband ASIP DSP and video ASIP DSP were briefly introduced. Readers are encouraged to get more ASIP design knowledge from [1]. ASIP design is so far experience based. If all requirements on performance, power consumption, and silicon cost cannot be reached by COTS, ASIP design is essential. ASIP enables performance computing and dramatically minimizes the cost of volume products.

This chapter is provided for readers who want to get fundamental knowledge in ASIP. For professional ASIP designers, more detailed information is provided in [1, 8], and [4].

References

1. D. Liu, *Embedded DSP Processor Design, Application Specific Instruction Set Processors*, Elsevier 2008 ISBN 9780123741233
2. D. Liu, A Nilsson, D Wu, J Eilert, and E Tell, *Bridging dream and reality: Programmable baseband processor for software-defined radio*, IEEE Communication Magazine, pp 134–140, September 2009
3. Coresonic Inc., http://www.coresonic.com/
4. P. Karlstrom, D. Liu, NoGAP a Micro Architecture Construction Framework, SAMOS IX: International Symposium on Systems, Architectures, MOdeling and Simulation, July 2009.
5. David Salomon, *Data Compression, The Complete Reference, 3rd Ed.*, Springer, ISBN: 9781846286025, 2006.
6. Berkeley Design Technology, Inc., http://www.bdti.com
7. Embedded Microprocessor Benchmarking Consortium, http://www.eembc.org
8. M. Hohenauer, H. Scharwaechter, K. Karuri, O. Wahlen, T. Kogel, R. Leupers, G. Ascheid, H. Meyr, G. Braun, *Compiler-in-loop Architecture Exploration for Efficient Application Specific Embedded Processor Design, magazine*, Design and Elektronik. WEKA, Verlag 2004.
9. D. Liu, et al, ePUMA, *Embedded Parallel DSP Processor with Unique Memory Access*, ICICS 2011, Singapore.
10. IBM, Cell Broadband Engine Programming Handbook, Version 1.11, May.2008.

FPGA-Based DSP

John McAllister

Abstract Field Programmable Gate Array (FPGA) offer an excellent platform for embedded DSP systems when real-time processing beyond that which multiprocessor platforms can achieve is required, and volumes are too small to justify the costs of developing a custom chip. This niche role is due to the ability of FPGA to host custom computing architectures, tailored to the application. Modern FPGAs play host to large quantities of heterogeneous logic, computational and memory components which can only be effectively exploited by heterogeneous processing architectures composed of microprocessors with custom co-processors, parallel software processor and dedicated hardware units. The complexity of these architectures, coupled with the need for frequent regeneration of the implementation with each new application makes FPGA system design a highly complex and unique design problem. The key to success in this process is the ability of the designer to best exploit the FPGA resources in a custom architecture, and the ability of design tools to quickly and efficiently generate these architectures. This chapter describes the state-of-the-art in FPGA device resources, computing architectures, and design tools which support the DSP system design process.

1 Introduction

In the mid-1980s, the pioneers of FPGA technology realised that whilst at that time transistors were scarce and as a result precious to circuit designers, Moore's Law would eventually make transistors so cheap that they could be used as a low cost collective in macro-blocks [30]. If these macro blocks permitted use as "*programmable*" logic, i.e. were functionally flexible to perform any simple logic

J. McAllister (✉)
Institute of Electronics, Communications and Information Technology (ECIT),
Queen's University Belfast, UK
e-mail: jp.mcallister@ieee.org

S.S. Bhattacharyya et al. (eds.), *Handbook of Signal Processing Systems*,
DOI 10.1007/978-1-4614-6859-2_22, © Springer Science+Business Media, LLC 2013

function, and could be arbitrarily interconnected, they could be used in networks to implement any digital logic functionality. This vision became incarnate in the Xilinx XC2064, the world's first FPGA device, in 1985. In the intervening period, the FPGA has changed markedly and dramatically from the simple "glue-logic" style devices of the original incarnations to domain-specific heterogeneous processing platforms comprising microprocessors, multi-gigabit serial transceivers, networked communications endpoints and vast levels of on-chip computation resource in the latest generations.

The key motivating factors for choosing FPGA as a target platform for DSP applications are flexibility, real-time performance and cost. Whilst software processors allow functional flexibility, the application's real-time performance requirements or physical constraints placed on the embedded realisation, for example in terms of size or power consumption, may be beyond that which these can achieve. In such a situation, unless volumes are sufficiently high, the Non-Recurring Engineering (NRE) costs associated with creating a customised Application Specific Integrated Circuit (ASIC) are such that this may not be commercially viable. For such high performance, low volume DSP systems, the ability of FPGA to host custom computing architectures tailored to the real-time requirements of the application and physical requirements of the operating environment, at relatively low cost, is a key advantage.

The historical perspective of FPGA is as a blank canvas "sea" of programmable logic ideal for hosting high performance custom circuit architectures. Indeed design tools which promote this kind of approach are prominent [1, 4, 6, 14, 43]. However, state-of-the-art FPGA are very different devices. They are massively complex combinations of heterogeneous processing, communication and memory resources capable of hosting very high performance DSP architectures. Typically they boast around 50 times the programmable Multiply Accumulate (MAC) processing resource of DSP processors[1] and host custom memory structures ranging from large capacity, low bandwidth off-chip storage to small, high bandwidth on-chip Static Random Access Memory (SRAM); combined with the almost limitless levels of data and task level parallelism available, their potential in place of fixed architecture, software programmable embedded devices is huge. However, this potential is only realised if the resources are properly harnessed in a manner which is easily accessible to application designers. Proper use of resources, in the context of modern FPGA, means creation of heterogeneous networks of microprocessors with optional accelerator co-processors [3, 37, 41] and parallel software processing architectures (i.e. multiple datapaths operating under the control of a single control unit) [7, 49] along with the classic dedicated hardware components.

The complexity of the architectures to be created, coupled with the requirement to create a new architecture for every new application makes the FPGA system design problem unique. The keys to success in this process are understanding the

---

[1] Based on comparison of Virtex®6 XC6VSX475T with assumed clock rate of 500 MHz and Texas Instruments TMS3206474 DSP.

requirements of the algorithm to be implemented, how to efficiently use the various resources of the target device, and having the ability to quickly generate and tune an efficient custom computing architecture. This chapter explores the capabilities of current FPGA for implementing DSP applications, the types of computing architectures which harness these capabilities best, and some examples of how these systems may be built quickly and efficiently. In Sects. 2 and 3, FPGA technology and the major processing resources available on state-of-the-art FPGA pertinent to DSP system implementation are described. Section 4 describes the classes of target architecture and their associated design processes for FPGA DSP, whilst Sect. 5 describes the characteristics of modern approaches to enabling higher productivity design processes for FPGA.

2 FPGA Technology

2.1 FPGA Fundamentals

The programmability of FPGA is based on two key principles: the use of programmable functional blocks, and programmable interconnect which allows multiple blocks to be connected to form more complex logical functions. The conceptual structure of a modern FPGA device, highlighting the key architectural components, is given in Fig. 1.

Modern FPGA are composed of four fundamental on-chip components:

1. Look Up Tables (LUTs)
2. DSP Units (DSPs)
3. Memories (RAMs)
4. Programmable Interconnect
5. Programmable Input/Output Pins

This current structure has grown from a classical FPGA structure composed of LUT, interconnect and input/output pin resources. A LUT is a Read-Only-Memory (ROM) which may be programmed to emulate logical functions by storing the relevant output in the memory location corresponding to the inputs which produce those outputs. Loading the data into each LUT on the chip is known as *configuring* the FPGA. By interconnecting these LUTs, which perform arbitrary logical functions, in an arbitrary manner using the programmable interconnect, networks of LUTs may be created to implement any logical functions. The resulting networks are then connected to the outside world via the programmable pins.

In the most recent generations of FPGA, the number of LUTs has grown steadily to very large numbers; in tandem, FPGA architectures have diversified to include on-chip units for high-performance arithmetic (DSPs) and memory storage (RAMs), creating a highly diverse set of resources with which designers may construct processing architectures. In the latest generations of FPGA, these resources place

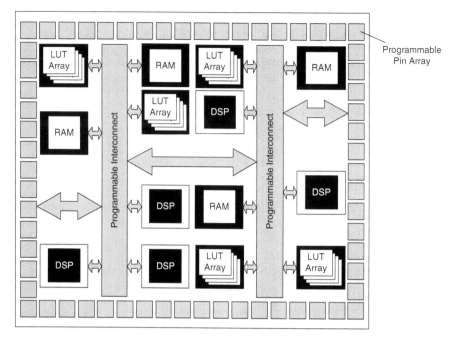

Fig. 1 Typical FPGA device architecture

unprecedented levels of computational capacity, memory and communication bandwidth at the fingertips of DSP system designers. The key enabling facility of FPGA, however, is parallelism: all these resources can be accessed concurrently, enabling parallel processing architectures limited only by the physical limits of the FPGA device avoiding the common communications bottlenecks apparent in software processing architectures.

2.2 Domain-Specific FPGA

The latest generations of FPGA mark a distinct divergence in device architecture, and on-chip processing architectures. On these devices, it became possible to implement multiprocessor architectures composed of units such as the Xilinx Microblaze™ [41] *softcore* processor (i.e. a processor implemented using the FPGA's programmable fabric) or *hardcore* processors embedded within the device silicon, distinct from the programmable fabric, such as the PowerPC processor on Xilinx Virtex®-II Pro [42]. Coupled with the introduction and expansion of other silicon embedded units, BRAMs, gigabit serial transceivers, DSP datapaths and network endpoints for rapid integration into computer networks via standards such as Ethernet or PCI Express, FPGA architectures have diversified to include a

Table 1 Current leading FPGA offerings

Vendor	Family	Target Applications
Xilinx	Virtex®-7	Highest bandwidth and performance logic
	Kintex®-7	DSP/Image processing
	Artix®-7	Low cost, low power, mixed-signal
Altera	Stratix®-5	Highest bandwidth applications
	Cyclone®-5	Mid range & SoC applications
	Arria®-5	Lowest cost, lowest power, high volume

wide range of heterogeneous components which may be used in different ways to implement DSP functions.

Furthermore, today's devices are not based on a single structure—each generation of devices is subdivided into families with each family exhibiting different features depending on the intended application domain. Consider the current mass range of FPGA offerings from the two market leaders, Xilinx Inc. and Altera Inc., as described in Table 1.

Both ranges show a three-tiered structure: high performance, high cost and power consumption devices (Xilinx Virtex®, Altera Stratix®), mid-range variants (Xilinx Kintex®, Altera Cyclone®), and low cost, low power devices (Xilinx Artix®, Altera Arria®). Whilst in general the trends are similar, the two ranges differentiate in specific ways, such as the mixed-signal focus of the Xilinx Artix® devices, and the SoC focus of the Altera Cyclone®[1].

2.3 FPGA for DSP

The level of computation resource enabled by the latest generations of FPGA device are unprecedented in any programmable embedded technology. Consider the computational capacity and memory bandwidth capabilities of the three variants of the Xilinx 7 Series FPGA technology, compared with the capabilities of the largest GPU (Graphics Processing Unit), the pinnacle of software-programmable embedded processing technology. These are summarised in Fig. 2.

As Fig. 2 shows, the levels of computational capacity and memory bandwidth resource available are unprecedented amongst other embedded processing technology. When compared with even the leading high performance computing technology (NVidia's Tesla M2090), the computational capacity and memory bandwidth of all the variants of Xilinx's 7 series FPGA is larger, and in the extreme case of the

---

[1]Xilinx have proposed a different, and specific kind of device, an Extensible Processing Platform, as their SoC variant. This is manifest in the Xilinx Zynq® product range, not considered in this document.

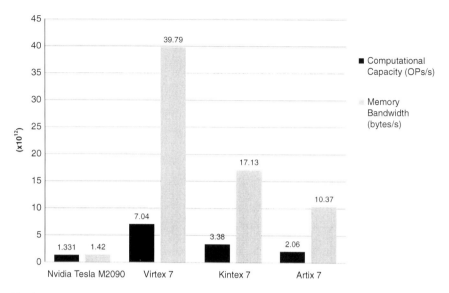

Fig. 2 FPGA capacity illustration

most capable Virtex®-7 technology the difference is close to an order of magnitude in terms of computational resource, and almost 50 times in terms of memory bandwidth.[2]

Given these massive levels of computational and memory bandwidth, the major issue facing designers of DSP systems is how best to harness them in a coherent processing architecture for high-end DSP applications. In Sect. 3 the nature of the processing resources in the top-end DSP FPGA (Virtex® 7 and Stratix® V) are described in detail.

3 State-of-the-Art FPGA Technology for DSP Applications

There are three types of resource on current FPGA most relevant to DSP system implementation: programmable logic, DSP-datapath, and memory. Sections 3.1–3.3 describe the characteristics of these features, and how they may be efficiently exploited for DSP system implementation.

---

[2]Metrics are approximated and assuming an FPGA operating frequency of 600 MHz.

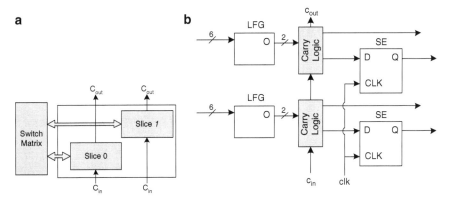

Fig. 3 Virtex®-7 FPGA programmable fabric architecture [47]. (**a**) Virtex®-7 CLB Structure, (**b**) Simplified Virtex®-7 Slice (*upper half*)

3.1 FPGA Programmable Logic Technology

3.1.1 Virtex®-7 Configurable Logic Block

Virtex® series FPGA are primarily constructed of *Configurable Logic Blocks* (CLBs). The structure of a CLB is outlined in Fig. 3a. As this shows, CLBs are composed of two *slices*, each of which contains four *Logic Function Generator* (LFG) units, as shown in Fig. 3b (only the upper two slices are shown, the given structure is replicated to create the full slice architecture). In Virtex®, Virtex®-II and Virtex®-II Pro technologies the LFG units had four inputs and one output, but this increased to five inputs and one output in Virtex®-4, and further to six inputs and two outputs in Virtex®-5 FPGA and beyond [47].

The basic slice architecture of Virtex®-7 contains four sets of: a single 6-input LFG, a storage element (SE) for registering the data emanating from the LFG, and a fast carry and switching logic which, coupled with the c_{out} and c_{in} ports on adjacent sets and slices can be used to cascade sets to implement high performance bit-parallel adders.

This relatively simple slice structure masks considerable complexity and diverse modes of operation. The LFGs can operate in one of four modes: as a single 6-input, 2-output LUT, as two 5-input, 1-output LUTs (assuming common inputs to the two input LUTs); as a 64 bit distributed RAM (DisRAM), or as a 32 element shift register (SRL). In DisRAM mode a single 64 bit RAM (Fig. 4a), may be realised with numerous such RAMs combined to implement larger or higher bandwidth RAMs if so desired. This is outlined in Fig. 4b which shows a 1 bit wide, 64 bit dual-port DisRAM. Larger bandwidth DisRAMs may be created by using numerous single-bit RAMs in parallel. In the SRL mode, each LFG may implement an addressable shift register of up to 32 elements in length, and multiple LFGs are cascadable for implementation of longer shift registers, as outlined in Fig. 5.

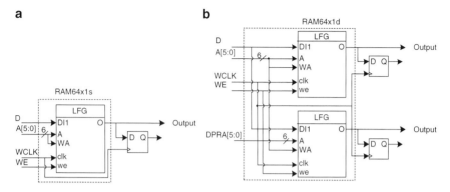

Fig. 4 Virtex®-7 FPGA LFG RAM configuration examples. (**a**) Virtex®-7 64 element DisRAM, (**b**) Virtex®-7 64 element Dual-Port DisRAM

Fig. 5 Virtex®-7 FPGA slice SRL

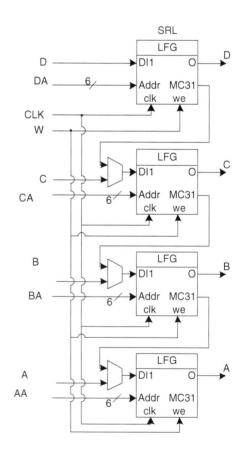

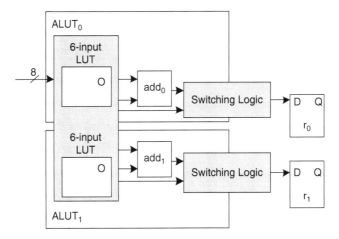

Fig. 6 Stratix®-V adaptive logic module (ALM)

3.1.2 Stratix®-V Adaptive Logic Module

Whilst the fundamental programmable component of the Xilinx Virtex® range of FPGA is the CLB, in the case of Altera Stratix® FPGA it is the *Logic Array Block* (LAB) [5]. Each of these is composed of ten *Adaptive Logic Modules* (ALMs). A simplified schematic of an ALM is shown in Fig. 6. As this shows, ALMs are based around 6-input LUT components, with a total of eight inputs to two LUTs. Each ALM can implement up to two combinational functions. An ALM also includes two programmable registers, two dedicated full adders, a carry chain, a shared arithmetic chain, and a register chain.

Each ALM can operate in four modes, each of which uses the variety of ALM resources differently.

1. **Normal Mode**: Suitable for general logic applications and combinational functions; eight data are input to the combinational logic, which implements two functions or a single function of up to six inputs.
2. **Extended LUT**: Suited to implementing HDL "if-else" conditional functions; a specific set of seven input functions may be implemented, where the seven inputs are composed of a 2-to-1 multiplexer fed by two arbitrary five input functions.
3. **Arithmetic**: The ALM is used to implement adders, counters, accumulators, wide parity functions and comparators; two sets of two 4-input LUTs are used, along with two dedicated full adders, which allow the LUTs to perform pre-adder logic.
4. **Shared Arithmetic**: Each ALM implements a 3-input addition, where each of two LUTs implement either the sum or the carry of three inputs; this mode is designed for multi-operand addition since it reduces the number of summation stages by increasing the width of each stage.

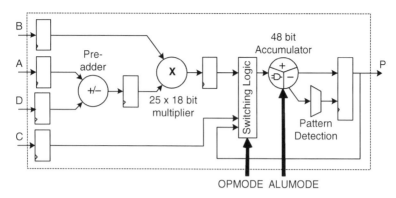

Fig. 7 Simplified Virtex®-7 DSP48E1 slice [45]

3.2 FPGA DSP: Datapath for DSP

The FPGA programmable fabric offers easy access to simple, high performance arithmetic, storage and logic components. However implementing more complex arithmetic operations (for example multipliers) in the programmable logic is much more resource expensive and consumes many LUTs and other resources. To address this, with the advent of more recent FPGA devices, such as Xilinx's Virtex®-II, hardcore versions of these were integrated on the FPGA silicon, and could be embedded almost seamlessly in circuit architectures along with programmable-logic based components. This marked a notable step forward in the use of FPGA for DSP since these previously relatively resource expensive operators were now available with higher performance and lower cost than before. In the latest generations of FPGA, this has gone a step farther with the integration of specialised DSP-datapath units.

3.2.1 Virtex®-7 DSP48E1 Slices

Virtex®-7 FPGA host multiple hardcore DSP-datapath units known as *DSP48E1 slices*. The architecture of one such slice is shown in Fig. 7—the Virtex®-7 range of FPGAs host between 1,260 and 3,600 of these components.

The basic architecture and operation of this component is obviously as a MAC. This fundamental DSP operation is popularly employed in the datapath of DSP processor architectures, and their performance is frequently benchmarked upon it [35]. However, the DSP48E1 has an array of associated operations which make it more capable:

- 25×18 bit pipelineable multiplication

  - 17 bit right shift for extended multiplier width implementation

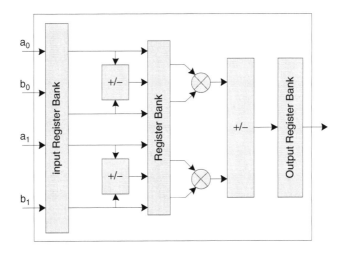

Fig. 8 Stratix®-V DSP module architecture

- Synchronous up/down counter operation
- 48 bit addition/subtraction with SIMD operation

  – Dual 24 bit addition/subtraction/accumulation
  – Quad 12 bit addition/subtraction/accumulation

- Pre-addition stage for efficient symmetric filter implementation
- Dynamic bitwise logic: AND, OR, NOT, NAND, NOr, XOR, XNOR
- Pattern detection for over/underflow support and rounding.

The DSP48E1 essentially operates as a programmable Arithmetic Logic Unit (ALU), with the *OPMODE* input controlling the overall operational mode of the slice, and the *ALUMODE* input controlling the operation of the add/bitwise logic component. The 48 bit addition is divisible into sub-operations (two 24 bit additions, or four 12 bit additions) which operate in parallel.

3.2.2 Stratix®-V DSP Module

The equivalent of a DSP48E1 in Stratix®-V is the *Variable Precision DSP Block* (DSPB). The overall structure of the DSP module is shown in Fig. 8. The unique aspect of these resources is their ability to work in a variety of distinct operating modes, on both real and complex-valued data. These modes are described in Table 2, which outlines the number of DSP blocks, and the equivalent number of independent 9, 12, 18 and 36 bit multiplications which may be realised using modules, for the Stratix®-V GT devices (the most computationally capable of the Stratix®-V family).

Table 2 Stratix®-V FPGA DSP block constitution

Device	DSPB	Multipliers			18×18 Multiply-Add	18×18 Multiply+ 36 bit Add
		9×9	16×16 18×18	27×27 36×18		
5GSD3	600	1,800	1,200	600	1,200	600
5GSD4	1,044	3,132	2,088	1,044	2,088	1,044
5GSD5	1,590	4,770	3,180	1,590	3,180	1,590
5GSD6	1,775	5,325	3,550	1,775	3,550	1,775
5GSD8	1,963	5,889	3,926	1,963	3,926	1,963

Each DSP block can operate in one of five basic modes, based around various configurations of the MAC function fundamental to DSP operations, the five classes being [5]:

1. **Independent Multiplier**: Parallel multiplication configurations, varying from three 9×9 multipliers, two 16×16 or one 27×27, 18×18 or 36×18 multiplier.
2. **Independent Complex Multiplier**: Exploits compositions of multiple DSPBs to achieve complex multiplication.
3. **Multiplier Adder Sum**: Cascaded compositions of DSPBs to achieve multiply-add-sum operation.
4. **Sum Of Square**: One DSP block can support a single sum-of-squares computation.
5. 18×18 **Multiplicand + 36 bit Summand**: One full resolution 18×18 multiply-add operation can be achieved in a single DSPB.
6. **Systolic FIR**: Cascaded DSPBs can achieve FIR filter operation for fully dynamic inputs, one dynamic input and one coefficient input, or one coefficient and one pre-adder output.

In addition, the Stratix®-V DSP block has dedicated rounding and saturation logic after the second stage adder, and the pipelining level of the device is configurable.

3.3 FPGA Memory Hierarchy in DSP System Implementation

Real-time implementation of DSP applications requires careful architecture design such that the datapath components, which perform the mathematical calculations, offer high enough computational capacity, and that these have sufficient input and output data bandwidth to keep them fed with data and continuously operating. This situation becomes particularly significant when implementing image processing applications on FPGA, where storage of entire frames of image data on-chip is sometimes not feasible [9], whilst storage in off-chip memory incurs significant data bandwidth and synchronisation problems. To help alleviate this issue, recent generations of FPGA device have exhibited a distinct hierarchy of memory storage, ranging from small, high bandwidth DisRAM-like units for data storage close to the datapath, to the low-bandwidth, bulk off-chip storage mechanisms. The general structures of these hierarchies in Virtex® and Stratix® FPGA are outlined in Sects. 3.3.1 and 3.3.2.

3.3.1 Virtex® Memory Hierarchy

In addition to the DisRAM configuration option of the Virtex® LFG outlined in Sect. 3.1.1, the latest generation of Xilinx FPGA have a three-tier memory hierarchy similar to that found in conventional microprocessor, except that in FPGA the absolute storage size and bandwidth at each level of the hierarchy is controlled by the designer, within physical capacity limits. On the chip, the DisRAM resource offered by the programmable logic is augmented by a series of dedicated, hardwired BRAM components on the chip [46].

In Virtex® FPGA these take the form of a number of 36 Kb RAMs, each of can be configured to implement a dual-port 36 Kb BRAM, or two dual-port 18 Kb BRAMs. Configuration of the memory may take either form without resorting to use of the programmable routing resource of the FPGA device; further, two BRAMs may be cascaded to implement a 64 Kb BRAM without accessing the FPGA routing resource. Virtex®-7 devices support between 795 and 1,880 of these BRAM components.

In addition, since the streaming nature of signal processing applications means that First-In-First-Out (FIFO) queues offer a simple, highly effective data buffering scheme, Virtex® BRAMs may be configured for use in this mode. They have extra logic integrated, including counters, comparators and status flag generation logic for this purpose. Each BRAM can implement one FIFO. The architecture of such a FIFO, which does not require programmable fabric resource, is shown in Fig. 9.

3.3.2 Stratix® Memory Hierarchy

The Stratix® FPGA family has a different approach to memory, employing a dedicated two-level hierarchy, composed of 640 bit *Memory Logic Array Blocks* (MLABs) and 20 Kbit *M20K* blocks. The structure of each level, and their intended uses are outlined in Table 3.

The MLAB resource is highly flexible, enabling customised balancing of capacity and bandwidth, with each offering storage of 64 8, 9 or 10 bit words, or 32 16, 18 or 20 bit words. The M20K is similarly flexible, trading off capacity and wordwidth between 512—16 K storage of 40 bit—1 bit words. Furthermore, each resource can operate in a number of modes:

- **Single Port**: 1 word read or written per cycle
- **Simple Dual Port**: Dual port behaviour, with one read and one write operation per cycle.

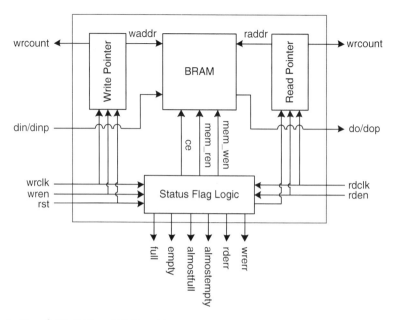

Fig. 9 Virtex® BRAM-based FIFO architecture [46]

Table 3 Stratix®-V memory hierarchy

| Resource | Bits/Block | Stratix V E Resource | | Use |
		5SEE9	5SEEB	
MLAB	640	15,850	2,640	FIFO buffers, filter delay lines
M20K	9,216	17,960	2,640	General purpose

- **True Dual Port**: Dual port behaviour with full read–write capability for each port per cycle—only support by M20K.
- **Shift Register**: Support for variable width, length and number of taps for filtering and correlation functions.
- **ROM**: Both MLAB and M20K can implement an intialisable ROM operation.
- **FIFO**: MLABs are ideal for small shallow FIFOs, with M20K enabling longer queues.

4 FPGA Design Processes

As outlined in Sect. 2, modern FPGA provide a diverse range of logic, computational and memory components for creation of custom DSP architectures. It is clear that strong similarities are evident across these devices in the form of distinct programmable logic, DSP-datapath and memory hierarchy resources in both Xilinx

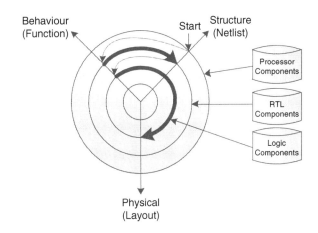

Fig. 10 FPGA design methodology Y-chart [11]

and Altera FPGAs. However to make effective use of these resources, suitable design approaches and supporting tools are required. Until recent generations of large capacity FPGA emerged offering these heterogeneous, the FPGA design process was one of creating a custom circuit architecture, predominately on the LUT-based FPGA programmable fabric. The methodology utilised to derive these architectures is summarised in Fig. 10 [11].

As Fig. 10 shows, the design process for an FPGA-based digital circuit typically commenced with a structural description of the system architecture, describing an interconnected set of processor components. This expression is subsequently refined to specify the behaviour of each component at the Register Transfer Level (RTL), and subsequently the system as a whole, before converting the behavioural description of each component to a structural one. Both these RTL expressions are typically described in a Hardware Description Language (HDL) such as VHDL or Verilog. Subsequently, the RTL structures are converted to gate level structural models using commercial synthesis tools, before generation of the final physical layout and programming file by vendor specific toolsets such as Xilinx ISE and Altera Quartus.

This approach has underpinned FPGA design processes for a lengthy period of time but it is becoming increasingly inadequate and unproductive with each new FPGA generation. The rate of increase of on-chip resources, and the associated increase in design complexity for the latest generations of FPGA technology requires much higher design productivity than RTL-level design of dedicated circuit architectures can provide. Hence the design demands of modern FPGA increasingly require state-of-the-art architectures and high productivity design processes and tools.

On first glance it appears the FPGA offer a "blank canvas" for implementation of DSP applications. However, to make best use of the resource, approaching the FPGA device from this viewpoint is not necessarily the best approach. Whilst highly flexible, to make most efficient use of the on-chip resources the various components should be used in specific ways. For instance, consider the use of DSP48E1 slices in Virtex®-7—these are highly efficient for 18×25 bit multiplication and 48 bit addition in a single slice, but once these wordsize bounds are exceeded, numerous DSP48E1 slices are required to account for the deficit [45], severely restricting how efficiently these may be implemented. Similarly, whilst the Virtex®-7 FPGA programmable fabric is highly efficient at implementing single port DisRAM components (64 bits/LUT), this efficiency is much reduced when multi-port RAMs are required. A similar scenario exists regarding the dual-port BRAM components, where unless the required memory bandwidth is less than dual-port, the efficiency of implementation is much reduced. To best exploit the fabric of the particular FPGA being targetted, these constraints must be explicitly taken into account during system architecture design [27].

At present, these is no single approach which is capable of addressing all these design complexity issues in a single toolset, or architecture design style—the choice of implementation strategy is subject to the real-time requirements of the application, and the relative constraints of the target FPGA device. For instance, the choice of implementation strategy for the hypothetical filter using programmable logic or DSP datapath in any particular system is based on whichever other components are present in the system and how they use the resources. If a 32,000 point FFT is present and is using up all the DSP48E1 resources then the choice of programmable fabric-based implementation is made, otherwise it may be more suitable to implement the FIR filter using DSP48E1; there is no general "best-fit" solution to implementing specific operations on an FPGA. However, a range of distinct design approaches, toolsets and architectural design styles are emerging. In Sects. 4.1 and 4.2 the two main architectural styles evident at present are described along with synthesis technology for these architectures, whilst Sect. 5 describes examples of emerging approaches which raise the level of design abstraction to the behavioural system level to enable exploration of the design space to explore these options (see Fig. 10).

4.1   Core-Based Implementation of FPGA DSP Systems

4.1.1   Strategy Overview

The programmable logic in FPGA offers the most mature facility for logic implementation, having gradually evolved from the LUT-based implementation philosophy on which FPGA was founded. As a result, strategies for creating dedicated processing architectures (known here as cores) to implement DSP functions is the most mature of any FPGA-based DSP implementation strategy. Given the abundance of register resource and the ability to easily implement simple arithmetic components in the programmable logic, it follows naturally that fine-grained pipelined core architectures are particularly good match with this host resource [40]. The availability of hardcore arithmetic components, such as multipliers, in later

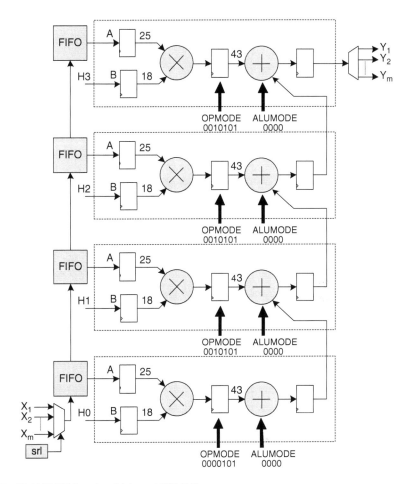

Fig. 11 DSP48E-based multichannel FIR [44]

generations of FPGA only serves to promote this style by offering high performance, low cost implementations of previously resource expensive mathematical operators.

The resulting abundance of bit-parallel pipelined MAC functionality enables high performance cores to be created for fundamental DSP operators such as FIR/IIR filters, FFTs or DCTs. This style of core design has become further ingrained with the advent of DSP48E1 in Virtex® and DSPBs in Stratix® FPGA, which offer another, even higher performance mechanism for building the same types of core architectures. Figure 11 shows how an array of DSP48E1 slices may be used to implement a multichannel FIR filter in Virtex®-5.

The longstanding suitability of FPGA to host such high performance architectures promoted much interest regarding its use as a host for systolic array type core architectures for FPGA, since these depend on fine-grained functional units with deep pipelines, a natural fit with FPGA's capability [40]. The variety of operations implemented in this fashion is huge, but mostly focuses on standard DSP operations such as filters and transforms, and extends to matrix computations, for example [21,50,51]. Furthermore, this style of architecture design enables powerful automatic core synthesis tools to be created. These convert Signal Flow Graph (SFG) algorithms automatically to systolic cores using graph transformation techniques such as folding and unfolding to balance throughput with resource [25,48].

Given FPGA's suitability as host to high performance core architectures, the device vendors themselves provide tools to automatically generate cores for varying real-time and resource requirements; tools such as Xilinx's Core Generator and Altera's MegaCore IP Library provide easy access to these cores, and promote a design reuse strategy for FPGA-based DSP system design. Further, this graph-based design style for rapid generation of custom core architectures is actively adopted in a number of commercial design tools, as outlined in Sect. 4.1.2.

4.1.2   Core-Based Implementation of DSP Applications from Simulink

Since simple cores are easily available in libraries such as Core Generator, the MegaCore IP Library or others, there is considerable potential for rapid creation of custom core-based architectures from these component parts. In particular, graph-based design tools for automatic composition of these to create more complex custom cores are widely evident. For instance, Synplicity Inc.'s Synplify DSP allows a user to automatically convert their DSP function, described as a MATLAB Simulink model, to a custom core. This tool can automatically pipeline and retime the graph to produce pipelined and multi-channel versions of a corresponding core. Xilinx and Altera use their System Generator and DSP Builder design tools respectively to enable a similar process, easing and promoting the uptake of their Core Generator and MegaCore IP library cores respectively. In both cases there is a tight coupling between the core behaviour, specified using a Simulink graph, and the resulting core architecture; a node on the Simulink graph translates to a component in the resulting circuit, implemented using a core exported from the vendor repository and cores are connected in a topology identical to that of the Simulink graph. A generalised version of the process used by both these tools is shown in Fig. 12 [4,43].

Behavioural models of the library cores are used to construct a Simulink.*mdl* model of the behaviour. After verificaton of the functionality, implementation-specific models of the cores addressing physical characteristics such as wordlengths and core timing are integrated into the Simulink model, which must then be manually manipulated to produce the same functionality taking account of these physical factors. This process allows verification that the functionality of the core matches that of the original algorithm. Once this is verified, this graph is exported to a RTL HDL description and passed to vendor-specific programming tools such as Xilinx ISE and Altera Quartus-II.

If a development platform exists, this may be integrated into the Simulink environment. A new Simulink graph which only creates the source stimulus for the

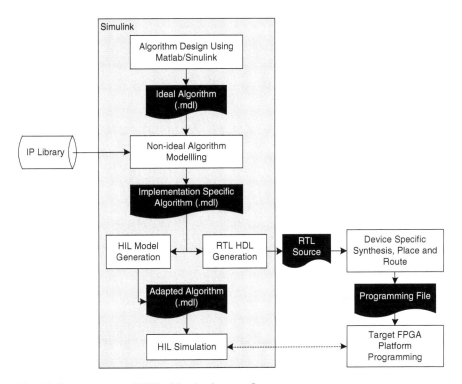

Fig. 12 System generator/DSP builder development flow

system implemented on the FPGA is then automatically generated. When the FPGA on the platform is programmed using the automatically generated programming file, Hardware-in-the-loop (HIL) testing ensues, where the host graph generates the stimulus data, communicates it to the FPGA, which processes it and passes the result back to the host, allowing the functional equivalence of the FPGA implementation with the original algorithm to be verified. This kind of tightly coupled approach allows rapid development of custom core architectures from Simulink graphs, and HIL testing of the core, but neither System Generator or DSP Builder exhibit advanced synthesis features such as auto-pipelining, retiming, multi-channelisation or graph folding apparent in other tools [48].

4.1.3 Electronic System Level (ESL)-Based Design of Core Based Architectures

The rapidly increasing scale of the design challenge facing FPGA designers is such that more radical, higher productivity design approaches have begun to emerge and stabilise. As a result, FPGA synthesis tools and techniques have diversified into Electronic System Level (ESL) design approaches, which commence at an

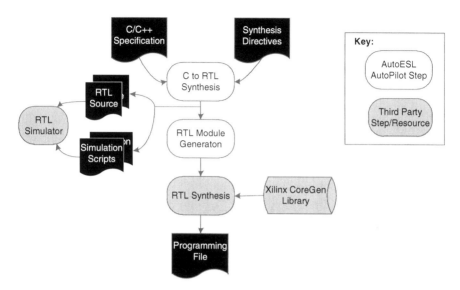

Fig. 13 AutoESL AutoPilot synthesis flow [8]

abstract specification format, such as a C program, and transform this into an FPGA processing architecture via a process which is at least semi-automated. Approaches such as Handel-C, Accelchip and PICO FPGA are typical of this approach.

In Handel-C [1], the parallel semantics of Communicating Sequential Processes [13] are combined with a subset of ANSI-C to allow application designers to describe the functionality of DSP operations with explicit parallelism, and translate this program to a gate-level netlist for implementation on FPGA. In Accelchip [6], a MATLAB description of the behaviour is automatically converted to a core, with the opportunity for the designer to optimise the implementation by applying loop transformations to the MATLAB source. Similar techniques are employed by PICO FPGA [14], which allows the user to specify the functionality in C and generate a selection of cores from which they may choose the most suitable option.

These approaches have all contributed significantly to the development of feasible ESL technology, but recent developments have pushed these closer to a tipping point of mass acceptance for FPGA. Consider the use of state-of-the-art synthesis tools, such as AutoESL. The design process exploited by this approach is illustrated in the flowchart of Fig. 13.

As Fig. 13 shows, the AutoESL AutoPilot synthesis environment converts modules described in a C/C++ specification format to an RTL architecture on FPGA, and integrates with third party RTL simulation, synthesis and component libraries for FPGA. When applied to the design of substantial DSP applications, including Sphere Decoding for Multiple-Input Multiple-Output (MIMO) communications channels, QR Decomposition and Matrix Multiplication, the resulting measures of productivity and implementation cost/performance are highly intriguing [8]. Whilst

C-based synthesis schemes were initially proposed to reduce the amount of time it takes to implement a given function on FPGA, it is apparent from the results quoted in [8] do not corroborate this outcome; rather these demonstrate a process which produces implementation with marginally reduced cost, as compared to those created via RTL-level design, but require the same design time. The value in this kind of approach comes from its ability to traverse the cost/performance trade-off with almost no subsequent redesign.

4.2   Programmable Processor-Based FPGA DSP System Design

Devices such as Virtex®-II Pro, which includes embedded PowerPC processor components, and the ability to use softcore processors such as Microblaze™ presents the possibility of creating Multiprocessor Programmable SoC (MPSoC) architectures on any FPGA [20]. Further, the unique ability of FPGA to allow such general purpose processors to be extended with custom datapath co-processors offers an excellent platform for developing application-specific MPSoCs, composed of multiple processors which maintain their software programmability, yet offer higher performance for domain specific applications by co-processor based extension [31,32]. Indeed, FPGA microprocessors have inbuilt features to allow such easy extension. Furthermore, since FPGA can easily host massively parallel processing architectures and have abundant on-chip programmable MAC datapath resources, they can play host to parallel programmable architectures, such as vector or SIMD processing styles [10] with the added ability to tailor the size and nature of the processor in an application-specific manner [7, 15, 28, 49].

All these options offer software programmability as a design approach for implementing DSP systems, rather than the dedicated hardware design styles described in Sect. 4.1. This section explores the various options for creating or extending software processing architectures for FPGA-based DSP; beginning by highlighting how the main commercial FPGA processors can be easily extended with custom datapath co-processors, a number of significant examples of this approach and massively parallel processor design are described.

4.2.1   Microblaze®-Centric Reconfigurable Processor Design

The Microblaze® softcore processor is a configurable Reduced Instruction Set Computer (RISC) for Xilinx FPGAs. The architecture of this processor is outlined in Fig. 14a. It is configurable in a number of ways with the configurable parts of the architecture shown in grey in Fig. 14a: the *Memory Management Unit* (MMU), *Instruction* and *Data Cache Units*, and a number of application specific datapath units.

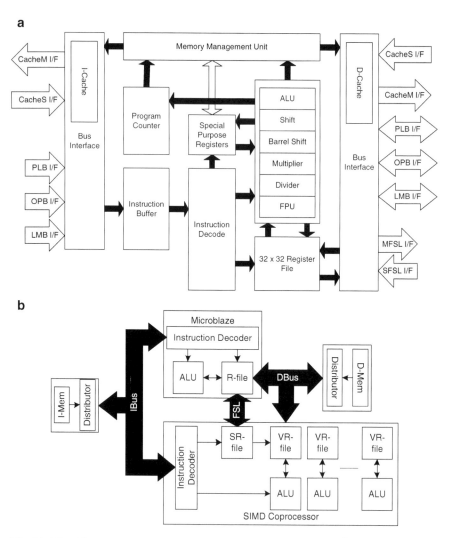

Fig. 14 Microblaze reconfigurable processing architecture. (**a**) Microblaze® Softcore Processor Architecture [41], (**b**) FSL-based Microblaze® Coprocessing [7]

Of particular significance is the configurable use of *Fast Simplex Links* (FSLs). A Microblaze® can host up to 16 FSLs, which are dedicated, point-to-point unidirectional data streaming interfaces, offering a low-latency interface directly into the processor pipeline for external co-processors. A simple schematic of how such a co-processor interfaces to the Microblaze using FSLs is given in Fig. 14b.

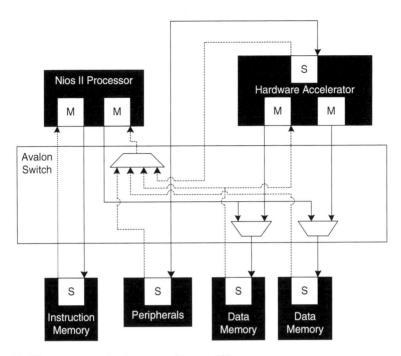

Fig. 15 Nios processor and subsystem architecture [3]

4.2.2 Nios-Centric Reconfigurable Processor Design

The Altera Nios II softcore processor exists as part of an extended subsystem architecture, which is shown in Fig. 15. As this shows, custom hardware accelerators may be integrated into the Nios memory subsystem using the Altera Avalon Switch Fabric [2].

Altera also offer a simple mechanism for quickly developing new co-processors which fit into this subsystem architecture. Altera's Nios II C2H Compiler [3] allows a user to describe the functionality of the system in Nios II compatible C code, and offload critical parts of the code to a co-processor, which is then automatically created and inserted into the subsystem of Fig. 15. A simple set of rules, outlined in Table 4, are used to convert C commands to a hardware architecture. This one-to-one mapping style makes the method of constructing the co-processor architecture explicit to the designer, allowing them to manipulate the C program to optimise it.

The C2H Compiler also provides rudimentary architectural optimisations; in particular automatically pipelining and scheduling loops and automatically removing Avalon master ports when several connect to the same memory block, merging the references and scheduling the accesses internally.

Table 4 Nios C2H synthesis rules

Command Type	Examples	Mapping
Arithmetic/Logical	`result += a + b;`	Hardware block
Iteration	`do(), for(),`	State machine + control
Selection	`if-else`	Multiplexer + control
Local Variables		Accelerator internal memory
Global Variables		Avalon MM Master port
Memory Accesses	`my *ptr = 8;`	Avalon MM Master port

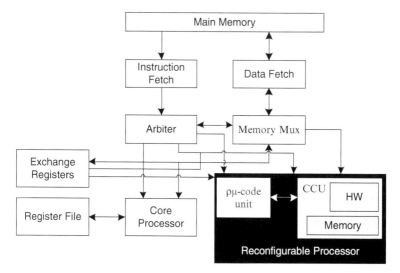

Fig. 16 Molen machine architecture

4.3 Reconfigurable Processor Design for FPGA

4.3.1 Molen

The *Molen* "polymorphic" processor [37] also focuses on accelerating critical tasks via the use of custom cores, but rather than attempt to couple the accelerator core into the pipeline of the General Purpose Processor (GPP), it considers the accelerator core to be a peer component in a larger computing machine composed of it and the GPP. An overview of the Molen machine architecture is given in Fig. 16.

Operating under a single thread of control, the core to be integrated (*HW* in Fig. 16) is wrapped in a *Control and Communications Unit* (CCU) wrapper, and a microcoded program is generated which fits a restricted instruction subset template defined for the Molen machine [37]. The *Arbiter* and *Memory Mux Units* arbitrate instruction and data fetches to/from the main memory between the GPP and the HW. All code is executed on the GPP, except operations tagged for implementation using the core, with the resulting communication between GPP and HW occurring via the exchange registers in Fig. 16.

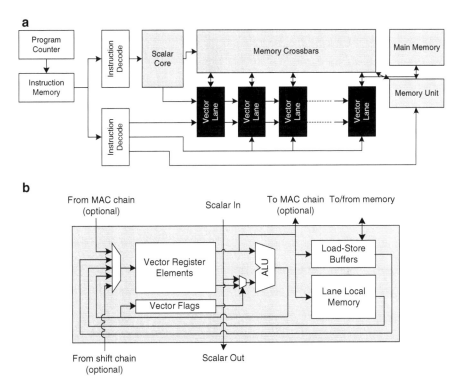

Fig. 17 VIPERS vector processor architecture [49]. (**a**) VIPERS Softcore vector processor, (**b**) VIPERS vector lane

In this arrangement, the design problem extends beyond using an extensible instruction set architecture for the GPP to include generation of custom instructions and wrapper logic for any arbitrary core such that it fits into the Molen architecture [22]. The Molen architecture reports impressive speed-up for DSP and image processing applications, reporting MPEG-2 application speedups by a factor of up to three for a Virtex®-II Pro FPGA using a PowerPC 403 GPP [22, 37].

4.3.2 VIPERS

The application-specific datapath route for accelerating DSP applications considered thus far is only one feasible option. The vast levels of simple on-chip programmable DSP-datapath (see Sect. 3.2) resource has driven attempts to achieve acceleration of critical functions by exploiting highly parallel software processing architectures using these units. VIPERS is one such approach in this vein [49]. VIPERS is a softcore vector processor, the architecture of which is outlined in Fig. 17a, with the architecture of each vector lane of the processor shown in Fig. 17b. Each vector lane contains an ALU (supporting arithmetic and logical operations,

Table 5 VIPERS processor configuration options

Parameter	Description	Range
NLane	Number of vector lanes	4–128
MVL	Maximum vector length	16–512
VPUW	Processor data width	8/16/32
MemWidth	Memory interface width	32–128
MemMinWidth	Minimum accessible data width in memory	8/16/32

maximum/minimum, merge, absolute value, absolute difference and comparison operations) as well as a barrel shifter and a DSPB-based multiplier. The architecture of VIPERS is configurable in a number of aspects, which are outlined in Table 5.

The ability to vary the number of independent vector lanes and the size of each lane (via the *MVL* parameter) provides scalable data level parallelism. Furthermore, the presence of optional MAC and shift chains in the architecture (the former of which exploits the DSPB resource) allows acceleration of critical functions in a manner which is the ideal use of the physical chip resource.

4.3.3 Discussion

The huge array of implementation options available to the DSP designer targetting FPGA provides the ability to implement a system by exploiting a variety of processing architectures and styles unparalleled elsewhere in the embedded computing domain. With options spanning between the two extremes of dedicated circuit implementation and MPSoC implementation perhaps the ideal scenario is to find the right balance between the two such that the high performance of the dedicated hardware and the flexibility of the software may be combined. To make best use of the FPGA resource, a balance must be struck between programmable logic, hardcore or softcore processors and DSP-datapath units based on the specific requirements of the application, and the constraints of the target device.

Modern FPGA are a diverse mix of processing resources, and as Sect. 4.1 and this one have described, harnessing these resources requires complex architectures consisting of multiple heterogeneous software processing architectures and cores. Accordingly, the process of DSP system design for FPGA is a complex hardware/-software cosynthesis, integration and design space exploration problem. This alone makes FPGA system design complex enough, but the situation is exacerbated since it is encountered *every* time a new DSP application is targeted to FPGA, since the entire motivation for choosing FPGA is for their ability to host custom computing architectures tailored to the application. As such, architecture synthesis for DSP applications on FPGA is a complex design problem encountered very frequently, a combination which places unique demands on Electronic System Level (ESL) design tools for embedded systems.

To maintain a productive design process in this environment it is vital that a designer can quickly evaluate a range of options for implementing their architecture, and choose the one which best matches their needs, applying some fine-tuning afterwards if required. This should all occur, preferably, without the designer having to resort to low abstraction, i.e. RTL level, manipulation of the architecture. To enable this kind of process a new generation of system design tools which allow rapid generation of FPGA architectures is beginning to emerge which help explore the trade-offs of resource, power, and real-time performance. A number of these are described in Sect. 5.

5   System Level Design for FPGA DSP

FPGA architecture development has historically been sustained by the use of RTL HDL-based coding, and implementation techniques similar to those employed for ASIC development [11]. However, as the complexity of ASICs increases along with the cost of fabricating a chip, the design and use of ASIC is increasingly confined to mass-market technologies such as mobile phones and PCs, all of which employ popular communications standards such as Bluetooth®, WiFi, and codecs such as MPEG. This has driven development of standards-based programmable SoC platforms which are highly efficient at implementing a set of standards in a particular area, for example 802.11 wireless communications protocols, or a variety of MPEG codecs, and programmable to realise specific standards from the target set [19, 39]. Consequently, the majority of SoC design tools are intended for a design process which may take a substantial period of time to build a highly efficient architecture once. This is distinct from FPGA-based design where architecture re-engineering is commonplace. A divergence in the nature of design tools supporting ASIC and FPGA design is a natural consequence.

Multiprocessor architectures, on the other hand, do not require any architectural manipulation, merely reprogramming. As such the requirement to frequently and rapidly generate custom computing architectures encountered in FPGA is unique. To maintain design productivity in this process, tools which enable this design process must quickly and automatically generate an architecture from an abstract representation of the algorithm functionality, and should be able to automatically explore the design space trading off performance with power and resource before automatically generating the implementation. A number of tools which enable this process for FPGA-based systems are beginning to emerge.

5.1   Daedalus

Daedalus is a system level design and rapid implementation toolset for DSP applications which has been used to generate FPGA-based MPSoC implementations of various image processing algorithms [23, 36]. A conglomerate of two previously

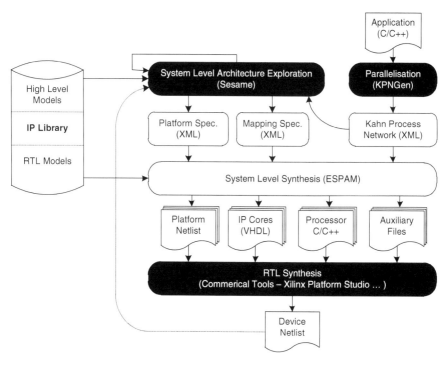

Fig. 18 Daedalus MPSoC design flow

existing tools, Sesame [26] and ESPAM [24], Daedalus enables a full synthesis flow from sequential application specification to implementation in a variety of MPSoC topologies, including point-to-point, bus-based or crossbar interconnect. The flow of data and operations in Daedalus is illustrated in Fig. 18.

The upper processes are mostly aimed at generating an allocation of processing resources, with the lower processes aimed at synthesis of the allocation. The functional specification of the system to be implemented, written as a Static Affine Nested Loop Program (SANLP) in a standard sequential language such as C++ or Matlab, is converted to a parallel Kahn Process Network form [16] by the *KPNGen* tool [38]. Daedalus accepts this specification and performs an exhaustive search of the design space to produce an MPSoC allocation which can support the real-time requirements of the system. The platform specification and mapping of the KPN tasks to distinct processors in the platform are then generated and passed to ESPAM for synthesis of the solution.

5.2   Compaan/LAURA

The Compaan toolchain is composed of the Compaan parallelising compiler and the LAURA architectural synthesis tool. It has predominately targetted core network implementations of Static Affine Nested Loop Programs (SANLPs) written in Matlab on FPGA [34]. Compaan acts as a parallelising compiler for the sequential specification, once again generating a KPN application specification. The LAURA tool then assumes either a fixed allocation of a one-to-one correspondence between the KPN structure and the target architecture [34]. Design space exploration of the implementation is affected by manipulating the KPN topology using transformations of the loops (e.g. skewing, unrolling) in the source SANLP [12, 33].

5.3   Koski

Koski is a Unified Modelling Language (UML)-based design environment for automatic programming, synthesis and exploration of MPSoC implementations of wireless sensor network applications [17]. The structure of the Koski design environment is outlined in Fig. 19. The final architecture is composed of multiple processors and cores connected to a HIBI network [29]. Again, application specification uses the KPN modelling domain. This specification, along with a specification of the architecture and of the implementation constraints, is processed by an architecture exploration toolset to generate the platform allocation. During *static* mapping, the allocation of processors, and mapping of tasks to processors is generated, with this allocation fine-tuned during a *dynamic* mapping phase. Upon finalisation of the mapping, the RTL and software synthesis technology of Koski is used to generate a prototype of the final implementation.

5.4   Open Issues in the Design of FPGA DSP Systems

The key to productive DSP system design on FPGA is the ability to quickly generate a wide range of implementations from which the designer can choose a suitable candidate. Given the range of techniques which may be used to implement DSP systems, as outlined in this chapter, highly promising technologies to fulfill this requirement are in evidence, but substantial problems remain.

Specifically, the heterogeneity of the resources on modern FPGA are significantly complicating the FPGA design process. For instance, whilst C-based synthesis tools such as AutoESL are capable of producing FPGA architectures to perform the computational requirements of a DSP application, they are much less well-developed in terms of synthesising complex memory structures for memory

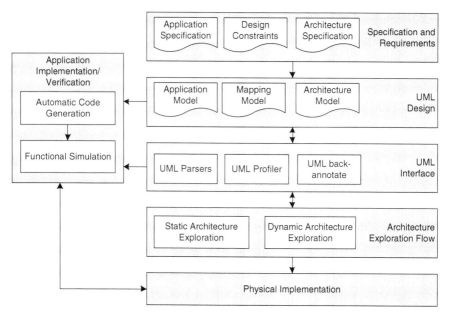

Fig. 19 Koski MPSoC design toolset

intensive applications, such as image processing operations [8]. Operations such as motion estimation or edge detection have extreme data capacity and bandwidth demands, necessitating detailed and convoluted design processes to derive complex, multi-level memory hierarchies to make such real-time realisations feasible. Whilst progress is being made in methodologies and automated solutions to this problem [9, 18, 27], automated creation of these multi-level hierarchies remains an open problem.

This memory problem illustrates only one aspect of the challenges facing FPGA DSP system designers, and vendors of tools to aid these designers—other open issues include synthesis of multi-FPGA architectures, automated resolution of off-chip communications, early estimation of implementation cost/performance, and reducing the run-times of low level mapping and place and route tasks for generating FPGA programming. Whilst these are taxing questions, the start made by the latest generation of FPGA design tools and technologies are highly encouraging. As the use of FPGA becomes more widespread to fill the widening gap created by the commercial justification required for the NRE costs of fabricating custom chips, and the power consumption and performance issues of multicore technology, the FPGA industry must, and will, establish system design approaches and supporting tools for its own unique needs.

6 Summary

FPGAs offer enormous potential for implementation of custom DSP computing architectures for ad-hoc, low volume DSP systems. The suitability of their programmable logic to host bit-parallel core architectures, and the advent of DSP-datapath units, along with hardcore and softcore processors has given designers to ability to harness a vast array of processing technologies with which to implement their systems. The options range from dedicated hardware architectures, to massively parallel software programmable architectures through GPP architectures with application specific co-processors, and everything inbetween.

Whilst this presents an enormous opportunity, it places significant stress on the FPGA system design process. Since the entire motivation for using FPGA is it's ability to host custom computer architectures tailored to the application, the complex heterogeneous system synthesis process required to generate an implementation of a DSP application is encountered frequently. As such, FPGA DSP system design presents an architecture synthesis challenge unencountered elsewhere in the embedded computing world. A new generation of design tools is starting to tackle and automate solutions to this system design problem, but many advances are yet to be made before it is solved.

References

1. Agility Design Solutions Inc.: Handel-C Language Reference Manual (2007). URL www.agilityds.com
2. Altera Corp.: Avalon Interface Specifications (2008). URL www.altera.com
3. Altera Corp.: Nios II C2H Compiler User Guide (2008). URL www.altera.com
4. Altera Corp.: DSP Design Flow User Guide (2009). URL www.altera.com
5. Altera Corp.: Stratix V Device Handbook (2012). URL www.altera.com
6. Banerjee, P., Haldar, M., Nayak, A., Kim, V., Saxena, V., Parkes, S., Bagchi, D., Pal, S., Tripathi, N., Zaretsky, D., Anderson, R., Uribe, J.: Overview of a compiler for synthesizing MATLAB programs onto FPGAs. IEEE Trans. VLSI Syst. **12**(3), 312–324 (2004)
7. Cho, J., Chang, H., Sung, W.: An FPGA-based simd processor with a vector memory unit. In: Proc. IEEE International Symposium on Circuits and Systems, pp. 525–528 (2006)
8. Cong, J., Liu, B., Neuendorffer, S., Noguera, J., Vissers, K., Zhang, Z.: High-level synthesis for FPGAs: From prototyping to deployment. IEEE Trans, on Computer-Aided Design of Integrated Circuits and Systems **30**(4), 473–491 (2011)
9. Fischaber, S., Woods, R., McAllister, J.: SoC memory hierarchy derivation from dataflow graphs. Journal of Signal Processing Systems (2009)
10. Gajski, D., Vahid, F., Narayan, S., Gong, J.: Specification and Design of Embedded Systems. Prentice Hall (1994)
11. Gajski, D.D., Abdi, S., Gerstlauer, A., Schirner, G.: Embedded System Design: Modeling, Synthesis and Verification. Springer, New York (2009)
12. Harriss, T., Walke, R., Kienhuis, B., Deprettere, E.F.: Compilation from matlab to process networks realised in FPGA. Design Automation for Embedded Systems **7**(4), 85–403 (2002)
13. Hoare, C.: Communicating Sequential Processes. Prentice Hall (1985)
14. Inc., S.: PICO FPGA Datasheet (2009). URL www.synfora.com

15. Jones, A., Hoare, R., Kusic, D., Fazekas, J., Foster, J.: An FPGA-based vliw processor with custom hardware execution. In: Proc. 13th International Symposium on Field-Programmable Gate Arrays, pp. 107–117 (2005)

16. Kahn, G.: The semantics of a simple language for parallel programming. In: Proc. IFIP Congress, pp. 71–475 (1974)

17. Kangas, T., Kukkala, P., Orsila, H., Salminen, E., Hannikainen, B., Hämäläinen, T., Riihmäki, J., Kuusilinna, K.: UML-based multiprocessor SoC design framework. ACM Transactions on Embedded Computing Systems **5**, 281–320 (2006)

18. Keinert, J., Teich, J.: Design of Image Processing Embedded Systems Using Multidimensional Data Flow. Springer, New York (2011)

19. Keutzer, K., Malik, S., Newton, R., Rabaey, J., Sangiovanni-Vincentelli, A.: System level design: Orthogonolization of concerns and platform-based design. IEEE Transactions on Computer-Aided Design of Integrated Circuits and Systems **19**(12), 1523–1543 (2000). URL http://www.gigascale.org/pubs/98.html

20. Kulmala, A., Salminen, E., Hämäläinen, T.: Instruction memory architecture evaluation on multiprocessor FPGA MPEG-4 encoder. In: IEEE Design and Diagnostics of Electronic Circuits and Systems, pp. 1–6 (2007)

21. Ma, L., Dickson, K., McAllister, J., McCanny, J.: QR decomposition-based matrix inversion for high performance embedded MIMO receivers. IEEE Transactions on Signal Processing **59**(4), 1858–1867 (2011)

22. Moscu Panainte, E., Bertels, K., Vassiliadis, S.: The MOLEN compiler for reconfigurable processors. ACM Transactions on Embedded Computing Systems **6**(1) (2007)

23. Nikolov, H.: System level design methodology for streaming multi-processor embedded systems. Ph.D. thesis, Leiden University, Netherlands (2009)

24. Nikolov, H., Stefanov, T., Deprettere, E.: Multi-processor system design with ESPAM. In: Proc. 4th International Conference Hardware/Software Codesign and System Synthesis, pp. 211–216 (2006)

25. Parhi, K.: VLSI Digital Signal Processing Systems : Design and Implementation. Wiley (1999)

26. Pimentel, A., Erbas, C., Polstra, S.: A systematic approach to exploring embedded system architectures at multiple abstraction levels. IEEE Transaction on Computers **55**(2), 1–14 (2006)

27. Qiang, L., Constantinides, G., Masselos, K., Cheung, P.: Combining data reuse with data-level parallelization for FPGA-targeted hardware compilation: A geometric programming framework. IEEE Trans. Computer Aided Design of Integrated Circuits and Systems **28**, 305–315 (2009)

28. Ravindran, K., Satish, N.R., Jin, Y., Keutzer, K.: An FPGA-based soft multiprocessor system for IPv4 packet forwarding. In: International Conference on Field Programmable Logic and Applications, pp. 487–492 (2005)

29. Salminen, E., Kangas, T., Hämäläinen, T., Riihimäki, J., Lahtinen, V., Kuusilinna, K.: HIBI communication network for system-on-chip. Journal of VLSI Signal Processing Systems **43**(2–3), 185–205 (2006)

30. Santo, B.: 25 microchips that shook the world. IEEE Spectrum **46**(5), 34–43 (2009)

31. Sheldon, D., Kumar, R., Lysecky, R., Vahid, F., Tullsen, D.: Application-specific customization of parameterized FPGA soft-core processors. In: Proc. IEEE/ACM International Conference on Computer-Aided Design, pp. 261–268 (2006)

32. Sheldon, D., Kumar, R., Vahid, F., Tullsen, D., Lysecky, R.: Conjoining soft-core FPGA processors. In: Proc. IEEE/ACM International Conference on Computer-Aided Design, pp. 694–701 (2006)

33. Stefanov, T., Kienhuis, B., Deprettere, E.: Algorithmic transformation techniques for efficient exploration of alternative application instances. In: Proc. 10th Int. Symp. on Hardware/Software Codesign, pp. 7–12 (2002)

34. Stefanov, T., Zissulescu, C., Turjan, A., Kienhuis, B., Deprettere, E.: System design using kahn process networks: The Compaan/Laura approach. In: Proc. Design Automation and Test in Europe (DATE) Conference, vol. 1, pp. 340–345 (2004)

35. Texas Instruments: TMS320C64x/C64x+ DSP CPU and Instruction Set Reference Guide (2008). URL http://www.ti.com
36. Thompson, M., Stefanov, T., Nikolov, H., Pimentel, A., Erbas, C., Polstra, E., Deprettere, E.: A framework for rapid system-level exploration, synthesis and programming of multimedia MP-SoCs. In: Proc. ACM/IEEE/IFIP Int. Conference on Hardware-Software Codesign and System Synthesis, pp. 9–14 (2007)
37. Vassiliadis, S., Wong, S., Gaydadjiev, G., Bertels, K., Kuzmanov, G., Moscu Panainte, E.: The MOLEN polymorphic processor. IEEE Transactions on Computers **53**(11), 1363–1375 (2004)
38. Verdoolaege, S., Nikolov, H., Stefanov, T.: pn: A tool for improved derivation of process networks. EURASIP Journal on Embedded Systems **2007**(1) (2007)
39. Wolf, W.: High Performance Embedded Computing - Architectures, Applications, and Methodologies. Morgan Kaufmann (2007)
40. Woods, R., McAllister, J., Yi, Y., Lightbody, G.: FPGA-based Implementation of Signal Processing Systems. Wiley (2008)
41. Xilinx Inc.: MicroBlaze Processor Reference Guide (2008). URL www.xilinx.com
42. Xilinx Inc.: Embedded Processor Block in Virtex-5 FPGAs (2009). URL www.xilinx.com
43. Xilinx Inc.: System Generator for DSP User Guide (2009). URL www.xilinx.com
44. Xilinx Inc.: Virtex-5 FPGA XtremeDSP Design Considerations (2009). URL www.xilinx.com
45. Xilinx Inc.: 7 Series DSP48E1 Slice User Guide (2012). URL www.xilinx.com
46. Xilinx Inc.: 7 Series FPGAs Memory Resources User Guide (2012). URL www.xilinx.com
47. Xilinx Inc.: 7 Series FPGAs Overview (2012). URL www.xilinx.com
48. Yi, Y., Woods, R.: Hierarchical synthesis of complex DSP functions using IRIS. IEEE Trans. Computer Aided Design of Integrated Circuits and Systems **25**(5), 806–820 (2006)
49. Yu, J., Eagleston, C., Chou, C., Perreault, M., Lemieux, G.: Vector processing as a soft processor accelerator. ACM Trans. Reconfigurable Technology and Systems **2**(2), 12:1–12:34 (2000)
50. Zhuo, L., Prasanna, V.: Scalable and modular algorithms for floating-point matrix multiplication on reconfigurable computing systems. IEEE Trans. Parallel and Distributed Systems **18**(4), 433–448 (2007)
51. Zhuo, L., Prasanna, V.: High-performance designs for linear algebra operations on reconfigurable hardware. IEEE Trans. Computers **57**(8), 1057–1071 (2008)

Application-Specific Accelerators
for Communications

Yang Sun, Kiarash Amiri, Michael Brogioli, and Joseph R. Cavallaro

Abstract For computation-intensive digital signal processing algorithms, complexity is exceeding the processing capabilities of general-purpose digital signal processors (DSPs). In some of these applications, DSP hardware accelerators have been widely used to off-load a variety of algorithms from the main DSP host, including FFT, FIR/IIR filters, multiple-input multiple-output (MIMO) detectors, and error correction codes (Viterbi, Turbo, LDPC) decoders. Given power and cost considerations, simply implementing these computationally complex parallel algorithms with high-speed general-purpose DSP processor is not very efficient. However, not all DSP algorithms are appropriate for off-loading to a hardware accelerator. First, these algorithms should have data-parallel computations and repeated operations that are amenable to hardware implementation. Second, these algorithms should have a deterministic dataflow graph that maps to parallel datapaths. The accelerators that we consider are mostly coarse grain to better deal with streaming data transfer for achieving both high performance and low power. In this chapter, we focus on some of the basic and advanced digital signal processing algorithms for communications and cover major examples of DSP accelerators for communications.

1 Introduction

In fourth-generation (4G) wireless systems, the signal processing algorithm complexity has far exceeded the processing capabilities of general-purpose digital signal processors (DSPs). With the inclusion of multiple-input multiple-output (MIMO) technology and advanced forward error correction coding technology in the 4G

Y. Sun (✉) • K. Amiri • M. Brogioli • J.R. Cavallaro
Rice University, 6100 Main St., Houston, TX 77005, USA
e-mail: ysunrice@gmail.com; kiaa@alumni.rice.edu; brogioli@alumni.rice.edu;
cavallar@rice.edu

S.S. Bhattacharyya et al. (eds.), *Handbook of Signal Processing Systems*,
DOI 10.1007/978-1-4614-6859-2_23, © Springer Science+Business Media, LLC 2013

wireless system, it is very important to develop area and power efficient 4G wireless receivers. Given area and power constraints for the mobile handsets one can not simply implement computation intensive DSP algorithms with gigahertz DSPs. Besides, it is also critical to reduce base station power consumption by utilizing optimized hardware accelerator design.

In this second edition, we will describe a few DSP algorithms which dominate the main computational complexity in a wireless receiver. These algorithms, including Viterbi decoding, Turbo decoding, LDPC decoding, MIMO detection, and channel equalization/FFT, need to be off-loaded to hardware coprocessors or accelerators, yielding high performance. These hardware accelerators are often integrated in the same die with DSP processors. In addition, it is also possible to leverage the field-programmable gate array (FPGA) to provide reconfigurable massive computation capabilities, as described in other chapter of this handbook [40].

DSP workloads are typically numerically intensive with large amounts of both instruction and data level parallelism. In order to exploit this parallelism with a programmable processor, most DSP architectures utilize Very Long Instruction Word, or VLIW architectures. VLIW architectures typically include one or more register files on the processor die, versus a single monolithic register file as is often the case in general-purpose computing. Examples of such architectures are the Freescale StarCore processor, the Texas Instruments TMS320C6x series DSPs as well as SHARC DSPs from Analog Devices, to name a few [3, 22, 63]. A comprehensive overview of the general-purpose DSP processors is given in Chapter of this handbook [58].

In some cases due to the idiosyncratic nature of many DSPs, and the implementation of some of the more powerful instructions in the DSP core, an optimizing compiler can not always target core functionality in an optimal manner. Examples of this include high performance fractional arithmetic instructions, for example, which may perform highly SIMD functionality which the compiler can not always deem safe at compile time.

While the aforementioned VLIW based DSP architectures provide increased parallelism and higher numerical throughput performance, this comes at a cost of ease in programmability. Typically such machines are dependent on advanced optimizing compilers that are capable of aggressively analyzing the instruction and data level parallelism in the target workloads, and mapping it onto the parallel hardware. Due to the large number of parallel functional units and deep pipeline depths, modern DSP are often difficult to hand program at the assembly level while achieving optimal results. As such, one technique used by the optimizing compiler is to vectorize much of the data level parallelism often found in DSP workloads. In doing this, the compiler can often fully exploit the single instruction multiple data, or SIMD functionality found in modern DSP instruction sets.

Despite such highly parallel programmable processor cores and advanced compiler technology, however, it is quite often the case that the amount of available instruction and data level parallelism in modern signal processing workloads far exceeds the limited resources available in a VLIW based programmable processor core. For example, the implementation complexity for a 40 Kbps DS-CDMA system would be 41.8 Gflops/s for 60 users [68], not to mention 100 Mbps+ 3GPP LTE system. This complexity largely exceeds the capability of nowadays DSP processors which typically can provide 1–5 Gflops performance, such as 1.5 Gflops TI C6711 DSP processor and 1.8 Gflops ADI TigerSHARC Processor. In other cases, the functionality required by the workload is not efficiently supported by more general-purpose instruction sets typically found in embedded systems. As such the need for acceleration at both the fine grain and coarse grain levels is often required, the former for instruction set architecture (ISA) like optimization, and the latter for task like optimization [17].

Additionally, wireless system designers often desire the programmability offered by software running on a DSP core versus a hardware based accelerator, to allow flexibility in various proprietary algorithms. Examples of this can be functionality such as channel estimation in baseband processing, for which a given vendor may want to use their own algorithm to handle various users in varying system conditions versus a pre-packaged solution. Typically these demands result in a heterogeneous system which may include one or more of the following: software programmable DSP cores for data processing, hardware based accelerator engines for data processing, and in some instances general-purpose processors or micro-controller type solutions for control processing.

The motivations for heterogeneous DSP system solutions including hardware acceleration stem from the tradeoffs between software programmability versus the performance gains of custom hardware acceleration in its various forms. There are a number of heterogenous accelerator based architectures currently available today, as well as various offerings and design solutions being offered by the research community.

There are a number of DSP architectures which include true hardware based accelerators which are not programmable by the end user. Examples of this include the Texas Instruments' C55x and C64x series of DSPs which include hardware based Viterbi or Turbo decoder accelerators for acceleration of wireless channel decoding [64, 65].

1.1 Coarse Grain Versus Fine Grain Accelerator Architectures

Coarse-grain accelerator based DSP systems entail a co-processor type design whereby larger amounts of work are run on the sometimes configurable co-processor device. Current technologies being offered in this area support offloading of functionality such as FFT and various matrix-like computations to the accelerator versus executing in software on the programmable DSP core.

As shown in Fig. 1, coarse grained heterogeneous architectures typically include a loosely coupled computational grid attached to the host processor. These types of architectures are sometimes built using an FPGA, ASIC, or vendor programmable acceleration engine for portions of the system. Tightly coupled loop nests or kernels are then offloaded from executing in software on the host processor to executing in hardware on the loosely coupled grid.

Fig. 1 Traditional coarse grained accelerator architecture [9]

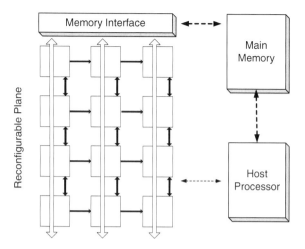

Fine-grain accelerator based architectures are the flip-side to the coarse grained accelerator mindset. Typically, ISAs provide primitives that allow low cost, low complexity implementations while still maintaining high performance for a broad range of input applications. In certain cases, however, it is often advantageous to offer instructions specialized to the computational needs of the application. Adding new instructions to the ISA, however, is a difficult decision to make. On the one hand they may provide significant performance increases for certain subsets of applications, but they must still be general enough such that they are useful across a much wider range of applications. Additionally, such instructions may become obsolete as software evolves and may complicate future hardware implementations of the ISA [76]. Vendors such as Tensilica, however, offer toolsets to produce configurable extensible processor architectures typically targeted at the embedded community [62]. These types of products typically allow the user to configure a predefined subset of processor components to fit the specific demands of the input application. Figure 2 shows the layout of a typical fine grained reconfigurable architecture whereby a custom ALU is coupled with the host processors pipeline.

In summary, both fine grained acceleration and coarse grained acceleration can be beneficial to the computational demands of DSP applications. Depending on the overall design constraints of the system, designers may chose a heterogeneous coarse grained acceleration system or a strictly software programmable DSP core system.

1.2 Hardware/Software Workload Partition Criteria

In partitioning any workload across a heterogeneous system comprised of reconfigurable computational accelerators, programmable DSPs or programmable host processors, and varied memory hierarchy, a number of criteria must be evaluated in

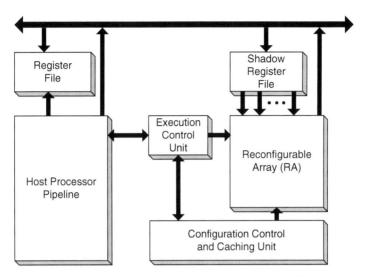

Fig. 2 Example fine grained reconfigurable architecture with customizable ALU for ISA extensions [9]

addition to application profile information to determine whether a given task should execute in software on the host processor or in hardware on FPGA or ASIC, as well where in the overall system topology each task should be mapped. It is these sets of criteria that typically mandate the software partitioning, and ultimately determine the topology and partitioning of the given system.

Spatial locality of data is one concern in partitioning a given task. In a typical software implementation running on a host processor, the ability to access data in a particular order efficiently is of great importance to performance. Issues such as latency to memory, data bus contention, data transfer times to local compute element such as accelerator local memory, as well as type and location of memory all need to be taken into consideration. In cases where data is misaligned, or not contiguous or uniformly strided in memory, additional overhead may be needed to arrange data before block DMA transfers can take place or data can efficiently be computed on. In cases where data is not aligned properly in memory, significant performance degradations can be seen due to decreased memory bandwidth when performing unaligned memory accesses on some architectures. When data is not uniformly strided, it may be difficult to burst transfer even single dimensional strips of memory via DMA engines. Consequently, with non-uniformly strided data it may be necessary to perform data transfers into local accelerator memory for computation via programmed I/O on the part of the host DSP. Inefficiencies in such methods of data transfer can easily overshadow any computational benefits achieved by compute acceleration of the FPGA. The finer the granularity of computation offloaded for acceleration in terms of compute time, quite often the more pronounced the side effects of data memory transfer to local accelerator memory.

Data level parallelism is another important criteria in determining the partitioning for a given application. Many applications targeted at VLIW-like architectures, especially signal processing applications, exhibit a large amount of both instruction and data level parallelism [32]. Many signal processing applications often contain enough data level parallelism to exceed the available functional units of a given architecture. FPGA fabrics and highly parallel ASIC implementations can exploit these computational bottlenecks in the input application by providing not only large numbers of functional units but also large amounts of local block data RAM to support very high levels of instruction and data parallelism, far beyond that of what a typical VLIW signal processing architecture can afford in terms of register file real estate. Furthermore, depending on the instruction set architecture of the host processor or DSP, performing sub-word or multiword operations may not be feasible given the host machine architecture. Most modern DSP architectures have fairly robust instruction sets that support fine grained multiword SIMD acceleration to a certain extent. It is often challenging, however, to efficiently load data from memory into the register files of a programmable SIMD style processor to be able to efficiently or optimally utilize the SIMD ISA in some cases.

Computational complexity of the application often bounds the programmable DSP core, creating a compute bottleneck in the system. Algorithms that are implemented in FPGA are often computationally intensive, exploiting greater amounts of instruction and data level parallelism than the host processor can afford, given the functional unit limitations and pipeline depth. By mapping computationally intense bottlenecks in the application from software implementation executing on host processor to hardware implementation in FPGA, one can effectively alleviate bottlenecks on the host processor and permit extra cycles for additional computation or algorithms to execute in parallel.

Task level parallelism in a portion of the application can play a role in the ideal partitioning as well. Quite often, embedded applications contain multiple tasks that can execute concurrently, but have a limited amount of instruction or data level parallelism within each unique task [69]. Applications in the networking space, and baseband processing at layers above the data plane typically need to deal with processing packets and traversing packet headers, data descriptors and multiple task queues. If the given task contains enough instruction and data level parallelism to exhaust the available host processor compute resources, it can be considered for partitioning to an accelerator. In many cases, it is possible to concurrently execute multiple of these tasks in parallel, either across multiple host processors or across both host processor and FPGA compute engine depending on data access patterns and cross task data dependencies. There are a number of architectures which have accelerated tasks in the control plane, versus data plane, in hardware. One example of this is the Freescale Semiconductor QorIQ platform which provides hardware acceleration for frame managers, queue managers, and buffer managers. In doing this, the architecture effectively frees the programmable processor cores from dealing with control plane management.

2   Hardware Accelerators for Communications

Processors in 3G and 4G cellular systems typically require high speed, throughput, and flexibility. In addition to this, computationally intensive algorithms are used to remove often high levels of multiuser interference especially in the presence of multiple transmit and receive antenna MIMO systems. Time varying wireless channel environments can also dramatically deteriorate the performance of the transmission, further requiring powerful channel equalization, detection, and decoding algorithms for different fading conditions at the mobile handset. In these types of environments, it is often the case that the amount of available parallel computation in a given application or kernel far exceeds the available functional units in the target processor. Even with modern VLIW style DSPs, the number of available functional units in a given clock cycle is limited and prevents full parallelization of the application for maximum performance. Further, the area and power constraints of mobile handsets make a software-only solution difficult to realize.

Figure 3 depicts a typical MIMO receiver model. Three major blocks, MIMO channel estimator & equalizer, MIMO detector, and channel decoder, determine the computation requirements of a MIMO receiver. Thus, it is natural to offload these very computational intensive tasks to hardware accelerators to support high data rate applications. Example of these applications include 3GPP LTE with 326 Mbps downlink peak data rate and IEEE 802.16e WiMax with 144 Mbps downlink peak data rate. Further, future standards such as LTE-Advance and WiMax-m are targeting over 1 Gbps speeds.

Data throughput is an important metric to consider when implementing a wireless receive. Table 1 summaries the data throughput performance for different MIMO wireless technologies as of 2009. Given the current DSP processing capabilities, it is very necessary to develop application-specific hardware accelerators for the complex MIMO algorithms.

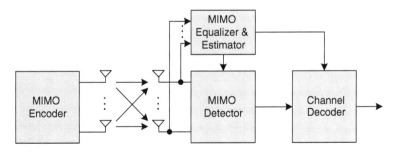

Fig. 3 Basic structure of an MIMO receiver

Table 1 Throughput performance of different MIMO systems

	HSDPA+ (2 × 2 MIMO)	LTE (2 × 2 MIMO)	LTE (4 × 4 MIMO)	WiMax Rel 1.5 (2 × 2 MIMO)	WiMax Rel 1.5 (4 × 4 MIMO)
Downlink	42 Mbps	173 Mbps	326 Mbps	144 Mbps	289 Mbps
Uplink	11.5 Mbps	58 Mbps	86 Mbps	69 Mbps	69 Mbps

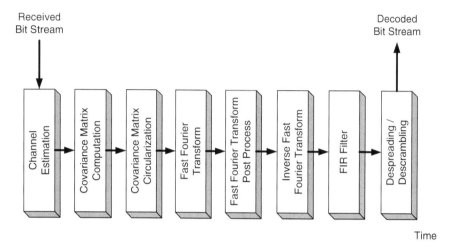

Fig. 4 Workload partition for a channel equalizer

2.1 MIMO Channel Equalization Accelerator

The total workload for a given channel equalizer performed as a baseband processing part on the mobile receiver can be decomposed into multiple tasks as
depicted in Fig. 4. This block diagram shows the various software processing
blocks, or kernels, that make up the channel equalizer firmware executing on the
digital signal processor of the mobile receiver. The tasks are: channel estimation
based on known pilot sequence, covariance computation (first row or column)
and circularization, FFT/IFFT (Fast Fourier transform and Inverse Fast Fourier
transform) post-processing for updating equalization coefficients, finite-impulse
response (FIR) filtering applied on the received samples (received frame), and user
detection (despreading–descrambling) for recovering the user information bits. The
computed data is shared between the various tasks in a pipeline fashion, in that the
output of covariance computation is used as the input to the matrix circularization
algorithm.

The computational complexity of various components of the workload vary with
the number of users in the system, as well as users entering and leaving the cell as
well as channel conditions. Regardless of this variance in the system conditions
at runtime, the dominant portions of the workload are the channel estimation,

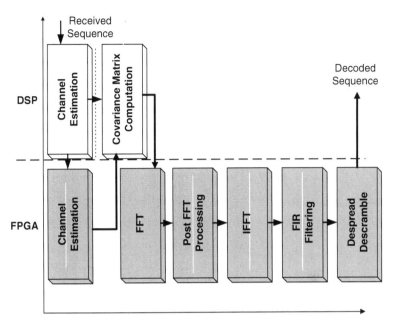

Fig. 5 Channel equalizer DSP/hardware accelerator partitioning

fast Fourier transform, inverse fast Fourier transform and FIR filtering as well as despreading and descrambling.

As an example, using the workload partition criteria for partitioning functionality between a programmable DSP core and system containing multiple hardware for a 3.5G HSDPA system, it has been shown that impressive performance results can be obtained. In studying the bottlenecks of such systems when implemented on a programmable DSP core in software, it has been found the key bottlenecks in the system to be the channel estimation, fast Fourier transform (FFT), inverse fast Fourier transform (IFFT), FIR filter, and to a lesser extent despreading and descrambling as illustrated in Fig. 4 [10]. By migrating the 3.5G implementation from a solely software based implementation executing on a TMS320C64x based programmable DSP core to a heterogeneous system containing not only programmable DSP cores but also distinct hardware acceleration for the various bottlenecks, the authors achieve almost an 11.2x speedup in the system [10]. Figure 5 illustrates the system partitioning between programmable DSP core and hardware (e.g. FPGA or ASIC) accelerator that resulted in load balancing the aforementioned bottlenecks.

The arrows in the diagram illustrate the data flow between local programmable DSP core on-chip data caches and the local RAM arrays. In the case of channel estimation, the work is performed in parallel between the programmable DSP core and hardware acceleration. Various other portions of the workload are offloaded to hardware based accelerators while the programmable DSP core performs the lighter weight signal processing code and book keeping.

Fig. 6 MIMO transmitter
and receiver

Despite the ability to achieve over 11x speedup in performance, it is important to note that the experimental setup used in these studies was purposely pessimistic. The various FFT, IFFT, etc. compute blocks in these studies were offloaded to discrete FPGA/ASIC accelerators. As such, data had to be transferred, for example, from local IFFT RAM cells to FIR filter RAM cells. This is pessimistic in terms of data communication time. In most cases the number of gates required for a given accelerator implemented in FPGA/ASIC was low enough that multiple accelerators could be implemented within a single FPGA/ASIC drastically reducing chip-to-chip communication time.

2.2   MIMO Detection Accelerators

MIMO systems, Fig. 6, have been shown to be able to greatly increase the reliability and data rate for point-to-point wireless communication [61]. Multiple-antenna systems can be used to improve the reliability and diversity in the receiver by providing the receiver with multiple copies of the transmitted information. This diversity gain is obtained by employing different kinds of space-time block code (STBC) [1, 59, 60]. In such cases, for a system with M transmit antennas and N receive antennas and over a time span of T time symbols, the system can be modeled as

$$\mathbf{Y} = \mathbf{HX} + \mathbf{N}, \tag{1}$$

where $\mathbf{H}$ is the $N \times M$ channel matrix. Moreover, $\mathbf{X}$ is the $M \times T$ space-time code matrix where its x_{ij} element is chosen from a complex-valued constellation Ω of the order $w = |\Omega|$ and corresponds to the complex symbol transmitted from the i-th antenna at the j-th time. The $\mathbf{Y}$ matrix is the received $N \times T$ matrix where y_{ij} is the perturbed received element at the i-th receive antenna at the j-th time. Finally, $\mathbf{N}$ is the additive white Gaussian noise matrix on the receive antennas at different time slots.

MIMO systems could also be used to further expand the transmit data rate using other space-time coding techniques, particularly layered space-time (LST) codes [21]. One of the most prominent examples of such space-time codes is Vertical Bell Laboratories Layered Space-Time (V-BLAST) [26], otherwise known as spatial multiplexing (SM). In the spatial multiplexing scheme, independent symbols are transmitted from different antennas at different time slots; hence, supporting even higher data rates compared to space-time block codes of lower data rate [1, 59].

The spatial multiplexing MIMO system can be modeled similar to Eq. (1) with $T = 1$ since there is no coding across the time domain:

$$\mathbf{y} = \mathbf{Hx} + \mathbf{n}, \tag{2}$$

where $\mathbf{H}$ is the $N \times M$ channel matrix, $\mathbf{x}$ is the M-element column vector where its x_i-th element corresponds to the complex symbol transmitted from the i-th antenna, and $\mathbf{y}$ is the received N-th element column vector where y_i is the perturbed received element at the i-th receive antenna. The additive white Gaussian noise vector on the receive antennas is denoted by $\mathbf{n}$.

While spatial multiplexing can support very high data rates, the complexity of the maximum-likelihood detector in the receiver increases exponentially with the number of transmit antennas. Thus, unlike the case in Eq. (1), the maximum-likelihood detector for Eq. (2) requires a complex architecture and can be very costly. In order to address this challenge, a range of detectors and solutions have been studied and implemented. In this section, we discuss some of the main algorithmic and architectural features of such detectors for spatial multiplexing MIMO systems.

2.2.1 Maximum-Likelihood (ML) Detection

The Maximum Likelihood (ML) or optimal detection of MIMO signals is known to be an NP-complete problem. The maximum-likelihood (ML) detector for Eq. (2) is found by minimizing the

$$\left\lVert \mathbf{y} - \mathbf{Hx} \right\rVert_2^2 \tag{3}$$

norm over all the possible choices of $\mathbf{x} \in \Omega^M$. This brute-force search can be a very complicated task, and as already discussed, incurs an exponential complexity in the number of antennas, in fact for M transmit antennas and modulation order of $w = |\Omega|$, the number of possible $\mathbf{x}$ vectors is w^M. Thus, unless for small dimension problems, it would be infeasible to implement it within a reasonable area-time constraint [11, 25].

2.2.2 Sphere Detection

Sphere detection can be used to achieve ML (or close-to-ML) with reduced complexity [19,31] compared to ML. In fact, while the norm minimization of Eq. (3) is exponential complexity, it has been shown that using the sphere detection method, the ML solution can be obtained with much lower complexity [19,30,70].

In order to avoid the significant overhead of the ML detection, the distance norm can be simplified [16] as follows:

$$D(\mathbf{s}) = \lVert \mathbf{y} - \mathbf{Hs} \rVert^2$$

$$= \lVert \mathbf{Q}^H\mathbf{y} - \mathbf{Rs} \rVert^2 = \sum_{i=M}^{1} \left| y_i' - \sum_{j=i}^{M} R_{i,j}s_j \right|^2, \tag{4}$$

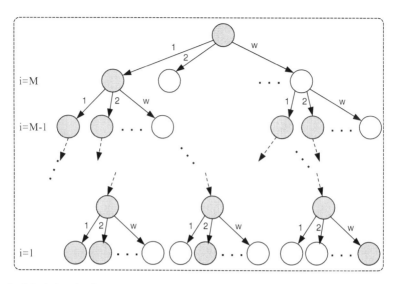

Fig. 7 Calculating the distances using a tree. Partial norms, *PNs*, of *dark nodes* are less than the threshold. *White nodes* are pruned out

where $\mathbf{H} = \mathbf{QR}$ represents the channel matrix QR decomposition, $\mathbf{R}$ is an upper triangular matrix, $\mathbf{QQ^H} = \mathbf{I}$ and $\mathbf{y}' = \mathbf{Q^H y}$.

The norm in Eq. (4) can be computed in M iterations starting with $i = M$. When $i = M$, i.e. the first iteration, the initial partial norm is set to zero, $T_{M+1}(\mathbf{s}^{(M+1)}) = 0$. Using the notation of [12], at each iteration the Partial Euclidean Distances (PEDs) at the next levels are given by

$$T_i(\mathbf{s}^{(i)}) = T_{i+1}(\mathbf{s}^{(i+1)}) + |e_i(\mathbf{s}^{(i)})|^2 \tag{5}$$

with $\mathbf{s}^{(i)} = [s_i, s_{i+1}, ..., s_M]^T$, and $i = M, M-1, ..., 1$, where

$$|e_i(\mathbf{s}^{(i)})|^2 = |y_i' - R_{i,i}s_i - \sum_{j=i+1}^{M} R_{i,j}s_j|^2. \tag{6}$$

One can envision this iterative algorithm as a tree traversal with each level of the tree corresponding to one i value or transmit antenna, and each node having w' children based on the modulation chosen.

The norm in Eq. (6) can be computed in M iterations starting with $i = M$, where M is the number of transmit antennas. At each iteration, partial (Euclidian) distances, $PD_i = |y_i' - \sum_{j=i}^{M} R_{i,j}s_j|^2$ corresponding to the i-th level, are calculated and added to the partial norm of the respective parent node in the $(i-1)$-th level, $PN_i = PN_{i-1} + PD_i$. When $i = M$, i.e. the first iteration, the initial partial norm is set to zero, $PN_{M+1} = 0$. Finishing the iterations gives the final value of the norm. As shown in Fig. 7, one can envision this iterative algorithm as a tree traversal problem where each level of the tree represents one i value, each node has its own PN, and w children, where w is the QAM modulation size. In order to reduce the search complexity, a threshold, C, can be set to discard the nodes with $PN > C$. Therefore, whenever a node k with a $PN_k > C$ is reached, any of its children will have $PN \geq PN_k > C$. Hence, not only the k-th node, but also its children, and all nodes lying beneath the children in the tree, can be pruned out.

There are different approaches to search the entire tree, mainly classified as depth-first search (DFS) approach and K-best approach, where the latter is based on breadth-first search (BFS) strategy. In DFS, the tree is traversed vertically [2, 12]; while in BFS [28, 72], the nodes are visited horizontally, i.e. level by level.

In the DFS approach, starting from the top level, one node is selected, the PNs of its children are calculated, and among those new computed PNs, one of them, e.g. the one with the least PN, is chosen, and that becomes the parent node for the next iteration. The PNs of its children are calculated, and the same procedure continues until a leaf is reached. At this point, the value of the global threshold is updated with the PN of the recently visited leaf. Then, the search continues with another node at a higher level, and the search controller traverses the tree down to another leaf. If a node is reached with a PN larger than the radius, i.e. the global threshold, then that node, along with all nodes lying beneath that, are pruned out, and the search continues with another node.

The tree traversal can be performed in a breadth-first manner. At each level, only the best K nodes, i.e. the K nodes with the smallest T_i, are chosen for expansion. This type of detector is generally known as the K-best detector. Note that such a detector requires sorting a list of size $K \times w'$ to find the best K candidates. For instance, for a 16-QAM system with $K = 10$, this requires sorting a list of size $K \times w' = 10 \times 4 = 40$ at most of the tree levels.

2.2.3 Computational Complexity of Sphere Detection

In this section, we derive and compare the complexity of the proposed techniques. The complexity in terms of number of arithmetic operations of a sphere detection operation is given by

$$J_{SD}(M,w) = \sum_{i=M}^{1} J_i E\{D_i\}, \tag{7}$$

where J_i is the number of operations per node in the i-th level. In order to compute J_i, we refer to the VLSI implementation of [12], and note that, for each node, one needs to compute the $R_{i,j}s_j$, multiplications, where, except for the diagonal element, $R_{i,i}$, the rest of the multiplications are complex valued. The expansion procedure, Eq. (4), requires computing $R_{i,j}s_j$ for $j = i+1, ..., M$, which would require $(M - i)$ complex multiplications, and also computing $R_{i,i}s_i$ for all the possible choices of $s_j \in \Omega$. Even though, there are w different s_js, there are only $(\frac{\sqrt{w}}{2} - 1)$ different multiplications required for QAM modulations. For instance, for a 16-QAM with $\{\pm 3 \pm 3j, \pm 1 \pm 1j, \pm 3 \pm 1j, \pm 1 \pm 3j\}$, computing only $(R_{i,j} \times 3)$ would be sufficient

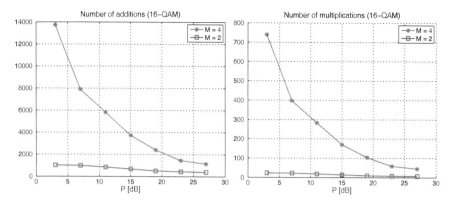

Fig. 8 Number of addition and multiplications operations for 16-QAM with different number of antennas, M

for all the choices of modulation points. Finally, computing the $\| \cdot \|^2$ requires a squarer or a multiplier, depending on the architecture and hardware availabilities.

In order to compute the number of adders for each norm expansion in (4), we note that there are $(M - i)$ complex valued adders required for $y_i' - \sum_{j=i+1}^{M} R_{i,j}s_j$, and w more complex adders to add the newly computed $R_{i,i}s_i$ values. Once the w different norms, $\left| y_i' - \sum_{j=i}^{M} R_{i,j}s_j \right|^2$, are computed, they need to be added to the partial distance coming from the higher level, which requires w more addition procedures. Finally, unless the search is happening at the end of the tree, the norms need to be sorted, which assuming a simple sorter, requires $w(w+1)/2$ compare-select operations.

Therefore, keeping in mind that each complex multiplier corresponds to four real-valued multipliers and two real-valued adders, and that every complex adder corresponds to two real-valued adders, J_i is calculated by

$$J_i(M,w) = J_{mult} + J_{add}(M,w)$$

$$J_{mult}(M,w) = ((\frac{\sqrt{w}}{2} - 1) + 4(M - i) + 1)$$

$$J_{add}(M,w) = (2(M - i) + 2w + w) + (w(w+1)/2) \cdot sign(i - 1),$$

where $sign(i - 1)$ is used to ensure sorting is counted only when the search has not reached the end of the tree, and is equal to:

$$sign(t) = \begin{cases} 1 & t \geq 1 \\ 0 & otherwise \end{cases}. \tag{8}$$

Moreover, we use θ, β and γ to represent the hardware-oriented costs for one adder, one compare-select unit, and one multiplication operation, respectively.

Figure 8 shows the number of addition and multiplication operations needed for a 16-QAM system with different number of antennas.

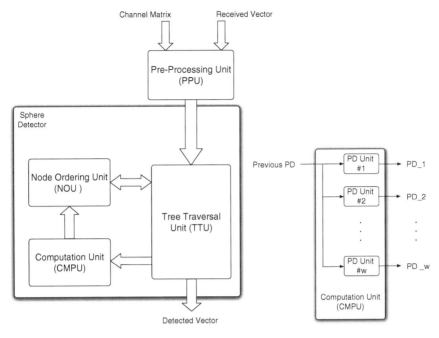

Fig. 9 Sphere Detector architecture with multiple PED function units

2.2.4 Depth-First Sphere Detector Architecture

The depth-first sphere detection algorithm [12, 19, 25, 31] traverses the tree in a depth-first manner: the detector visits the children of each node before visiting its siblings. A constraint, referred to as radius, is often set on the PED for each level of the tree. A generic depth-first sphere detector architecture is shown in Fig. 9. The Pre-Processing Unit (PPU) is used to compute the QR decomposition of the channel matrix as well as calculate $\mathbf{Q}^H\mathbf{y}$. The Tree Traversal Unit (TTU) is the controlling unit which decides in which direction and with which node to continue. Computation Unit (CMPU) computes the partial distances, based on (4), for w different s_j. Each PD unit computes $|y_i' - \sum_{j=i}^{M} R_{i,j} s_j|^2$ for each of the w children of a node. Finally, the Node Ordering Unit (NOU) is for finding the minimum and saving other legitimate candidates, i.e. those inside R_i, in the memory.

As an example to show the algorithm complexity, an FPGA implementation synthesis result for a 50 Mbps 4×4 16-QAM depth-first sphere detector is summarized in Table 2 [2].

2.2.5 K-Best Detector Architecture

K-best is another popular algorithm for implementing close-to-ML MIMO detection [15,28,72]. The performance of this scheme is suboptimal compared to ML and

Table 2 FPGA resource utilization for sphere detector

Device	Xilinx Virtex-4 xc4vfx 100-12ff1517
Number of Slices	4065/42176 (9%)
Number of FFs	3344/84352 (3%)
Number of Look-Up Tables	6457/84352 (7%)
Number of RAMB16	3/376 (1%)
Number of DSP48s	32/160 (20%)
Max. Freq.	125.7 MHz

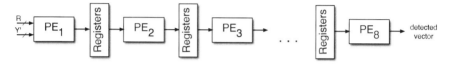

Fig. 10 The K-best MIMO detector architecture: the intermediate register banks contain the sorting information as well as the other values, i.e. **R** matrix

sphere detection. However, it has a fixed complexity and relatively straightforward architecture. In this section, we briefly introduce the architecture [28] to implement the K-best MIMO detector. As illustrated in Fig. 10, the PE elements at each stage compute the Euclidean norms of (6), and find the best K candidates, i.e. the K candidates with the smallest norms, and pass them as the surviving candidates to the next level. It should be pointed out that the Eq. (2) can be decomposed into separate real and imaginary parts [28], which would double the size of the matrices. While such decomposition reduces the complex-valued operations of nodes into real-valued operations, it doubles the number of levels of the tree. Therefore, as shown in Fig. 10, there are 8 K-best detection levels for the 4-antenna system. By selecting the proper K value, the real-value decomposition MIMO detection will not cause performance degradation compared to the complex-value MIMO detection [41].

In summary, both depth-first and K-best detectors have a regular and parallel data flow that can be efficiently mapped to hardware. The large amount of required multiplications makes the algorithm very difficult to be realized in a DSP processor. As the main task of the MIMO detector is to search for the best candidate in a very short time period, it would be more efficient to be mapped on a parallel hardware searcher with multiple processing elements. Thus, to sustain the high throughput MIMO detection, an MIMO hardware accelerator is necessary.

2.3 Channel Decoding Accelerators

Error correcting codes are widely used in digital transmission, especially in wireless communications, to combat the harsh wireless transmission medium. To achieve high throughput, researchers are investigating more and more advanced error correction codes. The most commonly used error correcting codes in modern systems are convolutional codes, Turbo codes, and low-density parity-check (LDPC) codes. As a core technology in wireless communications, FEC (forward error correction) coding has migrated from the basic 2G convolutional/block codes to more powerful 3G Turbo codes, and LDPC codes forecast for 4G systems.

As codes become more complicated, the implementation complexity, especially the decoder complexity, increases dramatically which largely exceeds the capability of the general-purpose DSP processor. Even the most capable DSPs today would need some types of acceleration coprocessor to offload the computation-intensive error correcting tasks. Moreover, it would be much more efficient to implement these decoding algorithms on dedicated hardware because typical error correction algorithms use special arithmetic and therefore are more suitable for ASICs or FPGAs. Bitwise operations, linear feedback shift registers, and complex look-up tables can be very efficiently realized with ASICs/FPGAs.

In this section, we will present some important error correction algorithms and their efficient hardware architectures. We will cover major error correction codes used in the current and next generation communication standards, such as 3GPP LTE, IEEE 802.11n Wireless LAN, IEEE 802.16e WiMax, etc.

2.3.1 Viterbi Decoder Accelerator Architecture

In telecommunications, convolutional codes are among the most popular error correction codes that are used to improve the performance of wireless links. For example, convolutional codes are used in the data channel of the second generation (2G) mobile phone system (e.g. GSM) and IEEE 802.11a/n wireless local area network (WLAN). Due to their good performance and efficient hardware architectures, convolutional codes continue to be used by the 3G/4G wireless systems for their control channels, such as 3GPP LTE and IEEE 802.16e WiMax.

A convolutional code is a type of error-correcting code in which each m-bit information symbol is transformed into an n-bit symbol, where m/n is called the code rate. The encoder is basically a finite state machine, where the state is defined as the contents of the memory of the encoder. Figure 11a, b show two examples of convolutional codes with constraint length $K = 3$, code rate $R = 1/3$ and constraint length $K = 7$, code rate $R = 1/2$, respectively.

The Viterbi algorithm is an optimal decoding algorithm for the decoding of convolutional codes [20, 67]. The Viterbi algorithm enumerates all the possible codewords and selects the most likely sequence. The most likely sequence is found by traversing a trellis. The trellis diagram for a $K = 3$ convolutional code (cf. Fig. 11a) is shown in Fig. 12.

In general, a Viterbi decoder contains four blocks: branch metric calculation (BMC) unit, add-compare-select (ACS) unit, survivor memory unit (SMU), and trace back (TB) unit as shown in Fig. 13. The decoder works as follows. BMC calculates all the possible branch metrics from the channel inputs. ACS unit recursively calculates the state metrics and the survivors are stored into a survivor

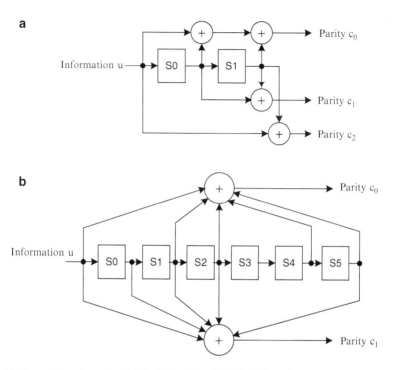

Fig. 11 Convolutional encoder: (**a**) K=3, R=1/3 and (**b**) K=7, R=1/2 encoder used for WLAN

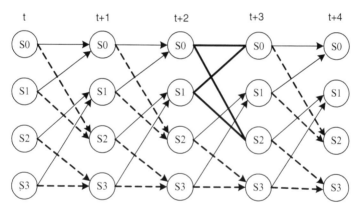

Fig. 12 A 4-state trellis diagram for the encoder in Fig. 11a. The *solid lines* indicate transitions for a "0" input and the *dashed lines* for a "1" input. Each stage of the trellis consists of 2^{K-2} Viterbi butterflies. One such butterfly is highlighted at step $t+2$

memory. The survivor paths contain state transitions to reconstruct a sequence of states by tracing back. This reconstructed sequence is then the most likely sequence sent by the transmitter. In order to reduce memory requirements and latency, Viterbi decoding can be sliced into blocks, which are often referred to as sliding windows.

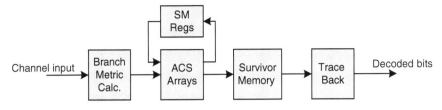

Fig. 13 Viterbi decoder architecture with parallel ACS function units

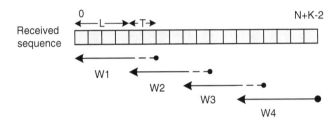

Fig. 14 Sliding window decoding with warmup. Decoding length $= L$. Warmup length $= T$

The sliding window principle is shown in Fig. 14. The ACS recursion is carried out for the entire code block. The trace back operation is performed on every sliding window. To improve the reliability of the trace back, the decisions for the last T (also referred to warmup window) steps will be discarded. Thus, after a fixed delay of $L + 2T$, the decoder begins to produce decoded bits on every clock cycle.

Like an FFT processor [5, 14, 42], a Viterbi decoder has a very regular data structure. Thus, it is very natural to implement a Viterbi decoder in hardware to support high speed applications, such as IEEE 802.11n wireless LAN with 300 Mbps peak data rate.

The most complex operations in the Viterbi algorithm is the ACS recursion. Figure 15a shows one ACS butterfly of the 4-state trellis described above (cf. Fig. 12). Each butterfly contains two ACS units. For each ACS unit, there are two branches leading from two states on the left, and going to a state on the right. A branch metric is computed for each branch of an ACS unit. Note that it is not necessary to calculate every branch metric for all 4 branches in an ACS butterfly, because some of them are identical depending on the trellis structure. Based on the old state metrics (SMs) and branch metrics (BMs), the new state metrics are updated as:

$$SM(0) = \min\left(SM(0) + BM(0,0), SM(1) + BM(1,0)\right) \tag{9}$$

$$SM(2) = \min\left(SM(0) + BM(0,2), SM(1) + BM(1,2)\right). \tag{10}$$

Given a constraint length of K convolutional code, 2^{K-2} ACS butterflies would be required for each step of the decoding. These butterflies can be implemented in serial

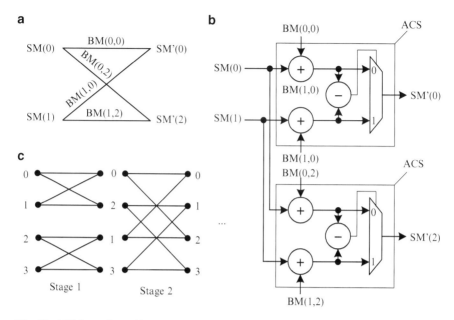

Fig. 15 ACS butterfly architecture: (**a**) Basic butterfly structure, (**b**) ACS butterfly hardware implementation, and (**c**) In-place butterfly structure for a 4-state trellis diagram

Fig. 16 Different radix trellis structure: (**a**) radix-2 trellis, (**b**) radix-4 trellis, and (**c**) radix-8 trellis

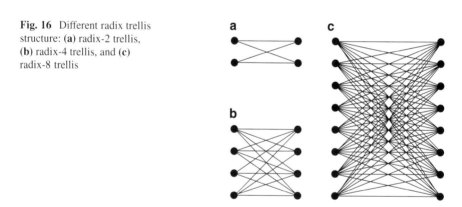

or parallel. To maximize the decoding throughput, a parallel implementation is often used. The basic parallel ACS architecture can process one bit of the message at each clock cycle. However, the processing speed can be possibly increased by N times by merging every N stages of the trellis into one high-radix stage with 2^N branches for every state. Figure 16 shows the radix-2, radix-4, and radix-8 trellis structures. Figure 17 shows a radix-8 ACS architecture which can process three message bits over three trellis path bits. Generally for a radix-N ACS architecture, it can process $\log_2 N$ message bits over $\log_2 N$ trellis path bits. For high speed applications, high radix ACS architectures are very common in a Viterbi decoder. The top level of a

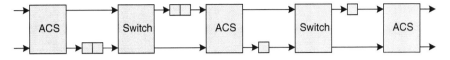

Fig. 17 Radix-8 ACS architecture

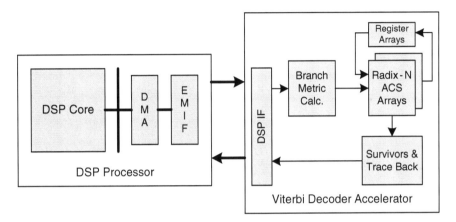

Fig. 18 A generic Viterbi decoder accelerator architecture. Data movement between DSP processor and accelerator is via DMA. Fully-parallel ACS function units are used to support high speed decoding

generic Viterbi decoder accelerator is shown in Fig. 18. Although a pure software approach is feasible for a modern DSP processor, it is much more cost effective to implement the Viterbi decoder with a hardware accelerator. The decoder can be memory mapped to the DSP external memory space so that the DMA transfer can be utilized without the intervention of the host DSP. Data is passed in and out is in a pipelined manner so that the decoding can be simultaneously performed with I/O operations.

2.3.2 Turbo Decoder Accelerator Architecture

Turbo codes are a class of high-performance capacity-approaching error-correcting codes [6]. As a break-through in coding theory, Turbo codes are widely used in many 3G/4G wireless standards such as CDMA2000, WCDMA/UMTS, 3GPP LTE, and IEEE 802.16e WiMax. However, the inherently large decoding latency and complex iterative decoding algorithm have made it rarely being implemented in a general-purpose DSP. For example, Texas Instruments' latest multi-core DSP processor TI C6474 employs a Turbo decoder accelerator to support 2 Mbps CDMA Turbo codes for the base station [65]. The decoding throughput requirement for 3GPP LTE Turbo codes is to be more than 80 Mbps in the uplink and 320 Mbps in the

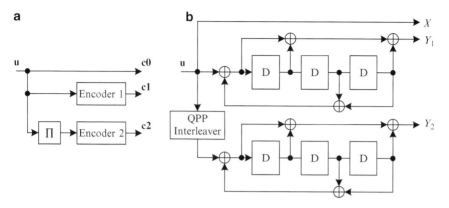

Fig. 19 Turbo encoder structure. (**a**) Basic structure. (**b**) Structure of Turbo encoder in 3GPP LTE

downlink. Because the Turbo codes used in many standards are very similar, e.g. the encoding polynomials are same for WCDMA/UMTS/LTE, the Turbo decoder is often accelerated by reconfigurable hardware.

A classic Turbo encoder structure is depicted in Fig. 19. The basic encoder consists of two systematic convolutional encoders and an interleaver. The information sequence **u** is encoded into three streams: systematic, parity 1, and parity 2. Here the interleaver is used to permute the information sequence into a second different sequence for encoder 2. The performance of a Turbo code depends critically on the interleaver structure [45].

The BCJR algorithm [4], also called forward–backward algorithm or Maximum *a posteriori* (MAP) algorithm, is the main component in the Turbo decoding process. The basic structure of Turbo decoding is functionally illustrated in Fig. 20. The decoding is based on the MAP algorithm. During the decoding process, each MAP decoder receives the channel data and *a priori* information from the other constituent MAP decoder through interleaving (π) or deinterleaving (π^{-1}), and produces extrinsic information at its output. The MAP algorithm is an optimal symbol decoding algorithm that minimizes the probability of a symbol error. It computes the *a posteriori* probabilities (APPs) of the information bits as follows:

$$\Lambda(\hat{u}_k) = \max_{\mathbf{u}:u_k=1}^{*} \left\{ \alpha_{k-1}(s_{k-1}) + \gamma_k(s_{k-1},s_k) + \beta_k(s_k)) \right\} \tag{11}$$

$$- \max_{\mathbf{u}:u_k=0}^{*} \left\{ \alpha_{k-1}(s_{k-1}) + \gamma_k(s_{k-1},s_k) + \beta_k(s_k)) \right\}, \tag{12}$$

where α_k and β_k denote the forward and backward state metrics, and are calculated as follows:

$$\alpha_k(s_k) = \max_{s_{k-1}}^{*} \left\{ \alpha_{k-1}(s_{k-1}) + \gamma_k(s_{k-1},s_k) \right\}, \tag{13}$$

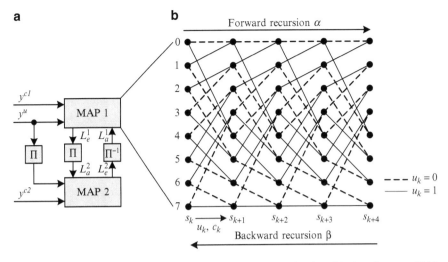

Fig. 20 Basic structure of an iterative Turbo decoder. (**a**) Iterative decoding based on two MAP decoders. (**b**) Forward/backward recursion on trellis diagram

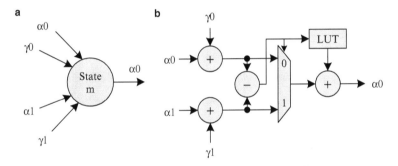

Fig. 21 ACSA structure. (**a**) Flow of state metric update. (**b**) Circuit implementation of an ACSA unit

$$\beta_k(s_k) = \max_{s_{k+1}}^*\{\beta_{k+1}(s_{k+1}) + \gamma_k(s_k, s_{k+1})\}. \qquad (14)$$

The γ_k term above is the branch transition probability that depends on the trellis diagram, and is usually referred to as a branch metric. The max star operator employed in the above descriptions is the core arithmetic computation that is required by the MAP decoding. It is defined as:

$$\overset{*}{\max}(a,b) = \log(e^a + e^b) = \max(a,b) + \log(1 + e^{-|a-b|}). \qquad (15)$$

A basic add-compare-select-add unit is shown in Fig. 21. This circuit can process one step of the trellis per cycle and is often referred to as Radix-2 ACSA unit.

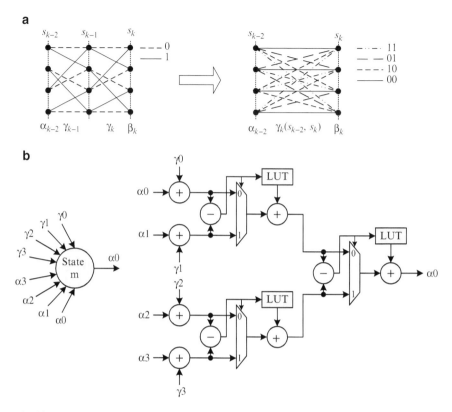

Fig. 22 (a) An example of radix-4 trellis. (b) Radix-4 ACSA circuit implementation

To increase the processing speed, the trellis can be transformed by merging every two stages into one radix-4 stage as shown in Fig. 22. Thus, the throughput can be doubled by applying this transform. For an N state Turbo codes, N such ACSA unit would be required in each step of the trellis processing. To maximize the decoding throughput, a parallel implementation is usually employed to compute all the N state metrics simultaneously.

In the original MAP algorithm, the entire set of forward metrics needs to be computed before the first soft log-likelihood ratio (LLR) output can be generated. This results in a large storage of K metrics for all N states, where K is the block length and N is the number of states in the trellis diagram. Similar to the Viterbi algorithm, a sliding window algorithm is often applied to the MAP algorithm to reduce the decoding latency and memory storage requirement. By selecting a proper length of the sliding window, e.g. 32 for a rate 1/3 code, there is nearly no bit error rate (BER) performance degradation. Figure 23a shows an example of the sliding window algorithm, where a dummy reverse metric calculation (RMC) is used to get the initial values for β metrics. The sliding window hardware architecture is shown in Fig. 23b. The decoding operation is based on three recursion units, two used for

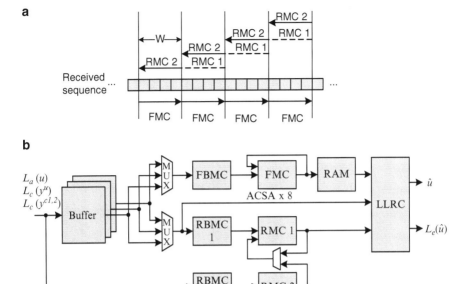

Fig. 23 Sliding window MAP decoder. (**a**) An example of sliding window MAP algorithm, where a dummy RMC is performed to achieve the initial β metrics. (**b**) MAP decoder hardware architecture

the reverse (or backward) recursions (dummy RMC 1 and effective RMC 2), and one for forward recursion (FMC). Each recursion unit contains parallel ACSA units. After a fixed latency, the decoder produces the soft LLR outputs on every clock cycle. To further increase the throughput, a parallel sliding window scheme [7, 13, 35, 36, 39, 43, 48, 53, 57, 66, 71, 75] is often applied as shown in Fig. 24.

Another key component of Turbo decoders is the interleaver. Generally, the interleaver is a device that takes its input bit sequence and produces an output sequence that is as uncorrelated as possible. Theoretically a random interleaver would have the best performance. But it is difficult to implement a random interleaver in hardware. Thus, researchers are investigating pseudo-random interleavers such as the row-column permutation interleaver for 3G Rel-99 Turbo coding as well as the new QPP interleaver [49] for 3G LTE Turbo coding. The main differences between these two types of pseudo-random interleavers is the capability to support parallel Turbo decoding. The drawback of the row-column permutation interleaver is that memory conflicts will occur when employing multiple MAP decoders for parallel decoding. Extra buffers are necessary to solve the memory conflicts caused by the row-column permutation interleaver [46]. To solve this problem, the new 3G LTE standard has adopted a new interleaver structure called QPP interleaver [49]. Given an information block length N, the x-th QPP interleaved output position is given by

$$\Pi(x) = (f_2 x^2 + f_1 x) \bmod N, 0 \leq x, f_1, f_2 < N. \tag{16}$$

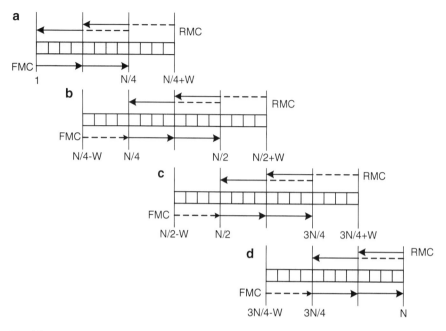

Fig. 24 An example of parallel sliding window decoding, where a decode block is sliced into 4 sections. The sub-blocks are overlapped by one sliding window length W in order to get the initial value for the boundary states

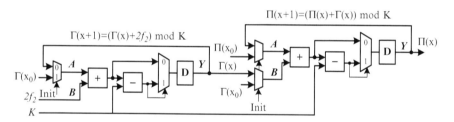

Fig. 25 A circuit implementation for the QPP interleaver $\pi(x) = (f_2 x^2 + f_1 x) \bmod K$ [57]

It has been shown in [49] that the QPP interleaver will not cause memory conflicts as long as the parallelism level is a factor of N. The simplest approach to implement an interleaver is to store all the interleaving patterns in non-violating memory such as ROM. However, this approach can become very expensive because it is necessary to store a large number of interleaving patterns to support decoding of multiple block size Turbo codes such as 3GPP LTE Turbo codes. Fortunately, there usually exists an efficient hardware implementation for the interleaver. For example, Fig. 25 shows a circuit implementation for the QPP interleaver in 3GPP LTE standard [57].

A basic Turbo accelerator architecture is shown in Fig. 26. The main difference between the Viterbi decoder and the Turbo decoder is that the Turbo decoder is

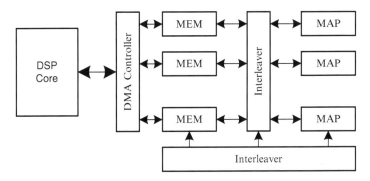

Fig. 26 Turbo decoder accelerator architecture. Multiple MAP decoders are used to support high throughput decoding of Turbo codes. Special function units such as interleavers are also implemented in hardware

based on the iterative message passing algorithms. Thus, a Turbo accelerator may need more communication and control coordination with the DSP host processor. For example, the interleaving addresses can be generated by the DSP processor and passed to the Turbo accelerator. The DSP can monitor the decoding process to decide when to terminate the decoding if there are no more decoding gains. Alternately, the Turbo accelerator can be configured to operate without DSP intervention. To support this feature, some special hardware such as interleavers have to be configurable via DSP control registers. To decrease the required bus bandwidth, intermediate results should not be passed back to the DSP processor. Only the successfully decoded bits need to be passed back to the DSP processor, e.g. via the DSP DMA controller. Further, to support multiple Turbo codes in different communication systems, a flexible MAP decoder is necessary. In fact, many standards employ similar Turbo code structures. For instance, CDMA, WCDMA, UMTS, and 3GPP LTE all use an eight-state binary Turbo code with polynomial (13, 15, 17). Although IEEE 802.16e WiMax and DVB-RCS standards use a different eight-state double binary Turbo code, the trellis structures of these Turbo codes are very similar as illustrated in Fig. 27. Thus, it is possible design multi-standard Turbo decoders based on flexible MAP decoder datapaths [38, 47, 57]. It has been shown in [57] that the area overhead to support multi-codes is only about 7%. In addition, when the throughput requirement is high, e.g. more than 20 Mbps, multiple MAP decoders can be activated to increase the throughput performance.

In summary, due to the iterative structures, a Turbo decoder needs more Gflops than what is available in a general-purpose DSP processor. For this reason, Texas Instruments' latest C64x DSP processor integrates a 2 Mbps 3G Turbo decoder accelerator in the same die [65]. Because of the parallel and recursive algorithms and special logarithmic arithmetics, it is more cost effective to realize a Turbo decoder in hardware.

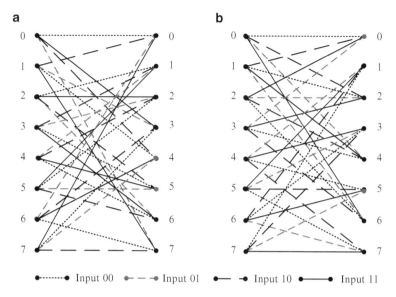

Fig. 27 Radix-4 trellis structures of (**a**) CDMA/WCDMA/UMTS/LTE Turbo codes and (**b**) WiMax/DVB-RCS Turbo codes

2.3.3 LDPC Decoder Accelerator Architecture

A low-density parity-check (LDPC) code [24] is another important error correcting code that is the among one of the most efficient coding schemes as of now. The remarkable error correction capabilities of LDPC codes have led to their recent adoption in many standards, such as IEEE 802.11n, IEEE 802.16e, and IEEE 802 10GBase-T. The huge computation and high throughput requirements make it very difficult to implement a high throughput LDPC decoder on a general-purpose DSP. For example, a 5.4 Mbps LDPC decoder was implemented on TMS320C64xx DSP running at 600 MHz [34]. This throughput performance is not enough to support high data rates defined in new wireless standards. Thus, it is important to develop area and power efficient hardware LDPC decoding accelerators.

A binary LDPC code is a linear block code specified by a very sparse binary $M \times N$ parity check matrix: $\mathbf{H} \cdot \mathbf{x}^T = 0$, where $\mathbf{x}$ is a codeword and $\mathbf{H}$ can be viewed as a bipartite graph where each column and row in $\mathbf{H}$ represent a variable node and a check node, respectively.

The decoding algorithm is based on the iterative message passing algorithm (also called belief propagation algorithm), which exchanges the messages between the variable nodes and check nodes on graph. The hardware implementation of LDPC decoders can be serial, semi-parallel, and fully-parallel as shown in Fig. 28. Fully-parallel implementation has the maximum processing elements to achieve very high throughput. Semi-parallel implementation, on the other hand, has a lesser number of processing elements that can be re-used, e.g. z number of processing elements

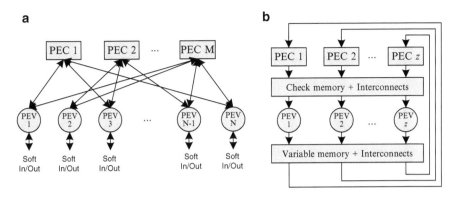

Fig. 28 Implementation of LDPC decoders, where PEC denotes processing element for check node and PEV denotes processing element for variable node: (**a**) fully-parallel and (**b**) semi-parallel

are employed in Fig. 28b. In a semi-parallel implementation, memories are usually required to store the temporary results. In many practical systems, semi-parallel implementations are often used to achieve 100 Mbps to 1 Gbps throughput with reasonable complexity [8, 27, 33, 44, 50–52, 54–56, 78].

In LDPC decoding, the main complexity comes from the check node processing. Each check node receives a set of variable node messages denoted as $\mathcal{N}_m$. Based on these data, check node messages are computed as

$$\Lambda_{mn} = \sum_{j \in \mathcal{N}_m \setminus n} \boxplus \lambda_{mj} = \left(\sum_{j \in \mathcal{N}_m} \boxplus \lambda_{mj} \right) \boxminus \lambda_{mn},$$

where Λ_{mn} and λ_{mn} denote the check node message and the variable node message, respectively. The special arithmetic operators $\boxplus$ and $\boxminus$ are defined as follows:

$$a \boxplus b \triangleq f(a,b) = \log \frac{1 + e^a e^b}{e^a + e^b}$$
$$= \text{sign}(a)\,\text{sign}(b) \left(\min(|a|, |b|) + \log(1 + e^{-(|a|+|b|)}) - \log(1 + e^{-\left||a|-|b|\right|}) \right),$$

$$a \boxminus b \triangleq g(a,b) = \log \frac{1 - e^a e^b}{e^a - e^b}$$
$$= \text{sign}(a)\,\text{sign}(b) \left(\min(|a|, |b|) + \log(1 - e^{-(|a|+|b|)}) - \log(1 - e^{-\left||a|-|b|\right|}) \right).$$

Figure 29 shows a hardware implementation from [51] to compute check node message Λ_{mn} for one check row m. Because multiple check rows can be processed simultaneously in the LDPC decoding algorithm, multiple such check node units can be used to increase decoding speed. As the number of ALU units in a general-purpose DSP processor is limited, it is difficult to achieve more than 10 Mbps throughput in a DSP implementation.

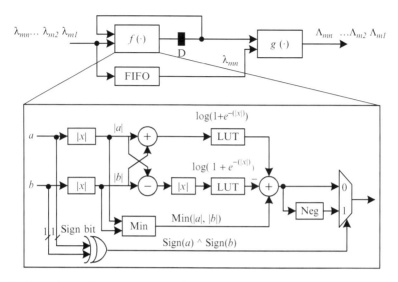

Fig. 29 Recursive architecture to compute check node messages [51]

Fig. 30 Structured LDPC parity check matrix with j block rows and k block columns. Each sub-matrix is a $z \times z$ identity shifted matrix

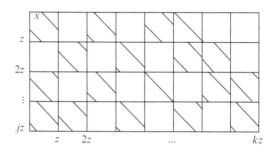

Table 3 Design parameters for **H** in standardized LDPC codes

	z	j	k	Check node degree	Variable node degree	Max. throughput
WLAN 802.11n	27–81	4–12	24	7–22	2–12	600 Mbps
WiMax 802.16e	24–96	4–12	24	6–20	2–6	144 Mbps

Given a random LDPC code, the main complexity comes not only from the complex check node processing, but also from the interconnection network between check nodes and variable nodes. To simplify the routing of the interconnection network, many practical standards usually employ structured LDPC codes, or quasi-cyclic LDPC (QC-LDPC) codes. The parity check matrix of a QC-LDPC code is shown in Fig. 30. Table 3 summaries the design parameters of the QC-LDPC codes for IEEE 802.11n WLAN and IEEE 802.16e WiMax wireless standards. As can be seen, many design parameters are in the same range for these two applications, thus it is possible to design a reconfigurable hardware to support multiple standards [51].

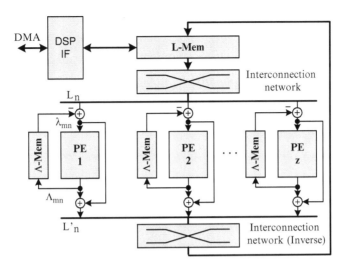

Fig. 31 Semi-parallel LDPC decoder accelerator architecture. Multiple PEs (number of z) are used to increase decoding speed. Variable messages are stored in L-memory and check messages are stored in Λ-memory. An interconnection network along with an inverse interconnection network are used to route data

As an example, a multi-standard semi-parallel LDPC decoder accelerator architecture is shown in Fig. 31 [51]. In order to support several hundreds Mbps data rate, multiple PEs are used to process multiple check rows simultaneously. As with Turbo decoding, LDPC decoding is also based on an iterative decoding algorithm. The iterative decoding flow is as follows: at each iteration, $1 \times z$ APP messages, denoted as L_n are fetched from the L-memory and passed through a permuter (e.g. barrel shifter) to be routed to z PEs (z is the parallelism level). The soft input information λ_{mn} is formed by subtracting the old extrinsic message Λ_{mn} from the APP message L_n. Then the PEs generate new extrinsic messages Λ_{mn} and APP messages L_n, and store them back to memory. The operation mode of the LDPC accelerator needs to be configured in the beginning of the decoding. After that, it should work without DSP intervention. Once it has finished decoding, the decoded bits are passed back to the DSP processor. Figure 32 shows the ASIC implementation result of this decoder (VLSI layout view) and its power consumption for different block sizes. As the block size increases, the number of active PEs increases, thus more power is consumed.

3 Summary

Digital signal processing complexity in high-speed wireless communications is driving a need for high performance heterogenous DSP systems with real-time processing. Many wireless algorithms, such as channel decoding and MIMO

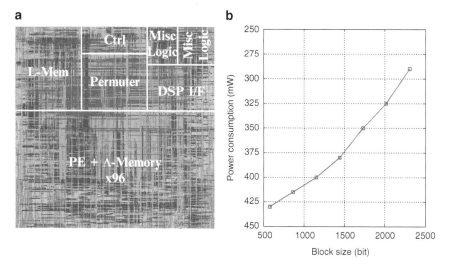

Fig. 32 An example of a LDPC decoder hardware accelerator [51]. (**a**) VLSI layout view (3.5 mm² area, 90 nm technology). (**b**) power consumptions for different block sizes

detection, demonstrate significant data parallelism. For this class of data-parallel algorithms, application specific DSP accelerators are necessary to meet real-time requirements while minimizing power consumption. Spatial locality of data, data level parallelism, computational complexity, and task level parallelism are four major criteria to identify which DSP algorithm should be off-loaded to an accelerator. Additional cost incurred from the data movement between DSP and hardware accelerator must be also considered.

There are a number of DSP architectures which include true hardware based accelerators. Examples of these include the Texas Instruments' C64x series of DSPs which include a 2 Mbps Turbo decoding accelerator [65], and Freescale Semiconductor's six core broadband wireless access DSP MSC8156 which includes a programmable 200 Mbps Turbo decoding accelerator (6 iterations), a 115 Mbps Viterbi decoding accelerator (K=9), an FFT/IFFT accelerator for sizes 128, 256, 512, 1,024 or 2,048 points at up to 350 million samples/s, and a DFT/IDFT for sizes up to 1,536 points at up to 175 million samples/s [23].

Relying on a single DSP processor for all signal processing tasks would be a clean solution. As a practical matter, however, multiple DSP processors are necessary for implementing a next generation wireless handset or base station. This means greater system cost, more board space, and more power consumption. Integrating hardware communication accelerators, such as MIMO detectors and channel decoders, into the DSP processor silicon can create an efficient System-on-Chip. This offers many advantages: the dedicated accelerators relieve the DSP processor of the parallel computation-intensive signal processing burden, freeing DSP processing capacity for other system control functions that more greatly benefit from programmability.

A new trend in DSP systems is to use GPGPU for signal processing. GPGPU stands for General-Purpose computation on Graphics Processing Units. Graphics Processing Units (GPUs) are high-performance many-core processors that can be used to accelerate a wide range of applications. For example, authors in [18] have demonstrated a GPU-based high speed LDPC decoder implementation. Authors in [73] and [74] have developed a high throughput MIMO detector and a 3GPP LTE Turbo decoder using GPU.

4 Further Reading

This chapter serves as a brief introduction to the application-specific accelerators for communications. For more detailed discussion on the VLSI signal processing system design and implementation, readers are encouraged to read the following book [42]. For more information on the software/hardware co-design as well as the hardware accelerators for 3G/4G wireless systems, one can read the following dissertations [9, 50]. Finally, major DSP processor vendors such as Texas Instruments, Analog Devices, and Freescale provide many application notes about their DSP hardware accelerators [3, 22, 63].

Readers are also advised to look at several other chapters of this handbook. For example, Gustafsson et al. [29] discusses the fundamental computer arithmetic, Takala [58] talks about the general-purpose DSP processors, Liu and Wang [37] introduces the application-specific instruction set DSP processors, and Zhang et al. [77] discusses the three-dimensional DSP systems.

Acknowledgements The authors at Rice University would like to thank Nokia, Nokia Siemens Networks (NSN), Xilinx, and US National Science Foundation (under grants CCF-0541363, CNS-0551692, CNS-0619767, EECS-0925942 and CNS-0923479) for their support of this work.

References

1. Alamouti, S.M.: A simple transmit diversity technique for wireless communications. IEEE Journal on Selected Areas in Communications **16**(8), 1451–1458 (1998)
2. Amiri, K., Cavallaro, J.R.: FPGA implementation of dynamic threshold sphere detection for MIMO systems. In: IEEE Asilomar Conf. on Signals, Syst. and Computers, pp. 94–98 (2006)
3. Analog Devices: The SHARC processor family.
   http://www.analog.com/en/embedded-processing-dsp/sharc/processors/index.html (2009)
4. Bahl, L., Cocke, J., Jelinek, F., Raviv, J.: Optimal decoding of linear codes for minimizing symbol error rate. IEEE Transactions on Information Theory **IT-20**, 284–287 (1974)
5. Bass, B.: A low-power, high-performance, 1024-point FFT processor. IEEE Journal of Solid-State Circuits **34**(3), 380–387 (1999)
6. Berrou, C., Glavieux, A., Thitimajshima, P.: Near Shannon limit error-correcting coding and decoding: Turbo-codes. In: IEEE Int. Conf. on Commun., pp. 1064–1070 (1993)

7. Bougard, B., Giulietti, A., Derudder, V., Weijers, J.W., Dupont, S., Hollevoet, L., Catthoor, F., Van der Perre, L., De Man, H., Lauwereins, R.: A scalable 8.7-nJ/bit 75.6-Mb/s parallel concatenated convolutional (turbo-) codec. In: IEEE International Solid-State Circuit Conference (ISSCC), vol. 1, pp. 152–484 (2003)

8. Brack, T., Alles, M., Lehnigk-Emden, T., Kienle, F., Wehn, N., Lapos, Insalata, N., Rossi, F., Rovini, M., Fanucci, L.: Low complexity LDPC code decoders for next generation standards. In: Design, Automation, and Test in Europe, pp. 1–6 (2007)

9. Brogioli, M.: Reconfigurable heterogeneous DSP/FPGA based embedded architectures for numerically intensive embedded computing workloads. Ph.D. thesis, Rice University, Houston, Texas, USA (2007)

10. Brogioli, M., Radosavljevic, P., Cavallaro, J.: A general hardware/software codesign methodology for embedded signal processing and multimedia workloads. In: IEEE Asilomar Conf. on Signals, Syst., and Computers, pp. 1486–1490 (2006)

11. Burg, A.: VLSI circuits for MIMO communication systems. Ph.D. thesis, Swiss Federal Institute of Technology, Zurich, Switzerland (2006)

12. Burg, A., Borgmann, M., Wenk, M., Zellweger, M., Fichtner, W., Bolcskei, H.: VLSI implementation of MIMO detection using the sphere decoding algorithm. IEEE Journal of Solid-State Circuits **40**(7), 1566–1577 (2005)

13. Cheng, C.C., Tsai, Y.M., Chen, L.G., Chandrakasan, A.: A 0.077 to 0.168 nJ/bit/iteration scalable 3GPP LTE turbo decoder with an adaptive sub-block parallel scheme and an embedded DVFS engine. In: IEEE Custom Integrated Circuits Conference, pp. 1–4 (2010)

14. Cooley, J.W., Tukey, J.W.: An algorithm for the machine calculation of complex Fourier series. Mathematics of Computation **19**(90), 297–301 (1965)

15. Cupaiuolo, T., Siti, M., Tomasoni, A.: Low-complexity high throughput VLSI architecture of soft-output ML MIMO detector. In: Design, Automation and Test in Europe Conference and Exhibition, pp. 1396–1401 (2010)

16. Damen, M.O., Gamal, H.E., Caire, G.: On maximum likelihood detection and the search for the closest lattice point. IEEE Transaction on Information Theory **49**(10), 2389–2402 (2003)

17. De Sutter, B., Raghavan, P., Lambrechts, A.: Coarse-grained reconfigurable array architectures. In: S.S. Bhattacharyya, E.F. Deprettere, R. Leupers, J. Takala (eds.) Handbook of Signal Processing Systems, second edn. Springer (2013)

18. Falcao, G., Silva, V., Sousa, L.: How GPUs can outperform ASICs for fast LDPC decoding. In: 23rd ACM International Conference on Supercomputing, pp. 390–399. ACM (2009)

19. Fincke, U., Pohst, M.: Improved methods for calculating vectors of short length in a lattice, including a complexity analysis. Mathematics of Computation **44**(170), 463–471 (1985)

20. Forney, G.D.: The Viterbi algorithm. Proceedings of the IEEE **61**(3), 268–278 (1973)

21. Foschini, G.: Layered space-time architecture for wireless communication in a fading environment when using multiple antennas. Bell Labs. Tech. Journal **2**, 41–59 (1996)

22. Freescale Semiconductor: Freescale Starcore architecture. www.freescale.com/starcore (2009)

23. Freescale Semiconductor: MSC8156 six core broadband wireless access DSP. www.freescale.com/starcore (2009)

24. Gallager, R.: Low-density parity-check codes. IEEE Transactions on Information Theory **IT-8**, 21–28 (1962)

25. Garrett, D., Davis, L., ten Brink, S., Hochwald, B., Knagge, G.: Silicon complexity for maximum likelihood MIMO detection using spherical decoding. IEEE Journal of Solid-State Circuits **39**(9), 1544–1552 (2004)

26. Golden, G., Foschini, G.J., Valenzuela, R.A., Wolniansky, P.W.: Detection algorithms and initial laboratory results using V-BLAST space-time communication architecture. Electronics Letters **35**(1), 14–15 (1999)

27. Gunnam, K., Choi, G.S., Yeary, M.B., Atiquzzaman, M.: VLSI architectures for layered decoding for irregular LDPC codes of WiMax. In: IEEE International Conference on Communications, pp. 4542–4547 (2007)

28. Guo, Z., Nilsson, P.: Algorithm and implementation of the K-best sphere decoding for MIMO detection. IEEE Journal on Selected Areas in Communications **24**(3), 491–503 (2006)

29. Gustafsson, O., Wanhammar, L.: Arithmetic. In: S.S. Bhattacharyya, E.F. Deprettere, R. Leupers, J. Takala (eds.) Handbook of Signal Processing Systems, second edn. Springer (2013)

30. Han, S., Tellambura, C.: A complexity-efficient sphere decoder for MIMO systems. In: IEEE International Conference on Communications, pp. 1–5 (2011)

31. Hassibi, B., Vikalo, H.: On the sphere-decoding algorithm I. expected complexity. IEEE Transaction On Signal Processing **53**(8), 2806–2818 (2005)

32. Hunter, H.C., Moreno, J.H.: A new look at exploiting data parallelism in embedded systems. In: Proceedings of the International Conference on Compilers, Architectures and Synthesis for Embedded Systems, pp. 159–169 (2003)

33. Jin, J., Tsui, C.: Low-complexity switch network for reconfigurable LDPC decoders. IEEE Transactions on Very Large Scale Integration (VLSI) Systems **18**(8), 1185–1195 (2010)

34. Lechner, G., Sayir, J., Rupp, M.: Efficient DSP implementation of an LDPC decoder. In: IEEE Int. Conf. on Acoustics, Speech, and Signal Processing, vol. 4, pp. 665–668 (2004)

35. Lee, S.J., Shanbhag, N.R., Singer, A.C.: Area-efficient high-throughput MAP decoder architectures. IEEE Transaction on VLSI Systems **13**(8), 921–933 (2005)

36. Lin, C.H., Chen, C.Y., Wu, A.Y.: Area-efficient scalable MAP processor design for high-throughput multistandard convolutional turbo decoding. IEEE Transactions on Very Large Scale Integration (VLSI) Systems **19**(2), 305–318 (2011)

37. Liu, D., Wang, J.: Application specific instruction set DSP processors. In: S.S. Bhattacharyya, E.F. Deprettere, R. Leupers, J. Takala (eds.) Handbook of Signal Processing Systems, second edn. Springer (2013)

38. Martina, M., Nicola, M., Masera, G.: A flexible UMTS-WiMax turbo decoder architecture. IEEE Transactions on Circuits and Systems II **55**(4), 369–273 (2008)

39. May, M., Ilnseher, T., Wehn, N., Raab, W.: A 150 Mbit/s 3GPP LTE turbo code decoder. In: IEEE Design, Automation & Test in Europe Conference & Exhibition, pp. 1420–1425 (2010)

40. McAllister, J.: FPGA-based DSP. In: S.S. Bhattacharyya, E.F. Deprettere, R. Leupers, J. Takala (eds.) Handbook of Signal Processing Systems, second edn. Springer (2013)

41. Myllylä, M., Silvola, P., Juntti, M., Cavallaro, J.R.: Comparison of two novel list sphere detector algorithms for MIMO-OFDM systems. In: IEEE International Symposium on Personal Indoor and Mobile Radio Communications, pp. 1–5 (2006)

42. Parhi, K.K.: VLSI digital signal processing systems design and implementation. Wiley (1999)

43. Prescher, G., Gemmeke, T., Noll, T.G.: A parametrizable low-power high-throughput turbo-decoder. In: IEEE Int. Conf. on Acoustics, Speech, and Signal Processing, vol. 5, pp. 25–28 (2005)

44. Rovini, M., Gentile, G., Rossi, F., Fanucci, L.: A scalable decoder architecture for IEEE 802.11n LDPC codes. In: IEEE Global Telecommunications Conference, pp. 3270–3274 (2007)

45. Sadjadpour, H., Sloane, N., Salehi, M., Nebe, G.: Interleaver design for turbo codes. IEEE Journal on Seleteced Areas in Communications **19**(5), 831–837 (2001)

46. Salmela, P., Gu, R., Bhattacharyya, S., Takala, J.: Efficient parallel memory organization for turbo decoders. In: Proc. European Signal Processing Conf., pp. 831–835 (2007)

47. Shin, M.C., Park, I.C.: A programmable turbo decoder for multiple 3G wireless standards. In: IEEE Solid-State Circuits Conference, vol. 1, pp. 154–484 (2003)

48. Studer, C., Benkeser, C., Belfanti, S., Huang, Q.: Design and implementation of a aarallel turbo-decoder ASIC for 3GPP-LTE. IEEE Journal of Solid-State Circuits **46**(1), 8–17 (2011)

49. Sun, J., Takeshita, O.: Interleavers for turbo codes using permutation polynomials over integer rings. IEEE Transaction on Information Theory **51**(1), 101–119 (2005)

50. Sun, Y.: Parallel VLSI architectures for multi-Gbps MIMO communication systems. Ph.D. thesis, Rice University, Houston, Texas, USA (2010)

51. Sun, Y., Cavallaro, J.R.: A low-power 1-Gbps reconfigurable LDPC decoder design for multiple 4G wireless standards. In: IEEE International SOC Conference, pp. 367–370 (2008)

52. Sun, Y., Cavallaro, J.R.: Scalable and low power LDPC decoder design using high level algorithmic synthesis. In: IEEE International SOC Conference (SoCC), pp. 267–270 (2009)

53. Sun, Y., Cavallaro, J.R.: Efficient hardware implementation of a highly-parallel 3GPP LTE, LTE-advance turbo decoder. Integration, the VLSI Journal, Special Issue on Hardware Architectures for Algebra, Cryptology and Number Theory **44**(4), 305–315 (2011)
54. Sun, Y., Cavallaro, J.R.: A flexible LDPC/turbo decoder architecture. Journal of Signal Processing System **64**(1), 1–16 (2011)
55. Sun, Y., Karkooti, M., Cavallaro, J.R.: VLSI decoder architecture for high throughput, variable block-size and multi-rate LDPC codes. In: IEEE International Symposium on Circuits and Systems (ISCAS), pp. 2104–2107 (2007)
56. Sun, Y., Wang, G., Cavallaro, J.R.: Multi-layer parallel decoding algorithm and VLSI architecture for quasi-cyclic LDPC codes. In: IEEE International Symposium on Circuits and Systems, pp. 1776–1779 (2011)
57. Sun, Y., Zhu, Y., Goel, M., Cavallaro, J.R.: Configurable and scalable high throughput turbo decoder architecture for multiple 4G wireless standards. In: IEEE International Conference on Application-Specific Systems, Architectures and Processors (ASAP), pp. 209–214 (2008)
58. Takala, J.: General-purpose DSP processors. In: S.S. Bhattacharyya, E.F. Deprettere, R. Leupers, J. Takala (eds.) Handbook of Signal Processing Systems, second edn. Springer (2013)
59. Tarokh, V., Jafarkhani, H., Calderbank, A.R.: Space-time block codes from orthogonal designs. IEEE Transactions on Information Theory **45**(5), 1456–1467 (1999)
60. Tarokh, V., Jafarkhani, H., Calderbank, A.R.: Space time block coding for wireless communications: Performance results. IEEE Journal on Selected Areas in Communications **17**(3), 451–460 (1999)
61. Telatar, I.E.: Capacity of multiantenna Gaussian channels. European Transaction on Telecommunications **10**, 585–595 (1999)
62. Tensilica Inc.: http://www.tensilica.com (2009)
63. Texas Instruments: TMS 320C6000 CPU and instruction set reference guide. http://dspvillage.ti.com (2001)
64. Texas Instruments: TMS 320C55x DSP CPU programmer's reference supplement. http://focus.ti.com/lit/ug/spru652g/spru652g.pdf (2005)
65. Texas Instruments: TMS320C6474 high performance multicore processor datasheet. http://focus.ti.com/docs/prod/folders/print/tms320c6474.html (2008)
66. Thul, M.J., Gilbert, F., Vogt, T., Kreiselmaier, G., Wehn, N.: A scalable system architecture for high-throughput turbo-decoders. The Journal of VLSI Signal Processing **39**(1–2), 63–77 (2005)
67. Viterbi, A.: Error bounds for convolutional coding and an asymptotically optimum decoding algorithm. IEEE Transactions on Information Theory **IT-13**, 260–269 (1967)
68. Wijting, C., Ojanperä, T., Juntti, M., Kansanen, K., Prasad, R.: Groupwise serial multiuser detectors for multirate DS-CDMA. In: IEEE Vehicular Technology Conference, vol. 1, pp. 836–840 (1999)
69. Willmann, P., Kim, H., Rixner, S., Pai, V.S.: An efficient programmable 10 gigabit Ethernet network interface card. In: ACM International Symposium on High-Performance Computer Architecture, pp. 85–86 (2006)
70. Witte, E., Borlenghi, F., Ascheid, G., Leupers, R., Meyr, H.: A scalable VLSI architecture for soft-input soft-output single tree-search sphere decoding. IEEE Trans. on Circuits and Systems II: Express Briefs **57**(9), 706–710 (2010)
71. Wong, C.C., Chang, H.C.: Reconfigurable turbo decoder with parallel architecture for 3GPP LTE system. IEEE Tran. on Circuits and Systems II: Express Briefs **57**(7), 566–570 (2010)
72. Wong, K., Tsui, C., Cheng, R.S., Mow, W.: A VLSI architecture of a K-best lattice decoding algorithm for MIMO channels. In: IEEE Internation Symposium on Circuits and Systems, vol. 3, pp. 273–276 (2002)
73. Wu, M., Sun, Y., Cavallaro, J.: Implementation of a 3GPP LTE turbo decoder accelerator on GPU. In: IEEE Workshop on Signal Processing Systems, pp. 192–197 (2010)
74. Wu, M., Sun, Y., Gupta, S., Cavallaro, J.R.: Implementation of a high throughput soft MIMO detector on GPU. Journal of Signal Processing System **64**(1), 123–136 (2011)

75. Wu, M., Sun, Y., Wang, G., Cavallaro, J.R.: Implementation of a high throughput 3GPP turbo decoder on GPU. Journal of Signal Processing System **Online First** (2011)

76. Ye, Z.A., Moshovos, A., Hauck, S., Banerjee, P.: CHIMAERA: A high performance architecture with a tightly coupled reconfigurable functional unit. In: Proceedings of the 27th Annual International Symposium on Computer Architecture, pp. 225–235 (2000)

77. Zhang, T., Pan, Y., Li, Y.: DSP systems using three-dimensional integration technology. In: S.S. Bhattacharyya, E.F. Deprettere, R. Leupers, J. Takala (eds.) Handbook of Signal Processing Systems, second edn. Springer (2013)

78. Zhong, H., Zhang, T.: Block-LDPC: a practical LDPC coding system design approach. IEEE Transactions on Circuits and Systems I **52**(4), 766–775 (2005)

General-Purpose DSP Processors

Jarmo Takala

Abstract Recently the border between DSP processors and general-purpose processors has been diminishing as general-purpose processors have obtained DSP features to support various multimedia applications. This chapter provides a view to general-purpose DSP processors by considering the characteristics of DSP algorithms and identifying important features in a processor architecture for efficient DSP algorithm implementations. Fixed-point and floating-point data paths are discussed. Memory architectures are considered from parallel access point of view and address computations are shortly discussed.

1 Introduction

Recently the border between DSP processors and general-purpose processors has been diminishing as general-purpose processors have obtained DSP features to support various multimedia applications. On the other hand, DSP processors, which used to be programmed with manual assembly, have nowadays incorporated features from general-purpose computers to support software development on high-level languages.

A DSP processor can be defined in a various ways and the simplest interpretation states that a DSP processor is any microprocessor that processes signals represented in digital form. As all programmable processors could be classified as DSP processors according to this definition, we would need to refine the definition. A more focused view on DSP processors can be obtained by considering the characteristics of digital signal processing. As DSP is application of mathematical algorithms to signals represented digitally and, on the other hand, DSP is applied in

J. Takala (✉)
Tampere University of Technology, Finland
e-mail: jarmo.takala@tut.fi

S.S. Bhattacharyya et al. (eds.), *Handbook of Signal Processing Systems*,
DOI 10.1007/978-1-4614-6859-2_24, © Springer Science+Business Media, LLC 2013

real-time systems where real-time constraint is determined by the repetition period of the algorithm, we can assume that DSP systems and, in particular, real-time DSP systems contain mainly repetitious application of data-driven behaviors defined by mathematical algorithms under strict timing constraints [29].

This implies that DSP processors are designed for repetitive, numerically intensive tasks. On the other hand, often DSP applications define two main requirements: timing and error. The timing requirement dictates that a sequence of operations defined by the algorithm in hand must be performed in a given time. In addition, error of the results must be less than specified, i.e., accuracy of computations must fulfill the requirements. Therefore, we can expect that DSP processors contain features to improve the accuracy and performance of computations according to DSP algorithms. In addition, as the processed signals often represent real world physical signals, there is need to interface the processor to various peripheral devices, e.g., A/D and D/A, to receive and send digital signals. In particular, in real-time systems, there is need to transfer data in and out with constant data rate requirements. Modern DSP processors often contain various peripheral devices for easy interfacing.

As many of the traditional DSP algorithms contain computation of sum of products, e.g., FIR filtering being the traditional example; filtered signal y_n is obtained with aid of an N-tap FIR filter as

$$y_n = \sum_{i=0}^{N-1} c_i x_{n-1} \qquad (1)$$

where x_n is the input sample at time instant n and c_i is the filter coefficient. As multiplication is often used, a fast multiplier has been an integral part of DSP processor architecture. Although the first commercially available DSP processor, Intel 2920 [20, 44], did not have a hardwired multiplier unit at all. The sum of products computation indicates accumulation, thus multiply-accumulate (MAC) is advantageous for DSP applications. In particular, single cycle multiply-accumulate instructions have been seen in DSP processors. Another important property is high memory bandwidth for feeding the arithmetic units with operands from memory. For this purpose, multiple-access memory architectures are used, which allow parallel instruction fetch and operand accesses. In addition, specialized addressing modes can improve the memory related performance as well as dedicated address generation units. DSP processors often contain specialized execution control mechanisms; in particular, efficient looping capabilities reduce the overhead due to repetitive execution. Specialized features to improve numerical accuracy are also present as accuracy of results is often one of the main criteria set by DSP applications. Finally, as DSP systems are often processing data representing real world signals, input/output interfaces and different peripherals are needed.

2   Arithmetic Type

DSP processors are divided as fixed-point and floating-point processors based on the type of arithmetic units in the processor. In the early days of DSP processors, fixed-point processors with 16-bit word width were used, e.g., Texas Instruments TMS32010 [32] and NEC μPD7720 [35]. While the 16-bit data was sufficient for speech applications, fairly soon it was realized that some applications require higher accuracy and 24-bit processors were introduced, e.g., Motorola DSP56000 [26]. In addition, floating-point DSP processors were introduced, e.g., Hitachi HD61810 [17], AT&T DSP32 [19], and NEC μPD77230 [34].

The floating-point processors contain more complex logic and, therefore, consume more power and are more expensive. However, the floating-point processors are easier to program as the dynamic range in floating-point representation is larger and there is no need to scale and optimize the signal levels during intermediate computations. Furthermore, high-level languages have floating-point data types while integers are the only supported fixed-point data types although signal processing calls for fractional data types for fixed-point arithmetic.

3   Data Path

The actual signal processing operations in a processor are carried out in various functional units, such as arithmetic logic units (ALU) or multipliers, and the collection these units is called as data path or data ALU. In order to store intermediate results, data path contains also accumulators, registers, or register files. The data path in DSP processors can be expected to be tailored for computations inherent in typical DSP algorithms. DSP processors may also be enhanced with special function units to improve performance for a group of applications. In the following sections, principal features of fixed-point data paths are studied and differences in floating-point data paths are discussed.

3.1   Fixed-Point Data Paths

A typical fixed-point data path includes multiplier, ALU, shifters, and registers and accumulators. Data path is used mainly for signal processing operations, while computations related to memory access, which call for integer arithmetic, are often performed in separate ALUs called as address generation units or address arithmetic units.

3.1.1   Multiplier and ALU

As traditional DSP processors are used for computing numerically intensive tasks a fast multiplier is an essential unit in a DSP processor. Although multipliers are included in general-purpose microprocessors, there is a major difference in multipliers in fixed-point DSP processors. In general-purpose fixed-point computations, integer data type is used and multipliers operate such that integer operands result in an integer product (b-bit operands produce b-bit product, i.e., the LSB of product is saved). In DSP processors, however, fractional data type is exploited, which implies that the LSB of the product is not sufficient, thus the product is obtained at full precision; multiplication of b-bit operands results in 2b-bit product (law of conservation of bits). This indicates that the integer multipliers present in standard microprocessors and microcontrollers is not well suited to signal processing. In some DSP processors, multipliers may produce narrower results for obtaining speed-up or smaller silicon area. Multiplier may also be pipelined implying latency, i.e., the product may not be available for the next instruction.

Multiplication is also involved in one of the most characteristic operations in DSP, multiply-accumulate and often DSP performance even characterized as MAC/s. This measure is often practical as DSP algorithms are, in general, data-independent and, therefore, deterministic in behavior. DSP processor may contain an additional adder to be used in MAC operation and these resources form a MAC unit. Processor may also contain parallel units to further boost the performance on DSP applications.

In similar fashion as in general-purpose processors, DSP processors contain arithmetic-logical unit, which performs the basic operations: addition, subtraction, increment, negate, and, or, not, etc. Often the addition in MAC operation is carried out in ALU. The ALU in DSP processor operates in a similar fashion as in general-purpose processors but the arithmetic operations are carried out with extended word width operands. This is due to the fact the consecutive MAC operations tend to increase the word width of the result. In order to avoid overflow in accumulation, additional guard bits are used, i.e., additional bits are used when performing the accumulation and storing the accumulation results. In general, $\log_2(N)$ additional bits are required for carrying out N additions without overflow. The more guard bits, the more headroom against the overflow.

In early DSP processors, multiplier was a separate unit, which stored its result in a specific "product" register as seen in Fig. 1. Such an arrangement implies that an additional instruction is needed to perform the accumulation and there is latency of one instruction in MAC operation. A similar behavior can be seen in Freescale DSP56300 illustrated in Fig. 2.

In order to increase the performance for DSP applications, processor can contain several multipliers. For example, TI TMS320C55x family of processors contain two MAC units [39]. The unit can perform multiplication with 17-bit operands and a 40-bit addition. TI TMS320C64x processors contain parallel functional units as shown in Fig. 3. Two of the eight function units can perform several type of multiplications: 32×32 multiplication, 16×16 multiplication, 16×32 multiplication, quad 8×8

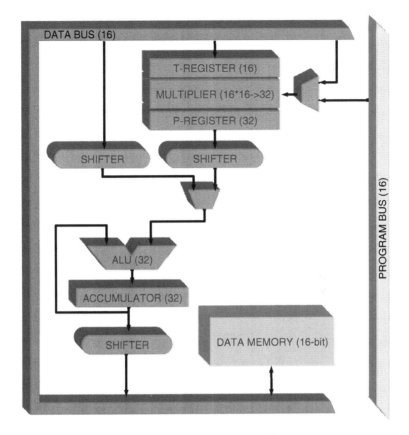

Fig. 1 Principal data path of TI TS320C25

multiplications, dual 16×16 multiplications, and Galois field multiplication. These two units support also dual 16×16 and quad 8×8 MAC operations. Freescale MSC8156 contains six SC3850 DSP cores and the data path of each core contains four ALUs, which can perform dual 16×16 MAC operations. The ALUs support also complex-valued multiplication of operands with 16-bit real and imaginary parts [14].

3.1.2 Registers

Early DSP processors were load-store architectures where operands in memory have to be moved to specific operand registers before the operand can be transferred to a functional unit. In a similar fashion, results from functional units were stored to specific registers or accumulators. In early DSP processor, the number of general-purpose registers was small indicating need to store intermediate results to memory. This represents clearly a performance bottleneck. Higher performance can be

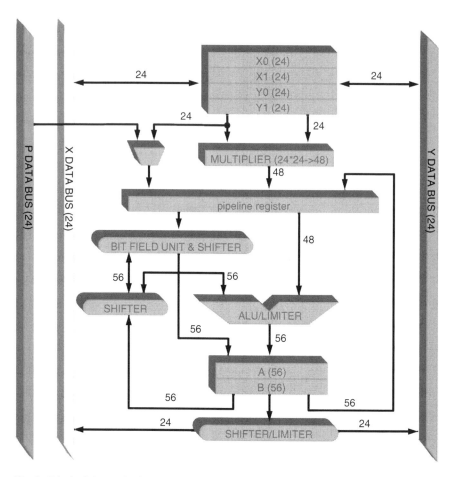

Fig. 2 Principal data path of Freescale DSP56300

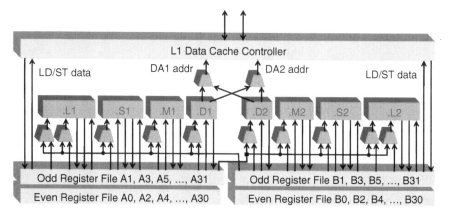

Fig. 3 Principal data path of TI TMS320C64x

expected when operands are fetched directly from the memory and with the aid of efficient addressing mechanisms. In addition, modern processors contain general-purpose registers and the number of registers is higher reducing the need for storing local data to memory.

As discussed in previous section, the width of accumulators and registers is often greater than the full precision result of multiplication as the additional bits (guard bits) in the most significant part provide headroom against an overflow when accumulating several products. Finally, several registers can be used as a combined extended precision register. For example, in Freescale DSP56300 in Fig. 2, 24-bit registers X1 and X2 can be used together as a 48-bit register. Arithmetic operations with extended precision can be performed by exploiting signed and unsigned versions of native add and multiply instructions and the final 96-bit result can be stored to accumulators A and B.

In general-purpose DSP processors, the trend has been towards larger number of general-purpose registers; e.g., TMS320C64x processors contain two register files, which each contain 32 registers as illustrated in Fig. 3. These 32-bit registers can be used to store data ranging from packed 8-bit to 64-bit fixed-point data. Word widths greater than 32 bits are stored by using to registers as a pair. In Freescale SC3850, 16 registers form a register file, which supports word widths up to 40 bits. Special packed data formats are supported such that two 20-bit operands can be packed into a single 40-bit register. This format is well suited for representing complex numbers.

3.1.3 Shifters

Shifters can be found in all general-purpose processors and the actual shifter unit is part of ALU. In general, two type of shift operations can be found: arithmetic and logical shift. The difference is in right shift: in logical shift, the vacant bit positions are filled with zeros while the sign-extension is used in arithmetic shift. In both cases, the vacant bits are filled with zeros in left shift. Yet another form of shift is circular shift (or rotation) where bits shifted out are moved in the vacant bit positions. Circular shifts are used in communications applications and some address computations.

The arithmetic shift is a useful operation in fixed-point processors as arithmetic shift b bits to the left corresponds to multiplication by a factor of 2^b and shift to the right is division by 2^b. In fixed-point DSP processors, shifters can be found in various locations as shifting is needed for different purpose. Shifting can be used to provide headroom against overflow in multiply-accumulation, i.e., if the product is shifted b bits to the right, 2^b accumulations can be performed without overflow. Such an organization was used in early DSP processors, e.g., in TI TMS320C25 depicted in Fig. 1, but the disadvantage is loss of precision as the least significant bits are lost in shifting. The shifter after the multiplier can still be useful when using fractional mode multiplication, i.e., product is shifted one bit to the left in order to remove the extra sign bit due to multiplication of fractional operands, e.g., in TI TMS320C54x [38].

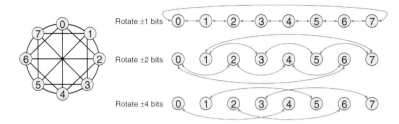

Fig. 4 Barrel shifter

Another use for shifters is scaling of operands from memory. When full precision products from fractional operands are used, the position of binary point in the extended precision result in accumulator is not the same as in the operands in the memory. For example, the b-bit operands in memory use representation with $b-1$ fractional bits, while full precision product in accumulators use representation with $2b-2$ fractional bits. Therefore, if we would like to add an operand from memory to the full precision result in the accumulator, the memory operand has to be multiplied by $2^{(b-1)}$, i.e., shifted $(b-1)$ bits to the left. Such an shift can be carried out with the top leftmost shifter connected to the data bus in Fig. 1. In a similar fashion, constants from instructions need to be shifted to the left and for this purpose, immediate field from program bus in DSP56300 in Fig. 2 has connection to the bit field unit & barrel shifter.

Finally, shifters are also exploited when storing extended precision results from wide accumulators or registers to native word width memory. This implies that a part of the bits need to be selected to be transferred to memory. For this purpose, the contents of the accumulator or register can be shifted such that the bits to be stored are in the LSB or MSB part and then stored to memory. For example, 16-bit operands with 15 fractional bits will produce a 32-bit full precision result with 30 fractional bits in the accumulator, thus the result should be shifted 15 bits to the right such that the 16 least significant bits can be stored to memory in the same representation as the original operands. Alternatively, if the 32-bit accumulator can be used as two 16-bit registers (upper and lower half), the product can be shifted one bit to the left and the upper half is stored to memory.

The shifters in DSP processors need to support shifts of variable number of bits in a single cycle unlike in low-cost microprocessors where shifts and rotations of only one bit to the left or right are typically supported. Therefore, DSP processors use barrel shifters (also known as plus-minus-2^i (PM2I) networks), which contains multiple layers for interconnections between register elements [23]. Each bit register in 2^n-bit barrel shifter is connected to registers, which are at distances $\pm 2^0, \pm 2^1, \ldots, \pm 2^{(n-1)}$ indicating that a shift of an arbitrary number of bits will take at most $(n-1)$ steps. This is illustrated in Fig. 4 where connections in an 8-bit barrel shift can be seen, i.e., each bit register has connections to five neighbors. The three different connection layers are shown in the right and it can be seen that any arbitrary shift will take at most two steps. Barrel shifters are implemented as cascade of parallel multiplexers. When the structure of barrel shifter is drawn by placing the register elements on a ring topology as in Fig. 4, the structure reminds barrel, hence the name.

3.1.4 Overflow Management

In fixed-point DSP processors, special care needs to be taken against the overflow. The simplest method to detect is overflow is the overflow bit in status register, which is set when ALU detects an overflow as a result of an arithmetic operation. Such a mechanism is available in most of the processors but it is not an efficient method for DSP applications as it requires software procedure to handle the error.

One method for overflow management, specific to DSP processors, is guard bits, i.e., oversized accumulators and registers with extended precision arithmetic units. The main purpose is to allow accumulation of products without overflow and the final extended precision results are quantized when stored to memory.

Another specific feature for overflow management are shifters. For example, to avoid extra overhead when quantizing extended precision results for memory, shifters between the accumulators or registers and buses are used. Another mechanism is to scale intermediate results to arrange headroom against the overflow. An example of such scaling is the shifter after the multiplier in Fig. 1. This approach reduces precision as some of the least significant bits of the products are lost during the right shift.

Yet another mechanism is dynamic scaling mode that is useful in block processing applications, e.g., discrete trigonometric transforms where block floating-point representation can be exploited. The principal idea is to emulate floating-point representation but to use a common exponent for a block of samples. The common exponent is defined by the largest magnitude in the block. In block processing, the word width increase of intermediate results is detected at each iteration and headroom against overflow is arranged with the aid of suitable down-scaling of operands during the next iteration. For example, in DSP56300 [13], such scaling is supported with a scaling bit in the status register. This bit is set if a result moved from an accumulator to memory requires scaling. This is a sticky bit, thus it can be used to detect need for scaling in a block of data. In the next iteration, the operands from memory can then be scaled towards zero.

The signal scaling will require careful planning from the programmer to maximize the intermediate signal levels in a proper manner. An automated method for avoiding overflow is to use saturation arithmetic or limiter as discussed in [16]. In such a case, the arithmetic unit limits the result to the largest positive or negative representable number in case of overflow as illustrated in Fig. 5 where behavior of limiter is shown when using 3-bit fractional representation. Overflow is also possible when storing results from oversized accumulators to memory, thus a limiter should be available not only in ALUs but between accumulators and memory buses as seen in Fig. 2. Saturation arithmetic is specified in various telecommunication standards and limiters would be useful is such applications.

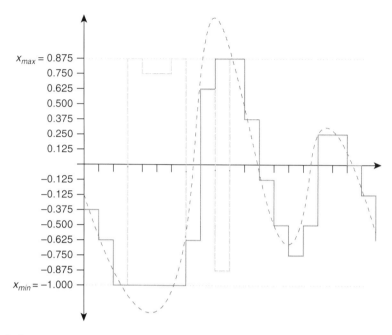

Fig. 5 Saturation arithmetic with 4-bit fractional representation (sign bit and three fraction bits), *thin dashed line*: full precision signal, *thick dashed line*: quantized overflown signal, *solid line*: quantized signal with saturation, x_{min} (x_{max}): minimum (maximum) representable number

3.2 Floating-Point Data Paths

The first floating-point DSP processors used proprietary number representations. In 1985, ANSI/IEEE standard 754-1985 [5] was introduced and soon floating-point arithmetic compatible with IEEE standard was introduced especially in general-purpose microprocessors. In modern floating-point DSP processors, the arithmetic units exploit IEEE standard.

In a floating-point arithmetic unit, the results are scaled automatically to maximize the precision of results as the number representation requires normalization of mantissa. This normalization results in the main advantage of floating-point representation: large dynamic range. For example, IEEE single-precision format defines 24-bit mantissa and 8-bit exponent where the minimum and maximum values for exponent are -126 and 127, respectively. Therefore, the dynamic range of the IEEE single-precision number system is

$$D_{sp} = \frac{\left(2 - 2^{-23}\right) \times 2^{127}}{1 \times 2^{-126}} = 2.89 \times 10^{76} \; ; \; 20\log_{10}(2.89 \times 10^{76})\mathrm{dB} \approx 1529 \, \mathrm{dB}.$$

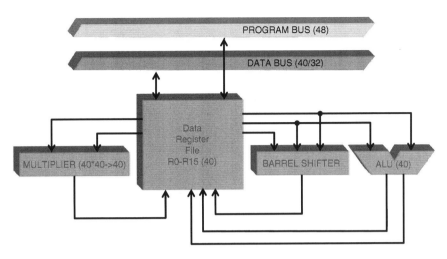

Fig. 6 Principal block diagram of ADSP2106X processors

Similar dynamic range with signed fixed-point number system requires word width of 255 bits:

$$D_{fxp} = \frac{1}{2^{-254}} = 2.89 \times 10^{76}.$$

The huge dynamic range implies that overflow is not a major concern like in fixed-point data paths. However, overflow is still possible and arithmetic units indicate such an exceptional result by setting the overflow bit in the status register and an error recovery routine provided by user or operating system is then executed.

The normalization of mantissa in a floating-point arithmetic units is carried out with the aid of a shifter. Such a shifter it is an integral part of floating-point arithmetic unit and, therefore, it is not visible for programmer unlike shifters in fixed-point DSP processors. Therefore, less shifters are seen in floating-point DSP processors. In addition, arithmetic units in floating-point DSP processors often support also fixed-point arithmetic.

The data paths of the first floating-point DSPs reminded the fixed-point data paths but soon floating-point processors were designed to be more flexible, e.g., general-purpose register files were used instead of operand registers and accumulators. An example of such a data path is Analog Devices ADSP-2106x [4] illustrated in Fig. 6. However, the data paths in modern VLIW DSP processors can be the same independent on the arithmetic type, e.g., TI TMS320C67x floating-point processors contain data path similar to TMS320C64x fixed-point processors illustrated in Fig. 3.

In floating-point processors, the multiplier does not produce product at full precision (as seen in Fig. 6) unlike in fixed-point DSP processors. Still quite often multipliers in floating-point DSP processors provide arithmetic results at extended precision unlike in general-purpose microprocessors. While 32-bit single-precision

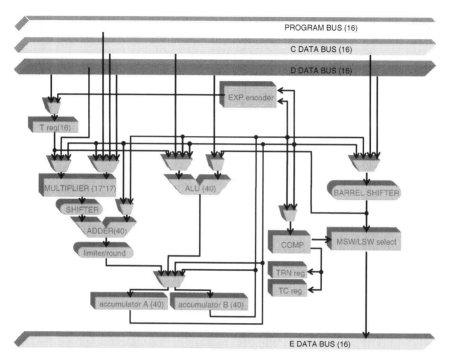

Fig. 7 Principal block diagram of TI TMS320C54x processors

IEEE format is popular there are processors supporting even higher precisions; e.g., TI TMS320C67x processors have hardware support for single-precision and double-precision instructions.

3.3 Special Function Units

In general, significant savings can be obtained if a system is tailored for the application at hand. This implies that efficiency can be improved by adding application-specific functions to the data path of a general-purpose DSP processor. An early example of such a customization is AT&T DSP1610, which included an additional bit manipulation unit. Another example is the compare-select-and-store unit integrated in the data path of TI TMS320C54x depicted in Fig. 7. This unit speeds up Viterbi decoder implementations. The data path has an additional unit for determining exponent value of operand in accumulator, which speeds up floating-point emulation. The previous processors were targeted to telecommunications applications and they may be called as domain-specific processors as they contain customized units to support a specific application domain, i.e., not only a single application but a group of applications having similar features. In a similar fashion,

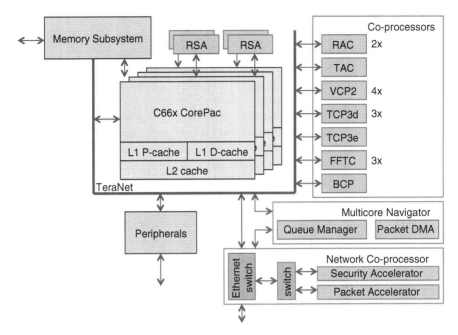

Fig. 8 Principal block diagram of TI TMS320C6670 processors

DSP processors may be targeted to video applications; Analog Devices ADSP-BF522 processor contains four 8-bit video ALUs, which support, e.g., average operation and subtract/absolute value/accumulate operation.

In modern DSP processors, the additional special units can also be rather complex, e.g., in ADSP-TS201S, 128-bit communications logic unit is used for trellis decoding and it executes complex correlations for CDMA communications applications [3]. TI TMS320C6455 [40] contains also special hardware for speeding up Viterbi decoding and turbo decoding but the units are not integrated in the data path; there are two co-processors for speeding up the tasks. Even larger number of co-processors can be found from TI TMS320C6670 [42] illustrated in Fig. 8; there are four rake/search accelerators (RSA) and 16 coprocessors. There are three co-processors for turbo decoding and one for turbo encoding, four for Viterbi decoding, two WCDMA receive and one transmit acceleration co-processors, three FFT co-processors, and a bit rate co-processor. The last co-processor carries out bit processing for up and down link, e.g., encoding, rate matching, segmentation and multiplexing. The system contains a network co-processor, which has accelerators for packet processing and encryption/decryption. The tasks and message passing between all the resources in the system are orchestrated with the aid of multicore Navigator and TeraNet switch fabric.

The main difference on supporting special functionalities is that the special function units (accelerators) integrated in the data path will be reflected by the instruction set, i.e., there are instructions to control the usage of special function units, while co-processors are mainly visible as memory or I/O mapped peripheral devices.

4   Memory Architecture

In order to obtain high performance in DSP, it is not enough to have high-performance arithmetic units in the data path. It is essential to be able to feed operands from memory to the data path; the memory bandwidth should match the performance of the data path. Quite often algorithms are applied to a large number of data structures such that there is not enough registers for all the operands, thus operands have to be obtained to memory.

In general, low-cost microprocessors use von Neumann architecture where instructions and data are both placed to the same storage structure implying that instruction fetches and data accesses are interleaved over common resources as illustrated in Fig. 9. This implies that several cycles are needed to complete, e.g., MAC operation needed for implementing one FIR tap in (1), which requires three memory accesses: fetching of instruction, coefficient, and sample.

Higher performance can be expected if memory accesses could be performed simultaneously. In first DSP processors, memory bandwidth was improved by exploiting Harvard architecture where instructions and data are stored in different independent memories as shown in Fig. 10a. This implies that while operands for current instruction are accessed, the next instruction can already be fetched. This approach doubles the memory bandwidth when one operand instructions are used as clearly seen by comparing the timing diagrams in Figs. 9 and 10a.

The speedup of two operand instructions is not doubled, thus one of the first modifications was to use repeat buffer where an instruction can be stored to avoid fetch from the memory thus the program bus is free for the data access. Such an approach is applicable mainly only in loops. In the first iteration, the instruction (I_1 in Fig. 10b) is fetched from program memory, thus the program bus is reserved and operand accesses are performed in the next cycle (O_{10} and O_{11}). In the next iterations, the instruction is fetched from the repeat buffer, thus the program bus is free for operand access and two operand accesses over two buses can be performed in parallel (O_{20} and O_{21} etc.).

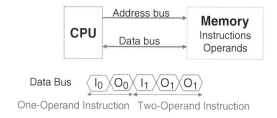

Fig. 9 von Neumann architecture. I_n: instruction access, O_n: Operand access

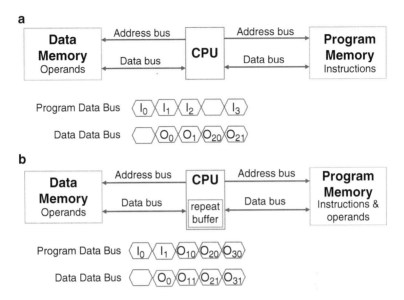

Fig. 10 Memory organizations with timing diagrams: (**a**) pure Harvard architecture and (**b**) Harvard architecture with repeat buffer. I_n: instruction fetch, O_n: Operand access

The memory bandwidth can be increased even further by adding memory modules and buses. In DSP56300, three memory modules are used as illustrated in Fig. 11. If the data is distributed between the two data memory modules, two operand instructions can be executed in one cycle. Even higher bandwidth can be obtained when more memory modules and buses are added. An example of such DSP processors is Hitachi DSPi, which contained six parallel memory modules [24]. However, in order to exploit the parallelism, data has to be carefully arranged between the memory modules to avoid access conflicts. The complexity of such data distribution task is illustrated with an example in the following.

Let us assume that an algorithm operating over a 4×4 matrix stored to four memory modules and it accesses elements in rows and columns. Here a question arises: how to distribute the data over the four memory modules such that four elements from memory can be accesses in parallel. The data could be stored to memories column by column, i.e., each column is stored to a single memory module as illustrated in Fig. 12a where the elements of matrix are numbered row by row, thus the first row of matrix contains elements $(0, 1, 2, 3)$. This data distribution method is called as low-order interleaving where the memory module address m and row address to memory r are obtained with the aid of address a as follows

$$m = a \bmod Q \tag{2}$$

$$r = \lfloor a/Q \rfloor \tag{3}$$

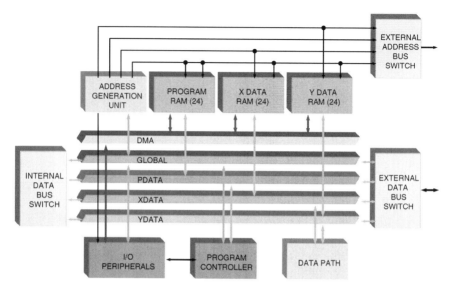

Fig. 11 Principal block diagram of DSP56300

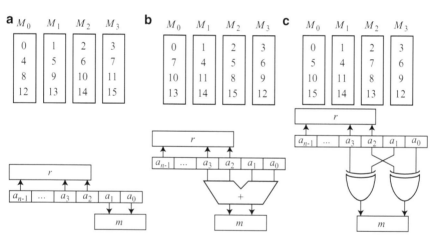

Fig. 12 Parallel access schemes: (**a**) low-order interleaving, (**b**) row rotation, and (**c**) linear transform

where Q denotes the number of memory modules and mod is modulus operation. When the number of memory modules is a power-of-two, the implementation is simple as illustrated in Fig. 12a. We can see that the elements in a row can be accessed in parallel as the elements are stored to separate memory modules $M_0,\ldots,M_3$. However, accessing elements in a column introduces access conflict as all the elements are located in the same memory module. The column access needs to be serialized slowing down the operation.

Row rotation scheme [8] can be used to allow conflict-free parallel access to rows and columns. The module and row address are now defined as

$$m = a \bmod Q \tag{4}$$

$$r = (a + \lfloor a/Q \rfloor) \bmod Q \tag{5}$$

where $\lfloor \ \rfloor$ denotes floor operation. The implementation in case the number of memory modules is a power-of-two requires an adder as illustrated in Fig. 12b. The storage of matrix elements is shown and it can be seen that the elements in each row and column are stored in different memory modules.

Another scheme is to apply linear transformations [18]. Here the computations of module and row address are performed at bit-level thus the module and row addresses are used as bit arrays $\mathbf{m}$ and $\mathbf{r}$ obtained from index address array $\mathbf{a}$ as follows

$$\mathbf{m} = \mathbf{Ta} \tag{6}$$

$$\mathbf{r} = \mathbf{Ka} \tag{7}$$

where $\mathbf{T}$ and $\mathbf{K}$ are binary matrices called row and module transformation matrices, respectively. These matrices can be designed in various way to obtain support for different access patterns. We can use the following matrices:

$$\mathbf{r} = \mathbf{Ka} = \begin{pmatrix} 1\ 0\ 0\ 0 \\ 0\ 1\ 0\ 0 \end{pmatrix} \begin{pmatrix} a_3 \\ a_2 \\ a_1 \\ a_0 \end{pmatrix} \; ; \; \mathbf{m} = \mathbf{Ta} = \begin{pmatrix} 1\ 0\ 1\ 0 \\ 0\ 1\ 0\ 1 \end{pmatrix} \begin{pmatrix} a_3 \\ a_2 \\ a_1 \\ a_0 \end{pmatrix} \tag{8}$$

and the corresponding matrix distribution and the implementation of module and row address computation can be seen in Fig. 12c. The computations in linear transformation schemes are based on modulo-2 arithmetic, thus the arithmetic is realized as bit-wise exclusive-OR operations and modulo operations and carry delay of adders are avoided. Linear transformation schemes are hence called as XOR schemes.

While the previous schemes have been applied in some DSP processors, a more popular scheme for parallel memory modules is high-order interleaving where the module address is defined by some most significant bits of the address as illustrated in Fig. 13a. The memory map in such a scheme will look like in Fig. 13b, i.e., one memory module is allocated to a block of addresses and the next module will reserve the next block of addresses. In such a case, a single data array is not distributed over the memory modules but several arrays are to be accessed in parallel and placed in different memory modules. For example, in FIR computations, the coefficients c_i in (1) and delay line containing the latest samples x_i are stored in different memory blocks, thus one coefficient and one sample from delay line can be accessed in

Fig. 13 Parallel access schemes: (**a**) high-order interleaving and memory maps with (**b**) high-order interleaving, and (**c**) low-order interleaving

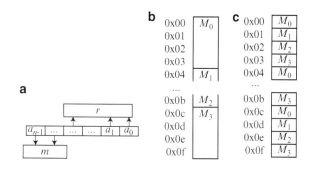

parallel. The memory map according to low-order interleaving is shown in Fig. 13c. In such a case, the idea is to access various elements from the same array in parallel rather than accessing elements from different arrays in parallel as it is often the case in modified Harvard architecture. This is also supported by the fact that in various processors based on Harvard architecture, the parallel memory modules reside in different memory spaces, e.g., in DSP65300 in Fig. 11, the parallel memory modules are in X and Y memory spaces, thus the module is determined during programming.

Harvard architecture is not the only method to increase the memory bandwidth. Similar performance can be obtained by exploiting multi-port memories. For example, dual-port memory with two buses provides the same memory bandwidth as two single port memories over to buses. The dual-port memory is, however, more expensive in terms of area, speed, and power. On the other hand, multi-port memory is more flexible as the same memory space is seen through the ports, thus there is no need to distribute the data between different modules. In LSI DSP16410, three-port memories are used such that one port is dedicated for instruction/coefficient memory space, the second port for data memory space, and the third port is left for DMA transfers [31].

Multiple access memories can also be used to increase the memory bandwidth, i.e., memory can be accessed multiple times in an instruction cycle. For example, TMS320C54x processors contain dual-access memory [38], which corresponds to a Harvard architecture with two data memories. Multiple access memory is also flexible as there is only one memory space, thus careful distribution of data is not needed like in Harvard architecture. However, multiple access memory may restrict the maximum clock frequency of the processor.

5 Address Generation

Intensive access to memory in DSP applications implies that address computations are performed frequently. As the data path is utilized by signal processing arithmetic, DSP processors often contain ALUs dedicated to memory address computations. Many of DSP applications operate over data in arrays, e.g., filters

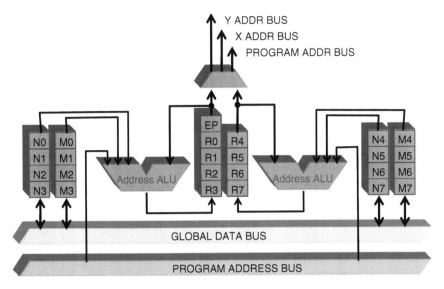

Fig. 14 Principal block diagram of address generation unit of DSP56300

exploit delay lines, as illustrated by FIR filter example in (1), and discrete transforms imply block processing. In early DSP processors, delay lines were implemented as a linear FIFO buffer in memory, which indicates that the samples need to be copied from a memory location to the next memory location at each iteration. This implies need for extra memory bandwidth. Similar behavior without the need for moving data is to exploit circular buffer or ring buffer, i.e., a memory array is managed with the aid of a pointer. Each time a new sample enters to buffer, the sample is stored to a memory location pointed by the pointer, and then the pointer is incremented. Once the buffer is filled and the pointer exceeds the reserved memory area, the pointer is initialized to point to the beginning of the buffer. Pointers can be maintained in specific address registers and updated with the aid of address ALUs. The pointer increment behaves as a modulo arithmetic, thus circular buffers are maintained with the aid of modulo addressing.

The previous features are used with the aid of addressing modes and the most powerful addressing mode in DSP processors is typically register indirect addressing with pre- or post-updates. The powerful addressing modes imply that the a single DSP instruction specifies several operations, thus often loop kernels in DSP processors are short compared to the corresponding kernels in general-purpose processors.

The address generation unit of DSP56300 illustrated in Fig. 14 is an example of an versatile and flexible address unit. Registers Rn, Nn, and Mn operate as triplets such that Rn contains the address that is used to compute the effective operand address. This register can be pre- or post-updated depending on the used selected addressing mode. Register Nn contains offset or index, which is added to the address in Rn. Register Mn is modifier, which defines the type of arithmetic used in the address computation: linear, modulo, or reverse-carry. Linear addressing is used in general-purpose addressing, modulo addressing is used with circular buffers, and reverse-carry addressing is useful on realizing the bit-reversed permutation found in radix-2 FFT. As there are eight register triplets, programmer can maintain eight different type of buffers simultaneously. The EP register is stack extension pointer and it is not used for operand addressing. Dedicated address generation units are used in conventional DSP processors, while multi-issue processors, e.g., VLIW processors, contain more parallel resources and arithmetic units can be used for computing both the addresses and signals.

6 Program Control

DSP algorithms are quite often data-independent and, therefore, computations are highly data-oriented in contrast to control-oriented algorithms containing data-dependent computations. In addition, algorithms often define tight loops, thus DSP processor should have efficient looping capabilities. If the loop kernel contains only few instructions, it is clear that the loop control overhead, i.e., instructions needed to decrement the loop counter, test against end condition, and branch, can be significant. DSP processors often have zero-overhead looping capability (also known as hardware looping), which means that processor has hardware support for the decrement-test-branch sequence, thus the control overhead of software looping can be avoided. Hardware looping, however, may introduce side effects or constraints, e.g., interrupts are disabled during the loop execution, branches or certain addressing modes are not allowed in loop kernels, or the number of instructions in loop kernel is limited.

In pipelined processors, branching is an expensive operation as the destination address of branching often is obtained in late pipeline stages implying that several instructions have already been fetched to the pipeline but due to the branch they should not be executed. Therefore, the pipeline needs to be flushed. This indicates that a standard branch consumes several cycles and is called as multi-cycle branch. One alternative method is to execute the instructions already fetched to the pipeline. This means that instructions following immediately the branch instruction will be executed regardless of the branch. Such a method is called as delayed branch and the number of instructions executed after branch instructions is called as delay slot.

In decision-intensive code, branching is needed often and it would be extremely expensive in architectures with deep pipelines. One approach to avoid multi-cycle branching in such cases is conditional execution. This means that a condition code is integrated to the instruction and the instruction is executed only when the condition is true. The instruction passes through the pipeline normally and only the execution stage is conditional.

7 Multicore DSP Processors

As the complexity of DSP applications has been increasing over the years, there has been trend to add parallelism to the processor's data path. For example, VLIW DSP processors were introduced to add more arithmetic units in the data path. However, performance requirements in many applications are beyond the capabilities of single processors and, therefore, multicore DSPs have been developed.

An example of such multicores is TI TMS320C6678 [43], which contains eight C66x VLIW cores [41] and provides peak performance of 160 GMAC/s for floating-point or 320 GMAC/s for fixed-point arithmetic with 1.25 GHz clock. The principal block diagram reminds the TMC320C6670 illustrated in Fig. 8. The eight cores, DMA transfer controllers, and various system peripherals are interconnected with TeraNet switch fabric. The cores have L1 caches for data and program, unified L2 cache, and shared memory. In a similar fashion, Freescale MCS8156 [15] contains Starcore SC3850 DSP cores interconnected with chip-level arbitration and switching system, which provides arbitration between the cores, system level shared memories, DDR SRAM controllers, and other system resources. Both these multicore DSPs represent homogeneous multiprocessor systems.

Even larger number of cores can be found in Tilera provides TILE-Gx, TILEPro, and TILE64 families of multicore processors, which contain 36 or 64 identical 64-bit processors [6]. These processors are interconnected with a 2D mesh network. HyperX HX3000 [22] by Coherent Logic contains a 10×10 array of processors connected with a 11×11 array of data management and routing units (DMR) as illustrated in Fig. 15. Each processor can access four local memories in it's neighbouring DMRs; one local memory block acts as a shared memory for four processors. Multicore chips by Mindspeed Technologies are based on picoArray, which

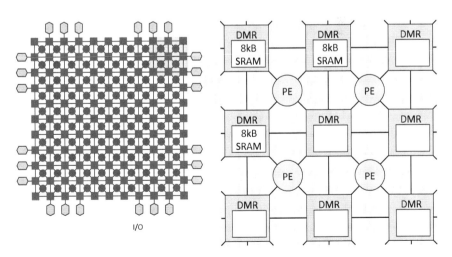

Fig. 15 Principal organization of HX3000. DMR: Data management and routing unit

consists 200–300 DSP cores interconnected with mesh type of interconnection. The cores are 16-bit 3-way VLIW processors [10]. Although in many of these multicore systems, there are shared memory resources, the programming model is based on message passing.

Yet another type of multicore organisation can be found from Sandbridge SB3500 chip [33, 36], which is heterogeneous system consisting of a general-purpose ARM processor and three multithreaded DSP cores. The processors use token triggering threading, thus threads are guaranteed to have a fixed number of cycles between issuing consecutive instructions. Each core has four hardware threads, which can execute 16-wide vector operations. the multithreading hides latencies in unconditional branches, interrupts, and memory accesses.

8   Conclusions

The DSP processor architectures have evolved over the time. Early DSP processors were programmed manually with assembly language and high performance was obtained with specialized function units operating in parallel. This implied non-orthogonal complex instruction set computer (CISC) type of instruction sets that are poor targets for compilers. Although reduced instruction set computer (RISC) and superscalar architectures have gained popularity in general-purpose computing, they have not been the mainstream in DSP processors. This is partly due to the fact that often DSP applications set hard real-time requirements indicating that the execution time of the software implementation should be deterministic. Typically the features used in superscalar processors introduce dynamic run-time behavior resulting in non-deterministic execution times.

Various methods to exploit parallelism has been used to improve the performance. Conventional DSP processor architectures have been enhanced with multiple MAC units, e.g., TMS320C55x, ADSP-BF5xx, or additional co-processors. Multiple function units are exploited with SIMD extensions, e.g., in SIMD mode in ADSP-21xxx, second data path executes the same operations as the instruction defines for the first data path but for a different register file. Multi-issue processors in form of VLIW machines have gained popularity in DSP applications.

9   Further Reading

Other chapters in the handbook cover topics related to DSP processors; Franke [11] discusses C compilers and code optimisation for DSP processors, Kessler [25] covers topics on compiling for VLIW DSPs, and Carro and Rutzig [9] introduces multicore system-on-chips. There are many textbooks discussing DSP processors and interested readers are advised to consider the following: [1, 21, 27–29]. In addition, Berkeley Design Technologies [7] carries out independent benchmarking of DSP processors and their web page contains useful information on various issues related to DSP implementations. Finally, DSP processor vendors provide application notes, which are a rich source for practical information, e.g., Texas Instruments [37], Freescale [12], Analog Devices [2], and LSI [30].

Acknowledgements This work has been supported by Academy of Finland through funding decision no. 253087.

References

1. Ackenhusen, J.G.: Real-Time Signal Processing: Design and Implementation of Signal Processing Systems. Prentice-Hall, Inc., Upper Saddle River, NJ (1999)
2. Analog Devices, Inc.: URL http://www.analog.com/
3. Analog Devices, Inc.: TigerSHARC Embedded Processor: ADSP-TS201S (2006)
4. Analog Devices, Inc.: SHARC Processor: ADSP-21060/ADSP-21060L/ADSP-21062/ADSP-21062L/ADSP-21060C/ADSP-21060LC (2008)
5. ANSI/IEEE Std 754-1985: IEEE standard for binary floating-point arithmetic. Standard, The Institute of Electrical and Electronics Engineers, Inc., New York, NY, U.S.A. (1985)
6. Bell, S., Edwards, B., Amann, J., Conlin, R., Joyce, K., Leung, V., MacKay, J., Reif, M., Bao, L., Brown, J., Mattina, M., Miao, C.C., Ramey, C., Wentzlaff, D., Anderson, W., Berger, E., Fairbanks, N., Khan, D., Montenegro, F., Stickney, J., Zook, J.: TILE64 - processor: A 64-core SoC with mesh interconnect. In: IEEE Int. Solid-State Circuits Conf., Digest of Technical Papers, pp. 88–598 (2008)
7. Berkeley Design Technologies, Inc.: URL www.bdti.com
8. Budnik, P., Kuck, D.: The organization and use of parallel memories. IEEE Trans. Comput. **20**(12), 1566–1569 (1971)
9. Carro, L., Rutzig, M.B.: Multicore systems on chip. In: S.S. Bhattacharyya, E.F. Deprettere, R. Leupers, J. Takala (eds.) Handbook of Signal Processing Systems, second edn. Springer (2013)
10. Duller, A., Panesar, G., Towner, D.: Parallel processing - the picoChip way. Communicating Processing Architectures **2003**, 125–138 (2003)
11. Franke, B.: C compilers and code optimization for DSPs. In: S.S. Bhattacharyya, E.F. Deprettere, R. Leupers, J. Takala (eds.) Handbook of Signal Processing Systems, second edn. Springer (2013)
12. Freescale Semiconductor, Inc.: URL http://www.freescale.com/
13. Freescale Semiconductor, Inc.: DSP56300 Family Manual: 24-Bit Digital Signal Processors (2005)
14. Freescale Semiconductor, Inc.: Differences Between the MSC8144 and the MSC815x DSPs (2009)
15. Freescale Semiconductor, Inc.: MSC8156 Six-Core Digital Signal Processor (2011)
16. Gustafsson, O., Wanhammar, L.: Arithmetic. In: S.S. Bhattacharyya, E.F. Deprettere, R. Leupers, J. Takala (eds.) Handbook of Signal Processing Systems, second edn. Springer (2013)
17. Hagiwara, Y., Kita, Y., Miyamoto, T., Toba, Y., Hara, H., Akazawa, T.: A single chip digital signal processor and its application to real-time speech analysis. IEEE Trans. Acoustics Speech Signal Process. **31**(1), 339–346 (1983)
18. Harper III, D.: Increased memory performance during vector accesses through the use of linear address transformations. IEEE Trans. Comput. **41**(2), 227–230 (1992)
19. Hayes, W., Kershaw, R., Bays, L., Boddie, J., Fields, E., Freyman, R., Garen, C., Hartung, J., Klinikowski, J., Miller, C., Mondal, K., Moscovitz, H., Rotblum, Y., Stocker, W., Tow, J., Tran, L.: A 32-bit VLSI digital signal processor. IEEE J. Solid-State Circuits **20**(5), 998–1004 (1985)

20. Hoff, M.E., Townsend, M.: An analog input/output microprocessor for signal processing. In: IEEE Int. Solid-State Circuits Conf., Digest Tech. Papers, p. 220 (1979)
21. Hu, Y.H.: Programmable Digital Signal Processors: Architecture, Programming, and Applications. Marcel Dekker, New York, NY (2002)
22. Humble, T., Mitra, P., Barhen, J., Schleck, B.: Real-time spatio-temporal twice whitening for MIMO energy detectors. In: Proc. CROWNCOM (2010)
23. Hwang, K., Briggs, F.A.: Computer Architecture and Parallel Processing. McGraw-Hill Book Co., Singapore (1985)
24. Kaneko, K., Nakagawa, T., Kiuchi, A., Hagiwara, Y., Ueda, H., Matsushima, H., Akazawa, T., Ishida, J.: A 50ns DSP with parallel processing architecture. In: IEEE Int. Solid-State Circuits Conf., Digest Tech. Papers, pp. 158–159 (1987)
25. Kessler, C.W.: Compiling for VLIW DSPs. In: S.S. Bhattacharyya, E.F. Deprettere, R. Leupers, J. Takala (eds.) Handbook of Signal Processing Systems, second edn. Springer (2013)
26. Kloker, K.: Motorola DSP56000 digital signal processor. IEEE Micro 6(6), 29–48 (1986)
27. Kuo, S.M., Gan, W.S.: Digital Signal Processors: Architectures, Implementations, and Applications. Pearson Education, Inc., Upper Saddle River, NJ (2005)
28. Kuo, S.M., Lee, B.H.: Real-Time Digital Signal Processing: Implementations, Applications, and Experiments with the TMS320C55x. Wiley, New York, NY (2001)
29. Lapsley, P.D., Bier, J., Shoham, A., Lee, E.A.: DSP Processor Fundamentals: Architectures and Features. Berkeley Design Technology, Inc., Fremont, CA (1996)
30. LSI Corp.: URL http://www.lsi.com/
31. LSI Corp.: DSP16410 Digital Signal Processor (2007)
32. Magar, S., Caudel, E., Leigh, A.: A microcomputer with digital signal processing capability. In: IEEE Int. Solid-State Circuits Conf., Digest Tech. Papers, vol. XXV, pp. 32–33 (1982)
33. Moudgill, M., Glossner, J., Agrawal, S., Nacer, G.: The Sandblaster 2.0 architecture and SB3500 implementation. In: Proc. Software Defined Radio Technical Forum. Washington, DC (2008)
34. Nishitani, T., Kuroda, I., Kawakami, Y., Tanaka, H., Nukiyama, T.: Advanced single-chip signal processor. In: Proc. IEEE Int. Conf. Acoustics, Speech, and Signal Process., vol. 11, pp. 409–412. Tokyo, Japan (1986)
35. Nishitani, T., Maruta, R., Kawakami, Y., Goto, H.: A single-chip digital signal processor for telecommunication applications. IEEE J. Solid-State Circuits 16(4), 372–376 (1981)
36. Surducan, V., Moudgill, M., Nacer, G., Surducan, E., Balzola, P., Glossner, J., Stanley, S., Yu, M., Iancu, D.: The Sandblaster software-defined radio platform for mobile 4G wireless communications. Int. J. Digital Multimedia Broadcasting 2009(Article ID 384507) (2009)
37. Texas Instruments, Inc.: URL http://www.ti.com/
38. Texas Instruments, Inc.: TMS320C54x DSP Functional Overview (2006)
39. Texas Instruments, Inc.: C55x v3.x CPU: Reference Guide (2009)
40. Texas Instruments, Inc.: TMS320C6455 Fixed-Point Digital Signal Processor (2009)
41. Texas Instruments, Inc.: TMS320C66x: DSP CorePac (2011)
42. Texas Instruments, Inc.: TMS320C6670: Multicore Fixed and Floating-Point System-on-Chip (2012)
43. Texas Instruments, Inc.: TMS320C6678: Multicore Fixed and Floating-Point Digital Signal Processor (2012)
44. Townsend, M., Hoff Jr., M.E., Holm, R.E.: An NMOS microprocessor for analog signal processing. IEEE Trans. Comput. 29(2), 97–102 (1980)

Mixed Signal Techniques

Olli Vainio

Abstract Mixed signal circuits include both analog and digital functions. Mixed signal techniques that are commonly encountered in signal processing systems are discussed in this chapter. First, the principles and general properties of sampling and analog to digital conversion are presented. The structure and operating principle of several widely used analog to digital converter architectures are then described, including both converters operating at the Nyquist rate and oversampled converters based on sigma–delta modulators. Next, different types of digital to analog converters are discussed. The basic features and building blocks of switched-capacitor circuits are then shown. Finally, mixed-signal techniques for frequency synthesis and clock synchronization are explained.

1 Introduction

Most physical quantities in the world are analog by nature, i.e., continuous in time and amplitude. Digital signal processing, on the other hand, is done with signal representations that are discrete in both time and amplitude. A signal processing system therefore often needs converters between the two domains, called A/D and D/A converters. For instance, a compact disc (CD) player includes a D/A converter for converting digitally encoded information to an audio signal [11]. There are also intermediate forms, especially sampled analog signals in switched-capacitor filters, which are discrete in time and continuous in amplitude. Presently a large portion of integrated circuits are mixed signal circuits, meaning that both analog and digital functions are included in the circuit.

O. Vainio
Tampere University of Technology, P. O. Box 553, FIN-33101 Tampere, Finland
e-mail: olli.vainio@tut.fi

S.S. Bhattacharyya et al. (eds.), *Handbook of Signal Processing Systems*,
DOI 10.1007/978-1-4614-6859-2_25, © Springer Science+Business Media, LLC 2013

This chapter discusses a few selected topics of mixed signal techniques that are common in signal processing systems. The chapter is organized as follows. Section 2 presents the basic principles of sampling, sample and hold circuits and performance parameters used to characterize data converters. In the next section, various analog to digital converter architectures are shown, including both Nyquist-rate and oversampled converters. Also the principles of sigma–delta modulators are explained. Section 4 describes digital to analog converters. Principles and basic building blocks of switched-capacitor circuits are discussed in Sect. 5. Finally, Sect. 6 presents techniques for frequency synthesis and clock synchronization.

2   General Properties of Data Converters

The block diagram of a mixed signal system, as commonly encountered in signal processing systems, is shown in Fig. 1. The continuous-time analog input signal $x(t)$, where t is the continuous time, originates from a transducer that converts some form of observable energy into voltage or current for electronic processing.

The discrete-time representation $x[n]$ of the analog signal $x(t)$ is typically obtained by periodic sampling according to the relation

$$x[n] = x(nT) \tag{1}$$

where n takes integer values and T is the sampling period. Its reciprocal, $f_s = 1/T$, is the *sampling frequency*.

Sampling can be regarded as the multiplication of the signal $x(t)$ by a sampling function, $s(t)$, which is a periodic impulse train

$$s(t) = \sum_{n=-\infty}^{\infty} \delta(t - nT) \tag{2}$$

where $\delta(t)$ is the unit impulse function or Dirac delta function. In the sampling process, the unit impulses are weighted by $x(t)$ such that the area of each impulse corresponds to the instantaneous amplitude value $x(nT)$ [20].

The frequency spectrum of a periodic impulse train is a periodic impulse train. Since multiplication in the time domain corresponds to convolution in the frequency domain, the spectrum of the sampled signal includes the original signal band and its replicas that repeat on the frequency axis. The replicas are called spectral images and they are centered at multiples of f_s. It can be concluded that if f_s is too small in

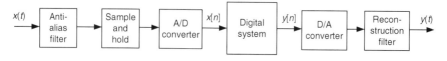

Fig. 1 A mixed signal system for digital processing of an analog input signal

relation to the bandwidth of $x(t)$, the images will overlap, and reconstruction of the original signal is no longer possible. Such a situation is called *aliasing*.

The *Nyquist sampling theorem* states that the signal $x(t)$ is uniquely determined by its samples $x[n]$ if

$$f_s \geq 2f_{max} \tag{3}$$

where f_{max} is the bandwidth of the signal. The frequency f_{max} is called the Nyquist frequency and the frequency $2f_{max}$ is the Nyquist rate. Usually anti-alias filtering is done before sampling in order to limit the signal bandwidth according to Eq. (3). In the discrete-time domain, the frequency is given as the angular frequency, $\omega = 2\pi f/f_s$, where f is the analog signal frequency in Hertz.

Analog to digital conversion is in principle the process of converting the weighted impulse train to a sequence of digitally coded samples in the discrete-time domain.

In practice, sampling using ideal impulses is not possible, and the sampling window has a finite width. The signal is therefore measured over a finite time interval and not instantaneously, giving rise to the *aperture effect*. The signal may be changing while it is being sampled, and for a specified amount of uncertainty, the slope or maximum frequency of the input signal is limited. Assuming a sinusoidal signal and a maximum allowed uncertainty of 1/2 LSB (least significant bit), the maximum frequency of a signal that can be converted to 1/2 LSB accuracy using a b-bit analog to digital converter (ADC) is

$$f_{max} = \frac{1}{\pi\tau 2^{b+1}} \tag{4}$$

where τ is the aperture time [21].

2.1   Sample and Hold Circuits

Commonly a sample and hold circuit (S/H, also known as zero-order hold) is used in connection with an ADC. The S/H output is kept constant during the conversion, and the aperture effect can be avoided. f_{max} then depends on f_s according to the Nyquist criterion in Eq. (3).

The basic sample and hold circuit is shown in Fig. 2a. When the switch is closed, the capacitor voltage follows the input voltage V_i. When the switch is open, C_H holds the voltage to which it was charged. Amplifiers may be used to buffer the input and output, as shown in Fig. 2a. The input buffer should have a large input impedance, small output impedance, and it should be stable for capacitive loads. The output buffer should have a large input impedance and a short settling time. An example of the behavior of V_i and V_o is depicted in Fig. 2b, where the switch is controlled by the signal S/H [21].

The Fourier transform of the sampling pulse is given by

$$H(f) = e^{-j\pi fT}\, T\, sinc(\pi fT). \tag{5}$$

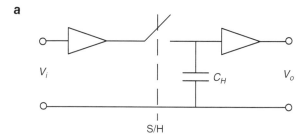

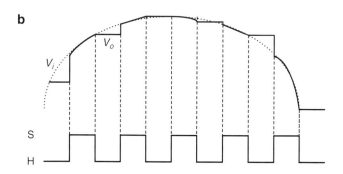

Fig. 2 (**a**) A sample and hold circuit. (**b**) Illustration of S/H operation

The S/H can be considered as an ideal impulse sampler followed by a filter with a frequency response of $sinc(\pi f/f_s)$, where $sinc(x) = \sin(x)/x$ [1]. The signal spectrum is modified because of the droop with frequency caused by the sinc response. This is illustrated in Fig. 3. At the Nyquist frequency $f_s/2$, the input signal is attenuated by the factor 0.64 or -3.9 dB. At the sampling frequency f_s the output of the ideal S/H is zero. The arrows in Fig. 3 represent the fundamental signal and the images of a sine wave of frequency $f_s/6$ Hz corresponding to $\omega = \pi/3$ in the discrete-time domain.

The circuit is also called a track and hold circuit when the control signal has such a duty cycle that the switch is closed for more time than it is open. Many integrated circuit ADCs include a S/H circuit internally. Such ADCs are also called sampling A/D converters.

Various performance parameters and nonideality measures are used to characterize S/H circuits [10, 21].

1. Tracking speed in the sample mode. There are both small-signal and large-signal limitations due to a finite bandwidth and limited slew rate.
2. Sample to hold transition error occurs when the S/H goes from sample mode to hold mode. Opening of the switch is not instantaneous, and the delay may be different for each switch operation. Charge injection during the sample to hold

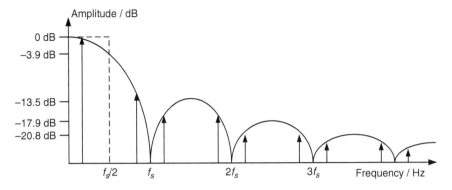

Fig. 3 Frequency response of the S/H circuit in relation to the signal band (*dashed line*)

transition results in a voltage error called the pedestal or step error. Because charge injection depends on the input voltage, nonlinear distortion may be caused.

3. Aperture time uncertainty or aperture jitter is caused by the effective sampling time changing from one sampling instance to the next.
4. Input signal feedthrough during the hold mode, typically caused by parasitic capacitive coupling.
5. In the hold mode, the voltage is slowly changing at a rate called the droop rate. The change is caused by charge leakage from the capacitor due to input bias current of the output amplifier, switch leakage and internal capacitor leakage.
6. General characteristics of analog circuits, such as dynamic range, linearity, gain and offset errors and power supply rejection ratio (PSRR).

2.2 Quantization

Assuming that a b-bit digital representation is used, the samples are assigned one of 2^b values by the ADC. Quantization causes quantization error to be superimposed to the signal. Quantization noise is commonly considered as zero-mean white noise that is uniformly distributed in the interval $\pm q/2$, where q is the quantization step size. If the input amplitude is normalized in the range ± 1, the quantization step is $q = 2^{1-b}$.

Denoting the quantization error as e, the quantization noise power can be calculated as

$$\sigma^2 = \int_{-q/2}^{q/2} e^2 de = q^2/12. \tag{6}$$

For a sine wave input, the signal to quantization noise power (SNR_q) in decibels for an ideal ADC is given by

$$SNR_q = 10\log(\frac{1/2}{q^2/12}) = 10\log(\frac{1/2}{(2^{1-b})^2/12}) \qquad (7)$$

yielding

$$SNR_q = 6.02b + 1.76 \, \text{dB}. \qquad (8)$$

Two conventions exist for relating the analog input voltages to the nominal transition points between code levels. We denote the most negative or negative full scale voltage by V_{min} and the most positive or positive full scale voltage by V_{max}. In the so-called *mid-tread* convention, the first transition is $q/2$ above V_{min} and the last transition is $3q/2$ below V_{max}. The midpoint of the voltage range, $(V_{min} + V_{max})/2$, is then located in the middle of a code bin. In the *mid-riser* convention, V_{min} is $-q$ from the first transition and V_{max} is q from the last transition. In this case, the midpoint, $(V_{min} + V_{max})/2$, is coincident with a code transition level [8].

The ADC output may be encoded, e.g., as unsigned binary numbers, two's complement representation or thermometer-code. The most common binary forms used for sample encoding in digital signal processing are two's complement fixed point and floating point formats [7,9].

The thermometer-code (also known as unary-weighted code) representation of a b-bit binary word has $2^b - 1$ bits. For instance, a 3-bit binary word 000, 001, ..., 111 is thermometer-coded as 0000000, 0000001, 0000011, ..., 1111111. Thus all the bits in this code are weighted equally. Converters between binary numbers and thermometer-code can be designed as combinational logic in a straightforward way [25].

2.3   Performance Specifications

Data converters are characterized by both static and dynamic performance parameters. The application determines which parameters are considered the most important.

2.3.1   Static Performance

The ADC produces one of the 2^b output codes for any analog input voltage. Thus the *dynamic range* (DR) of the converter is 2^b or expressed in decibels, DR = $20\log(2^b)$ = $6.02b$ dB. Ideally, a line connecting the output codes as a function of the input voltage is a straight line. There is *offset error,* if the ADC transfer function is shifted along the input voltage axis. Offset error is the value which should be added to the input values to either minimize the mean squared deviation from the output values, or to set the deviations to be zero at the first and last codes. *Gain error* is the value by which the input values should be multiplied to minimize the deviation. The *full-scale error* is a measure of a change of the slope of the ADC transfer function, including both offset and gain errors.

Each code step of the ADC should ideally have the same size in terms of the input voltage. Deviation in step size between the codes is called *differential nonlinearity* (DNL), defined as the difference between a specific code bin width and the average code bin width, divided by the average code bin width:

$$DNL(i) = \frac{V_{i+1} - V_i}{V_{LSB}} - 1 \tag{9}$$

where V_{i+1} and V_i are voltages corresponding to two successive code transitions and $V_{LSB} = (V_{max} - V_{min})/2^b$. The unit of DNL is LSB which is equal to one ideal code bin width.

Integral nonlinearity (INL) is the maximum difference between the ideal and actual code transition levels after correcting for gain and offset. The ideal transition levels are defined by a fitted straight line. There are two methods to specify the position of the straight line: either minimize the mean squared deviation from the output values or pass a straight line between the end points [8]. The INL corresponding to an output i can be calculated as

$$INL(i) = \frac{V_i - V_1}{V_{LSB}} - (i - 1) \tag{10}$$

where $0 < i < 2^b - 1$. V_i and V_1 are transition levels corresponding to codes i and one, respectively.

Standard ADC test methods for static linearity include feedback loops and histogram test methods. Direct implementations of standard test methods require test stimulus with much higher accuracy than the ADC under test. Recently, methods have been developed for testing without statistically known stimulus [12].

2.3.2 Dynamic Performance

The dynamic performance is characterized by investigating the frequency-domain characteristics of the ADC, typically calculating a fast Fourier transform (FFT) of the output for a sinusoidal input. The standard [8] assumes a pure sine wave input of specified amplitude and frequency. The output spectrum shows a peak for the fundamental frequency as well as components of harmonic distortion and noise caused by various sources.

The *signal-to-noise ratio* (SNR) is the ratio of the root mean square (RMS) amplitude of the ADC output signal to the RMS amplitude of the output noise, where harmonic distortion is not included. SNR is expressed in decibels, given by

$$SNR = 20\log(V_{OUT}^{rms}/V_{NOISE}^{rms}). \tag{11}$$

Harmonic distortion is caused by nonlinearities in the converter. The combined effect of the harmonic components is measured as the *total harmonic distortion* (THD), usually expressed as a decibel ratio:

$$THD = 20\log(\frac{\sqrt{V_2^2 + V_3^2 + \cdots + V_m^2}}{V_{SIGNAL}^{rms}}) \tag{12}$$

where $V_2, V_3, \cdots, V_m$ are the harmonic components including their aliases within the bandwidth of interest.

A parameter that combines the effects of distortion and noise is the *signal-to-noise and distortion* (SiNAD), calculated as the ratio of the RMS amplitude of the ADC output signal to the RMS amplitude of the output noise, where noise also includes nonlinear distortion and the effects of sampling time errors:

$$SiNAD = 20\log(V_{OUT}^{rms}/V_{NOISE+DISTORTION}^{rms}). \tag{13}$$

The *spurious-free dynamic range* (SFDR) is the ratio of the amplitude of the ADC output averaged spectral component at the input frequency to the amplitude of the highest spur, i.e., any unwanted signal in the spectrum. The spur may be a harmonic distortion component or at some point elevated noise floor. SFDR is usually measured in dBc (i.e., relative to the amplitude of the carrier frequency). It is calculated using RMS values as

$$SFDR(dBc) = signal(dB) - unwanted\,tone(dB). \tag{14}$$

SFDR is measured up to the Nyquist frequency or within a specified band [1].

3   Analog to Digital Converters

The selection of an ADC type depends on the application requirements, such as sampling rate, wordlength, power dissipation and any limitations on the possible nonidealities mentioned in the previous section. One way to classify the architectures is to make a distinction between Nyquist-rate converters and oversampled converters [10].

3.1   Nyquist-Rate A/D Converters

In Nyquist-rate ADCs the input signal bandwidth is close to that specified by the sampling theorem in Eq. (3). In practice, a somewhat higher sampling rate is often preferred to make it easier to construct anti-aliasing and reconstruction filters. In the following we mention some of the common Nyquist-rate ADC types.

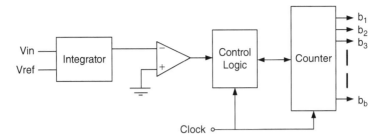

Fig. 4 A dual-slope ADC

3.1.1 Integrating Converters

Integrating ADCs are suitable for applications requiring high accuracy and low speed. The single-slope integrating ADC is based on using an analog integrator, a comparator and a digital counter. The integrator and comparator can be implemented using operational amplifiers (see Sect. 5.1). The integrator charges a capacitor with a constant slope while the counter is counting up. When the integrator voltage reaches the input signal voltage, the comparator output changes state, counting is stopped and the counter output gives the conversion result. The capacitor is then discharged and the counter is cleared for the next conversion cycle. The single-slope ADC suffers from drift with aging.

A preferable approach is the dual-slope ADC, shown in Fig. 4. The conversion cycle operates in two phases.

1. In phase 1 the counter is running for 2^b clock cycles. The integrator output is charged with a slope that is proportional to V_{in}.
2. In phase 2 the counter is reset and the integrator input is switched to reference voltage V_{ref}. The integrator output voltage is reduced with a constant slope. When it reaches zero, counting is stopped and the conversion result is ready.

The dual-slope converter avoids the drift problem of the single-slope ADC because both the integrator gain and the counting rate do not vary between the operation phases. The converter has high noise immunity, as the input voltage is practically averaged during phase 1. The signal is therefore effectively filtered using a sinc type filter response. By appropriately choosing the integration time, certain frequency components can be attenuated by aligning those with the transmission zeros of the sinc filter. This can be useful, for instance, to eliminate power line noise.

3.1.2 Successive-Approximation Converters

Successive-approximation converters use binary search to find the conversion result one bit at a time, starting from the most significant bit (MSB). It therefore takes b clock cycles to find the b-bit result. The block diagram of a successive-

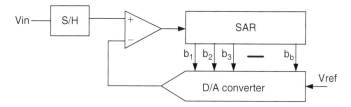

Fig. 5 A successive-approximation ADC

approximation ADC is shown in Fig. 5. It uses a successive-approximation register (SAR), including some control logic, to construct the result. During each iteration cycle, the SAR output word is given to a digital to analog converter (DAC), and its output is compared in an analog comparator with the input voltage from a S/H circuit. The S/H is needed to keep the comparator input stable during conversion. Successive-approximation converters are suitable for medium speed, moderate accuracy applications.

Converters can also be designed with nonlinear transfer functions for, e.g., data compressing applications [14]. Then the quantization steps are dense for small amplitudes and progressively larger for larger amplitudes, and the SNR is better for typical signal statistics than that achievable by linear quantization. For example, digital telephone systems use standardized A-law and μ-law algorithms for speech companding [4].

3.1.3 Flash Converters

Flash converters, also known as parallel converters, are used in very high speed applications. Flash ADCs are available for sampling rates exceeding 1 GHz. A disadvantage of the flash ADC architecture is that the size grows proportionally to 2^b. The resolution is typically limited to 8 bits and the power dissipation is quite high. A flash ADC architecture is shown in Fig. 6 for the example case of 3-bit output. The input voltage V_{in} is connected to a parallel array of comparators. Each comparator is also connected to a node of a resistor string which divides the reference voltage V_{ref} to a stack of fixed voltage levels for comparison with V_{in}. The comparator array generates a thermometer code representing the input voltage. One of the XOR gates gives a high output corresponding to the detected voltage level, and the 2^b to b encoder produces the output word.

By changing the resistor sizing it is possible to modify the transfer function of the flash ADC. It can be made nonlinear, e.g., logarithmic, or intentional 0.5 LSB offset may be included.

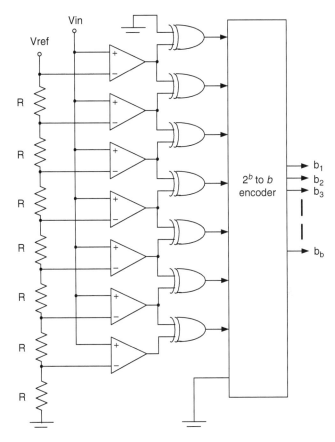

Fig. 6 A flash ADC

3.1.4 Sub-Ranging Converters

Sub-ranging or multi-step converters are popular in high-speed, medium-accuracy applications. The block diagram of an 8-bit subranging ADC is shown in Fig. 7. The conversion is done in two steps. First, the coarse ADC converts the 4 MSBs. This result is converted back to analog with a DAC and subtracted from the input voltage. The difference is converted digital with a fine ADC that produces the 4 LSBs. The subconverters are typically flash ADCs, needing altogether much less comparators than a full flash ADC of the same overall resolution. The fine ADC may be preceded by a gain amplifier with a gain of a power of two.

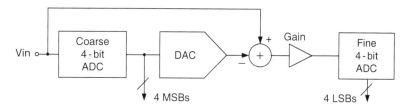

Fig. 7 A sub-ranging ADC with two 4-bit subconverters

3.2 Oversampled A/D Converters

Oversampled converters use a sampling rate that is much higher than the Nyquist rate. The oversampling ratio (OSR) is defined as the ratio of the actual sampling rate and the Nyquist rate. The OSR is typically between 8 and 512. Oversampled converters can utilize noise shaping, which means shaping the spectrum of the quantization noise so that most of the noise power is moved outside of the signal band. The sampling rate is subsequently lowered close to the Nyquist rate by decimation. This operation involves a so-called decimation filter, which is a lowpass filter that limits the signal bandwidth to avoid aliasing. The decimation filter also attenuates the out-of-band quantization noise.

3.2.1 Benefits of Oversampling

The quantization noise power, given by Eq. (6), is independent of the sampling frequency. Assuming that no noise shaping is done, the noise is evenly distributed in the frequency domain from zero to the Nyquist frequency. When oversampling is used, the portion of the quantization noise that is outside of the signal band can be filtered out, and the quantization noise power on the signal band is reduced to

$$\sigma_{OSR}^2 = \frac{q^2}{12}\left(\frac{1}{OSR}\right). \tag{15}$$

In the oversampled case the signal-to-noise ratio is therefore given by

$$SNR_{OSR} = 6.02b + 1.76 + 10\log(OSR) \text{ dB} \tag{16}$$

where the last term is the benefit achieved by oversampling. Thus, SNR is increased by 3 dB when the sampling rate is doubled, which corresponds to 0.5 bits increase in resolution.

Fig. 8 Block diagram of a
first-order sigma–delta
modulator

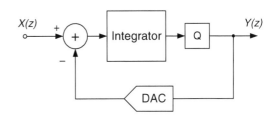

3.2.2 First-Order Sigma–Delta Modulator

Oversampled converters usually take advantage of noise shaping, that can be accomplished using sigma–delta (also called delta–sigma) modulators. Various modulator topologies exist. The block diagram of a first-order sigma–delta modulator for an ADC is shown in Fig. 8. The analog input voltage is added to the feedback signal and taken as input to the integrating filter. The integrator output is quantized by the quantizer Q. One-bit quantizers are most common but also multi-bit quantizers are used. The quantized output is the oversampled digital output from the modulator. It is also converted back to digital using a DAC and subtracted from the input in the addition node. In practice, a modulator is often implemented using switched-capacitor circuits (see Sect. 5). The one-bit quantizer is implemented with an analog comparator, and the one-bit DAC is simply a switch choosing one of two reference voltages.

The sigma–delta modulator can be analyzed using a linearized model, where the quantizer is modeled as an additive noise source. Denoting the input by $X(z)$, the noise by $N(z)$ and the integrator transfer function by $H(z)$, the output $Y(z)$ can be expressed as

$$Y(z) = STF(z)X(z) + NTF(z)N(z) \tag{17}$$

where the signal transfer function $STF(z)$ and the noise transfer function $NTF(z)$ are respectively given by

$$STF(z) = \frac{H(z)}{1 + H(z)} \tag{18}$$

$$NTF(z) = \frac{1}{1 + H(z)}. \tag{19}$$

When the integrator has a transfer function

$$H(z) = \frac{z^{-1}}{1 - z^{-1}} \tag{20}$$

Equation (17) becomes

$$Y(z) = z^{-1}X(z) + (1 - z^{-1})N(z). \tag{21}$$

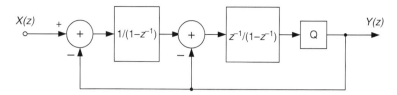

Fig. 9 Block diagram of a second-order sigma–delta modulator

We can see that the input passes through the modulator delayed by a unit delay, and the noise is filtered by a differentiator $1 - z^{-1}$. The differentiator has a zero at the zero frequency (DC, direct current) and the response rises towards high frequencies. The noise spectrum is therefore shaped by this transfer function. SNR again depends on the oversampling ratio and is given by

$$SNR_{SD1} = 6.02b - 3.41 + 30\log(OSR) \text{ dB.} \tag{22}$$

Doubling the OSR improves SNR by 9 dB, corresponding to 1.5 bits.

3.2.3 Higher-Order Sigma–Delta Modulators

The block diagram of a second-order sigma–delta modulator is shown in Fig. 9. It includes two integrators, one of which is nondelaying ($1/(1 - z^{-1})$) and the other one delaying ($z^{-1}/(1 - z^{-1})$). The output can be expressed as

$$Y(z) = z^{-1}X(z) + (1 - z^{-1})^2 N(z). \tag{23}$$

The noise is therefore filtered by a double differentiator. The SNR is given by

$$SNR_{SD2} = 6.02b - 11.14 + 50\log(OSR) \text{ dB.} \tag{24}$$

Doubling the OSR improves SNR by 15 dB, corresponding to 2.5 bits.

The topology shown in Fig. 9 can be generalized to create an Mth order sigma–delta modulator, for which the output can be expressed as

$$Y(z) = z^{-1}X(z) + (1 - z^{-1})^M N(z). \tag{25}$$

Thus M times differentiation of the noise is achieved. An Mth order sigma–delta modulator improves the SNR by $6M + 3$ dB for each doubling of the OSR, corresponding to $M + 0.5$ bits. However, such higher-order modulators are problematic from the stability point of view and alternative topologies have been developed, such as multiple feedback, feedforward and cascaded structures.

A common approach is to construct a higher-order modulator as a cascade of first- or second-order modulators. Such topologies are called multistage noise

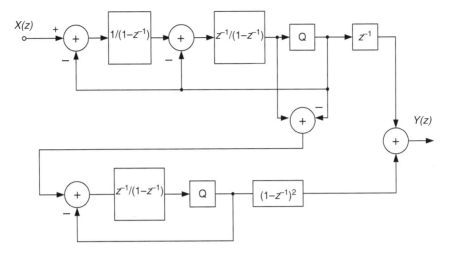

Fig. 10 Block diagram of a third-order (2-1) cascaded modulator

shaping or MASH modulators. An advantage of cascaded modulators is their stability. The common principle is to feed the quantization error of the first modulator to the input of another modulator. An example of cascaded modulators is shown in Fig. 10. This is a third-order modulator which consists of a second-order modulator followed by a first-order modulator.

The output of the 2-1 modulator is expressed as

$$Y(z) = z^{-2}X(z) + (1 - z^{-1})^3 N_2(z) \tag{26}$$

where $N_2(z)$ is the quantization noise of the second modulator.

3.2.4 Sigma–Delta A/D Converters

A sigma–delta ADC consists of a sigma–delta modulator and a digital decimation filter, followed by sampling rate reduction. The achievable resolution depends on the order of the sigma–delta modulator and the oversampling ratio. The decimation filter typically consists of at least two stages, where the first stage is made of one or more averaging (sinc) filters. A multistage decimation filter example is shown in Fig. 11 for decimation by the factor 32. $H_1(z)$ and $H_2(z)$ are optimized decimation filters and $H_3(z)$ is a droop correction filter, serving the purpose of equalizing the passband response [1, 13].

The best number of the sinc filters for an Mth order sigma–delta modulator is $M + 1$. Then only little aliased noise is passed, as the slope of attenuation of this lowpass filter at least equals the slope of the rising quantization noise [10].

Fig. 11 Multistage
decimation filter for
decimation by 32. The
downward array indicates
decimation, followed by the
decimation factor

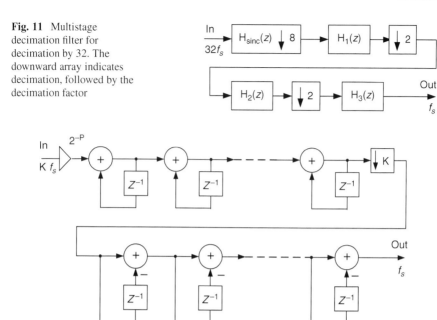

Fig. 12 Implementation of the averaging filters for decimation by factor K

An efficient way to implement the sinc filter is shown in Fig. 12. The filter has
the transfer function

$$H_{sinc}(z) = 2^{-P} \left(\frac{1 - z^{-K}}{1 - z^{-1}} \right)^{M+1} \qquad (27)$$

where K is the decimation ratio of the sinc filter. The scaling factor 2^{-P} has to satisfy
the condition

$$2^{-P} \le (1/K)^{M+1}. \qquad (28)$$

It should be noticed that although the structure of Fig. 12 may have occasional
overflows in the internal nodes, the result is still correct if proper scaling and two's
complement arithmetic is used.

4   Digital to Analog Converters

In principle, DAC operation could be achieved by generating a sequence of weighted
impulses from the digital samples and lowpass filtering to remove the image
frequencies. Such a filter is called a reconstruction filter. In practice, the DAC
output is usually taken from a S/H circuit. The sinc type response of the S/H circuit
causes a droop of about -3.9 dB at the Nyquist frequency, as illustrated in Fig. 3.

The reconstruction filter may be designed to compensate the droop, or compensation may be built into the digital signal processing algorithms. DAC architectures are based on either Nyquist-rate operation or oversampling. Important performance parameters for DACs (see Sect. 2) are: DNL, INL, SFDR, settling time, power dissipation, maximum clock frequency and output voltage swing [10].

4.1 Nyquist-Rate D/A Converters

Probably the simplest DAC principle is pulse width modulation (PWM) where the duty cycle of a binary voltage or current signal is altered according to the digital word. The output may be lowpass filtered with an analog filter. PWM is used widely, e.g., in electric motor control [18].

4.1.1 R-2R D/A Converters

The R-2R converters operate in either current mode or voltage mode. A current-mode R-2R DAC is shown in Fig. 13. Each switch is controlled by a bit of the word to be converted $b_1 \ldots b_b$. The resistor network creates a voltage division so that the branch currents have a binary-weighted relationship. The current through the 2R resistor at the kth level is

$$I_k = \frac{1}{2R} \frac{2^{k-1}}{2^b} (V_{ref+} - V_{ref-}), \tag{29}$$

$k = 1, \ldots, b$. The branch currents are added by the operational amplifier. The output voltage is

$$V_{out} = V_{ref-} \pm R \sum_{k=1}^{b} (b_k I_k) \tag{30}$$

with a plus or minus depending on whether $V_{ref+} < V_{ref-}$ or $V_{ref+} > V_{ref-}$, respectively. One drawback of this traditional current-mode R-2R DAC is that the output voltage swing is limited to one half of the supply voltage. Other R-2R ladder based structures are also used, each having its pros and cons [1].

4.1.2 Current-Steering D/A Converters

Current-steering DACs are suitable for very high speed applications, such as direct radio frequency (RF) synthesis. For example, 12-bit DACs are available for sample rates exceeding 4 GHz. High performance is due to the principle of current switching and there is no need to charge and discharge large internal capacitances. The basic

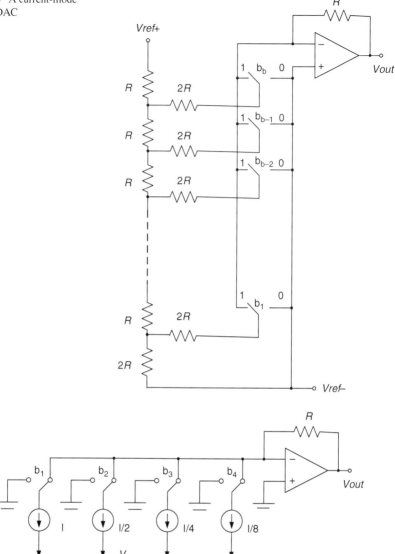

Fig. 13 A current-mode R-2R DAC

Fig. 14 A current-steering DAC

principle of a current-steering DAC is illustrated in Fig. 14. There is an array of binary-weighted current sources and each of the currents is directed to either the output or ground depending on the corresponding bit of the binary word. The currents are summed and converted to voltage by the resistor and the amplifier [10].

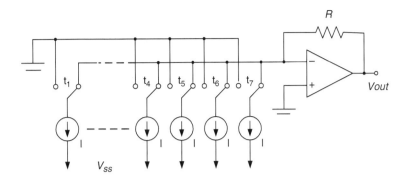

Fig. 15 A DAC using thermometer-code for current steering

4.1.3 Thermometer-Code D/A Converters

Thermometer-coding is often used in DACs. A current-steering DAC for 3-bit words could be constructed as shown in Fig. 15, where all the current sources are equally sized. One of the advantages of thermometer-code is that there is less glitching than in binary-weighted schemes.

4.1.4 Segmented D/A Converters

A popular approach to designing current-steering DACs is to use segmented converters. Segmentation means that unary weighting is used for some of the MSB current sources and binary weighting for the rest of the bits. Segmentation can improve converter throughput and reduce spurs compared to a fully binary-weighted scheme but adds to the power consumption and complexity of the DAC.

The architecture of a 10-bit segmented current-steering DAC is shown in Fig. 16 [3]. The four MSBs are thermometer-coded and the six LSBs are binary-weighted. The switching principle in a segmented DAC is illustrated in Fig. 17 using a 6-bit converter as an example. Here thermometer-coding is used for the two MSBs and this 3-bit code controls the three leftmost switches to steer the current to either the output or to the ground. The four LSBs are binary-coded and the corresponding switches are attached to a R-2R ladder that implements the binary weighting [10].

4.2 Oversampled D/A Converters

Oversampled D/A converters can be constructed using digital sigma–delta modulators. The block diagram of a sigma–delta modulator-based DAC is shown in Fig. 18. In many ways the system is a dual of a sigma–delta ADC. Modulator topologies such as those described in Sect. 3 can be used. The input is a digital word that is

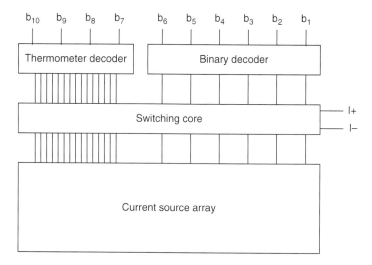

Fig. 16 Architecture of a 10-bit segmented current-steering DAC

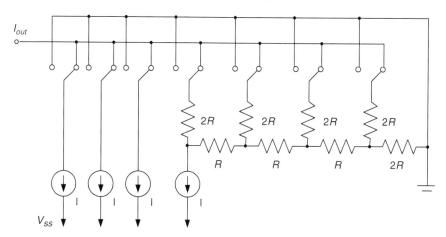

Fig. 17 A 6-bit segmented current-steering DAC

Fig. 18 Block diagram of a sigma–delta modulator-based DAC

sampled at the Nyquist rate (or a slightly higher rate). The signal is oversampled by the OSR factor, denoted by the arrow upwards. Oversampling is in principle carried out by inserting an appropriate number of zero-valued samples between each two original samples, and filtering the resulting sparse signal with a digital interpolation filter. The interpolated signal at the higher sampling rate occupies a bandwidth that is narrower by factor OSR than the original at the lower sampling rate. The signal is processed by a sigma–delta modulator which includes a quantizer and performs noise shaping, i.e., the spectrum of the quantization noise is shaped according to the order of the modulator. The quantized output of the modulator is converted analog with a 1-bit DAC. Such a DAC generating only two distinct output levels is inherently linear. Finally the output is filtered by an analog lowpass filter. Because of the oversampling, the specifications of the lowpass filter can be quite relaxed. However, the filter order should be at least one higher than the order of the sigma–delta modulator in order to match the rising spectrum of quantization noise.

Sigma–delta DACs are commonly used in applications requiring very high resolution, such as digital audio, because of the high linearity and low cost of such converters.

5 Switched-Capacitor Circuits

Switched-capacitor (SC) circuits are widely used to implement analog signal processing in integrated circuits. Several manufacturers offer SC filter components that can be configured by the user to implement a desired frequency response. The accuracy of SC circuits depends on capacitance ratios which can be well controlled in the manufacturing process, unlike absolute resistance values. That is the main reason to prefer SC circuits over active-RC implementations. However, active-RC filters are often used as prototypes when designing SC filters. SC circuits use analog voltage samples on a continuous amplitude scale. The time domain operation is discrete, and Z-domain synthesis techniques can be used to synthesize SC filters [5, 10].

The building blocks of SC circuits are operational amplifiers, switches and capacitors. The switches in CMOS integrated circuits are constructed as a parallel connection of an NMOS and a PMOS transistor in an arrangement called the transmission gate, in order to avoid the threshold voltage loss of a single transistor switch. A clock signal is needed to control the switches. Normally a 2-phase clock is used with nonoverlapping clock phases, denoted by ϕ_1 and ϕ_2 in the following.

5.1 Amplifiers

The choice of the amplifier type is affected by several parameters, such as open-loop gain, bandwidth, common-mode rejection ratio (CMRR), PSRR, noise, linearity and output dynamic range. Common amplifier types used in SC circuits are the internally compensated two-stage Miller operational amplifier and the operational transconductance amplifier (OTA), which is compensated by the load capacitance. In SC circuits, the amplifier only needs to drive capacitive loads and the output

Fig. 19 (**a**) A switched
capacitor. (**b**) Equivalent
resistor

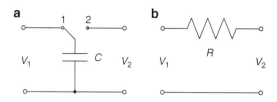

impedance can be high, which is characteristic for the OTA. This allows designing
the amplifier for higher speed and larger signal swing. If resistive loads need to
be driven, an output buffer is included in the amplifier to achieve a low output
impedance [10, 15].

In mixed-mode integrated circuits, the analog circuits need to operate together
with digital circuits, and low operation voltage is preferred because the supply
voltage affects power dissipation of the digital parts through a square relationship.
The design of analog circuits for low-voltage operation is challenging. However,
amplifiers can be implemented for sub-1 V operation with a gain that is large
enough, e.g., for headphone applications [17].

5.2   Switched Capacitors

The basic idea of SC circuits is to simulate the functionality of a resistor with
a switched capacitor [21]. Consider the circuit in Fig. 19a. When the switch is
in position 1, the capacitor is charged to voltage V_1, storing a charge $Q_1 = CV_1$.
If the switch is changed to position 2, the capacitor voltage changes to V_2 and the
amount of charge transferred from left to right is $\Delta Q = C(V_1 - V_2)$. Assuming that
the switch operates with a period T, the average current will be

$$I = \frac{C(V_1 - V_2)}{T}.$$

(31)

We may calculate the resistance R of an equivalent resistor as in Fig. 19b that passes
the same current by using the relation

$$I = \frac{V_1 - V_2}{R}$$

(32)

yielding

$$R = \frac{T}{C} = \frac{1}{f_s C}$$

(33)

where f_s is the switching frequency.

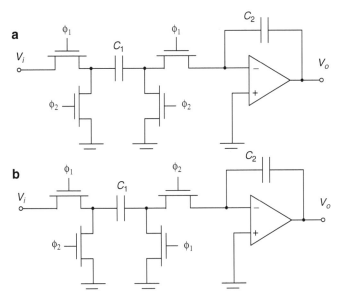

Fig. 20 (**a**) Inverting SC integrator. (**b**) Noninverting SC integrator

5.3 *Integrators*

Integrators are used as the basic building blocks in SC filters. Figure 20a shows an inverting and Fig. 20b shows a noninverting SC integrator circuit that form the basis of many SC networks. Notice that although the switches are drawn as single NMOS transistors, a PMOS transistor is usually connected in parallel. The transfer function of the inverting integrator is given by

$$H_i(z) = \frac{V_o(z)}{V_i(z)} = -\left(\frac{C_1}{C_2}\right)\frac{1}{1 - z^{-1}} \tag{34}$$

and the transfer function of the noninverting integrator is

$$H_n(z) = \frac{V_o(z)}{V_i(z)} = \left(\frac{C_1}{C_2}\right)\frac{z^{-1}}{1 - z^{-1}}. \tag{35}$$

The unit delay z^{-1} in the numerator is due to the fact that it takes the time of both clock phases for V_i to have an effect on V_o. Both of these integrator circuits are insensitive to stray capacitances that may be present at the capacitor plates.

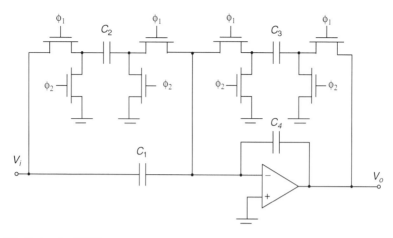

Fig. 21 First-order SC filter

5.4 First-Order Filter

A first-order SC filter based on an active-RC prototype can be constructed as shown in Fig. 21. The transfer function is

$$H_1(z) = \frac{V_o(z)}{V_i(z)} = -\frac{\frac{C_1}{C_4}(1 - z^{-1}) + \frac{C_2}{C_4}}{1 - z^{-1} + \frac{C_3}{C_4}}. \tag{36}$$

This transfer function indicates the benefit of SC filters that the poles and zeros are determined by capacitance ratios and the clock frequency rather than absolute component parameter values.

5.5 Second-Order Filter

A second-order filter is also known as the biquad. The structure of an SC biquad is shown in Fig. 22. The transfer function can be expressed as

$$H_2(z) = \frac{V_o(z)}{V_i(z)} = -\frac{a_2 + a_1 z^{-1} + a_0 z^{-2}}{b_2 + b_1 z^{-1} + z^{-2}} \tag{37}$$

where the relationships between the transfer function coefficients and the capacitances are

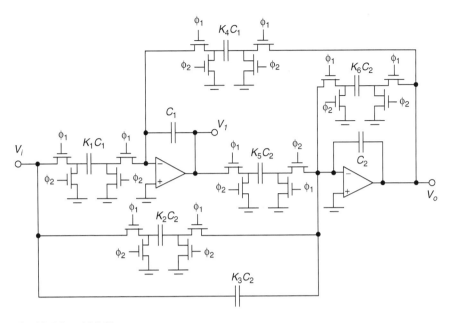

Fig. 22 Biquad SC filter

$$K_3 = a_0 \tag{38}$$

$$K_2 = a_2 - a_0 \tag{39}$$

$$K_1 K_5 = a_0 + a_1 + a_2 \tag{40}$$

$$K_6 = b_2 - 1 \tag{41}$$

$$K_4 K_5 = b_1 + b_2 + 1. \tag{42}$$

This structure is used for filters with a relatively low Q-value. Q in this case is the ratio of the pole's angular frequency and the 3 dB bandwidth [16]. Thus a large value of Q is related to a narrow bandwidth. In high-Q filters the structure of Fig. 22 tends to have a large spread of capacitance values, and alternative realizations have been developed that avoid this problem.

Higher-order SC filters can be built by cascading biquads. However, such cascade realizations are usually quite sensitive to component variations, especially in the filter passband. Less sensitive realizations can be obtained based on feedback structures developed for active-RC filters, such as the leap-frog or the follow-the-leader feedback structure. Alternatively, an SC circuit may simulate the operation of an LC ladder filter prototype [24].

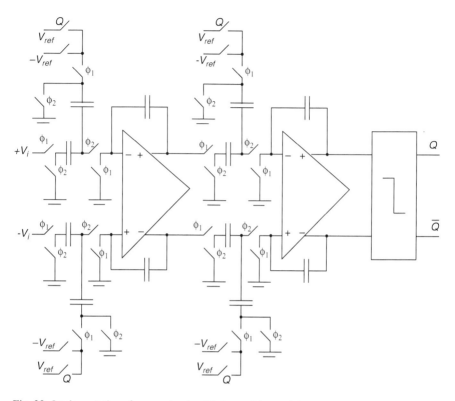

Fig. 23 Implementation of a second-order SC sigma–delta modulator

5.6 Differential SC Circuits

In many SC circuit applications, fully differential circuits are preferred over the single-ended counterparts. In a differential circuit the signal is carried by two wires and their voltage difference represents the signal value. Differential circuits are tolerant against power supply noise, clock feedthrough, switch charge injection errors and common-mode noise in general, because noise coupled to both wires does not affect the difference. A differential circuit also has a larger dynamic range of the signal. A drawback is larger circuit size, since the circuitry is doubled, except the amplifiers. As an example, Fig. 23 shows a second-order sigma–delta modulator implemented as a differential SC circuit. In this circuit the amplifiers are differential class AB OTA amplifiers [23].

5.7 Performance Limitations

SC filters can achieve a dynamic range of 80–90 dB. The useful dynamic range is limited by noise of the operational amplifier, consisting of 1/f and thermal noise, as well as the kT/C noise in the MOS transistor switches. The accuracy of the frequency response depends on several factors. Variation of the capacitance ratios is of the order of 0.1 % or higher, affecting the coefficients of the transfer function. If large capacitance ratios are needed, one of the capacitors is required to have a small size in order to keep the overall chip area reasonable, and its variations limit the accuracy. There are also parasitic capacitances between the source and drain diffusions of the transistors and the substrate in an integrated circuit.

The settling time of the operational amplifier is the main factor limiting the usable clock rate in a SC filter. The settling time is affected by the finite bandwidth and gain of the amplifier. The amplifier should be fast enough for the charge transfer to be completed during a clock phase. A highly selective filter, i.e., one with high-Q poles and a high order, suffers most response degradation due to residual settling errors. The error of the transient response is approximately of the form $e^{-t/2\tau_{op}}$, where t is time and τ_{op} is the time constant of the amplifier in the integrator configuration, being approximately the inverse of the amplifier unity-gain angular frequency. If 10 time constants are allowed for settling, the filter passband edge frequency should be at least 25 times smaller than the unity-gain frequency of the amplifier. Because of anti-aliasing considerations the clock frequency usually needs to be at least eight times the desired filter edge frequency [6].

6 Frequency Synthesis

The clock signal of an electronic system usually originates from a crystal oscillator. Crystal oscillators can generate the nominal frequency very accurately over a wide range of frequencies, from a few kilohertz up to several hundred megahertz. This is not always sufficient, as contemporary systems operate with clock rates in the gigahertz range. Mixed signal circuits can be used to multiply the clock frequency to a desired value.

6.1 Phase-Locked Loops

The block diagram of a phase-locked loop (PLL) is shown in Fig. 24. The input signal is a reference clock of frequency f_r, originating from an outside source such as a crystal oscillator. A part of the PLL is a voltage controlled oscillator (VCO) that oscillates at the desired frequency when the PLL is in locked mode. A VCO can be constructed, for instance, as a ring connection of an odd number of inverters,

Fig. 24 A phase-locked loop

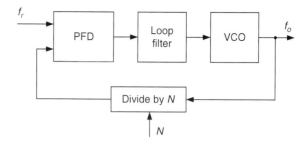

where at least one of the inverters is current-starved so that its propagation delay can be controlled by a control voltage [22]. The VCO output frequency is divided by an integer factor N and compared with the reference input in a phase-frequency detector (PFD) that generates a signal proportional to the phase and frequency difference between its inputs. This signal is lowpass filtered and used to control the VCO. In the locked mode, the output frequency is

$$f_o = N f_r. \tag{43}$$

PLL frequency synthesizers are commonly used as local oscillators in radio receivers and transmitters. A limitation of using only integer values of N is that only frequencies separated by integer multiples of f_r can be generated. This can be undesirable, e.g., if a radio standard requires narrow channel spacing.

6.2   Fractional-N Frequency Synthesis

A fractional division ratio in the PLL of Fig. 24 can be achieved by periodically switching the division ratio between two or more integer values. Suppose that the switching period has length q, and the division ratio is $N + 1$ for fraction p/q of the time and N for $(q - p)/q$ part of the time. The average output frequency is then given by

$$f_o = f_r(N + \frac{p}{q}). \tag{44}$$

However, this approach has the problem of producing unwanted spurious tones at the synthesizer output. A more advantageous method is shown in Fig. 25. A digital sigma–delta modulator is used to generate the fractional part of the division ratio and the integer part I is given separately. The sigma–delta modulator is clocked by the reference frequency. When the input of the sigma–delta modulator is $x(n)$ and the input wordlength b, the synthesizer output frequency is

$$f_o = N f_r = (I + \frac{x(n)}{2^b}) f_r. \tag{45}$$

Fig. 25 A fractional-*N*
frequency synthesizer with a
sigma–delta modulator

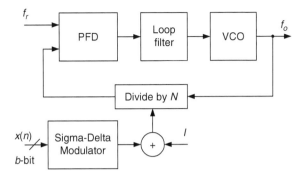

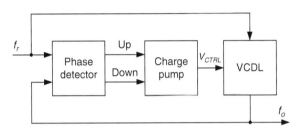

Fig. 26 A delay-locked loop

Therefore, the desired frequency can be approximated with an accuracy that depends
on the wordlength *b*. Techniques have been developed that even remove this
limitation, so that frequencies can be synthesized with perfect accuracy [2].

6.3 Delay-Locked Loops

A delay-locked loop (DLL) is a circuit that can be used for clock synchronization in
a multiple clock domain system. The block diagram of a DLL is shown in Fig. 26.
The input clock f_r goes into a voltage-controlled delay line (VCDL) and the local
clock f_o is taken from the output of the VCDL. Its phase is compared with the phase
of the input in a phase detector. Depending on the direction of the phase difference,
either an up or down signal is given to a charge pump that generates a control voltage
for the VCDL. The phases of f_r and f_o will therefore be matched. A VCDL can be
constructed, e.g., using current-starved inverters. A charge pump is built from two
current sources, one for charging and the other for discharging a capacitor according
to the state of the "Up" and "Down" signals [22].

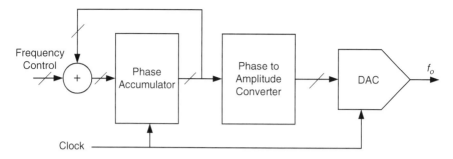

Fig. 27 A direct digital synthesis system

6.4 Direct Digital Synthesis

Direct digital synthesis (DDS) is widely used for frequency synthesis in communication systems. The DDS principle is illustrated in Fig. 27. On each clock cycle the phase accumulator adds to its contents the frequency control word from the input. The phase accumulator register is used to address a phase to amplitude converter, which is a look-up table holding samples of the waveform to be generated, typically a sinusoid. The frequency control word determines the step of the accumulator increments and therefore also the step size with which the waveform is sampled. The b-bit phase accumulator follows the modulo 2^b overflowing property. The rate of the overflows is the same as the output frequency

$$f_o = \frac{F f_s}{2^b} \qquad (46)$$

where F is the frequency control word and f_s is the clock frequency.

Typically the symmetry properties of the sine wave are exploited when constructing the table and only a quarter of the full wave is stored. Then the two MSBs are used to decode the quadrant of the sine wave and the remaining bits address a one-quadrant sine look-up table. A more efficient implementation can be achieved if, for instance, a second-degree polynomial approximation of the sine wave is used.

The DDS technique supports several modulation schemes. Frequency modulation is achieved by altering the input word. The frequency change is instantaneous and phase-continuous. Phase modulation can be accomplished by adding a programmable offset to the output of the phase accumulator. Depending on the wordlength, a very high resolution of the output frequency and phase tuning is possible using DDS. The frequency resolution is

$$\Delta f = \frac{f_s}{2^b}. \qquad (47)$$

A limit on the output frequency is set by the sampling theorem, i.e., $f_o \le f_s/2$.

The repetition period of the DDS output sequence in clock cycles is

$$P_D = \frac{2^b}{GCD(F, 2^b)} \tag{48}$$

where $GCD(F, 2^b)$ is the greatest common divisor of F and 2^b. The period is maximal if the LSB of the frequency control word is set to one. Then the errors caused by the quantized phase to amplitude converter samples are randomized, and the noise caused by the errors is whitened, at the expense of introducing an offset into the output frequency. Dithering is another method to randomize the quantization error in order to prevent idle tones [19].

References

1. R. J. Baker. CMOS Mixed-Signal Circuit Design. Wiley, Hoboken, NJ, 2009.
2. M. Borkowski. Digital Delta–Sigma Modulation: Variable Modulus and Tonal Behaviour in a Fixed-Point Digital Environment. Doctoral dissertation, University of Oulu, Acta Univ. Oul. C 306, 2008.
3. J. Deveugele and M. S. J. Steyaert. A 10-bit 250-MS/s binary-weighted current-steering DAC. IEEE J. Solid-State Circuits **41**, 320–329, 2006.
4. R. L. Freeman. Telecommunication System Engineering 3rd Ed. Wiley, New York, 1996.
5. R. L. Geiger, P. E. Allen and N. R. Strader. VLSI Design Techniques for Analog and Digital Circuits. McGraw-Hill, Singapore, 1990.
6. P. R. Gray and R. Castello. Performance limitations in switched-capacitor filters. In Y. Tsividis and P. Antognetti (eds). Design of MOS VLSI Circuits for Telecommunications, 314–333. Prentice-Hall, Englewood Cliffs, NJ, 1985.
7. O. Gustafsson and L. Wanhammar. Arithmetic. In S. S. Bhattacharyya, E. F. Deprettere, R. Leupers, and J. Takala, editors, Handbook of Signal Processing Systems. Springer, second edition, 2013.
8. IEEE Standard for Terminology and Test Methods for Analog-to-Digital Converters, IEEE Std 1241–2010. IEEE, New York, 2011.
9. E. C. Ifeachor and B. W. Jervis. Digital Signal Processing A Practical Approach. Addison-Wesley, Wokingham, England, 1993.
10. D. A. Johns and K. Martin. Analog Integrated Circuit Design. Wiley, New York, NY, 1997.
11. K. Konstantinides. Digital signal processing in home entertainment. In S. S. Bhattacharyya, E. F. Deprettere, R. Leupers, and J. Takala, editors, Handbook of Signal Processing Systems. Springer, second edition, 2013.
12. E. Korhonen. On-Chip Testing of A/D and D/A Converters: Static Linearity Testing Without Statistically Known Stimulus. Doctoral dissertation, University of Oulu, Acta Univ. Oul. C 365, 2010.
13. H. Lampinen. Studies and Implementations of Low-Voltage High-Speed Mixed Analog-Digital CMOS Integrated Circuits. Doctoral dissertation, Tampere University of Technology, Publications 362, 2002.
14. H. Lampinen and O. Vainio. A new dual-mode data compressing A/D converter. In Proc. IEEE Int. Symp. on Circuits and Systems, 429–432. Hong Kong, 1997.
15. H. Lampinen and O. Vainio. An optimization approach to designing OTAs for low-voltage sigma–delta modulators. IEEE Trans. Instrum. Meas. **50**, 1665–1671, 2001.

16. T. H. Lee. The Design of CMOS Radio-Frequency Integrated Circuits. Cambridge University Press, Cambridge, UK, 1998.

17. K. Lee, Q. Meng, T. Sugimoto, K. Hamashita, K. Takasuka, S. Takeuchi, U.-K. Moon and G. C. Temes. A 0.8 V, 2.6 mW, 88 dB dual-channel audio delta–sigma D/A converter with headphone driver. IEEE J. Solid-State Circuits **44**, 916–927, 2009.

18. W. S. Levine. Signal processing for control. In S. S. Bhattacharyya, E. F. Deprettere, R. Leupers, and J. Takala, editors, Handbook of Signal Processing Systems. Springer, second edition, 2013.

19. J. Lindeberg, J. Vankka, J. Sommarek and K. Halonen. A 1.5 V direct digital synthesizer with tunable delta–sigma modulator in 0.13-μm CMOS. IEEE J. Solid-State Circuits **40**, 1978–1982, 2005.

20. A. V. Oppenheim and R. W. Schafer. Discrete-Time Signal Processing. Prentice Hall, Englewood Cliffs, NJ, 1989.

21. R. Pallas-Areny and J. G. Webster. Analog Signal Processing. Wiley, New York, NY, 1999.

22. J. M. Rabaey, A. Chandrakasan and B. Nikolic. Digital Integrated Circuits: A Design Perspective, second edition. Prentice Hall, Upper Saddle River, NJ, 2003.

23. T. Ritoniemi, T. Karema, H. Tenhunen and M. Lindell. Fully differential CMOS sigma–delta modulator for high performance analog-to-digital conversion with 5 V operating voltage. In Proc. IEEE Int. Symp. on Circuits and Systems, 2321–2326. Espoo, Finland, 1988.

24. A. S. Sedra. Switched-capacitor filter synthesis. In Y. Tsividis and P. Antognetti (eds). Design of MOS VLSI Circuits for Telecommunications, 272–313. Prentice-Hall, Englewood Cliffs, NJ, 1985.

25. O. Vainio, D. Akopian and J. T. Astola. Systematic design of encoder and decoder networks with applications to high-speed signal processing. In Proc. IEEE Int. Symp. on Industrial Electronics, 172–175. Athens, Greece, 1995.

DSP Systems Using Three-Dimensional Integration Technology

Tong Zhang, Yangyang Pan, and Yiran Li

Abstract As three-dimensional (3D) integration technology becomes mature and starts to enter mainstream markets, it has attracted exploding interest from integrated circuit and system research community. This chapter discusses and demonstrates the exciting opportunities and potentials for digital signal processing (DSP) circuit and system designers to exploit 3D integration technology. In particular, this chapter advocates a 3D logic-DRAM integration design paradigm and discusses the use of 3D logic-memory integration in both programmable digital signal processors and application-specific digital signal processing circuits. To further demonstrate the potential, this chapter presents case studies on applying 3D logic-DRAM integration to clustered VLIW (very long instruction word) digital signal processors and application-specific video encoders. Since DSP systems using 3D integration technology is still in its research infancy, by presenting some first discussions and results, this chapter aims to motivate greater future efforts from DSP system research community to explore this new and rewarding research area.

1 Introduction

Three-dimensional (3D) integration has recently attracted tremendous interest from global semiconductor industries and is expected to lead to an industry paradigm shift [21,52]. The family of 3D integration technology includes packaging-based 3D integration such as system-in-package (SiP) and package-on-package (PoP), die-to-die and die-to-wafer 3D integration, and wafer-level back-end-of-the-line (BEOL)-compatible 3D integration. In general, all the 3D integration technologies would offer many potential benefits, including multi-functionality, increased performance, reduced power, small form factor, reduced packaging, flexible heterogeneous

T. Zhang (✉) • Y. Pan • Y. Li
Rensselaer Polytechnic Institute, Troy, NY, USA
e-mail: tzhang@ecse.rpi.edu; pany2@rpi.edu; liy16@rpi.edu

S.S. Bhattacharyya et al. (eds.), *Handbook of Signal Processing Systems*,
DOI 10.1007/978-1-4614-6859-2_26, © Springer Science+Business Media, LLC 2013

integration and reduced overall costs. As a result, 3D integration appears to be a viable enabling technology for future integrated circuits (ICs) and low-cost multi-functional heterogeneous integrated systems.

As 3D integration technology is quickly maturing and entering mainstream markets, it clearly provides a new spectrum of opportunities and challenges for circuit and system designers, and warrants significant rethinking and innovations from circuit and system design perspectives. Existing work mainly came from two research communities, i.e., electronic design automation (EDA) and computer architecture. In the context of EDA, a significant amount of research has been done in the areas of physical design (e.g., see [1, 16, 28, 58, 67]) and thermal analysis and management (e.g., see [13, 24, 36, 64, 66]). As pointed out in [5], architectural innovations will play an at least equally important role as EDA innovations in order to fully exploit the potential of 3D integration. At the architectural front, existing work predominantly focused on high-performance general-purpose microprocessors (e.g., see [7, 19, 31, 41, 49–51, 65, 76]), as 3D integration clearly provides a promising solution to address the looming memory wall problem [78] in computing systems.

Due to the demands of an ever-increasing number of signal processing applications that require high throughput at low power, digital signal processing (DSP) integrated circuits, including both programmable digital signal processors and application-specific DSP circuits, have become, and will continue to be, one of the main drivers for the semiconductor industry [69]. Nevertheless, little attention has been given to exploiting the potential of 3D integration to improve various programmable and/or application-specific DSP systems. This chapter represents our attempt to fill this missing link and, more importantly, motivate greater future efforts from DSP system research community to explore this new and rewarding research area. In this chapter, after briefly reviewing the basics of the 3D integration technology, we first discuss the rationale and opportunities for DSP systems, both programmable digital signal processors and application-specific signal processing circuits, to exploit the 3D integration technology. In particular, we advocate a 3D logic-DRAM integration paradigm, i.e., one or multiple DRAM dies are stacked with one logic die, for 3D DSP systems, and present an approach for 3D DRAM architecture design. Then we present two case studies on applying 3D logic-DRAM integration to clustered VLIW (very long instruction word) digital signal processors and application-specific video encoders in order to quantitatively demonstrate the promise of 3D DSP system design.

2   Overview of 3D Integration Technology

The objective of 3D integration is to enable electrical connectivity between multiple active device planes within the same package. Various 3D integration technologies are currently pursued and can be divided into the following three categories:

1. *3D packaging technology*: It is enabled by wire bonding, flip-chip bonding, and thinned die-to-die bonding [14]. As the most mature 3D integration technology,

it is already being used in many commercial products, noticeably in cell phones [26, 63]. Its major limitation is very low inter-die interconnect density (e.g., only few hundreds of inter-die bonding wires) compared to the other emerging 3D integration technologies.

2. *Transistor build-up 3D technology*: It forms transistors inside on-chip interconnect layer [3], on poly-silicon films [18], or on single-crystal silicon films [38, 39]. Although a drastically high vertical interconnect density can be realized, it is not readily compatible to existing fabrication processes and is subject to severe process temperature constraints that tend to dramatically degrade the circuit electrical performance.

3. *Monolithic, wafer-level, back-end-of-the-line (BEOL) compatible 3D technology*: It is enabled by wafer alignment, bonding, thinning and inter-wafer interconnections [53]. Realized by through silicon vias (TSVs), inter-die interconnects can have very high density. The TSVs can be formed before/during bonding (via-first) or after bonding (via-last) [52].

Among the above three different categories of 3D integration technologies, wafer-level BEOL-compatible 3D integration appears to be the most promising 3D integration for high-volume production and is quickly becoming mature because of tremendous recent research efforts from both academia and industries [8, 12, 33, 44, 47, 53, 56, 57, 61, 62]. For example, SEMATECH [68], the biggest semiconductor technology research and development consortium, launched its 3D program in early 2005. It has already generated a comprehensive cost-model and a draft of a 3D roadmap for the International Technology Roadmap for Semiconductors (ITRS) [34], and begun tool and process benchmarking.

From IC design perspective, vertical interconnects enabled by 3D integration technology, particularly wafer-level BEOL-compatible 3D integration, provide several distinct advantages in a straightforward manner, including:

- *Integration of disparate technologies*: Fabrication processing technologies specific to functions such as DRAM and RF circuits are incompatible with that of high performance logic CMOS devices. Converging these into a single 2D chip inevitably results in performance compromises (e.g., the density loss of embedded DRAM [54]). Obviously, 3D integration technology provides a natural vehicle to enable independent optimal fabrication and integration of these functions.

- *Massive bandwidth*: 3D integration can provide a massive vertical inter-die interconnect bandwidth. This can enable a very high degree of operational parallelism for improving the system performance and/or reducing power consumption. Meanwhile, this makes it possible to carry out a significant amount of cross-die co-design and co-optimization.

- *Wire-length reduction*: Expanding from 2D to 3D domain, we may possibly replace a lateral wire of tens or hundreds of microns with a 2–3 μm tall vertical via. Since interconnects play an important role in determining overall system speed and power consumption [15, 29], such wire-length reduction can significantly improve the IC system performance.

3 3D Logic-Memory Integration Paradigm for DSP Systems: Rationale and Opportunities

This section advocates a 3D logic-memory integrated paradigm, in particular the integration of a single logic die and multiple memory dies, for 3D integrated DSP systems. The motivation is twofold:

1. *Its great practical feasibility*: In spite of those appealing advantages of 3D integration as described above, 3D integration tends to have two apparent potential drawbacks: thermal effect and yield loss. Dies stacked in the middle are subject to bigger heat resistance and hence higher temperature, which could greatly degrade the circuit performance and reliability, particularly when the dies have very high activities and dissipate a large amount of heat. Although prior work from the EDA research community has developed techniques such as thermal via placement [17, 24] to mitigate the thermal issue to a certain degree, 3D stacked dies are fundamentally subject to the thermal problem, particularly for dies with high activity and hence high power consumption. Meanwhile, since 3D integration, especially wafer-level BEOL-compatible 3D integration, cannot use known good die (KGD), it is inherently subject to possible yield loss, which may result in overall cost increase. Moreover, the 3D integration process itself may introduce some defects, leading to an even larger yield loss.

   Compared with 3D logic–logic integration, 3D logic-memory integration could be much less vulnerable to the above two issues and hence has a much greater practical feasibility, at least in the foreseeable future. Compared with logic dies, digital memories tend to consume much less energy and hence dissipate much less heat. Meanwhile, because of the homogeneous functionality and regular architecture of digital memories, simple yet effective testing and fault tolerance techniques can be used to greatly improve yield.

2. *Its great potential to improve performance*: With the memory intensive nature, signal processing can greatly benefit from 3D memory stacking. Most signal processing functions can well leverage the large memory storage capacity and massive logic-memory interconnect bandwidth to straightforwardly improve the overall signal processing system performance. Moreover, due to the large algorithm and architecture design flexibility inherent in most signal processing functions, the 3D logic-memory integration paradigm directly enables a large space to pursue optimal signal processing algorithm/architecture design solutions and explore a wide spectrum of design trade-offs. This will be further demonstrated by two case studies presented later in this chapter.

Under a 3D logic-memory integration framework, we further discuss the opportunities and research directions that could potentially lead to effective explorations of 3D logic-memory integration in both programmable digital signal processors and application-specific signal processing ICs. Programmable digital signal processors are typically characterized with specialized hardware and instructions geared to efficiently executing common mathematical computations used in signal processing [45]. Being capable of well exploiting the instruction-level parallelism (ILP) abundant in many signal processing algorithms, VLIW (very long instruction word) architecture is being widely used to build modern programmable digital signal processors, although VLIW architecture tends to require more program memory. Meanwhile, enabled by continuous technology scaling, programmable digital signal processors are quickly adopting multi-core architecture in order to better handle various high-end communication and multimedia applications. As a result, similar to the memory wall problem in general purpose processors, programmable digital signal processors also experience an increasingly significant gap between on-chip processing speed and off-chip memory access speed/bandwidth. Hence, on-chip cache memory is also being used in programmable digital signal processors. However, because programmable digital signal processors are typically used in embedded systems with real-time constraints but the direct use of cache may introduce uncertainty of program execution time, design and use of on-chip cache memory in programmable digital signal processors can be very sophisticated and hence have been well studied (e.g., see [2, 20, 22, 23, 59, 77]).

It is very intuitive that 3D logic-memory integration provides many opportunities to greatly improve programmable digital signal processor performance, which include but certainly are not limited to:

- The 3D stacked memory can serve for the entire cache memory hierarchy with much reduced memory access latency. This can directly reduce the memory system design complexity and energy consumption induced by data movement among cache memory hierarchy.
- With the high-capacity memory storage, we could possibly design a cache memory system with much improved program execution time predictability. One possible solution is to use a coarse-grained caching of the program code, e.g., cache an entire sub-routine instead of fine-grained cache line as in current design practice, which can largely reduce the run-time cache misses and hence improve the execution time predictability by leveraging the large cache capacity enabled by 3D memory stacking.
- With sufficient 3D stacked memory storage capacity, signal processing data-flow diagram synthesis and scheduling may be potentially carried out in a much more efficient way. Modern data-flow diagram synthesis and scheduling techniques are typically memory constrained and inevitably involve certain design trade-off. 3D logic-memory integration may provide a unique opportunity to develop much better signal processing data-flow diagram synthesis and scheduling solutions.

In sharp contrast to programmable processors, application-specific signal processing IC design typically involves very close interaction between specific signal processing algorithm design and VLSI architecture design/optimization, which enables a much greater design flexibility and trade-off spectrum. Under the 3D logic-memory integration framework, the high memory storage capacity with massive logic-memory interconnect bandwidth naturally leads to much more space to explore signal processing algorithm and logic architecture design. Meanwhile,

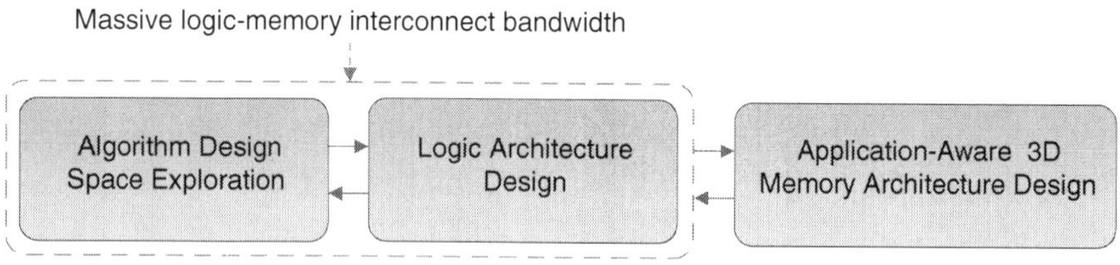

Massive logic-memory interconnect bandwidth

Fig. 1 Joint design methodology for 3D logic-memory integrated application-specific DSP ICs

leveraging the massive logic-memory interconnect bandwidth, we can customize the memory architecture and storage organization geared to the specific signal processing algorithm, leading to further potential to improve the overall system performance. Figure 1 illustrates the close coupling among signal processing algorithm, logic architecture, and 3D memory architecture design inherent in 3D logic-memory integrated application-specific DSP IC design.

Although specific materializations of the above joint design will certainly vary for different DSP algorithms, we believe that the following intuitive general guidelines can help designers to explore system design optimization opportunities:

- We should always fully analyze the memory access characteristics of the specific DSP algorithm, and try to accordingly customize the 3D memory architecture and logic-memory interconnect. This could help to greatly reduce the memory access energy consumption and/or maximize the overall signal processing system speed and parallelism.
- It is highly desirable to explore the potential of trading memory storage capacity for improving signal processing performance and/or computational efficiency. Some signal processing algorithms, which may be considered as infeasible due to their demand for large memory, could possibly become the most efficient options under the 3D logic-memory integration framework.
- Besides straightforward data storage, the 3D stacked memory could also be used to realize functions like table look-up or content addressable storage. For many signal processing algorithms, particularly those involving a significant amount of search operations, embedding table look-up or content address storage functions in the 3D memory could potentially achieve noticeable overall system performance.

To further demonstrate the design of 3D logic-memory integrated DSP systems, the remainder of this chapter presents two case studies in the context of both programmable digital signal processor and application-specific DSP IC. As the mainstream high capacity memory technology, DRAM appears to be the favorable option for the 3D logic-memory integration. Hence, in the following, we first present a practical 3D DRAM design strategy that could leverage 3D integration to improve DRAM design itself, then we will present the two case studies, including a 3D integrated VLIW digital signal processor and a 3D integrated H.264 video encoder.

4   3D DRAM Design

In current design practice, high capacity DRAMs have a hierarchical architecture that typically consists of banks, sub-banks, and sub-arrays. With its own address and data bus, each DRAM bank can be accessed independently from the other banks. Each bank is divided into sub-banks, and the data are read/written from/to one sub-bank during each memory access to one bank. Each sub-bank is further divided into sub-arrays, and each sub-array contains an indivisible array of DRAM cells surrounded by supporting peripheral circuits such as word-line decoders/drivers, sense amplifiers (SAs), and output drivers etc. Reference [35] gives detailed discussions on modern DRAM circuit and architecture design. DRAM cells, peripheral circuits, and interconnect, including H-tree outside/inside banks and sub-banks and the associated buffers, altogether determine the overall DRAM performance, including silicon area, access latency, and energy consumption.

The pitches of word-lines and bit-lines within each DRAM sub-array are in the range ~ 100 nm and below, while the pitch of TSVs for 3D integration is at least a few μm or tens of μm in order to allow low-cost fabrications of TSVs with high manufacturability. To well accommodate such significant pitch size mismatch, this section presents a practical 3D DRAM architecture design using a coarse-grained inter-sub-array 3D partitioning. The key is that only the address and/or data I/O wires of each memory sub-array associate with TSVs and the entire memory sub-array including cell array and peripheral circuits remain exactly the same as in 2D design. The top view of a 3D memory with inter-sub-array 3D partitioning looks exactly the same as its 2D counterpart, except that each leaf is a 3D sub-array set. Let n denote the number of memory dies being integrated. Each 3D sub-array set contains n individual 2D sub-arrays, and each 2D sub-array contains the memory cells and its own complete peripheral circuits such as decoders, drivers, and SAs. Within each 3D sub-array set, all the 2D sub-arrays only share address and data input/output through TSVs. The global address/data routing outside and inside each bank is simply distributed across all the n layers through TSVs, as illustrated in Fig. 2. Let N_{data} and N_{add} denote the data access and address input bandwidth of each 3D sub-array set, and recall that n denotes the number of memory layers. Without loss of generality, we assume that N_{data} and N_{add} are divisible by n. As illustrated in Fig. 2, the N_{data}-bit address bus and N_{add}-bit data bus are uniformly distributed across all the n layers and are shared by all the n 2D sub-arrays within the same 3D sub-array set through a TSVs bundle.

Compared with direct commodity DRAM die stacking, in the above presented design strategy, the DRAM global address/data routing is shared among DRAM dies to reduce the routing area overhead in each DRAM die. Since global routing tends to play an important role in determining the overall DRAM performance, this approach can achieve considerable performance gain. For the purpose of evaluation, we modified CACTI 5, a recent version of the widely used memory modeling tool CACTI [9], to support the above inter-sub-array 3D partitioning strategy. Using a 1 Gb DRAM as an example, we estimate the area, latency and energy of such 3D

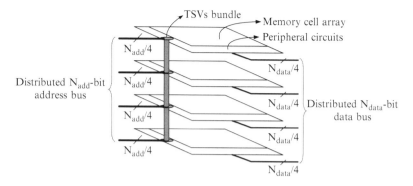

Fig. 2 Illustration of a 3D sub-array set with four layers

Table 1 Estimated results for an example 1 Gb DRAM using different design approaches at 65 nm technology node based on ITRS projection

		3D die packaging			Proposed 3D DRAM		
	2D DRAM	2-Layer	4-Layer	8-Layer	2-Layer	4-Layer	8-Layer
Footprint (mm²)	92	69	44	33	41	19	10
Access latency (ns)	26	23	19	18	20	16	14
Access energy (nJ)	2.20	1.75	1.50	1.40	1.50	1.40	1.35

DRAM at the 65-nm technology node. For the purpose of comparison, we further consider conventional 2D DRAM and 3D die packaging of separate DRAM dies without the use of TSVs (referred to as 3D die packaging). Table 1 shows the estimated results that clearly demonstrate the noticeable performance gain of such inter-sub-array 3D partitioning for 3D DRAM architecture design.

5 Case Study I: 3D Integrated VLIW Digital Signal Processor

The VLIW architecture is conventionally characterized by the fact that each instruction specifies several independent operations. This is compared to RISC (reduced instruction set computer) instructions that typically specify one operation and CISC (complex instruction set computer) instructions that typically specify several dependent operations. VLIW processors typically contain several identical processing clusters, where each cluster consists of a collection of execution units, such as integer ALUs, floating-point ALUs, load/store units, and branch units. All the clusters can execute several operations in one instruction at the same time.

Because VLIW processors can readily exploit the ILP and most signal processing applications have abundant ILP, most existing programmable digital signal processors inherently use VLIW architectures. VLIW processors rely on a compiler to

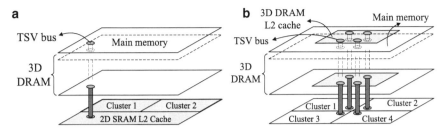

Fig. 3 Illustration of (**a**) straightforward 3D DRAM stacking, and (**b**) migrating on-chip L2 cache into 3D DRAM domain

explicitly exploit parallelism, which is different from RISC and CISC that rely on hardware to discover parallelism on-the-fly. The compiler schedules the operations based on expected architectural latencies, including the instruction execution time and the memory access time. Because of the data intensive nature of most signal processing applications, VLIW digital signal processors demand careful design and optimization of cache and memory hierarchy, which has been extensively studied (e.g., see [2, 20, 22, 23, 77]). It is very intuitive that VLIW digital signal processor and high-capacity DRAM integration enabled by 3D integration technologies can further improve the memory hierarchy performance and hence improve the overall system performance.

5.1 3D DRAM Stacking in VLIW Digital Signal Processors

In current design practice, a VLIW digital signal processor typically has a shared on-chip L2 cache and connects with an off-chip DRAM, where each L2 cache miss may result in significant penalty due to the long latency of off-chip data access. We have two obvious options to explore the design of VLIW digital signal processors with 3D DRAM stacking:

1. The most straightforward design option is to directly stack a VLIW digital signal processor die with 3D DRAM, as illustrated in Fig. 3a. This can be considered as moving the off-chip DRAM into the processor chip package. Clearly, such a straightforward design option can directly reduce the on-chip L2 cache miss penalty due to much less processor-DRAM access latency enabled by the 3D integration. Moreover, since 3D integration can enable a massive inter-die interconnect bandwidth through TSVs, the cache-memory communication bandwidth can be largely increased, which can further improve the overall system performance.
2. Beyond the above straightforward design option, we further study the potential of migrating the shared on-chip L2 cache into the 3D DRAM domain. As illustrated in Fig. 3b, this makes it possible to put more clusters on the VLIW digital signal

processor die, which can directly increase the system parallelism and potentially improve the overall computing system performance without increasing the chip footprint. In this context, the 3D DRAM has a heterogeneous structure and covers two levels of the entire memory hierarchy. Because L2 cache demands very short access latency, the 3D DRAM L2 cache should be particularly customized in order to achieve a comparable access latency as its on-chip SRAM counterpart.

5.2   3D DRAM L2 Cache

The 3D VLIW architecture configuration as shown in Fig. 3b migrates the on-chip L2 cache into the 3D DRAM domain. Since L2 cache access latency plays a critical role in determining the overall computing system performance, one may intuitively argue that, compared with on-chip SRAM L2 cache, 3D DRAM L2 cache may suffer from much longer access latency and hence result in significant performance degradation. In this section, we show that this intuitive argument may not necessarily hold true. In particular, as we increase the L2 cache capacity and the number of DRAM dies, the 3D DRAM L2 cache may have a latency comparable or even shorter than an SRAM L2 cache.

Commercial DRAM is typically much slower than SRAM mainly because, being commodity, DRAM has been always optimized for density and cost rather than speed. The speed of DRAM can be greatly improved by two approaches at the cost of density and cost, including:

1. We can reduce the size of each individual DRAM sub-array to reduce the memory access latency at the penalty of storage density. With shorter lengths of word-lines and bit-lines, a smaller DRAM sub-array can directly lead to reduced access latency because of the reduced load of the peripheral circuits.

2. We can adopt the multiple threshold voltage (multi-V_{th}) technique that has been widely used in logic circuit design [30], i.e., we still use high-V_{th} transistors in DRAM cells to maintain a sufficiently low DRAM cell leakage current, while using low-V_{th} transistors in peripheral circuits and H-tree buffers to reduce the latency. Such multi-V_{th} design is not typically used in commodity DRAM since it will increase leakage power consumption of peripheral circuits and, more importantly, will complicate the DRAM fabrication process and hence incur a higher cost.

Moreover, as we increase the L2 cache capacity, the global routing will play a bigger role in determining the overall L2 cache access latency. By using the above presented 3D DRAM design strategy, we can directly reduce the latency incurred by global routing, which will further contribute to reducing 3D DRAM L2 cache access latency compared with 2D SRAM L2 cache.

To evaluate the above arguments, Fig. 4 shows the comparison of access latency of 2D SRAM, single-V_{th} 2D DRAM, and multi-V_{th} 2D DRAM, under different L2 cache capacities, at 65 nm node. The results show that, as we increase the capacity of

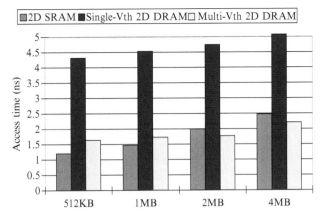

Fig. 4 Comparison of access latency of 2D SRAM, 2D single-V_{th} DRAM, and 2D multi-V_{th} DRAM at 65 nm node

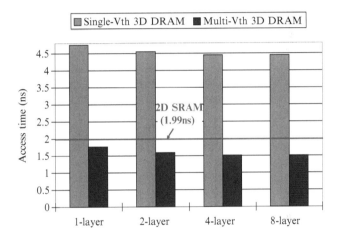

Fig. 5 Comparison of 2 MB L2 cache access latency as we increase the number of DRAM dies

L2 cache, multi-V_{th} DRAM L2 cache already excels its SRAM counterpart in terms of access latency, even without using 3D to further reduce global routing delay. This essentially agrees with the conclusions drawn from a recent work on embedded SOI DRAM design [4], which shows that embedded SOI DRAM can achieve shorter access latency than its SRAM counterpart at a capacity as small as 256 kB.

Figure 5 further shows that, when we use the 3D DRAM design strategy presented in Sect. 4 to implement a 2 MB L2 cache, the access latency advantage of multi-V_{th} 3D DRAM over 2D SRAM will further improve as we increase the number of DRAM dies. This is mainly because, as we stack more DRAM dies, the footprint and hence latency incurred by global routing will accordingly reduce.

Table 2 Configuration parameters used in Trimaran [10]

Frequency	200 MHz
Number of DRAM dies	6
Die size	58.5 mm^2
Parameter of one cluster	ALU number: 4; register file: 2 kB multi-ported; area size: 2.77 mm^2; technology: 90-nm
2D SRAM instruction cache (per cluster)	4 kB, 2 way, 16 byte blocks; access latency:1.26 ns
2D SRAM data cache (per cluster)	4 kB, 2 way, 16 byte blocks; access latency: 1.26 ns
Shared 2D SRAM L2 cache	256 kB, 2 way, 32 byte blocks; access latency: 2.43 ns; area: 5.69 mm^2; technology: 90-nm
Shared 3D DRAM L2 cache	512 kB, 2 way, 32 byte blocks; access latency: 4.12 ns
Shared main memory	1 GB, 4 kB page size; access latency: 22.5 ns; technology: 65-nm

5.3 Performance Evaluation

To evaluate the above presented two 3D VLIW digital signal processor architecture design options as illustrated in Fig. 3a,b, we use the Trimaran simulator [10] to carry out simulations over a wide range of signal processing benchmarks. Trimaran is an integrated compilation and performance monitoring infrastructure and covers HPL PlayDoh architecture [40]. We further integrate the memory subsystems of M5 [6] to strengthen the memory system simulation capability of Trimaran. For the 3D VLIW processor and DRAM integration, we assume that the VLIW processor and 3D DRAM are designed using 90- and 65-nm technologies, respectively. For the VLIW processor, each cluster contains 4 ALUs, a 2 kB multi-ported register file, a private 4 kB instruction cache and a 4 kB data cache. All the clusters share one 256 kB L2 cache.

We estimate the cluster silicon area based on [42] that reports a silicon implementation of a VLIW digital signal processor at 0.13-μm technology node similar to the architecture being considered in this work. In [42], one cluster, consisting of 6 ALUs and a 16 kB single-ported register file, occupies about 3.2 mm^2, and the entire VLIW digital signal processor contains 16 clusters and occupies 155 mm^2 (note that besides the 16 clusters, it contains two MIPS CPUs and a few other peripheral circuits). Accordingly, we estimate that one cluster with 4 ALUs and a 2 kB multi-ported register file is about 2.77 mm^2 at 90-nm technology node, the 256 kB L2 cache is 5.69 mm^2, and the entire VLIW processor die size is 58.5 mm^2. Therefore, when we move the 256 kB L2 cache into 3D DRAM domain, we can re-allocate the silicon area to two additional clusters. We have that 58.5 mm^2 die size with six stacked DRAM dies can achieve a total storage capacity of 1 GB with 128-bits output width at 65-nm node. Table 2 lists all the basic configuration parameters used in the simulations.

We use the conventional system architecture with off-chip DRAM as a baseline configuration, where the number of clusters is 2 and the L2 cache size is 256 KB.

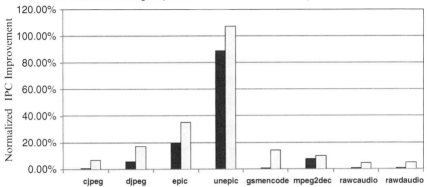

Fig. 6 Normalized IPC improvement of the two 3D VLIW digital signal processor design options over the baseline

The latency of off-chip main memory access is set to 500 cycles. For the first 3D VLIW digital signal processor design option as shown in Fig. 3a, the 1 GB main memory is directly stacked with the processor die. For the second 3D VLIW digital signal processor design option as shown in Fig. 3b, the 256 KB L2 cache is migrated into 3D DRAM and we put two more clusters into the processor die without increasing the die area. Using the Trimaran simulator and a variety of media benchmarks [46], we evaluate the effectiveness of these two 3D design options in terms of IPC (instructions per cycle). As shown in Fig. 6, the results clearly demonstrate the potential performance advantages of the 3D VLIW digital signal processor design options. The first design option with direct 3D DRAM stacking can lead to 10 %~80 % IPC improvement over the baseline scenario. Compared with direct 3D DRAM stacking, the second design option can further improve up to 10 % IPC by migrating the L2 cache into the 3D domain and increasing computational parallelism.

6 Case Study II: 3D Integrated H.264 Video Encoder

This section presents a case study on exploiting 3D memory stacking to improve application-specific video encoder design efficiency. The key and most resource demanding operation in video encoding is motion estimation, which has been widely studied over the past two decades and many motion estimation algorithms have been developed, covering a wide spectrum of trade-offs between motion estimation accuracy and computational complexity. Most existing algorithms perform block-based motion estimation, where the basic operations are calculating and comparing

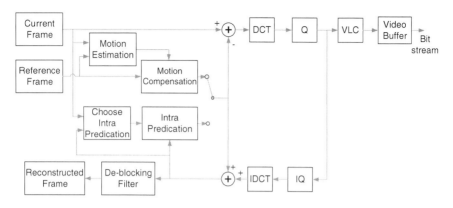

Fig. 7 H.264/AVC encoder structure

the matching functions between the current image macroblock (MB) and all the candidate MBs inside a confined area in the reference image frame(s). Hence, video encoding demands the simultaneous storage of several or many image frames in digital memory.

In current design practice, due to limited on-chip SRAM storage resources, all the current and reference image frames are stored in an off-chip high-capacity memory, most likely a DRAM, and on-chip SRAMs are used as buffers to streamline video encoding. Nevertheless, on-chip SRAM buffers still tend to occupy more than half of the silicon area of the entire encoder [32], most of which are used for motion estimation. As the video resolution and frame rate continue to increase and multi-frame-based motion estimation is being widely used, memory storage and access will inevitably continue to remain as the most critical issue in video encoder IC design.

It is very intuitive that 3D memory stacking can be the most viable option to address this issue and hold a great promise to largely improve overall video encoder performance in terms of silicon area and energy efficiency. In this section, we quantitatively evaluate the potential using H.264/AVC encoder design as a test vehicle.

6.1 H.264 Video Encoding Basics

As the latest video coding standard of ITU and ISO Moving Picture Experts Group (MPEG), H.264/AVC [75] has been widely adopted in real-life applications due to its excellent video compression efficiency. Figure 7 illustrates the typical H.264/AVC encoder structure. Each frame is processed in the unit of MB with a size of 16×16, each of which is encoded in either intra or inter mode. In the intra mode, the predicted MB is formed from pixels in the current frame that have been encoded, decoded and reconstructed. While in the inter mode, the predicted MB is formed by motion-compensated prediction from one or more reference frames. The prediction is subtracted from the current MB to produce a residual or difference MB, which is further transformed using DCT based on every 4×4 block and quantized to generate a sequence of coefficients. These coefficients are then reordered and entropy encoded, together with information about prediction mode, quantization step size, motion vector, etc, required to decode the MB, to form the compressed bitstream. Meanwhile, the quantized MB coefficients are decoded and reconstructed, which will be used in the encoding of subsequent MBs. Due to the nature of block transformation, blocking artifacts are inevitably introduced, and de-blocking filtering is used to reduce such block artifacts.

To achieve a higher compression efficiency than previous video coding standards, H.264/AVC adopted several more aggressive techniques, including intra prediction, 1/4-pixel motion estimation with multiple reference frames (MRF), variable block sizes (VBS), context-based adaptive variable length coding, rate-distortion opti-mization mode decision, etc. At the cost of a higher video compression efficiency, these techniques inevitably incur higher implementation complexity. In an H.264 encoder, almost every component has to access memory, and motion estimation has the most stringent memory resource demand in terms of memory storage and memory access bandwidth [74]. As the key operation in motion estimation, SAD (sum of absolute differences) computation is given by

$$SAD = \sum_{i=1}^{N} \sum_{j=1}^{N} \left| C(i,j) - R^{(m,n)}(i,j) \right|, \tag{1}$$

where C and $R^{(m,n)}$ denote the luminance intensity of current and reference MBs with the size of $N \times N$ pixels, and (m,n) presents the motion vector relevant to the current MB. Clearly, SAD computation involves two tasks, including image data retrieval and SAD arithmetic computation. Image data retrieval involves the fetch of current MBs and candidate MBs. Given a motion vector and the location of current MB, a candidate MB is formed within the corresponding search region in the reference frame. Different motion estimation algorithms have different rules on screening the spectrum of motion vectors to find an optimal motion vector. Hence, implementations of different motion estimation algorithms mainly differ in the sequence and number of candidate MBs that must be fetched from the reference frame(s), and accordingly they differ in the number of SAD computations.

6.2   3D DRAM Stacking in H.264 Video Encoder

Due to its memory intensive nature, a video encoder can very naturally benefit from 3D DRAM stacking. To exploit 3D memory stacking in video encoder design, the most straightforward option is to directly stack the existing video encoder and commodity DRAM die(s) without any modification or optimization. In this context, system performance gain only comes from using TSVs to replace chip-to-chip links, which can directly result in lower DRAM access energy consumption without demanding any changes on the design of both the video codec and DRAM. As discussed in [25, 55], for chip-to-chip data links, power is mainly dissipated on link termination resistors and drivers. In the context of 3D integration, the much shorter signal path within the range of tens of μm completely obviates the drivers and terminations. Since the capacitive load of TSV is much lower compared with that of chip-to-chip bus, CMOS buffers can be simply used to connect DRAM dies and logic die. Therefore, as demonstrated in [25], a typical DDR3 data link with an effective dynamic on-die termination (ODT) resistance of 25 Ω dissipates 15.9 mW, while its counterpart in 3D integration only consumes 0.38 mW, representing about 40× power reduction. Meanwhile, it is clear that the DRAM data fetch latency will also largely reduce in the 3D integration scenario.

Very intuitively, much higher performance gains should be expected if we could further optimize the video encoder and DRAM design to fully take advantage of the 3D integration implementation platform. This case study investigates the potential of customizing 3D stacked DRAM architecture geared to the most resource demanding motion estimation operation.

6.2.1 Architecting DRAM for Image Storage

It is feasible and certainly preferable to store both current and a few corresponding reference frames in 3D stacked DRAM. Due to the access latency and energy consumption overhead induced by DRAM row activation, it is always favorable to make as many consecutive reads/writes as possible on the same row before switching to another row. Therefore, if image blocks are linearly row-by-row (or column-by-column) mapped to the physical address space, a significant amount of row activations will occur, leading to high access latency and energy consumption. It is very straightforward that, instead of such a linear translated mapping, a MB-by-MB mapping geared to motion estimation should be used [37], i.e., each row stores the luminance intensity data of one or more MBs.

Usually a search region spans several MBs in reference frames. Let each MB be $N \times N$, and S_W and S_H represent the width and height of the search region, each search region spans $S_{BW} \times S_{BH}$ MBs in the reference frame, where

$$S_{BW} = 2 \cdot \left\lceil \frac{S_W - N}{2 \cdot N} \right\rceil + 1,$$

$$S_{BH} = 2 \cdot \left\lceil \frac{S_H - N}{2 \cdot N} \right\rceil + 1.$$

The most practically interesting scenario is the so-called *per-row DRAM-to-logic data delivery*, where the 3D stacked DRAM delivers a candidate MB row-by-row to the logic die, i.e., during each clock cycle, the 3D stacked DRAM delivers the

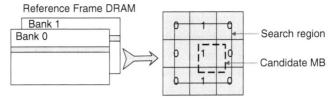

Total 9 MBs from the reference frame

Fig. 8 Frame memory organization for row-by-row data delivery when $S_{BW} = S_{HW} = 3$

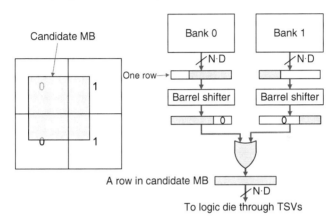

Fig. 9 Combine the data from two banks to form a row in candidate MB

luminance intensity data of one row within a candidate MB to the logic die. In current design practice, a DRAM chip usually consists of multiple banks, which can be accessed independently, in order to improve the data access parallelism. Since each candidate MB at most spans 2×2 MBs within the search region and each row in a candidate MB at most spans two MBs in reference frame, we store each image frame in two banks that alternatively store all the MBs row-by-row. For example, suppose each MB is 16×16 and the search region is 32×32, then we have $S_{BW} = S_{HW} = 3$, as illustrated in Fig. 8, where the numbers (i.e., either 0 or 1) within the nine MBs indicate the index of the two DRAM banks.

Given the current MB and motion vector, the candidate MB from the search region can be readily retrieved row-by-row. Since one DRAM word-line contains multiple consecutive rows within the same search region, once each DRAM word-line is activated and the data of the entire word-line are latched in the sense amplifiers, multiple rows within the search region can be sent to the logic die before we switch to another word-line. Assume luminance intensity of every pixel is represented by D bits, as shown in Fig. 9, each clock cycle two DRAM banks delivers two N-pixel rows within the search region, which are further shifted and combined to form one row in the candidate MB according to the present motion vector.

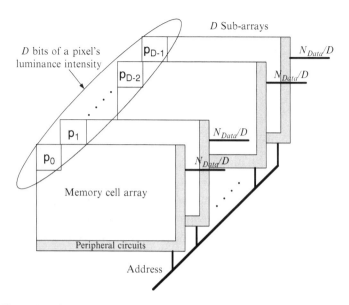

Fig. 10 Illustration of *bit-plane* data mapping strategy. All D sub-arrays share the same row and column address. N_{data}-bit data bus is uniformly distributed across D sub-arrays, where $N_{data} = N \cdot D$

We note that since each row contains several MBs and each MB may contain a few hundred bits (e.g., given $N = 16$ and $D = 8$, each MB contains 2,048 bits), a row within a bank should be further partitioned into several segments in order to maintain a reasonable access speed. Therefore, we follow the conventional DRAM design practice that employs a bank→sub-bank→sub-array hierarchy. Each DRAM bank consists of several identical sub-banks and at one time only one sub-bank can be accessed. By segmenting each row within one sub-bank, each sub-bank is further partitioned into a certain number of sub-arrays that collectively carry out DRAM access, and each sub-array contains an indivisible array of DRAM cells surrounded by supporting peripheral circuits such as address decoders/drivers, sense amplifiers and output drivers etc.

To map the MB luminance intensity data onto the sub-arrays within each sub-bank, we can use a *bit-plane* data mapping strategy to ensure the memory data access regularity across all the sub-arrays. Recall that the luminance intensity of one pixel uses D bits. We partition each sub-bank into D sub-arrays, $A_0, A_1, \ldots, A_{D-1}$, and each sub-array A_i stores the i-th bit of the luminance intensity data, as depicted in Fig. 10. Clearly, this bit-plane storage strategy ensures that all the sub-arrays within the same sub-bank can readily share the same row and column address and have a minimum sub-array data I/O routing overhead. In particular, each sub-array only needs to send N bits for realizing per-row DRAM-to-logic data delivery. In contrast, if we store the D bits for each pixel consecutively in the same sub-array, in stead of using such bit-plane storage strategy, different sub-arrays may contribute to the candidate MB data differently. As a result, the sub-arrays within the same sub-bank may not be able to fully share all the row and column address and each sub-array must have a data I/O width of $N \cdot D$ bits.

Another advantage of this bit-plane storage strategy is that it can easily support run-time graceful performance vs. energy trade-off, because the bit-plane structure makes it very easy to adjust the precision of luminance intensity data participating in motion estimation. It is well-known that appropriate pixel truncation [27] can lead to substantial reduction on computational complexity and power consumption without significantly affecting image quality. Such bit-plane memory structure can naturally support dynamic pixel truncation, which can meanwhile reduce the power consumption of memory data access. Given the D-bit full precision of luminance intensity data, if we only use $D_r < D$ bits in motion estimation, we can directly switch the $D - D_r$ sub-arrays, which store the lower $D - D_r$ bits for each pixel, into an idle mode to reduce the DRAM power consumption. It is intuitive that such lower-precision operation can be dynamically adjusted to allow more flexible performance vs. energy trade-off, e.g., we could first use low-precision data to calculate coarse SADs, and then run block matching with full precision in a small region around the candidate MB with the least coarse SAD.

It should be pointed out that, unlike conventional design solutions, the above presented design strategy under the 3D logic-DRAM integrated system framework can realize any arbitrary and discontinuous motion vector search and hence seamlessly support most existing motion estimation algorithms. Finally, we note that, although the above discussion only focuses on data storage for motion estimation, the same DRAM storage approaches can be used to facilitate the motion compensation as well in both video encoders and decoders.

6.2.2   Motion Estimation Memory Access

With the above DRAM architecture design strategy, the motion estimation engine on the logic die can access the 3D DRAM to directly fetch the current MB and candidate MB through a simple interface. Assume that the video encoder should support a multi-frame motion estimation with up-to m reference frames. In order to seamlessly support multi-frame motion estimation while maintaining the same video encoding throughput, we store all these m reference frames separately, each reference frame is stored in two banks. The motion estimation engine can access all the m reference frames simultaneously, i.e., the motion estimation engine contains m parallel SAD computation units, each unit carries out motion estimation based upon one reference frame. We denote the MB at the top-left corner of each frame with a 2D position index of $(0, 0)$. Assuming that each frame contains $F_W \times F_H$ MBs, the MB at the bottom-right corner of each frame has a 2D position index of $(F_W - 1, F_H - 1)$. Assuming that each word-line in one bank stores s MBs, we store all the MBs row-by-row. Hence, given the MB index (x, y), we first can identify its bank index as $x\%2$, where $\%$ is the modulo operator that finds the remainder of division. Then we can derive the corresponding DRAM row address as

$$A_{row} = \left\lfloor \frac{y \cdot F_W + x}{s} \right\rfloor, \tag{2}$$

where $\lfloor \cdot \rfloor$ denotes the floor operation that returns the biggest integer that is less than or equal to the operand. Since each address location in one bank corresponds to one row in one MB and each MB has N rows, each MB spans N consecutive column addresses. The first column address is obtained as

$$A_{col} = ((y \cdot F_W + x)\%s) \cdot N, \tag{3}$$

and the N consecutive column addresses are $\{A_{col}, A_{col} + 1, \cdots, A_{col} + N - 1\}$.

For each computation unit to access one reference frame, it only needs to provides two sets of parameters, 2D position index of current MB (x_{MB}, y_{MB}) and motion vector (x_{MV}, y_{MV}), which determine the location of the candidate MB in the reference frame. Given these inputs, DRAM will deliver the corresponding candidate MB row-by-row within N cycles. Due to the specific DRAM architecture described above, each 2-bank DRAM frame storage can easily derive its own internal row and column address and the corresponding configuration parameters for the barrel shifters, which is described as follows.

As pointed out earlier, each candidate MB may span at most 4 adjacent MBs in the reference frame as illustrated in Fig. 9. Given the 2D position index of current MB (x_{MB}, y_{MB}) and motion vector (x_{MV}, y_{MV}), we can directly derive the 2D position indexes of these up-to 4 adjacent MBs as

$$\left(x_{MB} + \text{sign}(x_{MV}) \cdot \left\lceil \frac{|x_{MV}|}{N} \right\rceil, \ y_{MB} + \text{sign}(y_{MV}) \cdot \left\lceil \frac{|y_{MV}|}{N} \right\rceil \right), \tag{4}$$

$$\left(x_{MB} + \text{sign}(x_{MV}) \cdot \left\lceil \frac{|x_{MV}|}{N} \right\rceil, \ y_{MB} + \text{sign}(y_{MV}) \cdot \left\lfloor \frac{|y_{MV}|}{N} \right\rfloor \right), \tag{5}$$

$$\left(x_{MB} + \text{sign}(x_{MV}) \cdot \left\lfloor \frac{|x_{MV}|}{N} \right\rfloor, \ y_{MB} + \text{sign}(y_{MV}) \cdot \left\lceil \frac{|y_{MV}|}{N} \right\rceil \right), \tag{6}$$

$$\left(x_{MB} + \text{sign}(x_{MV}) \cdot \left\lfloor \frac{|x_{MV}|}{N} \right\rfloor, \ y_{MB} + \text{sign}(y_{MV}) \cdot \left\lfloor \frac{|y_{MV}|}{N} \right\rfloor \right), \tag{7}$$

where $\text{sign}(\cdot)$ represents the sign the of operand. Clearly, dependent upon the specific values of x_{MV} and y_{MV}, we may have 1, 2, or 4 distinctive indexes, i.e., the candidate MB may span 1, 2, or 4 adjacent MBs in the reference frame. For the general case, let us consider the scenario when one candidate MB spans 4 adjacent MBs in the reference frame as illustrated in Fig. 11. After we obtain the indexes of the 4 reference MBs, we can directly calculate the corresponding DRAM row addresses. In order to determine the range of the column addresses, we calculate

$$y_{off-set} = \begin{cases} |y_{MV}|\%N, & y_{MV} \geq 0 \\ N - |y_{MV}|\%N, & y_{MV} < 0 \end{cases} \tag{8}$$

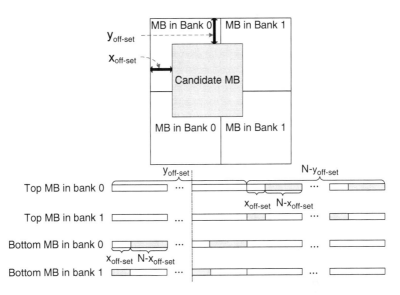

Fig. 11 An example to illustrate the calculation of memory access address to fetch one candidate MB

The candidate MB spans the last $N - y_{off-set}$ rows in the two top MBs, and the first $y_{off-set}$ rows in the two bottom MBs, as illustrated in Fig. 11. Hence, we can correspondingly determine the column addresses. Finally, we should determine the configuration data for the two barrel shifters (as shown in Fig. 9) that forms one row in the candidate MB at a time. In this context, we can calculate

$$x_{off-set} = \begin{cases} |x_{MV}| \% N, & x_{MV} \geq 0 \\ N - |x_{MV}| \% N, & x_{MV} < 0 \end{cases} \qquad (9)$$

For each row read from the reference MBs on the left as shown in Fig. 11, the barrel shifter will shift the data towards left by $x_{off-set}$ pixels; for each row read from the reference MBs on the right as shown in Fig. 11, the barrel shifter will shift the data towards right by $N - x_{off-set}$ pixels. By implementing the above simple calculations in the DRAM domain, the 3D stacked DRAM can readily deliver the candidate MB to the motion estimation engine on the logic die, while the motion estimation engine only needs to supply the current motion vector and the position of the current MB.

6.3 Performance Evaluation

In the H.264/AVC coding standard, the size of each MB is 16×16 and the search region is set 32×32 for HDTV1080p ($1,920 \times 1,080$). Assuming that the luminance density of each pixel uses 8 bits, each 2-bank frame DRAM has a data I/O bandwidth

Table 3 Estimated results for each 2 MByte 2-bank frame storage DRAM block

Number of sub-banks	1 DRAM layer					4 DRAM layers				
	1	2	4	8	16	1	2	4	8	16
Access time (non-burst) (ns)	9.49	7.37	6.65	6.57	7.36	9.36	7.25	6.53	6.44	6.67
Burst access time (ns)	7.17	4.83	4.11	3.91	4.39	7.09	4.84	4.03	3.83	3.93
Energy per access(non-burst) (nJ)	0.71	1.00	1.29	1.61	1.99	0.93	1.22	1.28	1.59	1.00
Energy per burst access (nJ)	0.15	0.45	0.76	1.08	1.46	0.13	0.43	0.74	1.06	1.42
Footprint (mm^2)	1.10	2.08	3.12	4.79	7.49	0.22	0.34	0.47	0.69	1.13

of 128 bits. We assume that the encoder must be able to support multi-frame motion estimation with up-to five reference frames. Hence, we need six 2-bank frame 3D DRAM blocks to store the current frame and five reference frames in the stacked 3D DRAM. This leads to an aggregate data I/O bandwidth of $128 \times 6 = 768$ bits, corresponding to 768 TSVs for logic-DRAM data interconnect. We use the inter-sub-array 3D partitioning strategy presented in Sect. 4 to estimate 3D DRAM performance. For the target HDTV1080p resolution, each image frame needs about 2MByte storage. Table 3 shows the estimated 3D DRAM results for each 2 MByte 2-bank frame storage DRAM block at 65-nm node. Since each sub-bank always has eight sub-arrays, we explore the 3D DRAM design space by varying the size of each sub-array and the number of sub-banks. In this study, the number of bit-lines in each sub-array is fixed as 512.

Table 3 clearly shows a trade-off: as we increase the number of sub-banks by reducing the size of each sub-array, we could directly reduce the access latency, while the access energy consumption and DRAM footprint will increase. We considered both 1-layer and 4-layer 3D DRAM stacking, and the results clearly show the advantages of 4-layer 3D DRAM stacking. As pointed out in the above, this proposed 3D DRAM attempts to access the data on the same word-line as much as possible. Conventionally, access to the data on the same word-line is denoted as burst access. Table 3 shows the difference between burst access and non-burst access, which clearly shows that burst access is much more preferable.

The above presented image storage architecture can seamlessly support any arbitrary motion vector search pattern, hence can naturally support various motion estimation algorithms. In this case study we considered the following popular algorithms: exhaustive full search (FS), three step search (TSS) [11], new three step search (NTSS) [48], four step search (FSS) [60, 73], and diamond search (DS) [70, 71]. We apply these algorithms to two widely used HDTV 1080p video sequences *Tractor* and *Rush hour* [72], where 15 frames are extracted and analyzed in each video sequence. Figure 12 shows the peak signal-to-noise ratio (PSNR) vs. average memory energy consumption for processing each image frame without using on-chip SRAM buffer. Each curve contains five points, corresponding to the scenarios using 1, 2, 3, 4, and 5 reference frames, respectively. We note that, due to the very regular memory access pattern in full search, explicit memory access can be greatly reduced by data reuse in the motion estimation engine. In this study, we

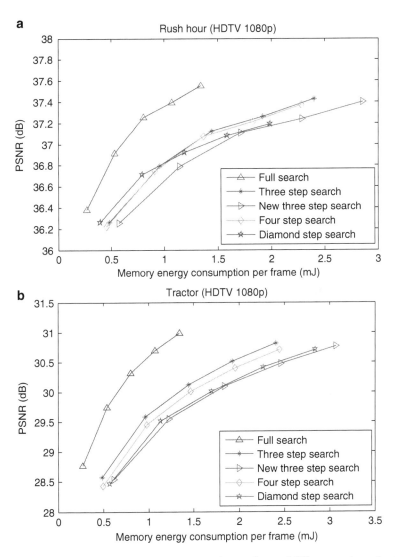

Fig. 12 PSNR vs. average memory energy consumption per frame of different motion estimation algorithms without using on-chip SRAM buffer, where the five points on each curve correspond to the use of 1, 2, 3, 4, and 5 reference frames

assume the use of the full search motion estimation design approach with excellent data reuse presented in [43]. As clearly shown in Fig. 12, full search tends to consume the minimal memory access energy when using the above proposed 3D DRAM image storage architecture.

Based upon the above 3D DRAM parameters, we estimate the overall memory sub-system power consumption and obtain the comparison with conventional

Table 4 Comparison between current design practice and the design using 3D stacked DRAM with full search

			Design using 3D stacked DRAM	
		Current practice	(w/o on-chip SRAM)	(w/ on-chip SRAM)
Off-chip DRAM	Capacity (Gb)	4	4	4
	Access power (mW)	274.56	21.90	21.90
	I/O Power (mW)	124.02	9.89	9.89
3D DRAM	Capacity (Mb)	N/A	100	100
	Access power (mW)	N/A	910.93	18.50
	Leakage power (mW)	N/A	31.03	31.03
	Barrel shifter power (mW)	N/A	12.61	0
	I/O power (mW)	N/A	146.03	2.96
On-chip SRAM	Capacity (Kb)	250	N/A	250
	Footprint (mm^2)	0.91	N/A	0.91
	Access power (mW)	187.12	N/A	187.12
Total memory access power (mW)		585.70	1135.99	271.39

design practice. Regardless to whether 3D DRAM stacking is used or not, we assume the image frames are originally stored in an off-chip 4 Gb commodity DRAM, which also stores data and instructions for other cores in the envisioned heterogeneous multi-core systems. In the case of conventional design practice, the motion estimation accelerator contains an on-chip SRAM buffer to store the current MB and a 80×80 search region in each reference frame (i.e., a total 250 Kb when using five reference frames). In the case of 3D-based design, we consider two scenarios: (i) we eliminate the on-chip SRAM buffer in order to reduce the logic die area, where motion estimation computation blocks directly obtain image data from 3D stacked DRAM, and (ii) we still keep the on-chip SRAM buffer in order to reduce the frequency of 3D stacked DRAM access. The estimation and comparison results are list in Tables 4 and 5, in which full search and three step search algorithms are used, respectively. The access power listed in the table is the power consumed for memory read/write under the real-time constraint (i.e., 30 frames per second for HDTV1080p), including address decoder, sense amplifiers, repeaters, routing, etc. The I/O power of DRAM and SRAM is aggregated and listed in the table. It clearly shows the energy efficiency advantages when 3D stacked DRAM is being used, especially in three step search case, mainly because the use of 3D stacked DRAM can largely reduce the frequency of off-chip 4 Gb commodity DRAM access and data access to the small 3D stacked DRAM dedicated for motion estimation is much more energy-efficient than access to off-chip 4 Gb commodity DRAM. Nevertheless, accessing 3D DRAM directly without on-chip SRAM consumes a large amount of power if full search is used. It is because a large search region imposes more frequent data access from 3D DRAM. Moreover, the above two different scenarios (i.e., with and without on-chip SRAM buffer when using 3D

Table 5 Comparison between current design practice and the design using 3D stacked DRAM with three step search

			Design using 3D stacked DRAM	
		Current practice	(w/o on-chip SRAM)	(w/ on-chip SRAM)
Off-chip DRAM	Capacity (Gb)	4	4	4
	Access power (mW)	274.56	21.90	21.90
	I/O power (mW)	124.02	9.89	9.89
3D DRAM	Capacity (Mb)	N/A	100	100
	Access power (mW)	N/A	146.49	18.50
	Leakage power(mW)	N/A	31.03	31.03
	Barrel shifter power (mW)	N/A	2.61	0
	I/O power (mW)	N/A	23.48	2.96
On-chip SRAM	Capacity (Kb)	250	N/A	250
	Footprint (mm^2)	0.91	N/A	0.91
	Access power (mW)	29.97	N/A	29.97
Total memory access power (mW)		428.55	235.40	114.24

stacked DRAM) essentially represent different trade-offs between the logic die area reduction and total memory access power consumption.

Using Synopsys tool sets and a 65 nm CMOS standard cell library, we synthesize parallel motion estimation computation engines and estimate the power consumption. Combining the above memory sub-system power consumption results, we estimate the overall average motion estimation power consumption as shown in Fig. 13, where we consider both with and without on-chip SRAM buffer. Because the search steps of NTSS, FS-TSS and DS algorithms may vary during the run time, we show the average values obtained from computer simulations. The lower and upper parts of each bar in the figures represent memory access and computation logic power consumption respectively. The power consumption is obtained at the clock frequencies that are just fast enough for real-time processing in all the studies.

7 Conclusion

Motivated by recent significant developments of 3D integration technology towards high manufacturability, this chapter studied the exciting opportunities enabled by 3D integration for digital signal processing circuits and systems design. Because of the memory intensive nature of most signal processing systems, 3D integration of a single signal processing logic die and multiple high density memory dies appears to be a very natural vehicle to exploit 3D integration technology in DSP systems. The key of such 3D integrated DSP system design is how to most effectively leverage the massive logic-memory interconnect bandwidth and very short logic-memory communication latency to push the overall system performance and efficiency

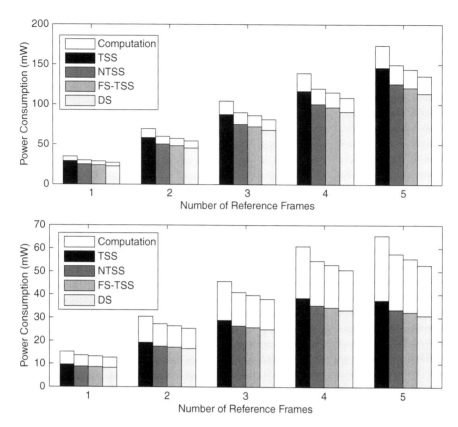

Fig. 13 Average power consumption of entire motion estimation using advanced algorithms (**a**) without on-chip SRAM buffer, and (**b**) with on-chip SRAM buffer

beyond what is achievable today. This chapter discussed the use of 3D logic-memory integration in both programmable digital signal processors and application-specific signal processing ICs. Focusing on 3D logic-DRAM integration, this chapter discussed 3D DRAM architecture design and presented two case studies on applying 3D logic-DRAM integration to programmable VLIW processors and application-specific video encoders.

References

1. Ababei, C., Feng, Y., Goplen, B., Mogal, H., Zhang, T., Bazargan, K., Sapatnekar, S.: Placement and routing in 3D integrated circuits. IEEE Design & Test of Computers **22**, 520–531 (2005)
2. Abraham, S.G., Mahlke, S.A.: Automatic and efficient evaluation of memory hierarchies for embedded systems. In: 32nd Annual International Symposium on Microarchitecture(MICRO-32), pp. 114–125 (1999)

3. Banerjee, K., Souri, S., Kapur, P., Saraswat, K.: 3-d ics: a novel chip design for improving deep-submicrometer interconnect performance and systems-on-chip integration. Proceedings of the IEEE **89**, 602–633 (2001)
4. Barth, J., Reohr, W., Parries, P., Fredeman, G., Golz, J., Schuster, S., Matick, R., Hunter, H., Tanner, C., Harig, J., Hoki, K., Khan, B., Griesemer, J., Havreluk, R., Yanagisawa, K., Kirihata, T., Iyer, S.: A 500 MHz random cycle, 1.5 ns latency, SOI embedded DRAM macro featuring a three-transistor micro sense amplifier. IEEE Journal of Solid-State Circuits **43**, 86–95 (2008)
5. Bernstein, K., Andry, P., Cann, J., Emma, P., Greenberg, D., Haensch, W., Ignatowski, M., Koester, S., Magerlein, J., Puri, R., Young, A.: Interconnects in the third dimension: Design challenges for 3D ICs. In: Proc. of ACM/IEEE Design Automation Conference (DAC), pp. 562–567 (2007)
6. Binkert, N.L., Dreslinski, R.G., Hsu, L.R., Lim, K.T., Saidi, A.G., Reinhardt, S.K.: The m5 simulator: Modeling networked systems. IEEE Micro **26**, 2006 (2006)
7. Black, B., Annavaram, M., Brekelbaum, N., DeVale, J., Jiang, L., Loh, G.H.: Die Stacking (3D) Microarchitecture. In: Proc. of IEEE/ACM International Symposium on Microarchitecture (Micro), pp. 469–479 (2006)
8. Burns, J., Aull, B., Chen, C., Chen, C., Keast, C., Knecht, J., Suntharalingam, V., Warner, K., Wyatt, P., Yost, D.: A Wafer-Scale 3-D Circuit Integration Technology. IEEE Trans. Electron Devices **53**, 2507–2516 (2006)
9. CACTI: An integrated cache and memory access time, cycle time, area, leakage, and dynamic power model. http://www.hpl.hp.com/research/cacti/
10. Chakrapani, L.N., Gyllenhaal, J., Hwu, W.W., Mahlke, S.A., Palem, K.V., Rabbah, R.M.: Trimaran: An infrastructure for research. In: in Instruction-Level Parallelism. Lecture Notes in Computer Science, p. 2005 (2004)
11. Chau, L.P., Jing, X.: Efficient three-step search algorithm for block motion estimation video coding. In: Proc. of IEEE International Conference on Acoustics, Speech, and Signal Processing, pp. 421–424 (2003)
12. Chen, K.N., Lee, S., Andry, P., Tsang, C., Topol, A., Lin, Y., Lu, J., Young, A., Ieong, M., Haensch, W.: Structure Design and Process Control for Cu Bonded Interconnects in 3D Integrated Circuits. In: Technical Digest of IEEE International Electron Devices Meeting (IEDM), pp. 367–370 (2006)
13. Chen, M., Chen, E., Lai, J.Y., Wang, Y.P.: Thermal Investigation for Multiple Chips 3D Packages. In: Proc. of Electronics Packaging Technology Conference, pp. 559–564 (2008)
14. Claasen, T.: An Industry Perspective on Current and Future State of the Art in System-on-Chip (SoC) Technology. Proceedings of the IEEE **94**, 1121–1137 (2006)
15. Cong, J.: An interconnect-centric design flow for nanometer technologies. Proceedings of the IEEE **89**, 505–528 (2001)
16. Cong, J., Luo, G.: A multilevel analytical placement for 3D ICs. In: Proc. of Asia and South Pacific Design Automation Conference, pp. 361–366 (2009)
17. Cong, J., Zhang, Y.: Thermal Via Planning for 3-D ICs. In: Proc. of IEEE/ACM International Conference on Computer-Aided Design (ICCAD), pp. 745–752 (2005)
18. Crowley, M., Al-Shamma, A., Bosch, D., Farmwald, M., Fasoli, L.: 512 Mb PROM with 8 layers of antifuse/diode cells. In: IEEE Intl. Solid-State Circuit Conf. (ISSCC), p. 284 (2003)
19. Emma, P., Kursun, E.: Is 3D chip technology the next growth engine for performance improvement? IBM Journal of Research and Development **32**(6), 541–552 (2008)
20. Fisher, J.A., Faraboschi, P., Young, C.: Embedded Computing: A VLIW Approach to Architecture, Compilers and Tools. Morgan Kaufmann (2004)
21. Garrou, P., Bower, C., Ramm, P.: Handbook of 3D Integration: Technology and Applications of 3D Integrated Circuits. Wiley (2008)
22. Gibert, E., Sanchez, J., Gonzales, A.: Effective Instruction Scheduling Techniques for an Interleaved Cache Clustered VLIW Processor. In: Proceedings of the 35 th Annual IEEE/ACM International Symposium on Microarchitecture, pp. 123–133 (2002)

23. Gibert, E., Sanchez, J., Gonzales, A.: Local Scheduling Techniques for Memory Coherence in a Clustered VLIW Processor with a Distributed Data Cache. In: Proceedings of the International Symposium on Code Generation and Optimization, pp. 193–203 (2003)
24. Goplen, B., Sapatnekar, S.: Placement of thermal vias in 3-D ICs using various thermal objectives. IEEE Transactions on Computer-Aided Design of Integrated Circuits and Systems **25**, 692–709 (2006)
25. Gu, S., Marchal, P., Facchini, M., Wang, F., Suh, M., Lisk, D., Nowak, M.: Stackable memory of 3D chip integration for mobile applications. In: Proc. of IEEE International Electron Devices Meeting (IEDM), pp. 1–4 (2008)
26. Harrer, H., Katopis, G., Becker, W.: From chips to systems via packaging - A comparison of IBM's mainframe servers. IEEE Circuits and Systems Magazine **6**, 32–41 (2006)
27. He, Z.L., Tsui, C.Y., Chan, K.K., Liou, M.: Low-power vlsi design for motion estimation using adaptive pixel truncation. IEEE Trans. on Circuits and Systems for Video Technology **10**(5), 669–678 (2000)
28. Healy, M., Vittes, M., Ekpanyapong, M., Ballapuram, C.S., Lim, S.K., Lee, H.H.S., Loh, G.H.: Multiobjective Microarchitectural Floorplanning for 2-D and 3-D ICs. IEEE Transactions on Computer-Aided Design of Integrated Circuits and Systems **26**, 38–52 (2007)
29. Ho, R., Mai, K.W., Horowitz, M.A.: The future of wires. Proceedings of the IEEE **89**, 490–504 (2001)
30. Horowitz, M., Stark, D., Alon, E.: Digital circuit design trends. IEEE Journal of Solid-State Circuits **43**, 757–761 (2008)
31. Huang, W., Allen-Ware, M., Carter, J., Cheng, E., Skadron, K., Stan, M.: Temperature-aware architecture: Lessons and opportunities. IEEE Micro **31**(3), 82–86 (2011)
32. Huang, Y., Chen, T.C., Tsai, C.H., Chen, C.Y., Chen, T.W., Chen, C.S., Shen, C.F., Ma, S.Y., Wang, T.C., Hsieh, B.Y., Fang, H.C., Chen, L.G.: A 1.3TOPS H.264/AVC single-chip encoder for HDTV applications. In: International Solid-State Circuit Conference, pp. 128–129. San Francisco (2005)
33. Hwang, C.G.: New paradigms in the silicon industry. In: Proc. of Technical Digest of IEEE International Electron Devices Meeting (IEDM), pp. 19–26 (2006)
34. International Technology Roadmap for Semiconductors(ITRS): http://www.itrs.net
35. Itoh, K.: VLSI Memory Chip Design. Springer (2001)
36. Jain, A., Jones, R., Chatterjee, R., Pozder, S.: Analytical and numerical modeling of the thermal performance of three-dimensional integrated circuits. IEEE Transactions on Components and Packaging Technologies **33**(1), 56–63 (2010)
37. Jaspers, E., de With, P.: Bandwidth reduction for video processing in consumer systems. IEEE Trans. on Consumer Electronics **47**(4), 885–894 (2001)
38. Jung, S.M., Jang, J., Cho, W., Moon, J., Kwak, K.: The Revolutionary and Truly 3-Dimensional 25F2 SRAM Technology with the Smallest S3 (Stacked Single-Crystal Si) cell, $0.16\mu m^2$, and SSTFT (Stacked Single-Crystal Thin Film Transistor) for Ultra High Density SRAM. In: Proc. of Symposium on VLSI Technology, pp. 228–229 (2004)
39. Jung, S.M., Jang, J., Kim, K.: Three Dimensionally Stacked NAND Flash Memory Technology Using Stacked Single Crystal Si Layers in ILD and TANOS Structure for Beyond 30 nm Node. In: Technical Digest of IEEE International Electron Devices Meeting (IEDM), pp. 37–40 (2006)
40. Kathail, V., Schlansker, M.S., Rau, B.R.: Hpl-pd architecture specification: Version 1.1. Tech. rep., Hewlett-Packard Company (2000)
41. Kgil, T., D'Souza, S., Saidi, A., Binkert, N., Dreslinski, R., Reinhardt, S., Flautner, K., Mudge, T.: PicoServer: Using 3D Stacking Technology To Enable A Compact Energy Efficient Chip Multiprocessor. In: Proc. of 12th International Conference on Architectural Support for Programming Languages and Operating Systems (ASPLOS) (2006)
42. Khailany, K., Williams, T., Lin, J., Long, E.P., Rygh, M., W.Tovey, D., J.Dally, W.: A Programmable 512 GOPS Stream Processor for Signal, Image, and Video Processing. IEEE JOURNAL OF SOLID-STATE CIRCUITS **43** (2008)

43. Kim, M., Hwang, I., Chae, S.I.: A fast VLSI architecture for full-search variable block size motion estimation in MPEG-4 AVC/H.264. In: Proc. Asia and South Pacific Design Automation Conference, pp. 631–634 (2005)

44. Koester, S.J., Young, A.M., Yu, R.R., Purushothaman, S., Chen, K.N., La Tulipe, D.C., Rana, N., Shi, L., Wordeman, M.R., Sprogis, E.J.: Wafer-level 3D integration technology. IBM Journal of Research and Development **52**(6), 583–597 (2008)

45. Lapsley, P., Bier, J., Shoham, A., Lee, E.A.: DSP Processor Fundamentals: Architectures and Features. IEEE Press (1997)

46. Lee, C., Potkonjak, M., Mangione-Smith, W.: MediaBench: a tool for evaluating and synthesizing multimedia and communications systems. In: Proc. of IEEE/ACM International Symposium on Microarchitecture, pp. 330–335 (1997)

47. Lee, S., Chen, K.N., Lu, J.Q.: Wafer-to-wafer alignment for three-dimensional integration: A review. Journal of Microelectromechanical Systems **20**(4), 885–898 (2011)

48. Li, R., Zeng, B., Liou, M.L.: A new three-step search algorithm for block motion estimation. IEEE Trans. on Circuits and Systems for Video Technology **4**, 438–442 (Aug.)

49. Liu, C.C., Ganusov, I., Burtscher, M., Tiwari, S.: Bridging the Processor-Memory Performance Gap with 3D IC Technology. IEEE Design and Test of Computers **22**, 556?64 (2005)

50. Loh, G.: 3D-stacked memory architecture for multi-core processors. In: Proceedings of the 35th ACM/IEEE Intl. Conf. on Computer Architecture (2008)

51. Loh, G., Xie, Y., Black, B.: Processor Design in 3D Die-Stacking Technologies. IEEE Micro **27**, 31–48 (2007)

52. Lu, J.Q.: 3-D Hyperintegration and Packaging Technologies for Micro-Nano Systems. Proceedings of the IEEE **97**, 18–30 (2009)

53. Lu, J.Q., Cale, T., Gutmann, R.: Wafer-level three-dimensional hyper-integration technology using dielectric adhesive wafer bonding. Materials for Information Technology: Devices, Interconnects and Packaging (Eds. E. Zschech, C. Whelan, T. Mikolajick) pp. 386–397 (Springer-Verlag (London) Ltd, August 2005)

54. Matick, R., Schuster, S.: Logic-based eDRAM: Origins and rationale for use. IBM J. Res. & Dev. **49**, 145–165 (2005)

55. M.Facchini, T.Carlson, A.Vignon, M.Palkovic, F.Catthoor, L.Benini, P.Marchal: System-level power/performance evaluation of 3d stacked drams for mobile applications. In: Proc. of Design, Automation & Test in Europe Conference & Exhibition, pp. 923–928 (2009)

56. Moor, P.D., Ruythooren, W., Soussan, P., Swinnen, B., Baert, K., Hoof, C.V., Beyne, E.: Recent advances in 3d integration at imec. Enabling Technologies for 3-D Integration (edited by C.A. Bower, P.E. Garrou, P. Ramm, and K. Takahashi) (2006)

57. Morrow, P., Black, B., Kobrinsky, M., Muthukumar, S., Nelson, D., Park, C.M., Webb, C.: Design and fabrication of 3d microprocessors. Enabling Technologies for 3-D Integration (edited by C.A. Bower, P.E. Garrou, P. Ramm, and K. Takahashi) (2006)

58. Nain, R., Chrzanowska-Jeske, M.: Fast Placement-Aware 3-D Floorplanning Using Vertical Constraints on Sequence Pairs. IEEE Transactions on Very Large Scale Integration (VLSI) Systems **19**(9), 1667–1680 (2011)

59. Panda, P.R., Catthoor, F., Dutt, N.D., Danckaert, K., Brockmeyer, E., Kulkarni, C., Kjeldsberg, P.G.: Data and memory optimization techniques for embedded systems. ACM Transactions on Design Automation of Electronic Systems **6**, 149–206 (2001)

60. Po, L.M., Ma, W.C.: A novel four-step search algorithm for fast block motion estimation. IEEE Trans. on Circuits and Systems for Video Technology **6**, 313–317 (1996)

61. Pozder, S., Chatterjee, R., Jain, A., Huang, Z., Jones, R., Acosta, E.: Progress of 3D Integration Technologies and 3D Interconnects. In: Proc. of IEEE International Interconnect Technology Conference, pp. 213–215 (2007)

62. Pozder, S., Jones, R., Adams, V., Li, H.F., Canonico, M., Zollner, S., Lee, S., Gutmann, R., Lu, J.Q.: Exploration of the scaling limits of 3d integration. Enabling Technologies for 3-D Integration (edited by C.A. Bower, P.E. Garrou, P. Ramm, and K. Takahashi) (2006)

63. Rickert, P., Krenik, W.: Cell phone integration: SiP, SoC, and PoP. IEEE Design & Test of Computers **23**, 188–195 (2006)

64. Ryu, S.K., Lu, K.H., Zhang, X., Im, J.H., Ho, P., Huang, R.: Impact of near-surface thermal stresses on interfacial reliability of through-silicon vias for 3-D interconnects. IEEE Transactions on Device and Materials Reliability **11**(1), 35–43 (2011)
65. Saen, M., Osada, K., Okuma, Y., Niitsu, K., Shimazaki, Y., Sugimori, Y., Kohama, Y., Kasuga, K., Nonomura, I., Irie, N., Hattori, T., Hasegawa, A., Kuroda, T.: 3-D System Integration of Processor and Multi-Stacked SRAMs Using Inductive-Coupling Link. IEEE Journal of Solid-State Circuits **45**(4), 856–862 (2010)
66. Sapatnekar, S.: Addressing thermal and power delivery bottlenecks in 3D circuits. In: Proc. of Asia and South Pacific Design Automation Conference, pp. 423–428 (2009)
67. Seiculescu, C., Murali, S., Benini, L., De Micheli, G.: SunFloor 3D: A Tool for Networks on Chip Topology Synthesis for 3-D Systems on Chips. IEEE Transactions on Computer-Aided Design of Integrated Circuits and Systems **29**(12), 1987–2000 (2010)
68. SEMATECH Consortium. http://www.sematech.org
69. Strauss, W.: The real DSP chip market. IEEE Signal Processing Magazine **20**, 83 (2003)
70. Tham, J.Y., Ranganath, S., Ranganath, M., Kassim, A.A.: A novel unrestricted center-biased diamond search algorithms for block motion estimation. IEEE Trans. on Circuits and Systems for Video Technology **8**, 369–377 (1998)
71. Tham, J.Y., Ranganath, S., Ranganath, M., Kassim, A.A.: A new diamond search algorithm for fast block-matching motion estimation. IEEE Trans. on Image processing **9**, 287–290 (2000)
72. Sveriges Television (SVT). http://www.svt.se
73. Wang, K.T., Chen, O.C.: Motion estimation using an efficient four-step search method. In: Proc. of IEEE International Symposium on Circuits and Systems, pp. 217–220 (1998)
74. Wang, R., Li, J., Huang, C.: Motion compensation memory access optimization strategies for H.264/AVC decoder. In: Proc. of IEEE International Conference on Acoustics, Speech, and Signal Processing, pp. 97–100 (2005)
75. Wiegand, T., Sullivan, G.J., Bjontegaard, G., Luthra, A.: Overview of the H.264/AVC video coding standard. IEEE Trans. on Circuits and Systems on Video Technology **13**(7), 560–576 (2003)
76. Wu, Q., Zhang, T.: Design Techniques to Facilitate Processor Power Delivery in 3-D Processor-DRAM Integrated Systems. IEEE Transactions on Very Large Scale Integration (VLSI) Systems **19**(9), 1655–1666 (2011)
77. Wu, Z., Wolf, W.: Design study of shared memory in vliw video signal processors. In: Proceedings of the 1998 International Conference onParallel Architectures and Compilation Techniques, pp. 52–59 (1998)
78. Wulf, W.A., McKee, S.A.: Hitting the MemoryWall: Implications of the Obvious. Computer Architecture News **23**, 20–24 (1995)

Part III
Design Methods and Tools

Methods and Tools for Mapping Process Networks onto Multi-Processor Systems-On-Chip

Iuliana Bacivarov, Wolfgang Haid, Kai Huang, and Lothar Thiele

Abstract Applications based on the Kahn process network (KPN) model of computation are determinate, modular, and based on FIFO communication for inter-process communication. While these properties allow KPN applications to efficiently execute on multi-processor systems-on-chip (MPSoC), they also enable the automation of the design process. This chapter focuses on the second aspect and gives an overview of methods for automating the design process of KPN applications implemented on MPSoCs. Whereas previous chapters mainly introduced techniques that apply to restricted classes of process networks, this overview will be dealing with general Kahn process networks.

1 Introduction

Multi-processor system-on-chip (MPSoC) is one of the most promising and solid paradigm for implementing embedded systems for signal processing in communication, medical, and multi-media applications. MPSoC platforms are heterogeneous by nature as they use multiple computation, communication, memory, and peripheral resources. They allow the parallel execution of (multiple) applications and, at the same time, they offer the flexibility to optimize performance, energy consumption, or cost of the system. Nevertheless, to optimize an MPSoC in the presence of tight time-to-market and budget constraints, a systematic design flow is required.

To deal with this challenge, Kienhuis et al. [36] suggested to structure the design flow in a certain manner, now commonly referred to as the Y-chart approach. It is a systematic methodology for selecting an embedded system implementation from

I. Bacivarov (✉) • W. Haid • K. Huang • L. Thiele
Computer Engineering and Networks Laboratory, ETH Zurich, Switzerland
e-mail: iuliana.bacivarov@tik.ee.ethz.ch; wolfgang.haid@tik.ee.ethz.ch;
kai.huang@tik.ee.ethz.ch; lothar.thiele@tik.ee.ethz.ch

S.S. Bhattacharyya et al. (eds.), *Handbook of Signal Processing Systems*, 867
DOI 10.1007/978-1-4614-6859-2_27, © Springer Science+Business Media, LLC 2013

Fig. 1 Y-chart approach for designing MPSoC

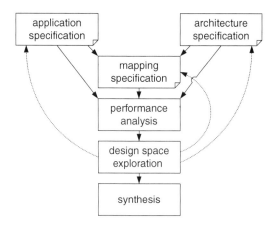

a set of alternatives, a process often denoted as design space exploration. One key idea underlying this approach is to explicitly separate application and architecture specifications. A separate mapping specification describes how the application is spatially (binding) and temporally (scheduling) executed on the architecture. Design space exploration is then performed by iteratively analyzing and optimizing the application, the structure of the underlying (hardware) architecture as well as candidate mappings, as shown in Fig. 1.

Many design flows implementing the Y-chart approach have been proposed. For a review, see [11]. These flows have in common that they impose a set of system-level concepts to facilitate design space exploration, such as the use of a formal model of computation, providing restrictions on the set of scheduling policies, and relying on modular specifications. For instance, the application may be formally specified as a data flow model, a synchronous model, or a discrete event model, in order to enable automated performance analysis. In a similar way, resource sharing policies may be limited to an event-triggered or a time-triggered policy, to prune the design space. Finally, using modular system-level specifications will enable quick system modifications concerning the application, architecture, and mapping.

In the context of (array) signal processing applications executing on MPSoC, the Kahn process network (KPN) model of computation [33] is frequently used. Assuming a network of autonomous, concurrently executing processes that communicate point-to-point via unbounded FIFO channels, the KPN model has additional favorable properties. The KPN model is determinate, i.e. the functional behavior is independent on the scheduling of processes. The inter-process communication via FIFO channels using blocking read semantics can be efficiently implemented either in software, hardware, or in heterogeneous HW/SW systems. Computation and control are completely distributed, requiring no global synchronization, communication, or memory. The resulting modularity allows applications to be scaled easily and opens up many degrees of freedom for implementing a system.

Due to these properties, the KPN model of computation is "compatible" with the Y-chart approach and has led to numerous design flows. Although they share the same model of computation, these design flows consider different design objectives, they focus on different aspects, and leverage different properties of the KPN model or one of its subclasses. In this chapter, an overview of KPN-based design flows is given, emphasizing both, similarities and differences in these flows. The following section reviews existing design flows and the way they relate to the Y-chart. Afterwards, a closer look at individual steps in the design flow is taken and several methods to tackle them are presented. Finally, for exemplification, a specific design flow is considered in detail.

2   KPN Design Flows for Multiprocessor Systems

Several design flows based on the Y-chart approach and the KPN model have been developed. Table 1 shows a (non-exhaustive) list of design flows targeted at the implementation of KPN applications on MPSoC platforms. In addition to the listed design flows, there are other approaches related to the KPN model with different aims. Ptolemy [57] and Metropolis [3] allow the analysis and simulation of applications specified as KPNs, among other models of computation. However, they are targeted more towards hardware/software codesign and in particular towards the system synthesis and verification. The design space exploration is not the main focus of these frameworks. The Mathworks Real-Time Workshop [46] and the National Instruments LabVIEW Microprocessor SDK [48] target the implementation of signal processing applications on single-processor systems. SystemCoDesigner [35] and PeaCE [21] are HW/SW codesign flows based on a model of computation that combines the KPN model with finite state machines, see chapter [15]. Note that even though the focus of this chapter is on the design flows for MPSoC listed in Table 1, many of the presented ideas also apply to the other mentioned design flows.

Table 1 KPN design flows for MPSoCs

Design flow	Web page
Artemis [55]	http://daedalus.liacs.nl
Distributed Operation Layer (DOL) [62]	http://www.tik.ee.ethz.ch/~shapes/dol.html
Embedded System-Level Platform Synthesis and Application Mapping (ESPAM) [49]	http://daedalus.liacs.nl
Koski [34]	Not available online
Multiapplication and Multiprocessor Synthesis (MAMPS) [41]	http://www.es.ele.tue.nl/mamps
Open Dataflow (OpenDF) [6]	http://opendf.sourceforge.net
Software/Hardware Integration Medium (SHIM) [13]	Not available online
StreamIt [19]	http://www.cag.lcs.mit.edu/streamit

Generally, KPN design flows for MPSoCs respect the four design phases of the Y-chart: system specification, performance analysis, design space exploration, and system synthesis, as shown in Fig. 1.

Based on these four phases, the design process can be described as follows: The starting point of the design flow is a parallelized KPN specification of the application. In this specification, the coarse-grain data and functional parallelism of the application is made explicit. Fine-grained word or instruction-level parallelism can effectively be handled by today's compilers. Usually, the KPN is manually specified by the programmer. There are, however, also tools available that allow deriving a KPN from sequential programs, such as the Compaan [37] and pn [67] tools. KPN design flows usually provide a functional simulation capability that enables the execution of KPN specification on a standard single-processor machine in a multi-tasking environment. Due to the determinacy of KPNs, the timing-independent functionality of the application can be validated this way.

Second, the architecture needs to be specified. This is frequently done in form of a system-level specification describing the architectural resources, such as processors, memories, interconnects, and I/O devices. This specification can either describe a fixed MPSoC or the template of a configurable MPSoC platform. In both cases, the architecture specification needs to contain all the information required for design space exploration and performance analysis. In the case of a configurable platform, the architecture specification is also the basis for the synthesis of the final target platform later in the design flow. Hence, it needs to contain information required by the RTL synthesis tool, such as references to VHDL or Verilog code of hardware components, complete IP blocks, and configuration files.

The application and architecture specification phase is followed by defining a mapping of the application onto the architecture. In this step, processes are bound to processors and channels are bound to communication paths containing memories and interconnects. In addition, the scheduling and arbitration policies for shared resources are defined.

Usually, the final mapping is the result of a design space exploration, which is done based on the system performance analysis. The methods applied for performance analysis range from simple back-of-the-envelope calculations to formal analysis methods, simulations, and measurements. In KPN design flows, performance analysis during design space exploration is possible and is usually done at a rather high level of abstraction. As shown in the next section, different methods targeted towards KPN applications have been proposed in this context that achieve high accuracy within short analysis times. Being able to defer the use of simulation or measurements until late in the design cycle is one of the key advantages of KPN design flows.

After manual or automated design space exploration, the system is finally implemented by making use of appropriate synthesis techniques. For this purpose, KPN design flows feature powerful synthesis tools that implement a system based on the application, architecture, and mapping specification in software, hardware, or both hardware and software. Clearly, this is a key advantage of KPN design flows because the pitfalls of implementing a parallel system, such as hardware-software interface generation, deadlocks, starvation, and data races are handled in an automated way.

The design flows listed in Table 1 implement this basic Y-chart approach in different ways: On the one hand, the methods that are applied in each of the four phases differ between the design flows, as discussed in the next section. On the other hand, the scope (set of optimization variables) of design space exploration is different. Basically, one can distinguish between software design flows, where the target platform is fixed, and hardware/software co-design flows, where a template of a target platform is given and the instantiation of a specific platform is part of the design space exploration. This is shown in Table 2 where a few case studies are summarized that have been performed using the design flows listed in Table 1. DOL, SHIM, and StreamIt assume fixed hardware platforms, whereas the scope of the other design flows encompasses the implementation of the target platform on FPGAs.

3 Methods

KPN design flows attempt to assist a system designer in implementing an application as a hardware/software system by offering support for several activities such as:

- System specification.
- System synthesis.
- Performance analysis.
- Design space exploration.

For each of these activities, methods have been proposed that differ in goal, scope, degree of automation, and complexity. In the previous chapters, mainly methods for subclasses of KPNs have been discussed. In this chapter, we give an overview of methods that are applicable to general KPNs in the context of MPSoCs. For each of the activities mentioned above, we discuss the challenges and proposed solutions.

3.1 System Specification

Developing applications that run correctly and efficiently on MPSoCs is challenging. The difficulty consists in finding an appropriate level of abstraction that balances the conflicting goals of (a) developing applications in a productive manner and of (b) enabling efficient automated implementation. While productivity is usually achieved by programming at a high abstraction level, efficiency is usually achieved by optimizing code at a low abstraction level. Many case studies provide evidence that for streaming applications, the KPN model of computation achieves a good trade-off between these two goals. On the one hand, streaming applications

Table 2 Case studies performed with various design flows

Design flow	Case study Application	Target Platform	Performance Analysis	Exploration Method
Artemis [54]	Motion-JPEG encoder	Molen architecture on Xilinx Virtex-II Pro FPGA	Trace-driven simulation	Evolutionary algorithm
DOL [62]	MPEG-2 decoder	Atmel DIOPSIS 940	Real-time analytic model	Evolutionary algorithm
ESPAM [49]	Motion-JPEG encoder	Multi-MicroBlaze on Xilinx Virtex-II Pro FPGA	Measurement	Exhaustive search
Koski [34]	WLAN terminal	Multi-NIOS on Altera Stratix-II FPGA	High-level simulation	Simulated annealing
MAMPS [41]	H263 and JPEG decoders	Multi-MicroBlaze on Xilinx Virtex-II Pro FPGA	High-level simulation	Dedicated heuristic
OpenDF [32]	MPEG-4 SP decoder	FPGA (no particular type specified)	Not applicable	Not applicable
SHIM [66]	JPEG decoder	Sony/Toshiba/IBM Cell BE	Not applicable	Not applicable
StreamIt [19]	12 streaming applications	RAW architecture	SDF analytic model	Simulated annealing

can often naturally be modeled as a KPN which promotes productivity. On the other hand, runtime environments have been developed that efficiently implement processes and channels.

Specifically, the KPN model can be seen as a *coordination* model [17] which considers the programming of a distributed system as the combination of two distinct activities: the actual *computing* part comprising a number of processes involved in manipulating data and a *coordination* part reflecting the communication and cooperation between processes. The coordination model allows reuse of components because the application programmer can easily build new algorithms by a new composition of existing processes. Furthermore the coordination model allows applications to be ported to different target architectures because usually only the glue-code that implements the coordination part is architecture dependent.

Due to these reasons, KPN applications are usually specified in a way that reflects the coordination model. Two different approaches can be distinguished, namely specification using a host as well as a coordination language, and specification using a domain-specific language. When using distinguished host and coordination languages, the KPN processes are specified in a host language (often in C or C++) whereas the coordination part is specified separately using a coordination language (often in XML or UML). This is, for instance, the approach taken in the Artemis and DOL design flows, where C and XML are used. When using a domain-specific language, computation and coordination are expressed in a single language that provides constructs for both parts. OpenDF and StreamIt, for instance, are based on domain-specific languages.

In both cases, applications are usually expressed based on the principles of encapsulation and explicit concurrency: Each process completely encapsulates its own state together with the code that operates on it and operates independently from other processes in the system, except for the data dependencies that are made explicit by channels. This allows for modular, scalable, and platform-independent application specifications.

System specification for KPNs is thus different from two other frequently used approaches for MPSoC software development, namely specification based on a board support package and specification based on a high-level application programming interface (API). When developing an application based on the board support package that is usually shipped with an MPSoC, the abstraction level is rather low. The focus is thus often on correctly implementing an application using low-level primitives for initialization, communication, or synchronization, rather than on optimizing an application. When using a high-level API the designer is relieved from dealing with low-level details (provided that the API has been ported to the target MPSoC). Compared to the KPN based approach, however, automatically optimizing programs written using a high-level API, such as MPI or OpenMP, is more difficult: Due to the lack of an underlying model of computation, the basis for automatically analyzing and optimizing a program is essentially missing.

3.2   System Synthesis

The Y-chart approach opens a gap between the system-level specification and the actual implementation of the design, sometimes referred to as the implementation gap. The challenge in bridging this gap is to preserve the KPN semantics on the one hand and achieve the desired performance on the other hand. Also, the pitfalls of parallel programming, such as deadlocks, starvation, and data races need to be handled. This is the task of system synthesis.

Different approaches for the synthesis of KPNs for software, hardware, and in combined hardware/software platforms have been proposed. The target architectures have been comprised of single-processors, multi-processors, and FPGAs. In all cases, system synthesis deals with the implementation of processes and channels as well as the arbitration of resources in case that processes and channels are mapped to shared resources. If not all parameters of an implementation are fixed before synthesis, the remaining degrees of freedom need to be exploited during system synthesis. In that case, system synthesis is often considered as an optimization problem where frequently considered optimization goals are the minimization of code size, the minimization of buffer requirements, or finding the schedules that minimize delays and maximize system throughput.

While many of these problems can be solved only for restricted subclasses, a few observations apply to general KPNs:

*   First, KPN applications can be efficiently implemented on architectures with different processor, interconnect, and memory configurations, as shown in Table 2. As an example, KPN applications can be implemented on distributed memory (message-passing) architectures as well as shared memory architectures. The FIFO communication can be implemented using dedicated hardware FIFOs or buses, but also more complex communication topologies, such as hierarchical buses or networks-on-chip.
*   Second, KPN applications can be executed in a purely data-driven manner based on their determinacy. This means that resources can operate independently from each other without any global synchronization. Pair-wise synchronization is only needed between processes that are directly connected by channels. From another perspective, this means that KPN applications can be scheduled with any scheduling policy that prevents deadlocks, i.e., preemptive, non-preemptive, or cooperative scheduling could be used. Due to these very relaxed requirements, KPN applications can usually be implemented easily on top of existing (real-time) operating systems. On the other hand, implementing a runtime-environment for a new platform from scratch is also possible because not many services need to be provided by the runtime-environment.
*   Third, KPN applications can be easily partitioned into processes running in hardware and processes running in software. This is due to the parallel specification of the KPN application on the one hand, and due to the simple interaction of processes over FIFO channels on the other hand which facilitates the synthesis of the HW/SW interface.

The observations above indicate that synthesizing a KPN is conceptually not a very difficult task. Implementing a KPN based on a multi-processor operating system, for instance, is rather simple: Processes can be implemented as operating system processes or threads, and channels can be implemented using existing inter-process communication schemes. The difficulties in KPN synthesis origin from optimizing an implementation by minimizing the overhead for FIFO communication and the runtime environment. This can be achieved by considering low-level details of an implementation, for instance by efficiently using the hardware communication infrastructure (e.g. DMA engines) or by efficiently using the memory hierarchy (e.g. caches or scratchpad memories). On the other hand, optimizations can also be done at a high level, for instance, by (automatically) adjusting the granularity and topology of a KPN to the target architecture. This includes the replication of processes to increase the parallelism in a KPN or the merging of processes to reduce inter-process communication. We refrain from giving further details here and refer to the previous chapters for details on applicable techniques.

Finally, a further problem needs to be considered in the synthesis of KPNs: The denotational semantics of the Kahn model is based on FIFO channels with unbounded capacity. Since unbounded channels cannot be realized in physical implementations, however, KPNs need to be transformed in a way that allows for an implementation on channels with finite capacity. It can be shown that an operational semantics of KPNs based on channels with finite capacity matches the denotational semantics when artificial deadlocks can be avoided. An artificial deadlock is a deadlock caused by one or more channels having insufficient capacity. Due to the Turing-completeness of KPNs, it is in general not possible to determine sufficient channel sizes at design time, however. One possibility to deal with this situation are runtime approaches that detect and resolve artificial deadlocks during execution [51]. Another possibility is to restrict the communication behavior of the processes such that the channels become amenable to analysis at design time.

3.3 Performance Analysis

During the design process, a designer is typically faced with questions such as whether the timing properties of a certain design will meet the design requirements, what architectural element will act as a bottleneck, or what the memory requirements will be. Consequently, one of the major challenges in the design process is to quantitatively analyze specific characteristics of a system, such as end-to-end delays, buffer requirements, throughput, energy consumption, or temperature rises due to application activities. We refer to this analysis as performance analysis.

The performance analysis of KPNs executing on MPSoCs poses a major challenge due to multiple and heterogeneous hardware resources, the distributed execution of the application, and the interaction of computation and communication on shared resources. To deal with these challenges, multiple methods have been

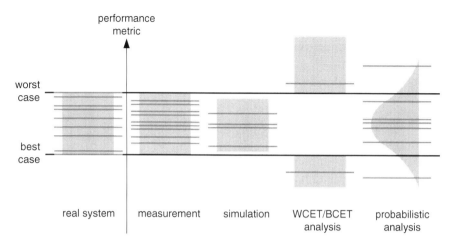

Fig. 2 Scope of different performance analysis methods for MPSoC

successfully used in the context of KPN design flows. These methods differ in accuracy, evaluation time, set-up effort, and scope.

In Fig. 2, the scope of different performance analysis methods is compared. Leftmost, the interval of values for a performance metric as occurring in the real system is shown. This performance metric could be the end-to-end delay of a system, the utilization of a computation or communication resource, or the occupation of a channel buffer, for instance. Different performance analysis methods now differ regarding the values that can be obtained.

When taking measurements of the real system, the measured values only represent a subset of all possible values. Most likely, due to insufficient coverage of corner cases and the limited number of measurement samples, the interval bounds can only be estimated based on the measurements. This observation applies to simulation as well. Best-case and worst-case analysis methods take a different approach by providing safe results about the interval bounds, i.e. upper and lower bounds on the worst-case and best-case behavior, respectively. On the other hand, usually not all parts of a system can be accurately modeled. In that case, (safe) optimistic and pessimistic assumptions need to be made, leading to bounds on system performance measures that are not tight. Finally, also probabilistic methods are used to provide quantitative statements about system behavior. In the following, we take a closer look at simulation and best-case/worst-case analysis due to their frequent application in KPN design flows.

Simulation is presumably the most frequently used method for performance analysis. This is reflected by the availability of a wide range of simulation tools that are applicable to different levels of abstraction. The most accurate but also slowest class are cycle-accurate simulators. Instruction-accurate simulators (also referred to as instruction-set simulators or virtual platforms) provide a good trade-off between speed and accuracy which allows entire MPSoCs to be modeled and simulated.

An example is the so-called full system simulator of the Cell Broadband Engine which also allows switching between different simulation modes with different accuracies [31]. Besides performance analysis, virtual platforms can also be used for software development and debugging. For this purpose, the full system simulator of the Cell Broadband Engine provides a fast, purely functional simulation mode in which timing is not considered.

At higher levels of abstraction, also other kinds of simulation are used for performance analysis. One example is trace-based simulation in the Artemis design flow [55] or in DOL [29], for instance. In trace-based simulation, first an untimed execution trace of the application is recorded that contains computation and communication events of processes and channels. Based on an architecture description, the mapping of the application onto the architecture, and estimates about the time to process events, this trace is refined towards timing behavior. This technique allows designers to estimate the system performance. Depending on the level of detail in the trace and the modeling of the execution platform, estimation errors of less than 5% have been reported with a significantly reduced simulation time compared to instruction-accurate simulation.

For the design of hard real-time systems, worst-case guarantees on the system timing need to be given. As stated above, worst-case bounds are difficult to obtain from simulation due to insufficient corner case coverage and often prohibitively long execution times of a simulation run. Therefore, analytic methods appear to be a promising method for providing worst-case guarantees even in the case of complex and large-scale MPSoC implementations. Prominent methods for analytic performance analysis are listed in the following.

- *Holistic Methods*: Holistic analysis is a collection of techniques for the analysis of distributed systems. The principle is to extend concepts of classical single-processor scheduling theory to distributed systems, integrating the analysis of computation and communication resource scheduling. Several holistic analysis techniques have been aggregated in the modeling and analysis suite for real-time applications (MAST) [18].
- *Compositional Performance Analysis Methods*: The basic idea of compositional performance analysis methods is to construct an analysis model of small components and propagate timing information between these components. Typical components model the execution of processes on a processor, the transmission of data packets on interconnects, or traffic shapers. Timing information is described by event models, such as periodic, periodic with jitter and bursts, or more general models in terms of arrival curves. Prominent methods of compositional performance analysis are modular performance analysis (MPA) [8] and symbolic timing analysis for systems (SymTA/S) [27]. Both methods support a rich set of scheduling policies, such as preemptive and non-preemptive fixed priority scheduling, earliest deadline first scheduling, or time division multiple access. MPA is used in the DOL design flow, for instance.
- *Automata Based Analysis*: Performance analysis of MPSoCs has also been tackled using state-based formalisms. One example are timed automata [1]:

Fig. 3 Application, architecture, and mapping design space

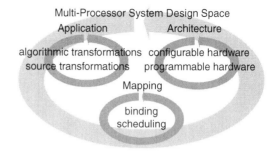

The approach is to model a system as a network of interacting timed automata and formally verify its behavior by means of reachability analysis using the Uppaal model checker [26].

A comparison of these performance analysis methods is provided in [52]. Note that beside being suited for the analysis of real-time systems, analytic models are often used as the basis for performing system optimization, such as scheduling parameter optimization [69] or robustness optimization [25].

Finally, one can observe that none of the methods shown in Fig. 2 can fulfill all the requirements concerning accuracy, scope, and set-up effort. Therefore, combinations of the different methods have been proposed: Simulation has been coupled with native execution on the target platform to reduce simulation time [2, 40]. Different analytic methods have been coupled to broaden the analysis scope [42, 60]. Subsystems in simulation have been replaced by analytic models to reduce simulation time and eliminate the need to generate a detailed simulation model of a component [43]. In these efforts, the modularity of KPNs is often leveraged by using the FIFO channels as the interface between the different performance analysis methods.

3.4 Design Space Exploration

Designers of MPSoCs face a large design space due to the combinatorial explosion related to the available degrees of freedom. At several points in the design flow and at various levels of abstraction, they need to decide between design alternatives. Specifically, the design space of MPSoCs can be roughly divided into three domains: the application design space, the architecture design space, and the mapping design space. These three domains can be further split up, as shown in Fig. 3.

Exploration of the *application design space* can be split up into two main kinds of transformations, namely algorithmic and source transformations. Algorithmic transformations make explicit the coarse-grained parallelism in a sequential application by transforming it into a KPN. Given a KPN application, source transformations split and merge processes to trade-off parallelism and communication overhead.

Exploration of the *architecture design space* attempts to find an optimized architecture for a given application. The goal is to instantiate programmable and (re-)configurable hardware components that allow an efficient implementation of an application.

Exploration of the *mapping design space* is the last step in the design space exploration. Given a KPN application and an architecture, processes and channels of the KPN are bound to processors and interconnects in the architecture, and scheduling policies are defined on shared communication or computation resources.

Usually, design space exploration is a multi-objective optimization problem. The goal is thus to find a set of Pareto-optimal designs which represent solutions with different trade-offs between the optimization goals such as performance, cost, energy consumption, or peak temperature. The final choice is left to the designer who needs to decide which of the Pareto-optimal designs to implement or to refine to the next level of abstraction.

Available approaches to the exploration of design spaces can be characterized as follows. Gries [20] presents a more detailed survey of automated design space exploration and performance analysis in different design flows.

- *Manual Exploration*: The selection of design points is done by the designer. When taking this approach, the advantage of using a KPN design flow lies in efficient performance analysis and automated synthesis of selected designs.
- *Exhaustive Search*: All design points in a specified region of the design parameters are evaluated. Very often, this approach is combined with local optimization in one or several design parameters in order to reduce the size of the design space. Due to the availability of fast performance analysis techniques for KPNs, exhaustive search is a realistic option if the design space is limited (or can be pruned) to roughly a few thousand designs.
- *Reduction to Single Objective*: For design space exploration with multiple conflicting criteria, there are several approaches available that reduce the problem to a single criterion optimization. For example, manual or exhaustive sampling is done in one (or several) directions of the search space and a constraint optimization, e.g. iterative improvement or analytic methods is done in the other. One may also combine the various objectives to a single criterion by means of a weighted sum where the weights express the preferences of the designer.
- *Black-box Randomized Search*: The design space is sampled and searched via a black-box optimization approach, i.e. new design points are generated based on the information gathered so far and by defining an appropriate neighborhood function (variation operator). The properties of these new design points are estimated which increases the available information about the design space. Examples of sampling and search strategies are Pareto simulated annealing, Pareto tabu search, or evolutionary multi-objective optimization. These black box optimization methods are often combined with local search methods that optimize certain design parameters or structures. This approach is most frequently used in KPN design flows, as illustrated in Table 2.

- *Problem-Dependent Approaches*: In addition to the above methods, one can find also a close integration of the exploration with a problem-dependent performance analysis of implementations. This approach is often used in design flows that are based on subclasses of KPNs. The StreamIt and MAMPS design flow, for instance, are based on SDF (Synchronous Data Flow) graphs and use adopted techniques for design space exploration, see chapter [22].

4 Specification, Synthesis, Analysis, and Optimization in DOL

Until now, this chapter introduced KPN design flows and the corresponding main design activities. This section will provide additional technical details by means of a concrete example of a typical design flow: the Distributed Operation Layer (DOL) [30, 62]. The underlying concepts for system specification, synthesis, performance analysis, and design space exploration will be considered, as well as a few typical experimental results for the size of the implementation, the runtime, and accuracy of the applied methods. DOL is currently being extended towards scenario-based design flow [10], and supports the design, optimization, and simultaneous execution of multiple dynamic applications on a MPSoC starting with a similar programming model as [16]. However, this section does not discuss these extensions, focusing on the typical design flow for single Kahn process networks.

4.1 Distributed Operation Layer

The distributed operation layer (DOL) [30, 62] is a platform independent MPSoC design flow based on the Kahn process network (KPN) model of computation [33] and targeted at real-time multimedia and (array) signal processing applications.

The DOL design cycle, as shown in Fig. 4, follows the Y-chart approach in which the application specification is platform-independent and needs to be related to a concrete architecture by means of an explicit mapping. As usual, the design starts with the specification of the application and architecture (and sometimes even a mapping). Then, code for the functional simulation of the application is automatically generated for testing and debugging the parallel application code with standard debugging tools on a standard PC/workstation.

Once the application is functionally correct, it can be mapped onto the target architecture. Based on the architecture and the mapping specification, the system is synthesized by generating the corresponding binaries. Note that, here, system synthesis refers to software synthesis only as the architecture specification is considered to be unaltered during the exploration phase. Then, the synthesis involves the generation of the mapping-dependent source code for processors, the

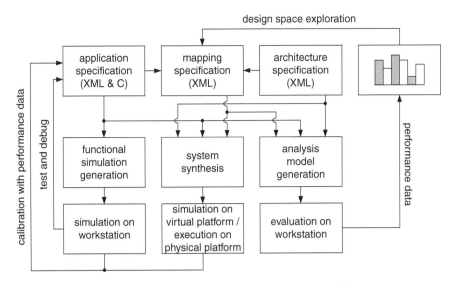

Fig. 4 Overview of the DOL design flow

compilation, and the linking to platform specific libraries as well as to the run-time environment. Generated binaries can either be executed on a simulator of the target platform or on the real MPSoC. Both, the functional and the low-level simulation provide performance figures that will enrich the application specification. This information will be used in later phases for the calibration of the analysis model.

The design flow described so far is typical for MPSoC design and very similar to the other design flows listed in Table 1 and explained in the previous chapters. What is different in DOL is its focus on the design and analysis of real-time signal processing applications. To this end, an analytic worst-/best-case performance analysis method has been embedded into the design flow. Besides enabling the analysis of real-time systems, using an analytic method for performance analysis facilitates rapid design space exploration due to short analysis times. The resulting performance data are embedded in a design space exploration loop in search of the optimal mapping.

4.2 System Specification

For designing the specification format of an MPSoC, one has to consider three criteria. First, the specification format should be expressive enough to represent the class of envisioned applications, i.e. (real-time) signal processing applications. Second, the specification should facilitate automation of system synthesis and analysis. The third criterion is the possibility of mapping an application in different ways onto an architecture. In the DOL framework, these criteria are met by

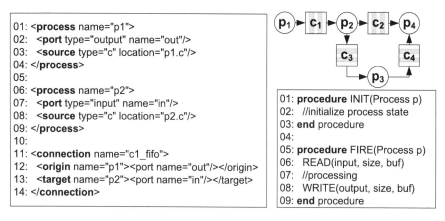

```
01: <process name="p1">
02:   <port type="output" name="out"/>
03:   <source type="c" location="p1.c"/>
04: </process>
05:
06: <process name="p2">
07:   <port type="input" name="in"/>
08:   <source type="c" location="p2.c"/>
09: </process>
10:
11: <connection name="c1_fifo">
12:   <origin name="p1"><port name="out"/></origin>
13:   <target name="p2"><port name="in"/></target>
14: </connection>
```

```
01: procedure INIT(Process p)
02:   //initialize process state
03: end procedure
04:
05: procedure FIRE(Process p)
06:   READ(input, size, buf)
07:   //processing
08:   WRITE(output, size, buf)
09: end procedure
```

Fig. 5 Kahn process network model. *Left*: XML description of the process network structure. *Right top*: example of a process network. *Right bottom*: C code of individual processes

specifying the application as a Kahn Process Network [33] and by specifying the application independently of architecture.

When designing parallel applications irrespective of architectures, an important feature is the ability to specify different topologies of the process network with different degrees of parallelism. For this reason, the KPN coordination part is kept separately (described in XML) from the source code of the individual processes (described in C/C++), see Fig. 5. Similar hybrid XML/C formats are employed by other frameworks as well (e.g. in Artemis [55], ESPAM [49], and MAMPS [41]).

While the syntax of the XML file is specified using an XML schema, the C code is based on a simple API. As shown in Fig. 5, this API basically consists of four functions, two of which concern computation, namely INIT and FIRE, and two of which concern communication inside the FIRE procedure, namely READ and WRITE:

- INIT contains the code that is executed once at start-up to initialize a process.
- FIRE contains the code that is repeatedly called by the scheduler.
- READ implements the blocking read from a FIFO channel.
- WRITE implements the blocking write to a FIFO channel.

A similar API is defined by Y-API [38], for instance, a library for specifying and executing Kahn process networks.

The architecture model in DOL is an abstract representation of the underlying execution platform. Its purpose is to determine at a system-level the consequences of the application mapping. This abstract architecture models the topology (i.e. the set of processors and communication paths between processors) and includes performance figures of the underlying platform useful for performance analysis, e.g. the clock frequency and throughput of architectural resources. The architecture model is a structural description that does not express the functional behavior, and

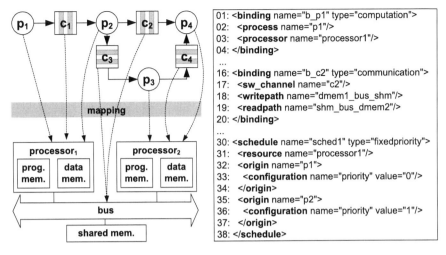

```
01: <binding name="b_p1" type="computation">
02:   <process name="p1"/>
03:   <processor name="processor1"/>
04: </binding>
...
16: <binding name="b_c2" type="communication">
17:   <sw_channel name="c2"/>
18:   <writepath name="dmem1_bus_shm"/>
19:   <readpath name="shm_bus_dmem2"/>
20: </binding>
...
30: <schedule name="sched1" type="fixedpriority">
31:   <resource name="processor1"/>
32:   <origin name="p1">
33:     <configuration name="priority" value="0"/>
34:   </origin>
35:   <origin name="p2">
36:     <configuration name="priority" value="1"/>
37:   </origin>
38: </schedule>
```

Fig. 6 Mapping of a Kahn process network onto a two-processors architecture and an example of a corresponding mapping XML file

which is specified in XML, similar to the application model. This XML architecture representation is not specific to DOL, but also encountered in other frameworks, such as Artemis and MAMPS.

The application model is brought in correspondence to the architecture model by a mapping (see Fig. 6) which can be either established manually by an experienced designer or generated automatically by design space exploration. This mapping fixes the allocation of hardware resources, the binding of the application elements onto these resources, and the scheduling on shared resources. For the mapping specification, once more, the XML format is used. The mapping XML serves as intermediate format and interface between tools, i.e. the design space exploration tool generates a mapping XML as an output, which is the input for the software synthesis tool.

The application XML, the architecture XML, and the mapping XML are the basis for the following DOL synthesis steps, i.e. for the functional simulation and the implementation of the final MPSoC, but also for the generation of the analytic performance analysis model (see Fig. 4).

4.3 System Synthesis

Similar to other frameworks, an application specified in DOL cannot be directly executed by just compiling the provided source code of the processes. A synthesis step is required that generates the "glue code" implementing the processes and channels, the bootstrapping and the scheduling of the application. Specifically, synthesis is done first for a standard PC/workstation to support the functional verification and debugging of the application (in which case it should be rather termed functional simulation generation) and second for the target MPSoC (which is properly known as system synthesis). However, due to similarities, the two steps are treated together in the following subsections as facets of system synthesis in the DOL design flow. Note that when the behavior of a part of an application can be restricted to a subclass of KPN, the general approach described below could be combined with one of the corresponding synthesis techniques described in the previous chapters.

4.3.1 Functional Simulation Generation

The purpose of providing a functional simulation that can be executed on a standard PC/workstation is to provide the application developer with a convenient approach to test and debug the application. Specifically, functional bugs within the application can be exposed and debugged by running a functional simulation on a standard PC and using standard debugging tools, e.g. the GNU debugger gdb.

A second role of the functional simulation is to obtain architecture-independent application parameters for performance analysis, such as the amount of data transferred between processes or the number of activations of processes. These parameters can easily be obtained from functional simulation by monitoring the calls of the READ, WRITE, and FIRE methods. By back-annotating these parameters to the application specification as shown in Fig. 4, they can be referred to during performance analysis, as explained later in this section.

Using the DOL design flow, a functional simulation can be automatically generated according to the application specification: For each process, an execution thread is instantiated. To implement software channels, inter-thread communication channels are used. The execution of the application is then controlled by a simple data-driven scheduler. Since Kahn process networks are determinate, this is a viable possibility because the scheduling does not influence the input/output behavior of the application.

In DOL, the functional simulation is based on the SystemC library. Therefore, processes can be implemented as user-space threads which incurs less runtime overhead compared to using an operating system thread library, such as the pthread library. Figure 7 shows the software architecture of the functional simulation based on SystemC: Each Kahn process is embedded into a SystemC thread, whereas each Kahn software channel is implemented as a SystemC channel. Moreover, the main file that bootstraps the process network and implements the scheduler to coordinate the quasi-parallel execution of processes is generated automatically as well.

Another frequently chosen library is the pthreads library. On multicore multiprocessors where single operating system threads can be executed on different cores, a functional simulation based on pthreads can even achieve a speed-up compared to the sequential version of the application. In [14], speeds-ups of more than 3 have been reported for executing applications specified in SHIM on a quad-core Intel Xeon processor.

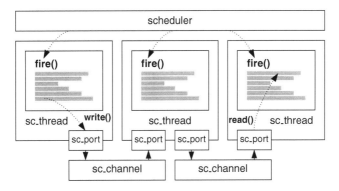

Fig. 7 Software architecture of the functional simulation of a KPN application based on SystemC

4.3.2 Software Synthesis

After the application has been functionally verified by functional simulation, it is ported to the target platform. This requires an architecture dependent runtime environment in which the application is executed. The role of the runtime environment is to hide architectural details of the MPSoC platform by providing a set of high-level services enabling the execution of an application on the platform, such as task scheduling, inter-process communication, or inter-processor communication. Depending on the target platform, developing (parts of) the runtime environment might be necessary to create the basis for software synthesis.

In case of the DOL design flow, different hardware MPSoC platforms are supported:

- *Cell Broadband Engine [53]*: MPSoC consisting of a PowerPC-based Power Processor Element and eight DSP-like Synergistic Processing Elements interconnected via a ring bus.
- *Atmel Diopsis 940 [50]*: Tile-based MPSoC, where a single tile is composed of an ARM9 processor and a DSP interconnected by an AMBA bus; up to eight tiles are interconnected via a network-on-chip.
- *MPARM [4]*: Homogeneous MPSoC consisting of identical ARM7 processors connected by an AMBA bus.
- *Intel SCC [47]*: Many-core homogeneous architecture with 48 cores organized in 24 tiles, each tile embedding two cores. Tiles are connected via a mesh on-chip network, and each tile has also a message passing buffer.

Figure 8 depicts a block diagram of the MPARM architecture. Software synthesis for MPARM is based on the RTEMS (Real-Time Executive for Multi-processor Systems) [59] operating system. Basic services provided by RTEMS are the scheduling of processes, device drivers for inter-process communication, and device drivers for system input/output. Based on these services, it is rather simple to bootstrap and execute a process network. As an example, Listing 1 illustrates parts of the code for bootstrapping a process network based on the RTEMS API. Software

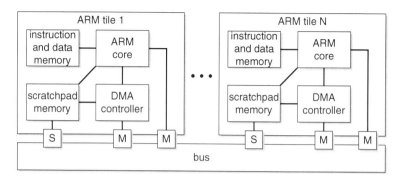

Fig. 8 Block diagram of the MPARM architecture

synthesis for the Cell Broadband Engine is described in [24] in the context of the
DOL design flow, and in [66] in the context of the SHIM design flow, for instance.

Listing 1 shows parts of a main file for a producer-consumer type application
running on MPARM. In lines 2–3, memory is allocated for the local data of the
producer and consumer processes. In lines 6–9, two tasks are created for the
processes by allocating a task control block, by assigning a task name and a task
ID, by allocating a stack, and by setting initial attributes like the task priority and
the task mode. Lines 11–12 show the creation of a message queue. In lines 14–18,
the rtems_task_start directive puts the tasks into the ready state, enabling the
scheduler to execute them. Finally, the initialization tasks deletes itself (line 20).

Listing 1 RTEMS initialization task in which two tasks are bootstrapped to run a producer and
consumer process of a process network

```
 1   rtems_task  Init(rtems_task_argument  arg)  {
 2      producer_wrapper  ←  malloc(sizeof(RtemsProcessWrapper));
 3      consumer_wrapper  ←  malloc(sizeof(RtemsProcessWrapper));
 4
 5      for  (j  ←  0;  j < 2;  j++)  {
 6         status  ←  rtems_task_create(j + 1, 128,
 7                       RTEMS_MINIMUM_STACK_SIZE,  RTEMS_DEFAULT_MODES,
 8                       RTEMS_DEFAULT_ATTRIBUTES,  &(task_id[j]));
 9      }
10
11      status  ←  rtems_message_queue_create(1, 10, 1,
12                    RTEMS_DEFAULT_ATTRIBUTES,  &queue_id[0]);
13
14      status  ←  rtems_task_start(task_id[1], producer_task,
15                    (rtems_task_argument)producer_wrapper);
16
17      status  ←  rtems_task_start(task_id[2], consumer_task,
18                    (rtems_task_argument)consumer_wrapper);
19
20      rtems_task_delete(RTEMS_SELF);
21   }
```

For all the mentioned platforms, the main challenge is to provide an efficient FIFO channel implementation that allows overlapping computation and communication in order to reduce the runtime overhead as much as possible. Aspects that play an important role in this context are the size and location of channel buffers, the efficient use of DMA controllers for data transfers between processors, and the minimization of synchronization messages.

4.4 Performance Analysis

The DOL design flow is targeted towards the design of real-time multi-media and signal processing applications. These systems must meet real-time constraints, which means that not only the correctness and performance of a system are of major concern but also the timeliness of the computed results. Typical questions in this context are:

- What is the response time to certain events? Is this response time within the required real-time limits?
- Can the system accept additional load and still meet the quality-of-service and real-time constraints?
- Is a system schedulable, that is, are all real-time constraints met?

To be able to answer these questions, a suitable combination of system design and performance analysis is required. To this end, it is essential that the architecture, application, and runtime-environment of a system are amenable to formal analysis, because simulation or measurements are not able to provide guarantees about timing properties. On the other hand, performance analysis methods with a reasonable scope and accuracy need to be employed such that effects occurring in the system implementation can be faithfully modeled. For MPSoC applications, this includes the modeling of heterogeneous resources and their sharing, the modeling of complex timing behavior arising from variable execution demands and interference on shared resources, or the modeling of different processing semantics.

Many approaches have been proposed to solve this problem, see [12] for an overview. Frequently used approaches are time-triggered and synchronous approaches, for instance. In purely time-triggered approaches, such as the time-triggered architecture [39] or Giotto [28], processing time of resources is allocated to tasks in fixed time slots. This fixed allocation facilitates analysis, but dimensioning of the slots turns out to be difficult for varying workloads. For instance, using the worst-case workload for setting the slot sizes might lead to over-dimensioned systems. Purely synchronous approaches implemented in synchronous languages, such as Esterel, Lustre, and Signal, rely on a global clock that divides the execution of a system into a sequence of atomic processing steps [5]. While synchronous approaches are successfully used for single-processor systems, applying them to MPSoCs is difficult because MPSoCs are usually split up into different (asynchronous) clock domains such that the synchronous assumption does not hold.

The approach taken in DOL relies on using compositional performance analysis where a system is modeled as a set of processing components that interact via event streams. This is a good match for MPSoCs as well as for KPNs. Contrary to other approaches, the approach is rather flexible in that it is not limited to a certain system architecture, scheduling policy, or execution semantics. Specifically, Modular Performance Analysis (MPA) is integrated into the DOL design flow. In the following, the basic concepts of MPA are reviewed. Afterwards, it is summarized how MPA is integrated into the DOL design flow.

4.4.1 Modular Performance Analysis

Modular Performance Analysis (MPA) [8, 71] is an analytical approach for the analysis of real-time systems. It is based on real-time calculus [64] which has its foundations in network calculus, a method for worst-case analysis of communication networks [9, 44]. With MPA, hard upper and lower bound for performance metrics of a distributed real-time system can be computed. As shown in Fig. 9, the performance model of a system is decomposed into a network of abstract processing components that model the computation and communication in a system. These processing components are connected by abstract event streams that model the timing properties of the data streams flowing through the system. Finally, resources are modeled by resource streams that model the availability of processing resources to computation and communication tasks. Different scheduling policies can be modeled by differently connecting processing components and resource streams. As an example, Fig. 9 illustrates fixed priority (FP) scheduling on processors and time-division multiple-access (TDMA) scheduling on the bus.

The processing components modeling computation and communication are characterized by the worst-case/best-case execution demand and the minimal and maximal token size, respectively. Event streams are characterized by so-called arrival curves and resource streams by service curves. Summarizing, these abstractions allow the modeling of computation and communication on heterogeneous resources in a unified manner.

Based on these abstractions, the system is analyzed by consecutive propagation of event streams between components. Depending on the mapping of processing components to a resource and its scheduling/arbitration, the timing properties of the streams change. System properties such as resource utilization, system throughput, end-to-end delays, or buffer sizes can be derived this way. Tool support for actually performing the analysis is provided by a freely available Matlab toolbox [70] that implements the underlying algebraic operations.

4.4.2 Integration of MPA into the DOL Design Flow

It has been mentioned that the goal of system synthesis is to bridge the implementation gap, that is, to refine a high-level system specification into an actual

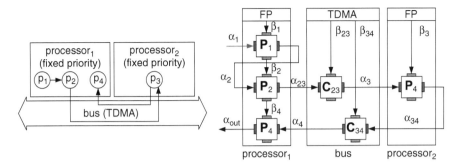

Fig. 9 MPA model of a system with two processors connected by a bus on which a KPN with four processes is executing. In the MPA model, horizontal edges represent event streams whereas vertical edges represent resource streams

system implementation. Similarly, there is an "abstraction gap" between an MPA model and the implementation: The execution of sequential processes on a processor is modeled by an abstract processing component, the availability of resources is abstracted by service curves, and the dataflow through the systems by arrival curves. Bridging this abstraction gap, that is, creating an analysis model that correctly models the implementation is a non-trivial task. On the one hand, the high-level system specification is conceptually similar to the analysis model but does not contain all the parameters required to generate an MPA model, such as best-case/worst-case task execution times or token sizes. The implementation, on the other hand, implicitly contains this information, but extracting the information is not straightforward.

In the DOL design flow, the abstraction gap is bridged by analysis model generation and calibration. In model generation, the high-level system specification is translated into a corresponding MPA model. In model calibration, the required model parameters are obtained. In both steps, the modular structure of the application and architecture specification is leveraged. The basic approach is depicted in Fig. 10. The analysis model generation represents a branch in the design flow that is parallel to the system synthesis. The analysis model calibration makes use of this feature later on in order to build a database with necessary performance data for the formal model. In the following, the basic approach is described. Further details are provided in [23].

The goal of model generation is to translate the high-level application, architecture, and mapping specification into a corresponding MPA model implemented as a Matlab script. Due to the modular specification of the application that is made explicit in the process network XML description, this is straightforward: Each process is simply modeled as an abstract processing component. Similarly, the communication channels between processes are modeled as abstract communication components and connected to the processing components according to the topology of the KPN.

Fig. 10 Analysis model
generation and calibration
in the DOL flow

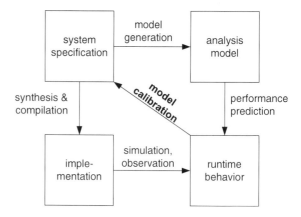

The aim of model calibration is to obtain the quantities to parameterize the generated model such that it correctly models the implementation. Basically, three different types of parameters can be distinguished:

• First, there are the application parameters that are architecture and timing independent. An example is the minimal and maximal size of tokens transmitted over each channel.

• Second, there are the parameters that depend only on the architecture and the runtime environment. Examples are the throughput of the different communication resources or the context switch time of the runtime environment.

• Third, there are the application parameters that depend on the architecture and the mapping. Basically, whenever the architecture or mapping changes, new parameters need to be determined. An example is the worst-case/best-case execution time of a process on a processor.

Depending on the parameter type, there are different ways to obtain them. Timing independent parameters can be obtained from the functional simulation, due to the determinism of KPN applications. Architecture dependent parameters need to be obtained once a new hardware architecture or runtime environment is employed for realizing a system. The parameters of the third category, i.e. application parameters that depend on the architecture and the mapping, are more difficult to determine. Similar to system-level performance analysis, different methods for worst-case/best-case execution time analysis have been proposed, for instance [72]. In the DOL design flow, timed simulation on a virtual platform is employed. Note that compared to formal methods, this approach is not suitable for the calibration of hard real-time system models unless complete coverage of corner cases is exhibited in the calibration simulation runs. One can observe that similar approaches are taken in other design flows. In the Artemis design flow, for instance, model generation and calibration is used to create a model for trace-based simulation [56].

Finally, the DOL framework have been extended with capabilities for worst-case thermal analysis, as nowadays providing guarantees on maximum temperature is as important as functional correctness and timeliness. Aware of the performance-temperature correlation, DOL is optimizing the system design with respect to both worst-case performance and worst-case temperature, analyzed in the same MPA framework. The basic worst-case peak temperature analysis method in MPA for a single processor under a broad range of uncertainties in terms of task execution times, task invocation periods, and jitter in task arrivals is described in [58]. Extensions are then proposed in [61] for analysis of MPSoC platforms by considering both the self-heating of the processor and the heat transfer between neighboring processors. In the same manner as it is done for timing, thermal analysis models are automatically generated from the same set of specifications as used for software synthesis. To increase the model accuracy, both analysis models are calibrated with data corresponding to real system parameters obtained in an automatic manner, prior to design space exploration. The calibration tool-chain for the thermal model is described in [65].

4.5 Design Space Exploration

The final piece of the DOL flow is design space exploration, built on top of analysis and synthesis tools to find an optimal mapping. In general, the problem of optimally mapping an application to a heterogeneous distributed architecture is known to be NP-complete. Even for systems of modest complexity, one thus needs to resort to heuristics to solve the problem. In addition, the mapping problem is usually multi-objective such that there is no single optimal solution but a set of Pareto-optimal solutions constituting a so-called Pareto-front.

In DOL, the aim of the design space exploration is to compute the set of Pareto-optimal solutions representing different trade-offs in the design space. Based on the (approximated) Pareto-front, the designer chooses the final solution to implement. Therefore, the mapping problem is specified as a multi-objective optimization problem.

Formally, a multi-objective optimization problem is defined on the decision space X which contains all possible design decisions, i.e. architectures, applications and mappings. To each implementation $x \in X$ there is associated an objective vector f in the objective space Z that consists of n objectives $f = (f_1, \ldots, f_n)$ which should be minimized (or maximized). An order relation $\leq$ is defined on the objective space Z, which induces a preference relation $\preceq$ on the decision space X: $x_1 \preceq x_2 \Leftrightarrow f(x_1) \leq f(x_2)$, for $x_1, x_2 \in X$. In other words, for the mapping problem for instance, if mapping x_1 is better (minimal) in all objectives than mapping x_2, the optimization algorithm will "preferentially" select mapping x_1.

The design search space, symbolized with Ψ, is the set of all subsets of X, i.e. it includes all possible solution sets $A \subseteq X$. The final goal is to determine an optimal element of Ψ, i.e. an optimal subset of all possible implementations X. This subset should reflect all trade-offs induced by the multiple objectives.

Algorithm 1 Main optimization function
___
 1: randomly choose $A \in \Psi$ ▷ generate initial set P of size m
 2: set $P \leftarrow A$
 3:
 4: **while** termination criterion not fulfilled **do**   ▷ main optimization loop
 5:     $P' \leftarrow heuristicSetMutation(P)$
 6:     **if** $P' \preccurlyeq P$ **then**
 7:         $P \leftarrow P'$
 8:     **end if**
 9: **end while**
10: **return** P

Algorithm 2 Heuristic set mutation function
___
 1: **function** HEURISTICSETMUTATION(()P)
 2:     generate $\{r1,\ldots,rk\} \in X$ based on P
 3:     $P' \leftarrow P \cup \{r1,\ldots,rk\}$
 4:     **while** $|P'| > m$ **do**
 5:         choose $p \in P'$ with $fitness(p) = min_{a \in P}\{fitness(a)\}$
 6:         $P' \leftarrow P' \setminus \{p\}$
 7:     **end while**
 8:     **return** P'
 9: **end function**

The preference relation $\preceq$ on X that has been defined above can now be used to define a corresponding set preference relation, symbolized with $\preccurlyeq$, on the search space Ψ.

This set preference relation provides the information on the basis of which two candidate Pareto sets can be compared: $A \preccurlyeq B \Leftrightarrow \forall b \in B, \exists a \in A : a \preceq b$. This property reflects the concept of Pareto-dominance: A design point dominates another one if it is equal or better in all criteria and strictly better in at least one. Moreover, the search in the design space will be pursued until a good Pareto-optimal set approximation $A \in \Psi$ is found.

In DOL, evolutionary algorithms are used to solve the mapping optimization problem. Evolutionary algorithms find solutions to a given problem by iterating two main steps [74]: (1) selection of promising candidates, based on an a-priori evaluation of candidate solutions and (2) generation of new candidates by variation of previously selected candidates. The principle of the selection in evolutionary multi-objective optimization is sketched in Algorithm 1. For a complete description, we refer to [75]. The starting point is a randomly generated population $P \in \Psi$ of size m. During optimization, a heuristic mutation operator based on selection and variation generates another set $P' \in \Psi$, which is wanted to be better than P in the context of the predefined set preference relation $\preccurlyeq$, i.e. $P' \preccurlyeq P$. Finally, P is replaced by P', if the later is preferable to the former (i.e., $P' \preccurlyeq P$), or P it remains unchanged in the opposite case.

The heuristic set mutation operator is detailed in Algorithm 2. First, k new solutions are created based on P, after an appropriate selection and variation

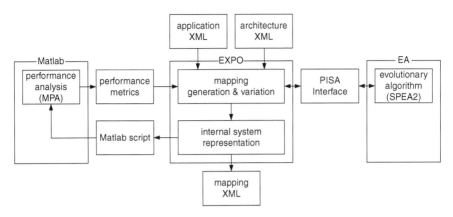

Fig. 11 Design space exploration in the DOL framework

operation. While the variation is problem-specific, the selection is independent of the problem, using either an uniform random selection or a fitness-based selection. Then, the k new solutions are added to P, resulting in a set P' of size $m + k$. P' is iteratively truncated to size m by removing the solutions with worst fitness values. Note that the fitness values are associated in a performance evaluation process, i.e., in DOL we use the MPA framework as described in Sect. 4.4.

While the selection algorithm described is domain independent, specific methods are used to include domain specific knowledge into the search process and select "best" solutions among a population. These are the domain representation (i.e., the system mapping), the evaluation of designs (i.e., the MPA analysis), and the variation of a population of solutions by mutation and cross-over operations.

Although standard variation schemes exist for mutation and crossover, their implementation is strongly dependent on system properties. The mutation generates a local neighborhood of selected design points. In the DOL context, the mutation affects the mapping solutions; for instance, different mappings can be generated with different bindings of processes onto processors. The crossover recombines two selected solutions to generate a new one. Note that during mutation or crossover, infeasible (mapping) solutions can be generated. In this situation, a repair strategy is invoked, which, in conformity with the evolutionary algorithm principle, attempts to maintain a high diversity in the population. An example could be the rerouting of inter-process communication, when during re-mapping a process was bound to a new location.

In DOL, the design space exploration framework includes several tools, as shown in Fig. 11. In particular, the EXPO [63] tool is the central module of the framework. As underlying multi-objective search algorithm, Strength Pareto Evolutionary Algorithm (SPEA2) [73] is used that communicates with EXPO via the PISA [7] interface. Similarly, the design space exploration framework in Artemis [55] is also based on PISA and SPEA2.

Using the frameworks of EXPO and PISA relieves the designer from implementing those parts of the design space exploration that are independent of the actual optimization problem. For example, the selection may be handled inside the multi-objective optimizer SPEA2. The designer just needs to focus on the problem specific parts, that is, the generation, variation, and evaluation of solutions. The implementation of problem specific parts starts with the specification, where the application and architecture (and later on, the mapping) are automatically extracted from the corresponding XML files and represented in the design space exploration framework. Then, candidate mappings are inspected (as described above) based on the provided variation methods. Finally, during design space exploration, the objective values of all candidate mappings are computed by generating the corresponding Matlab MPA scripts and interfacing Matlab for their evaluation. In DOL, all these operations are automatically parameterized using the application and architecture specification. Note that the approach described above is a heuristic search procedure. Therefore, it does not guarantee the optimality of the final solution, i.e., the final set of solutions. However, in our experiments we have identified that after several design space exploration cycles, the found solutions are close to optimal even for large problem complexities.

4.6 Results of the DOL Framework

In this section, a few results are highlighted that have been obtained by applying the DOL design flow described above. Specifically, the design and analysis of a Motion-JPEG (MJPEG) decoder [68] running on MPARM [4] is considered. For the execution of the system, we used a 31-frame input bitstream encoded using the QVGA (320×240) YUV 444 format.

The MJPEG decoder decompresses a sequence of frames by applying JPEG decompression to each frame. Because of the inherent parallelism in the MJPEG algorithm, the decoding is done in a pipeline with five stages, each stage being implemented as a Kahn process. The first and last stages are the splitting of streams into frames (ss) and the merging of frames back to streams (ms). The variable length decoding and the splitting of frames into macroblocks form the second stage (sf). The zigzag scan, inverse quantization and the inverse discrete cosine transform form the third (zii), while combining macroblocks back to frames forms the fourth stage (mf).

Using the design space exploration framework of DOL based on the PISA interface and SPEA2, one can compute the Pareto optimal mappings of the MJPEG application onto an MPARM system with a variable number of processors. The mapping has been optimized in conformity with two design objectives: (1) end-to-end delay in the system computed as the result of MPA analysis, which is an upper bound of the actual end-to-end delay and (2) the cost of the system evaluated as a sum of costs associated with the used processors, memories, and the bus. In the experiments, a population size of 60 individuals has been chosen and the algorithm

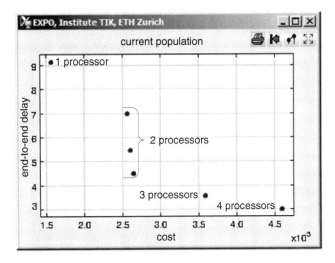

Fig. 12 Pareto optimal solutions resulted after the design space exploration (screenshot of the EXPO tool)

has been executed for 50 generations. These parameters generally depend on the complexity of the problem to solve. The obtained Pareto front is shown in Fig. 12, consisting of 6 mapping solutions onto a different number of processors. The search in this design space took about 2 h.

For illustration purposes only, we employ a simple configuration with a small number of processes that can be mapped in different ways onto the architecture and can communicate via different hardware communication paths. However, a more efficient implementation can be obtained if the same application is specified with a scalable number of processes processing data in parallel. This would enlarge the parallelism of the design but also the dimensionality of the design space. A more complex design space exploration with the DOL framework is shown in [62], where a scaled version of an MPEG-2 decoder has been mapped onto a tile-based heterogeneous architecture.

Other KPN flows, like Sesame in the Artemis project [55] that use exactly the same optimization frameworks of PISA and SPEA2, report comparable parameters and results for the design space exploration. However, their design space exploration is considerably shorter, i.e. 5 s for a design with 8 processes, because they use a much simpler additive performance model. Of course, for larger problem sizes all the parameters will scale and the design space exploration can take much longer if more accurate analysis methods, like MPA or simulation, are used. However, this is an acceptable cost since designers are exploring the entire design space only once.

In the remainder of this section, a mapping of the MJPEG application onto a 3-tile MPARM system, that is, three ARM processors interconnected via a shared AMBA bus, is considered. The resulting MPA model is shown in Fig. 13.

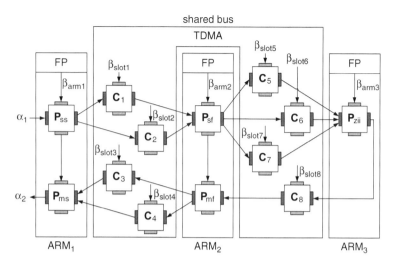

Fig. 13 MPA model of MJPEG application mapped onto a 3-tile MPARM platform. P_x are the five processes of the MJPEG decoding pipeline that communicate via the C_y software channels

Table 3 Duration of different design steps in the MJPEG design, measured on a 1.86 GHz Intel Pentium Mobile machine with 1 GB of RAM. The simulations were executed to decode a 31-frame input bitstream encoded using the QVGA (320×240) YUV 444 format

Step		Duration
Model calibration (one-time effort)	Functional simulation generation	42 s
	Functional simulation	3.6 s
	Synthesis (generation of binary)	4 s
	Simulation on MPARM	13,550 s
	Log-file analysis and back-annotation	12 s
Model generation		1 s
Performance analysis based on generated model		2.5 s

To evaluate the efficiency of the design flow (i.e., the time spent in obtaining results), Table 3 lists the durations of the different design steps for performance analysis of the different design solutions. Several conclusions can be drawn:

- Automated software synthesis can be done fast. Actually, most of the time required to generate the functional simulation or the binaries for the target platform is required to compile the generated source code rather than to generate this code.
- The table shows that timed simulation on the virtual platform is the most time-consuming step in the design flow. Minimization of simulation time is thus paramount and actually possible in a systematic design flow, as has been shown above. Conversely, simulation time can become a major bottleneck in MPSoC design when following a less systematic design flow requiring many design iterations involving timed simulation.

- The generation and analysis of a system's MPA model is a matter of seconds. Note that similar times have been reported for alternative performance analysis methods like trace-based simulation in the Artemis design flow, for instance. While further reducing this time is desirable, it is a reasonable time frame for performance analysis within a design space exploration loop.
- The one-time calibration to obtain the parameters for the MPA model takes several seconds albeit being completely automated. Extracting these parameters manually would be a major effort.

In order to evaluate the accuracy of MPA estimations, the performance bounds computed with MPA are compared to actual (average-case behavior) quantities observed during system simulation. The differences are in a range of 10–20%, which is typical for a compositional performance analysis. Differences in the same range have been observed for several systems in [52], for instance. There are two main reasons for these differences. First, several operators in the formal performance analysis do not yield tight bounds. Second, the simulation of a complex system in general cannot determine the actual worst-case and best-case behavior. The simulations on the system level do not use exhaustive test patterns and do not cover all possible corner cases in the interference through joint resources.

Moreover, to illustrate the connection between the worst-case chip temperature and worst-case latency, we represent eight selected mapping configurations of the MJPEG decoder application together with their worst-case chip temperature and worst-case latency calculated in MPA. Interesting here is the effect of the physical placement that cannot be ignored anymore. So even if the mapping is already defined, the system designer might still optimize the system (i.e., reduce the temperature) by selecting an appropriate physical placement. This is highlighted by solution pairs where only the placement of the processing components has changed but temperature differences of 8K can still appear [45].

Finally, the DOL framework itself is evaluated in terms of code size of the prototype implementation. The DOL design flow and the associated tools are implemented in Java (Fig. 14). To give an indication about the size of the implementation, Table 4 shows the code size of different parts of the design flow (excluding the plug-ins for design space exploration and thermal analysis). One can see that apart from the tool-internal representations of the system specification, the largest part is the MPA code generator for performance analysis. The software synthesizers and the monitoring for the MPA model calibration are comparatively small. Similar observations can be made for other design flows, as well.

5 Concluding Remarks

The mapping of process networks onto multi-processor systems requires a systematic and automated design methodology. This chapter provides an overview over different existing methods and tools, which are all starting from a general

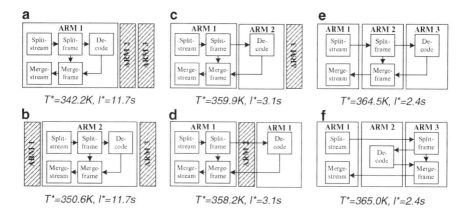

Fig. 14 Worst-case latency versus worst-case peak temperature for similar bindings but different placements, of an MJPEG decoder evaluated on MPARM platform [4]

Table 4 Java code size of different parts of the DOL design flow

Part of design flow	Lines of code
DOL representation of system specifications	6,200
Functional simulation generator	4,100
MPARM code generator	2,100
MPA Matlab code generator	5,000
Log-file analysis of functional and timed simulation	1,200

Kahn process network (KPN) model of computation and are implementing the established Y-chart approach. Due to fundamental properties of the Kahn model, many problems in the design process can be solved in an automated manner. Thus, the system specification, synthesis, performance analysis, and design space exploration can be implemented in a fully automated way.

After an overview over all these activities, this chapter provides a practical illustration of their implementation in the distributed operation layer (DOL) framework. The design steps followed by DOL are somewhat common to all the KPN flows. What is typical to the DOL framework is the embedding of an accurate formal performance analysis model into the design flow. This presents a clear advantage over the standard simulation-based approaches employed for performance analysis, which typically take more time to execute than a formal model and cannot offer guarantees for (hard) real-time signal-processing applications, due to the incomplete coverage of the design space.

Another key point is the need for a scalable design flow which allows to design large and complex MPSOC systems, which can clearly be noticed from Table 2. As soon as we are faced with more complex MPSoCs, this forthcoming difficulty needs to be considered in all steps of the design trajectory. In particular, it will have an impact at the system-level, where basic design decisions are taken. In this sense, the Kahn model and design methods based on it are promising candidates due to the modular system specification. It offers a great potential for compositional (and fast) performance analysis and design space exploration. By taking a closer look at the DOL framework, it can be observed that it features a specification format that can easily be scaled (i.e. provided by the XML and C basis), it includes a compositional formal performance analysis in the design, and the optimization is done with the support of modular tools such as EXPO and PISA. These features provide the basis for scalable mappings and mapping optimizations.

References

1. Alur, R., Dill, D.L.: A theory of timed automata. Theoretical Computer Science **126**(2), 183–235 (1994)
2. Bacivarov, I., Bouchhima, A., Yoo, S., Jerraya, A.A.: ChronoSym: A new approach for fast and accurate SoC cosimulation. International Journal of Embedded Systems (IJES) **1**(1/2), 103–111 (2005). DOI http://dx.doi.org/10.1504/IJES.2005.008812
3. Balarin, F., Watanabe, Y., Hsieh, H., Lavagno, L., Passerone, C., Sangiovanni-Vincentelli, A.: Metropolis: An integrated electronic system design environment. Computer **36**(4), 45–52 (2003). DOI http://dx.doi.org/10.1109/MC.2003.1193228
4. Benini, L., Bertozzi, D., Alessandro, B., Menichelli, F., Olivieri, M.: MPARM: Exploring the multi-processor SoC design space with SystemC. The Journal of VLSI Signal Processing **41**, 169–182(14) (2005). DOI doi:10.1007/s11265-005-6648-1
5. Benveniste, A., Caspi, P., Edwards, S.A., Halbwachs, N., Le Guernic, P., De Simone, R.: The synchronous languages 12 years later. Proceedings of the IEEE **91**(1), 64–83 (2003)
6. Bhattacharyya, S.S., Brebner, G., Janneck, J.W., Eker, J., von Platen, C., Mattavelli, M., Raulet, M.: OpenDF — a dataflow toolset for reconfigurable hardware and multicore systems. In: First Swedish Workshop on Multi-Core Computing (MCC). Uppsala, Sweden (2008)
7. Bleuler, S., Laumanns, M., Thiele, L., Zitzler, E.: PISA – a platform and programming language independent interface for search algorithms. In: Int'l Conf. on Evolutionary Multi-Criterion Optimization (EMO), pp. 494–508. Faro, Portugal (2003)
8. Chakraborty, S., Künzli, S., Thiele, L.: A general framework for analyzing system properties in platform-based embedded system design. In: Proc. Design, Automation and Test in Europe (DATE), pp. 190–195. Munich, Germany (2003)
9. Cruz, R.L.: A Calculus for Network Delay, Part I: Network Elements in Isolation. IEEE Trans. Inf. Theory **37**(1), 114–131 (1991)
10. Distributed Application Layer. URL http://www.tik.ee.ethz.ch/~euretile
11. Densmore, D., Sangiovanni-Vincentelli, A., Passerone, R.: A platform-based taxonomy for ESL design. IEEE Design & Test of Computers **23**(5), 359–374 (2006)
12. Edwards, S., Lavagno, L., Lee, E.A., Sangiovanni-Vincentelli, A.: Design of embedded systems: Formal models, validation, and synthesis. Proceedings of the IEEE **85**(3), 366–390 (1997)
13. Edwards, S.A., Tardieu, O.: SHIM: A Deterministic Model for Heterogeneous Embedded Systems. IEEE Trans. on VLSI Systems **14**(8), 854–867 (2006)
14. Edwards, S.A., Vadudevan, N., Tardieu, O.: Programming shared memory multiprocessors with deterministic message-passing concurrency: Compiling SHIM to Pthreads. In: Proc. Design, Automation and Test in Europe (DATE), pp. 1498–1503. Munich, Germany (2008)
15. Falk, J., Haubelt, C., Zebelein, C., Teich, J.: Integrated modeling using finite state machines and dataflow graphs. In: S.S. Bhattacharyya, E.F. Deprettere, R. Leupers, J. Takala (eds.) Handbook of Signal Processing Systems, second edn. Springer (2013)

16. Geilen, M., Basten, T.: Kahn process networks and a reactive extension. In: S.S. Bhattacharyya, E.F. Deprettere, R. Leupers, J. Takala (eds.) Handbook of Signal Processing Systems, second edn. Springer (2013)
17. Gelernter, D., Carriero, N.: Coordination languages and their significance. Commun. ACM **35**(2), 97–107 (1992)
18. González Harbour, M., Gutiérrez García, J.J., Palencia Gutiérrez, J.C., Drake Moyano, J.M.: MAST: Modeling and analysis suite for real time applications. In: Proc. Euromicro Conference on Real-Time Systems, pp. 125–134. Delft, The Netherlands (2001)
19. Gordon, M.I., Thies, W., Amarasinghe, S.: Exploiting coarse-grained task, data, and pipeline parallelism in stream programs. In: Proc. Int'l Conf. on Architectural Support for Programming Languages and Operating Systems (ASPLOS), pp. 151–162. San Jose, CA, USA (2006)
20. Gries, M.: Methods for evaluating and covering the design space during early design development. Integration, the VLSI Journal **38**(2), 131–183 (2004)
21. Ha, S., Kim, S., Lee, C., , Yi, Y., Kwon, S., Joo, Y.P.: PeaCE: A hardware-software codesign environment for multimedia embedded systems. ACM Trans. Design Automation of Electronic Systems **12**(3), 1–25 (2007)
22. Ha, S., Oh, H.: Decidable dataflow models for signal processing: Synchronous dataflow and its extensions. In: S.S. Bhattacharyya, E.F. Deprettere, R. Leupers, J. Takala (eds.) Handbook of Signal Processing Systems, second edn. Springer (2013)
23. Haid, W., Keller, M., Huang, K., Bacivarov, I., Thiele, L.: Generation and calibration of compositional performance analysis models for multi-processor systems. In: Proc. Int'l Conf. on Systems, Architectures, Modeling and Simulation (IC-SAMOS), pp. 92–99. Samos, Greece (2009)
24. Haid, W., Schor, L., Huang, K., Bacivarov, I., Thiele, L.: Efficient execution of Kahn process networks on multi-processor systems using protothreads and windowed FIFOs. In: Proc. IEEE Workshop on Embedded Systems for Real-Time Multimedia (ESTIMedia), pp. 35–44. Grenoble, France (2009)
25. Hamann, A., Racu, R., Ernst, R.: Multi-dimensional robustness optimization in heterogeneous distributed embedded systems. In: Proc. Real Time and Embedded Technology and Applications Symposium (RTAS), pp. 269–280. Bellevue, WA, United States (2007)
26. Hendriks, M., Verhoef, M.: Timed automata based analysis of embedded system architectures. In: Workshop on Parallel and Distributed Real-Time Systems. Rhodes, Greece (2006)
27. Henia, R., Hamann, A., Jersak, M., Racu, R., Richter, K., Ernst, R.: System level performance analysis — the SymTA/S approach. IEE Proceedings Computers and Digital Techniques **152**(2), 148–166 (2005)
28. Henzinger, T.A., Horowitz, B., Kirsch, C.M.: Giotto: A time-triggered language for embedded programming. Proceedings of the IEEE **91**(1), 84–99 (2003)
29. Huang, K., Bacivarov, I., Liu, J., Haid, W.: A modular fast simulation framework for stream-oriented MPSoC. In: IEEE Symposium on Industrial Embedded Systems (SIES), pp. 74–81. Lausanne, Switzerland (2009)
30. Huang, K., Haid, W., Bacivarov, I., Keller, M., Thiele, L.: Embedding formal performance analysis into the design cycle of MPSoCs for real-time streaming applications. ACM Transactions in Embedded Computing Systems (TECS) (2012)
31. IBM SDK for multicore acceleration: http://www-128.ibm.com/developerworks/power/cell/
32. Janneck, J.W., Miller, I.D., Parlour, D.B., Roquier, G., Wipliez, M., Raulet, M.: Synthesizing hardware from dataflow programs: An MPEG-4 simple profile decoder case study. In: IEEE Workshop on Signal Processing Systems (SiPS), pp. 287–292. Washington, D.C., USA (2008)
33. Kahn, G.: The semantics of a simple language for parallel programming. In: Proc. IFIP Congress, pp. 471–475. Stockholm, Sweden (1974)
34. Kangas, T., Kukkala, P., Orsila, H., Salminen, E., Hännikäinen, M., Hämäläinen, T.D.: UML-based multiprocessor SoC design framework. ACM Trans. on Embedded Computing Systems **5**(2), 281–320 (2006)

35. Keinert, J., Streubühr, M., Schlichter, T., Falk, J., Gladigau, J., Haubelt, C., Teich, J., Meredith, M.: SystemCoDesigner — an automatic ESL synthesis approach by design space exploration and behavioral synthesis for streaming applications. ACM Trans. on Design Automation of Electronic Systems **14**(1), 1:1–1:23 (2009)

36. Kienhuis, B., Deprettere, E., Vissers, K., van der Wolf, P.: An approach for quantitative analysis of application-specific dataflow architectures. In: Proc. Int'l Conf. on Application-Specific Systems, Architectures and Processors (ASAP), pp. 338–349. Washington, DC, USA (1997)

37. Kienhuis, B., Rijpkema, E., Deprettere, E.: Compaan: Deriving process networks from Matlab for embedded signal processing architectures. In: Proc. of the Int'l Workshop on Hardware/Software Codesign (CODES), pp. 13–17. San Diego, CA, USA (2000)

38. de Kock, E.A., Essink, G., Smits, W.J.M., van der Wolf, P., Brunel, J.Y., Kruijtzer, W.M., Lieverse, P., Vissers, K.A.: YAPI: Application modeling for signal processing systems. In: Proc. Design Automation Conference (DAC), pp. 402–405. Los Angeles, CA, USA (2000)

39. Kopetz, H., Bauer, G.: The time-triggered architecture. Proceedings of the IEEE **91**(1), 112–126 (2003)

40. Kraemer, S., Gao, L., Weinstock, J., Leupers, R., Ascheid, G., Meyr, H.: HySim: A fast simulation framework for embedded software development. In: Proc. Int'l Conf. on Hardware/Software Codesign and System Synthesis (CODES+ISSS), pp. 75–80. Salzburg, Austria (2007)

41. Kumar, A., Fernando, S., Ha, Y., Mesman, B., Corporaal, H.: Multiprocessor systems synthesis for multiple use-cases of multiple applications on FPGA. ACM Trans. on Design Automation of Electronic Systems **31**(3), 40:1–40:27 (2008)

42. Künzli, S., Hamann, A., Ernst, R., Thiele, L.: Combined approach to system level performance analysis of embedded systems. In: Proc. Int'l Conf. on Hardware/Software Codesign and System Synthesis (CODES/ISSS), pp. 63–68. Salzburg, Austria (2007)

43. Künzli, S., Poletti, F., Benini, L., Thiele, L.: Combining simulation and formal methods for system-level performance analysis. In: Proc. Design, Automation and Test in Europe (DATE), pp. 236–241 (2006)

44. Le Boudec, J.Y., Thiran, P.: Network Calculus — A Theory of Deterministic Queuing Systems for the Internet, *Lecture Notes in Computer Science*, vol. 2050. Springer Verlag (2001)

45. Marwedel, P., Teich, J., Kouveli, G., Bacivarov, I., Thiele, L., Ha, S., Lee, C., Xu, Q., Huang, L.: Mapping of applications to MPSoCs. In: CODES+ISSS. October 9–14, 2011, Taipei, Taiwan. (2011)

46. MathWorks Real-Time Workshop: http://www.mathworks.com/products/rtw/

47. Mattson, T.G., Riepen, M., Lehnig, T., Brett, P., Haas, W., Kennedy, P., Howard, J., Vangal, S., Borkar, N., Ruhl, G., Dighe, S.: The 48-core SCC processor: The programmer's view. In: Proceedings of the 2010 ACM/IEEE International Conference for High Performance Computing, Networking, Storage and Analysis, SC '10, pp. 1–11. IEEE Computer Society, Washington, DC, USA (2010). DOI 10.1109/SC.2010.53. URL http://dx.doi.org/10.1109/SC.2010.53

48. NI LabVIEW Microprocessor SDK: http://www.ni.com/labview/microprocessor_sdk.htm

49. Nikolov, H., Stefanov, T., Deprettere, E.: Systematic and automated multiprocessor system design, programming, and implementation. IEEE Trans. on Computer-Aided Design of Integrated Circuits and Systems **27**(3), 542–555 (2008)

50. Paolucci, P.S., Jerraya, A.A., Leupers, R., Thiele, L., Vicini, P.: SHAPES: A tiled scalable software hardware architecture platform for embedded systems. In: Proc. Int'l Conf. on Hardware/Software Codesign and System Synthesis (CODES+ISSS), pp. 167–172. Seoul, South Korea (2006)

51. Parks, T.M.: Bounded Scheduling of Process Networks. Ph.D. thesis, University of California, Berkeley (1995)

52. Perathoner, S., Wandeler, E., Thiele, L., Hamann, A., Schliecker, S., Henia, R., Racu, R., Ernst, R., González Harbour, M.: Influence of different system abstractions on the performance analysis of distributed real-time systems. In: Proc. Int'l Conf. on Embedded Software (EMSOFT), pp. 193–202. Salzburg, Austria (2007). DOI http://doi.acm.org/10.1145/1289927.1289959

53. Pham, D.C., Aipperspach, T., Boerstler, D., Bolliger, M., Chaudhry, R., Cox, D., Harvey, P., Harvey, P.M., Hofstee, H.P., Johns, C., Kahle, J., Kameyama, A., Keaty, J., Masubuchi, Y., Pham, M., Pille, J., Posluszny, S., Riley, M., Stasiak, D.L., Suzuoki, M., Takahashi, O., Warnock, J., Weitzel, S., Wendel, D., Yazawa, K.: Overview of the architecture, circuit design, and physical implementation of a first-generation Cell processor. IEEE Journal of Solid-State Circuits **41**(1), 179–196 (2006)
54. Pimentel, A.D.: The Artemis workbench for system-level performance evaluation of embedded systems. Int. J. Embedded Systems **3**(3), 181–196 (2008)
55. Pimentel, A.D., Erbas, C., Polstra, S.: A systematic approach to exploring embedded system architectures at multiple abstraction levels. IEEE Trans. on Computers **55**(2), 99–112 (2006)
56. Pimentel, A.D., Thompson, M., Polstra, S., Erbas, C.: Calibration of abstract performance models for system system-level design space exploration. Journal of Signal Processing Systems **50**(2), 99–114 (2008)
57. Ptolemy web site: http://ptolemy.eecs.berkeley.edu
58. Rai, D., Yang, H., Bacivarov, I., Chen, J.J., Thiele, L.: Worst-case temperature analysis for real-time systems. In: Proc. Design, Automation and Test in Europe (DATE'11). Grenoble, France (2011)
59. RTEMS web page: http://www.rtems.com
60. Schliecker, S., Stein, S., Ernst, R.: Performance analysis of complex systems by integration of dataflow graphs and compositional performance analysis. In: Proc. Design, Automation and Test in Europe (DATE), pp. 273–278 (2007)
61. Schor, L., Bacivarov, I., Yang, H., Thiele, L.: Worst-case temperature guarantees for real-time applications on multi-core systems. In: Proc. IEEE Real-Time and Embedded Technology and Applications Symposium (RTAS). Beijing, China (2012)
62. Thiele, L., Bacivarov, I., Haid, W., Huang, K.: Mapping applications to tiled multiprocessor embedded systems. In: Proc. Int'l Conf. on Application of Concurrency to System Design (ACSD), pp. 29–40. Bratislava, Slovak Republic (2007)
63. Thiele, L., Chakraborty, S., Gries, M., Künzli, S.: A framework for evaluating design tradeoffs in packet processing architectures. In: Proc. Design Automation Conference (DAC), pp. 880–885. New Orleans, LA, USA (2002)
64. Thiele, L., Chakraborty, S., Naedele, M.: Real-time calculus for scheduling hard real-time systems. In: Proc. Int'l Symposium on Circuits and Systems (ISCAS), vol. 4, pp. 101–104. Geneva, Switzerland (2000)
65. Thiele, L., Schor, L., Yang, H., Bacivarov, I.: Thermal-aware system analysis and software synthesis for embedded multi-processors. In: Proc. Design Automation Conference (DAC), pp. 268 – 273. ACM, San Diego, California, USA (2011)
66. Vasudevan, N., Edwards, S.A.: Celling SHIM: Compiling deterministic concurrency to a heterogeneous multicore. In: Proc. ACM Symposium on Applied Computing (SAC), pp. 1626–1631. Honolulu, HI, USA (2009)
67. Verdoolaege, S., Nikolov, H., Stefanov, T.: pn: A tool for improved derivation of process networks. EURASIP Journal on Embedded Systems **2007** (2007)
68. Wallace, G.K.: The JPEG still picture compression standard. IEEE Trans. on Consumer Electronics **38**(1), 18–34 (1992). DOI 10.1109/30.125072
69. Wandeler, E., Thiele, L.: Optimal TDMA time slot and cycle length allocation for hard real-time systems. In: Proc. Asia and South Pacific Conf. on Design Automation (ASP-DAC), pp. 479–484. Yokohama, Japan (2006)
70. Wandeler, E., Thiele, L.: Real-Time Calculus (RTC) Toolbox. http://www.mpa.ethz.ch/ Rtctoolbox (2006). URL http://www.mpa.ethz.ch/Rtctoolbox
71. Wandeler, E., Thiele, L., Verhoef, M., Lieverse, P.: System architecture evaluation using modular performance analysis: A case study. Int'l Journal on Software Tools for Technology Transfer (STTT) **8**(6), 649–667 (2006)
72. Wilhelm, R., Engblom, J., Ermedahl, A., Holsti, N., Thesing, S., Whalley, D., Bernat, G., Ferdinand, C., Heckmann, R., Mitra, T., Mueller, F., Puaut, I., Puschner, P., Staschulat, J., Stenström, P.: The worst-case execution time problem — overview of methods and survey of tools. ACM Trans. on Embedded Computing Systems **7**(3), 36:1–36:53 (2008)

73. Zitzler, E., Laumanns, M., Thiele, L.: SPEA2: Improving the strength pareto evolutionary algorithm for multiobjective optimization. In: Proc. Evolutionary Methods for Design, Optimisation, and Control (EUROGEN), pp. 95–100. Athens, Greece (2001)
74. Zitzler, E., Thiele, L.: Multiobjective Evolutionary Algorithms: A Comparative Case Study and the Strength Pareto Approach. IEEE Trans. on Evolutionary Computation **3**(4), 257–271 (1999)
75. Zitzler, E., Thiele, L., Bader, J.: SPAM: Set preference algorithm for multiobjective optimization. In: Conf. on Parallel Problem Solving From Nature (PPSN), pp. 847–858. Dortmund, Germany (2008)

Dynamic Dataflow Graphs

Shuvra S. Bhattacharyya, Ed F. Deprettere, and Bart D. Theelen

Abstract Much of the work to date on dataflow models for signal processing system design has focused on decidable dataflow models that are best suited for one-dimensional signal processing. This chapter reviews more general dataflow modeling techniques that are targeted to applications that include multidimensional signal processing and dynamic dataflow behavior. As dataflow techniques are applied to signal processing systems that are more complex, and demand increasing degrees of agility and flexibility, these classes of more general dataflow models are of correspondingly increasing interest. We first provide a motivation for dynamic dataflow models of computation, and review a number of specific methods that have emerged in this class of models. Our coverage of dynamic dataflow models in this chapter includes Boolean dataflow, CAL, parameterized dataflow, enable-invoke dataflow, dynamic polyhedral process networks, scenario aware dataflow, and a stream-based function actor model.

1 Motivation for Dynamic DSP-Oriented Dataflow Models

The decidable dataflow models covered in [31] are useful for their predictability, strong formal properties, and amenability to powerful optimization techniques. However, for many signal processing applications, it is not possible to represent

S.S. Bhattacharyya (✉)
University of Maryland, College Park, MD, USA
e-mail: ssb@umd.edu

E.F. Deprettere
Leiden University, Leiden, The Netherlands
e-mail: edd@liacs.nl

B.D. Theelen
Embedded Systems Innovation by TNO, Eindhoven, The Netherlands
e-mail: bart.theelen@tno.nl

S.S. Bhattacharyya et al. (eds.), *Handbook of Signal Processing Systems*,
DOI 10.1007/978-1-4614-6859-2_28, © Springer Science+Business Media, LLC 2013

all of the functionality in terms of purely decidable dataflow representations. For example, functionality that involves conditional execution of dataflow subsystems or actors with dynamically varying production and consumption rates generally cannot be expressed in decidable dataflow models.

The need for expressive power beyond that provided by decidable dataflow techniques is becoming increasingly important in design and implementation signal processing systems. This is due to the increasing levels of application dynamics that must be supported in such systems, such as the need to support multi-standard and other forms of multi-mode signal processing operation; variable data rate processing; and complex forms of adaptive signal processing behaviors.

Intuitively, *dynamic dataflow* models can be viewed as dataflow modeling techniques in which the production and consumption rates of actors can vary in ways that are not entirely predictable at compile time. It is possible to define dynamic dataflow modeling formats that are decidable. For example, by restricting the types of dynamic dataflow actors, and by restricting the usage of such actors to a small set of graph patterns or "schemas", Gao, Govindarajan, and Panangaden defined the class of *well-behaved dataflow graphs*, which provides a dynamic dataflow modeling environment that is amenable to compile-time bounded memory verification [19].

However, most existing DSP-oriented dynamic dataflow modeling techniques do not provide decidable dataflow modeling capabilities. In other words, in exchange for the increased modeling flexibility (expressive power) provided by such techniques, one must typically give up guarantees on compile-time buffer underflow (deadlock) and overflow validation. In dynamic dataflow environments, analysis techniques may succeed in guaranteeing avoidance of buffer underflow and overflow for a significant subset of specifications, but, in general, specifications may arise that "break" these analysis techniques—i.e., that result in inconclusive results from compile-time analysis.

Dynamic dataflow techniques can be divided into two general classes: (1) those that are formulated explicitly in terms of interacting combinations of state machines and dataflow graphs, where the dataflow dynamics are represented directly in terms of transitions within one or more underlying state machines; and (2) those where the dataflow dynamics are represented using alternative means. The separation in this dichotomy can become somewhat blurry for models that have a well-defined state structure governing the dataflow dynamics, but whose design interface does not expose this structure directly to the programmer. Dynamic dataflow techniques in the first category described above are covered in [31]—in particular, those based on explicit interactions between dataflow graphs and finite state machines. This chapter focusses on the second category.[1] Specifically, dynamic dataflow modeling techniques that involve different kinds of modeling abstractions, apart from state transitions, as the key mechanisms for capturing dataflow behaviors and their potential for run-time variation.

---

[1] Except for the Scenario Aware Dataflow model in Sect. 6.

Numerous dynamic dataflow modeling techniques have evolved over the past couple of decades. A comprehensive coverage of these techniques, even after excluding the "state-centric" ones, is out of the scope this chapter. The objective is to provide a representative cross-section of relevant dynamic dataflow techniques, with emphasis on techniques for which useful forms of compile time analysis have been developed. Such techniques can be important for exploiting the specialized properties exposed by these models, and improving predictability and efficiency when deriving simulations or implementations.

2 Boolean Dataflow

The Boolean dataflow (BDF) model of computation extends synchronous dataflow with a class of dynamic dataflow actors in which production and consumption rates on actor ports can vary as two-valued functions of *control tokens*, which are consumed from or produced onto designated *control ports* of dynamic dataflow actors. An actor input port is referred to as a *conditional input port* if its consumption rate can vary in such a way, and similarly an output port with a dynamically varying production rate under this model is referred to as a *conditional output port*.

Given a conditional input port p of a BDF actor A, there is a corresponding input port C_p, called the *control input* for p, such that the consumption rate on C_p is statically fixed at one token per invocation of A, and the number of tokens consumed from p during a given invocation of A is a two-valued function of the data value that is consumed from C_p during the same invocation.

The dynamic dataflow behavior for a conditional output port is characterized in a similar way, except that the number of tokens produced on such a port can be a two-valued function of a token that is consumed from a control input port or of a token that is produced onto a *control output* port. If a conditional output port q is controlled by a control output port C_q, then the production rate on the control output is statically fixed at one token per actor invocation, and the number of tokens produced on q during a given invocation of the enclosing actor is a two-valued function of the data value that is produced onto C_q during the same invocation of the actor.

Two fundamental dynamic dataflow actors in BDF are the *switch* and *select* actors, which are illustrated in Fig. 1a. The switch actor has two input ports, a control input port w_c and a *data* input port w_d, and two output ports w_x and w_y. The port w_c accepts Boolean valued tokens, and the consumption rate on w_d is statically fixed at one token per actor invocation. On a given invocation of a switch actor, the data value consumed from w_d is copied to a token that is produced on either w_x or w_y depending on the Boolean value consumed from w_c. If this Boolean value is true, then the value from the data input is routed to w_x, and no token is produced on w_y. Conversely if the control token value is false, then the value from w_d is routed to w_y with no token produced on w_x.

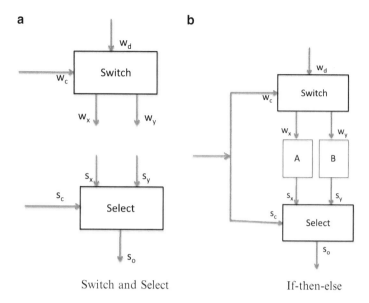

Fig. 1 (a) Switch and select actors in Boolean dataflow, and (b) an if-then-else construct expressed
in terms of Boolean dataflow

A BDF select actor has a single control input port s_c; two additional input ports
(*data input ports*) s_x and s_y; and a single output port s_o. Similar to the control port of
the switch actor, the s_c port accepts Boolean valued tokens, and the production rate
on the s_o port is statically fixed at one token per invocation. On each invocation of
the select actor data is copied from a single token from either s_x or s_y to s_o depending
on whether the corresponding control token value is true or false respectively.

Switch and select actors can be integrated along with other actors in various ways
to express different kinds of control constructs. For example, Fig. 1b illustrates an
if-then-else construct, where the actors A and B are applied conditionally based on a
stream of control tokens. Here A and B are synchronous dataflow (SDF) actors that
each consume one token and produce one token on each invocation.

Buck has developed scheduling techniques to automatically derive efficient
control structures from BDF graphs under certain conditions [11]. Buck has also
shown that BDF is Turing complete, and furthermore, that SDF augmented with just
switch and select (and no other dynamic dataflow actors) is also Turing complete.
This latter result provides a convenient framework with which one can demonstrate
Turing completeness for other kinds of dynamic dataflow models, such as the
enable-invoke dataflow model described in Sect. 5. In particular, if a given model
of computation can express all SDF actors as well as the functionality associated
with the BDF switch and select actors, then such a model can be shown to be Turing
complete.

3 CAL

In addition to providing a dynamic dataflow model of computation that is suitable for signal processing system design, CAL provides a complete programming language and is supported by a growing family of development tools for hardware and software implementation. The name "CAL" is derived as a self-referential acronym for the *CAL actor language*. CAL was developed by Eker and Janneck at U.C. Berkeley [14], and has since evolved into an actively-developed, widely-investigated language for design and implementation of embedded software and field programmable gate array applications (e.g., see [30, 56, 77]). One of the most notable developments to date in the evolution of CAL has been its adoption as part of the recent MPEG standard for reconfigurable video coding (RVC) [8].

A CAL program is specified as a network of CAL actors, where each actor is a dataflow component that is expressed in terms of a general underlying form of dataflow. This general form of dataflow admits both static and dynamic behaviors, and even non-deterministic behaviors.

Like typical actors in any dataflow programming environment, a CAL actor in general has a set of input ports and a set of output ports that define interfaces to the enclosing dataflow graph. A CAL actor also encapsulates its own private state, which can be modified by the actor as it executes but cannot be modified directly by other actors.

The functional specification of a CAL actor is decomposed into a set of *actions*, where each action can be viewed as a template for a specific class of firings or invocations of the actor. Each firing of an actor corresponds to a specific action and executes based on the code that is associated with that action. The core functionality of actors therefore is embedded within the code of the actions. Actions can in general consume tokens from actor input ports, produce tokens on output ports, modify the actor state, and perform computation in terms of the actor state and the data obtained from any consumed tokens.

The number of tokens produced and consumed by each action with respect to each actor output and input port, respectively, is declared up front as part of the declaration of the action. An action need not consume data from all input ports nor must it produce data on all output ports, but ports with which the action exchanges data, and the associated rates of production and consumption must be constant for the action. Across different actions, however, there is no restriction of uniformity in production and consumption rates, and this flexibility enables the modeling of dynamic dataflow in CAL.

A CAL actor A can be represented as a sequence of four elements

$$\sigma_0(A), \Sigma(A), \Gamma(A), pri(A), \tag{1}$$

where $\Sigma(A)$ represents the set of all possible values that the state of A can take on; $\sigma_0(A) \in \Sigma(A)$ represents the initial state of the actor, before any actor in the enclosing dataflow graph has started execution; $\Gamma(A)$ represents the set of actions of A; and $pri(A)$ is a partial order relation, called the *priority relation* of A, on $\Gamma(A)$ that specifies relative priorities between actions.

Actions execute based on associated *guard conditions* as well as the priority relation of the enclosing actor. More specifically, each actor has an associated guard condition, which can be viewed as a Boolean expression in terms of the values of actor input tokens and actor state. An actor A can execute whenever its associated guard condition is satisfied (*true*-valued), and no higher-priority action (based on the priority relation $pri(A)$) has a guard condition that is also satisfied.

In summary, CAL is a language for describing dataflow actors in terms of ports, actions (firing templates), guards, priorities, and state. This finer, *intra-actor* granularity of formal modeling within CAL allows for novel forms of automated analysis for extracting restricted forms of dataflow structure. Such restricted forms of structure can be exploited with specialized techniques for verification or synthesis to derive more predictable or efficient implementations.

An example of this capability for *specialized region detection* in CAL programs is the technique of deriving and exploiting so-called *statically schedulable regions* (SSRs) [30]. Intuitively, an SSR is a collection of CAL actions and ports that can be scheduled and optimized statically using the full power of static dataflow techniques, such as those available for SDF, and integrated into the schedule for the overall CAL program through a top-level dynamic scheduling interface.

SSRs can be derived through a series of transformations that are applied on intermediate graph representations. These representations capture detailed relationships among actor ports and actions, and provide a framework for effective *quasi-static scheduling* of CAL-based dynamic dataflow representations. Quasi-static scheduling is the construction of dataflow graph schedules in which a significant proportion of overall schedule structure is fixed at compile-time. Quasi-static scheduling has the potential to significantly improve predictability, reduce run-time scheduling overhead, and as discussed above, expose subsystems whose internal schedules can be generated using purely static dataflow scheduling techniques.

Further discussion about CAL can be found in [42], which discusses the application of CAL to reconfigurable video coding.

4 Parameterized Dataflow

Parameterized dataflow is a meta-modeling approach for integrating dynamic parameters and run-time adaptation of parameters in a structured way into a certain class of dataflow models of computations, in particular, those models that have a well-defined concept of a graph *iteration* [6]. For example, SDF and cycle-static SDF (CSDF), which are discussed in [31], and multidimensional SDF (MDSDF), which is discussed in [37], have well defined concepts of iterations based on solutions to the associated forms of balance equations. Each of these models can be integrated with parameterized dataflow to provide a dynamically parameterizable form of the original model.

When parameterized dataflow is applied in this way to generalize a specialized dataflow model such as SDF, CSDF, or MDSDF, the specialized model is referred to as the *base model*, and the resulting, dynamically parameterizable form of the base model is referred to as *parameterized XYZ*, where *XYZ* is the name of the base model. For example, when parameterized dataflow is applied to SDF as the base model, the resulting model of computation, called *parameterized synchronous dataflow* (*PSDF*), is significantly more flexible than SDF as it allows arbitrary parameters of SDF graphs to be modified at run-time. Furthermore, PSDF provides a useful framework for quasi-static scheduling, where fixed-iteration looped schedules, such as single appearance schedules [7], for SDF graphs can be replaced by *parameterized looped schedules* [6, 40] in which loop iteration counts are represented as symbolic expressions in terms of variables whose values can be adapted dynamically through computations that are derived from the enclosing PSDF specification.

Intuitively, parameterized dataflow allows arbitrary attributes of a dataflow graph to be parameterized, with each parameter characterized by an associated domain of admissible values that the parameter can take on at any given time. Graph attributes that can be parameterized include scalar or vector attributes of individual actors, such as the coefficients of a finite impulse response filter or the block size associated with an FFT; edge attributes, such as the delay of an edge or the data type associated with tokens that are transferred across the edge; and graph attributes, such as those related to numeric precision, which may be passed down to selected subsets of actors and edges within the given graph.

The parameterized dataflow representation of a computation involves three cooperating dataflow graphs, which are referred to as the *body* graph, the *subinit* graph, and the *init* graph. The body graph typically represents the functional "core" of the overall computation, while the subinit and init graphs are dedicated to managing the parameters of the body graph. In particular, each output port of the subinit graph is associated with a body graph parameter such that data values produced at the output port are propagated as new parameter values of the associated parameter. Similarly, output ports of the init graph are associated with parameter values in the subinit and body graphs.

Changes to body graph parameters, which occur based on new parameter values computed by the init and subinit graphs, cannot occur at arbitrary points in time. Instead, once the body graph begins execution it continues uninterrupted through a graph iteration, where the specific notion of an iteration in this context can be specified by the user in an application-specific way. For example, in PSDF, the most natural, general definition for a body graph iteration would be a single *SDF iteration* of the body graph, as defined by the SDF repetitions vector [31].

However, an iteration of the body graph can also be defined as some constant number of iterations, for example, the number of iterations required to process a fixed-size block of input data samples. Furthermore, parameters that define the body graph iteration can be used to parameterize the body graph or the enclosing PSDF specification at higher levels of the model hierarchy, and in this way, the processing that is defined by a graph iteration can itself be dynamically adapted as the application executes. For example, the duration (or block length) for fixed-parameter processing may be based on the size of a related sequence of contiguous network

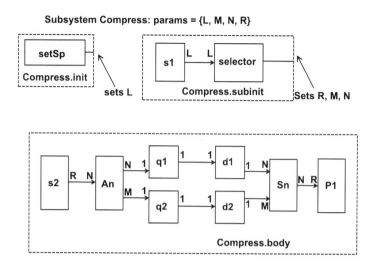

Fig. 2 An illustration of a speech compression system that is modeled using PSDF semantics. This illustration is adapted from [6]

packets, where the sequence size determines the extent of the associated graph iteration.

Body graph iterations can even be defined to correspond to individual actor invocations. This can be achieved by defining an individual actor as the body graph of a parameterized dataflow specification, or by simply defining the notion of iteration for an arbitrary body graph to correspond to the *next actor firing* in the graph execution. Thus, when modeling applications with parameterized dataflow, designers have significant flexibility to control the windows of execution that define the boundaries at which graph parameters can be changed.

A combination of cooperating body, init, and subinit graphs is referred to as a *PSDF specification*. PSDF specifications can be abstracted as PSDF actors in higher level PSDF graphs, and in this way, PSDF specifications can be integrated hierarchically.

Figure 2 illustrates a PSDF specification for a speech compression system. This illustration is adapted from [6]. Here *setSp* ("set speech") is an actor that reads a header packet from a stream of speech data, and configures L, which is a parameter that represents the length of the next speech instance to process. The *s1* and *s2* actors are input interfaces that inject successive samples of the current speech instance into the dataflow graph. The actor *s2* zero-pads each speech instance to a length R ($R \geq L$) so that the resulting length is divisible by N, which is the speech segment size. The *An* ("analyze") actor performs linear prediction on speech segments, and produces corresponding auto-regressive (AR) coefficients (in blocks of M samples), and residual error signals (in blocks of N samples) on its output edges. The actors *q1* and *q2* represent quantizers, and complete the modeling of the transmitter component of the body graph.

Receiver side functionality is then modeled in the body graph starting with the actors *d1* and *d2*, which represent dequantizers. The actor *Sn* ("synthesize") then reconstructs speech instances using corresponding blocks of AR coefficients and error signals. The actor *P1* ("play") represents an output interface for playing or storing the resulting speech instances.

The model order (number of AR coefficients) M, speech segment size N, and zero-padded speech segment length R are determined on a per-segment basis by the *selector* actor in the subinit graph. Existing techniques, such as the Burg segment size selection algorithm and AIC order selection criterion [32] can be used for this purpose.

The model of Fig. 2 can be optimized to eliminate the zero padding overhead (modeled by the parameter R). This optimization can be performed by converting the design to a parameterized cyclo-static dataflow (PCSDF) representation. In PCSDF, the parameterized dataflow meta model is integrated with CSDF as the base model instead of SDF.

For further details on this speech compression application and its representations in PSDF and PCSDF, the semantics of parameterized dataflow and PSDF, and quasi-static scheduling techniques for PSDF, see [6].

Parameterized cyclo-static dataflow (PCSDF), the integration of parameterized dataflow meta-modeling with cyclo-static dataflow, is explored further in [57]. The exploration of different models of computation, including PSDF and PCSDF, for the modeling of software defined radio systems is explored in [5]. In [36], Kee et al. explore the application of PSDF techniques to field programmable gate array implementation of the physical layer for 3GPP-Long Term Evolution (LTE). The integration of concepts related to parameterized dataflow in language extensions for embedded streaming systems is explored in [41]. General techniques for analysis and verification of hierarchically reconfigurable dataflow graphs are explored in [46].

5 Enable-Invoke Dataflow

Enable-invoke dataflow (EIDF) is another DSP-oriented dynamic dataflow modeling technique [51]. The utility of EIDF has been demonstrated in the context of behavioral simulation, FPGA implementation, and prototyping of different scheduling strategies [49–51]. This latter capability—prototyping of scheduling strategies—is particularly important in analyzing and optimizing embedded software. The importance and complexity of carefully analyzing scheduling strategies are high even for the restricted SDF model, where scheduling decisions have a major impact on most key implementation metrics [9]. The incorporation of dynamic dataflow features makes the scheduling problem more critical since application behaviors are less predictable, and more difficult to understand through analytical methods.

EIDF is based on a formalism in which actors execute through dynamic transitions among modes, where each mode has "synchronous" (constant production/consumption rate behavior), but different modes can have differing dataflow rates. Unlike other forms of mode-oriented dataflow specification, such as stream-based functions (see Sect. 8), SDF-integrated starcharts (see [15]), SysteMoc (see [15]), and CAL (see Sect. 3), EIDF imposes a strict separation between fireability checking (checking whether or not the next mode has sufficient data to execute), and mode execution (carrying out the execution of a given mode). This allows for lightweight fireability checking, since the checking is completely separate from mode execution. Furthermore, the approach improves the predictability of mode executions since there is no waiting for data (blocking reads)—the time required to access input data is not affected by scheduling decisions or global dataflow graph state.

For a given EIDF actor, the specification for each mode of the actor includes the number of tokens that is consumed on each input port throughout the mode, the number of tokens that is produced on each output port, and the computation (the *invoke function*) that is to be performed when the actor is invoked in the given mode. The specified computation must produce the designated number of tokens on each output port, and it must also produce a value for the *next mode* of the actor, which determines the number of input tokens required for and the computation to be performed during the next actor invocation. The next mode can in general depend on the current mode as well as the input data that is consumed as the mode executes.

At any given time between mode executions (actor invocations), an enclosing scheduler can query the actor using the *enable function* of the actor. The enable function can only examine the number of tokens on each input port (without consuming any data), and based on these "token populations", the function returns a Boolean value indicating whether or not the next mode has enough data to execute to completion without waiting for data on any port.

The set of possible next modes for a given actor at a given point in time can in general be empty or contain one or multiple elements. If the next mode set is empty (the next mode is null), then the actor cannot be invoked again before it is somehow reset or re-initialized from environment that controls the enclosing dataflow graph. A null next mode is therefore equivalent to a transition to a mode that requires an infinite number of tokens on an input port. The provision for multi-element sets of next modes allows for natural representation of non-determinism in EIDF specifications.

When the set of next modes for a given actor mode is restricted to have at most one element, the resulting model of computation, called *core functional dataflow* (*CFDF*), is a deterministic, Turing complete model [51]. CFDF semantics underlie the *functional DIF* simulation environment for behavioral simulation of signal processing applications. Functional DIF integrates CFDF-based dataflow graph specification using the *dataflow interchange format* (*DIF*), a textual language for representing DSP-oriented dataflow graphs, and Java-based specification of intra-actor functionality, including specification of enable functions, invoke functions, and next mode computations [51].

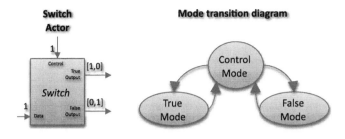

Fig. 3 An illustration of the design of a `switch` actor in CFDF

Figures 3 and 4 illustrate, respectively, the design of a CFDF actor and its implementation in functional DIF. This actor provides functionality that is equivalent to the Boolean dataflow switch actor described in Sect. 2.

6 Scenario Aware Dataflow

This section discusses Scenario-Aware Dataflow (SADF), which is a generalization of dataflow models with strict periodic behavior. Like most dataflow models, SADF is primarily a coordination language that highlights how actors (which are potentially executed in parallel) interact. To express dynamism, SADF distinguishes data and control explicitly, where the control-related coherency between the behavior (and hence, the resource requirements) of different parts of a signal processing application can be captured with so-called *scenarios* [26]. The scenarios commonly coincide with dissimilar (but within themselves more static) modes of operation originating, for example, from different parameter settings, sample rate conversion factors, or the signal processing operations to perform. Scenarios are typically defined by clustering operational situations with similar resource requirements [26]. The scenario-concept in SADF allows for more precise (quantitative) analysis results compared to applying traditional SDF-based analysis techniques. Still, common subclasses of SADF can be synthesized into efficient implementations [65].

Constructor

(modes and dataflow behavior defined)

```
public CFSwitch() {
      _control       = addMode("control");
      _control_true  = addMode("control_true");
      _control_false = addMode("control_false");

      _data_in     = addInput("data_in");
      _control_in  = addInput("control_in");
      _true_out    = addOutput("true_out");
      _false_out   = addOutput("false_out");.

      _control.setConsumption(_control_in, 1);
      _control_true.setConsumption(_data_in, 1);
      _control_true.setProduction(_true_out, 1);
      _control_false.setConsumption(_data_in, 1);
      _control_false.setProduction(_false_out, 1);
}
```

Invoke Function

(performs action and determines next mode)

```
public CoreFunctionMode
                  invoke(CoreFunctionMode mode){
      if (_init==mode) {
         return _control;
      }
      if(_control==mode) {
           if((Boolean)pullToken(_control_in)) {
              return _control_true;
           } else {
              return _control_false;
           }
      }
      if(_control_true==mode) {
           Object obj = pullToken(_data_in);

           pushToken(_true_out, obj);
           return _control;
      }
      if(_control_false==mode) {
           Object obj = pullToken(_data_in);

           pushToken(_false_out, obj);
           return _control;
      }
}
```

Enable Function

(determines if firing condition is met)

```
public boolean enable(CoreFunctionMode mode){
      if (_control==mode) {
          if(peek(_control_in)>0) {
             return true;
          }
          return false;
      }else if (_control_true==mode){
          if(peek(_data_in)>0) {
             return true;
          }
          return false;
      } else if (_control_false==mode) {
          if(peek(_data_in)>0) {
             return true;
          }
          return false;
      }
      return false;
}
```

Fig. 4 An implementation of the `switch` actor design of Fig. 3 in the functional DIF environment

6.1 SADF Graphs

In this subsection SADF is introduced by some examples from the multi-media domain. Consider the MPEG-4 video decoder for the Simple Profile from [66, 70]. It supports video streams consisting of intra (I) and predicted (P) frames. For an image size of 176×144 pixels (QCIF), there are 99 macro blocks to decode for I frames and no motion vectors. For P frames, such motion vectors determine the new position of certain macro blocks relative to the previous frame. The number of motion vectors and macro blocks to process for P frames ranges between 0 and 99. The MPEG-4 decoder clearly shows variations in the functionality to perform and in the amount of data to communicate between the operations. This leads to large fluctuations in resource requirements [52]. The order in which the different situations occur strongly depends on the video content and is generally not periodic.

Figure 5 depicts an SADF graph for the MPEG-4 decoder in which nine different scenarios are identified. SADF distinguishes two types of actors: *kernels* (solid vertices) model the data processing parts, whereas *detectors* (dashed vertices)

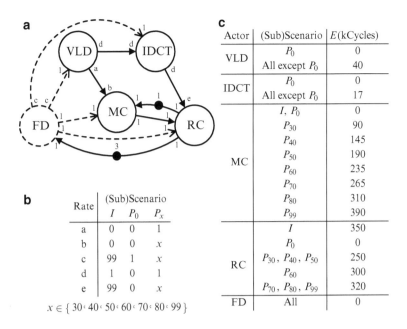

b

Rate	(Sub)Scenario		
	I	P_0	P_x
a	0	0	1
b	0	0	x
c	99	1	x
d	1	0	1
e	99	0	x

$x \in \{\, 30 \trianglelefteq 40 \trianglelefteq 50 \trianglelefteq 60 \trianglelefteq 70 \trianglelefteq 80 \trianglelefteq 99 \,\}$

c

Actor	(Sub)Scenario	E(kCycles)
VLD	P_0	0
	All except P_0	40
IDCT	P_0	0
	All except P_0	17
MC	I, P_0	0
	P_{30}	90
	P_{40}	145
	P_{50}	190
	P_{60}	235
	P_{70}	265
	P_{80}	310
	P_{99}	390
RC	I	350
	P_0	0
	P_{30}, P_{40}, P_{50}	250
	P_{60}	300
	P_{70}, P_{80}, P_{99}	320
FD	All	0

Fig. 5 Modeling the MPEG-4 decoder with SADF. (**a**) Actors and channels; (**b**) parameterized rates; (**c**) worst-case execution times

control the behavior of actors through scenarios.[2] Moreover, *data* channels (solid edges) and *control* channels (dashed edges) are distinguished. Control channels communicate scenario-valued tokens that influence the control flow. Data tokens do not influence the control flow. The availability of tokens in channels is shown with a dot. Here, such dots are labeled with the number of tokens in the channel. The start and end points of channels are labeled with *production* and *consumption rates* respectively. They refer to the number of tokens atomically produced respectively consumed by the connected actor upon its *firing*. The rates can be fixed or scenario-dependent, similar as in PSDF. Fixed rates are positive integers. Parameterized rates are valued with non-negative integers that depend on the scenario. The parameterized rates for the MPEG-4 decoder are listed in Fig. 5b. A value of 0 expresses that data dependencies are absent or that certain operations are not performed in those scenarios. Studying Fig. 5b reveals that for any given scenario, the rate values yield a consistent SDF graph. In each of these scenario graphs, detector FD has a repetition vector entry of 1 [70], which means that scenario changes as prescribed by the behavior of detectors may occur only at iteration boundaries of each such scenario graph. This is not necessarily true for SADF in general as discussed below.

---

[2]In case of one detector, SADF literature may not show the detector and control channels explicitly.

SADF specifies execution times of actors (from a selected time domain, see Sect. 6.2) per scenario. Figure 5c lists the worst-case execution times of the MPEG-4 decoder for an ARM7TDMI processor. Figure 5b, c show that the worst-case communication requirements occur for scenario P_{99}, in which all actors are active and production/consumption rates are maximal. Scenario P_{99} also requires maximal execution times for VLD, IDCT, and MC, while for RC it is the scenario I in which the worst-case execution time occurs. Traditional SDF-based approaches need to combine these worst-case requirements into one (unrealistically) conservative model, which yields too pessimistic analysis results.

An important aspect of SADF is that sequences of scenarios are made explicit by associating *state machines* to detectors. The dynamics of the MPEG-4 decoder originate from control-flow code that (implicitly or explicitly) represents a state-machine with video stream content dependent guards on the transitions between states. One can think of if-statements that distinguish processing I frames from processing P frames. For the purpose of compile-time analysis, SADF abstracts from the content of data tokens (similar to SDF and CSDF) and therefore also from the concrete conditions in control-flow code. Different types of state machines can be used to model the occurrences of scenarios, depending on the compile-time analysis needs as presented in Sect. 6.2. The dynamics of the MPEG-4 decoder can be captured by a state-machine of nine states (one per scenario) associated to detector FD.

The operational behavior of actors in SADF follows two steps, similar to the *switch* and *select* actors in BDF. The first step covers the control part which establishes the mode of operation. The second step is like the traditional data flow behavior of SDF actors[3] in which data is consumed and produced. Kernels establish their scenario in the first step when a scenario-valued token is available on their control inputs. The operation mode of detectors is established based on external and internal forces. *Subscenario* denotes the result of the internal forces affecting the operation mode. External forces are the scenario-valued tokens available on control inputs (similar as for kernels). The combination of tokens on control inputs for a detector determine its scenario,[4] which (deterministically) selects a corresponding state-machine. A transition is made in the selected state machine, which establishes the subscenario. Where the scenario determines values for parameterized rates and execution time details for kernels, it is the subscenario that determines these aspects for detectors. Tokens produced by detectors onto control channels are scenario-valued to coherently affect the behavior of controlled actors, which is a key feature of SADF. Actor firings in SADF block until sufficient tokens are available. As a result, the execution of different scenarios can overlap in a pipelined fashion. For example, in the MPEG-4 decoder, IDCT is always ready to be executed immediately after VLD, which may already have accepted a control token with a

---

[3]Execution of the reflected function or program is enabled when sufficient tokens are available on all (data) inputs, and finalizes (after a certain execution time) with producing tokens on the outputs.

[4]If a detector has no control inputs, it operates in a default scenario ε and has one state machine.

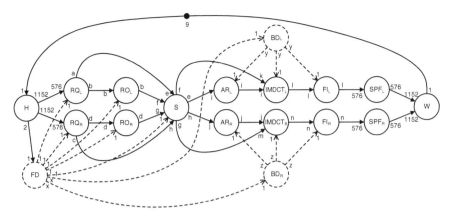

Fig. 6 Modeling an MP3 decoder with SADF using hierarchical control

different scenario value from FD. The ability to express such so-called *pipelined reconfiguration* is another key feature of SADF.

Now, consider the MP3 audio decoder example taken from [66] depicted in Fig. 6. It illustrates that SADF graphs can contain multiple detectors, which may even control each other's behavior. MP3 decoding transforms a compressed audio bitstream into pulse code modulated data. The stream is partitioned into frames of 1,152 mono or stereo frequency components, which are divided into two granules of 576 components structured in blocks [58]. MP3 distinguishes three frame types: Long (*L*), Short (*S*) and Mixed (*M*) and two block types: Long (*BL*) and Short (*BS*). A Long block contains 18 frequency components, while Short blocks include only 6 components. Long frames consist of 32 Long blocks, Short frames of 96 Short blocks and Mixed frames are composed of 2 Long blocks, succeeded by 90 Short blocks. The frame type and block type together determine the operation mode. Neglecting that the frame types and specific block type sequences are correlated leads to unrealistic models. The sequences of block types is dependent on the frame type, as is reflected in the structure of source code of the MP3 audio decoder. SADF supports *hierarchical control* to intuitively express this kind of correlation between different aspects that determine the scenario.

Figure 7a lists the parameterized rates for the MP3 decoder. Only five combinations of frame types occur for the two audio channels combined. A two-letter abbreviation is used to indicate the combined fame type for the left and right audio channel, respectively: *LL*, *SS*, *LS* and *SL*. Mixed frame type *M* covers both audio channels simultaneously. Detector FD determines the frame type with a state machine of five states, each uniquely identify a subscenario in {*LL*, *SS*, *LS*, *SL*, *M*}. The operation mode of kernel S depends on the frame types for both audio channels together and therefore it operates according to a scenario from this same set. The scenario of kernels RQ_L, RO_L and RQ_R, RO_R is only determined by the frame type for either the left or right audio channel. They operate in scenario *S*, *M* or *L* by receiving control tokens from FD, valued with either the left or right letter in *LL*, *SS*, *LS*, *SL* or with *M*.

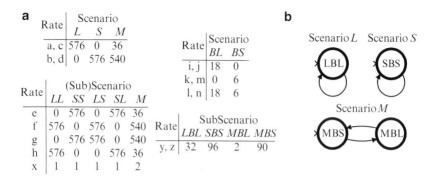

a

Rate	Scenario		
	L	S	M
a, c	576	0	36
b, d	0	576	540

Rate	(Sub)Scenario				
	LL	SS	LS	SL	M
e	0	576	0	576	36
f	576	0	576	0	540
g	0	576	576	0	540
h	576	0	0	576	36
x	1	1	1	1	2

Rate	Scenario	
	BL	BS
i, j	18	0
k, m	0	6
l, n	18	6

Rate	SubScenario			
	LBL	SBS	MBL	MBS
y, z	32	96	2	90

b

Fig. 7 Properties of the MP3 decoder model. (**a**) Parameterized rates; (**b**) state machines for BD_L and BD_R

Detectors BD_L and BD_R identify the appropriate number and order of Short and Long blocks based on the frame scenario, which they receive from FD as control tokens valued L, S or M. From the perspective of BD_L and BD_R, block types BL and BS are refinements (subscenarios) of the scenarios L, S and M. Figure 7b shows the three state machines associated with BD_L as well as BD_R. Each of their states implies one of the possible subscenarios in $\{LBL, SBS, MBL, MBS\}$. The value of the control tokens produced by BD_L and BD_R to kernels AR_L, $IMDCT_L$, FI_L and AR_R, $IMDCT_R$, FI_R in each of the four possible subscenarios matches the last two letters of the subscenario name (i.e., BL or BS). Although subscenarios LBL and MBL both send control tokens valued BL, the difference between them is the number of such tokens (similarly for subscenarios SBS and MBS).

Consider decoding of a Mixed frame. It implies the production of two M-valued tokens on the control port of detector BD_L. By interpreting each of these tokens, the state machine for scenario M in Fig. 7b makes one transition. Hence, BD_L uses subscenario MBL for its first firing and subscenario MBS for its second firing. In subscenario MBL, BD_L sends 2 BL-valued tokens to kernels AR_L, $IMDCT_L$ and SPF_L, while 90 BS-valued tokens are produced in subscenario MBS. As a result, AR_L, $IMDCT_L$ and SPF_L first process 2 Long blocks and subsequently 90 Short blocks as required for Mixed frames.

The example of Mixed frames highlights a unique feature of SADF: reconfigurations may occur *during* an iteration. An iteration of the MP3 decoder corresponds to processing frames, while block type dependent variations occur during processing Mixed frames. Supporting reconfiguration within iterations is fundamentally different from assumptions underlying other dynamic dataflow models, including for example PSDF. The concept is orthogonal to hierarchical control. Hierarchical control is also different from other dataflow models with hierarchy such as heterochronous dataflow [27]. SADF allows *pipelined* execution of the controlling and controlled behavior together, while other approaches commonly prescribe that the controlled behavior must first finish completely before the controlling behavior may continue.

6.2 *Analysis*

Various analysis techniques exist for SADF, allowing the evaluation of both qualitative properties (such as consistency and absence of deadlock) and best/worst-case and average-case quantitative properties (like minimal and average throughput).

Consistency of SADF graphs is briefly discussed now. The MPEG-4 decoder is an example of a class of SADF graphs where each scenario is like a consistent SDF graph and scenario changes occur at iteration boundaries of these scenario graphs (but still pipelined). Such SADF graphs are said to be *strongly consistent* [70], which is easy to check as it results from structural properties only. The SADF graph of the MP3 decoder does not satisfy these structural properties (for Mixed frames), but it can still be implemented in bounded memory. The required consistency property is called *weak consistency* [66]. Checking weak consistency requires taking the possible (sub)scenario sequences as captured by the state machines associated to detectors into account, which complicates a consistency check considerably.

Analysis of quantitative properties and the efficiency of the underlying techniques depend on the selected type of state machine associated to detectors as well as the chosen time model. For example, one possibility is to use non-deterministic state machines, which merely specify what sequences of (sub)scenarios *can* occur but not how often. This is typically used for best/worst-case analysis. Applying the techniques in [20, 23, 24] then allows computing that a throughput of processing 0.253 frames per kCycle can be guaranteed for the MPEG-4 decoder. An alternative is to use probabilistic state machines (i.e., Markov chains), which additionally capture the occurrence probabilities of the (sub)scenario sequences to allow for average-case analysis as well. Assuming that scenarios I, P_0, P_{30}, P_{40}, P_{50}, P_{60}, P_{70}, P_{80} and P_{99} of the MPEG-4 decoder may occur in any order and with probabilities 0.12, 0.02, 0.05, 0.25, 0.25, 0.09, 0.09, 0.09 and 0.04 respectively, the techniques in [67] allow computing that the MPEG-4 decoder processes on average 0.426 frames per kCycle. The techniques presented in [71] combine the association of Markov chains to detectors with exponentially distributed execution times to analyze the response time distribution of the MPEG-4 decoder for completing the first frame.

The semantics of SADF graphs where Markov chains are associated to detectors while assuming generic discrete execution time distributions[5] has been defined in [66] using Timed Probabilistic Systems (TPS). Such transition systems operationalize the behavior with states and guarded transitions that capture events like the begin and end of each of the two steps in firing actors and progress of time. In case an SADF graph yields a TPS with finite state space, it is amenable to analysis techniques for (Priced) Timed Automata or Markov Decision Processes and Markov Chains by defining reward structures as also used in (probabilistic) model checking. Theelen et al. [67] discusses that specific properties of dataflow models in

---

[5]This covers the case of constant execution times as so-called point distributions [66, 67].

general and SADF in particular allow for substantial state-space reductions during such analysis. The underlying techniques have been implemented in the SDF[3] tool kit [62], covering the computation of worst/best-case and average-case properties for SADF including throughput and various forms of latency and buffer occupancy metrics [68]. In case such exact computation is hampered by state-space explosion, [68, 70] exploit an automated translation into process algebraic models expressed in the Parallel Object-Oriented Specification Language (POOSL) [69], which allows for statistical model checking (simulation-based estimation) of the properties. The combination of Markov chains and exponentially distributed execution times has been studied in [71], using a process algebraic semantics based on Interactive Markov Chains [33] to apply a general-purpose model checker for analyzing response time distributions.

In case we abstract from the stochastic aspects of execution times and scenario occurrences, SADF is still amenable to worst/best-case analysis. Since SADF graphs are timed dataflow graphs, they exhibit *linear timing behavior* [20, 43, 76], which facilitates network-level worst/best-case analysis by considering the worst/best-case execution times for individual actors. For linear timed systems this is know to lead to the overall worst/best-case performance. For the class of strongly-consistent SADF graphs with a single detector (also called *FSM-based SADF*), very efficient performance analysis can be done based on a $(\max, +)$-algebraic interpretation of the operational semantics. It allows for worst-case throughput analysis, some latency analysis and can find critical scenario sequences without exploring the state-space of the underlying TPS. Instead, the analysis is performed by means of state-space analysis and maximum-cycle ratio analysis of the equivalent $(\max, +)$-automaton [20, 23, 24]. Geilen et al. [23] shows how this analysis can be extended for the case that scenario behaviors are not complete iterations of the scenario SDF graphs.

6.3   Synthesis

FSM-based SADF graphs have been extensively studied for implementation on (heterogeneous) multi-processor platforms [64]. Variations in resource requirements need to be exploited to limit resource usage without violating any timing require-ments. The result of the design flow for FSM-based SADF implemented in the SDF[3] tool kit [62] is a set of Pareto optimal mappings that provide a trade-off in valid resource usages. For certain mappings, the application may use many computational resources and few storage resources, whereas an opposite situation may exist for other mappings. At run-time, the most suitable mapping is selected based on the available resources not used by concurrently running applications [59].

There are two key aspects of the design flow of [62, 64]. The first concerns mapping channels onto (possibly shared) storage resources. Like other dataflow models, SADF associates unbounded buffers with channels, but a complete graph may still be implemented in bounded memory. FSM-based SADF allows for

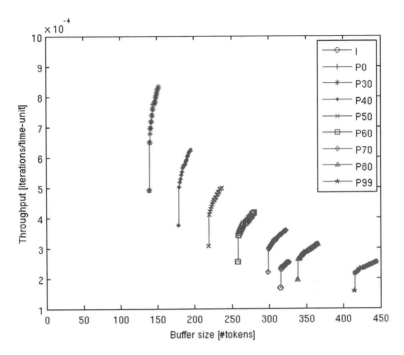

Fig. 8 Throughput/buffer size trade-off space for the MPEG-4 decoder

efficient compile-time analysis of the impact that certain buffer sizes have on the timing of the application. Hence, a synthesized implementation does not require run-time buffer management, thereby making it easier to guarantee timing. The design flow in [64] dimensions the buffer sizes of all individual channels in the graph sufficiently large to ensure that timing (i.e., throughput) constraints are met but also as small as possible to save memory and energy. It exploits the techniques of [63] to analyze the trade-off between buffer sizes and throughput for each individual scenario in the FSM-based SADF graph. After computing the trade-off space for all individual scenarios, a unified trade-off space for all scenarios is created. The same buffer size is assigned to a channel in all scenarios. Combining the individual spaces is done using Pareto algebra [22] by taking the free product of all trade-off spaces and selecting only the Pareto optimal points in the resulting space. Figure 8 shows the trade-off space for the individual scenarios in the MPEG-4 decoder. In this application, the set of Pareto points that describe the trade-off between throughput and buffer size in scenario P_{99} dominate the trade-off points of all other scenarios. Unifying the trade-off spaces of the individual scenarios therefore results in the trade-off space corresponding to scenario P_{99}. After computing the unified throughput/buffer trade-off space, the synthesis process in [64] selects a Pareto point with the smallest buffer size assignment that satisfies the throughput constraint as a means to allocate the required memory resources in the multiprocessor platform.

A second key aspect of the synthesis process is the fact that actors of the same or different applications may share resources. The set of concurrently active applications is typically unknown at compile-time. It is therefore not possible to construct a single static-order schedule for actors of different applications. The design flow in [64] uses static-order schedules for actors of the same application, but sharing of resources between different applications is handled by run-time schedulers with TDMA policies. It uses a binary search algorithm to compute the minimal TDMA time slices ensuring that the throughput constraint of an application is met. By minimizing the TDMA time slices, resources are saved for other applications. Identification of the minimal TDMA time slices works as follows. In [3], it is shown that the timing impact of a TDMA scheduler can be modeled into the execution time of actors. This approach is used to model the TDMA time slice allocation it computes. Throughput analysis is then performed on the modified FSM-based SADF graph. When the throughput constraint is met, the TDMA time slice allocation can be decreased. Otherwise it needs to be increased. This process continues until the minimal TDMA time slice allocation satisfying the throughput constraint is found.

7 Dynamic Polyhedral Process Networks

The chapter on *polyhedral process networks* (PPN) [73] deals with the automatic derivation of certain dataflow networks from *static affine nested loop programs* (SANLP). An SANLP is a nested loop program in which loop bounds, conditions and variable index expressions are (quasi-)affine expressions in the iterators of enclosing loops and static parameters.[6] Because many signal processing applications are not static, there is a need to consider *dynamic affine nested loop programs* (DANLP) which differ from SANLPs in that they can contain

1. *If-the-else* constructs with no restrictions on the condition [60].
2. *Loops* with no condition on the bounds [44].
3. *While* statements other than `while(1)` [45].
4. Dynamic parameters [78].

Remark. In all DANLP programs presented in subsequent subsections, arrays are indexed by affine functions of static parameters and enclosing for-loop iterators. This is why the *A* is still in the name.

---

[6]The corresponding tool is called PNgen [74], and is part of the `Daedalus` design framework [48], http://daedalus.liacs.nl.

```
1   %parameter N 8 16;        7   for i = 1:1:N,
2                             8    if t(i) <= 0,
3   for i = 1:1:N,            9     [x(i)] = F2( x(i) );
4    [x(i), t(i)] = F1(...);  10   end
5   end                       11   [...] = F3( x(i) );
6                             12  end
```

Fig. 9 Pseudo code of a simple weakly dynamic program

```
1   %parameter N 8 16;         16    [out_0] = F2(in_0);
2                              17    [x_2(i)] = opd (out_0);
3   for i = 1:1:N,             18    [ ctrl(i) ] = opd ( i );
4    ctrl(i) = N+1;            19   end
5   end                        20
6   for i = 1:1:N,             21    C = ipd( ctrl(i) );
7    [out_0, out_1] = F1(...); 22   if i = C,
8    [x_1(i)] = opd (out_0);   23    [in_0] = ipd (x_2(C));
9    [t_1(i)] = opd (out_1);   24   else
10  end                        25    [in_0] = ipd (x_1(i));
11                             26   end
12  for i = 1:1:N,             27
13   [t_1(i)] = ipd (t_1(i));  28    [out_0] = F3(in_0);
14   if t_1(i) <= 0,           29    [...] = opd (out_0);
15    [in_0] = ipd (x_1(i));   30   end
```

Fig. 10 Example of dynamic single assignment code

7.1 Weakly Dynamic Programs

While in a SANLP condition statements must be affine in static parameters and iterators of enclosing loops, if conditions can be anything in a DANLP. Such programs have been called *weakly dynamic programs* (WDP) in [60]. A simple example of a WDP is shown in Fig. 12.

The question of course is whether the argument of function $F3$ originates from the output of function $F2$ or function $F1$.

In the case of a SANLP, the input–output equivalent PPN is obtained by (1) converting the SANLP—by means of an *array analysis* [16, 17]—to a *single assignment code* (SAC) used in the compiler community and the systolic array community (see [34]); (2) deriving from the SAC a *polyhedral reduced dependence graph* [55] (PRDG); and (3) constructing the PPN from the PRDG [13, 39, 55].

While in a SAC every variable is written *only once*, in a *dynamic single assignment code* (dSAC) every variable is written *at most once*. For some variables, it is not known whether or not they will be read or written at compile time. For a WDP, however, not all dependences are known at compile time and, therefore, the analysis must be based on the so-called *fuzzy array dataflow analysis* [18]. This approach allows the conversion of a WDP to a dSAC. The procedure to generate the dSAC is out of the scope of this chapter. The dSAC for the WDP in Fig. 9 is shown in Fig. 10.

C in the dSAC shown in Fig. 10 is a parameter emerging from the *if*-statement in line 8 of the original program shown in Fig. 9. This *if*-statement also appears in the dSAC in line 14. The dynamic change of the value of C is accomplished by the lines 18 and 21 in Fig. 10. The control variable ctrl(i) in line 18 stores the iterations for which the data dependent condition that introduces C is true. Also, the variable ctrl(i) is used in line 21 to assign the correct value to C for the current iteration. See [60] for more details.

The dSAC can now be converted to two graph structures, namely the *Approximate reduced dependence graph* (ADG), and the *Schedule tree* (STree). The ADG is the dynamic counterpart of the static PRDG. Both the PRDG and the ADG are composed of processes N, input ports IP, output ports OP, and edges E [13, 55]. They contain all information related to the data dependencies between functions in the SAC and the dSAC, respectively. However, in a WDP some dependencies are not known at compile time, hence the name *approximate*. Because of this, the ADG has the additional notion of *linearly bounded set*, as follows.

Let be given four sets of functions
$S1 = \{f_x^1(i) \mid x = 1..|S1|, \ i \in Z^n\}$, $S2 = \{f_x^2(i) \mid x = 1..|S2|, \ i \in Z^n\}$, $S3 = \{f_x^3(i) \mid x = 1..|S3|, \ i \in Z^n\}$, $S4 = \{f_x^4(i) \mid x = 1..|S4|, \ i \in Z^n\}$, an integral $m \times n$ matrix A and an integral n-vector b. A *linearly bounded set* (LBS) is a set of points $LBS = \{ \ i \in Z^n \mid A.i \geq b,$

$$if \ S1 \not\equiv \emptyset \Rightarrow \forall_{x=1..|S1|}, \ f_x^1(i) \geq 0,$$
$$if \ S2 \not\equiv \emptyset \Rightarrow \forall_{x=1..|S2|}, \ f_x^2(i) \leq 0,$$
$$if \ S3 \not\equiv \emptyset \Rightarrow \forall_{x=1..|S3|}, \ f_x^3(i) > 0,$$
$$if \ S4 \not\equiv \emptyset \Rightarrow \forall_{x=1..|S4|}, \ f_x^4(i) < 0 \ \}.$$

The set of points $B = \{ \ i \in Z^n \mid A.i \geq b \ \}$ is called *linear bound* of the LBS and the set $S = S1 \cup S2 \cup S3 \cup S4$ is called *filtering set*. Every $f_x^j(i) \in S$ can be an arbitrary function of i.

Consider the dSAC shown in Fig. 10. The exact iterations i are not known at compile time because of the dynamic condition at line 14 in the dSAC (Fig. 10). That is why the notion of linearly bounded set is introduced, by which the unknown iterations i are approximated. So, ND_{N2} is the following LBS: $ND_{N2} = \{i \in Z \mid 1 \leq i \leq N \wedge 8 \leq N \leq 16, \ t\_1(i) \leq 0\}$. The linear bound of this LBS is the polytope $B = \{1 \leq i \leq N \wedge 8 \leq N \leq 16\}$ that captures the information known at compile time about the bounds of the iterations i. The variable $t\_1(i)$ is interpreted as an unknown function of i called filtering function whose output is determined at run time.

The STree contains all information about the execution order amongst the functions in the dSAC. The STree represents one valid schedule between all these functions called *global schedule*. From the STree a local schedule between any arbitrary set of the functions in the dSAC can be obtained by pruning operations on the STree. Such a local schedule may for example be needed when two or more processes are merged [61]. The STres is obtained by converting the dSAC to a syntax tree using a standard syntax parser, after which all the nodes and edges that are not related to nodes Fi (nodes $F1$, $F2$, and $F3$ in Fig. 10). See [60] for further details. A summary is depicted in Fig. 11.

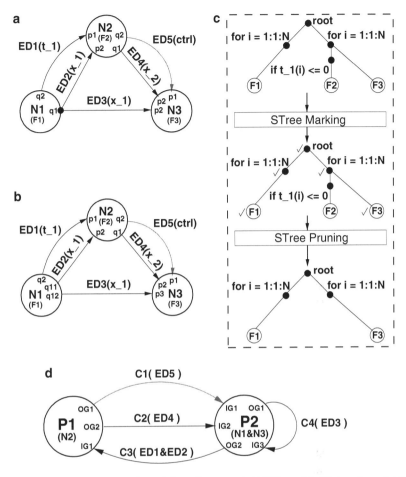

Fig. 11 Examples of (**a**) approximated dependence graph (ADG) model; (**b**) transformed ADG; (**c**) schedule tree and transformations; (**d**) process network model

The difference between the ADG in Fig. 11a and the transformed ADG in Fig. 11b is that an ADG may have several input ports connected to a single output port whilst in the transformed ADG every input port is connected to only one single output port (in accordance with the Kahn Process Network semantics [35]).

Parsing the STree in Fig. 11c top-down from left to right generates a program that gives a valid execution order (global schedule) among the functions $F1$, $F2$ and $F3$ which is the original order given by the dSAC.

The process network in Fig. 11d may be the result of a design space exploration, and some optimizations. For example, process $P2$ is constructed by grouping nodes $N1$ and $N3$ in the ADG in Fig. 11b. Because the behavior of process $P2$ is sequential (by default), it has to execute the functionality of nodes $N1$ and $N3$ in sequential order. This order is obtained from the STree in Fig. 11c. See [60] for details.

```
1   %parameter N 1 10;
2
3   for j = 1 to N,
4     X[j] = f(...)
5     for i = 1 to max_f,
6       if i <= X[j],
7       y[i] = F1()
8       end
9     end
10  end
11  [] = F2( y[5] )
```

```
1   %parameter N 1 10;
2
3   for j = 1 to N,
4     for i = 1 to f(...),
5     y[ i ] = F1()
6     end
7   end
8   [...] = F2( y[5] ),
```

An example of a `Dynloop` program.

An equivalent Weakly
Dynamic Program.

Fig. 12 A `Dynloop` program and its equivalent WDP program

In a (static) PPN, there are two models of FIFO communication [72], namely *in-order communication* and *out-of-order communication*. In the first model, the order in which tokens are read from a FIFO channel is the same as the order in which they have been written to the channel. In the second model that order is different. In a PPN that is input–output equivalent to a WDP, there are two more FIFO communication models, namely *in-order with coloring* and *out-of-order with coloring*. This is necessary because the number of tokens that will be written to a channel and read from that channel is not known at compile time. See [60] for details.

Buffer sizes can be determined using the procedure given in [73] and in [74], except that a conservative strategy (over-estimation) is needed due to the fact that the rate and the exact amount of data tokens that will be transferred over a particular data channel is unknown at compile-time. This can be done by modifying the iteration domains of all input/output ports, such that all dynamic *if*-conditions defining any of these iteration domains evaluate always to `true`.

7.2 Dynamic Loop-Bounds

Whereas in a SANLP loop bounds have to be affine functions of iterators of enclosing loops and static parameters, loop bounds in a DANLP program can be dynamic. Such programs have been called `Dynloop` programs in [44]. A simple example of a `Dynloop` program is shown at the left side in Fig. 12

A `Dynloop` program can be cast in the form of a WDP. See Sect. 7.1. The WDP corresponding to the `Dynloop` program at the left in Fig. 12 is shown at the right in Fig. 12.

```
1   %parameter N 1 10;          14  if max_f >= 5,
                                 15     c1 = ctrl_c1_1[N, 5]
2   for j = 1 to N,             16     c2 = ctrl_c2_1[N, 5]
3     X[j] = f()                17  else
4     for i = 1 to max_f,       18     c1 = N + 1
5       if i <= X[j],           19     c2 = max_f + 1
6         y_1[j,i] = F1()       20  end
7         ctrl_c1[i] = j        21  if c1 <= N & c2 == 5,
8         ctrl_c2[i] = i        22    in_0 = y_1[c1,c2]
9       end                     23  else
10        ctrl_c1_1[j,i] = ctrl_c1[i]   24    in_0 = 0
11        ctrl_c2_1[j,i] = ctrl_c2[i]   25  end
12    end                       26  [...] = F2( in_0 )
13  end
```

Fig. 13 Final dSAC

The maximum value of $f()$, denoted by max_f, see line 5 at the right in Fig. 12 is substituted for the upper bound of the loop at line 4 at the left in Fig. 12. The value of max_f can be determined by studying the range of function $f()$.[7]

As in Sect. 7.1, a dynamic single assignment code (dSAC) can now be obtained by means of a fuzzy array dataflow analysis (FADA) [18]. This analysis introduces parameters to deal with the dynamic structure in the WDP. The values of these parameters have to be changed dynamically. This is done by introducing for every such parameter a control variable that stores the correct value of the parameter for every iteration. However, the straightforward introduction of control values as done in Sect. 7.1 violates the dSAC condition that every control variable is written *at most once*. To obtain a valid dSAC, an additional dataflow analysis for the control variables is necessary, resulting in additional control variables. See [44] for details.

The final dSAC is shown in Fig. 13 where it has been assumed that the variable y(5) has been initialized to zero.

The control variables must be initialized with values that are greater than the maximum value of the corresponding parameters. For the example at hand, parameter $c1 \in [1..N]$, and $c2 \in [1..\text{max\_f}]$. Therefore, the corresponding control variables are initialized as follows:

$$\forall i : 1 \leq i \leq \text{max\_f} : \text{ctrl\_c1}[i] = N+1,$$
$$\text{ctrl\_c2}[i] = \text{max\_f}+1.$$

This initialization is not shown in Fig. 13 for the sake of brevity.

After applying the standard *linearization* [72], and its extension described in Sect. 7.1, and estimating buffer sizes as described in that same subsection, the resulting PPN is as shown in Fig. 14.

---

[7]If that is not possible, then an alternative way to estimate max_f is given in [44].

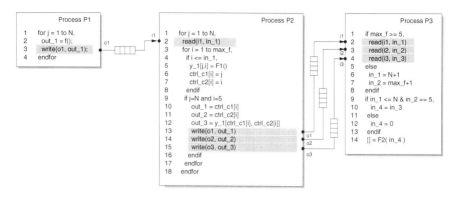

Fig. 14 The final PPN derived from the program in Fig. 13

7.3 Dynamic While-Loops

Whereas in a SANLP program only `while(1)` loops are allowed, in a DANLP program any while-loop is acceptable. Such DANLP programs have been called *while-loop affine programs* (WLAP) in [45].

There are a number of publications that address the problem of while loops parallelization [4, 10, 12, 25, 28, 29, 53, 54]. The approach presented here has the advantage that it

- Supports both task-level and data-level parallelism.
- Generates also parallel code for multi-processor systems having distributed memory.
- Provides an automatic data-dependence analysis procedure.
- Exposes and utilizes all available parallelism.

An example is shown at the left side in Fig. 15.

Again, the question is from where, say, function $F7$ gets its scalar argument x. Because this is not known at compile-time, a *fuzzy array dataflow analysis* (FADA) [18] is necessary to find all data dependencies.

The approach to convert a WLAP program to an input–output equivalent *polyhedral process network* (PPN) goes in four steps. First, all data-dependency relations in the initial WLAP program have to be found by applying the FADA analysis on it. Recall that the result of the analysis is approximated, i.e., it depends on parameters which values are determined at run-time. Second, based on the results of the analysis, the initial WLAP is transformed into a *dynamic Single Assignment Code* (dSAC) representation. See Sect. 7.1. Parameters that are introduced by the FADA appear in the dSAC, and their values are assigned using control variables. Third, the control variables are generated in a way that extends the methods in Sects. 7.1 and 7.2 to be applicable for WLAP programs as well, see [45]. Fourth, the topology of the corresponding PPN is derived as well as the code to be executed in the processes of the PPN.

```
1  &parameter EPS 0.005

2  for i = 1 to N,
3    y[i] = F1()
4    x = F2( y[i] )
5    while ( x >= EPS )
6      x = F3()
7      for j = i+1 to N+1,
8        y[j] = F4( y[j-1] )
9        x = F5( x, y[j] )
10     end
11     y[i] = F6( x )
12   end
13   out = F7( x )
14 end
```

An example of a WLAP program

```
1  %parameter EPS 0.005

2  w = 0
3  ctrl_x_5 = (N+1,0)
4  for i = 1 to N,
5    y_1[i] = F1()
6    in_2 = y_1[i]
7    x_2[i] = F2( in_2 )
8    while (in_w = σ_x(⟨W^c(i^c w)⟩)) >= EPS),
9      w = w + 1
11     x_3[i,w] = F3()
11     for j = i+1 to N+1,
12       in_4 = σ_y(⟨S_4^c(i^c w^c j)⟩)
13       y_4[i,w,j] = F4( in_4 )
14       in_5_x = σ_x(⟨S_5^c(i^c w^c j)⟩)
15       in_5_y = y_4[i,w,j]
16       x_5[i,w,j] = F5( in_5_x, in_5_y )
17       ctrl_x_5 = (i,w)
18     end
19     in_6 = σ_x(⟨S_6^c(i^c w)⟩)
20     y_6[i,w] = F6( in_6 )
21   end
22   ctrl_x_5_[i] = ctrl_x_5
23   (α^c β) = ctrl_x_5_[i]
24   in_7 = σ_x(⟨S_7^c(i^c α^c β)⟩)
25   out = F7( in_7 )
26 end
```

The corresponding final dSAC

Fig. 15 An example of a while-loop affine program and its corresponding dynamic single assignment program

The iterator w is associated with the while loop and is initialized with value 0, meaning that the while loop has never been executed. The parameter α captures the value of the for-loop iterator in the enclosing while-loop and is initialized to $N + 1$. The parameter β is the upper bound of the while-loop iterator w. Because $\alpha \in [1..N]$ and $\beta \geq 1$, the above initializations satisfy the condition that their values are never taken by the corresponding parameters. From line 23 at the right side in Fig. 15, it follows that the control variable ctrl_x_5 is initialized to ctrl_x_5 = (N+1,0) at line 3 at the right side in Fig. 15. Where does the control variable ctrl_x_5 come from? It comes from the construction of the dSAC. The procedure to derive the final dSAC is largely based on [18] and its extension in Sect. 7.2. The problem is again that the dSAC resulting from the FADA analysis is not a proper dSAC because it violates the property that every variable is written *at most once*. The relation between writing to and reading from the control variables must be identified by performing a dataflow analysis for the control variables, where the writings to them occur inside a while-loop. To that end, an additional control variable ctrl_x_5_ is introduced right *after* the while-loop, see line 22 at the right in Fig. 15. The new control variable is written at every iteration of *for*-loop i and takes the value either of control variable ctrl_x_5 assigned on the last iteration of the while-loop, or its initial value, if the while-loop is not executed. A *static* exact

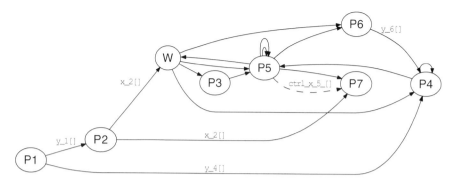

Fig. 16 The PPN for the program in Fig. 15

```
 1  %parameter EPS 0.005

 2  w = 0
·3  for i = 1 to N,
 4    while(1),
 5      w = w + 1
 6      if (w > 2) then w = 2
 7      if (w == 1),
 8        read(P2, 1, in_w)
 9      else
10        read(P5, 2, in_w)
11      end
12      out_w = (in_w >= EPS)
13      write(P3, 3, out_w)
14      write(P4, 4, out_w)
15      write(P5, 5, out_w)
16      write(P6, 6, out_w)
17      if (!out_w) <break>
18    end
19  end
```
Code of process W

```
 1  w = 0
 2  ctrl_x_5 = (N+1,0)
 3  for i = 1 to N,
 4    while(1),
 5      w = w + 1
 6      if (w > 2) then w = 2
 7      read(W, 1, in_w)
 8      if (!in_w) <break>
 9      for j = i+1 to N+1,
10        if (j == i+1),
11          if (w == 1),
12            read(P3, 2, in_5_x)
13          else
14            read(P5, 3, in_5_x)
15          en
16        else
17          read(P5, 4, in_5_x)
18        end
19        read(P4,5, in_5_y)
20        out_5 = F5( in_5_x, in_5_y )
21        ctrl_x_5 = (i,w)
22        if (j == N+1),
23          write(P5, 6, out_5)
24        else
25          write(P5, 7, out_5)
26        endif
27      end
28    end
29    out_5_c = ctrl_x_5
30    out_5_x = out_5
31    write(P7, 8, out_5_c)
32    write(P7, 9, out_5_x)
33  end
```
Code of process P5

```
 1  w = 0
 2  for i = 1 to N,
 3    read(P5, 1, in_c)
 4    if (in_c.β>=1 && 1<= in_c.α <= i),
 5      read(P5, 2, in_7)
 6    else
 7      read(P2, 3, in_7)
 8    end
 9    out = F7( in_7 )
10  end
```
Code of process P7

Fig. 17 Processes W, $P5$, and $P7$ after linearization

array dataflow analysis (EADA) [16] can be performed on this new control variable `ctrl_x_5_`. This is possible because the new control variable is not surrounded by the dynamic while-loop, i.e., it is outside the while loop.

The PPN that corresponds to the final dSAC in Fig. 15 is depicted in Fig. 16.

This PPN consists of 8 processes and 18 channels. The processes $P1$–$P7$ correspond to the functions $F1$–$F7$ in Fig. 15. Process W corresponds to the while condition at line 8 of the final dSAC in Fig. 15

The code for processes W, $P5$, and $P7$ is shown in Fig. 17. Process W is an example of a process detecting the termination of the while-loop at line 5 at the left in Fig. 15. Process $P5$ is an example of a process executing a function enclosed in the while-loop. Process $P7$ is an example of a process that runs a function *outside* the while-loop, and has a data dependency with a function *inside* the while-loop.

7.4 Parameterized Polyhedral Process Networks

Parameters that appear in a SANLP program are static. In a DANLP, parameters can be dynamic. A polyhedral process network [73] that is input–output equivalent to such a DANLP program is, then, a *parameterized polyhedral process network* called P^3N in [78].

Remark. There are two assumptions here. First, dynamic conditions, dynamic loop bounds and dynamic while-loops are left out to focus only on dynamic parameters. Second, values of the dynamic parameters are obtained from the environment.

The formal definition of a P^3N is given in [78], and is only slightly different from the definition given in [73]. Although the consistency of a P^3N has to be checked at run-time, still some analysis can be done at compile-time. A simple example of a P^3N is shown in Fig. 18.

Figure 18a is a static PPN, process *P3* of which is shown in Fig. 18b. Figure 18c is a P^3N version of the PPN in Fig. 18a. Process *P3* of the P^3N in Fig. 18c is shown in Fig. 18d. The PPN and the P^3N have the same *dataflow* topology. Processes *P2* and *P3* in the P^3N in Fig. 18c are reconfigured by two parameters *M* and *N* whose values are updated from *the environment* at run-time using process *Ctrl* and FIFO channels *ch7*, *ch8*, and *ch9*. The P^3N shown in Fig. 18c may be derived from a sequential program, yet it can also be constructed from library elements as in [31].

Recall from [73] that a parametric polyhedron $\mathscr{P}(\mathbf{p})$ is defined as $\mathscr{P}(\mathbf{p}) = \{(w, x_1, \ldots, x_d) \in \mathbb{Q}^{d+1} \mid A \cdot (w, x_1, \ldots, x_d)^T \geq B \cdot \mathbf{p} + b\}$ with $A \in \mathbb{Z}^{m \times d}, B \in \mathbb{Z}^{m \times n}$ and $c \in \mathbb{Z}^m$. For nested loop programs, w is to be interpreted as the one-dimensional while(1) index, and d as the depth of a loop nest. For a particular value of w the polyhedron gets closed, i.e., it becomes a polytope. The parameter vector $\mathbf{p}$ is bounded by a polytope $\mathscr{P}_{\mathbf{p}} = \{\mathbf{p} \in \mathbb{Q}^n \mid C \cdot \mathbf{p} \geq d\}$.

The domain D_P of a process is defined as the set of all integral points in its underlying parametric polyhedron, i.e., $D_P = \mathscr{P}_P(\mathbf{p}) \cap \mathbb{Z}^{d+1}$. The domains D_{IP} and D_{OP} of an input port *IP* and an output port *OP*, respectively, of a process are subdomains of the domain of that process.

The following four notions play a role in the operational semantics of a P^3N:

- Process iteration.
- Process cycle.
- Process execution.
- Quiescent point.

A **process iteration** of process P is a point $(w, x_1, \ldots, x_d) \in D_P$, where the following operations are performed sequentially: reading a token from each IP for which $(w, x_1, \ldots, x_d) \in D_{IP}$, executing process function F_P, and writing a token to each OP for which $(w, x_1, \ldots, x_d) \in D_{OP}$.

A **process cycle** $CYC_P(\mathscr{S}, \mathbf{p}) \subset D_P$ is the set of lexicographically ordered points $\in D_P$ for a particular value of $w = \mathscr{S} \in \mathbb{Z}^+$. The lexical ordering is typically imposed by a loop nest.

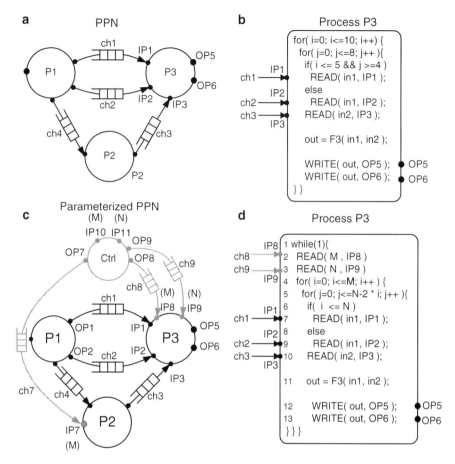

Fig. 18 (**a**) An example of a PPN, (**b**) process *P3* in the PPN, (**c**) an example of a P³N, and (**d**) process *P3* in the P³N

A **Process execution** E_P is a sequence of process cycles denoted by $CYC_P(1, \mathbf{p}_1) \rightarrow CYC_P(2, \mathbf{p}_2) \rightarrow \ldots \rightarrow CYC_P(k, \mathbf{p}_k)$, where $k \rightarrow \infty$.

A point $Q_P(\mathscr{S}, \mathbf{p}_i) \in CYC_P(\mathscr{S}, \mathbf{p}_i)$ of process P is a **quiescent point** if $CYC_P(\mathscr{S}, \mathbf{p}_i) \in E_P$ and $\neg(\exists (w, x_1, \ldots, x_d) \in CYC_P(\mathscr{S}, \mathbf{p}_i) : (w, x_1, \ldots, x_d) \prec Q_P(\mathscr{S}, \mathbf{p}_{\mathscr{S}}))$.

Thus, process P can change parameter values at the first process iteration of any process cycle during the execution. The notion of quiescent points as being the points at which values of the parameters **p** can change appears also in [47]. The behavior of the control process *Ctrl* is given in Fig. 19a.

Process *Ctrl* starts with at least one valid parameter combination (lines 1--2) and then reads parameters from the environment (lines 3--4) every pre-specified time interval. For every incoming parameter combination, the process function Eval (line 5) checks whether the combination of parameter values is valid. The

a Process Ctrl

1 M_new = M_init
2 N_new = N_init
   while(1){
IP10 ◆3   READ_PARM (M, IP10)
IP11 ◆4   READ_PARM (N, IP11)

5   [M_new, N_new] =
      Eval(M, N, M_new, N_new)

6   WRITE_PARM (M_new, OP7) |OP7| → ch7
7   WRITE_PARM (M_new, OP8) |OP8| → ch8
8   WRITE_PA M (N_new, OP9) |OP9| → ch9
   }

b Process Function Eval

[M_new, N_new]
Eval(M, N, M_old, N_old){

// checking parameters
par_ok = Check(M, N);

if(par_ok){
   return (M, N)
else {
   return (M_old, N_old)
} }

Fig. 19 (**a**) Control process *Ctrl* and (**b**) process function *Eval*

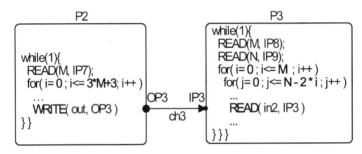

P2

while(1){
  READ(M, IP7);
  for(i= 0 ; i<= 3*M+3; i++)
  ...
   WRITE(out, OP3)
}}

|OP3| IP3|
   ch3

P3

while(1){
  READ(M, IP8);
  READ(N, IP9);
  for(i= 0 ; i<= M ; i++)
   for(j= 0 ; j<= N - 2 * i ; j++)
   ...
   READ(in2, IP3)
   ...
}}}

Fig. 20 Which combinations (M, N) do ensure consistency of P^3N?

implementation of function Eval is given in Fig. 19b. If the combination is valid, then function Eval returns the current parameter values (M, N). Otherwise, the last valid parameters combination (propagated through M_new, N_new in this example) is returned. After the evaluation of the parameters combination, process *Ctrl* writes the parameter values to output ports (lines 6–8) when all channels *ch7*, *ch8*, and *ch9* have at least one buffer place available. When at least one channel buffer is full, the incoming parameters combination is discarded and the control process continues to read the next parameters combination from the environment. Furthermore, the depth of the FIFOs of the control channels determines how many process cycles of the dataflow processes are allowed to overlap.

Valid parameter values lead to the consistent execution of a P^3N, i.e., without deadlocks and with bounded memory (FIFOs with finite capacity). To illustrate the problem, consider channel *ch3* connecting processes *P2* and *P3* of the P^3N given in Fig. 18c.

The access of processes *P2* and *P3* to channel *ch3* is depicted in Fig. 20.

Consistency requires that, for each corresponding process cycle of both processes $CYC_{P2}(i,M_i)$ and $CYC_{P3}(i,M_i,N_i)$, the number of tokens produced by process $P2$ to channel $ch3$ must be equal to the number of tokens consumed by process $P3$ from channel $ch3$. For example, if $(M,N) = (7,8)$, $P2$ produces 25 tokens to $ch3$ and $P3$ consumes 25 tokens from the same channel after one corresponding process cycle of both processes. It can be verified that $P2$ produces 13 tokens to $ch3$ while $P3$ requires 20 tokens from it in a corresponding process cycle when $(M,N) = (3,7)$. Thereby, in order to complete one execution cycle of $P3$ in this case, it will read data from $ch3$ which will be produced during the next execution cycle of $P2$. Evidently this leads to an incorrect execution of the P^3N. From this example, it is clearly seen that the incoming values of (M,N) must satisfy certain relation to ensure the consistent execution of the P^3N.

Although the consistency of a P^3N has to be checked at run-time, still some analysis can be done at design-time. This is because input ports and output ports of a process cycle are parametric polytopes. The number of points in a port domain equals the number of tokens that will be written to a channel or read from a channel depending on whether the port is an output port or an input port, respectively. The condition $|D_{OP}^{CYC}| = |D_{IP}^{CYC}|$ can be checked by comparing the number of points in both port domains. The counting problem can be solved in polynomial time using the *Barvinok* library [73, 75]. In general the number of points in domain $D_X = \mathscr{P}_X(\mathbf{p}) \cap \mathbb{Z}^{d+1}$, where X stand for either a process P, an input port IP, or an output port OP, is a set of quasi-polynomials [73].

For the example shown in Fig. 20, the difference $|D_{OP}^{CYC}| - |D_{IP}^{CYC}|$ is,

$$
\begin{cases}
(1+N+N\cdot M-M^2)-(3M+4)=0 & \text{if } (M,N) \in C1 \\
(1+\frac{3}{4}N+\frac{1}{4}N^2+\frac{1}{4}N-\frac{1}{4}\cdot\{0,1\}_N)-(3M+4)=0 & \text{if } (M,N) \in C2
\end{cases}
$$

where $C1 = \{(M,N) \in \mathbb{Z}^2 \mid M \le N \wedge 2M \ge 1+N\}$, $C2 = \{(M,N) \in \mathbb{Z}^2 \mid 2M \le N\}$, and $\{0,1\}_N$ is a periodic coefficient with period 2.[8] In this example, if the range of the parameters is $0 \le M,N \le 100$, then there are only ten valid parameter combinations. If $0 \le M,N \le 1,000$, then there are 34 valid parameter combinations, and if $0 \le M,N \le 10,000$, then the number of valid combinations is 114.

The symbolic subtraction of the quasi-polynomials can result in constant zero, non-zero constant, or a quasi-polynomial. In the first case, consistency is always preserved for all parameters within the range. In the second case, all parameters within the range are invalid, because they violate the consistency condition. In the third case, a quasi-polynomial remains, and only some parameter combinations within the range are valid for the consistency condition. The equations can be solved at design time, and all valid parameter combinations are put in a table which is stored in a function *Check*. At run-time, the control process only propagates those incoming parameter combinations that match an entry in the table. Alternatively, function *Check* evaluates the difference between the two quasi-polynomials against

---

[8]$\{0,1\}_N$ is 0 or 1 depending on whether N is even or odd, respectively.

zero with incoming parameter values at run-time. When using a table, the execution time of the P^3N is almost equal to the execution time of the corresponding PPN. On the other hand, evaluation the polynomials at run-time overlaps the dataflow processing. For medium and high workloads (execution latency of the processes) the overhead is negligible. See [78] for further details.

8 The Stream-Based Function Model

The active entities in dataflow graphs are pure functions, embedded in *Actors*. In dataflow (polyhedral and parameterized polyhedral) process networks, the active entities encompass state and pure functions. They are called *Processes*. Actors and processes are specified in a *host language* like C, C++ or Java, by convention.

Remark. For reasons of convenience, the active entities in dataflow graphs and dataflow networks will in this section be called *actors*, whether they are void of state, encompass a single thread of control, or are processes. Thus static dataflow graphs [31], dynamic dataflow graphs, polyhedral process [73] and Sect. 7.4, and Kahn process networks [21] will collectively be referred to as *Dataflow Actor Networks* (DFAN). They obey the Kahn coordination semantics, possibly augmented with actor firing rules, annotated with deadlock free minimal FIFO buffer capacities, and one or more global schedules associated with them.

This section deals with a parallel model for actors, called *Stream-based function* (SBF) in [38] SBF is appealing when it comes to implementing a DFAN in a heterogeneous multiprocessor execution platform that consists of a number of computational elements (processors), and a communication, synchronization, and storage infrastructure. This requires a *mapping* that relates an untimed application model and a timed architecture model together: Actors are assigned to processors (both of type ISA and dedicated), FIFO channels are logically embedded in— often distributed—memory, and DFAN communication primitives and protocols are transformed to platform specific communication primitives and protocols. The specification of platform processors and DFAN actors may be quite different.

The SBF can serve as an intermediate specification between the conventional DFAN actor specification, and a variety of computational elements in the execution platform.

8.1 The Stream-Based Function Actor

The SBF is composed of a *set of functions*, called *function repertoire*, a *transition and selection function*, called *controller*, and a combined function and data *state*, called *private memory*. The controller selects a function from the function repertoire that is associated with a *current function state*, and makes a transition to the *next*

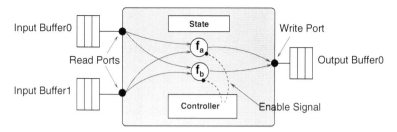

Fig. 21 A simple SBF actor

function state. A selected function is *enabled* when all its input arguments can be read, and all its results can be written; it is blocked otherwise. A selected non-blocked function must evaluate or *fire*. Arguments and results are called *input tokens* and *output tokens*, respectively.

An actor operates on sequences of tokens, streams or signals, as a result of a repetitive enabling and firing of functions from the function repertoire. Tokens are *read* from external ports and/or loaded from private memory, and are *written* to external ports and/or stored in private memory. The external ports connect to FIFO channels through which actors communicate point-to-point.

Figure 21 depicts an illustrative stream-based function (SBF) actor. Its function repertoire is $P = \{f_{init}, f_a, f_b\}$. The two read ports, and the single write port connect to two input and one output FIFO channels, respectively. P contains at least an initialization function f_{init}, and is a finite set of functions. The functions in P must be selected in a mutual exclusive order. That order depends on the controller's selection and transition function. The private memory consists of a function state part, and a data state part that do not intersect.

8.2 The Formal Model

If C denotes the SBF actor's function state space, and if D denotes its data state space, then the actor's state space S is the Cartesian product of C and D,

$$S = C \times D, \qquad C \cap D = \emptyset. \tag{2}$$

Let $c \in C$ be the current function state. From c, the controller selects a function and makes a function state transition by activating its selection function μ and transition function ω as follows,

$$\mu : C \to P, \quad \mu(c) = f, \tag{3}$$

$$\omega : C \times D \to C, \qquad \omega(c,d) = c'. \tag{4}$$

The evaluation of the combined functions μ and ω is instantaneous. The controller cannot change the content of C; it can only observe it. The transition function allows for a dynamic behavior as it involves the data state space D. When the transition function is only a map from C to C, then the trajectory of selected functions will be static (see [31, 73]). Assuming that the DFAN does not deadlock, and that its input streams are not finite, the controller will repeatedly invoke functions from the function repertoire P in an endless firing of functions,

$$f_{init} \xrightarrow{\mu(\omega(S))} f_a \xrightarrow{\mu(\omega(S))} f_b \xrightarrow{\mu(\omega(S))} \ldots f_x \xrightarrow{\mu(\omega(S))} \ldots \tag{5}$$

Such a behavior is called a *Fire-and-Exit* behavior. It is quite different from a threaded approach in that all synchronization points are explicit. The role of the function f_{init} in (5) is crucial. This function has to provide the starting current function state $c_{init} \in C$

$$c_{init} = f_{init}(channel_a \ldots channel_z). \tag{6}$$

It evaluates first and only once, and it may read tokens from one or more external ports to return the starting current function state.

The SBF model is reminiscent of the *Applicative State Transition* (AST) node in the systolic array model of computation [34] introduced by Annevelink [1], after the *Applicative State Transition* programming model proposed by Backus [2]. The AST node, like the SBF actor, comprises a function repertoire, a selection function μ, and a transition function ω. However, the AST node does not have a private memory, and the initial function state c_{init} is read from a unique external channel. Moreover, AST nodes communicate through register channels. The SBF actor model is also reminiscent of the CAL, which is discussed in Sect. 3.

Clearly, the actors in decidable dataflow models [31] and the multidimensional dataflow models [37], and the processes in polyhedral precess networks [73] and Kahn process networks [21] are special cases of the SBF actor. Analysis is still possible in case the SBF-based DFAN originates from *weakly dynamic* nested loop programs, see Sect. 7.1, Dynloop programs, see Sect. 7.2, WLAP programs, see Sect. 7.3, and some parameterized polyhedral process networks, see Sect. 7.4.

Actors in dataflow actor networks communicate point to point over FIFO buffered channels. Conceptually, buffer capacities are unlimited, and actors synchronize by means of a *blocking read* protocol: an actor that attempts to read tokens from a specific channel will block whenever that channel is empty. Writing tokens will always proceed. Of course, channel FIFO capacities are not unlimited in practice, so that a mapped DFAN does have a *blocking write* protocol as well: an actor that attempts to write tokens to a specific channel will block whenever that channel is full. A function $[c, d] = f(a, b)$ has to bind to input ports p_x and p_y, respectively, when the function has to receive its arguments a and b, in this order. Similarly, that function has to bind to ports q_x and q_y, respectively, when ports q_x and q_y are to receive results c and d, in this order.

In the SBF actor, the controller's selection function μ selects both a function from the function repertoire and the corresponding input and output ports. Note that a function that is selected from the function repertoire may read arguments and write results from non-unique input ports and to non-unique output ports, respectively, as is also the case with polyhedral process networks [73].

This allows to separate function selection and binding, so that reading, executing, and writing can proceed in a pipelined fashion. Although standard blocking read and blocking write synchronization is possible, the SBF actor allows for a more general deterministic dynamic approach. In this approach, the actor behavior is divided into a *channel checking* part and a *scheduling* part. See also Sect. 5. In the channel checking part, channels are visited *without empty or full channel blocking*. A visited input channel C_{in} returns a $C_{in}.1$ signal when the channel is not empty, and a $C_{in}.0$ signal when the channel is empty. And similarly for output channels. These signals indicate whether or not a particular function from the function repertoire can fire. In the scheduling part, a function can only be invoked when the channel checking signals allow it to fire. If not, then the function will not be invoked. As a consequence, the actor is blocked. Clearly channel checking and scheduling can proceed in parallel.

9   Summary

This chapter, has reviewed several DSP-oriented dataflow models of computation that are oriented towards representing dynamic dataflow behavior. As signal processing systems are developed and deployed for more complex applications, exploration of such generalized dataflow modeling techniques is of increasing importance. This chapter has complemented the discussion in [31], which focuses on the relatively mature class of decidable dataflow modeling techniques, and builds on the dynamic dataflow principles introduced in certain specific forms [15, 21].

Acknowledgements   In this work, Bhattacharyya has been supported in part by the US Air Force Office of Scientific Research. The authors also thank Marc Geilen (m.c.w.geilen@tue.nl) and Sander Stuijk (s.stuijk@tue.nl), both from the Eindhoven University of Technology, for their contribution to Sect. 6.

References

1. Annevelink, J.: HIFI: A design method for implementing signal processing algorithms on VLSI processor arrays. Ph.D. thesis, Delft University of Technology, Department of Electical Engineering, Delft, The Netherlands (1988)
2. Backus, J.: Can programming be liberated from the von Neumann style? A functional style and its algebra of programs. Communications of the ACM **21**(8), 613–641 (1978)

3. Bekooij, M., Hoes, R., Moreira, O., Poplavko, P., Pastrnak, M., Mesman, B., Mol, J., Stuijk, S., Gheorghita, V., van Meerbergen, J.: Dataflow analysis for real-time embedded multiprocessor system design. In: P. van der Stok (ed.) Dynamic and Robust Streaming in and between Connected Consumer-Electronic Devices, pp. 81–108. Springer (2005)
4. Benabderrahmane, M.W., Pouchet, L.N., Cohen, A., Bastoul, C.: The polyhedral model is more widely applicable than you think. In: Proc. International Conference on Compiler Construction (ETAPS CC'10). Paphos, Cyprus (2010)
5. Berg, H., Brunelli, C., Lucking, U.: Analyzing models of computation for software defined radio applications. In: Proceedings of the International Symposium on System-on-Chip (2008)
6. Bhattacharya, B., Bhattacharyya, S.S.: Parameterized dataflow modeling for DSP systems. IEEE Transactions on Signal Processing **49**(10), 2408–2421 (2001)
7. Bhattacharyya, S.S., Buck, J.T., Ha, S., Lee, E.A.: Generating compact code from dataflow specifications of multirate signal processing algorithms. IEEE Transactions on Circuits and Systems — I: Fundamental Theory and Applications **42**(3), 138–150 (1995)
8. Bhattacharyya, S.S., Eker, J., Janneck, J.W., Lucarz, C., Mattavelli, M., Raulet, M.: Overview of the MPEG reconfigurable video coding framework. Journal of Signal Processing Systems (2010). DOI:10.1007/s11265-009-0399-3
9. Bhattacharyya, S.S., Leupers, R., Marwedel, P.: Software synthesis and code generation for DSP. IEEE Transactions on Circuits and Systems — II: Analog and Digital Signal Processing **47**(9), 849–875 (2000)
10. Bijlsma, T., Bekooij, M.J.G., Smit, G.J.M.: Inter-task communication via overlapping read and write windows for deadlock-free execution of cyclic task graphs. In: Proceedings SAMOS'09, pp. 140–148. Samos, Greece (2009)
11. Buck, J.T.: Scheduling dynamic dataflow graphs with bounded memory using the token flow model. Ph.D. thesis, Department of Electrical Engineering and Computer Sciences, University of California at Berkeley (1993)
12. Collard, J.F.: Automatic parallelization of while-loops using speculative execution. Int. J. Parallel Program. **23**(2), 191–219 (1995)
13. Deprettere, E.F., Rijpkema, E., Kienhuis, B.: Translating imperative affine nested loop programs to process networks. In: E.F. Deprettere, J. Teich, S. Vassiliadis (eds.) Embedded Processor Design Challenges, LNCS 2268, pp. 89–111. Springer, Berlin (2002)
14. Eker, J., Janneck, J.W.: CAL language report, language version 1.0 — document edition 1. Tech. Rep. UCB/ERL M03/48, Electronics Research Laboratory, University of California at Berkeley (2003)
15. Falk, J., Haubelt, C., Zebelein, C., Teich, J.: Integrated modeling using finite state machines and dataflow graphs. In: S.S. Bhattacharyya, E.F. Deprettere, R. Leupers, J. Takala (eds.) Handbook of Signal Processing Systems, second edn. Springer (2013)
16. Feautrier, P.: Dataflow analysis of scalar and array references. Int. Journal of Parallel Programming **20**(1), 23–53 (1991)
17. Feautrier, P.: Automatic parallelization in the polytope model. In: The Data Parallel Programming Model, pp. 79–103 (1996)
18. Feautrier, P., Collard, J.F.: Fuzzy array dataflow analysis. Tech. rep., Ecole Normale Superieure de Lyon (1994). ENS-Lyon/LIP N° 94-21
19. Gao, G.R., Govindarajan, R., Panangaden, P.: Well-behaved programs for DSP computation. In: Proceedings of the International Conference on Acoustics, Speech, and Signal Processing (1992)
20. Geilen, M.: Synchronous dataflow scenarios. ACM Trans. Embed. Comput. Syst. **10**(2), 16:1–16:31 (2011)
21. Geilen, M., Basten, T.: Kahn process networks and a reactive extension. In: S.S. Bhattacharyya, E.F. Deprettere, R. Leupers, J. Takala (eds.) Handbook of Signal Processing Systems, second edn. Springer (2013)
22. Geilen, M.C.W., Basten, T., Theelen, B.D., Otten, R.J.H.M.: An algebra of pareto points. Fundamenta Informaticae **78**(1), 35–74 (2007)

23. Geilen, M.C.W., Falk, J., Haubelt, C., Basten, T., Theelen, B.D., Stuijk, S.: Performance analysis of weakly-consistent scenario-aware dataflow graphs. Tech. Rep. ESR-2011-03, Eindhoven University of Technology (2011)
24. Geilen, M., Stuijk, S.: Worst-case performance analysis of synchronous dataflow scenarios. In: Proceedings of the eighth IEEE/ACM/IFIP international conference on hardware/software codesign and system synthesis, CODES/ISSS '10, pp. 125–134. ACM, New York, NY, USA (2010)
25. Geuns, S., Bijlsma, T., Corporaal, H., Bekooij, M.: Parallelization of while loops in nested loop programs for shared-memory multiprocessor systems. In: Proc. Int. Conf. Design, Automation and Test in Europe (DATE'11). Grenoble, France (2011)
26. Gheorghita, S.V., Stuijk, S., Basten, T., Corporaal, H.: Automatic scenario detection for improved WCET estimation. In: Proceedings of the 42nd annual Design Automation Conference, DAC '05, pp. 101–104. ACM, New York, NY, USA (2005)
27. Girault, A., Lee, B., Lee, E.: Hierarchical finite state machines with multiple concurrency models. Computer-Aided Design of Integrated Circuits and Systems, IEEE Transactions on **18**(6), 742 –760 (1999)
28. Griebl, M., Collard, J.F.: Generation of Synchronous Code for Automatic Parallelization of while-loops. EURO-PAR'95, Springer-Verlag LNCS, number 966, pp. 315–326 (1995)
29. Griebl, M., Lengauer, C.: A communication scheme for the distributed execution of loop nests with while loops. Int. J. Parallel Programming **23** (1995)
30. Gu, R., Janneck, J., Raulet, M., Bhattacharyya, S.S.: Exploiting statically schedulable regions in dataflow programs. In: Proceedings of the International Conference on Acoustics, Speech, and Signal Processing, pp. 565–568. Taipei, Taiwan (2009)
31. Ha, S., Oh, H.: Decidable dataflow models for signal processing: Synchronous dataflow and its extensions. In: S.S. Bhattacharyya, E.F. Deprettere, R. Leupers, J. Takala (eds.) Handbook of Signal Processing Systems, second edn. Springer (2013)
32. Haykin, S.: Adaptive Filter Theory. Prentice Hall (1996)
33. Hermanns, H.: Interactive Markov chains: and the quest for quantified quality. Springer-Verlag, Berlin, Heidelberg (2002)
34. Hu, Y.H., Kung, S.Y.: Systolic arrays. In: S.S. Bhattacharyya, E.F. Deprettere, R. Leupers, J. Takala (eds.) Handbook of Signal Processing Systems, second edn. Springer (2013)
35. Kahn, G.: The semantics of a simple language for parallel programming. In: Proc. of Information Processing (1974)
36. Kee, H., Wong, I., Rao, Y., Bhattacharyya, S.S.: FPGA-based design and implementation of the 3GPP-LTE physical layer using parameterized synchronous dataflow techniques. In: Proceedings of the International Conference on Acoustics, Speech, and Signal Processing, pp. 1510–1513. Dallas, Texas (2010)
37. Keinert, J., Deprettere, E.F.: Multidimensional dataflow graphs. In: S.S. Bhattacharyya, E.F. Deprettere, R. Leupers, J. Takala (eds.) Handbook of Signal Processing Systems, second edn. Springer (2013)
38. Kienhuis, B., Deprettere, E.F.: Modeling stream-based applications using the SBF model of computation. Journal of Signal Processing Systems **34**(3), 291–299 (2003)
39. Kienhuis, B., Rijpkema, E., Deprettere, E.F.: Compaan: Deriving Process Networks from Matlab for Embedded Signal Processing Architectures. In: Proc. 8th International Workshop on Hardware/Software Codesign (CODES'2000). San Diego, CA, USA (2000)
40. Ko, M., Zissulescu, C., Puthenpurayil, S., Bhattacharyya, S.S., Kienhuis, B., Deprettere, E.: Parameterized looped schedules for compact representation of execution sequences in DSP hardware and software implementation. IEEE Transactions on Signal Processing **55**(6), 3126–3138 (2007)
41. Lin, Y., Choi, Y., Mahlke, S., Mudge, T., Chakrabarti, C.: A parameterized dataflow language extension for embedded streaming systems. In: Proceedings of the International Symposium on Systems, Architectures, Modeling and Simulation, pp. 10–17 (2008)
42. Mattavelli, M., Raulet, M., Janneck, J.W.: MPEG reconfigurable video coding. In: S.S. Bhattacharyya, E.F. Deprettere, R. Leupers, J. Takala (eds.) Handbook of Signal Processing Systems, second edn. Springer (2013)

43. Moreira, O.: Temporal analysis and scheduling of hard real-time radios running on a multi-processor. Ph.D. thesis, Eindhoven University of Technology (2012)
44. Nadezhkin, D., Nikolov, H., Stefanov, T.: Translating affine nested-loop programs with dynamic loop bounds into polyhedral process networks. In: ESTImedia, pp. 21–30 (2010)
45. Nadezhkin, D., Stefanov, T.: Automatic derivation of polyhedral process networks from while-loop affine programs. In: ESTImedia, pp. 102–111 (2011)
46. Neuendorffer, S., Lee, E.: Hierarchical reconfiguration of dataflow models. In: Proceedings of the International Conference on Formal Methods and Models for Codesign (2004)
47. Neuendorffer, S., Lee, E.: Hierarchical reconfiguration of dataflow models. In: Proc. of MEMOCODE, pp. 179–188 (2004)
48. Nikolov, H., Stefanov, T., Deprettere, E.: Systematic and automated multi-processor system design, programming, and implementation. IEEE Transactions on Computer-Aided Design **27**(3), 542–555 (2008)
49. Plishker, W., Sane, N., Bhattacharyya, S.S.: A generalized scheduling approach for dynamic dataflow applications. In: Proceedings of the Design, Automation and Test in Europe Conference and Exhibition, pp. 111–116. Nice, France (2009)
50. Plishker, W., Sane, N., Bhattacharyya, S.S.: Mode grouping for more effective generalized scheduling of dynamic dataflow applications. In: Proceedings of the Design Automation Conference, pp. 923–926. San Francisco (2009)
51. Plishker, W., Sane, N., Kiemb, M., Anand, K., Bhattacharyya, S.S.: Functional DIF for rapid prototyping. In: Proceedings of the International Symposium on Rapid System Prototyping, pp. 17–23. Monterey, California (2008)
52. Poplavko, P., Basten, T., van Meerbergen, J.L.: Execution-time prediction for dynamic streaming applications with task-level parallelism. In: DSD, pp. 228–235 (2007)
53. Raman, E., Ottoni, G., Raman, A., Bridges, M.J., August, D.I.: Parallel-stage decoupled software pipelining. In: Proc. 6th annual IEEE/ACM international symposium on Code generation and optimization, pp. 114–123 (2008)
54. Rauchwerger, L., Padua, D.: Parallelizing while loops for multiprocessor systems. In: In Proceedings of the 9th International Parallel Processing Symposium (1995)
55. Rijpkema, E., Deprettere, E., Kienhuis, B.: Deriving process networks from nested loop algorithms. Parallel Processing Letters **10**(2), 165–176 (2000)
56. Roquier, G., Wipliez, M., Raulet, M., Janneck, J.W., Miller, I.D., Parlour, D.B.: Automatic software synthesis of dataflow program: An MPEG-4 simple profile decoder case study. In: Proceedings of the IEEE Workshop on Signal Processing Systems (2008)
57. Saha, S., Puthenpurayil, S., Bhattacharyya, S.S.: Dataflow transformations in high-level DSP system design. In: Proceedings of the International Symposium on System-on-Chip, pp. 131–136. Tampere, Finland (2006)
58. Shlien, S.: Guide to MPEG-1 audio standard. Broadcasting, IEEE Transactions on **40**(4), 206–218 (1994)
59. Shojaei, H., Ghamarian, A., Basten, T., Geilen, M., Stuijk, S., Hoes, R.: A parameterized compositional multi-dimensional multiple-choice knapsack heuristic for CMP run-time management. In: Design Automation Conf., DAC 09, Proc., pp. 917–922. ACM (2009)
60. Stefanov, T., Deprettere, E.: Deriving process networks from weakly dynamic applications in system-level design. In: Proc. IEEE-ACM-IFIP International Conference on Hardware/-Software Codesign and System Synthesis (CODES+ISSS'03), pp. 90–96. Newport Beach, California, USA (2003)
61. Stefanov, T., Kienhuis, B., Deprettere, E.: Algorithmic transformation techniques for efficient exploration of alternative application instances. In: Proc. 10th Int. Symposium on Hardware/-Software Codesign (CODES'02), pp. 7–12. Estes Park CO, USA (2002)
62. Stuijk, S., Geilen, M., Basten, T.: SDF^3: SDF For Free. In: Application of Concurrency to System Design, 6th International Conference, ACSD 2006, Proceedings, pp. 276–278. IEEE Computer Society Press, Los Alamitos, CA, USA (2006). DOI 10.1109/ACSD.2006.23. URL http://www.es.ele.tue.nl/sdf3

63. Stuijk, S., Geilen, M., Basten, T.: Throughput-buffering trade-off exploration for cyclo-static and synchronous dataflow graphs. IEEE Trans. on Computers **57**(10), 1331–1345 (2008)
64. Stuijk, S., Geilen, M., Basten, T.: A predictable multiprocessor design flow for streaming applications with dynamic behaviour. In: Proceedings of the Conference on Digital System Design, DSD '10, pp. 548–555. IEEE (2010). DOI 10.1109/DSD.2010.31
65. Stuijk, S., Geilen, M.C.W., Theelen, B.D., Basten, T.: Scenario-aware dataflow: Modeling, analysis and implementation of dynamic applications. In: ICSAMOS, pp. 404–411 (2011)
66. Theelen, B.D., Geilen, M.C.W., Stuijk, S., Gheorghita, S.V., Basten, T., Voeten, J.P.M., Ghamarian, A.: Scenario-aware dataflow. Tech. Rep. ESR-2008-08, Eindhoven University of Technology (2008)
67. Theelen, B.D., Geilen, M.C.W., Voeten, J.P.M.: Performance model checking scenario-aware dataflow. In: Proceedings of the 9th international conference on Formal modeling and analysis of timed systems, FORMATS'11, pp. 43–59. Springer-Verlag, Berlin, Heidelberg (2011)
68. Theelen, B.D.: A performance analysis tool for scenario-aware streaming applications. In: QEST, pp. 269–270 (2007)
69. Theelen, B.D., Florescu, O., Geilen, M.C.W., Huang, J., van der Putten, P.H.A., Voeten, J.P.M.: Software/hardware engineering with the parallel object-oriented specification language. In: Proceedings of the 5th IEEE/ACM International Conference on Formal Methods and Models for Codesign, MEMOCODE '07, pp. 139–148. IEEE Computer Society, Washington, DC, USA (2007)
70. Theelen, B.D., Geilen, M.C.W., Basten, T., Voeten, J.P.M., Gheorghita, S.V., Stuijk, S.: A scenario-aware data flow model for combined long-run average and worst-case performance analysis. In: Proceedings of MEMOCODE, pp 185194, pp. 185–194. IEEE Computer Society Press (2006)
71. Theelen, B.D., Katoen, J.P., Wu, H.: Model checking of scenario-aware dataflow with CADP. In: DATE, pp. 653–658 (2012)
72. Turjan, A., Kienhuis, B., Deprettere, E.: Realizations of the Extended Linearization Model. in Domain-Specific Embedded Multiprocessors (Chapter 9), Marcel Dekker, Inc. (2003)
73. Verdoolaege, S.: Polyhedral process networks. In: S.S. Bhattacharyya, E.F. Deprettere, R. Leupers, J. Takala (eds.) Handbook of Signal Processing Systems, second edn. Springer (2013)
74. Verdoolaege, S., Nikolov, H., Stefanov, T.: pn: a tool for improved derivation of process networks. EURASIP J. Embedded Syst. (2007)
75. Verdoolaege, S., Seghir, R., Beyls, K., Loechner, V., Bruynooghe, M.: Counting integer points in parametric polytopes using Barvinok's rational functions. Algorithmica (2007)
76. Wiggers, M.: Aperiodic multiprocessor scheduling. Ph.D. thesis, University of Twente (2009)
77. Willink, E.D., Eker, J., Janneck, J.W.: Programming specifications in CAL. In: Proceedings of the OOPSLA Workshop on Generative Techniques in the context of Model Driven Architecture (2002)
78. Zhai, J.T., Nikolov, H., Stefanov, T.: Modeling adaptive streaming applications with parameterized polyhedral process networks. In: Proceedings of the 48th Design Automation Conference, DAC '11, pp. 116–121. ACM, New York, NY, USA (2011)

DSP Instruction Set Simulation

Florian Brandner, Nigel Horspool, and Andreas Krall

Abstract An instruction set simulator is an important tool for system architects and for software developers. However, when implementing a simulator, there are many choices which can be made and that have an effect on the speed and the accuracy of the simulation. They are especially relevant to DSP simulation. This chapter explains the different strategies for implementing a simulator.

1 Introduction

The *instruction set architecture* (ISA) of a computer comprises a complete definition of the instructions, the way those instruction are represented by bit patterns in the computer memory, and the semantics of each instruction when executed. An ISA is normally implemented by the physical hardware of the computer, using hardware resources such as the register file, the arithmetic and logic unit, the bus, and so on. However it is equally possible for the ISA to be implemented entirely in software. In this situation, we are using the computer which runs the software (the *host computer*) to emulate the ISA of a different computer (the *guest computer*). The software application which implements the ISA is commonly known as an *instruction set simulator* [66, Chap. 2]. It should be noted that some

F. Brandner (✉)
Department of Applied Mathematics and Computer Science,
Technical University of Denmark, Kongens Lyngby, Denmark
e-mail: flbr@imm.dtu.dk

N. Horspool
Department of Computer Science, University of Victoria, Victoria, BC, Canada
e-mail: nigelh@cs.uvic.ca

A. Krall
Institut für Computersprachen, Technische Universität Wien, Vienna, Austria
e-mail: andi@complang.tuwien.ac.at

S.S. Bhattacharyya et al. (eds.), *Handbook of Signal Processing Systems*,
DOI 10.1007/978-1-4614-6859-2_29, © Springer Science+Business Media, LLC 2013

implementations of instruction set simulators make use of extra hardware resources to boost their performance, and some are implemented entirely in hardware (in which case, the correct name for the technology is *instruction set emulation*).

Instruction set simulators have several important uses. The three main ones, and ones which will be briefly explained below, are:

- To implement a virtual machine.
- To emulate a computer architecture using a different architecture.
- To perform low-level monitoring of software execution.

A virtual machine is used for the purpose of providing a managed code environment. For example, Java source code is compiled to Java bytecode, which are instructions for the *Java virtual machine* (JVM). The JVM has an ISA which is perhaps too complex to be implemented fully in hardware but is easily implementable in software. (The Java code is *managed* in the sense that the bytecode instructions are verified by the JVM to ensure that unsafe usage cannot occur.) Note that this usage of the term "virtual machine" is distinct from the notion of virtual machine when used in the context of *virtualization*. The purpose of virtualization is to allow multiple operating systems to share a single hardware platform through the use of virtual hardware. The instruction sets of the virtual machines would, in the context of virtualization, be identical to the instruction set of the hardware platform and there would thus be no place for instruction set simulation.

The second use is where one computer emulates another computer with a different ISA. It is particularly useful when software applications have been developed for one architecture and need to be executed on another architecture. This, for example, was the technique by which Apple eased the transition for its users when the architecture of the Macintosh computers was changed from the Motorola 68000-series to the PowerPC and then, again, to the Intel x86. Having one computer emulate another computer architecture is also useful when a new computer is under development and not yet available to the software developers.

The third use arises because a software implementation of an ISA can easily be augmented with additional software which makes various performance measurements. It can measure quantities such as the number of instructions executed, the numbers of conditional branches taken and not taken, the number of stalls in a simulated pipeline, the hit rate of a simulated cache, the number of simulated clock cycles per instruction, and so on. Such measurements can be very useful to a system designer who has to choose a system configuration, such as choosing between a unified L1 cache or a split instruction cache and data cache, and choosing the sizes of the caches and the number of cache levels, etc. These measurements can also be useful to a software developer who is fine-tuning code to make it execute as efficiently as possible. In such a scenario, it makes perfect sense to execute an x86 simulator using an x86 platform, and similarly for other platforms.

It is the second and third uses which are the most relevant to DSP instruction set simulation. Using a simulator which provides detailed performance measurements, a systems engineer can configure the system to optimize the trade-offs between factors such as cost, power consumption, and speed. Meanwhile, a software developer can use the simulator to begin testing and debugging code written for the DSP system before that system has been deployed.

There are two levels of implementation for an instruction set simulator: *full-system level* and *CPU-level* [74].

- Full-system simulation is the most complete and most detailed level: the whole computer system including the I/O bus and the peripherals are simulated in software. Thus, for example, if the code being simulated generates a simulated interrupt then the simulator proceeds by *executing* the instructions of the interrupt service routine, and so on. Similarly, instructions which generate system calls would cause the simulation to continue by a change in execution mode and subsequent execution of code that belongs to an operating system. This would be the level of implementation needed by a system designer or by a software developer producing kernel-level code for an operating system.
- CPU-level emulation, in contrast, requires simulating only the CPU as it executes a single application program. All system calls, e.g. for performing an I/O operation, would be mapped by the simulator into equivalent calls to services provided by the host operating system. There would be no attempt to continue the execution of instructions into operating system code. This would be the level of simulation appropriate for implementing a virtual machine, or for allowing a software developer to produce application programs for the emulated architecture.

There are two main approaches for implementing an ISA simulator in software. One approach is to use an interpreter, and the other is to use code translation. The first choice has several variations, as does the second. These choices and their variations, as well as the rôle of hardware to assist in the simulation, are explained in more detail in the remainder of this chapter. A final possibility of using special-purpose hardware to emulate the ISA will not be covered here.

2 Interpretive Simulation

The oldest and simplest software implementation of an ISA simulator is an instruction interpreter modeled after the *fetch-decode-execute* cycle of a von Neumann style computer. The straightforward implementation of an ISA interpreter [37] is sometimes called a "classic interpreter" and is explained below. However, there are some obvious inefficiencies which affect the speed of a classic interpreter. Improvements to the interpreter structure which address those inefficiencies give rise to some variations which are also described below.

```
// load instructions and data into MEM array
PC = start_address ;
while (isRunning) {
    instr = MEM[PC/4];   // fetch the next instruction
    PC += 4;             // increment PC by 4 bytes
    opcode = decode(instr);    // extract the opcode
    switch( opcode ) {
        ...
        case ADDI:   // add-immediate instruction
            C  = instr & 0xFFFF;
            Rs = (instr >> 21) & 0x1F;
            Rt = (instr >> 16) & 0x1F;
            sum = IntegerAdd(REG[Rs], C);
            if (Rt != 0) REG[Rt] = sum;
            break;
        ...
        case SW:  // store word instruction
            C  = instr & 0xFFFF;
            Rx = (instr >> 21) & 0x1F;
            Ry = (instr >> 16) & 0x1F;
            MEM[(REG[Ry]+C)/4] = REG[Rx];
            break;
        ...
    }
}
```

Fig. 1 Structure of an ISA interpreter for the MIPS architecture

2.1 The Classic Interpreter

The interpreter has an array MEM which models the RAM memory of the architecture being simulated, and it has a variable PC which models the program counter register of that architecture. If the computer has a set of numbered registers, an array REG can be used to model those registers. A small portion of possible C code for an interpreter of the MIPS instruction set is shown in Fig. 1. It uses an array REG to mimic a register file, it has a variable which holds the current state of the condition code register, and its MEM array holds an exact replica of what would be in the main memory of the simulated computer.

Two of the simpler instructions have been picked for the example code; even so, there are some hidden complexities. For example, overflow by the add-immediate instruction must be detected and must be handled in the same way as the simulated architecture. This tricky detail is assumed to be implemented by the IntegerAdd function.

The interpreter code can exactly mimic the hardware resources of the computer being simulated. If caches, or other components of the system, need to be simulated, the interpreter code can be augmented in a straightforward manner. An important point is that additional code to monitor execution performance can easily be added too.

However, there is a significant performance penalty when a simple interpreter is used. If we compare the speed of the actual computer against the speed of an interpreter simulating that computer when executed on that same computer, there is typically a slowdown of a few orders of magnitude. The speed which can be achieved depends very much on the complexity of the ISA. Much effort has been put into making interpreters work faster, but the modifications made to the interpreter scheme can interfere with the correctness or the completeness of the emulation. These problems will be discussed below.

It should also be observed that a significant mismatch between the characteristics of the guest system and the host system can incur a big performance penalty. For example, attempting to simulate a system with 40 bit words (which are quite common for fixed-point arithmetic DSPs) on a host computer which has 32 bit words can cause severe complications. Of course, performance penalties due to such mismatches will not be limited to just interpretive techniques; they are likely to be observed no matter what simulation strategy is employed.

Instead of directly interpreting the binary representation, interpretive simulators often store the program in an intermediate representation which is more efficient to handle. Such an example is the well known SPIM simulator for the MIPS processor [41]. Decoding of the instructions is only done once when the binary representation is converted into the intermediate representation. When a new processor is under development, the binary representation is usually specified as the last step. Therefore, instruction set simulators for design space exploration often use an intermediate representation for the simulated program. When the size and the encoding of the instructions is not yet determined, the instruction cache can only be approximated.

2.2 Threaded Code

The outermost control structure of a classical interpreter is a switch statement inside a loop. Although it is the code in each case of the switch statement which performs the useful work, there is significant overhead associated with transferring control from the end of one case to the start of the case needed to simulate the next instruction. This overhead can be greatly reduced if we pre-translate the code to be interpreted into a different format. One such format is *threaded code* [6].

For example, if the program being interpreted is an assembled version of the MIPS code shown in Fig. 2a, then the interpreter could begin by filling a data array with the information shown in Fig. 2b.

Now, this TCODE array can be used to implement direct transfers of control from the code for one MIPS instruction to the next MIPS instruction, and so on. The interpreter code of Fig. 1 simply needs to be modified to have the structure shown in Fig. 3.

Fig. 2 Threaded code
translation example.
(**a**) MIPS assembler code
(**b**) Threaded code array

```
a
@ MIPS Assembler

...
addi 1,2, 29
addi 1,1, 1
sw  1,24(17)
addi 17,17, 4
sw  1,24(17)
...
```

```
b
void *TCODE[] = {
    ...
    &ADDI,
    &ADDI,
    &SW,
    &ADDI,
    &SW,
    ...
};
```

```
        // load instructions and data into Mem array
        PC = start_address ;

        TPC = &TCODE[PC/4];
        goto *TPC++;  // jump to first instruction

            ...

ADDI:   // add-immediate instruction
        instr = MEM[PC/4];
        PC = PC + 4;
        C  = instr & 0xFFFF;
        Rs = (instr >> 21) & 0x1F;
        Rt = (instr >> 16) & 0x1F;
        sum = IntegerAdd(REG[Rs], C);
        if (Rt != 0) REG[Rt] = sum;
        goto *TPC++;  // transfer to next instruction

SW:     // store word instruction
        instr = MEM[PC/4];
        PC = PC + 4;
        C  = instr & 0xFFFF;
        Rx = (instr >> 21) & 0x1F;
        Ry = (instr >> 16) & 0x1F;
        MEM[(REG[Ry]+C)/4] = REG[Rx];
        goto *TPC++;  // transfer to next instruction

            ...
```

Fig. 3 Threaded code interpreter for the MIPS architecture

Threaded code is an excellent implementation strategy for a virtual machine,
such as the JVM. In this situation, instructions can be identified in advance and
translated to entries in the TCODE array. The instruction operands can also be
interspersed between the label addresses, so that all information needed by the code
which implements an instruction can be extracted from the TCODE array. The PC
variable is then redundant and can be eliminated—it is replaced by the TPC variable.
In effect, the virtual machine code has been translated into a less compact format
which is much more amenable to execution by an interpreter.

For a real machine, one usually cannot perform such a simple translation. In particular, it may be impossible to identify which locations in the guest computer's memory hold instructions and which hold data. Even if the guest computer's address space has been divided up into a text region (which holds instructions) and a data region, it is still quite possible that the text region holds some data (especially read-only data). If the guest computer code is *self-modifying code*, there is a bigger problem because the TCODE array may lose its correspondence with the instructions in the guest computer's memory.

2.3   Interpreter Optimizations

Particular combinations of instructions tend to recur in assembler code. An interpreter can take advantage of this by creating new instructions which combine frequently occurring n-tuples of instructions into single *superinstructions* which are also known as *superoperators* [53]. If the n-tuples have a high static frequency of occurrence, the memory footprint of the program is reduced. If they have a high dynamic frequency, the interpretation speed is increased. Ideally, we can find n-tuples which have both high static and dynamic frequencies.

The improved interpretation speed is obtained because (a) there are fewer fetch-decode-execute cycles being performed, and (b) the combined actions of the opcodes involved in the n-tuple can be simplified. For example, if the two MIPS instructions:

```
addi    $3,$2,4
add     $3,$3,$3
```

are combined into a single instruction NewOp1 $3,$2,4 then the intermediate update of register 3 can be eliminated as unnecessary.

If a threaded code interpreter is in use, there is effectively no limit to the number of superoperator combinations which can be supported. There is a trade-off between the memory footprint of the interpreter and interpretation speed.

3   Compiled Simulation

Interpretive simulation is quite slow. Unnecessary operations like instruction decoding and all details of instruction fetch are repeated again and again. These details can be optimized away with compiled simulation. The idea of compiled simulation is to take an application program (usually represented in assembly language source form or as a binary file) and generate a specialized simulator which executes just this one application. This technique is due to Mills et al. [45].

Fig. 4 Simulator generation
for compiled simulation

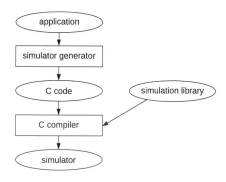

3.1 Basic Principles

All compiled simulators are based on the following principle; see Fig. 4 for an overview. The binary or the assembly language source of the application to be simulated is read by the simulator generator. The generator then translates the application to a high-level language, such as C or C++. The generated source code is translated to machine code by an optimizing high-level language compiler of the host computer and linked against a generic simulator library. The resulting program is a simulator, which emulates the behavior of the original application when executed. In addition to the optimizations available to the host's high-level language compiler, optimizations might be applicable during the source code generation by the simulator generator, e.g., the generator may monitor execution modes of the processor and thus omit certain simulation overhead.

Compiled simulators are distinguished by the way in which sequences of instructions are translated to high-level language snippets. Mills [45] translates the whole application available in assembly language source to a single function of the C programming language. Each assembly language instruction is represented by a C macro which emulates the behavior of the particular instruction. The macro uses the original address of the instruction as a label of a C `switch` statement and is followed by C statements which emulate the instruction. The macros

```
#define SUB(addr)      case (addr): AC -= *SP++;
#define BNE(addr, to)  case (addr): if (AC) {PC = (to); break;}
```

define subtract and branch instructions for an accumulator architecture. The variable PC corresponds to the program counter of the emulated architecture. The macros are used in a giant C `switch` statement which represents the whole instruction memory of the emulated architecture (see Fig. 5). The program counter has to be set and evaluated in the switch statement only when a branch instruction is emulated. The macros read like the original program.

Mills' method works very well for medium sized programs. The whole program is translated to a single large C function. The advantage is that this function can be compiled to very efficient machine code. Local variables of the function will

Fig. 5 Mills represents instructions as C macros within a `switch` statement, where the labels correspond to the instruction addresses

```
PC = 0x1000;
while (isRunning) {
    switch(PC) {
        SUB(0x1000)
        BNE(0x1002, 0x1000)
        ...
    }
}
```

Fig. 6 Large simulation functions are avoided by partitioning of the address space of the simulated program

```
PC = 0x1000;
while (isRunning) {
    switch(PC>>10) {
        case 0: iblock0(); break;
        case 1: iblock1(); break;
        case 2: iblock2(); break;
        ...
    }
}

void iblock0() {
    while (isRunning) {
        switch(PC) {
            SUB(0x1000)
            BNE(0x1002, 0x1000)
            ...
            default: return;
        }
    }
}
```

be mapped to registers of the host architecture and local optimizations are applied. This advantage is lost, however, when the program becomes too large. The size of the function may exceed limits imposed by the compiler, or non-linear algorithms employed by the compiler may lead to drastically increased translation times. In either case, the approach of generating a single C function becomes impractical. Large programs are thus partitioned into aligned blocks of instructions with a size that is a power of two [5], where each block is assigned a dedicated simulation function. Due to the use of multiple simulation functions, most of the processor's state has to be stored in global variables, and this interferes with optimizations that would otherwise be performed by the C compiler. Figure 6 shows the generated C code for a partitioned compiled simulator.

For large programs it is advantageous to translate *basic blocks*. These are linear sequences of instructions with at most one branch at the end. Each basic block is translated to a C function. During simulation, each C function returns the address of the next function to be executed. The sequence of function calls is managed by a generic simulation loop. This approach works particularly well for functional

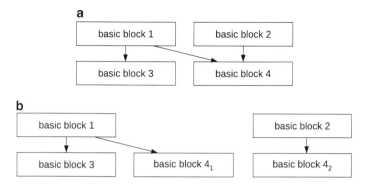

Fig. 7 Simulation of processors with long pipelines across basic block boundaries using block duplication. (**a**) Original control flow, (**b**) Control flow with duplicated blocks

simulators and architectures where the simulation of instructions does not cross the boundaries of the simulation functions.

Compiled simulation is generally very similar to *reverse compilation*. However, reverse compilation becomes more difficult when the architecture supports delay slots for branch and memory access instructions. An example is the C6x series of DSP processors from Texas Instruments. Bermudo et al. [7] show that control flow graph reconstruction can be simplified by duplication of basic blocks.

3.2 Simulation of Pipelined Processors

The problem with compiled simulation for pipelined processors is that the execution of instructions is partitioned into subtasks which are executed during different cycles. Each of these subtasks must be scheduled statically to give correct behavior and high potential for optimization. This is simple when the processor is an in-order architecture and the instructions do not cross basic block boundaries. When instructions cross basic block boundaries duplication of basic blocks is necessary.

Take, for example, the basic blocks of the control flow in Fig. 7a. Basic block 3 has only basic block 1 as predecessor. Therefore, instructions started in basic block 1 can be extended to basic block 3. Basic block 4 has to be duplicated, as depicted by Fig. 7b, because instructions started in basic blocks 1 and 2 can both be continued in basic block 4.

Determination of basic blocks can be impossible when the program contains computed indirect branches. The problem lies in determining all possible targets for the indirect branch, and this is in general an incomputable problem [34]. To handle the situation, it is necessary to combine the compiled simulator with an interpretive simulator. Whenever a branch target is computed which was not known to be the start of a basic block, the simulator switches into interpretive mode.

Interpretation continues until the end of the basic block is reached. The interpretive mode can also be used to support debugging features, such as single stepping.

3.3 Static Binary Translation

Static binary translation is similar to compiled simulation and it was used earlier than compiled simulation [44, 65]. The only difference is that instead of generating a simulator in a high-level language, machine code for the host architecture is generated. The advantage is that no code size restrictions or constraints on the organization of the machine code exist. The removal of these constraints usually leads to faster simulation. Additionally the translation time is reduced, since generation and parsing of the high-level language version of the simulator is eliminated. A disadvantage is that all optimizations, which are usually done by the high-level language compiler, have to be carried out by the binary translator. The use of a compiler framework greatly reduces the development time of a simulator based on static binary translation.

4 Dynamically Compiled Simulation

In the case of instruction set simulation based on dynamic binary translation, code generation is performed on-the-fly at simulation time. This approach was first used for execution profiling in the Shade system [18]. The advantage is that more information, such as the targets of indirect branches, is available and that it is possible to handle self-modifying code. A disadvantage is that the translation is repeated for every simulation run, and the overall simulation time has to include that translation time.

4.1 Basic Simulation Principles

To keep the translation overhead small, simulation usually starts in interpretive mode. During interpretation, information on basic block execution frequencies is collected. When a predetermined execution threshold for a basic block is exceeded, this basic block is translated to machine code. Some systems use *traces* as the basic unit of compilation. A trace is a cycle-free linear sequence of instructions possibly containing multiple branches into and out of the trace.

The translated code blocks or translated traces are typically kept in a *translation cache*. If, during the simulation, a matching code fragment is found in the translation cache, the simulator jumps to the cached machine code and executes it, otherwise simulation continues using interpretation. When the end of the translated code is reached, control is returned to the simulator to resume the interpretive mode of execution. To avoid unnecessary switches back into interpretive mode, branches in already translated code are updated whenever a new block is translated. In the case of self-modifying code or dynamic code loading/generation within the simulated program, the translation cache has to be invalidated accordingly. Detecting memory updates that overlap with translated code regions may cause significant overhead.

4.2 Optimizations

Since compilation time is included in the simulation time, only a few important optimizations are performed during dynamic code generation. The most important is register allocation. Registers and other hardware resources of the simulated architecture are usually stored in memory. At the beginning of a basic block or a trace the state of the simulated processor is partially loaded into registers. Similarly, modified values are saved upon leaving the compiled code. The register allocation strategy typically follows some simple heuristics. For example, some registers of the simulated processor may be mapped to registers of the host computer without performing any analysis of the code in order to reduce compilation overhead.

Processor simulation often involves the update of status registers or other information which are later overwritten by subsequent simulated instructions without any intervening uses. Liveness analysis can detect whether such computed values are actually used by a subsequent instruction. For basic blocks and traces, this analysis is very fast. The gathered information may be used to avoid useless computations and should lead to significant performance improvements.

Better code can be generated when regions of code which contain loops are optimized and translated to machine code. However, due to the more complex control flow, optimizing these regions requires more time. Therefore, only the most frequently executed code regions are considered for this translation method. For example, a mixed-mode simulator can be used with three different levels of optimization. Seldomly executed code is interpreted, moderately executed code is optimized at a basic block level, and very frequently executed code is optimized at a region level [10].

5 Parallel Simulation

The advent of modern multi-core processors with 8, 16 or even more computing cores poses several interesting questions as to how an instruction set should be simulated. One question concerns how the increased computing power of additional cores should be exploited to speed up the time-consuming simulation. In addition, DSP cores are more and more integrated with other computing resources, forming complex multi-core architectures or even *Multi-Processor System-On-Chips* (MP-

SoC). Accurately and efficiently capturing the parallel processing of these devices during simulation is another major challenge—in particular, as the number of the involved computing elements grows. In the following sections, we point to some recent developments in modeling and exploiting parallel computers for instruction set simulation.

5.1 Parallelizing Simulation

The strategies to parallelize and distribute the work of simulating a single DSP core on several processors strongly depend on the simulation technique being used. A common problem, however, is the need to synchronize the parallel simulation steps and guarantee a consistent global state. This synchronization may impose a severe penalty in terms of execution time. The level of granularity at which the simulation is parallelized thus becomes crucial. For instance, off-loading simulation tasks to other processor cores or accelerators, such as the graphics card (GPGPU) or an FPGA, inevitably leads to additional communication delays induced by data transfers between the main processor and the accelerator. The solution here is to split the simulation tasks in such a way that the amount of communication and synchronization is minimized. For example, a good design clusters interdependent simulation tasks in order to reduce the number of the synchronization steps. A good strategy, for instance, is to keep interdependent components of the simulator *close* to each other, i.e., off-load them to the same accelerator or place them on the same processor core. Another strategy is to partition the simulation along simulation functions, for example, by splitting the simulator into a component that focuses on functional simulation only and a second component that provides accurate information on timing and the state of the simulated hardware.

Performing bookkeeping tasks separately from the actual simulation is another source of parallelism. For instance, interactions with external systems and the end-user as well as collecting and processing statistics can be done in parallel to the actual simulation run. In addition, pre-processing steps can be performed in parallel to speed-up the execution by pre-decoding instructions within an interpreter or by pre-generating code during dynamic binary translation. The advantage here is that the actual simulation is freed from performing those bookkeeping tasks and thus may proceed faster. Care has to be taken, though, that the simulation is not slowed down by synchronization primitives protecting shared data structures between the main simulation thread and the other threads. Also the extent of the pre-processing and the number of threads reserved for performing these bookkeeping tasks has to be chosen carefully in order to avoid excessive memory consumption and contention of computational resources.

5.2   Simulation of Parallelism

Future computing platforms are expected to provide an abundance of computing cores that are expected to be integrated on a single chip, sharing on-chip inter-connection networks, buses, I/O interfaces, and memory resources. Large scale simulation of all these computing cores is a challenging problem that requires highly efficient simulators that have to exploit parallelism at all levels in order to keep simulation times to an acceptable level.

A first insight when designing a highly parallel simulator is that most computing cores in the system actually operate independently of each other *most* of the time. It thus does not appear to be necessary to keep the simulation of all these cores fully synchronized at all times. Synchronization is only needed when the simulated application reaches a synchronization point on its own, e.g., it acquires a lock, performs an atomic memory update, or executes some other form of explicit synchronization primitive. If the simulator is able to recognize all of the synchronization primitives possibly used by the application and if the application correctly uses those primitives, it is guaranteed that the simulation will yield correct results without any additional synchronization between the concurrently operating simulation threads.

This approach basically trades accuracy for simulation speed, since simulation of the interactions among the independently operating computing cores is kept to a minimum. Depending on the accuracy requirements, this approach might be an acceptable trade-off. A more serious problem arises when the functional correctness of the simulation depends on the interleaving of the execution of two or more computing cores in the simulated system. For instance, this might arise when memory space is shared without the use of explicit synchronization primitives—note that this might be intentional, e.g., an optimization, or it might be unintentional, e.g., a data race. The question, whether the simulation is guaranteed to be correct or not, depends on the memory model of the simulated multi-core platform, i.e., a set of rules that define if and how a consistent global view of shared memory resources is achieved. Most modern multi-core systems assume *weak* memory models that usually only guarantee that the system will eventually reach a consistent view but does not provide any guarantees on when and how this state is reached. Stronger memory models usually define an order on events in the multi-core system and a means to enforce that all cores in the system observe the same order. The strongest form of guarantee is given by a model ensuring *sequential consistency*, i.e., that a strict linear ordering of all events in the system exists. Clearly, strong memory models demand additional synchronization to ensure that the behavior of the modeled system is simulated correctly, while weaker models generally do not require such synchronizations. Note that similar guarantees also have to be defined and enforced for other forms of shared resources. An important detail that is often overlooked is that, whatever the memory model of the simulated platform, the host platform running the simulation needs to be accounted for when discussing consistency guarantees. Implementing a stronger model on top of a platform providing only weak guarantees will require a careful design of the simulator which inevitably causes additional overhead, while the converse situation will cause much less of a headache.

6 Hardware Supported Simulation

Although today's workstation computers provide tremendous performance, the speed of an instruction set simulator may be slow and often unsatisfactory. It is also likely that the simulation overhead for complex *Systems-On-Chip* (SoC) and multi-core devices will further increase in the future, and thus widen the gap between simulation time and the actual execution time on a real system. Even though instruction set simulation inherently offers a high degree of parallelism, it cannot be fully exploited due to communication and synchronization overhead. Dedicated hardware provides a promising approach to overcome this difficulty by leveraging fine-grained parallelism. This section is primarily focussed on simulation platforms that use *field programmable gate arrays* (FPGAs) in order to connect the simulator to external hardware, or to perform simulation tasks within the FPGA itself. Other approaches that rely on custom ASIC designs [20, 22, 32] also show impressive results, but appear to be less applicable than general solutions.

6.1 Interfacing Simulators with Hardware

In the simplest case, instruction set simulation is performed entirely in software by a workstation computer, while other devices are realized using FPGAs or real hardware as depicted in Fig. 8. This approach is particularly suited to system

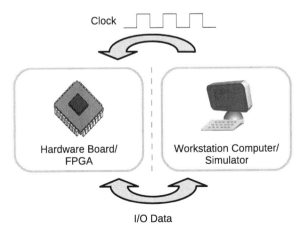

Fig. 8 External hardware can interface with an instruction set simulator in software by driving the hardware clock from within the simulator. The transmission of I/O data is effectively synchronized and can be controlled entirely in software

integration and verification, but usually offers little or no improvement in terms of simulation speed.

Keeping the hardware and software models synchronized is a major problem in such systems. There are problems both with simulation performance and with correctness and accuracy of the performance measurements. A typical solution is to control the hardware clock from within the software simulation [48, 64], e.g., by toggling the value of an I/O pin. This allows the models to be kept in perfect synchronization on a cycle-per-cycle basis. However, depending on the interconnect, accessing the I/O pin might be costly. For example, initiating a bus transfer on every simulated cycle is likely to dominate the overall execution time on the simulator's host machine, as bus interfaces usually operate only at a fraction of the clock frequency of modern microprocessors. Transaction-based synchronization [39, 63] can be used to reduce the number of synchronization operations, but also impacts the accuracy of the simulation results.

6.2 Simulation in Hardware

Integrating the simulation tools with external devices that are implemented in hardware is a valuable approach, but does not improve the performance of the instruction set simulation itself. However, since a tight coupling between the software model and the external hardware has already been established it is now possible to implement parts of the instruction set simulator within the hardware [67], for example by using an FPGA. This approach is especially interesting for architectures that are highly parallel and contain features that are hard to handle using software. Digital signal processors typically fall into this category, and may thus draw significant benefit from hardware supported simulation (Fig. 9).

In contrast to *prototyping* [19], the primary goal is not to implement the architecture design using an FPGA, but to implement a hardware structure that models the behavior at a certain level of granularity, e.g., on the cycle- or transaction-level. For example, executing a single cycle of the modeled architecture might be realized using several cycles in the FPGA [28]. The number of cycles in the FPGA might also vary depending on the complexity of the currently active instructions. This also implies that signals, registers, functional units, and other hardware components of the original architecture are not necessarily represented in the simulation model as such. The designer can thus choose to implement the simulation model using software techniques, or a basic hardware implementation, or by use of highly optimized hardware structures in order to achieve a suitable trade-off between simulation speed, design complexity, and flexibility.

The observation that functional simulation usually can be performed efficiently on modern workstation computers, while the modeling of dynamic behavior and timing is usually overly complex and slow, led to the approach adopted in several simulation systems [13, 28, 52]. The simulation is split into two separate parts that

Execution Trace

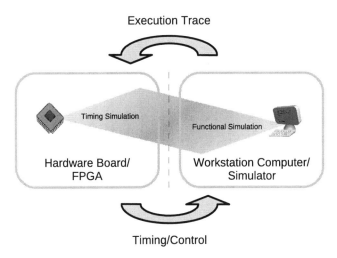

Timing/Control

Fig. 9 Simulation can be performed in part using software techniques and hardware components. An efficient partitioning of such a simulation system is to perform the complex timing emulation within an FPGA, while a purely functional simulation is executed on the host computer

operate in parallel: the functional model and the timing model. In the case of *trace-driven* simulation, sometimes also referred to as *functional-first* simulation, the simulation process is controlled by the functional model. On the other hand, in *timing-directed* simulation, also called *timing-first* simulation, the timing model controls and drives progress. In the former case, the functional model executes the program without considering dynamic effects, such as variable latency instructions, interlocking, branch prediction, or cache and memory delays. If, at some point, the timing model detects a mismatch between the instruction stream executed by the functional model and the instructions executed on a real system, e.g., due to a mispredicted branch, the simulation has to be reverted and restarted in order to correct the mismatch. In trace-driven systems, the timing model has control and triggers the execution of the functional model at the appropriate times. Both approaches share the advantage that the functional model can be implemented using software techniques, and are thus more flexible and easier to extend than hardware models. However, there is no reason why the functional part could not execute within the FPGA [52]. In addition, reuse is enabled by combining different timing models with the same functional model [52].

Another approach is to divide the simulation tasks between hardware and software depending on execution frequency and implementation complexity. Certain simulation tasks are selectively promoted to hardware if they dominate the performance of the system and if their implementation in hardware is cheap in terms of implementation effort and hardware resources.

7   Generation of Instruction Set Simulators

Designing and developing an instruction set simulator for a new architecture or
processor design is a cumbersome and error-prone task. Especially during the
early stages of the design process, the instruction set and its implementation
may change frequently and thus complicate the task of simulator development.
Processor description languages [35, 46] provide a promising approach in which
the specification of the instruction set and implementation details of a new design
are specified using a formal language. These languages provide helpful abstractions,
leading to processor models that are easy to maintain and extend, and thus reduce
development time and cost. Furthermore, these languages allow for *design space
exploration*, a technique for exploring different processor designs in order to find
the best solution for a particular task. The resulting *application-specific instruction
processors* (ASIP) combine superior performance and reduced power consumption
while still providing a certain degree of flexibility in comparison to pure hardware
solutions.

7.1   *Processor Description Languages*

The development of a specialized DSP architecture is typically constrained by
development cost, time to market, and the production cost per unit. Efficient and
accurate simulation tools play a central role in exploring and evaluating design
alternatives. In addition, these tools allow software development to start during the
early stages of the design process. This not only leads to a shorter development
cycle, but also lets developers collect data and perform various measurements
that are crucial for tuning the design to the requirements of the application
domain in question. However, developing a new instruction set simulator, or even
customizing existing tools, is an error-prone and time-consuming task. Extensible
and maintainable models of the processor design are thus desperately needed.

Processor description languages and related tools provide a promising solution
to this problem. In recent years, these languages have evolved from being mere
research projects into products that are actively used and successfully adopted by
leading companies in order to develop and design highly specialized and tuned
architectures [30]. A processor model typically consists of several layers that
include information on the hardware organization, the instruction set, instruction
semantics and timing. Meta information such as assembly syntax, *application
binary interface* (ABI) conventions, etc., can be specified in most languages.
Structuring the models, for example by instruction classes, enables code reuse
across different instructions and leads to concise and compact models. In this way,
adding new instructions or adapting existing instructions is often only a matter of
a few lines of code. Processor description languages, sometimes also referred to as

Architecture Description Languages, can roughly be categorized into three distinct groups:

1. *Structural* languages offer primitives that directly match abstractions typically found in hardware design. The processor model thus closely resembles the structure of the original hardware design.
2. *Behavioral* languages on the other hand primarily focus on the instruction set architecture and typically provide some means to structure and order instruction variants and meta information such as assembly syntax, binary encoding, and abstract instruction semantics.
3. *Mixed* approaches combine the structural and behavioral view of the processor and provide mechanisms to model both the hardware structure and the instruction set architecture. Usually these languages also provide a mechanism to specify a mapping between the behavioral and the structural description that allows one to relate the two models to each other.

The information that is available in these processor specifications can be used to (semi-)automatically generate software tools that are customized for the particular processor. This includes standard development tools such as a compiler, an assembler, and a linker. However it also includes instruction set simulators and hardware models in a general purpose *hardware description language* (HDL), such as VHDL or Verilog. The various flavors of languages are usually geared towards a particular application. Behavioral approaches, for example, typically support the generation of compiler back-ends and related development tools, such as assemblers and linkers, very well. However they lack information needed for accurate simulation and automatic generation of hardware models. Structural languages are usually well suited for these *low-level* tasks, but in turn lack abstract semantic models of instructions. Generating a compiler is thus more complicated in these systems. To overcome such limitations, many systems have progressively adopted features and ideas from the other style. Most contemporary languages thus follow the mixed approach that supports both application fields equally well.

Figure 10 depicts an example workflow for the design of a new architecture using a processor description language. The designers iteratively extend and adapt the architecture model in order to reach the best possible solution. During each iteration step, a series of simulation runs is performed to evaluate the modification and expose bottlenecks that have to be dealt with during the next iteration. This process is often referred to as *design space exploration* (DSE). Although DSE can be applied in a more general setting even without the use of processor description languages, these languages effectively reduce the time spent on exploring and evaluating design alternatives.

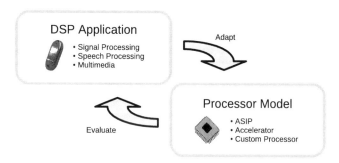

Fig. 10 During design space exploration the processor model, specified using a processor description language, is iteratively adopted in order to achieve the best possible performance, power, and area trade-off for a given DSP application

7.2 Retargetable Simulation

A key component that comes with virtually every processor description language is a simulation framework that can be customized and adapted quickly and easily to a new architecture model. This is usually achieved by generating the processor specific portions of the simulator, e.g., a software instruction decoder and appropriate simulation functions that model the behavior of individual instruction. The simulation techniques employed by these *retargetable* simulation engines do not differ from those found in hand-coded simulators, and range from simple functional interpreters and highly sophisticated binary translators to highly precise event-driven simulators. However, these engines potentially have to cope with a large range of different architectures, making optimizations found in hand-written simulators impossible in these engines for the sake of generality. It is thus not uncommon for retargetable simulation engines to restrict the range of supported architectures to certain architecture styles. In-order pipelined architectures, for example, are typically well supported, while less attention is typically paid to dynamic features of modern superscalar processors. It is important to note that the execution model of the processor description language and the techniques used to implement the execution model within the retargetable simulation framework may impact the accuracy of the simulation results. Designers thus need to evaluate whether the processor description language and its accompanying tools are suited for the particular task at hand.

As with hand-coded frameworks, the simplest form of retargetable simulation is interpretation. The semantic information that is available in the processor models is grouped into simple simulation functions that are invoked from within the interpreter loop. Each function performs an atomic simulation step that is indivisible for external observers during simulation. In the case of functional simulation, a single function is generated that captures the behavior of the instruction completely. For transaction-based or even cycle-accurate simulation, multiple simulation functions are emitted. The simple interpreter loop is usually also extended using state machines that control which functions need to be invoked. This is particularly important for handling various hazards, interlocks, and stalls that may emerge during the simulation. These techniques allow a very detailed simulation of the dynamic behavior of a processor; the generated simulators thus achieve very accurate results. However, simulation speed is usually not satisfactory.

Retargetable simulators based on compiled simulation and dynamic binary translation can be used to circumvent the problem of unsatisfactory speed. In a similar manner to the interpreter approach, simulation functions are generated. However, these functions do not emulate the behavior of the instruction itself but instead generate code that is later executed. In the case of static compiled simulation, code fragments of a high-level language, such as C, are emitted by these functions that precisely emulate the instruction's behavior. In the case of dynamic compilation and binary translation, machine code fragments or fragments of an intermediate representation that can be compiled to machine code are emitted.

Integrating retargetable simulation engines with other simulation tools and external components is often a major problem. This is in part caused by the very nature of these engines. The generic architecture independent parts are incomplete without the generated portions of the simulator. Extending them is complex and often impossible without considering the processor model. The generated portions on the other hand can not be modified directly without losing the advantage of retargetability. Leveraging existing general-purpose modeling frameworks such as SystemC [68] is a viable option that allows interfacing with a large number of existing models that come with these frameworks. Some processor description languages take this idea even a step further and do not supply a specialized language to specify the instruction semantics. Instead, a general purpose modeling language can be used, allowing the processor model to directly interact with external components and models.

8   Related Work on Simulation Approaches

Interpretive Simulation

Klint describes the classic implementation of an ISA interpreter [37]. In 1990 James Larus made SPIM available [41]. SPIM is an interpretive simulator for the MIPS architecture which is commonly used for teaching. SPIM takes assembly language programs as input which are stored in an intermediate representation. Magnusson shows how instruction cache simulation and execution profiling can be done efficiently with a threaded-code interpreter [42]. Processors with out-of-order execution usually are simulated using interpretive simulation since static modeling of the dynamic behavior of the processor is nearly impossible. A famous interpretive simulator for such kind of processors is SimpleScalar [3].

Compiled Simulation

Mills et al. were the first to use compiled simulation [45]. They presented a simple and efficient method to translate medium sized applications into a fast compiled simulator. Their technique inspired all the later work.

Reshadi et al. compile only the instruction decoding part to achieve the flexibility of an interpretive simulator while reaching a simulation speed which is close to compiled simulation [58].

SyntSim is a generator for functional simulators [11]. Using profile information, parts of the executable are compiled. The performance is between a factor of 2 and 16 slower than native code.

Errico et al. generate interpretive and both static and dynamic compiled simulators from a simulator specification [25]. The dynamic simulator generates C code at run time which is compiled by GCC and dynamically loaded. Translation of aligned code pages is done identically both for the static and the dynamic simulator.

Dynamically Compiled Simulation

SimOS [61] is a full-system simulator that offers various simulators including Embra [73], a fast dynamic translator. Embra focuses on the efficient simulation of the memory hierarchy, in particular the efficient simulation of the memory address translation and memory protection mechanisms, as well as caches. Embra follows a compile-only approach, i.e., simulated instructions are always translated to native code and then executed. The translated code fragments for basic blocks are stored in a translation cache to speed up the look-up of native code blocks during the simulation. The basic simulator can be extended and adapted using customized translations. For example, different cache configurations and coherency protocols are realized using these customized translations.

Shade [18] is another dynamically compiling simulator that aims primarily at fast execution tracing. It offers a rich interface to trace and process events during simulation. Similar to the customized translations in Embra, Shade allows user-defined as well as pre-defined code to collect trace data. Trace collection is controlled by *analyzers* that specify whether information should be considered during tracing on a per-opcode or per-instruction basis. The tracing level, i.e., the amount of data collected during simulation can be varied at runtime. In this way, only critical portions of the program execution need be executed with full tracing.

Ebcioğlu et al. present sophisticated code generation techniques to efficiently generate code for parallel processors at runtime [21, 23]. Several code generator optimizations are presented and combined with runtime statistics collection. For example, instructions of the target processor are initially interpreted, and compilation of *tree regions* is only triggered for hot paths of the simulated program. Instruction-level parallelism is further improved by aggressive instruction scheduling techniques based on these tree regions. The simulation of synchronous exceptions, i.e., irregular control flow due to page faults, division by zero, etc., is optimized by code motion techniques [31]. However, only the behavior of the target processor is modeled. The simulation neither includes a timing model nor a model of the target processor's memory organization.

Jones et al. use *large translation units* (LTUs) [36] to speed up the simulation of the EmCore processor, a processor which is compatible with the ARC architecture. Several basic blocks are translated at once within an LTU, leading to reduced compilation overhead and improved simulation speed. The simulator supports two operation modes: (1) a fast cycle-approximate functional simulation, and (2) a cycle-accurate timing model. Similar to typical static compiled simulation, code generation is performed using the C programming language in combination with a standard C compiler. Shared libraries are generated from the C source code and dynamically loaded at runtime. The generated code is stored permanently and can thus be reused across several simulation runs to improve the simulation speed further.

Parallel Simulation

Parallel Embra [40] combines binary translation with loose timing constraints. It relies on the underlying shared memory system for event ordering and synchronization when distributing the simulation up to 64 cores. Similarly Parallel Mambo [71] distributes the simulation to multiple cores in functional simulation. COREMU [72] uses a thin library layer to handle the communication between multiple instances of sequential QEMU based emulators with negligible uniprocessor emulation overhead.

Almer et al. [1] showed scalable multi-core simulation using parallel dynamic binary translation. They simulated up to 2,048 cores of the ARCompact instruction set architecture on a 32 core x86 host machine reaching simulation speeds of up to 25,307 MIPS. Raghav et al. [55] developed an interpreting simulator for thousand core architectures running on a general purpose graphic processor unit (GPGPU). For instruction decoding, they used look-up tables which were mapped to the texture memory of the GPGPU. The experimental evaluation was done for ARM and x86 multicores with 32 to 1,024 cores using an Nvidia GeForce GTX 295 graphics card. When simulating 1,024 cores executing the same application program, up to 1,000 MIPS could be emulated; when the application programs were different, the simulation speed dropped to 5 MIPS.

Böhm et al. [8] use decoupled dynamic binary translation to distribute the translation to multiple cores of the host machine. Over a large set of benchmarks they achieve 11.5% speedup on average on a quad core Intel Xeon processor. Qin et al. [54] translate frequently interpreted code pages to C++ and use the compiler of the host machine to generate a dynamically linked library. The translation process is distributed to multiple cores to speed up the simulation.

Retargetable Simulation

Several retargetable simulation frameworks are publicly available from open source projects and universities, others are offered by commercial software vendors. SimpleScalar is a retargetable simulator based on interpretation [3]. New processors are described using *DEF* files, which contain a set of C macro definitions and plain C code. A similar environment is offered by the commercial SimiCS simulator [43], but instead of plain C code a specialized device modeling language (DLM) is offered. The Liberty Simulation Environment (LSE) [70] supports very detailed event-based simulation. Hardware structures are described using the Liberty Structural Specification Language (LSS) [69]. Another approach is taken by the Unisim project,[1] it relies on the SystemC language and its tools [68] to provide an execution environment for the simulation of processors [2]. The project offers several components, such as memories, caches, buses, and even complete processors, with well defined interfaces that can quickly be integrated with other components.

The listed simulation environments offer very powerful support for implementing efficient and accurate simulation tools. However, the focus of these systems is targeted towards simulation only. For design space exploration of application-specific processors, an efficient simulation engine alone is not enough. Processor description languages usually support the automatic generation of all the tools needed for DSE. A book by Mishra and Dutt [46] provides an excellent introduction to this topic and covers most of the contemporary languages.

LISA by CoWare Inc. is one of the most successful processor description languages. The system provides powerful tools to automatically derive an efficient simulator based on compiled simulation [50] or interpretation from a processor specification. Nohl et al. propose the JIT-CC approach to speed up the simulation of LISA models [49] by caching information on previously decoded instructions. The LISA simulators can also be enhanced with hybrid simulation so that irrelevant parts of the program can be fast-forwarded using native execution [29]. Time-consuming accurate simulation is performed only where required.

A similarly mature processor modeling environment is offered by Target Compiler Technologies. Based on the *nML* processor description language [27], a complete tool chain for the modeled processor can be derived [30], including the retargetable simulator CHECKERS. Both the LISA and nML languages have successfully been used during the design of commercial processors in the digital signal processing domain, e.g., the CoolFlux DSP [60].

The *EXPRESSION* language [33] has been developed at the University of California, Irvine.[2] Efficient simulators based on compiled simulation [57, 58] and interpretive simulation can be derived from EXPRESSION models. The interpreter

---

[1] http://unisim.org/site/.

[2] http://www.ics.uci.edu/~express/.

is based on Reduced Colored Petri Nets (RCPNs) and supports efficient simulation of out-of-order execution [56]. The system also supports a hybrid form of static and dynamic compiled simulation [59] in order to reduce the compilation overhead at runtime.

Like the Unisim framework, the *ArchC* processor description language[3] is based on SystemC [4]. It follows the SystemC TLM standard (described in more detail in another chapter of this book). It is thus possible to integrate the derived simulators with external SystemC modules.

Farfeleder et al. present a very efficient compiled simulator based on a processor description language. Individual basic blocks of the simulated program are compiled to C functions [26, 38]. A technique called basic block duplication is used for efficient simulation of pipelined architectures.

A similar approach is taken in [10]. Instead of static compiled simulation via C code, machine code is generated directly using the LLVM compiler infrastructure.[4] In addition to simple basic blocks, hot *regions* of code are recognized and recompiled using more aggressive optimizations. These regions may contain arbitrary control flow, including loops. The system is based on a structural processor description language and supports the efficient simulation of external interrupts using a rollback mechanism [9].

Hardware Supported Simulation

IBM has pioneered the use of dedicated hardware resources to speed up binary translation. The *DAISY* processor [22] and, later, BOA [62] were optimized to support the efficient execution of PowerPC programs. The VLIW processors offered dedicated registers to hold the processor state of the simulated CPU, additional shadow registers to support rollbacks if speculative execution fails, and extensions to support the efficient processing of external interrupts. Transmeta [20] has adopted this approach to simulate X86 processors. Both approaches are strictly functional in that timing is not considered.

Schnerr et al. [63] use an off-the-shelf TI C6x VLIW processor to improve the simulation speed of embedded SoC systems. Cycle-accurate instruction set simulation is performed very efficiently using the VLIW processor. The simulation interfaces to the hardware components of the SoC using a synchronization protocol either on a per-cycle basis or based on transactions [64].

A similar synchronization scheme was proposed by Nakamura. A processor simulation in software is synchronized with hardware components in an FPGA using shared registers [48]. In addition, a processor emulation platform based on FPGAs is available [47].

---

[3]http://archc.sourceforge.net/.

[4]http://www.llvm.org/.

The *ReSim* simulation engine [28] is based on *traces*. Traces are collected using a simple simulator, e.g., a functional SimpleScalar model, and later processed by a dedicated timing simulator implemented in an FPGA. This splitting lowers the complexity of the FPGA implementation, improves speed, and can easily be replayed using different hardware configurations.

Splitting the simulation task into a functional and timing partition was first proposed by Chiou et al. for the *FAST* simulator [12–14]. In contrast to ReSim, the simulation is not based on traces, instead the functional simulator runs in parallel either on a host computer or a general purpose processor that is integrated with the FPGA. In both cases the functional and the timing partition are tightly coupled. For example, the timing partition informs the functional simulator in the case of mispredictions.

A different splitting of the simulation tasks is used by the *ProtoFlex* environment [15–17], which mainly targets the efficient simulation of multi-processor systems. Individual processor components and instructions are promoted to the more efficient FPGA emulation, depending on execution frequency and complexity trade-offs. The FPGA resources are shared among the individual emulated processors.

The *ASim* simulation framework [24] was also extended to support partial execution in FPGAs. *HASim* [51,52] supports implementation of individual modules of a processor using either software or via an FPGA.

References

1. Almer, O., Böhm, I., von Koch, T.J.K.E., Franke, B., Kyle, S.C., Seeker, V., Thompson, C., Topham, N.P.: Scalable multi-core simulation using parallel dynamic binary translation. In: International Conference on Embedded Computer Systems: Architectures, Modeling, and Simulation, ICSAMOS '11, pp. 190–199. IEEE (2011)
2. August, D., Chang, J., Girbal, S., Gracia-Perez, D., Mouchard, G., Penry, D.A., Temam, O., Vachharajani, N.: UNISIM: An open simulation environment and library for complex architecture design and collaborative development. IEEE Computer Architecture Letters 6(2), 45–48 (2007)
3. Austin, T., Larson, E., Ernst, D.: SimpleScalar: An infrastructure for computer system modeling. Computer 35(2), 59–67 (2002)
4. Azevedo, R., Rigo, S., Bartholomeu, M., Araujo, G., Araujo, C., Barros, E.: The ArchC architecture description language and tools. Int. J. Parallel Program. 33(5), 453–484 (2005). DOI http://dx.doi.org/10.1007/s10766-005-7301-0
5. Bartholomeu, M., Azevedo, R., Rigo, S., Araujo, G.: Optimizations for compiled simulation using instruction type information. In: Symposium on Computer Architecture and High Performance Computing (SBAC-PAD 2004), pp. 74–81 (2004). DOI http://doi.ieeecomputersociety.org/10.1109/CAHPC.2004.28
6. Bell, J.R.: Threaded code. Commun. ACM 16(6), 370–372 (1973). DOI http://doi.acm.org/10.1145/362248.362270
7. Bermudo, N., Horspool, N., Krall, A.: Control flow graph reconstruction for reverse compilation of assembly language programs with delayed instructions. In: SCAM'05: Proceedings of the Fifth International Workshop on Source Code Analysis and Manipulation, pp. 107–116 (2005)

8. Böhm, I., von Koch, T.J.K.E., Kyle, S.C., Franke, B., Topham, N.: Generalized just-in-time trace compilation using a parallel task farm in a dynamic binary translator. ACM SIGPLAN Notices **46**(6), 74–85 (2011). DOI http://dx.doi.org/10.1145/1993316.1993508

9. Brandner, F.: Precise simulation of interrupts using a rollback mechanism. In: SCOPES '09: Proceedings of the 12th International Workshop on Software and Compilers for Embedded Systems, pp. 71–80 (2009)

10. Brandner, F., Fellnhofer, A., Krall, A., Riegler, D.: Fast and accurate simulation using the LLVM compiler framework. In: RAPIDO '09: 1st Workshop on Rapid Simulation and Performance Evaluation: Methods and Tools (2009)

11. Burtscher, M., Ganusov, I.: Automatic synthesis of high-speed processor simulators. In: MICRO 37: Proceedings of the 37th annual IEEE/ACM International Symposium on Microarchitecture, pp. 55–66 (2004). DOI http://dx.doi.org/10.1109/MICRO.2004.7

12. Chiou, D., Sanjeliwala, H., Sunwoo, D., Xu, J.Z., Patil, N.: FPGA-based fast, cycle-accurate, full-system simulators. In: WARFP'06: Proceedings of the second Workshop on Architecture Research using FPGA Platforms (2006)

13. Chiou, D., Sunwoo, D., Kim, J., Patil, N., Reinhart, W.H., Johnson, D.E., Xu, Z.: The FAST methodology for high-speed SoC/computer simulation. In: ICCAD '07: Proceedings of the 2007 IEEE/ACM International Conference on Computer-Aided Design, pp. 295–302 (2007)

14. Chiou, D., Sunwoo, D., Kim, J., Patil, N.A., Reinhart, W., Johnson, D.E., Keefe, J., Angepat, H.: FPGA-accelerated simulation technologies (FAST): Fast, full-system, cycle-accurate simulators. In: MICRO '07: Proceedings of the 40th Annual IEEE/ACM International Symposium on Microarchitecture, pp. 249–261 (2007). DOI http://dx.doi.org/10.1109/MICRO.2007.16

15. Chung, E.S., Hoe, J.C., Falsafi, B.: ProtoFlex: Co-simulation for component-wise FPGA emulator development. In: WARFP '06: In Proceedings of the 2nd Workshop on Architecture Research using FPGA Platforms (2006)

16. Chung, E.S., Nurvitadhi, E., Hoe, J.C., Falsafi, B., Mai, K.: A complexity-effective architecture for accelerating full-system multiprocessor simulations using FPGAs. In: FPGA '08: Proceedings of the 16th International ACM/SIGDA Symposium on Field Programmable Gate Arrays, pp. 77–86 (2008). DOI http://doi.acm.org/10.1145/1344671.1344684

17. Chung, E.S., Papamichael, M.K., Nurvitadhi, E., Hoe, J.C., Mai, K., Falsafi, B.: ProtoFlex: Towards scalable, full-system multiprocessor simulations using FPGAs. ACM Transactions on Reconfigurable Technology and Systems (TRETS **2**(2), 1–32 (2009)

18. Cmelik, B., Keppel, D.: Shade: A fast instruction-set simulator for execution profiling. In: SIGMETRICS '94: Proceedings of the 1994 ACM SIGMETRICS Conference on Measurement and Modeling of Computer Systems, pp. 128–137 (1994)

19. Cofer, R.C., Harding, B.: Rapid System Prototyping with FPGAs: Accelerating the Design Process. Newnes (2005)

20. Dehnert, J.C., Grant, B.K., Banning, J.P., Johnson, R., Kistler, T., Klaiber, A., Mattson, J.: The Transmeta Code Morphing™ software: Using speculation, recovery, and adaptive retranslation to address real-life challenges. In: CGO '03: Proceedings of the International Symposium on Code Generation and Optimization, pp. 15–24 (2003)

21. Ebcioğlu, K., Altman, E., Gschwind, M., Sathaye, S.: Dynamic binary translation and optimization. IEEE Transactions on Computers **50**(6), 529–548 (2001)

22. Ebcioğlu, K., Altman, E.R.: DAISY: Dynamic compilation for 100% architectural compatibility. In: ISCA '97: Proceedings of the 24th International Symposium on Computer Architecture, pp. 26–37 (1997)

23. Ebcioğlu, K., Altman, E.R., Gschwind, M., Sathaye, S.: Optimizations and oracle parallelism with dynamic translation. In: MICRO 32: Proceedings of the 32nd annual ACM/IEEE International Symposium on Microarchitecture, pp. 284–295 (1999)

24. Emer, J., Ahuja, P., Borch, E., Klauser, A., Luk, C.K., Manne, S., Mukherjee, S.S., Patil, H., Wallace, S., Binkert, N., Espasa, R., Juan, T.: Asim: A performance model framework. Computer **35**(2), 68–76 (2002). DOI http://dx.doi.org/10.1109/2.982918

25. Errico, J.D., Qin, W.: Constructing portable compiled instruction-set simulators - an ADL-driven approach. In: DATE '06: Proceedings of the Conference on Design, Automation and Test in Europe, pp. 112–117 (2006)
26. Farfeleder, S., Krall, A., Horspool, N.: Ultra fast cycle-accurate compiled emulation of inorder pipelined architectures. EUROMICRO Journal of Systems Architecture **53**(8), 501–510 (2007)
27. Fauth, A., Praet, J.V., Freericks, M.: Describing instruction set processors using nML. In: EDTC '95: Proceedings of the 1995 European Conference on Design and Test, pp. 503–507 (1995)
28. Fytraki, S., Pnevmatikatos, D.: ReSim, a trace-driven, reconfigurable ILP processor simulator. In: DATE '09: Proceedings of Design, Automation and Test in Europe 2009 (2009)
29. Gao, L., Kraemer, S., Leupers, R., Ascheid, G., Meyr, H.: A fast and generic hybrid simulation approach using C virtual machine. In: CASES '07: Proceedings of the 2007 International Conference on Compilers, Architecture, and Synthesis for Embedded Systems, pp. 3–12 (2007)
30. Goossens, G., Lanneer, D., Geurts, W., Praet, J.V.: Design of ASIPs in multi-processor SoCs using the Chess/Checkers retargetable tool suite. In: International Symposium on System-on-Chip, pp. 1–4 (2006). DOI 10.1109/ISSOC.2006.321968
31. Gschwind, M., Altman, E.: Optimization and precise exceptions in dynamic compilation. ACM SIGARCH Computer Architecture News **29**(1), 66–74 (2001)
32. Gschwind, M., Altman, E.R., Sathaye, S., Ledak, P., Appenzeller, D.: Dynamic and transparent binary translation. Computer **33**(3), 54–59 (2000). DOI http://dx.doi.org/10.1109/2.825696
33. Halambi, A., Grun, P., Ganesh, V., Khare, A., Dutt, N., Nicolau, A.: EXPRESSION: A language for architecture exploration through compiler/simulator retargetability. In: DATE '99: Proceedings of the Conference on Design, Automation and Test in Europe, pp. 485–490 (1999). DOI http://doi.acm.org/10.1145/307418.307549
34. Horspool, R.N., Marovac, N.: An approach to the problem of detranslation of computer programs. Comput. J. **23**(3), 223–229 (1980)
35. Ienne, P., Leupers, R.: Customizable Embedded Processors: Design Technologies and Applications (Systems on Silicon). Morgan Kaufmann Publishers Inc., San Francisco, CA, USA (2006)
36. Jones, D., Topham, N.P.: High speed CPU simulation using LTU dynamic binary translation. In: HiPEAC'09: Proceedings of the 4th International Conference on High Performance Embedded Architectures and Compilers, pp. 50–64 (2009)
37. Klint, P.: Interpretation techniques. Software: Practice and Experience **11**(9), 963 – 973 (1981)
38. Krall, A., Farfeleder, S., Horspool, N.: Ultra fast cycle-accurate compiled emulation of inorder pipelined architectures. In: SAMOS '05: Proceedings of the International Workshop on Systems, Architectures, Modeling, and Simulation, LNCS 3553, pp. 222–231 (2005)
39. Kudlugi, M., Hassoun, S., Selvidge, C., Pryor, D.: A transaction-based unified simulation/emulation architecture for functional verification. In: DAC '01: Proceedings of the 38th Conference on Design Automation, pp. 623–628 (2001). DOI http://doi.acm.org/10.1145/378239.379036
40. Lantz, R.E.: Fast functional simulation with parallel Embra. In: 4th Annual Workshop on Modeling, Benchmarking and Simulation, MOBS'08 (2008)
41. Larus, J.: Assemblers, linkers and the SPIM simulator. In: D.A. Patterson, J.L. Hennessy (eds.) Computer Organization and Design: The Hardware/software Interface. Morgan Kaufmann (2005)
42. Magnusson, P.S.: Efficient instruction cache simulation and execution profiling with a threaded-code interpreter. In: WSC '97: Proceedings of the 29th Conference on Winter Simulation, pp. 1093–1100 (1997). DOI http://doi.acm.org/10.1145/268437.268745
43. Magnusson, P.S., Christensson, M., Eskilson, J., Forsgren, D., Hållberg, G., Högberg, J., Larsson, F., Moestedt, A., Werner, B.: Simics: A full system simulation platform. Computer **35**(2), 50–58 (2002)
44. May, C.: Mimic: a fast System/370 simulator. In: Symposium on Interpreters and Interpretive Techniques, pp. 1–13 (1987). DOI http://doi.acm.org/10.1145/29650.29651
45. Mills, C., Ahalt, S.C., Fowler, J.: Compiled instruction set simulation. Software: Practice and Experience **21**(8), 877–889 (1991)

46. Mishra, P., Dutt, N.: Processor Description Languages, Volume 1. Morgan Kaufmann Publishers Inc., San Francisco, CA, USA (2008)
47. Nakamura, Y., Hosokawa, K.: Fast FPGA-emulation-based simulation environment for custom processors. IEICE Transactions on Fundamentals of Electronics, Communications and Computer Sciences **E89-A**(12), 3464–3470 (2006)
48. Nakamura, Y., Hosokawa, K., Kuroda, I., Yoshikawa, K., Yoshimura, T.: A fast hardware/-software co-verification method for system-on-a-chip by using a C/C++ simulator and FPGA emulator with shared register communication. In: DAC '04: Proceedings of the 41st annual Conference on Design Automation, pp. 299–304 (2004). DOI http://doi.acm.org/10.1145/996566.996655
49. Nohl, A., Braun, G., Schliebusch, O., Leupers, R., Meyr, H., Hoffmann, A.: A universal technique for fast and flexible instruction-set architecture simulation. In: DAC '02: Proceedings of the 39th Conference on Design Automation, pp. 22–27 (2002)
50. Pees, S., Hoffmann, A., Meyr, H.: Retargetable compiled simulation of embedded processors using a machine description language. ACM Transactions on Design Automation of Electronic Systems. **5**(4), 815–834 (2000)
51. Pellauer, M., Vijayaraghavan, M., Adler, M., Arvind, Emer, J.: A-Ports: An efficient abstraction for cycle-accurate performance models on FPGAs. In: FPGA '08: Proceedings of the 16th International ACM/SIGDA Symposium on Field Programmable Gate Arrays, pp. 87–96 (2008). DOI http://doi.acm.org/10.1145/1344671.1344685
52. Pellauer, M., Vijayaraghavan, M., Adler, M., Arvind, Emer, J.: Quick performance models quickly: Closely-coupled partitioned simulation on FPGAs. In: ISPASS '08: IEEE International Symposium on Performance Analysis of Systems and Software, pp. 1–10 (2008). DOI 10.1109/ISPASS.2008.4510733
53. Proebsting, T.A.: Optimizing an ANSI C interpreter with superoperators. In: POPL '95: Proceedings of the 22nd ACM SIGPLAN-SIGACT Symposium on Principles of Programming Languages, pp. 322–332 (1995). DOI http://doi.acm.org/10.1145/199448.199526
54. Qin, W., D'Errico, J., Zhu, X.: A multiprocessing approach to accelerate retargetable and portable dynamic-compiled instruction-set simulation. In: Proceedings of the 4th international conference on Hardware/software codesign and system synthesis, CODES+ISSS '06, pp. 193–198. ACM, New York, NY, USA (2006). DOI http://doi.acm.org/10.1145/1176254.1176302. URL http://doi.acm.org/10.1145/1176254.1176302
55. Raghav, S., Ruggiero, M., Atienza, D., Pinto, C., Marongiu, A., Benini, L.: Scalable instruction set simulator for thousand-core architectures running on gpgpus. In: International Conference on High Performance Computing and Simulation, HPCS '10, pp. 459–466. IEEE (2010). DOI 10.1109/HPCS.2010.5547092
56. Reshadi, M., Dutt, N.: Generic pipelined processor modeling and high performance cycle-accurate simulator generation. In: DATE '05: Proceedings of the Conference on Design, Automation and Test in Europe, pp. 786–791 (2005). DOI http://dx.doi.org/10.1109/DATE.2005.166
57. Reshadi, M., Dutt, N., Mishra, P.: A retargetable framework for instruction-set architecture simulation. ACM Transactions on Embedded Computing Systems (TECS) **5**(2), 431–452 (2006)
58. Reshadi, M., Mishra, P., Dutt, N.: Instruction set compiled simulation: A technique for fast and flexible instruction set simulation. In: Proceedings of the 40th Conference on Design Automation, pp. 758–763 (2003). DOI http://doi.acm.org/10.1145/775832.776026
59. Reshadi, M., Mishra, P., Dutt, N.: Hybrid-compiled simulation: An efficient technique for instruction-set architecture simulation. ACM Transactions on Embedded Computing Systems (TECS) **8**(3), 1–27 (2009)
60. Roeven, H., Coninx, J., Ade, M.: CoolFlux DSP: The embedded ultra low power C-programmable DSP core. In: GSPx'04: International Signal Processing Conference, pp. 1–7 (2004)
61. Rosenblum, M., Herrod, S.A., Witchel, E., Gupta, A.: Complete computer system simulation: The SimOS approach. IEEE Parallel & Distributed Technology **3**(4), 34–43 (1995)

62. Sathaye, S., Ledak, P., Leblanc, J., Kosonocky, S., Gschwind, M., Fritts, J., Bright, A., Altman, E., Agricola, C.: BOA: Targeting multi-gigahertz with binary translation. In: In Proceedings of the 1999 Workshop on Binary Translation, pp. 2–11 (1999)

63. Schnerr, J., Bringmann, O., Rosenstiel, W.: Cycle accurate binary translation for simulation acceleration in rapid prototyping of SoCs. In: DATE '05: Proceedings of the Conference on Design, Automation and Test in Europe, pp. 792–797 (2005). DOI http://dx.doi.org/10.1109/DATE.2005.106

64. Schnerr, J., Haug, G., Rosenstiel, W.: Instruction set emulation for rapid prototyping of SoCs. In: DATE '03: Proceedings of the Conference on Design, Automation and Test in Europe, pp. 562–567 (2003)

65. Sites, R.L., Chernoff, A., Kirk, M.B., Marks, M.P., Robinson, S.G.: Binary translation. Communications of the ACM 36(2), 69–81 (1993). DOI http://doi.acm.org/10.1145/151220.151227

66. Smith, J.E., Nair, R.: Virtual Machines. Morgan Kaufman (2005)

67. Suh, T., Lee, H.H.S., Lu, S.L., Shen, J.: Initial observations of hardware/software co-simulation using FPGA in architectural research. In: WARFP'06: In Proceedings of the 2nd Workshop on Architecture Research using FPGA Platforms (2006)

68. Open SystemC Initiative. http://www.systemc.org/home

69. Vachharajani, M., Vachharajani, N., August, D.I.: The Liberty Structural Specification Language: A high-level modeling language for component reuse. In: PLDI '04: Proceedings of the ACM SIGPLAN 2004 Conference on Programming Language Design and Implementation, pp. 195–206 (2004)

70. Vachharajani, M., Vachharajani, N., Penry, D.A., Blome, J.A., Malik, S., August, D.I.: The Liberty Simulation Environment: A deliberate approach to high-level system modeling. ACM Transactions on Computer Systems 24(3), 211–249 (2006)

71. Wang, K., Zhang, Y., Wang, H., Shen, X.: Parallelization of IBM Mambo system simulator in functional modes. SIGOPS Oper. Syst. Rev. 42, 71–76 (2008). DOI http://doi.acm.org/10.1145/1341312.1341325. URL http://doi.acm.org/10.1145/1341312.1341325

72. Wang, Z., Liu, R., YufeiChen, Wu, X., Chen, H., Zhang, W., Zang, B.: COREMU: a scalable and portable parallel full-system emulator. In: Proceedings of the 16th ACM symposium on Principles and Practice of Parallel Programming, pp. 213–222. ACM (2011). URL http://doi.acm.org/10.1145/1941553.1941583

73. Witchel, E., Rosenblum, M.: Embra: Fast and flexible machine simulation. In: SIGMETRICS '96: Proceedings of the 1996 ACM SIGMETRICS International Conference on Measurement and Modeling of Computer Systems, pp. 68–79 (1996)

74. Yi, J.J., Lilja, D.J.: Simulation of computer architectures: Simulators, benchmarks, methodologies, and recommendations. IEEE Transactions on Computers 55(3), 268–280 (2006)

Integrated Modeling Using Finite State Machines and Dataflow Graphs

Joachim Falk, Christian Haubelt, Christian Zebelein, and Jürgen Teich

Abstract In this chapter, different application modeling approaches based on the integration of finite state machines with dataflow models are reviewed. Restricted Models of Computation (MoC) may be exploited in design methodologies to generate optimized hardware/software implementations from a given application model. A particular focus is put on the analyzability of these models with respect to schedulability and the generation of efficient schedule implementations. In this purpose, clustering methods for model refinement and schedule optimization are of particular interest.

1 Introduction

Dataflow graphs are widely accepted for modeling DSP algorithms, e.g., multimedia and signal processing applications. On the one hand, known optimization strategies are well known for static dataflow graphs, while on the other hand, modeling complex multimedia applications using only static dataflow models is a challenging task. Often, these problems can be circumvented by using dynamic dataflow models which are a good fit for multimedia and signal processing applications. However, the ad-hoc representation of control flow in dynamic dataflow actors has lead to problems concerning adequate information extraction from these models to serve as a basis for the analysis of these graphs. Hence, neither static nor dynamic

J. Falk (✉) • J. Teich
Hardware-Software-Co-Design, University of Erlangen-Nuremberg,
Am Weichselgarten 3, 91058 Erlangen, Germany
e-mail: falk@cs.fau.de; teich@cs.fau.de

C. Zebelein • C. Haubelt
Applied Microelectronics and Computer Engineering, University of Rostock,
Richard-Wagner-Str. 31, 18119 Rostock-Warnemünde, Germany
e-mail: christian.zebelein@uni-rostock.de; christian.haubelt@uni-rostock.de

S.S. Bhattacharyya et al. (eds.), *Handbook of Signal Processing Systems*,
DOI 10.1007/978-1-4614-6859-2_30, © Springer Science+Business Media, LLC 2013

dataflow graphs seem to be the best choice to establish a design methodology upon it. One of the key problems is due to the fact that dataflow actors often show control dominant behavior which cannot adequately be expressed in many dataflow models. Consequently, different modeling approaches integrating *finite state machines* (*FSMs*) with dataflow graphs have been proposed in the past. However, this often comes with the drawback of decreased analysis capabilities. Nevertheless, several successful academic system-level design methodologies have emerged recently based on integrated modeling approaches.

In this chapter, we will present different integrated modeling approaches (Sect. 2). Section 3 is devoted to the problem of identifying static dataflow actors within a general dataflow graph. Different design methodologies, namely Ptolemy II [28], OpenDF [4], and SystemCoDesigner [27], will be presented in Sect. 4. We will focus particularly on the analyzability of the application model and the exploitation of these analysis capabilities in the corresponding design methodology. In Sect. 5, clustering approaches for static dataflow graphs embedded in more general dataflow graphs will be presented. Finally, in Sect. 6, we recap the important points of this chapter.

2   Modeling Approaches

In this section, we will discuss modeling approaches combining dataflow graphs with Finite State Machines (*FSMs*). Starting with the recapitulation of some fundamental dataflow models like *Synchronous Dataflow* (*SDF*) [29] and *Dynamic Dataflow* (*DDF*), the *charts* (pronounced "star charts") [18] formalism is introduced. The *charts* formalism was one of the first integrated modeling approaches implemented in the Ptolemy project [12]. Furthermore, we will compare *enable-invoke* [34] and *core functional dataflow* [34] with the *charts* approach. Later on, the CAL Actor Language (*CAL*) [11], Extended CoDesign Finite State Machines (*ECFSMs*) [35], and SYSTEMOC [13] will be discussed.

2.1   Dataflow Graphs

In *synchronous dataflow* (*SDF*) [29] graphs, the firing of an actor is an atomic computation that consumes a fixed number of tokens from each incoming edge and produces a fixed number of tokens on each outgoing edge. The most restrictive subclass of *SDF* graphs are Homogeneous *SDF* (*HSDF*) graphs, where each actor exactly consumes and produces a single token from each incoming and on each outgoing edge, respectively. Edges are conceptionally unbounded FIFO queues representing a possibly infinite stream of data. The consumption and production rates can be used to unambiguously define an iteration that returns the queues to

Fig. 1 A state transition
graph representing an *FSM*

their original state. This can be done by solving the balance equation $\gamma_i p_i = \gamma_j c_j$ where the edge (i,j) is assumed to be directed from actor i to actor j. p_i is the number of tokens produced by actor i onto this particular edge, whereas c_j is the number of tokens consumed by actor j. γ_i and γ_j denote the number of actor firings of actor i and j, respectively, to return the edge (i,j) to its original state. Beside the trivial solution $\gamma_i = \gamma_j = 0$, the balance equation may also be satisfied by a finite number of firings.

Considering a network of N actors with M edges, there will be M equations with N unknowns. For connected graphs, there exists either the only solution $\gamma_i = 0, \forall 1 \le i \le N$ or a unique smallest solution, called the *minimal repetition vector* γ, with $\gamma_i > 0, \forall i$. A more thorough treatment of the Synchronous Dataflow *MoC* can be found in [21].

In *dynamic dataflow (DDF)*, actors consume and produce a variable number of tokens on each firing. It is assumed that a *DDF* actor must assert, prior to any firing, the required number of tokens on each input. If the requirements are met, the *DDF* actor can fire. A more detailed discussion of the *dynamic dataflow MoC* can be found in [6].

2.2 *charts

One of the first modeling approaches integrating Finite State Machines (*FSMs*) with dataflow models is *charts (pronounced "star charts") [18]. The concurrency model in *charts is not restricted to be dataflow: Other choices are discrete event models or synchronous/reactive models. However, in the scope of this chapter, we will limit our discussion to the *FSM*/dataflow integration.

We start by formally defining deterministic finite state machines:

Definition 1 (*FSM*). A deterministic *FSM* is a five tuple $(Q, \Sigma, \Delta, \sigma, q_0)$, where Q is a finite set of states, Σ is a set of symbols denoting the possible inputs, Δ is a set of symbols denoting possible outputs, $\sigma : Q \times \Sigma \to Q \times \Delta$ is the transition mapping function, and $q_0 \in Q$ is the initial state.

In one reaction, an *FSM* maps its current state $q \in Q$ and an input symbol $a \in \Sigma$ to a next state $q' \in Q$ and an output symbol $b \in \Delta$, where $(q', b) = \sigma(q, a)$. Given a sequence of input symbols, an *FSM* performs a sequence of reactions starting in its initial state q_0. Thereby, a sequence of output symbols in Δ is produced.

FSMs are often represented by *state transition diagrams*. In *state transition diagrams*, vertices correspond to states and edges model state transitions, see Fig. 1. Edges are labeled by *guard/action* pairs, where *guard* $\in \Sigma$ specifies the input symbol triggering the corresponding state transition and *action* $\in \Delta$ specifies the output symbol to be produced by the *FSM* reaction. The edge without source state points to the initial state.

One reaction of the *FSM* is given by triggering a single enabled state transition. An enabled state transition is an outgoing transition from the current state where the guard matches the current input symbol. Thereby, the *FSM* changes to the destination state of the transition and produces the output symbol specified by the action. To simplify the notation, each state is assumed to have implicit self transitions for input symbols not used as guards for any explicit transition. The action in such cases is empty or the output symbol $\varepsilon \in \Delta$ indicating the absence of an output value.

FSMs as presented in Definition 1 can be easily extended to multiple input signals and multiple output signals. In such a scenario, input and output symbols are defined by subdomains, i.e., $\Sigma = \Sigma_1 \times \cdots \times \Sigma_n$ and $\Delta = \Delta_1 \times \cdots \times \Delta_m$.

The *charts approach allows nesting of dataflow graphs and *FSMs* by allowing the refinement of a dataflow actor via *FSMs* and allowing an *FSM* state to be refined via a dataflow graph. We first discuss how *FSMs* are used to refine dataflow actors.

2.2.1 Refining Dataflow Actors via *FSMs*

To use an *FSM* as a refinement for a dataflow actor, the consumption and production rates have to be determined from the *FSM* for the refined actor. However, the derivation of the consumption and production rates depends on the Model of Computation (*MoC*) of the dataflow graph containing the actor. The tokens on the channels connected to the refined dataflow actor are mapped to the input alphabet Σ of the refining *FSM*. In the guard of the *FSM*, we will use $i[n]$ to refer to the nth token on the channel connected to the input port i. In the action of the *FSM*, we will use $o[n]$ to refer to the nth token to produce on the channel connected to the output port o. Each unique $i[n]$ has a symbol subdomain Σ_m associated with it and the cross product of these subdomains form the input alphabet to the *FSM*. The same is true for $o[n]$ and the output alphabet Δ, thus perfectly matching the *FSM* model (see Definition 1). In each reaction of the *FSM*, the corresponding action emits a symbol from the output alphabet of the *FSM*, which is used to derive a token value to be produced on each outgoing edge of the actor.

FSM in Dynamic Dataflow: If an *FSM* refines a *DDF* actor, then each state of the *FSM* is associated with a number of tokens that will be consumed by the next firing of the actor. The semantics is as follows: For a given state q of the *FSM*, the number of consumed tokens is determined by examining each outgoing transition of the state. For each input port i of the actor, the maximum index n is determined with which the input port i occurs in any of the transitions leaving the state q. If the input port i does not occur in any transition, an index of -1 is assumed. Then, for the given state q the consumption rate for each input port i is the maximum index n incremented by one. Note that this derivation of the consumption rates implicitly produces actors which are continuous in Kahn's sense [26]. An introduction to Kahn Process Networks (*KPN*) can be found in [17].

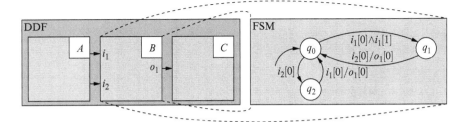

Fig. 2 Refinement of a *DDF* actor by an *FSMs* in *charts

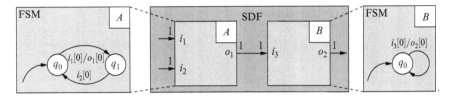

Fig. 3 Refinement of *SDF* actors by *FSMs* in *charts

To illustrate the above concept, consider state q_0 in the *FSM* depicted in Fig. 2 refining the dynamic actor B. In this case, regardless if the transition to q_1 or q_2 is taken, the actor will consume exactly two tokens from i_1 and one token from i_2.

FSM in Synchronous Dataflow: When an *FSM* refines an *SDF* actor, it must externally obey *SDF* semantics, i.e., it must consume and produce a fixed number of tokens in each firing. Consider Fig. 3 taken from [18] as an example. The top level *HSDF* model consists of two actors connected by a single edge. The minimal repetition vector is $\gamma_A = \gamma_B = 1$. Since the edge (A, B) contains no initial tokens, actor A fires before actor B in each iteration. In the first iteration, actor A fires if at least one token is available on each input edge. Afterwards, *HSDF* actor B is enabled in the first iteration but also *HSDF* actor A might be enabled in the second iteration assuming sufficient input data is available. Hence, *FSM A* and *FSM B* can execute concurrently. If sufficient input data is available for actor A.

We now consider the refinement of *SDF* actors by *FSMs*. In the simplest case, the consumption and production rate of each state of the *FSM* are equal. Then, the *FSM* can be used unmodified as an *SDF* actor where the consumption and production rate of the actor are rates which hold in each state of the *FSM*. An example of such an *FSM* can be seen for the refinement of the *SDF* actor B in Fig. 3. The *FSM* neatly corresponds to the *SDF* model consuming one token and producing one token in each reaction.

However, the *FSM* for actor A from Fig. 3 does not correspond to the simple case. In this case the consumption and production rates of the refined *SDF* actor are the maximum of the consumption and production rates of each state of the *FSM*. If consumed tokens are not present in the guards leaving a state, they are simply ignored. If values of produced tokens are not specified in the action of an enabled state transition, the values are interpreted as ε values indicating the absence of values. Note that even an ε value produces a token in the enclosing dataflow graph. This solution is a little unsatisfactory and indeed there is another approach to handle this case. The *FSM* can be embedded into a *heterochronous dataflow graph*.

FSM in Heterochronous Dataflow: The idea of *heterochronous dataflow* (*HDF*) is similar to parameterized dataflow modeling [3], that is dynamic behavior is allowed and is represented via *FSMs*. However, all *FSMs* in the *heterochronous dataflow graph* are forced to only change state once the *heterochronous dataflow graph* has executed a full iteration. This constraint ensures that the consumption and production rates of the *HDF* actor does not change while the *HDF* graph executes its iteration. However, after the iteration is finished, the *HDF* actors are free to update their state leading to new consumption and production rates for the *HDF* actors in the system. With these new consumption and production rates, a new balance equation is solved and a new repetition vector calculated which is executed in the next iteration. For the duration of this next iteration, all *HDF* actors have to keep their consumption and production rates unmodified.

Let us consider again Fig. 3. But now instead of an *SDF* domain, we use an *HDF* domain. In the case that actor A is in state q_0, to execute a full iteration of the *HDF* graph, the actors A and B are executed exactly once. Note that while actor A is executed, it remains in state q_0 regardless of the value $i_1[0]$ of the first token (note that tokens carrying ε values are still tokens and not absence of tokens) on input port i_1. After the full iteration of the *HDF* graph ($\gamma_A = \gamma_B = 1$) has finished, the *FSM* of actor A may change its state to q_1 depending on the value $i_1[0]$ of the first token on input port i_1. In the case that actor A is now in state q_1, a full iteration of the *HDF* graph corresponds to the sole execution of actor A. After the full iteration of the *HDF* graph has finished, the *FSM* of actor A may change its state to q_0 depending on the value $i_2[0]$ of the first token on input port i_2.

2.2.2 Refining *FSM* States via Dataflow Graphs

Previously, we have seen how an *FSM* can be used to refine a dataflow actor. On the opposite side, an *FSM* can be used to coordinate between multiple dataflow graphs. This coordination is achieved by refining *FSM* states by dataflow graphs. The dataflow graph is composed into a single actor which is executed if the *FSM* is in the refined state. To refine a state by a *dataflow graph*, a notion of *iteration* is necessary as the execution of one reaction of the *FSM* has to terminate. An *iteration* has been chosen as a natural boundary to stop the execution of the embedded dataflow graph. However, the existence of a finite iteration is undecidable for general dataflow graphs. Hence, the application of refinements of states to dataflow graphs is restricted in *charts to certain subclasses of dataflow, e.g., *synchronous dataflow graphs* (*SDF*) [29] and *cyclo-static dataflow graphs* (*CSDF*) [7], which provide such a notion of iteration naturally. Moreover, combining the actors in a subgraph of a dataflow graph into a single actor, which will execute an iteration for the subgraph, is not always possible. The problem is treated in detail in Sect. 5.

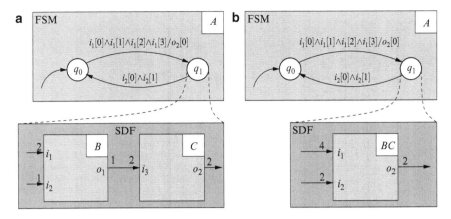

Fig. 4 (**a**) *SDF* graph refining a state (**b**) *SDF* graph is substituted by a single *SDF* actor

As an example, consider Fig. 4a, which is taken from [18]. The top level *FSM A* consists of two states q_0 and q_1. If the current state of the *FSM* is not refined, e.g., q_0, the *FSM* reacts like an ordinary *FSM*. State q_1 is refined by an *SDF* graph consisting of two actors B and C which are connected by a single edge. The minimal repetition vector may be determined as $\gamma_B = 2, \gamma_C = 1$, i.e., the repetition vector is satisfied by two firings of actor B followed by one firing of actor C implementing one iteration of the *SDF* graph. Assuming the above behavior, the *SDF* graph can be substituted by a single *SDF* actor as shown in Fig. 4b. If a dataflow graph refines a state of an *FSM*, the firing rules of the dataflow graph are exported to the environment of the *FSM*, i.e., if the *FSM* is in the state which is refined by the dataflow graph, then for a reaction of the *FSM* exactly as many tokens will be required as for the execution of the *iteration* of the embedded dataflow graph.

After the *embedded dataflow graph* has finished its iteration, the *FSM* will execute a transition. If the *FSM* is embedded in the *DDF* or *SDF* domain, then the *FSM* executes a transition after the *corresponding dataflow graph* has finished its iteration. If the *FSM* is embedded in the *HDF* domain, then this transition is delayed until all parent graphs have finished their iteration.

Assuming that *FSM A* is in state q_1, the *FSM* is not embedded in an *HDF* graph, and the tokens $i_2[0]$ and $i_2[1]$ have both the presence value (note that tokens carrying ε values are still tokens and not absence of tokens), then the next activation of actor A will execute the embedded *SDF* graph for one iteration followed by a state transition to state q_0.

2.3 Enable-Invoke Dataflow

The *enable-invoke dataflow (EIDF)* [34] derives its name from the requirement that each actor invocation is separated into an *enable function* ν and an *invoke function* κ.

These functions correspond to testing if sufficient tokens are present on the input channels (*enable function* ν), therefore enabling the *invoke function* κ which executes one firing of the actor. On execution, the invoke function κ consumes and produces tokens (whose presence has been guaranteed by the previous call to the *enable function* ν). For an efficient usage of the *enable-invoke dataflow* model in a dynamic scheduling environment, the *enable function* and *invoke function* are designed in such a way that there are no redundant computations between the *enable function* and the *invoke function*.

More formally, the *enable function* $\nu : \mathbb{N}_0^{|I|} \times M \to \{\mathbf{t}, \mathbf{f}\}$ maps a vector $\mathbb{N}_0^{|I|}$ containing the number of tokens present on the actor input ports I and the current mode $m \in M$ to a Boolean value (**t**)rue or (**f**)false. The *invoke function* can only be executed if the *enable function* evaluates to **t** for a mode m in the set of current modes and the number of tokens present on the actor input ports. The *invoke function* $\kappa : \Sigma \times M \to \Delta \times 2^M$ maps a vector Σ of token values from the actor input ports and a mode $m \in M$ to a vector of token values Δ to be produced on the actor output ports and a set $M_{\text{active}} \subseteq M$ of new modes.[1] If the set of new modes returned by the *invoke function* is empty, then the actor will be disabled for the remainder of the system execution.

As seen from the above definitions, the actor can be simultaneously in multiple modes at once. As each mode can have different consumption and production rates, the behavior of the dataflow graph containing such an actor may differ depending on the arrival times of the tokens on the actor input ports. Hence, the *enable-invoke dataflow* model is able to describe actors which are not continuous in Kahn's sense [26]. We will call such actors *non-deterministic*.

A subset of *enable-invoke dataflow* is *core functional dataflow* (*CFDF*) which constrains the *invoke function* of *enable-invoke dataflow* to return exactly one mode in the set of active modes M_{active}. This constraint is sufficient to again guarantee actors to be continuous in Kahn's sense [26] and therefore to be *deterministic*. The behavior of dataflow graphs containing only such deterministic actors is independent of the scheduling algorithm chosen to schedule the graph. To exemplify this, we consider the example from Fig. 5, which is taken from [33].

The depicted dataflow graph contains four *SDF* actors A, B, C, D, and a *core functional dataflow* actor E implementing the *switch actor* known from *Boolean dataflow* (*BDF*) [9]. The *switch actor* consumes a control token from input port i_1 and either goes to the $m_{\mathbf{t}}$ mode if the control token $i_0[0]$ was **t** or otherwise to the $m_{\mathbf{f}}$ mode. In the $m_{\mathbf{t}}$ mode, a token is forwarded from input port i_2 to the output port o_1, while the $m_{\mathbf{f}}$ mode forwards the token from input port i_2 to the output port o_2.

As can be seen from Fig. 5, the model for *core functional dataflow* is very similar to the **charts* approach when *FSMs* are embedded in the *SDF* domain. However, **charts* can interpret them in three different ways: (i) The *FSM* actor is embedded in the *SDF* domain, and the actor consumes/produces for every actor invocation on each port the maximum number of tokens the *FSM* consumes/produces in any of its

---

[1] We use 2^M to denote the power set of the set of modes M.

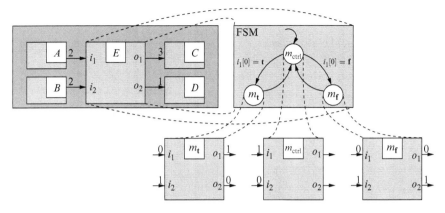

Fig. 5 Example of a dataflow graph containing four *SDF* actors A, B, C, D, and a *core functional dataflow* actor E implementing the *switch actor* known from *Boolean dataflow* [9]

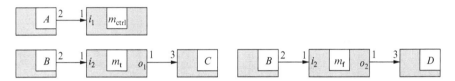

Fig. 6 Example of the static dataflow graphs which can be combined from the three modes which are present in *core functional dataflow* actor E with the four *SDF* actors A, B, C, D

states. (ii) The *FSM* actor is embedded in the *HDF* domain, and the *FSM* is only allowed to perform state transitions if a full iteration of the graph has been executed. (iii) The *FSM* actor is embedded in the *DDF* domain, and the *FSM* performs a state transition after each firing of the actor. The *core functional dataflow* domain behaves like *FSM* actors embedded in the *DDF* domain in *charts*, that is after each firing of a *core functional dataflow* actor, the mode of the actor may change potentially modifying the consumption and production rates on the ports.

Plishker et al. [33] have presented an analysis method which still can exploit the static information present in the *core functional dataflow* domain to improve scheduler generation compared to a simple *round robin scheduler*. The idea is based on finding sets of modes, where each mode in the set belongs to a distinct *CFDF* actor. The set of modes is constraint to correspond to a consistent *SDF* graph. A consistent *SDF* graph is a graph which has a non-trivial solution for its balance equations. Hence, $\gamma_a > 0, \forall$ actors a. A detailed explanation of consistency can be found in [21].

The *SDF* actors in Fig. 5 are simply considered *CFDF* actors with exactly one mode. Three consistent *SDF* graphs (cf. Fig. 6) can be constructed for the *CFDF* graph from Fig. 5.

Having such consistent *SDF* graphs, two approaches can be taken: (i) The *heterochronous dataflow* approach only switches modes if the graph has finished an iteration. This has the advantage that all channel fill sizes are known at compile-time, as the consistent *SDF* graph starts from a known initial marking of the channels. This information allows the synthesis step for the consistent *SDF* graph to calculate a *static schedule*. The *static schedule* can eliminate the checking for the presence of tokens. Furthermore, other transformations which depend on knowledge of the channel fill sizes for each actor execution in the static schedule can also be performed, e.g., lowering of FIFOs to registers if applicable, etc. The disadvantage is the constraint that the *FSM* can only change its state at the end of an iteration. This makes the model *non-compositional* as the points in time for which transitions in an *FSM* of a refined dataflow actor *A* are allowed depend on the consumption and production rates of other dataflow actors in the same *HDF* domain. (ii) The approach presented in [33] takes the opposite view. The mode transitions can be taken at any time. This implies that no knowledge of the channel fill sizes is present when a schedule for a consistent *SDF* graph is executed. Therefore, each actor in the schedule has to test for presence of tokens, i.e., the scheduler has to execute the *enable function* to decide if the actor can be fired via its *invoke function*. Furthermore, optimizations depending on knowledge of the channel fill sizes for each actor execution in the static schedule cannot be performed. However, a speedup compared to the *round robin scheduler* can still be obtained for *multirate dataflow graphs*. The speedup stems from the fact that the generated schedules queries the *enable function* of the contained actors in the correct proportion to each other. The simple round robin scheduler would check the actors *A* to *E* for executability in sequence. However, this constant checking is wasteful as *B* needs only be checked once per two executions of *E* and *C* only once per three executions of *E*.

2.4 CAL Actor Language

The CAL Actor Language (*CAL*) is a programming language which was first invented [11] to describe actors and their behavior only. Later on, the language was extended to allow specification of dataflow graphs containing these actors.

We will first briefly review the syntax of *CAL* actors and the accompanying formal model. For the mathematical model, we assume that all data values processed by *CAL* actors are from the universal set of values *V*. The *CAL* language itself is primarily concerned with the definition of the actors themselves and assumes an environment which provides the concrete dataflow graph and channel types in which the actors are instantiated. Data types which are underspecified in the *CAL* actor description are assumed to be inferred from the types used in the concrete channels to which the actor will be connected in the environment. For an in-depth review of the type system used in *CAL*, we refer the interested reader to the language manual [11].

a

```
1 actor Add() T t1, T t2 ==> T s:
2  action [a], [b] ==> [sum]
3  var sum do sum := a + b; end
4 end
```

b

```
1 actor Select() S,A,B ==> O:
2  action S:[s],A:[v] ==> [v]
3  guard s end
4
5  action S:[s],B:[v] ==> [v]
6  guard not s end
7 end
```

Fig. 7 Two simple *CAL* actor examples an adder and the *BDF* [9] Select actor. (**a**) Addition actor consuming one token from each of its input ports t1 and t2 producing one token on the output port s containing the summation of the input tokens. (**b**) The Select actor forwarding tokens from either port A or port B depending on a control token on port S

CAL actors are described by atomic firing steps [10], where each firing step consumes a finite (possibly empty) sequence of tokens from each input port of the actor and produces a finite (possibly empty) sequence of tokens on each output port. A simple example of a *CAL* actor can be seen in Fig. 7a which consumes one token from each input port t1 and t2, calculates the sum of the data values in these tokens, and produces an output token on the output port s containing this data value.

The names of the input and output ports as well as optional type declarations for the tokens are specified by the port signature of the actor, e.g., as seen in Fig. 7a on Line 1 where the two input ports t1 and t2 as well as the output port s are constrained to have the same token type T inferred from the environment. The computation executed by an actor specified in *CAL* is performed by its actions, e.g., as depicted in the Lines 2–3 where the summation operation of the Add actor is performed. This action is guarded by an *input/output pattern* (Line 2) encoding the requirements on the number of available tokens on input ports, on the number of free token places on the output ports.

Furthermore, the actions can be guarded by general conditions depending on the values of the input tokens as used in modeling of the Select actor known from Boolean dataflow (*BDF*) [9]. The Select actor, depicted in Fig. 7b, forwards a token from either its input port A or B to its sole output port O depending on a control input port S. If the token on S carries a true value, then the token from input port A is consumed and forwarded to the output port (Lines 2–3). Otherwise, the token from input port B is forwarded to the output port (Lines 5–6). In both cases, the control token from port S is consumed.

Moreover, *CAL* actors themselves can have a state $\sigma \in \Sigma$, e.g., representing state variables or states from a *schedule FSM* controlling action execution. An example of an actor using state variables is depicted in Fig. 8a where the state variable is defined on Line 2 and carries the accumulated integration value. Furthermore, there can be additional conditions depending on these state variables guarding action execution. An example for such conditions can be found in the *schedule FSM* of the PingPong actor depicted in Fig. 8b. The execution of action FA is guarded with the condition that the *schedule FSM state* is equal to s0 and the action FB is guarded with condition that the *schedule FSM state* is equal to s1. In this

a

```
1 actor Integrate() T t ==> T s:
2   T sum := 0;
3
4   action [x] ==> [sum]
5   do sum := sum + x; end
6 end
```

b

```
1 actor PingPong() A,B ==> O:
2 FA: action A:[v] ==> [v] end
3 FB: action B:[v] ==> [v] end
4     schedule fsm s0:
5       s0 (FA) --> s1;
6       s1 (FB) --> s0;
7     end
8 end
```

Fig. 8 Two simple stateful *CAL* actors one with a state variable the other one with a *schedule FSM* controlling action activation. (**a**) Integration actor using the state variable sum to keep the current integration value. (**b**) Example of a simple merge actor merging from port A then B repeated ad infinitum

a

```
1 actor NDMerge() A,B ==> O:
2   action A:[v] ==> [v] end
3   action B:[v] ==> [v] end
4 end
```

b

```
1 actor PMerge() A,B ==> O :
2 FA: action A:[v] ==> [v] end
3 FB: action B:[v] ==> [v] end
4     priority FA > FB end
5 end
```

Fig. 9 Two *CAL* actor using different strategies for selection of the action to execute in the presence of multiple enabled actions. (**a**) Example of a simple non-deterministic merge actor specified in *CAL*. (**b**) Example of a merge actor preferring tokens from input port A over tokens from B

sense, schedule *FSMs* only represent syntactic sugar of the *CAL* language as they could be replaced with an explicit actor state variable and guard conditions on this state variable.[2] However, explicitly using schedule *FSMs* clearly separates the actor data manipulation calculating the token values produced and the communication behavior controlling the amount of tokens consumed and produced.

Every atomic firing of an action is executed in a state σ, consumes a finite sequence of input tokens $s \in \mathbb{S}$ from each input port, produces a finite sequence of output tokens $s' \in \mathbb{S}$, and determines a new actor state σ'.[3] The action can only be executed if its guard conditions are satisfied. More formally, the actions can be seen as transitions $\sigma \xrightarrow{s \to s'} \sigma'$ in a transition relation $\tau \subseteq \Sigma \times \mathbb{S}^n \times \mathbb{S}^m \times \Sigma$ where n and m denote the number of input and output ports, respectively.

If multiple transitions $\sigma \xrightarrow{s \to s'} \sigma' \in \tau$ are enabled simultaneously, the *CAL* environment can select one of these transitions *non-deterministically* for execution. An example of such a *CAL* actor is depicted in Fig. 9a which describes a non-deterministic merge actor where the environment can select either the token on

---

[2]The *CAL* language even allows to specify schedule *FSMs* by a regular expression.

[3]We use $\mathbb{S}$ to denote the finite sequence of input tokens from the universal set of values V.

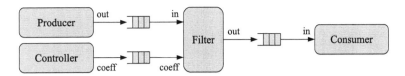

Fig. 10 Filter example dataflow graph: The *ACFSMs* (Producer, Controller, Filter and Consumer) are connected via FIFO channels. The Filter actor has two inputs (in and coeff) from which tokens are consumed and one output (out) onto which tokens are produced

port A or on port B to forward to the output port O. If this decision should be encoded in the actor instead, *CAL* provides a priority mechanism which creates a non-reflexive, anti-symmetric, and transitive partial order relation on the transitions. Actions which are lower in this partial order cannot be executed if there is at least one action enabled which is higher in the partial order. To exemplify this, we can look at the priority merge actor from Fig. 9b which prioritizes the action FA before action FB. Therefore, action FB can never be executed if there is a token present on input port A.

With the priority extension, the formal model of a *CAL* actor consists of a triple $(\sigma_0, \tau, \succ)$ containing its initial actor state $\sigma_0 \in \Sigma$, its transition relation $\tau \subseteq \Sigma \times \mathbb{S}^n \times \mathbb{S}^m \times \Sigma$ encoding actions and guards, and the partial order $\succ \subset \tau \times \tau$ encoding the priority relationship of the actions.

2.5 Extended Codesign Finite State Machines

The Extended Codesign Finite State Machine (*ECFSM*) Model of Computation (*MoC*) has been presented in [35]. As will be seen, the original Codesign Finite State Machine (*CFSM*) MoC used in POLIS [2] is a special case of the *ECFSM* *MoC*. Both models, however, are refinements of the Abstract Codesign Finite State Machine (*ACFSM*) Model, also presented in [35].

In this section, we will therefore first introduce the *ACFSM* Model of Computation, followed by the discussion of the refinement step which transforms a given *ACFSM* into an *ECFSM*. An *ACFSM* is a formal model consisting of a network of *FSMs* (in the following called actors) connected via unbounded *FIFO channels*. As in CAL, tokens (also called *events*) are transmitted across the channels carrying data values of some data type. An example of such a dataflow network consisting of several *ACFSMs* in Fig. 10 is shown.

Briefly, a single *ACFSM* consists of an *FSM* controlling the communication behavior of the actor, transforming a finite sequence of input tokens into a finite sequence of output tokens.

In order to formally describe the behavior of a single *ACFSM*, we will first look at the behavior of the Filter actor from Fig. 10. Its task is to filter a sequence of pixels from the Producer actor by multiplying all pixels of a frame by a

Fig. 11 Behavior of
the `Filter` actor shown
in Fig. 10

```
1 module Filter {
2   input byte in, coeff;
3   output byte out;
4   int nLines, nPix, line, pix;
5   byte k;
6   forever {
7     (nLines, nPix) = read(in, 2);
8     k = read(coeff, 1);
9     for(line = 0; line < nLines; ++line) {
10      if(present(coeff, 1))
11        k = read(coeff, 1);
12      for(pix = 0; pix < nPix; ++pix) {
13        write(out, read(in, 1) * k, 1);
14      }
15    }
16  }
17 }
```

coefficient (provided by the `Controller` actor). At the beginning of each frame,
the `Producer` actor writes two tokens onto the output `out` containing the number
of lines per frame and the number of pixels per line, respectively. These two values
are read by the `Filter` actor from the input `in`, whereas the initial multiplication
coefficient is read from the input `coeff`. For each line, the coefficient values may
change, depending on whether or not a token is available on the input `coeff`.
Subsequently, the filtered pixels are written onto output `out`.

The behavior of the `Filter` actor can be represented by C-like pseudocode as
shown in Fig. 11. There exist three primitives which can be used in order to access
the FIFO channels connected to the inputs and outputs of the actor: `read(in,
n)` and `write(out, data, n)` consume and produce n tokens from input `in`
and output `out`, respectively. Note that whereas `read` blocks until enough tokens
are available, `write` never blocks, as the FIFO channels are unbounded. The third
primitive `present(in, n)` returns true if at least n tokens are available on the
input `in`.

In order to transform the `Filter` pseudocode into an *ACFSM*, we will now
discuss the formal definition of an *ACFSM*:

Definition 2. An *ACFSM* is a triple $a = (I, O, T)$ consisting of a finite set of *inputs*
$I = \{i_1, i_2, \ldots, i_n\}$, a finite set of *outputs* $O = \{o_1, o_2, \ldots, o_n\}$, and a finite set of
transitions T. In the following, we will use $i[k]$ to indicate the token on the k-th
position on the FIFO channel connected to input $i \in I$ (as seen from the *ACFSM*).
Analogously, we will use $o[k]$ to indicate the token produced by a transition onto the
k-th position on the FIFO channel connected to output $o \in O$.

Definition 3. A transition of an *ACFSM* is a tuple $T = (req, cons, prod, f_{\text{guard}}, f_{\text{action}})$: The *input enabling rate* $req : I \rightarrow \mathbb{N}_0$ maps each input $i \in I$ to the number
of tokens which must be available on the channel connected to i in order to enable
the transition. The *input consumption rate cons* $: I \rightarrow \mathbb{N}_0 \cup \{ALL\}$ specifies for each
input $i \in I$ the number of tokens which will be consumed from the channel connected

to i when the transition is executed. Specifying $cons(i) = ALL$ for a given input i *resets* the channel connected to i, i.e., all tokens currently stored in the channel are consumed. Otherwise, a transition is not allowed to consume more tokens than requested, i.e., $cons(i) \leq req(i)$. Analogously, the *output production rate prod* $: O \rightarrow \mathbb{N}_0$ specifies for each output $o \in O$ the number of tokens which will be produced on the channel connected to o when the transition is executed. The *guard function* f_{guard} is a boolean-valued expression over the values of the tokens required on the inputs, i.e., $\{i[k] \mid i \in I \wedge 1 \leq k \leq req(i)\}$. Note that only if both the input enabling rate *req* and the guard function f_{guard} are satisfied, a transition can be executed. Finally, the *action function* f_{action} determines the values of the tokens which are to be produced on the outputs, i.e., $\{o[k] \mid o \in O \wedge 1 \leq k \leq prod(o)\}$, when the transition is executed.

In contrast to CAL, *ACFSMs* cannot have explicit state variables. Nevertheless, these can be modeled by *state feedback channels*, i.e., by an input/output pair of an *ACFSM* which is connected to the same channel containing a certain state variable. If a transition wants to update the state variable, it must consume the token containing the old value, and produce a token containing the new value.

The execution semantics of an *ACFSM* can be described as follows: Initially, an *ACFSM* is *idle*, i.e., waiting for input tokens. An *ACFSM* is *ready* if at least one transition $t \in T$ is enabled, i.e., if both the input consumption rate $t.req$ and the guard function $t.f_{\text{guard}}$ are satisfied. Subsequently, a single enabled transition t will be chosen, transitioning the *ACFSM* into the *executing* state. During the *executing* state, the *ACFSM* *atomically* consumes tokens from the inputs according to $t.cons$, performs the action function $t.f_{\text{action}}$, and produces tokens on the outputs according to $t.prod$.

It should be noted that due to the non-blocking reads which can be implemented by *ACFSMs*, they are not continuous in Kahn's sense [26], i.e., the arrival of tokens at different times may change the behavior of the dataflow graph.

In [35], the authors claim that "*ACFSM* transitions [...] can be conditioned to the *absence* of an event over a signal". In principle, there are two possibilities how this could be modeled:

- Provide a special syntax for the input enabling rate such that also checks for empty channels can be expressed. However, the input enabling rate as given in Definition 3 only permits the expression of *minimum conditions*. Therefore, an input i with $req(i) = 0$ could lead to the execution of the corresponding transition even if the channel connected to i is *not* empty.
- Specify priorities for the transitions of an *ACFSM* such that transitions requiring more tokens can have a higher priority than transitions requiring less tokens. However, in the original paper [35] it is also stated that "if several transitions can be executed [...], the *ACFSM* is non-deterministic and can execute any of the matching transitions", i.e., there seems to be no concept of priorities for transitions.

For example, lines 10–11 in Fig. 11 cannot be adequately modeled without checking for the absence of tokens: One transition is needed which checks if at least one token is available on input `coeff`, i.e., $req(\text{coeff}) = 1$. However, as we do not want to block execution if there is no token available, a second transition is needed with $req(\text{coeff}) = 0$. As discussed above, the second transition could be executed even if the corresponding channel is not empty. Without the means of expressing a check for an empty FIFO channel, we can only assume that the first transition has a higher priority than the second, allowing us to check for the absence of a new multiplication coefficient.

The behavior of the `Filter` actor given in Fig. 11 can be transformed into an *ACFSM* as follows: First, feedback channels are added for the state variables `nLines`, `nPix`, `line`, `pix` and `k`. Additionally, a feedback channel is needed for storing the current state of the *ACFSM*, `state`. The corresponding input/output pairs will be named accordingly, resulting in

- the set of inputs $I = \{\text{in}, \text{coeff}, \text{nLines}, \text{nPix}, \text{line}, \text{pix}, \text{k}, \text{state}\}$ and
- the set of outputs $O = \{\text{out}, \text{nLines}, \text{nPix}, \text{line}, \text{pix}, \text{k}, \text{state}\}$.

We assume that an initial token with a value of 0 is placed on each state feedback channel. The resulting transitions are shown in Table 1. For example, the first transition can be enabled if two tokens are available on input `in`, one token on `coeff`, and one token on each input connected to a state feedback channel (except `pix`). Additionally, the current state of the *ACFSM* must be the initial state, i.e., the value of the (single) token on input `state` must represent the initial state, i.e., `state[1] = 0`. If these requirements are satisfied, the transition will be executed, and the specified output tokens produced: For example, the value of the first token consumed from input `in` is used to produce the single token on output `nLines`. Note that the value of the current state is set to 1, allowing other transitions to be executed.

In conclusion, *ACFSMs* specify the *topology* of the dataflow graph and the *functional* behavior of each module. However, the communication over unbounded channels with FIFO semantics needs to be refined in order to be implementable with a finite amount of resources. This refinement from infinite-sized queues to implementable finite-sized queues is exactly the transformation from an Abstract *CFSM* into an Extended *CFSM*, and will be briefly discussed in the following.

Write operations carried out by the transitions of an *ACFSM* are always *non-blocking*, as the corresponding channels have an infinite capacity. For a given finite-sized channel, however, if write operations should be non-blocking and the channel in question is full, new data arriving on the channel will overwrite old data stored in the channel. If this behavior is not desired, different approaches may be applied. First, it is often sufficient to increase the channel's capacity. This approach fails, however, if the average production rate of the producer is greater then the average consumption rate of the consumer. In general, calculating sufficient channel capacities is undecidable. Therefore, *ECFSMs* introduce blocking behavior to prevent data loss. Finally, traditional *CFSMs* are a special case of *ECFSMs*, where all channels have a capacity of one.

Table 1 Corresponding *ACFSM* of the `Filter` actor shown in Fig. 11 consisting of six transitions

$(req(i), cons(i)) / f_{\text{guard}}$								$prod(o) / f_{\text{action}}$						
in	coeff	nLines	nPix	line	pix	k	state	out	nLines	nPix	line	pix	k	state
2	1	1	1	1	0	1	1	0	1	1	1	0	1	1
$state[1] = 0$								$nLines[1] \Leftarrow in[1]$						
								$nPix[1] \Leftarrow in[2]$						
								$line[1] \Leftarrow 0$						
								$k[1] \Leftarrow coeff[1]$						
								$state[1] \Leftarrow 1$						
0	0	(1,0)	0	(1,0)	0	0	1	0	0	0	0	0	0	1
$state[1] = 1 \wedge line[1] = nLines[1]$								$state[1] \Leftarrow 0$						
0	1	(1,0)	0	(1,0)	1	1	1	0	0	0	0	1	1	1
$state[1] = 1 \wedge line < nLines$								$k[1] \Leftarrow coeff[1]$						
								$pix[1] \Leftarrow 0$						
								$state[1] \Leftarrow 2$						
0	0	(1,0)	0	(1,0)	1	0	1	0	0	0	0	1	0	
$state[1] = 1 \wedge line[1] < nLines[1]$								$pix[1] \Leftarrow 0$						
								$state[1] \Leftarrow 2$						
0	0	0	(1,0)	1	(1,0)	0	1	0	0	0	1	0	0	1
$state[1] = 2 \wedge pix[1] = nPix[1]$								$line[1] \Leftarrow line[1] + 1$						
								$state[1] \Leftarrow 1$						
1	0	0	(1,0)	0	1	(1,0)	(1,0)	1	0	0	0	1	0	0
$state[1] = 2 \wedge pix[1] < nPix[1]$								$out[1] \Leftarrow in[1] * k[1]$						
								$pix[1] \Leftarrow pix[1] + 1$						

In order to adequately model the multiplication coefficient update (Lines 10–11 in Fig. 11), we assume that the transitions are assigned priorities according to their ordering, with higher priorities first. Note also that for reasons of readability, $cons(i)$ is omitted if $req(i) = cons(i)$

2.6 SysteMoC

SYSTEMOC [13] extends the notion of *ECFSM* by actor states and hierarchy. At a glance, the notation extends conventional dataflow graph models by finite state machines controlling the consumption and production of tokens by actors. SYSTEMOC itself is an extension library for *SystemC* [1, 19], which is itself a C++ class library. We will first introduce the non-hierarchical SYSTEMOC model. The vertices of the graph correspond to actors, that are more formally defined below:

Definition 4 (Actor). An actor is a tuple $a = (I, O, \mathscr{F}_{\text{func}}, \mathscr{R})$ containing *actor ports* partitioned into the set of *actor input ports* I and *actor output ports* O, the *actor functionality* $\mathscr{F}_{\text{func}}$ and the *actor FSM* $\mathscr{R}$.

In contrast to *CAL* and *ECFSM*, we have split the state of the actor into two parts. The *actor functionality state* which is manipulated by the functions $f \in \mathscr{F}_{\text{func}}$ from the *actor functionality* and the *actor FSM state* contained in $\mathscr{R}$ which is controlled by the *actor FSM*. This distinction enables the mathematical expression of the

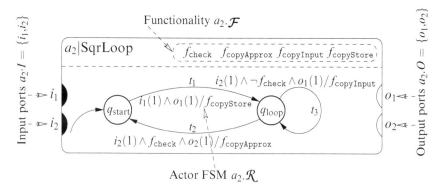

Fig. 12 Visual representation of a SqrLoop actor a_2 as used in the network graph $g_{\text{sqr.flat}}$ displayed in Fig. 13a. The SqrLoop actor is composed of *input ports I* and *output ports O*, its *functionality* $\mathcal{F}_{\text{func}}$, and the *actor FSM* $\mathcal{R}$ determining the communication behavior of the actor. The condition $i_x(n)/o_y(m)$ on an *actor FSM transition* denotes a predicate that tests the availability of at least n tokens/m free places on the actor input port i_x/output port o_y.

separation of the actor data manipulation calculating the token values produced, e.g., $f_{\text{copyApprox}}$ as shown in Fig. 12, and the *communication behavior* controlling the amount of tokens consumed and produced. Subsequent analysis steps will abstract from the state of the actor functionality and only consider the *actor FSM state*. More intuitively, one can think of the state of the actor functionality as state variables of the actor, which are primarily modified by the data path of the actor modeled by functions, e.g., $f_{\text{copyApprox}} \in \mathcal{F}_{\text{func}}$.

Furthermore, we have extended the well known dataflow graph definition with a notion of ports, i.e., channels no longer connect only actors but actor ports as depicted in Fig. 13a. We will call such an extended representation a *network graph*, formally defined below:

Definition 5 (Non-Hierarchical Network Graph). A *non-hierarchical network graph* is a directed graph $g_n = (A, C, P)$ containing a set of *actors A*, a set of *channels* $C \subseteq A.O \times A.I$, and a channel parameter function $P : C \to \mathbb{N}_\infty \times V^*$ which associates with each channel $c \in C$ its buffer size $n \in \mathbb{N}_\infty = \{1, 2, 3, \ldots \infty\}$ and also a possibly non-empty sequence $v \in V^*$ of initial tokens.[4] Furthermore, we require that each actor port $p \in A.I \cup A.O$ is connected to exactly one channel c.

The execution of a SYSTEMOC model can be divided into three phases: (i) checking for enabled transitions for each actor, i.e., sufficient tokens and free places

---

[4] We use the "."-operator, e.g., $a.I$, for member access of tuples whose members have been explicitly named in their definition, e.g., member I of actor $a \in A$ from Definition 4. Moreover, this member access operator has a trivial extension to sets of tuples, e.g., $A.I = \bigcup_{a \in A} a.I$, which is also used throughout this document. We use V^* to denote the set of all possible *finite sequences* of tokens $v \in V$, i.e., $V^* = \bigcup_{n \in \{0,1,\ldots\}} V^n$.

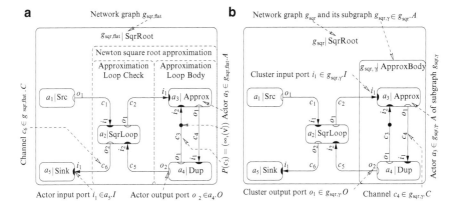

Fig. 13 The *network graphs* displayed above implement Newton's iterative algorithm for calculating the square roots of an infinite input sequence of numbers generated by the Src actor a_1. The square root values are generated by Newton's iterative algorithm SqrLoop actor a_2 for the error bound checking and a_3–a_4 to perform an approximation step. After satisfying the error bound, the result is transported to the Sink actor a_5. (**a**) Example of a non-hierarchical network graph $g_{sqr,flat}$, (**b**) And a corresponding hierarchical version g_{sqr}

on input and output ports and guard functions evaluating to true, (ii) selecting and executing one enabled transition per actor, and (iii) consuming and producing tokens as specified by the transition. Note that step (iii) might enable new transitions. Hence, SYSTEMOC is similar in this regard to *ECFSMs* where actors are blocked as long as the output buffers cannot store the results. More formally, the *actor FSM* is defined as follows:

Definition 6 (Actor FSM). The *actor FSM* of an actor $a \in A$ is a tuple $a.\mathscr{R} = (T, Q, q_0)$ containing a finite set of *transitions* T, a finite set of *states* Q and an *initial state* $q_0 \in Q$. A *transition* $t = (q, req, cons, prod, f_{guard}, f_{action}, q') \in T$ containing a current state $q \in Q$ and a next state $q' \in Q$. The *enabling rate req* : $I \cup O \to \mathbb{N}_0$, *input consumption rate cons* : $I \to \mathbb{N}_0$, and the *output production rate prod* : $O \to \mathbb{N}_0$ map each input/output to the required number of tokens/places to fire the transition, the number of tokens consumed if the transition is fired, and the number of tokens produced if the transition is fired, respectively. Furthermore, a transition is not allowed to consume more tokens than requested, i.e., $cons(i) \leq req(i)$, or produce more tokens than reserved, i.e., $prod(o) \leq req(o)$. The *guard function* f_{guard} is a boolean-valued expression over the values of the tokens required on the inputs, i.e., $\{i[k] \mid i \in I \land 0 \leq k < req(i)\}$. Note that only if both the input enabling rate *req* and the guard function f_{guard} are satisfied, a transition can be executed. Finally, the *action function* f_{action} determines the values of the tokens which are to be produced on the outputs, i.e., $\{o[k] \mid o \in O \land 0 \leq k < prod(o)\}$, when the transition is executed.

Next, we introduce the notion of *clusters* which we will be used to introduce hierarchy into the non-hierarchical network graph model by allowing actors in a *network graph* or *cluster* to be either a *proper actor* or a *cluster*. This will be used in Sect. 5.1 to encode a *quasi-static schedule* for a set of actors. These actors will be put into a cluster and the resulting quasi-static schedule will be annotated to the cluster by means of a *cluster FSM*. A cluster itself is simply a *network graph* extended by the notion of ports.

Definition 7 (Cluster). A *cluster* is a graph $g_\gamma = (I, O, A, C, P)$ containing *cluster ports* partitioned into *cluster input ports* $I \subseteq A.I$ and *cluster output ports* $O \subseteq A.O$, a set of *actors* A, a set of *channels* $C \subseteq A.O \times A.I$, and a channel parameter function $P : C \to \mathbb{N}_\infty \times V^*$ which associates with each channel $c \in C$ its buffer size $n \in \mathbb{N}_\infty = \{1, 2, 3, \ldots \infty\}$ and also a possibly non-empty sequence $v \in V^*$ of initial tokens. Hence, an input or output port of an actor of the cluster is either connected to a channel inside the cluster or is a cluster input or output port.

As can be seen from Definitions 5 and 7, a non-hierarchical network graph g_n is simply a cluster without cluster ports, i.e., $g_n.I = g_n.O = \emptyset$, where all actor $g_\gamma.A$ are proper actors, i.e., $g_\gamma.A \cap G_\gamma = \emptyset$.[5] We will call a cluster (network graph) g_γ to be non-hierarchical if it contains no subgraph, i.e., $g_\gamma.A \cap G_\gamma = \emptyset$, and hierarchical otherwise. More formally, we define a hierarchical network graph as follows:

Definition 8 (Hierarchical Network Graph). A *hierarchical network graph* g_n is a *cluster without cluster ports*, i.e., $g_n.I = g_n.O = \emptyset$.

The transformation of a possibly non-hierarchical original cluster into a hierarchical cluster will be denoted by usage of the $\Gamma : G_\gamma \times 2^A \to G_\gamma \times G_\gamma$ *cluster operator*. This operation takes an original cluster and a set of actors $A_{\text{to cluster}}$ to generate a clustered network graph containing a subgraph which replaces the set of clustered actors $A_{\text{to cluster}}$. An example of such a cluster operation can be seen in Fig. 13a depicting the original network graph $g_{\text{sqr.flat}}$ and Fig. 13b showing the clustered network graph g_{sqr} containing the subgraph $g_{\text{sqr.}\gamma}$ which replaces the clustered actors, i.e., $\Gamma(g_{\text{sqr.flat}}, A_{\text{to cluster}}) = (g_{\text{sqr}}, g_{\text{sqr.}\gamma})$.

The semantics of a hierarchical network graph g_n is defined by the semantics of its corresponding non-hierarchical network graph g'_n. This non-hierarchical network graph is obtained by recursively applying the dissolve operation Γ^{-1} until the resulting network graph no longer contains any subgraphs. The dissolve operation Γ^{-1} is the inverse of the cluster operation.

Hence, so far we only introduced *structural hierarchy*. To allow arbitrary nesting of dataflow graphs and *FSMs*, we will finally introduce the notion of *cluster FSM* $g_\gamma.\mathscr{R}$ associated with a cluster g_γ responsible for the scheduling of the dataflow graph contained in g_γ.

---

[5]We use G_γ to denote the set of all possible clusters.

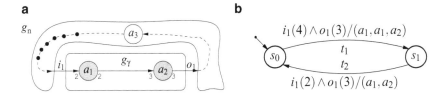

Fig. 14 Example of a cluster and its cluster FSM. (**a**) Data flow graph g_n with a cluster g_γ, (**b**) Cluster FSM $g_\gamma.\mathscr{R}$ belonging to cluster g_γ

Definition 9 (Cluster *FSM*). The *cluster FSM* of a cluster g_γ is a tuple $g_\gamma.\mathscr{R} = (T, Q, q_0)$ containing a finite set of *transitions* $T \ni t = (q, req, cons, prod, f_{\text{sched}}, q')$, a finite set of *states* Q, and an *initial state* $q_0 \in Q$. Cluster and Actor *FSMs* differ in their possible actions $f_{\text{action}}/f_{\text{sched}}$ which are selected from the actor functionality $\mathscr{F}_{\text{func}}$ in the case of *actor FSMs* and are scheduling sequences $\mathscr{F}_{\text{sched}}$ in the case of *cluster FSMs*. A scheduling sequence $f_{\text{sched}} \in \mathscr{F}_{\text{sched}}$ is a finite sequence of actor/cluster firings from the actors/clusters contained in the dataflow graph g_γ, i.e., $\mathscr{F}_{\text{sched}} = g_\gamma.A^*.$[6]

In summary, a *cluster FSM* is an *actor FSM* which simply has a different mechanism controlling the execution of the contained actors in the cluster. To represent *structural hierarchy* a cluster without a *cluster FSM* is generated. In this case the actors contained in the cluster are *self-scheduled*, i.e., they fire as soon as sufficient tokens and free places permit it. To exemplify, we consider the network graph depicted in Fig. 14a. It contains a cluster g_γ with actors a_1 and a_2. The scheduling of these actors is given by the cluster FSM depicted in Fig. 14b. The actions of this cluster FSM are sequences of actor firings, e.g., (a_1, a_1, a_2) from transition t_1 which specified to fire actor a_1 twice followed by a firing of actor a_2. Without a cluster FSM the cluster g_γ is self-scheduled, e.g., the actor a_1 might fire a third time before actor a_2 is fired.

3   Automatic Model Extraction

In order to be able to apply MoC specific analysis methods such as static schedule generation or deadlock detection, it is important to recognize data flow models of computation such as *SDF* and *CSDF*. We will show that this can be accomplished by

---

[6] We use the "."-operator, e.g., $g_\gamma.A$, for member access of tuples whose members have been explicitly named in their definition, e.g., member A of cluster g_γ from Definition 7. We use A^* to denote the set of all possible *finite sequences* of actors/clusters $a \in A$, i.e., $A^* = \bigcup_{n \in \{0,1,\dots\}} A^n$. An element of this set can be interpreted as a static schedule of actors/clusters which can be fired one after the other.

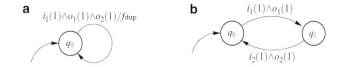

Fig. 15 Basic representation of an *SDF* actor (**a**) and a *CSDF* actor (**b**) by actor *FSMs*. (**a**) Basic representation of an *SDF* actor via an *actor FSM*. The depicted *FSM* belongs to the Dup actor a_4 depicted in Fig. 13, (**b**) Basic representation of an *CSDF* actor via an *actor FSM*

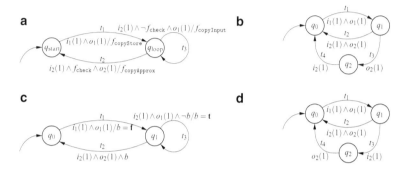

Fig. 16 Various actor *FSMs* which do not fit the basic representation for static actors as exemplified in Fig. 15. (**a**) The *actor FSM* of the SqrLoop actor a_2 which is obviously not a static actor, (**b**) An *actor FSM* that cannot be converted to a basic representation for static actors, (**c**) An *actor FSM*, which seems to belong to the *dynamic domain*, but exhibits *CSDF* behavior. This is due to the manipulation of the boolean variable b which leads to a cyclic execution of the transitions t_1, t_3, and t_2, (**d**) An *actor FSM* that can be converted to a basic representation for static actors

the analysis of the *actor FSM* of a SysteMoC actor. The most basic representation of static actors in our general data flow notation can be seen in Fig. 15.

The *SDF* actor depicted in Fig. 15a corresponds to the Dup actor a_4 as displayed in Fig. 13. The depicted *actor FSM* contains only one transition which consumes one token from the input port i_1 and duplicates this token on its output ports o_1 and o_2. Clearly, the actor exhibits a static communication behavior corresponding to the *SDF model of computation*. The idea of the classification algorithm is to check if the communication behavior of a given actor can be reduced to a basic representation, which can be easily classified into the *SDF* or *CSDF* model of computation. We first note that basic representation for both *SDF* and *CSDF* models are circular *FSMs*, e.g., as depicted for the *CSDF* actor in Fig. 15b.

Not all *actor FSMs* exhibit such a regular pattern (cf. Fig. 16). However, some of them, e.g., Fig. 16c,d, still represent static actors. It can be distinguished if analysis of the *actor functionality state* (cf. Definition 4) is required to recognize static actors, e.g., Fig. 16c, or not, e.g., Fig. 16d. In the following we will not consider the *actor functionality state*. Therefore, the presented algorithm will fail to classify Fig. 16c as a static actor. In that sense, the algorithm only provides a *sufficient* criterion. A *sufficient* and *necessary* criterion cannot be given as the problem is undecidable in the general case.

The algorithm starts by trying to determine a *classification candidate* from a *breadth-first search* of the *FSM*. This candidate is later checked for consistency with the entire *FSM* state space.

Definition 10 (Classification Candidate). A possible *CSDF* behavior of the actor is captured by a *classification candidate* $c = (\rho_0, \rho_1, \ldots \rho_{\tau-1})$ where each $\rho = (cons, prod)$ represents a phase of the *CSDF* behavior and τ is the number of phases.

To exemplify, we consider Fig. 15b. In this cases we have two phases, i.e., $\tau = 2$, where the first phase (ρ_0) consumes one token from port i_1 and produces one token on port o_1 while the second phase (ρ_1) consumes one token from port i_2 and produces one token on port o_2.

It can be observed that all paths starting from the initial *actor FSM* state q_0 must comply with the *classification candidate*. Furthermore, as *CSDF* exhibits cyclic behavior, the paths must also contain a cycle. We now search for such a path $p = (t_1, t_2, \ldots, t_n)$ of transitions t_i in the *actor FSM* starting from the initial state q_0 via a *breadth-first search*. The path can be decomposed into an acyclic prefix path p_a and a cyclic path p_c such that $p = p_a \frown p_c$, i.e., $p_a = (t_1, t_2, \ldots t_{i-1})$ being the prefix and $p_c = (t_i, t_{i+1}, \ldots t_n)$ being the cyclic part. The *breadth-first search* finds the shortest path which loops back onto itself.

After such a path p has been found, it will be contracted by unifying adjacent transitions t_i, t_{i+1} if t_i only consumes tokens or t_{i+1} only produces tokens. For clarification, we look at Fig. 16b,d. In both depicted *FSMs* the transition t_2 consumes and produces exactly as much tokens as the transitions sequence (t_3, t_4). However, the transition sequence (t_3, t_4) in Fig. 16d is equivalent to the single transition t_2 while in Fig. 16b the transition sequence cannot be contracted due to the changed order between the consumption of tokens on i_2 and the production of tokens on o_2.

For further illustration, both *FSMs* have been used in a dataflow graph depicted in Fig. 17a. The resulting dependencies between the transitions in a transition trace of the actors a_2 and a_3 are shown in Fig. 17b. As can be seen in Fig. 17c if the transition sequence (t_3, t_4) of actor a_3 is contracted into a single transition t_c, then the resulting dependencies of t_c are exactly the same as for transition t_2. This is the reason why the *FSM* from Fig. 16d can be classified into a *CSDF* actor with an *actor FSM* as depicted in Fig. 15b. Furthermore, the contraction is a valid transformation as it does not change the data dependencies in the dataflow graph. This can be seen by comparing the transition dependencies depicted in Fig. 17b,c. Compacting the transition sequence (t_3, t_4) of actor a_3 generates the transition t_c, which has a data dependency from the consumption of a token on i_2 to the production of a token on o_2. However, the previous transition sequence (t_3, t_4) also induces this data dependency as t_4 can only be executed after t_3.

In contrast to this, the contraction of the transition sequence (t_3, t_4) of actor a_2 into a transition t_d does introduce a new erroneous data dependency from the consumption of a token on i_2 to the production of a token on o_2. The data

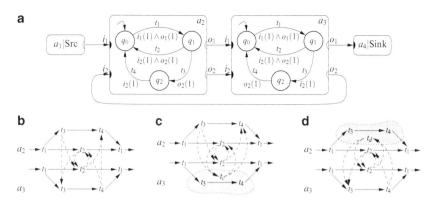

Fig. 17 Dataflow graph containing *FSMs* from Fig. 16b,d and the dependencies in their transition traces if various transition sequences are contracted. (**a**) Dataflow graph containing *FSMs* from Fig. 16b (in actor a_2) and Fig. 16d (in actor a_3). Due to data dependencies actor a_2 will never execute transition t_2. On the other hand, actor a_3 is free to chose either the transition sequence (t_3, t_4) or the transition t_2 in its execution trace, (**b**) Dependencies in the transition trace of actor a_2 and a_3 from (**a**). *Dashed lines* represent data dependencies. *Solid lines* represent dependencies induced by the sequential nature of the *FSM*, (**c**) Dependencies in the transition trace if transition sequence (t_3, t_4) of actor a_3 has been contracted into the transition t_c. Note that transition t_2 and the contracted transition t_c of actor a_3 have the same data dependencies, (**d**) Dependencies in the transition trace if transition sequence (t_3, t_4) of actor a_2 has been contracted into the transition t_d inducing two erroneous dependency cycles $a_2.t_d \rightarrow a_3.t_2 \rightarrow a_2.t_d$ and $a_2.t_d \rightarrow a_3.t_3 \rightarrow a_3.t_4 \rightarrow a_2.t_d$

dependency is erroneous as the original transition sequence (t_3, t_4) has no such dependency as it first produces the token on o_2 before trying to consume a token on i_2. Indeed, the erroneous contraction will introduce a deadlock into the system as can be seen in Fig. 17d where the transition t_d is part of two dependency cycles $a_2.t_d \rightarrow a_3.t_2 \rightarrow a_2.t_d$ and $a_2.t_d \rightarrow a_3.t_3 \rightarrow a_3.t_4 \rightarrow a_2.t_d$ which are not present in the original dependency structure depicted in Fig. 17b.

After the *classification candidate* has been derived from the *FSM*, it has to be validated if it holds for all states of the *FSM*. However, due to the existence of contracted transitions in the *classification candidate* as well as in transition sequences of the *FSM*, a simple matching of phases ρ_i to *FSM* states is infeasible. Instead, we match a *FSM* state q_n with a tuple $r_n = (i, cons, prod)$ where i denotes the *CSDF* phase ρ_i from which it was initially derived and *cons* and *prod* are the tokens which are still to be consumed and produced to complete the consumption and production of the phase ρ_i. The annotation of tuples r instead of *CSDF* phase ρ allows us to match multiple *FSM* state for a *CSDF* phase.

The validation algorithm starts with annotating the tuple $r_0 = (0, \rho_0.cons, \rho_0.prod)$ to the initial state q_0.[7] Starting from the initial state q_0 and the initial tuple

---

[7] We use the "."-operator, e.g., $\rho.cons$, for member access of tuples whose members have been explicitly named in their definition, e.g., member *cons* of *CSDF* phase ρ from Definition 10.

r_0 in a *breadth-first search* over the *state transition diagram* further tuples r_n are annotated to states. The *breadth-first search* stops for a state q if a tuple r is already annotated. The *classification candidate* is rejected if different visits of states lead to contradicting annotations. The *classification candidate* is accepted if all states have tuples r annotated and the *breadth-first search* returns. Furthermore, not all transitions leaving a state q_n are valid for an annotated tuple r_n. For a transition to be valid it has to conform to both *transition criteria* as given below:

Definition 11 (Transition Criteria I). Each outgoing transition $t \in T_n = \{t \in T \mid t.q = q_n\}$ of a visited actor state $q_n \in Q$ must consume and produce less or equal tokens than specified by the annotated tuple r_n, i.e., $\forall t \in T_n : r_n.cons \geq t.cons \land r_n.prod \geq t.prod$. Furthermore, the transition t has to request exactly as much tokens on the input ports as it consumes, i.e., $t.cons = t.req$. Otherwise, the annotated tuple r_n is invalid and the *classification candidate* will be rejected.

Definition 12 (Transition Criteria II). Each outgoing transition $t \in T_n = \{t \in T \mid t.q = q_n\}$ of a visited actor state $q_n \in Q$ must not produce tokens if there are still tokens to be consumed in that phase after the execution of the transition t, i.e., $\forall t \in T_n : r_n.cons = t.cons \lor t.prod = \mathbf{0}$. Otherwise, the annotated tuple r_n is invalid and the *classification candidate* will be rejected.

The *transition criteria* defined above ensure that a *CSDF* phase ρ can only be matched with a transition sequence which induces the same data dependencies as ρ. If a state q_n with annotated tuple r_n has a transition t which does not conform to both *transition criteria* the *classification candidate* is invalid and will be rejected. If the transition t from q_n with annotated tuple $r_n = (i, cons, prod)$ to q_{n+1} is valid then the new tuple r_{n+1} will be calculated as follows:

$$
\begin{aligned}
cons_{\text{left}} &= r_n.cons - t.cons \\
prod_{\text{left}} &= r_n.prod - t.prod \\
i' &= (i+1) \bmod \tau \\
r_{n+1} &= \begin{cases} (i, cons_{\text{left}}, prod_{\text{left}}) & \text{if } cons_{\text{left}} \neq \mathbf{0} \land prod_{\text{left}} \neq \mathbf{0} \\ (i', \rho_{i'}.cons, \rho_{i'}.prod) & \text{otherwise} \end{cases}
\end{aligned}
$$

Where $cons_{\text{left}}$ and $prod_{\text{left}}$ are the consumption and production which remain to be done for the *CSDF* phase ρ_i. If the remaining consumption and production of tokens are both zero then the new tuple r_{n+1} will start with the new *CSDF* phase $\rho_{i'}$ which is calculated from $i + 1$ modulo the number of *CSDF* phase τ of the *classification candidate*. Putting it all together, the algorithm to validate a *classification candidate* is given in Fig. 18. In Sect. 5.1 we will show how detected static actors can be automatically *clustered* to reduce scheduling overhead.

```
 1 FUN validateClassCand
 2   IN: c  // candidate
 3   IN: R // actor FSM
 4 BEGIN
 5   create a queue Q
 6   enqueue q_0 on to Q
 7   annotate r_0 to q_0
 8   WHILE Q is not empty
 9     dequeue an item from Q into q_n
10     FOREACH transition t ∈ {t ∈ T | t.q = q_n}
11       RETURN f if t does not conform to both transition criteria
12       calculate r_{n+1} from r_n annotated to q_n and t
13       IF q_{n+1} = t.q' has no tuple r' annotated
14         annotate r_{n+1} to q_{n+1}
15         enqueue q_{n+1} onto Q
16       ELSE
17         RETURN f if r' ≠ r_{n+1}
18   RETURN t
19 END
```

Fig. 18 Algorithm to validate a *classification candidate c* for the *actor FSM R*

4 Design Methodologies

In this section, we will discuss design flows based on integrated Finite State Machines (*FSM*) and dataflow modeling. A focus is thereby given to the exploitation of knowledge about models of computation for analyzability and, hence enabling subsequent model refinements and optimization of the implementations.

4.1 Ptolemy II

The Ptolemy II project [12] started in 1996 as the successor of the Ptolemy classic project [8] by the University of California, Berkeley. Ptolemy II is a software infrastructure for modeling, analysis, and simulation of embedded systems. The project focuses on the integration of different Models of Computation (*MoC*) by so-called *hierarchical heterogeneity*. Currently, the supported *MoCs* are continuous time, discrete event, synchronous dataflow, finite state machines, concurrent sequential processes, process networks, etc. Ptolemy II supports *domain polymorphism*, i.e., reuse of actors in different MoCs. The coupling of different MoCs is realized via the *charts approach, as presented in Sect. 2.2. Ptolemy II therefore uses an abstract syntax based on the concept of so-called *clustered graphs*. A clustered graph consists of entities and relations. Entities have ports and relations connect to ports. In general, entities correspond to actors and relations to edges. This enables the designer to model, analyze, or simulate heterogeneous systems. Each hierarchy

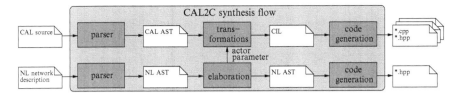

Fig. 19 Design flow using of the *CAL2C* utility compiling *CAL* into *C++* code

level of the *clustered graphs* is associated with a certain MoC. For the analysis of such a hierarchy level, the classical approaches known for the corresponding MoC are employed.

However, the decomposition of an application into hierarchy levels with explicit MoC assignments can be problematic for the designer. Therefore, a tendency to select the most general MoC, e.g., process networks instead of synchronous dataflow for the hierarchy levels can be observed. Furthermore, Ptolemy II only allows simulation and target code generation. Also, there is currently no support for mixed hardware/software implementations.

In the following, we will discuss *OpenDF*, a design framework based on *CAL*, and *SystemCoDesigner*, a design methodology based on SYSTEMOC. Both of these frameworks support the synthesis of integrated *FSM* and dataflow models into hardware/software implementations.

4.2   The OpenDF Design Flow

The *OpenDF* project [4] provides design entry, simulation, and debugging support for the *CAL* language [11]. Dataflow graphs are specified via network language (*NL*) which instantiates the *CAL* actors. The actor parameters derived from the instantiation of the actors in the *NL* file are inserted and the *CAL* source code is parsed and transformed into an *XML* intermediate representation. This representation is used for various transformations, e.g., type inference, type checking, constant propagation, and dataflow analysis. The result is an intermediate representation of the *CAL* source code in static single assignment form (*SSA*). From this result, either a simulation can be executed or two different synthesis backends generating *C++* or *verilog* source code can be started.[8]

The software synthesis tool *CAL2C* [24] supports two synthesis targets one targeting *POSIX* threads and software *FIFOs*, the other one targeting the SystemC environment, using it as a process network (*PN*) simulation framework. The synthesis process, also depicted in Fig. 19, is done by (i) translating the *NL* dataflow

---

[8]The *CAL* simulator component of the *OpenDF* environment can be downloaded from sourceforge [23].

description file into a header instantiating all actors and FIFOs and their initial token valuation, (2) translating the *CAL* actor itself by emitting all the actions and internal state variables as *C* code by an intermediate step via the *C Intermediate Language* [31] and its pretty printer, and (3) by emitting code corresponding to the action schedule responsible for the activation of the actions in the functionality. This code corresponds to the guards, the schedule *FSM*, and the priority specification of the *CAL* actor.

The hardware synthesis tool *CAL2HDL* [25] starts by evaluating the *NL* dataflow description deriving the set of actors to synthesize and their parameters. Each actor in this set is synthesized separately, i.e., no scheduling strategy optimization is applied, resulting in purely self-scheduled actor system. The actor synthesis proceeds from the *XML* intermediate representation by inserting the actor parameters from the *NL* dataflow description as constants into the actors performing constant propagation and inlining all functions into actions. The resulting intermediate representation is self-contained, i.e., it is no longer referencing any external functions. After this, the intermediate representation is transformed to *SSA* form. The next step is the generation of a corresponding hardware module for each action by translating the *SSA* representation to *RTL* via behavioral compilation, i.e., generating multiplex/demultiplex logic for register access, generating control logic for control flow elements (if-then-else, looping constructs, ...), and static scheduling of operators where possible. Furthermore, a central scheduler hardware module is instantiated, responsible for executing the schedule *FSM* by evaluating the guards and selecting a highest priority transition to execute from the priority specification of the *CAL* actor. All these modules communicate via usage of *RTL* signals. Finally, network synthesis is performed by a straightforward transformation of the dataflow graph from the *NL* dataflow description into a corresponding hardware structure. Edges in the dataflow graph are translated to FIFOs with the sizes specified in the *NL* file. Furthermore, actors instantiations in *NL* may have clock domain annotations. Therefore, actors in the same clock domain are connected via usage of synchronous FIFOs while actors in different domains are connected via asynchronous FIFOs.

4.3   SystemCoDesigner

SYSTEMCODESIGNER [27] is a hardware/software codesign framework supporting the synthesis of implementations from SYSTEMOC [13] models. It provides timing simulation for possible architecture mappings as well as software and hardware synthesis backends for SYSTEMOC.

4.3.1   Overview

The overall design flow of the SYSTEMCODESIGNER design methodology is based on (i) a behavioral SYSTEMOC model of an application, (ii) generation of hardware accelerators from SYSTEMOC modules using behavioral synthe-

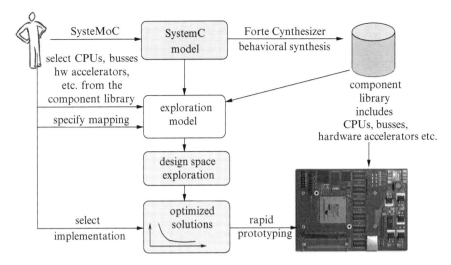

Fig. 20 Design flow using SYSTEMCODESIGNER: The application is given by a SYSTEMoC behavioral model. Forte Cynthesizer [16] is used to automatically synthesize hardware accelerators. Design space exploration using multi-objective evolutionary algorithms automatically searches for the best architecture candidates. The entire hardware/software implementation is prototyped automatically for FPGA-based platforms. The SYSTEMoC behavioral model of the application is used for both automatic design space exploration and creation of hardware accelerators allowing for rapid prototyping

sis, (iii) determination of their performance parameters like required hardware resources, throughput and latency or estimated software execution times, (iv) design space exploration for finding the best candidate architectures, and (v) rapid prototype generation for FPGA platforms. The design flow implemented by SYSTEMCODESIGNER is shown in Fig. 20.

The first step in the SYSTEMCODESIGNER design flow is to describe the application in form of a SYSTEMoC model. Each SYSTEMoC actor can then be transformed and synthesized into both hardware accelerators and software modules. Whereas the latter one is achieved by simple code transformations, the hardware accelerators are built by high level synthesis using Forte Cynthesizer [16]. This allows for quick extraction of important performance parameters like the achieved throughput and the required area by a particular hardware accelerator. In case of Xilinx FPGAs used as target platform, hardware resources in form of flip flops, look-up tables, and block RAMs are estimated. These values can be used to evaluate different solutions found during automatic design space exploration.

The performance information together with the executable specification and a so-called *architecture template* serve as the input model for design space exploration (cf. Fig. 20). The architecture template is represented by a graph which contains all possible hardware accelerators, processors, memories, and the communication infrastructure from which the design space exploration has to select those ones which are necessary in order to fulfill the user requirements in terms of overall throughput and chip size. The fewer components are allocated, the fewer hardware resources are required. This, however, in general comes along with a reduction in throughput, thus leading to tradeoffs between execution speed and implementation costs. Multi-objective optimization together with symbolic optimization techniques [30] are used to find sets of optimized solutions.

From this set of optimized solutions, the designer selects a hardware/software implementation best suited for his needs. Once this decision has been made, the last step of the proposed ESL design flow is the rapid prototyping of the corresponding FPGA-based implementation. The implementation is assembled by interconnecting the hardware accelerators and processors of the selected implementation with special communication modules provided in a component library. Furthermore, the program code for all microprocessors is generated by compiling source code derived from the SYSTEMOC model and linking it with a run-time library for the microprocessor. Finally, the entire platform is compiled into an FPGA bit stream using, e.g., the Xilinx Embedded Development Kit (EDK) [36] tool chain.

5 Exploiting Static MoCs For Scheduling

In this section, novel methods for exploiting static MoCs embedded in more general dataflow graphs will be presented. In this context, clustering methods for the purpose of multi-processor scheduling are of particular interest.

5.1 SysteMoC Clustering

Generating *program code* from a dataflow graph for microprocessor target requires a scheduling strategy for the actors mapped to this microprocessor. While *static scheduling* [5, 22] is used for models with limited expressiveness, e.g., static dataflow graphs. However, real world designs also contain dynamic actors. The most basic strategy used is to postpone all scheduling decisions to run-time (*dynamic scheduling*) with the resulting significant scheduling overhead and, hence, a reduced system performance. However, this strategy is suboptimal if it is known that some of the actors exhibit regular communication behavior like *SDF* and *CSDF* actors. The scheduling overhead problem could be mended by coding the actors of the application at an appropriate level of granularity, i.e., combining as much functionality into one actor that the computation costs dominate the scheduling overhead. This can be thought of as the designer manually making clustering decisions. However, the appropriate level of granularity is chosen by the designer by trading schedule efficiency improving bandwidth against resource consumption (e.g., buffer capacities), and scheduling flexibility improving latency. Aggregation of functionality into one actor can also mean that more data has to be produced and consumed atomically which requires larger buffer capacities and may delay data unnecessarily, therefore degrading the latency if multiple computational resources are available for execution of the application. Furthermore, if the mapping of actors to resources itself is part of the synthesis step, as it is in case of design space exploration, an appropriate level of granularity can no longer be chosen in advance by the designer as it depends on the mapping itself.

In order to obtain efficient schedules in terms of latency and throughput, subgraphs of static actors can be clustered and thereby prescheduled at compile-time with the scheduling decisions which can be performed at this time while scheduling decisions not amenable to compile-time analysis are deferred to run-time. Unfortunately, existing algorithms [29, 32] might result in infeasible schedules or greatly restrict the clustering design space by considering only *SDF* subgraphs that can be clustered into monolithic *SDF* actors without introducing deadlocks. The problem stems from the fact that static scheduling adds constraints to the possible schedules of a system. Imposing too strict constraints on the possible schedules excludes all feasible schedules deadlocking the system. As a remedy a generalized clustering algorithm is presented in [14] which annotates a *quasi-static schedule (QSS)* via the *cluster FSM* formalism given in Definition 9. Intuitively, a *QSS* denotes a schedule in which a relatively large proportion of scheduling decisions have been made at compile-time. Advantages of *QSSs* include lower run-time overhead and improved predictability compared to general dynamic schedules. Clustering decisions, e.g., the subgraph which should be clustered, can be driven by the SYSTEMCODESIGNER design methodology presented in the previous section. In the following, we will also call such a scheduled cluster a *composite actor*, as it can be viewed as a black box where only the *cluster FSM* influences the communication behavior of such a cluster. To exemplify the computation of a quasi-static schedule, we consider the cluster $g_{sqr,\gamma}$ from Fig. 13b implementing the `ApproxBody` of Newton's iterative square root algorithm. This cluster contains two *SDF* actors $a_3|$`Approx` (doing the actual approximation) and $a_4|$`Dup` (a simple token dupli-cator). A valid static schedule for the cluster is $a_3.i_1(1)\&a_4.o_2(1)/(a_3,a_4)$ which checks the prerequisite[9] of the availability of one token on the input $a_3.i_1$ and one free place on the output $a_4.o_2$ and fires actor a_3 followed by actor a_4 if the prerequisite is satisfied. [10] This schedule can be expressed by a *cluster FSM* with a single state with the $a_3.i_1(i)\&a_4.o_2(1)/(a_3,a_4)$ transition as a self-loop.

However, the presented static schedule corresponds to a conversion of the cluster into an *SDF* composite actor. This type of conversion can only express fully static schedules which is insufficient for replacing general static clusters. We will show

---

[9]Note that we use $p(n)$ to denote that at least n tokens/free space must be available on the channel connected to the actor port p.

[10]In general, the longer the static scheduling sequence which can be executed by checking a single prerequisite, the less schedule overhead is imposed by this schedule.

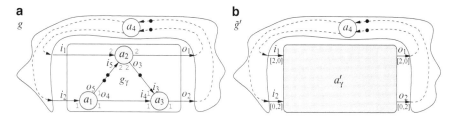

Fig. 21 An example with two feedback loops $o_1 \rightarrow i_2$ and $o_2 \rightarrow i_1$ over the dynamic dataflow actor a_4 and a *CSDF* composite actor a'_γ replacement candidate for cluster g_γ. (**a**) Data flow graph g with a cluster g_γ, (**b**) Composite actor a'_γ replacing g_γ

why composite actors only exhibiting *SDF* or even *CSDF* semantics are insufficient to replace a general static cluster as they may introduce deadlocks in the presence of a dynamic dataflow environment.

To exemplify this, we consider Fig. 21a with the cluster g_γ containing the actors a_1, a_2 and a_3 where i_1, i_2 and o_1, o_2 are the input and output ports of the cluster. The simplest possible case is the static schedule $i_1(2)\&i_2(2)\&o_1(2)\&o_2(2)/ (a_1,a_1,a_2,a_3,a_3)$, i.e., which checks the prerequisite of two tokens on port i_1 and i_2 as well as two free places on both output ports o_1 and o_2, and, if the prerequisite is satisfied, fire actor a_1 twice, then a_2, and finally a_3 twice. This schedule corresponds to the static schedule of the *SDF* graph contained in the cluster. Clearly, assuming actor a_4 only routes tokens, i.e., the sum of all tokens on channels connected to actor a_4 is the same before and after a firing of actor a_4, this schedule is infeasible as always two tokens are missing to activate the cluster, therefore introducing a deadlock into the system.

One might be tempted to solve the problem by converting the cluster g_γ into a *CSDF* composite actor a'_γ as shown in Fig. 21b. The *CSDF* composite actor has two phases and thus implements a quasi-static schedule. In the first phase, a positive check for two tokens on port i_1 is followed by the execution of the static schedule (a_2) while the second *CSDF* phase checks for two tokens on port i_2 and execute the static schedule (a_1,a_3,a_1,a_3). However, also this schedule fails in the general case if we assume that the dynamic dataflow actor a_4 randomly switches between two forwarding modes: Forwarding mode (i) forwards two tokens to port i_1 followed by two tokens to port i_2. On the other hand, forwarding mode (ii) forwards one token to port i_2, two tokens to port i_1 followed finally by one token to port i_2. Forwarding mode (i) can be satisfied with the suggested *CSDF* schedule. However, forwarding mode (ii) will deadlock with the same schedule. We could try to solve this problem by substituting another *CSDF* composite actor. In the first phase, this actor would check for one token on port i_2 and execute the schedule (a_1,a_3). In the second phase, it checks for two tokens on port i_1 and one token on port i_2 and execute the schedule (a_1,a_2,a_3). However, this *CSDF* actor would fail for forwarding mode (i).

The above examples demonstrate that in the general case, a static cluster cannot be converted into an *SDF* nor *CSDF* composite actor without possibly introducing

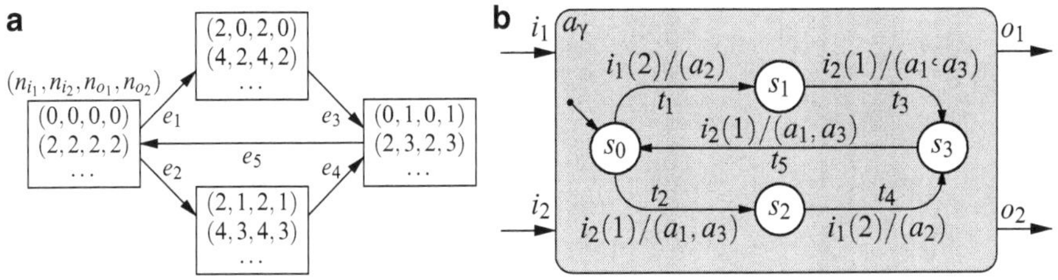

Fig. 22 The composite actor a_γ replacing cluster g_γ can be represented via a *cluster Finite State Machine*. The *cluster FSM* is derived from the Hasse diagram of the input/output states of the g_γ cluster. The transitions are annotated with preconditions and the static scheduling sequences to execute if a transition is taken. (**a**) Hasse diagram of input/output states, (**b**) Composite actor a_γ replacing g_γ

a deadlock into the transformed dataflow graph. However, via the *cluster FSM* representation as introduced in Sect. 2.6, it is possible to represent a quasi-static schedule for all clusters where no unbounded accumulation of tokens on the internal FIFOs in the cluster can result if unbounded output FIFOs are connected to the cluster output ports.[11] Via the *cluster FSM* notation, the cluster g_γ can be replaced by a composite actor a_γ which can be expressed as shown in Fig. 22b. As can be seen, two transitions t_1 and t_2 are leaving the start state q_0. Whereas t_1 requires at least two tokens on input port i_1 (denoted by the precondition $i_1(2)$) and execute the static schedule (a_2), t_2 requires at least one input token on input port i_2 (denoted by $i_2(1)$) and execute the static schedule (a_1, a_3).

The key idea for producing a deadlock free *QSS* is to assume the so-called "worst case" environment for the cluster. This is the case when, each output port $o \in g_\gamma.O$ is part of a feedback loop to each input port $i \in g_\gamma.I$ and any produced token on an output port $o \in g_\gamma.O$ is needed for the activation of an actor $a \in g_\gamma.A$ in the same cluster through these feedback loops. In particular, postponing the production of an output token results in a deadlock of the entire system. Hence, the *QSS* determined by the clustering algorithm guarantees the production of a maximum number of output tokens from the consumption of a minimal number of input tokens.

We will represent the production of a maximal number of output tokens with a minimal number of input tokens by so-called *input/output states* of the static cluster. For the cluster g_γ, this relation is given in Table 2 which depicts the input/output states of the cluster g_γ. For an efficient computation of this table, we refer the interested reader to [14].

We first note that this table is infinite but periodic as can be seen from the depiction of the input/output states in the corresponding Hasse diagram (cf. Fig. 22a). Equivalent values, i.e., input/output states which can be reached by firing all actors of the cluster according to the *repetition vector* of the cluster, are aggregated into one vertex in the Hasse diagram. This is also the reason why the Hasse diagram

---

[11] A more formal definition of this condition can be found in [14].

Table 2 Minimal number of input tokens n_{i_1}/n_{i_2} necessary on cluster g_γ input port i_1/i_2 and maximal number of output tokens n_{o_1}/n_{o_2} producible on the cluster output ports o_1/o_2 with these input tokens

$(n_{i_1}, n_{i_2},$	(0,0,	(0,1,	(2,0,	(2,1,	(2,2,	(2,3,	(4,2,	(4,3,	(4,4,	⋯
$n_{o_1}, n_{o_2})$	0,0)	0,1)	2,0)	2,1)	2,2)	2,3)	4,2)	4,3)	4,4)	⋯

For example, to produce two tokens on output port o_2, two tokens are required from both input ports i_1 and i_2. The maximal number of output tokens producible from these input tokens are two output tokens on o_1 and o_2

a

```
1 actor a₁ () i₁.i₂.i₃.i₄==>o₁.o₂ :
2   action i₁:[u],i₂:[v]==>o₁:[u+v]
3   end
4   action i₃:[u],i₄:[v]==>o₂:[u-v]
5   end
6 end
```

b

```
1 actor Add() i₁.i₂==>o₁ :
2   action i₁:[u],i₂:[v]==>o₁:[u+v]
3   end
4 end
5 actor Sub() i₃.i₄==>o₂ :
6   action i₃:[u],i₄:[v]==>o₂:[u-v]
7   end
8 end
```

Fig. 23 A *CAL* actor and its decomposition into two *SDF* actors. (**a**) A *CAL* actor which combines the function of both an addition as well as a subtraction actor, (**b**) Decomposition of the actor from (**a**) into two actors one doing addition and the other one subtraction

contains cycles, as normally, these values would all be depicted as unique vertices. As can be seen from Fig. 22a,b there is a one-to-one correspondence between the vertices and edges in the Hasse diagram and the states and transitions in the *cluster FSM*. For a more technical explanation of deriving the *cluster FSM* from the Hasse diagram we again refer the reader to [14]. Furthermore, there exists a more advanced representation of the *cluster FSM* via so-called *Rules* [15] which allows a more compact representation, thus reducing the code size required to represent the *quasi-static schedule*.

5.2   Statically Schedulable Regions in CAL

While the clustering algorithm presented in Sect. 5.1 assumes that the static subgraph which will be clustered consists of pure *SDF* or *CSDF* actors, Janneck et al. [20] have presented a methodology to derive static parts from dynamic *CAL* actors. To exemplify, we consider the *CAL* actor depicted in Fig. 23a. This actor does not exhibit *SDF* semantics, but it can be decomposed into the two *SDF* actors as depicted in Fig. 23b. The idea from [20] presents a systematic algorithm to decompose a dynamic *CAL* actor into *SDF* actors and a dynamic residual actor.

The analysis determines a *statically related group* of ports for each actor. Each such group of ports is later transformed into an *SDF* actor. Connected subgraphs of these derived *SDF* actors can then be scheduled quasi-statically.

To begin with, we will first introduce some notation. We use $ports(a)$ to denote the set of actor input and output ports of the *CAL* actor a. Furthermore, we will use $\Gamma(a)$ to denote the set of actions of the actor a. To refer to the set of ports which are modified by an action $l \in \Gamma(a)$, that is, the action will consume or produce tokens on that port, the notation $ports(a)_l$ is used. The number of tokens which are transferred on a port p by an action l will be denoted by $\text{trans}(a)_{l,p}$. Furthermore, *CAL* actors themselves can have a state $\sigma \in \Sigma$, e.g., representing state variables or states from a *schedule FSM* controlling action execution. We will call an action l *state-guarded* if its execution depends on the value of the actor state σ. Moreover, an action l is called *state-changing* if the execution of the action may modify the state σ of the *CAL* actor.

With the previous, notation we can define a *statically related group* of ports Z of an actor a. This set of ports $Z \subseteq ports(a)$ is a maximal subset of the ports of a which conforms to the following conditions:

$$\forall l \in \Gamma(a) : (Z \subseteq ports(a)_l) \vee (Z \cap ports(a)_l = \emptyset) \tag{1}$$

$$\forall p \in Z : \forall l, m \in \Gamma(a), p \in ports(a)_l \cap ports(a)_m : \text{trans}(a)_{l,p} = \text{trans}(a)_{m,p}$$

$$\text{and} \text{trans}(a)_{l,p} \text{is also a constant integer} \tag{2}$$

The conditions ensure that (1) all the ports in the *statically related group* are processed in an atomic action, e.g., like *SDF* actors consume and produce tokens atomically, and if multiple actions are working on these ports that (2) all these actions produce and consume the same number of tokens. Indeed, there may be multiple *statically related groups* of ports Z. For example, consider the actor a_1 depicted in Fig. 23a. This actor has two *statically related groups* of ports $Z_1 = \{i_1, i_2, o_1\}$ and $Z_2 = \{i_3, i_4, o_2\}$. This is also the port partitioning along which the actor has been split into its two constituent *SDF* actors as depicted in Fig. 23b.

However, the conditions are not able to recognize *CSDF* actors nor *SDF* actors where the consumption and production of the actor is distributed over more than one action. For example, consider actor a_2 as depicted in Fig. 24a. The actor is undoubtedly a *CSDF* actor but no *statically related group* of ports can be found as both ports violate Eq. (2). This stems from the fact that the analysis to derive *statically related groups* does not consider the actor state σ. Therefore, the analysis cannot recognize the cyclic action execution of a_2. This is also the reason why in contrast to the analysis presented in Sect. 3 the algorithm to derive *statically related groups* cannot recognize *SDF* actors where the consumption and production of the actor is distributed over more than one action.

We now consider the dataflow graph depicted in Fig. 25b. The dataflow graph has seven *statically related groups* Z_1–Z_7 which can be transformed into *SDF* actors for quasi-static scheduling. The *statically related groups* induce two subgraphs (in fact there are three but one subgraph consists only of one actor alone). We will denote the subgraph containing the *statically related groups* Z_1, Z_3, and Z_5 with g_1 and the other subgraph consisting of the groups Z_2, Z_4 and Z_6 with g_2. Scheduling of subgraph g_2

a

```
1 actor a₂() i₁==>o₁:
2 Boolean t := false;
3 action i₁:[v] ==> o₁:[v]
4 guard not t
5 do t := true; end
6 action i₁:[u,v] ==> o₁:[u,v]
7 guard t
8 do t := false; end
9 end
```

b

```
1 actor a₃() i₁.i₂.i₃==>o₁.o₂:
2 Boolean t := false;
3 action i₁:[v]==>o₁:[v]
4 guard t
5 do t := false; end
6 action i₂:[v]==>o₂:[v]
7 guard not t end
8 action i₃:[v]==>[]
9 do t := v; end
10 end
```

Fig. 24 Examples of a *CSDF* actor in *CAL* which is not recognized by the quasi-static scheduling algorithm presented in [20]. (**a**) A *CSDF* actor which forwards first one token from i_1 to o_1 and then atomically two tokens from i_1 to o_1, (**b**) A forward actor a_3 which depending on the state variable t forwards either from i_1 to o_1 or from i_2 to o_2. The state variable can be reassigned by a token on i_3

a

```
1 actor a₄() i₁.i₂.i₃==>o₁.o₂:
2 action i₁:[u,v,w]==>o₁:[u,v,w]
3 end
4 action i₂:[v]==>o₂:[v]
5 end
6 end
```

b

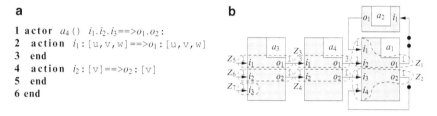

Fig. 25 A last *CAL* actor and a dataflow graph containing the presented *CAL* actors. The dataflow graph has seven *statically related groups* Z_1–Z_7 which are denoted with a *dashed border*. (**a**) A forward actor a_4 which forwards three tokens atomically from i_1 to o_1 or one token from i_2 to o_2, (**b**) A dataflow graph containing the *CAL* actors a_1 to a_4 their ports and the partitioning of these ports into *statically related groups*

is straightforward and can be done fully statically. Subgraph g_1 is a more interesting case. The repetition vector for g_1 is $3 \times Z_1, 1 \times Z_3, 3 \times Z_5$. However, the action for Z_5 as depicted in Fig. 24b Lines 3–5 does disable itself after one firing. Therefore, the quasi-static schedule for g_1 must be able to stall until Z_5 is enabled again. A similar problem exists due to the feedback loop over actor a_4. This problem could be handled by quasi-static scheduling as proposed in Sect. 5.1. Another approach would use the approach of Plishker et al. as described in Sect. 2.3 to obtain a speedup for *enable-invoke* dataflow.

6 Conclusions

Due to the complexity of current dataflow applications their modeling requires dynamic behavior. However, these applications still contain large isles of static actors, which are amenable to compile time analysis. In this chapter, we presented several models of computation, which are able to integrate static dataflow with a dynamic dataflow environment while still preserving the analyzability of the static parts of the dataflow graph. The usage of *finite state machines* has proven invaluable for the coordination between the static and the dynamic parts within dataflow graphs. As an example for such a coordination, we presented quasi-static scheduling, which reduces the scheduling overhead for one processor of an MPSoC while still accommodating a worst-case environment of the cluster.

References

1. Baird, M. (ed.): IEEE Standard 1666-2005 SystemC Language Reference Manual. IEEE Standards Association, New Jersey, USA (2005)
2. Balarin, F., Giusto, P., Jurecska, A., Passerone, C., Sentovich, E., Tabbara, B., Chiodo, M., Hsieh, H., Lavagno, L., Sangiovanni-Vincentelli, A., Suzuki, K.: Hardware-Software Co-Design of Embedded Systems: The POLIS Approach. Kluwer Academic Publishers (1997)
3. Bhattacharya, B., Bhattacharyya, S.: Parameterized dataflow modeling for DSP systems. Signal Processing, IEEE Transactions on 49(10), 2408–2421 (2001)
4. Bhattacharyya, S., Brebner, G., Eker, J., Mattavelli, M., Raulet, M.: OpenDF – A dataflow toolset for reconfigurable hardware and multicore systems (2008). First Swedish Workshop on Multi-Core Computing, MCC, Ronneby, Sweden, November 27–28, 2008
5. Bhattacharyya, S.S., Buck, J.T., Ha, S., Lee, E.A.: Generating compact code from dataflow specifications of multirate signal processing algorithms. IEEE Transactions on Circuits and Systems I: Fundamental Theory and Applications 42(3), 138–150 (1995)
6. Bhattacharyya, S.S., Deprettere, E.F., Theelen, B.: Dynamic dataflow graphs. In: S.S. Bhattacharyya, E.F. Deprettere, R. Leupers, J. Takala (eds.) Handbook of Signal Processing Systems, second edn. Springer (2013)
7. Bilsen, G., Engels, M., Lauwereins, R., Peperstraete, J.: Cyclo-static dataflow. IEEE Transaction on Signal Processing 44(2), 397–408 (1996)
8. Buck, J., Ha, S., Lee, E.A., Messerschmitt, D.G.: Ptolemy: A framework for simulating and prototyping heterogenous systems. International Journal in Computer Simulation 4(2), 155–182 (1994)
9. Buck, J.T.: Scheduling dynamic dataflow graphs with bounded memory using the token flow model. Ph.D. thesis, Dept. of EECS, UC Berkeley, Berkeley, CA 94720, U.S.A. (1993)
10. Dennis, J.B.: First version of a data flow procedure language. In: Programming Symposium, Proceedings Colloque sur la Programmation, pp. 362–376. Springer-Verlag, London, UK (1974)
11. Eker, J., Janneck, J.W.: CAL language report – language version 1.0. Tech. rep., University of California at Berkeley (2003)
12. Eker, J., Janneck, J.W., Lee, E.A., Liu, J., Liu, X., Ludvig, J., Neuendorffer, S., Sachs, S., Xiong, Y.: Taming heterogeneity - the Ptolemy approach. Proceedings of the IEEE 91(1), 127–144 (2003)
13. Falk, J., Haubelt, C., Teich, J.: Efficient representation and simulation of model-based designs in SystemC. In: Proc. FDL'06, Forum on Design Languages 2006, pp. 129–134. Darmstadt, Germany (2006)
14. Falk, J., Keinert, J., Haubelt, C., Teich, J., Bhattacharyya, S.: A generalized static data flow clustering algorithm for MPSoC scheduling of multimedia applications. In: EMSOFT'08: Proceedings of the 8th ACM international conference on Embedded software (2008)

15. Falk, J., Zebelein, C., Haubelt, C., Teich, J.: A rule-based static dataflow clustering algorithm for efficient embedded software synthesis. In: Proceedings of Design, Automation and Test in Europe (DATE'11) (2011)

16. http://www.forteds.com (2011)

17. Geilen, M., Basten, T.: Kahn process networks and a reactive extension. In: S.S. Bhattacharyya, E.F. Deprettere, R. Leupers, J. Takala (eds.) Handbook of Signal Processing Systems, second edn. Springer (2013)

18. Girault, A., Lee, B., Lee, E.: Hierarchical finite state machines with multiple concurrency models. Computer-Aided Design of Integrated Circuits and Systems, IEEE Transactions on **18**(6), 742–760 (1999)

19. Grötker, T., Liao, S., Martin, G., Swan, S.: System Design with SystemC. Kluwer Academic Publishers (2002)

20. Gu, R., Janneck, J.W., Raulet, M., Bhattacharyya, S.S.: Exploiting statically schedulable regions in dataflow programs. J. Signal Processing Systems **63**(1), 129–142 (2011)

21. Ha, S., Oh, H.: Decidable dataflow models for signal processing: Synchronous dataflow and its extensions. In: S.S. Bhattacharyya, E.F. Deprettere, R. Leupers, J. Takala (eds.) Handbook of Signal Processing Systems, second edn. Springer (2013)

22. Hsu, C., Bhattacharyya, S.S.: Cycle-breaking techniques for scheduling synchronous dataflow graphs. Tech. Rep. UMIACS-TR-2007-12, Institute for Advanced Computer Studies, University of Maryland at College Park (2007). URL http://hdl.handle.net/1903/4328

23. Janneck, J.W.: The open dataflow project. http://opendf.sourceforge.net (2009)

24. Janneck, J.W., Miller, I.D., Parlour, D.B., Roquier, G., Wipliez, M., Raulet, M.: Automatic software synthesis of dataflow program: An MPEG-4 simple profile decoder case study. In: Proc. of the IEEE Workshop on Signal Processing Systems (SiPS'08), pp. 281–286 (2008)

25. Janneck, J.W., Miller, I.D., Parlour, D.B., Roquier, G., Wipliez, M., Raulet, M.: Synthesizing hardware from dataflow programs: An MPEG-4 simple profile decoder case study. In: Proc. of the IEEE Workshop on Signal Processing Systems (SiPS'08), pp. 287–292 (2008)

26. Kahn, G.: The semantics of simple language for parallel programming. In: IFIP Congress, pp. 471–475 (1974)

27. Keinert, J., Streubühr, M., Schlichter, T., Falk, J., Gladigau, J., Haubelt, C., Teich, J., Meredith, M.: SYSTEMCODESIGNER - An automatic ESL synthesis approach by design space exploration and behavioral synthesis for streaming applications. Transactions on Design Automation of Electronic Systems **14**(1), 1–23 (2009)

28. Lee, E.A.: Overview of the Ptolemy project. Tech. Rep. UCB/ERL M03/25, Department of Electrical Engineering and Computer Sciences, University of California, Berkeley, CA, 94720, USA (2004)

29. Lee, E.A., Messerschmitt, D.G.: Synchronous data flow. Proceedings of the IEEE **75**(9), 1235–1245 (1987)

30. Lukasiewycz, M., Glaß, M., Haubelt, C., Teich, J., Regler, R., Lang, B.: Concurrent topology and routing optimization in automotive network integration. In: Proceedings of the 2008 ACM/EDAC/IEEE Design Automation Conference (DAC'08), pp. 626–629. Anaheim, USA (2008)

31. Necula, G.C., McPeak, S., Rahul, S.P., Weimer, W.: CIL: Intermediate language and tools for analysis and transformation of C programs. In: R.N. Horspool (ed.) Compiler Construction, *Lecture Notes in Computer Science*, vol. 2304, pp. 209–265. Springer (2002)

32. Pino, J., Bhattacharyya, S., Lee, E.: A hierarchical multiprocessor scheduling system for DSP applications. In: Signals, Systems and Computers, 1995. 1995 Conference Record of the Twenty-Ninth Asilomar Conference on, vol. 1, pp. 122–126 (1995)

33. Plishker, W., Sane, N., Bhattacharyya, S.: A generalized scheduling approach for dynamic dataflow applications. In: Design, Automation Test in Europe Conference Exhibition, 2009. DATE '09., pp. 111–116 (2009)

34. Plishker, W., Sane, N., Kiemb, M., Bhattacharyya, S.S.: Heterogeneous design in functional DIF. In: Proceedings of the 8th international workshop on Embedded Computer Systems: Architectures, Modeling, and Simulation, SAMOS '08, pp. 157–166. Springer-Verlag, Berlin, Heidelberg (2008)
35. Sangiovanni-Vincentelli, A.L., Sgroi, M., Lavagno, L.: Formal models for communication-based design. In: Proceedings of the 11th International Conference on Concurrency Theory, CONCUR '00, pp. 29–47. Springer-Verlag, London, UK (2000)
36. XILINX: Embedded SystemTools Reference Manual - Embedded Development Kit EDK 8.1ia (2005)

C Compilers and Code Optimization for DSPs

Björn Franke

Abstract Compilers take a central role in the software development tool chain for any processor and enable high-level programming. Hence, they increase programmer productivity and code portability while reducing time-to-market. The responsibilities of a C compiler go far beyond the translation of the source code into an executable binary and comprise additional code optimization for high performance and low memory footprint. However, traditional optimizations are typically oriented towards RISC architectures that differ significantly from most digital signal processors. In this chapter we provide an overview of the challenges faced by compilers for DSPs and outline some of the code optimization techniques specifically developed to address the architectural idiosyncrasies of the most prevalent digital signal processors on the market.

1 Introduction

Compilers translate the source code written by the programmer into an executable program for the particular target machine. While compilers generally aid the programmer to work more productively through higher levels of abstraction compilers have to meet many additional—and sometimes conflicting—requirements. For performance-critical DSP applications highest efficiency of the generated code is expected and only small overhead compared to manually written assembly code is tolerated. At the same time, high code density is paramount for memory constrained and cost-sensitive DSP applications. Optimized code should be predictable, correct and amendable to debugging, and generally "well-behaved" irrespective of coding

B. Franke (✉)
University of Edinburgh, School of Informatics, Informatics Forum, 10 Crichton Street, Edinburgh EH8 9AB, Scotland, UK
e-mail: bfranke@inf.ed.ac.uk

S.S. Bhattacharyya et al. (eds.), *Handbook of Signal Processing Systems*,
DOI 10.1007/978-1-4614-6859-2_31, © Springer Science+Business Media, LLC 2013

style and idiosyncrasies of the target architecture. In this chapter we explore some of the challenges faced by C compilers for DSPs and take a look into the optimizations that are applied to meet the strict requirements on performance and code size.

2   C Compilers for DSPs

C compilers for digital signal processors are faced with a number of challenges arising from idiosyncratic architectural features and the users' demand for near-optimal code quality. At the same time, C compilers for DSPs can be tuned for a their particular digital signal processing application domain. Advantage can be taken of recurrent algorithmic features and their mapping onto the underlying hardware can be improved over time. Before we look into these code optimizations in the next section we present a brief overview of the industrial context in which DSP compiler development takes place and of some DSP-specific compiler and C language extensions. Finally, we introduce a number of DSP benchmarks suitable for compiler and platform performance evaluation.

2.1   *Industrial Context*

Compilers are rarely developed from scratch, but most frequently an existing compiler is ported to a new processor or a compiler development toolkit is used to generate the processor-specific parts of a compiler from high-level machine descriptions. While this approach ensures rapid success in deploying a working compiler it may also have some shortcomings:

- Depending on the chosen compiler framework it may be difficult or costly to fit in a new, particular optimization critical to achieve a certain performance level for the key application of customer *XYZ*.
- A set of optimizations might improve the *average* performance on some DSP benchmark suite, however, individual benchmarks might show a performance degradation unacceptable to some customers.

A compiler might also correctly implement the ISO-C standard, but behave differently to the "unofficial standard" set by the GCC compiler and customers complain about this apparent "bug". In addition, some users might want extended control over the compiler optimizations, either to fine-tune the optimizations to their application or to prevent the compiler from performing optimizations that might interfere with some manually applied code optimizations.

In either case, compilers rarely meet the requirements of all the users and often this results in a pragmatic solution where a compiler is extended with additional switches controlled by the user, e.g. platform-specific `#pragma`'s or compiler-known functions, or DSP-specific extensions to the C programming language are used.

2.2   Compiler and Language Extensions

In the performance-critical DSP environment both the compiler and the C programming language itself are frequently extended to narrow the gap between the
high-level language and the target processor and, thus, enabling the compiler to
generate "better" code. Programmers can use these extensions to convey extra
information to the compiler. Thereby the compiler is directed to make explicit
use of specific processor features such as specialized instructions or dual memory
banks. *Compiler-known functions* extend the compiler with special functions and
are available to the programmer as convenient shortcuts to custom DSP operations
provided by the specific target processor. On the other hand, *DSP-C* and *Embedded
C* are extensions to the ISO-C language standard [20] and provide the programmer
with new, DSP-specific language keywords. This later approach maintains a certain
degree of portability and readability of the code while making use of DSP-specific
hardware extensions to increase code efficiency.

2.2.1   Compiler-Known Functions

Compiler-known functions (CKFs), sometimes also called intrinsic functions, are
function calls that are not mapped to regular function calls by the compiler, but
are directly replaced by one or more assembly instructions. CKFs enhance the
readability of the C code while at the same time they provide a convenient way
of directing the compiler to explicitly emit a particular sequence of instructions.
CKFs are useful in cases where the compiler is known to not generate the desired
instruction, e.g. because its pattern is not tree shaped and does not fit into the usual
tree covering based instruction selection algorithm employed by the compiler, or the
programmer wants to exercise specific control over the code generation process for
performance reasons.

For instance, the compiler for the NXP TriMedia TM5250 processor defines a
CKF `ifir16` for computing the signed sum of products of signed 16-bit halfwords.
This operation is useful in implementing complex arithmetic, but results in several
assembly instructions when implemented in "plain" C rather than a CKF. The NXP
compiler, however, has knowledge about this special function and maps what looks
like an ordinary function call to `ifir16` directly to a single assembly instruction
of the same name.

CKFs require a detailed understanding of the target processor by the programmer and code written using CKFs is machine-dependent. Portability can be
re-established, however, by providing a suitable set of simulation functions (written
in "plain" ISO-C) for the CKFs that are less efficient, but functionally equivalent to
their machine-specific counterparts.

2.2.2   DSP-C and Embedded C

DSP-C [3] and *Embedded C* [12] are DSP-specific extensions to the ISO C language
enable the programmer to gain control over hardware resources typically not
available on general-purpose computing systems, but commonly found in DSPs,
while at the same time writing portable code. In particular, support for fixed point
data types, multiple memories and circular buffers are provided. These extensions
serve a dual purpose: First, they enhance the expressiveness of the C programming
language and provide the programmer with constructs that target idioms (fixed point
data types, circular buffers) frequently encountered in DSP algorithms. Second, they
shift the responsibility for the efficient use of non-standard hardware resources (dual
memory banks) from the compiler to the programmer and, thus, help alleviate the
compiler of a"hard" data partitioning and memory assignment problems. A more
comprehensive discussion of the DSP-C standard can be found in [3].

2.3   Compiler Benchmarks

Measuring the success of a compiler to generate dense and efficient code and
the ability to compare the results across platforms or to hand-coded assembly
implementations largely depends on standardized benchmark applications.

Early, but now largely outdated DSP benchmarks were developed in academia
and aimed at supporting research in DSP code generation. For example, DSP-
stone [49] and UTDSP [24] benchmark suites have been used in a large number of
compiler research projects because they comprise compact, but compute-intensive
DSP kernels written in C and operate on small data sets. The workload presented by
these benchmarks, however, is not representative for modern DSP applications and
the lack of a clearly defined benchmarking methodology make them less suitable
for meaningful performance evaluations.

More recently, the Embedded Microprocessor Benchmark Consortium (EEMBC)
has developed a set of performance benchmarks for the hardware and software
used in embedded systems. The EEMBC benchmarks include, but are not restricted
to, DSP applications and cover different embedded application domains including
automotive, consumer, digital entertainment, networking, office automation and
telecom. The primary goal of EEMBC is to provide industry with standardized
benchmarks and, thus, enabling processor vendors and their customers to compare
processor performance for their chosen application domain. In addition, the EEMBC
benchmarks provide an additional benefit in that performance figures for compiler
generated code ("out-of-the-box") and manually tuned algorithms ("optimized") are
quoted. These figures give an interesting insight to the compilers' code generation
abilities as for some platforms the difference between the two performance figures
can be quite significant. Unlike DSPstone and UTDSP the EEMBC benchmarks
come closer to representing real-world applications, however, their greater "realism"
comes at the cost of increased code size and complexity that sometimes make it
more difficult for the compiler engineer to review the generated code.

3 Code Optimizations

With efficiency being one of the most important design goals of digital signal processors it is critical for a compiler to generate the most efficient code for any DSP application. Hence, most DSP compilers perform routinely optimizations such as:

- Rearranging code to combine blocks and minimize branching
- Eliminating unnecessary recalculations of values
- Combining equivalent constants
- Eliminating unused code and storage of unreferenced values
- Global register allocation
- Inlining or replacing function and library calls with program code
- Substituting operations to conserve resources
- Eliminating ambiguous pointer references when possible
- Many more

Some of the more idiosyncratic architectural features such as complex instruction patterns targeting frequently occurring DSP operations may still not be fully exploited by conventional compilers. Much of the early work on code optimization for embedded DSPs has therefore investigated improved code generation approaches for selecting machine-specific instructions [30], scheduling with register constraints [10], and register allocation [29]. For DSPs with heterogeneous datapaths traditional code generation approaches fail to produce efficient code and compiler writers have developed more advanced, graph-based code generation techniques [28]. However, most modern DSP architectures have become more "compiler-friendly" and more conventional code generation techniques can be applied.

In the following paragraphs we discuss a number of code optimization techniques specific to digital signal processors. We cover address code and control flow optimizations, loop and memory optimizations, and, eventually, optimizations for reduced code size.

Conversion from floating-point to fixed-point implementation of an algorithm is usually not performed by a compiler. While the conversion can be automated [31] additional application knowledge about the expected dynamic range of the processed signals is required.

A more comprehensive discussion of code optimization techniques for a wider range of embedded processors, including DSPs and multimedia processors, is presented in [25]. A general survey of compiler transformations for high-performance computing can be found in [4]. An excellent textbook introduction to the design and implementation of optimizing compilers is [33].

3.1   Address Code Optimizations

Typical code contains a large number of instruction for the computation of addresses of data stored in memory. These auxiliary computations are necessary, but compete for processing resources with the computations on actual user data. Hence, address code optimizations have been developed that aim at making address computations more efficient and, thus, increasing overall performance. In the following two paragraphs we discuss two of these optimizations that support the compiler in making better use of the address generation units common to many DSPs.

3.1.1   Pointer-to-Array Conversion

Many DSP architectures have specialized Address Generation Units (AGUs) (see [45]), but early compilers were unable to generate efficient code for them, especially in programs containing explicit array references. Programmers, therefore, used pointer-based accesses and pointer arithmetic within their programs in order to give "hints" to the early compiler on how and when to use post-increment/decrement addressing modes available in AGUs. For instance, consider example in Fig. 1a, a kernel loop of the DSPstone benchmark *matrix2*. Here the pointer accesses "encourage" the compiler to utilize the post-increment address modes of the AGU of a DSP.

If, however, further analysis and optimization is needed before code generation, then such a formulation is problematic as such techniques often rely on explicit array index representations and cannot cope with pointer references. In order to maintain semantic correctness compilers use conservative optimization strategies, i.e. many possible array access optimizations are not applied in the presence of pointers. Obviously, this limits the maximal performance of the produced code. It is highly desirable to overcome this drawback, without adversely affecting AGU utilization.

Although general array access and pointer analysis are without further restrictions equally powerful and intractable, it is easier to find suitable restrictions of the array data dependence problem while keeping the resulting algorithm applicable to real-world programs. Furthermore, as array-based analysis is more mature than pointer-based analysis within available commercial compilers, programs containing arrays rather than pointers are more likely to be efficiently implemented.

Pointer-to-Array Conversions [15] collects information from pointer-based code in order to regenerate the original accesses with explicit indexes that are suitable for further analyses. Example 1(b) shows the loop with explicit array indexes that is semantically equivalent to the previous loop in Example 1(a). Not only it is easier to read and understand for a human reader, but it is amendable to existing array data-flow analyses. The data-flow information collected by these analyses can be used for e.g. redundant load/store eliminations, software-pipelining and loop parallelization.

a
```
int *p_a = &A[0] ;
int *p_b = &B[0] ;
int *p_c = &C[0] ;

for (k = 0 ; k < Z ; k++) {
  p_a = &A[0] ;
  for (i = 0 ; i < X; i++) {
    p_b = &B[k*Y] ;
    *p_c = *p_a++ * *p_b++ ;
    for (f = 0 ; f < Y-2; f++) {
      *p_c += *p_a++ * *p_b++ ;
    }
    *p_c++ += *p_a++ * *p_b++ ;
  }
}
```

b
```
for (k = 0 ; k < Z ; k++) {
  for (i = 0 ; i < X; i++) {
    C[X*k+i] = A[Y*i] * B[Y*k];
    for (f = 0 ; f < Y-2; f++) {
      C[X*k+i] +=
          A[Y*i+f+1] * B[Y*k+f+1];
    }
    C[X*k+i] +=
        A[Y*i+Y-1] * B[Y*k+Y-1];
  }
}
```

Fig. 1 Example: Pointer-to-Array conversion. (**a**) Original loop. (**b**) Loop after pointer-to-array conversion

Fig. 2 Example:
Pointer-to-Array conversion
and delinearization

```
for (k = 0 ; k < Z ; k++) {

  for (i = 0 ; i < X; i++) {

    C[k][i] = A[i][0] * B[k][0];
    for (f = 0 ; f < Y-2; f++) {
      C[k][i] += A[i][f+1] * B[k][f+1];

    }
    C[k][i] += A[i][Y-1] * B[k][Y-1];

  }
}
```

A further step towards regaining a high-level representation that can be analyzed by existing formal methods is the application of de-linearisation methods. De-linearisation is the transformation of one-dimensional arrays and their accesses into other shapes, in particular, into multi-dimensional arrays. The code in Fig. 1b shows the example loop after application of clean-up conversion and de-linearisation. The arrays *A*, *B* and *C* are no longer linear arrays, but have been transformed into matrices. Such a representation enables more aggressive compiler optimizations such as data layout optimizations. Later phases in the compiler can easily linearize the arrays for the automatic generation of efficient memory accesses (Fig. 2).

3.1.2   Offset Assignment

Address generation units (AGU) perform address calculations in parallel to other on-going computations of the DSP. Efficient utilization of these dedicated functional units depends on their effective use by the compiler. For example, an AGU might compute the address of the next data item to be accessed by adding a constant (small) modifier (e.g. $+1$ or -1) to the current data address and keeping the new address ready for the next memory operation to come. This, however, is only possible if the next data item can be placed within this range.

Placing variables in memory in such a way that for a given sequence of memory accesses the best use of parallel address computations can be made is one of the possible compiler optimizations targeting AGUs. The goal of *offset assignment* is then "to generate a data memory layout such that the number of address offsets of two consecutive memory accesses that are outside the auto-modify range is minimized" [47].

For example, consider the following sequence of accesses to the variables a, b, c, d, e, and f: $S = \{a,b,c,d,e,f,a,d,a,d,a,c,d,f,a,d\}$. Assuming an auto-modify range of ± 1, i.e. address calculations refering the data item stored at the address immediately before or after the current item are free, and a single index register we compute the addressing costs for two different memory layout schemes in Fig. 3. Each time the index register needs to be written because of an access to an element not neighboring the current position a cost of 1 is incurred (= dotted arrows). For the naïve scheme shown in the upper half of the diagram the total addressing costs are 9 whereas in the improved memory layout shown in the bottom half of the diagram the total costs have been reduced to 4.

In general, the offset assignment problem has a large solution space and can be optimally solved for small instances only. Several heuristics that have been developed are based on finding a maximum-weighted Hamiltonian path in the access graph by using a greedy edge selection strategy [47].

3.2   Control Flow Optimizations

Control flow optimizations such as the elimination of branches to unconditional branches, branches to the next basic block, branches to the same target and many more are part of every optimizing compiler. However, some particular architectural features of digital signal processors such as zero overhead loops and predicated instructions require specialized control flow optimizations.

3.2.1   Zero Overhead Loops

Zero overhead loops are an architectural feature found in many DSPs (see [45]) that help implementing efficient code for loops with iteration counts known at

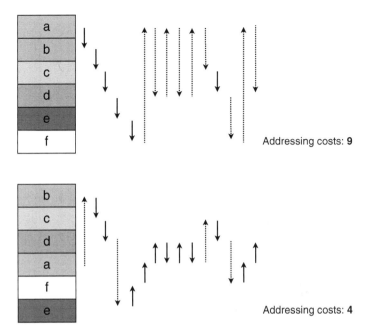

Fig. 3 Example: addressing costs of different memory layout schemes

compiler-time. They do not incur the overhead for increment and branch instructions for each iteration, but instead a zero overhead loop is initialized once and subsequently a fixed number of instructions is executed a specified number of times. Constraints regarding the number of instructions in the loop body, whether or not control flow instructions are permissible inside a zero overhead loop, the nesting of zero overhead loops, early exits or function calls complicate code generation for zero overhead loops because a number of checks need to be performed [40, 46].

A conventional loop implementation with an explicit termination test is shown in Fig. 4a. For each of the ten iterations a sequence of increment, comparison and branch instructions needs to be executed. Compare this to the code fragment in Fig. 4b where a zero overhead loop is used and no additional instructions are required after the initial loop setup.

While zero overhead loops are conceptually simple to exploit within a compiler, it is the number and type of constraints that sometimes make it difficult to generate code for zero overhead loops for all applicable loops. For example, many compiler frameworks generate the loop control code before code for the loop body is generated. This is problematic if the architecture places constraints on the maximum size of the loop body of a zero overhead loop. Possible solutions include to defer the decision on zero overhead loop code generation if the compiler framework is flexible enough to allow this, or to rely on a conservative heuristic concerning the ratio of intermediate operations to machine instructions.

a
```
         mov.d    #1,d10
LSTART:  /* loop body */
         . . .
         . . .
LTEST:   mov.d #1,d11
         add.d d10,d11,d10
         mov.d #10,d12
         slt.d d10,d12,d13
         bnz.d d13,LSTART
```

b
```
              do #10,LEND
              /* loop body */
              . . .
              . . .
LEND:
              . . .
```

Fig. 4 Example: code generation for zero-overhead loops. (**a**) Loop with explicit test. (**b**) Zero-overhead loop

a
```
      if (cond) Branch L1
      r2 = MEM[A]
      r1 = r2 + 1
      r0 = MEM[r1]
L1:   r9 = r3 + r4
```

b
```
      p1, p2   = cond
      r2 = MEM[A] <p2>
      r1 = r2 + 1 <p2>
      r0 = MEM[r1] <p2>
L1:   r9 = r3 + r4
```

Fig. 5 Example: If-Conversion (**a**) Conditional branch. (**b**) After If-Conversion ($p1$ denotes the predicate for the true condition and $p2$ is its inverse, i.e. $p1 = \neg p2$)

The use of zero overhead loops not only increases performance, but code size is reduced at the same time (see Sect. 3.5). In particular, zero overhead loops compare favorably in terms of code size to *loop unrolling* (see Sect. 3.3.1) while often providing a similar or better performance gain.

3.2.2   If-Conversion

Predicated execution is a standard feature in many modern DSPs, and is supported by compilers through *If-Conversion* [13], i.e. a transformation which converts control dependencies into data dependencies. A program containing conditional branches is transformed into predicated code in such a way that instructions previously located in either of the two paths of a conditional branch are guarded by predicates that evaluate to either *true* (= instruction enabled) or *false* (= instruction disabled).

A simple piece of code containing a conditional branch is shown in Fig. 5a. The branch condition is evaluated and only one of the control-dependent paths is executed. In the code after If-Conversion shown in Fig. 5b the conditional branch has been eliminated and a predicate $p2$ is computed. Subsequently, instructions from *both* control flow paths are executed and the predicate controls which results are used. Note that the conditional instructions now depend on the predicate $p2$. The code after If-Conversion does not contain any control dependences, but these have been converted into data dependences.

If-Conversion is a useful optimization as it avoids pipeline stalls due to mispredicted branches. However, If-Conversion does not guarantee performance improvements in all cases. For large conditionally executed code blocks, for example, conventional branches work better. In such cases many instructions need to be fetched and decoded by the processor that turn out to be disabled and, thus, waste valuable CPU time. In addition, deeply nested predicates quickly become impractical. For small, balanced branches, however, If-Conversion effectively increases processor pipeline utilization through reduced number of stalls and higher flexibility in instruction scheduling.

3.3   Loop Optimizations

Most execution time of DSP applications is spent on loops. Hence, effective code optimization of loop structures promises significant overall performance improvements. Unfortunately, the optimization of loops is a much "harder" problem than optimizing straight-line code. This is because the individual iterations of a loop may not be independent of each other, but data dependences might exist, i.e. data generated in one iteration of the loop gets carried forward to another iteration where it is consumed. Clearly, any loop restructuring transformation must obey these dependences in order to preserve correctness.

In the following paragraphs we present a number of loop transformations that are of particular importance to DSP kernels and applications. A more comprehensive list of loop optimizations can be found in e.g. [48] or [4].

Within the compiler a suitable framework is used for representing and systematically manipulating loop structures. The two most popular frameworks for this purpose are *unimodular transformations* and the *polyhedral model*. The approach based on unimodular transformations represents statements of n-deep loop nests as integer points in an n-dimensional space which can be captured in matrix notation upon which transformations described by unimodular matrices can be applied. These matrices can describe the *loop reversal*, *loop interchange*, and *loop skewing* transformations and combinations thereof. The polyhedral model goes beyond this and describes the statements contained in a loop nest as unions of sets of polytopes. The polytope model can represent imperfectly nested loops and affine transformations. Data dependences and boundaries of the polytopes are expressed as additional constraints. Both unimodular transformations and the polytope model are excellent formalisms to capture loop transformations. Still, the key problem remains to *identify* a suitable sequence of loop transformations to eventually improve the overall performance of the loop under consideration [39].

a
```
for (i = 0; i < 16; i++) {
  /* i = 0, 1, 2, ... */
  c += a[i] * b[i];
}
```

b
```
for (i = 0; i < 16; i += 2) {
  /* i = 0, 2, 4, ... */
  c += a[i] * b[i];
  /* i = 1, 3, 5, ... */
  c += a[i+1] * b[i+1];
}
```

Fig. 6 Example: loop unrolling. (**a**) Original loop. (**b**) Loop unrolled twofold

3.3.1 Loop Unrolling

Loop unrolling is probably the single-most important code transformation for any DSP application [42]. As the name suggests this transformation unrolls a loop, i.e. the loop body is replicated one or more times and the step of the loop iteration variable is adjusted accordingly. A simple example where the loop body is duplicated and the loop step doubled is shown in Fig. 6.

DSP applications spent most of their time in small, but compute-intensive loops. To improve overall performance, efficient hardware utilization is paramount and one way of achieving this is to merge consecutive loop iterations. The main benefits from loop unrolling arise from a reduced number of control flow instructions (conditional branches) due to fewer iterations of the unrolled loop and better scheduling opportunities due to the larger loop body. For example, the number of iterations of the unrolled loop in Fig. 6b is half of that of the original loop in Fig. 6a. As a consequence the overhead associated with the comparison with the upper loop bound and the increment of the loop iteration variable is reduced by a factor of 2.[1] At the same time the size of the loop body is doubled. This provides the instruction scheduler with more instructions to hide the latencies of e.g. memory accesses and to better exploit the available instruction level parallelism.

For small loops *complete* unrolling can be beneficial, i.e. the entire surrounding loop is flattened. For larger loops—with either a large number of iterations or a large number of instructions contained in the loop body—this is not practical because of the resulting code bloat.

Determining the optimal loop unroll factor is difficult [32, 44], in particular, if the target machine comprises an instruction cache. Code expansion resulting from loop unrolling can degrade the performance of the instruction cache. The added scheduling freedom can result in additional register pressure due to increased live ranges of variables. The resulting memory spills and reloads may negate the benefits of unrolling. In addition, control flow instructions in the loop body complicate unrolling decisions for example, if the compiler cannot determine that a loop may take an early exit.

---

[1]This reduction may be less if the original loop is implemented using *Zero-Overhead Loops*, see [45].

In cases where the loop iteration count is not divisible without remainder by the unroll factor *epilogue code* containing the remaining few iterations needs to be generated. This further increases code size in addition to the already duplicated loop body.

3.3.2 Loop Tiling

Loop tiling [37] is a compiler optimization that helps improving data cache performance by partitioning the loop iteration space into smaller blocks (also called *tiles*). By adjusting the size of the working set to the size of the data cache data reuse of array elements is maximized within each tile. The example in Fig. 7 illustrates the application of loop tiling for a simple matrix multiplication kernel.

The original code in Fig. 7a contains a nested loop with the innermost loop operating on the data arrays a, b and c. The arrays a and b are accessed in row and column order, respectively. In particular, different iterations of the j-loop operate on the same elements of the array a, and the accesses to array b do not depend on the outer i loop. This means, data elements from both a and b are reused for the computations of different elements $c[i][j]$. If the data arrays are large and the caches of the target machine small the data cache has not enough capacity to hold previously accessed data until its reused, thus a larger number of cache misses than necessary will occur. Loop tiling reduces the size of the "working set", thus increasing data locality and making sure that data will not get evicted from the cache until its reuse in a later loop iteration. The code fragment in Fig. 7b shows the matrix multiplication loop after loop tiling where three new loop levels (*ii*, *jj*, and *kk*) have been introduced. Each of the new, inner loops operates on a small tile of size t of its "parent" loop (e.g. between i and $i + t$), whereas the outer loops now iterate over these tiles taking larger steps of size t.

To achieve the best effect of loop tiling this code transformation is frequently combined with *array padding* and *loop interchange*. Array padding adds a small numbers of "dummy" elements to an array. This can lead to better aligned data placement in memory and helps further reducing the number of cache misses. Loop interchange exchanges the order of two or more nested loops if it is safe to do so. This may further increase cache efficiency due to better data locality.

Determining the optimal loop tiling factor to make the best use of the available caches is difficult and both analytical [1] and empirical [11] approaches have been developed.

3.3.3 Loop Reversal

Loop reversal is a simple loop transformation that "runs a loop backwards". In most cases loop reversal does not itself improve the generated code, but it acts as an enabler to better code generation, e.g. for zero-overhead loops, or to other optimizations such as *loop fusion*.

a

```
for (i = 0; i < N; i++) {
  for (j = 0; j < N; j++) {
    c[i][j] = 0.0;
    for (k = 0; k < N; k++) {
      c[i][j] += a[i][k] * b[k][j];
    }
  }
}
```

b

```
for (i = 0; i < N; i += t) {
  for (j = 0; j < N; j += t) {
    for (k = 0; k < N; k += t) {
      for (int ii = i; ii < i + t; ii++)
        for (int jj = j; jj < j + t; jj++) {
          c[ii][jj] = 0.0;
          for (int kk = k; kk < k + t; kk++)
            c[ii][jj] += a[ii][kk] * a[kk][jj];
        }
    }
  }
}
```

Fig. 7 Example: loop tiling. (**a**) Original loop. (**b**) After loop tiling (tile size $= t$)

a
```
for (i = N-1; i > 0; i--) {
  a[i] = 0;
}
```

b
```
for (i = 0; i < N; i++) {
  a[i] = 0;
}
```

Fig. 8 Example: loop reversal. (**a**) Original loop. (**b**) Loop after loop reversal

Consider the example in Fig. 8. The original loop in Fig. 8a initializes the elements of array *a* with a constant value, starting with the last element and working towards the beginning. If the target platform supports zero-overhead loops (see [45]) and the compiler is able to generate appropriate code (see Sect. 3.2.1) it still might be that the particular shape of the presented loop defeats the pattern recognition for zero-overhead loop code generation. In this case the reversal of the loop such as shown in Fig. 8b can help creating a canonical form more readily amendable to zero-overhead loop code generation. This is an example of a loop transformation that does not immediately improve code quality, but enables other parts of the compiler to produce better code. Loop reversal is only legal for loops that have no loop carried dependences.

a
```
void  f ( short * A , short * B ,
            short * C)
{ int  i ;
   for  ( i = 0;  i < N;  i += 2)
      { A[ i ] = B[ i ] + C[ i ];
        A[ i +1] = B[ i +1] + C[ i +1];
      }
}
```

b
```
void  f ( short * A , short * B ,
            short * C)
{ int  i ;
   for  ( i = 0;  i < N;  i += 2)
      { VADD_2(A , B , C);

      }
}
```

Fig. 9 Example: loop vectorization. (**a**) Original loop. (**b**) Loop after loop vectorization

3.3.4 Loop Vectorization

Most media processors and some DSPs such as the Analog Devices SHARC and TigerSHARC processors provide short vector (or SIMD) instructions, capable of executing one operation on multiple data items in parallel (see [45]). This parallelism, however, defeats the standard code generation methodology based on tree pattern matching with dynamic programming where intermediate representation data flow trees are covered with instruction patterns. Taking full advantage of SIMD instructions requires a different approach where the compiler operates on full data flow graphs.

Consider the example in Fig. 9. The original loop in Fig. 9a operates on 16-bit data stored in the arrays B and C. The original loop has already been unrolled twofold (see Sect. 3.3.1) in order to prepare the loop for vectorization. It simply adds the corresponding elements of the two arrays and stores the computed sum in the elements of array A. The iterations of the loop are completely independent of each other and pairs of successive iterations can be summarized and executed simultaneously by a single vector add instruction as shown in Fig. 9b.

A compiler targeting SIMD operations must ensure that the way data is stored in memory matches the data layout constraints for the operands of the vector operations. Furthermore, the compiler must be capable of identifying suitable operations in a larger data flow graph that can replaced with a single SIMD instruction. Due to their parallel nature these operations might be partially or fully independent of each other.

Code selection for short vector instructions can be accomplished in a standalone preprocessor [38] or fully integrated in the code generation stage of the compiler [19, 26, 36]. Another approach is taken in [14] where high-level algorithmic specifications of DSP algorithms are translated to C code and vectorization is considered already in this automated C code generation stage.

a
```
for ( i = 0; i < M; i++) {

    for ( j = 0; j < M; j++) {
      a[i][j] = 5 * j;
    }
}
```

b
```
for ( i = 0; i < M; i++) {

    for ( j = 0; j < M; j++) {
      *(a+M*i+j) = 5 * j;
    }
}
```

Fig. 10 Example: loop invariant code motion, Part I. (**a**) Original code. (**b**) Code after lowering

a
```
for ( i = 0; i < M; i++) {
    pa = a+M*i;

    for ( j = 0; j < M; j++) {
      *(pa+j) = 5 * j;

    }

}
```

b
```
ivar_pa = a;
for ( i = 0; i < M; i++) {
    ivar1_j = ivar_pa;
    ivar2_j = 0;
    for ( j = 0; j < M; j++) {
      *(ivar1_j) = ivar2_j;
      ivar1_j++;
      ivar2_j += 5;
    }
    ivar_pa += M;
}
```

Fig. 11 Example: loop invariant code motion, Part II. (**a**) After loop invariant code motion. (**b**) After induction variable creation

3.3.5 Loop Invariant Code Motion

Loop Invariant Code Motion [40] moves constant expressions out of the loop body and, hence, reduces the number of instructions executed per loop iteration. An example of this transformation is shown in Figs. 10 and 11. After the original loop in Fig. 10a has been lowered address computations for the array access in the loop body become explicit (see Fig. 10b). The first part of the address expression, however, is only dependent on the outer loop iterator i, but is constant for any value of the inner loop iterator j. "Hoisting" this constant subexpression out of the j-loop results in the code shown in Fig. 11a. Not only does this code avoid the repeated computation of a constant expression, it is also amendable to further optimization. *Strength Reduction* (see also Sect. 3.5.1) can be applied to replace the costly multiplication operation with a cheaper addition, and *Induction Variable Creation* can eliminate the use of the loop iterator j in favor of an induction variable incremented in each iteration that can be mapped efficiently the available post-increment addressing modes (see [45]) offered by many DSPs.

Before any loop-invariant computation can be moved out of a loop the compiler needs to verify this property. This is easy for genuine constants. For general expression it must be shown that all reaching definitions of the arguments of an expression are outside the loop. After that a new block of code containing the hoisted, loop-invariant computation is inserted as a *pre-header* to the loop.

Sequential Execution

1. Iteration	2. Iteration	3. Iteration	4. Iteration	5. Iteration

Software Pipelining

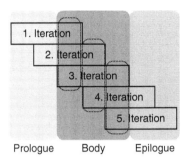

Prologue   Body   Epilogue

Time ⟶

Fig. 12 Software pipelining

3.3.6 Software Pipelining

Software Pipelining [2] is a low-level loop optimization and VLIW scheduling technique. It aims at exploiting parallelism at the instruction level by overlapping consecutive iterations of a loop so that a faster execution rate is realized (see Fig. 12). Instead of executing loop iterations in sequential order software pipelining partially overlaps loop iterations. After a number of initial iterations that are summarized in a loop prologue a stable and repetitive scheduling pattern for the loop body emerges. In the example instructions from three loop iterations are executed simultaneously. Eventually, the epilogue code executes the remaining instructions that do not belong to this repeated pattern towards the end of the loop.

Software pipelining is highly important for VLIW processors. More details can be found in [21].

3.4 Memory Optimizations

Memory optimizations play an important role when memory latency or bandwidth form a bottleneck. In this section we discuss one memory optimization of particular importance to DSPs as it deals with dual memory banks and the partitioning and assignment of variables to these memories.

```
void lmsfir(float input[], float output[],
            float expected[], float coefficient[],
            float gain)
{
    /* Variable declarations omitted. */

    sum = 0.0;
    for (i = 0; i < NTAPS; ++i){
        sum += input[i] * coefficient[i];
    }
    output[0] = sum;
    error = (expected[0] - sum) * gain;
    for (i = 0; i < NTAPS-1; ++i){
        coefficient[i] += input[i] * error;
    }
    coefficient[NTAPS-1] = coefficient[NTAPS-2] +
        (input[NTAPS-1]* error);
}
```

Fig. 13 Example: *lmsfir* function

3.4.1 Dual Memory Bank Assignment

Due to their streaming nature memory bandwidth is critical for most digital signal processing applications. To accommodate for these bandwidth requirements digital signal processors are typically equipped with dual memory banks (see [45]) that enable simultaneous access to two operands if the data is partitioned appropriately.

Efficient assignments of variables to memory banks can have a significant beneficial performance impact, but are difficult to determine. For instance, consider the example shown in Fig. 13. This shows the *lmsfir* function from the UTDSP *lmsfir_8_1* benchmark. The function has five parameters that can be allocated to two different banks. Local variables are stack allocated and outside the scope of explicit memory bank assignment as the stack sits on a fixed memory bank on most target architectures. In Fig. 14 four of the possible legal assignments are shown. In the first case in Fig. 14a all data is placed in the X memory bank. This is the default case for many compilers where no explicit memory bank assignment is specified. Clearly, no advantage of dual memory banks can be realized and this assignment results in an execution time of 100 cycles for e.g. an Analog Devices TigerSHARC TS-101 platform. The best possible assignment is shown in Fig. 14b, where input and gain are placed in X memory and output, expected, and coefficient in Y memory. Simultaneous accesses to the input and coefficient arrays have been enabled and, consequently, this assignment reduces the execution time to 96 cycles. Interestingly, an "equivalent" assignment scheme as shown in Fig. 14c that simply swaps the assignment between the two memory banks does not perform as well. In fact, the "inverted" scheme derived from the best assignment results in an execution time of 104 cycles, a 3.8 % slowdown over the baseline due to additional register-to-register moves introduced by the compiler. The worst possible assignment scheme is

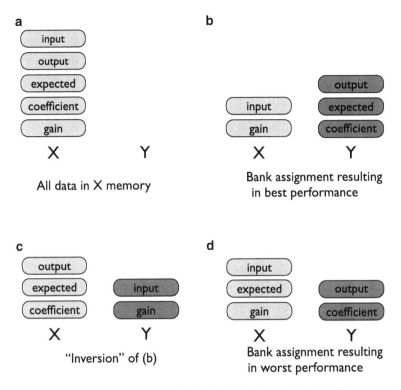

Fig. 14 Four memory bank assignments for the *lmsfir* function resulting in different execution times

shown in Fig. 14d. Still, input and coefficient are placed in different banks enabling parallel loads, but this scheme takes 110 cycles to execute, a 9.1 % slowdown over the baseline.

An early attempt to solve the dual memory bank assignment problem is described in [41]. They produced a low-level solution that performs a greedy minimum-cost partitioning of the variables using the loop-nest depth of each interference as a priority heuristic. The problem was formulated as an interference graph where two nodes interfere if they represent a potentially parallel access in a basic block. This is a very intuitive representation of the problem and has been used in many other solutions since.

Another approach [22] uses synchronous data flow specifications and the simple conflict graphs that accompany such programs. Three techniques were proposed, a traditional coloring algorithm, an integer linear programming based algorithm and a low complexity greedy heuristic using variable size as a priority metric.

Another integer linear programming solution prior to [22] is described in [27]. This technique runs after the compiler back-end has generated object code and lets it consider spill code and ignore accesses which do not reach memory. The assignment problem is modeled as an interference graph where two variables interfere if they are in the same basic block and there is no dependence between them. The interference is weighted according to the number of potentially parallel memory accesses.

More recently a more accurate integer linear programming model for DSP memory assignment has been presented [17]. The model described here is considerably more complicated than the one previously presented in [27] but provides larger improvements.

Finally, a technique that operates at a higher level than the other methods is described in [43]. It performs memory assignment on the high-level IR, thus allowing the coloring method to be used with each of the back-ends within the compiler. The problem is modeled as an independence graph and the weights between variables take account of both execution frequency and how close the two accesses are in the code.

Source-level transformations targeting DSP-C as an output language [35] and the use of machine-learning techniques [34] for dual memory bank assignment have been proposed in recent years.

3.5   Optimizations for Code Size

Most digital signal processors comprise fast, but small on-chip memories that store both data and the program code. While it is possible to provide additional, external memory this may not be desirable for reasons of cost and higher PCB integration. In such a situation the size of available on-chip memories place hard constraints on the code size which as an optimization goal becomes at least as important as performance. Incidentally, smaller code may also be faster code, especially if an instruction cache is present. Lower instruction count and a reduced number of instruction cache misses contribute to higher performance. However, there is no strict correlation between code size and performance. For example, some optimizations for performance such as loop unrolling (see Sect. 3.3.1) *increase* code size whereas more code-size aware code optimization such as redundant code elimination may lead to *lower* performance. Eventually, it will be the responsibility of the application design team to trade-off the (real-time) performance requirements of their application and the memory constraints set by their chosen processor. In the following paragraph we present a number of compiler-driven optimizations for code size that are applicable to most DSPs and do not require any additional hardware modules e.g. for code decompression. A comprehensive survey of code-size reductions methods can be found in [5].

3.5.1   Generic Optimizations for Code Compaction

In this paragraph we discuss a number of code optimizations that are routinely applied by most compilers as part of their effort to eliminate code redundancies.

```
a x = 3;                  b x = 3;                  c x = 3;
   y = 2;                    y = 2;                    y = 2;

   if (x+1 > y) {            if (4 > 2) {
      r = x;                    r = x;                  r = x;
   } else {                  } else {
      /* Unreachable */         /* Unreachable */       /* Unreachable code
      r = y;                    r = y;                     eliminated. */
   }                         }
```

Fig. 15 Example: unreachable code elimination. (**a**) Original code. (**b**) Code after constant propagation and folding. (**c**) Code after unreachable code elimination

Eliminating redundant or unreachable code certainly reduces code size without much effort and should be considered as a first measure if code size is a concern.

Redundant Code Elimination aims at eliminating those computations in a program that at a given point are guaranteed to be available as a results of an earlier computation. If such a computation can be identified then the redundant code fragment can be eliminated and the results of the previous computation used instead. One common variant of redundant code elimination is *Common Subexpression Elimination* where the identification and substitution of redundant code is restricted to subexpressions of a generally larger expression. While the behavior of the code does not change it is important to note that the expected performance impact depends on a number of factors. On the one hand, fewer instructions are issued due to the elimination of redundant computations. This has a positive impact on performance. On the other hand, register pressure is increased as a previously computed results needs to be stored for its later use and the increased register use may possibly lead to register spilling to memory. If this is the case, any performance and even code size benefit may get negated. In general, great care needs to be taken when applying redundant code elimination to make sure the desired code size reduction effect is achieved.

Unreachable Code Elimination detects and eliminates those computations that cannot be reached on any control flow path from the rest of the program. Unreachable code fragments can exist for a number of reasons. First, the user may have defined functions that are not used anywhere in the program. Second, the user may use conditional statements to enable/disable certain code fragments, e.g. for debugging purposes, rather than using the more appropriate C pre-processor macros for conditional compilation. Finally, unreachable code can result from the earlier application of other optimizations. Consider the example in Fig. 15a. The code in the *else* block is not reachable because the conditional expression of the *if*-statement will *always* be true. However, at this stage the compiler cannot detect this. After constant propagation and constant folding (see Fig. 15b) the code the conditional expression only makes use of constant values rather than variables and $4 > 2$ can be replaced with 1 (= *true*). Subsequently, the now provably unreachable code in the *else* block of the *if*-statement can be eliminated as shown in Fig. 15c.

Care needs to be taken when applying unreachable code elimination because it is not always possible to accurately work out the control flow relations in a program. For example, if a program makes use of both the C and assembly languages certain C routines that appear to be unreachable may get called from the assembly code. The compiler needs to be conservative to ensure correctness and cannot make any assumptions about e.g. functions with external linkage.

Dead Code Elimination targets computations whose results are never used. Note that a code fragment may reachable, but still dead if the results computed by this fragment are not used (or output) at any later point in the program. "Results" of a computation do include any side effects such as exceptions generated or signals raised by this piece of code and that affect other parts of the program under optimization. Only if it can be guaranteed that a fragment of code does not produce any direct or indirect results in this strict sense (e.g. it cannot produce any *Division-by-Zero Exceptions* or similar) that are consumed elsewhere it can be eliminated.

Strength Reduction [7] replaces a costly sequence of instructions with an equivalent, but cheaper (= faster or shorter) instruction sequence. For example, an optimizing compiler may replace the operation $2 \times x$ with either $x + x$ or $x \ll 1$. Strength reduction is most widely known for use inside loops where repeated multiplications are replaced with cheaper additions. In addition to the obvious performance improvements from replacing a costly operation with a faster one, strength reduction often decreases the total number of instructions in a loop and, thus, contributes to code size reductions. If applied in combinations with other optimizations such as algebraic reassociation, strength reduction further reduces the number of operations. On the other hand, strength reduction may increase the number of additions as the number of multiplies is decreased.

Certainly, optimizations for code size interfere with each other as much as optimizations for performance. Hence, it is a difficult problem to select the most appropriate set of code transformations (and their order of application) to achieve a maximum code size reduction effect. An iterative approach to compiler tuning where many different versions of the same code are tested and statistical analysis is employed to identify the compiler options that have the largest effect on code size is presented in [18]. The use of genetic algorithms to find optimization sequences that generate small object codes is subject of [8]. Code compaction in a binary rewriting tool is described in [9]. For architectures that provide an additional narrow instruction set the generation of mixed-mode instruction width code is discussed in e.g. [23].

3.5.2   Coalescing of Repeated Instruction Sequences

In addition to the generic optimizations discussed in the previous paragraph that target the obvious sources of code size reduction such as dead or unreachable code, there exist further optimizations that aim at specifically generating more compact code, possibly at the cost of decreased performance. One of these techniques [6] tries to identify repeated code fragments that are subsequently replaced by a single

a
```
/* Region 1 */
...
x = a + b;
y = c - d;
z = x * y;
...
...
/* Region 2 */
r = n + m;
s = o - p;
t = r * s;
...
```

b
```
/* Region 1 */
...
f(&x,&y,&z,
  a,b,c,d);

...

...
/* Region 2 */
f(&r,&s,&t,
  n,m,o,p);
...
```

c
```
/* New function */
void f(&x,&y,&z,
       a,b,c,d)
{
  x = a + b;
  y = c - d;
  z = x * y;
}
```

Fig. 16 Example: procedural abstraction. (**a**) Original code. (**b**) Code after procedural abstraction. (**c**) Newly created function

instance and control transfer instructions (jumps or function calls) to this single block of code. This optimization effectively "outlines" repeated fragments of code and replaces the original, redundant occurrences with function calls (jumps) to the newly created function (block). This code transformation is also known as *procedural abstraction* if a single, new function is created, or *cross-jumping* or *tail merging* if a region of code is replaced with a direct jump to a single instance of an identical code block. An example of procedural abstraction is illustrated in Fig. 16.

Coalescing of repeated instructions sequences, either through procedural abstraction or cross-jumping, involves a number steps. Initially, repeated instructions sequences need to be identified using a *suffix tree* [16] from which a *repeat table* is constructed. Repeated instruction sequences identified during these stages might still contain *unsafe* operations that interfere with code size reducing code transformations. These *control hazards*, e.g. jumps into or out of the code fragment, must be avoided, for example, by splitting these repeats appropriately. Each of the smaller fragments can then be dealt with separately. Once the compiler has identified a set of repeats suitable for compaction it will determine if coalescing of repeated instruction sequences is profitable and, if so, either apply the procedural abstraction or cross-jumping transformation. The former requires that the block of code to be "outlined" is *single-entry, single-exit*, i.e. all internal jumps must be within the body of the region. For cross-jumping all of the control-flow successors must match for each repeated region.

If repeated code patterns occur frequently and are large enough the overall benefit from procedural abstraction or cross-jumping outweighs their code size and performance overheads. On the other hand, for each code region replaced with a function call there is a function call overhead resulting in additional instructions possibly outweighing the benefit, especially for very short code fragments. Also, the additional control flow instructions may introduce a significant performance overhead when compared with the original straight-line code.

Within an optimizing compiler the coalescing of repeated instruction sequences is typically performed on a low-level intermediate representation.[2] When applied at a lower level, possibly after register allocation, the identification of repeated instruction sequences should not be restricted to exact, instruction-by-instruction matches, but allow for some abstraction of branches and registers. Two regions may only differ in the use of labels or register names and by treating these similar regions as matches the overall effectiveness of code compression can be increased.

References

1. Jaume Abella and Antonio Gonzalez and Josep Llosa and Xavier Vera: Near-Optimal Loop Tiling by Means of Cache Miss Equations and Genetic Algorithms. In: Proceedings of the 2002 International Conference on Parallel Processing Workshops (ICPPW '02), Washington, DC, USA, 2002.
2. Allan, Vicki H. and Jones, Reese B. and Lee, Randall M. and Allan, Stephen J.: Software pipelining. ACM Computing Surveys, Vol. 27, No. 3, pp. 367–432, 1995.
3. ACE Associated Compiler Experts bv: DSP-C – An extension to ISO/IEC IE 9899:1990. Netherlands, 1998.
4. David F. Bacon, Susan L. Graham, and Oliver J. Sharp: Compiler Transformations for High-Performance Computing. ACM Computing Surveys, Vol. 26, No. 4, December 1994.
5. Árpád Beszédes, Rudolf Ferenc, Tibor Gyimóthy, André Dolenc, and Karsisto, Konsta: Survey of code-size reduction methods. ACM Computing Surveys, Vol. 35, No. 3, pp. 223–267, 2003.
6. Keith D. Cooper and Nathaniel McIntosh: Enhanced Code Compression for Embedded RISC Processors. In: Proceedings of the SIGPLAN Conference on Programming Language Design and Implementation, pp. 139–149, 1999.
7. Keith D. Cooper, L. Taylor Simpson, and Christopher A. Vick: Operator Strength Reduction. ACM Transactions on Programming Languages and Systems, Vol. 23, Issue 5, pp. 603–625, Sept. 2001.
8. Keith D. Cooper, Philip J. Schielke, and Devika Subramanian; Optimizing for Reduced Code Space using Genetic Algorithms. In: Proceedings of the ACM SIGPLAN 1999 W orkshop on Languages, Compilers, and Tools for Embedded Systems, pp. 1–9, Atlanta, Georgia, United States, 1999.
9. Sauma K. Debray, William Evans, Robert Muth, and Bjorn De Sutter: Compiler Techniques for Code Compaction. ACM Transactions on Programming Languages and Systems, Vol. 22, No. 2, March 2000, pp. 378–415.
10. Depuydt, Francis and Goossens, Gert and De Man, Hugo: Scheduling with register constraints for DSP architectures. Integration, the VLSI Journal, pp. 95–120, Vol. 18, No. 1, 1994.
11. Dubach, Christophe and Cavazos, John and Franke, Björn and Fursin, Grigori and O'Boyle, Michael F.P. and Temam, Olivier: Fast compiler optimisation evaluation using code-feature based performance prediction. In: Proceedings of the 4th International Conference on Computing Frontiers (CF '07), Ischia, Italy, 2007.
12. ISO/IEC JTC1 SC22 WG14 N1169: Programming languages - C - Extensions to support embedded processors. Technical Report ISO/IEC TR 18037, 2006.
13. Fang, Jesse Zhixi: Compiler Algorithms on If-Conversion, Speculative Predicates Assignment and Predicated Code Optimizations. In: Proceedings of the 9th International Workshop on Languages and Compilers for Parallel Computing (LCPC '96), pp. 135–153, London, UK, 1997.

---

[2] A high-level representation is used in Fig. 16 to preserve clarity and brevity of the example.

14. Franz Franchetti and Markus Püschel: Short Vector Code Generation and Adaptation for DSP Algorithms. In: Proceedings of the International Conference on Acoustics, Speech, and Signal Processing (ICASSP), pp. 537–540, 2003.

15. Franke, Björn and O'Boyle, Michael: Array recovery and high-level transformations for DSP applications. ACM Transactions on Embedded Computing Systems, Vol.2 , No. 2, pp. 132–162, 2003.

16. C.W. Fraser, E.W. Myers, and A.L. Wendt: Analyzing and compressing assembly code. SIGPLAN Notices, Vol. 19, No. 6, pp. 117–121, June 1984.

17. G. Gréwal and T. Wilson and A. Morton: An EGA Approach to the Compile-Time Assignment of Data to Multiple Memories in Digital-Signal Processors, SIGARCH Computer Architecture News, Vol. 31, No. 1, pp. 49–59, 2003.

18. Masayo Haneda, Peter M.W. Knijnenburg, and Harry A.G. Wijshoff: Code Size Reduction by Compiler Tuning. In: S. Vassiliadis et al. (Eds.): SAMOS 2006, LNCS 4017, pp. 186–195, Springer-Verlag, 2006.

19. Hohenauer, Manuel and Engel, Felix and Leupers, Rainer and Ascheid, Gerd and Meyr, Heinrich: A SIMD optimization framework for retargetable compilers. ACM Transactions on Architecture and Code Optimization, Vol. 6, No. 1, pp. 1–27, 2009.

20. ISO/IEC JTC1/SC22/WG14 working group. ISO/IEC 9899:1999 – Programming languages – C, 1999.

21. Kessler, C.: Compiling for VLIW DSPs. In: S.S. Bhattacharyya, E.F. Deprettere, R. Leupers, J. Takala (eds.) Handbook of Signal Processing Systems, second edn. Springer (2013)

22. M-Y. Ko and S. S. Bhattacharyya: Data Partioning for DSP Software Synthesis. In: Proceedings of the International Workshop on Software and Compilers for Embedded Systems (SCOPES '03), pp. 344–358, Vienna, Austria, 2003

23. Arvind Krishnaswamy and Rajiv Gupta: Mixed-width instruction sets. Communications of the ACM, Vol. 46, No. 8, pp. 47–52, 2003.

24. Lee, C.G.: UTDSP Benchmark Suite. http://www.eecg.toronto.edu/~corinna/DSP/ infrastructure/UTDSP.html.Cited1July2009

25. Leupers, R.: Code Optimization Techniques for Embedded Processors: Methods, Algorithms, and Tools. Kluwer Academic Publishers, Norwell, MA, USA, 2000.

26. Leupers, Rainer: Code selection for media processors with SIMD instructions. In: Proceedings of the Conference on Design, Automation and Test in Europe (DATE '00), Paris, France, pp. 4–8, 2000.

27. R. Leupers and D. Kotte: Variable Partitioning for Dual Memory Bank DSPs. In: Proceedings of the IEEE International Conference on Acoustics, Speech and Signal Processing (ICASSP '01), pp. 1121–1124, Salt Lake City, USA, 2001.

28. Leupers, Rainer and Bashford, Steven: Graph-based code selection techniques for embedded processors. ACM Transactions on Design Automation of Electronic Systems, pp. 794–814, Vol. 5, No. 4, 2000.

29. Leupers, Rainer: Register allocation for common subexpressions in DSP data paths, In: Proceedings of the 2000 Asia and South Pacific Design Automation Conference (ASP-DAC '00), pp. 235–240, Yokohama, Japan, 2000.

30. Liao, Stan and Devadas, Srinivas and Keutzer, Kurt and Tjiang, Steve and Wang, Albert: Code optimization techniques for embedded DSP microprocessors. In: Proceedings of the 32nd Annual ACM/IEEE Design Automation Conference (DAC '95), pp. 599–604, San Francisco, California, United States, 1995.

31. Menard, Daniel and Chillet, Daniel and Charot, François and Sentieys, Olivier: Automatic floating-point to fixed-point conversion for DSP code generation. In: Proceedings of the 2002 International Conference on Compilers, Architecture, and Synthesis for Embedded Systems (CASES '02), pp. 270–276, Grenoble, France, 2002.

32. Monsifrot, Antoine and Bodin, François and Quiniou, Rene: A Machine Learning Approach to Automatic Production of Compiler Heuristics. In: Proceedings of the 10th International Conference on Artificial Intelligence: Methodology, Systems, and Applications (AIMSA '02), pp. 41–50, London, UK, 2002.

33. Stephen S. Muchnick: Advanced compiler design and implementation. Morgan Kaufmann Publishers Inc., San Francisco, CA, USA, 1997.
34. Alastair Murray and Björn Franke: Using Genetic Programming for Source-Level Data Assignment to Dual Memory Banks. In: Proceedings of the 3rd Workshop on Statistical and Machine Learning Approaches to Architecture and Compilation (SMART09), Paphos, Cyprus, 2009.
35. Alastair Murray and Björn Franke: Fast Source-Level Data Assignment to Dual Memory Banks. In: Proceedings of the Workshop on Software & Compilers for Embedded Systems (SCOPES 2008), March 2008, Munich, Germany.
36. Nuzman, Dorit and Rosen, Ira and Zaks, Ayal: Auto-vectorization of interleaved data for SIMD. ACM SIGPLAN Notices, Vol. 41, No. 6, pp. 132–143, 2006.
37. Preeti Ranjan Panda, Hiroshi Nakamura, Nikil D. Dutt, Alexandru Nicolau: Augmenting Loop Tiling with Data Alignment for Improved Cache Performance. IEEE Transactions on Computers, Vol. 48, No. 2, pp. 142–149, February, 1999.
38. Pokam, Gilles and Bihan, Stéphane and Simonnet, Julien and Bodin, François: SWARP: A retargetable preprocessor for multimedia instructions. Concurrency and Computation: Practice & Experience, Volume 16, Issue 2–3, pp. 303–318, January 2004.
39. Pouchet, Louis-Noël and Bastoul, Cédric and Cohen, Albert and Cavazos, John: Iterative Optimization in the Polyhedral Model: Part II, Multidimensional Time. In: Proceedings of the 2008 ACM SIGPLAN Conference on Programming Language Design and Implementation (PLDI '08), pp. 90–100, New York, NY, USA, 2008.
40. Pujare, S., Lee, C.G., and Chow, P.: Machine-independent compiler optimizations for the UofT DSP architecture. In: Proceedings of the 6th International Conference on Signal Processing Applications and Technology (ICSPAT), pp. 860–865, 1995.
41. M.A.R. Saghir, P.Chow and C.G. Lee: Exploiting Dual Data-Memory Banks in Digital Signal Processors. In: Proceedings of the 7th International Conference on Architectural Support for Programming Languages and Operating Systems (ASPLOS-VII), pp. 234–243, Cambridge, Massachusetts, USA, 1996.
42. Sair, S., Kaeli, D., Meleis, W.: A study of loop unrolling for VLIW-based DSP processors. In: Proceedings of the IEEE Workshop on Signal Processing Systems, 1998.
43. V. Sipková: Efficient Variable Allocation to Dual Memory Banks of DSPs. In: Proceedings of the 7th International Workshop on Software and Compilers for Embedded Systems (SCOPES '03), pp. 359–372, Vienna, Austria, 2003.
44. Mark Stephenson and Saman Amarasinghe: Predicting Unroll Factors Using Supervised Classification. In: Proceedings of the International Symposium on Code Generation and Optimization (CGO '05), pp. 123–134, Washington, DC, USA, 2005.
45. Takala, J.: General-purpose DSP processors. In: S.S. Bhattacharyya, E.F. Deprettere, R. Leupers, J. Takala (eds.) Handbook of Signal Processing Systems, second edn. Springer (2013)
46. Uh, Gang-Ryung; Wang, Yuhong; Whalley, David B.; Jinturkar, Sanjay; Burns, Chris and Cao, Vincent: Techniques for Effectively Exploiting a Zero Overhead Loop Buffer. In: Proceedings of the 9th International Conference on Compiler Construction (CC '00), pp. 157–172, London, UK, 2000.
47. Vienna University of Technology, Institute of Communications and Radio-Frequency Engineering: Address Code Optimization. www.address-code-optimization.org, retrieved August 2009.
48. Michael Joseph Wolfe: Optimizing Supercompilers for Supercomputers. MIT Press, Cambridge, MA, USA, 1990.
49. Zivojnovic, V., Martinez, J., Schläger, C. & Meyr, H.: DSPstone: A DSP-Oriented Benchmarking Methodology. In: Proceedings of the International Conference on Signal Processing Applications and Technology (ICSPAT'94), Dallas, Oct. 1994.

Kahn Process Networks and a Reactive Extension

Marc Geilen and Twan Basten

Abstract Kahn and MacQueen have introduced a generic class of determinate asynchronous data-flow applications, called Kahn Process Networks (KPNs) with an elegant mathematical model and semantics in terms of Scott-continuous functions on data streams together with an implementation model of independent asynchronous sequential programs communicating through FIFO buffers with blocking read and non-blocking write operations. The two are related by the Kahn Principle which states that a realization according to the implementation model behaves as predicted by the mathematical function. Additional steps are required to arrive at an actual implementation of a KPN to take care of scheduling of independent processes on a single processor and to manage communication buffers. Because of the expressiveness of the KPN model, buffer sizes and schedules cannot be determined at design time in general and require dynamic run-time system support. Constraints are discussed that need to be placed on such system support so as to maintain the Kahn Principle. We then discuss a possible extension of the KPN model to include the possibility for sporadic, reactive behavior which is not possible in the standard model. The extended model is called Reactive Process Networks. We introduce its semantics, look at analyzability and at more constrained data-flow models combined with reactive behavior.

M. Geilen (✉)
Eindhoven University of Technology, Den Dolech 2, Eindhoven, The Netherlands
e-mail: m.c.w.geilen@tue.nl

T. Basten
Eindhoven University of Technology and TNO-ESI,
Den Dolech 2, Eindhoven, The Netherlands
e-mail: a.a.basten@tue.nl

S.S. Bhattacharyya et al. (eds.), *Handbook of Signal Processing Systems*,
DOI 10.1007/978-1-4614-6859-2_32, © Springer Science+Business Media, LLC 2013

1 Introduction

Process networks are a popular model to express behavior of data-flow and streaming nature. This includes audio, video and 3D multimedia applications such as encoding and decoding of MPEG video streams. Using process networks, an application is modeled as a collection of concurrent processes communicating streams of data through FIFO channels. Process networks make task-level parallelism and communication explicit, have a simple semantics, are compositional and allow for efficient implementations without time-consuming synchronizations. There are several variants of process networks. One of the most general forms are Kahn process networks [29, 30], where the nodes are arbitrary sequential programs, that communicate via channels of the process network with blocking read and non-blocking write operations. Although harder to analyze than more restricted models, such as Synchronous Data Flow networks [34], the added flexibility makes KPNs a popular programming model. Where synchronous data-flow models can be statically scheduled at compile time, KPNs must be scheduled dynamically in general, because their expressive power does not allow them to be statically analyzed. A run-time system is required to schedule the execution of processes and to manage memory usage for the channels. On a heterogeneous implementation platform, a KPN may be distributed over several components with individual scheduling and memory management domains. In this case, the execution of the process network has to be coordinated in a distributed fashion.

A process network is *determinate* if its input/output behavior can be expressed as a function. Kahn Process Networks represent the *largest class* of determinate data-flow process networks if we compare them based on the input/output functions they can express. The model abstracts from timing behavior and focusses on functional input/output behavior of a network of parallel processes.

In this chapter we discuss the syntax and operational semantics of the model, a denotational semantics and the Kahn Principle, which relates them. The denotational semantics of KPNs of [29] is attractive from a mathematical point of view, because of its abstractness and compositionality. In a realization of a process network, FIFO sizes and contents play an important role and are influenced by a run-time environment that governs the execution of the network. It is for this reason that we present a simple operational semantics of process networks, similar to [17, 48]. We study and prove properties of the resulting transition system and we illustrate a simple proof of the Kahn Principle. Not because these are new results, but we believe that they provide insight in the fundamental properties and limitation of the Kahn model. We look at methods and requirements for directly implementing Kahn Process Networks and we look at possible extensions of Kahn Process Networks, in particular with the ability to express reactive behavior.

Fig. 1 Example: a Kahn
Process Network computing
the Fibonacci sequence

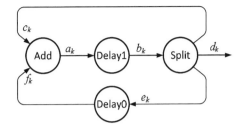

1.1 Example

Figure 1 shows an example of a Kahn Process Network. It consists of four *processes* which compute deterministic functions. Processes have *inputs* and *outputs*. A KPN has *unbounded* FIFO channels connecting an output of a process to an input of a process. Unconnected channels represent inputs or outputs of the KPN. The network in Fig. 1 has no inputs and one output (d_k). The process Add reads one integer number from each input (c_k and f_k, $k \geq 0$) and writes the sum of both numbers to its output ($a_k = c_k + f_k$). The processes Delay n first write the value n to their output and subsequently start copying their input to their output. The fourth process, Split, copies its input to three separate outputs, implying $c_k = d_k = e_k = b_k$. If the network executes, Delay0 and Delay1 write respectively a value 0 and 1 to their outputs. The value 1 is copied by the Split actor giving $d_0 = 1$. Now the Add process has the values 0 and 1 on its inputs, adds them up and writes the value 1 to its output. This value is copied again by Delay1 and Split and the value 1 ($= d_1$) is produced again on the output. It is easy to see that $d_k = a_{k-1}$ for $k \geq 1$ and $f_k = d_{k-1} = a_{k-2}$ for $k \geq 2$. Thus, $d_k = d_{k-1} + d_{k-2}$ with $d_0 = d_1 = 1$ and the recurrence equation corresponds to the Fibonacci sequence and because of the proper initialization by the Delay n processes, the KPN starts producing the Fibonacci sequence.

A more practical example of a KPN is shown in Fig. 2. It shows a part of a pipeline of a JPEG decoder. Input stream to the network is a stream of compressed data, the Variable Length Decoder function turns it into a stream of macro-blocks of 8 by 8 pixels. Those blocks are subsequently passed through three functions transforming those blocks, undoing the quantization, the zig-zag reordering of data and the discrete cosine transformation respectively. The corresponding mathematical functions are in practice specified by code as illustrated for the Inverse Zig Zag function (see Sect. 6). In this case the function is an infinite loop because the functions operates block by block on a stream of blocks. It uses a `read` operation to read a macro block from its input `in`, then does some processing on it and at the end uses a `write` operation to write the result to its output.

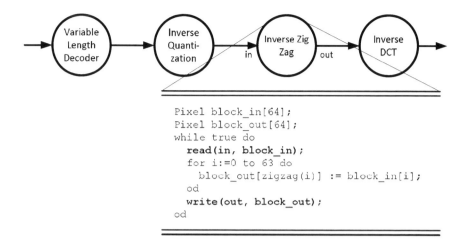

```
Pixel block_in[64];
Pixel block_out[64];
while true do
    read(in, block_in);
    for i:=0 to 63 do
        block_out[zigzag(i)] := block_in[i];
    od
    write(out, block_out);
od
```

Fig. 2 KPN of JPEG decoder

1.2 Preliminaries

We introduce some mathematical notation and preliminaries. We assume the reader is familiar with the concept of a complete partial order (CPO) (see for instance [14]). We use $(X, \sqsubseteq)$ to denote a CPO on the set X with partial order relation $\sqsubseteq \subseteq X \times X$ to denote the corresponding partial order relation. We use $\sqcup D$ to denote the least upper bound of a directed subset D of X. For convenience, we assume a universal, countable, set *Chan* of channels and for every channel $c \in$ *Chan* a corresponding finite channel alphabet Σ_c. We use Σ to denote the union of all channel alphabets, and A^* (A^ω) to denote the set of all finite (and infinite) strings over alphabet A and $A^{*,\omega} = A^* \cup A^\omega$. $\sqsubseteq$ denotes the prefix relation on strings (a complete partial order, see for instance [14]). If σ and τ are strings, $\sigma \cdot \tau$ denotes the usual concatenation of the strings. If $\sigma \sqsubseteq \tau$ then $\tau - \sigma$ denotes the string τ without its prefix σ.

A *history* of a channel denotes the sequence of data elements communicated along a channel [46], for instance the sequence of Fibonacci numbers. A history h of a set C of channels is a mapping from channels $c \in C$ to strings over Σ_c. The set of all histories of C is denoted as $H(C)$. If $h \in H(C)$ and $D \subseteq C$, then $h|D$ denotes the history obtained from h by restricting the domain to D. If h_1 is a history of C_1 and h_2 is a history of C_2, we write $h_1 \sqsubseteq h_2$ if $C_1 \subseteq C_2$ and for every $c \in C_1$, $h_1(c) \sqsubseteq h_2(c)$. The set of histories together with the relation $\sqsubseteq$ on histories form a complete partial order with as bottom element the empty history $\emptyset$. If $h_1 \sqsubseteq h_2$, then $h_2 - h_1$ denotes the history which maps a channel $c \in C_1$ to $h_2(c) - h_1(c)$. Histories h_1 and h_2 are called *consistent* if they share an upper bound, i.e., if there exists some history h_3 such that $h_1 \sqsubseteq h_3$ and $h_2 \sqsubseteq h_3$. If $h_1, h_2 \in H(C)$, then the concatenation $h_1 \cdot h_2$ is the history such that $h_1 \cdot h_2(c) = h_1(c) \cdot h_2(c)$ for all $c \in C$. A history is called finite if it maps every channel in its domain to a finite string and only a finite number of channels to a non-empty string. A finite history is a finite element of the CPO of histories. (Recall from [14] that an element $k \in X$ of a CPO $(X, \sqsubseteq)$ is a finite element if for every chain $D \subseteq X$, if $k \sqsubseteq \sqcup D$ then there is some $d \in D$ such that $k \sqsubseteq d$.)

2   Denotational Semantics

We discuss the formal definition of the semantics, i.e., the behavior, of a KPN. Traditionally, the focus is on the functional input/output behavior of the network, abstracting from the timing. More precisely, we are interested in a *function* relating the output produced by the network to the input provided to the network. For the JPEG decoder for instance, the sequence of decoded output blocks is a function of the compressed input data. We first concentrate on a *denotational* semantics, focussing on *what* input/output relation a KPN computes and later, in Sect. 3 on an *operational* semantics, which also captures *how* that output is computed.

The denotational semantics of KPNs [29] defines the behavior of a KPN as a *Scott-continuous input/output function* on the input or output stream histories. Assuming that the (Scott-continuous) input/output functions of the individual processes are known, then the input/output function of the KPN as a whole can be defined as the least fixed-point of a collection of equations specifying the relations between channel histories based on the processes. For the JPEG example, the overall function is straightforwardly obtained by function composition. A more challenging situation arises when there is feedback, such as in the Fibonacci example.

A KPN *process* with m inputs and n outputs is a function $f : (\Sigma^{*,\omega})^m \to (\Sigma^{*,\omega})^n$ with the characteristic (Scott-continuity) such that it *preserves least upper bounds*: $f(\sqcup i_k) = \sqcup f(i_k)$. This implies that such a processes is *monotone*: if $i \sqsubseteq j$ then $f(i) \sqsubseteq f(j)$; when additional input is provided, additional output may be produced, but existing output cannot be changed.

Because processes may exhibit memory (the current output may depend on input from the past), the function is defined in terms of the sequences representing the entire input *history* of the channel.

The functions of the individual processes of the Fibonacci example can be defined as follows. Delayn $: \mathbb{N}^{*,\omega} \to \mathbb{N}^{*,\omega}; i \in \mathbb{N}^{*,\omega}; n \in \{0,1\}$:

$$\mathsf{Delayn}(i) = n \cdot i \tag{1}$$

Add $: (\mathbb{N}^{*,\omega})^2 \to \mathbb{N}^{*,\omega}$ is defined inductively by the following equations $(i, j \in \mathbb{N}^{*,\omega};$ $x, y \in \mathbb{N}, \varepsilon$ is the empty sequence):

$$\begin{cases} \mathsf{Add}(i, \varepsilon) = \varepsilon \\ \mathsf{Add}(\varepsilon, i) = \varepsilon \\ \mathsf{Add}(x \cdot i, y \cdot j) = (x + y) \cdot \mathsf{Add}(i, j) \end{cases} \tag{2}$$

The FIFO channels in a KPN connect output streams of processes to input streams of other processes, creating relationships between the functions of the processes. In the example, the streams a, b, c, d, e and f are governed by the joint equations of the processes:

$$\begin{cases} b = \mathsf{Delay1}(a) \\ f = \mathsf{Delay0}(e) \\ a = \mathsf{Add}(c, f) \\ c = b, \;\; d = b, \;\; e = b \end{cases} \tag{3}$$

One can show that this set of equations has a single, unique, solution, namely the Fibonacci numbers, for its output sequence d.

In general, the semantics of a KPN are defined in a similar way as illustrated for the Fibonacci example above. In [29], Kahn presented the denotational semantics of process networks as the solution to a set of equations that capture the input/output relations of the individual processes and the way they are connected in the network. These equations are called the *network equations*.

A KPN consists of a set P of processes and a set S of streams, partitioned into *inputs I, outputs O* and *internal channels C*.

The processes p, with inputs I_p and outputs O_p, that constitute the network are described by (Scott-continuous) functions $f_p : H(I_p) \to H(O_p)$ that map input histories to output histories of the processes. The network as a whole can be described as a function, defined by Kahn's network equations as follows. For $h \in H(C \cup I \cup O)$ to be a valid history describing a behavior of the whole KPN (by describing the data communicated on each of its channels), it must satisfy the *network equations* derived from the processes $p \in P$:

$$h|O_p = f_p(h|I_p) \text{for all} p \in P$$

It simply expresses that if we take from h the channels that are the inputs and outputs of p, then these must be related corresponding to the function f_p.

To characterize the behavior of the KPN as a function f that maps an input history i to a history of *all channels* (including the output channels, which we are after) in the KPN, $f : H(I) \to H(I \cup C \cup O)$, we can derive the following equations (substituting $f(i)$ for h and adding an equation for the input channels).

$$\begin{cases} f(i)|I = i \\ f(i)|O_p = f_p(f(i)|I_p) \text{for all} p \in P \end{cases} \tag{4}$$

Note that every internal or output channel of the KPN is an output channel of one of the processes of the KPN. Hence, all channels are defined by these equations. Equation (4) is a recursive equation and in general, there may be many functions f that satisfy this equation, but only one that corresponds to an actual, causal behavior respecting causality (one where, intuitively speaking, symbols are produced before

Table 1 Kleene iteration of the Fibonacci example

k	0	1	2	3	4	5	6	7	8	9	10	11	12
$a(k)$	ε	ε	ε	1	1	1	1.2	1.2	1.2	1.2.3	1.2.3	1.2.3	1.2.3.5
$b(k)$	ε	1	1	1	1.1	1.1	1.1	1.1.2	1.1.2	1.1.2	1.1.2.3	1.1.2.3	1.1.2.3
$c(k)$	ε	ε	1	1	1	1.1	1.1	1.1	1.1.2	1.1.2	1.1.2	1.1.2.3	1.1.2.3
$d(k)$	ε	ε	1	1	1	1.1	1.1	1.1	1.1.2	1.1.2	1.1.2	1.1.2.3	1.1.2.3
$e(k)$	ε	ε	1	1	1	1.1	1.1	1.1	1.1.2	1.1.2	1.1.2	1.1.2.3	1.1.2.3
$f(k)$	ε	0	0	0.1	0.1	0.1	0.1.1	0.1.1	0.1.1	0.1.1.2	0.1.1.2	0.1.1.2	0.1.1.2.3

they are consumed). The causal solution is the *smallest* function f that satisfies the network equations. Technically, this solution can be obtained as the least fixed-point of an appropriate functional. From Eq. (4) we derive the functional

$$\Phi : (H(I) \rightarrow H(I \cup C \cup O)) \rightarrow (H(I) \rightarrow H(I \cup C \cup O))$$

defined as (compare Eq. (4)):

$$\begin{cases} \Phi(f)(i)|I = i \\ \Phi(f)(i)|O_p = f_p(f(i)|I_p) \text{for all} p \in P \end{cases} \tag{5}$$

Clearly, a fixed-point of Φ satisfies Eq. (4). Moreover, Φ is a continuous function on a CPO and according to Kleene's fixed-point theorem [14] it has a least-fixed point, the limit of the ascending Kleene chain (Kleene iteration), $\bigsqcup \{\Phi^k(h_\varepsilon) \mid k \geq 0\}$, where h_ε denotes the empty history associating an empty string with every channel. Therefore, Kleene iteration provides a way of constructively computing the network behavior, by updating the output of a process whenever its input has changed. If this procedure is repeated, then either it stops and the complete output has been computed or it continues forever and the infinite output equals the limit of the sequence of outputs computed during the procedure. Kleene iteration for the Fibonacci example (up to 12 steps, although the process continues ad infinitum) is illustrated in Table 1 and illustrates how it converges to the infinite Fibonacci sequence.

3 Operational Semantics

The KPN denotational semantics specifies the input/output behavior of a KPN in an elegant, abstract way as a mathematical function. This description however is far away from the actual operation or implementation of a KPN. Formal reasoning about implementations or run-time systems for KPNs can be easier based on an *operational* semantics. For instance, the denotational semantics provides no information to reason about buffer sizes required for the FIFO communication, or reasoning about potential deadlocks. In this section, we present a compositional operational semantics to hierarchical KPNs.

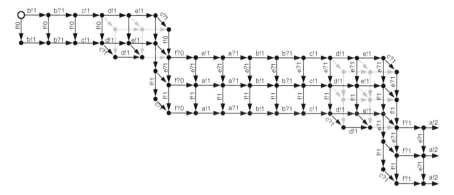

Fig. 3 Operational semantics of the Fibonacci example

3.1 Labeled Transition Systems

We give an operational semantics to KPNs in the form of a labeled transition system (LTS). We use, more specifically, an LTS with an initial state, with designated input and output actions in the form of reads and writes of symbols on channels, as well as internal actions. We illustrate this with the LTS of the Fibonacci example (Fig. 3).

Definition 1. (LTS) An LTS is a tuple $(S, s_0, I, O, Act, \rightarrow)$ consisting of a (countable) set S of states, an initial state $s_0 \in S$, a set $I \subseteq Chan$ of input channels, a set $O \subseteq Chan$ (disjoint from I) of output channels, a set Act of actions consisting of input (read) actions $\{c?a \mid c \in I, a \in \Sigma_c\} \subseteq Act$, output (write) actions $\{c!a \mid c \in O, a \in \Sigma_c\} \subseteq Act$ and (possibly) internal actions (all other actions), and a labeled transition relation $\rightarrow \subseteq S \times Act \times S$ describing possible transitions between states.

Thus, $c!a$ is a write action to channel c with symbol a; $c?a$ models passing of a symbol from input channel c to the LTS. The initial state in Fig. 3 is indicated with the larger open circle. From this state, two write actions are possible, $b!1$ and $f!0$, leading to different, new states. We write $s_1 \xrightarrow{\alpha} s_2$ if $(s_1, \alpha, s_2) \in \rightarrow$ and $s_1 \xrightarrow{\alpha}$ if there is some $s_2 \in S$ such that $s_1 \xrightarrow{\alpha} s_2$.

With a write operation, the symbol on the output channel is determined by the LTS. With a read operation, the symbol that appears on the input channel is determined by the environment of the LTS. Therefore, a read operation is modeled with a set of input actions that provides a transition for every possible symbol of the alphabet. (Note that the Fibonacci example KPN has no inputs to the network, only an output d.)

If Act is a set of actions and $C \subseteq Chan$ a set of channels, we write $Act|C$ to denote $\{c!a, c?a \in Act \mid c \in C\}$, i.e., actions of Act on channels in C. An execution σ is a path through the transition system starting from the initial state, a sequence $s_0 \xrightarrow{\alpha_0} s_1 \xrightarrow{\alpha_1} \dots$ of states $s_i \in S$ and actions $\alpha_i \in Act$, such that $s_i \xrightarrow{\alpha_i} s_{i+1}$ for all

$i \geq 0$ (up to the length of the execution). If σ is such an execution, then we use $|\sigma| \in \mathbb{N} \cup \{\infty\}$ to denote the length of the execution. For $k \leq |\sigma|$, we use σ^k to denote the prefix of the execution up to and including state k. If $s_0 \xrightarrow{\alpha_0} s_1 \xrightarrow{\alpha_1} \ldots \xrightarrow{\alpha_{n-1}} s_n$, we write $s_0 \xRightarrow{a} s_n$, where $a = \alpha_0 \cdot \alpha_1 \cdot \ldots \cdot \alpha_n$. From a given execution σ with actions $a = \alpha_0 \cdot \alpha_1 \cdot \ldots$, we extract the consumed input and the produced output on a set $D \subseteq Chan$ of channels as follows. For a channel $c \in D$, $a?c$ is a (finite or infinite) string over Σ_c that results from projecting a onto *read* actions on c. $a!c$ is a (finite or infinite) string over Σ_c that results from projecting a onto *write* actions on c. Input history $a?D = \{(c,a?c) \mid c \in D\}$ and output history $a!D = \{(c,a!c) \mid c \in D\}$.

Furthermore, we use the same notation for executions σ with actions a: $\sigma?c = a?c$, $\sigma!c = a!c$, $\sigma?D = a?D$ and $\sigma!D = a!D$. Thus $\sigma?I$ denotes the input consumed by the network in execution σ and $\sigma!O$ denotes the output produced by the network. The *I/O-history* $h(\sigma)$ of an execution σ is $\sigma?I \cup \sigma!O$. To reason about the input *offered* to the network (consumed or not consumed), we say that σ is an execution with input $i : I \to \Sigma^{*,\omega}$ if $\sigma?I \sqsubseteq i$ (the consumed data is consistent with i, but need not be all of i). If $i \sqsubseteq j$ then σ is also an execution with input j.

Executions in general may be only partially completed or they may be unrealistic or wrong because certain actions are systematically being ignored. To be able to exclude such executions when necessary, we need the notions of *maximality* and *fairness*. Let $\sigma = s_0 \xrightarrow{\alpha_0} s_1 \xrightarrow{\alpha_1} \ldots$ be an execution of the LTS.

- (MAXIMALITY) Execution σ with input i is called maximal if it does not stop prematurely, if it is infinite or in its last state only read actions on input channels from which all input of i has been consumed are possible, i.e., if $|\sigma| = n$ and $s_n \xrightarrow{\alpha}$ then $\alpha = c?a$ for some $c \in I$ and $a \in \Sigma_c$ and $\sigma?c = i(c)$.
- (FAIRNESS) Execution σ is fair with input i if it is finite, or it is infinite and if at some point an action is enabled, it is eventually executed or disabled (the latter does not occur for KPNs), i.e.,

  - if for some $n \in \mathbb{N}$ and internal or output action α, $s_n \xrightarrow{\alpha}$, then there is some $k \geq n$ such that $\alpha_k = \alpha$ or $s_k \not\xrightarrow{\alpha}$;
  - if for some $n \in \mathbb{N}$, $c \in I$ and $a \in \Sigma_c$, $s_n \xrightarrow{c?a}$, and $i(c) = (\sigma^n?c)a\tau$ for some $\tau \in \Sigma_c^{*,\omega}$, then there is some $k \geq n$ such that $\alpha_k = c?a$ or $s_k \not\xrightarrow{c?a}$;

Every finite execution with input i of an LTS can be extended to a fair and maximal execution with input i. We can describe the externally observable behavior of a labeled transition system by relating the output actions the LTS produces with the input actions provided to the network. In general this gives a relation between input histories and output histories. We restrict the attention to "proper" executions in the sense that only maximal and fair executions are taken into account.

Definition 2. (INPUT/OUTPUT RELATION) The input/output relation IO of an LTS is the relation $\{(i, \sigma!O) \mid \sigma \text{ is a maximal and fair execution with input } i\}$.

3.2 Determinacy

In general, the input/output relation is too abstract to adequately characterize the behavior of an LTS. If it is non-deterministic, the order in which input is consumed or output is produced can be relevant (see Sect. 7). Kahn process networks do not exhibit such non-determinism. Transitions are deterministic and if multiple actions are available at the same time, then they are independent (i.e., they can be executed in any order with an identical result). This leads to a special type of LTS, which we call *determinate*. Recall the LTS in Fig. 3, showing the beginning of the LTS of the Fibonacci example. Although there are very many different paths, there is only very limited actual choice in choosing a path from the initial top-left state. Any path we choose executes exactly the same actions in a slightly modified order caused by concurrency in the process network. We give a precise definition of this type of LTS and summarize their properties.

Definition 3. (DETERMINACY) LTS $(S, s_0, I, O, Act, \rightarrow)$ is determinate if for any $s, s_1, s_2 \in S$, $\alpha_1, \alpha_2 \in Act$, if $s \xrightarrow{\alpha_1} s_1$ and $s \xrightarrow{\alpha_2} s_2$, the following hold:

1. (DETERMINISM) if $\alpha_1 = \alpha_2$ is some input or output action, then $s_1 = s_2$, i.e., executing a particular action has a unique deterministic result;
2. (CONFLUENCE) if α_1 and α_2 are not two input actions on the same channel (i.e., instances of the same read operation), then there is some s_3 such that $s_1 \xrightarrow{\alpha_2} s_3$ and $s_2 \xrightarrow{\alpha_1} s_3$. Figure 3 shows many instances of this structure, where if from some state multiple transitions are possible, the actions that are not chosen remain enabled and taking them in different orders leads to the same state.
3. (INPUT COMPLETENESS) if $\alpha_1 = c?a$ for some $c \in I$, then for every $a' \in \Sigma_c$, $s \xrightarrow{c?a'}$, i.e., input symbols are completely defined by the environment, the LTS cannot be selective in the choice of symbols it accepts;
4. (OUTPUT UNIQUENESS) if $\alpha_1 = c!a$ and $\alpha_2 = c!a'$ for some $c \in O$, then $a = a'$, i.e., output symbols are completely determined by the transition system. Here, the environment cannot be selective in its choice of symbol to receive.

A *sequential* LTS is a determinate LTS with the additional property that

5. (SEQUENTIALITY) if $\alpha_1 = c \sharp a$ ($\sharp \in \{!, ?\}$) (some read or write operation), then $\alpha_2 = c \sharp a'$ for some $a' \in \Sigma_c$ and $c \in I \cup O$, i.e., the LTS accepts at most one input/output operation at any point in time and no other (for instance internal) actions.

Although a determinate LTS may have multiple actions enabled at the same time. The order in which they are taken has no influence on the consumed input or produced output of the network in a fair and maximal execution. Two different executions are essentially the same, because the actions of one can be reordered without changing the input or output histories, to obtain the other execution (they are equivalent Mazurkiewicz traces).

Proposition 1. *The input/output relation of a determinate labeled transition system is a continuous function.*

A detailed proof can be found in [18]. If λ is a labeled transition system, we use f_λ to denote its I/O relation. In particular, if λ is determinate, this is an I/O *function*.

An important property in process networks is a deadlock condition. We can define a deadlock as the possibility to reach a state from which no further transitions are possible, a finite maximal execution. It is usually called a deadlock only if this is an undesirable situation. An important corollary from the analysis above is that if a determinate LTS has some execution which leads to a deadlock state, then *all of its executions* lead to the same deadlock state. In other words, the deadlock cannot be avoided by a smarter scheduling strategy, it is inherent to the determinate LTS specification and as we will see, therefore also holds for KPN.

3.3   Operational Semantics

We can now formalize an operational semantics of a KPN as a determinate LTS.

Definition 4. (KAHN PROCESS NETWORK) A Kahn process network is a tuple $(P, C, I, O, Act, \{\lambda_p \mid p \in P\})$ that consists of the following elements:

- A finite set P of *processes*.
- A finite set $C \subseteq Chan$ of *internal channels*, a finite set $I \subseteq Chan$ of input channels and a finite set $O \subseteq Chan$ of output channels, all distinct.
- Every constituent process $p \in P$ is itself defined by a determinate labeled transition system $\lambda_p = (S_p, s_{p,0}, I_p, O_p, Act_p, \rightarrow)$, with $I_p \subseteq I \cup C$ and $O_p \subseteq O \cup C$. The sets $Act_p \backslash (Act_p | (I_p \cup O_p))$ of internal actions of the processes are disjoint.
- The set Act of *actions* consisting of the actions of the constituent processes: $Act = \bigcup_{p \in P} Act_p$.
- For every channel $c \in C \cup I$, there is exactly one process $p \in P$ that reads from it ($c \in I_p$) and for every channel $c \in C \cup O$, there is exactly one process $p \in P$ that writes to it ($c \in O_p$).

To define the operational semantics of a KPN, we need a notion of *global state* of the network; this state is composed of the individual states of the processes and the current contents of the internal channels. A configuration of the process network is a pair (π, γ) consisting of a process state π and a channel state γ, where

- a *process state* $\pi : P \rightarrow S = \bigcup_{p \in P} S_p$ is a function that maps every process $p \in P$ on a local state $\pi(p) \in S_p$ of its transition system;
- a *channel state* $\gamma : C \rightarrow \Sigma^*$ is a history function that maps every internal channel $c \in C$ on a *finite* string $\gamma(c)$ over Σ_c.

The set of all configurations is denoted by *Confs* and there is a designated initial configuration $c_0 = (\pi_0, \gamma_0)$, where π_0 maps every process $p \in P$ to its initial state

$s_{p,0}$ and γ_0 maps every channel $c \in C$ to the empty string ε. We assign to a KPN $\kappa = (P, C, I, O, Act, \{\lambda_p \mid p \in P\})$, an operational semantics in the form of an LTS $(Confs, c_0, I, O, Act, \rightarrow)$. The labeled transition relation $\rightarrow$ is inductively defined by the following five rules (given in Plotkin style [44] inference rules; if the condition above the rule is satisfied, the conclusion below is also valid). For reading from and writing to internal channels by processes we have the following two rules respectively:

$$\frac{\pi(p) \xrightarrow{c?a}_p s, \gamma(c) = a\sigma, c \in C}{(\pi, \gamma) \xrightarrow{c?a} (\pi\{s/p\}, \gamma\{\sigma/c\})} \qquad \frac{\pi(p) \xrightarrow{c!a}_p s, \gamma(c) = \sigma, c \in C}{(\pi, \gamma) \xrightarrow{c!a} (\pi\{s/p\}, \gamma\{\sigma \cdot a/c\})}$$

Input channels and output channels are open to the environment:

$$\frac{\pi(p) \xrightarrow{c?a}_p s, c \in I}{(\pi, \gamma) \xrightarrow{c?a} (\pi\{s/p\}, \gamma)} \qquad \frac{\pi(p) \xrightarrow{c!a}_p s, c \in O}{(\pi, \gamma) \xrightarrow{c!a} (\pi\{s/p\}, \gamma)}$$

Individual processes may perform internal actions:

$$\frac{\pi(p) \xrightarrow{\alpha}_p s, \alpha \notin Act_p|(I_p \cup O_p)}{(\pi, \gamma) \xrightarrow{\alpha} (\pi\{s/p\}, \gamma)}$$

This LTS is denoted as $\Lambda(\kappa)$.

The labeled transition system of a KPN is determinate. We check the four properties of a determinate LTS.

- Determinism follows from determinism of the process that accepts or produces the input or output action respectively.
- Confluence. If both actions originate from different processes, it can be checked that they cannot disable each other. If they originate from the same process, it follows from confluence of that process.
- Input completeness follows immediately from input completeness of the constituent processes.
- Output uniqueness similarly follows directly from output uniqueness of the process delivering the output.

The LTS of the KPN has the same property (determinacy) as the individual processes. Therefore, the presented semantics of KPN is *compositional*, a determinate process network is constructed from individual determinate processes, and this means it can itself be used as a process in a larger KPN. This way, we can hierarchically construct larger KPNs. At the lowest level we can start with primitive processes, for instance sequential processes which are typically implemented by sequential programs with read and write operations.

If κ is a KPN, then we use f_κ to denote the I/O function realized by that KPN. In the remainder, we assume that $(P, C, I, O, Act, \{\lambda_p \mid p \in P\})$ is a Kahn Process Network with LTS $(Confs, c_0, I, O, Act, \rightarrow)$.

Using the operational semantics of KPN, we can now reason about resources such as buffer capacities for the FIFOs connecting the processes. Abstracting from specific buffer sizes, we can also consider whether finite buffer capacities are sufficient. In the Fibonacci example of Fig. 3, there exist executions which require a buffer size of 2 on channel e, for instance if we take the path according to the upper envelope of the picture, two write actions occur in the channel e before the first read action. The path along the bottom envelope of the picture requires only a buffer size of 1 for channel e because the first read occurs before the second write. An execution σ is *bounded* if there exists a mapping $B : C \to \mathbb{N}$ such that at any state (π, γ) of σ, $|\gamma(c)| \leq B(c)$ for all $c \in C$. Not every KPN allows a bounded execution. Some KPNs may have both bounded and unbounded executions. This is relevant for realizations of KPNs, as discussed in Sect. 6.

4   The Kahn Principle

The operational semantics given in the previous section is a model closer to a realization of a KPN than the denotational semantics of Sect. 2. The denotational semantics defines behavior as the least solution to a set of network equations [29]. The correspondence between both semantics, demonstrating that they are consistent, is referred to as the *Kahn Principle*. It was stated convincingly, but without proof, by Kahn in [29] and was later proved by Faustini [17] for an operational model of process networks, in [48] for an operational model of concurrent transition systems and in [38] for an operational characterization using I/O automata.

Based on the operational semantics, a determinate LTS, a functional relation is obtained between inputs and outputs. This function is shown to correspond to the least solution of Kahn's network equations.

The proof presented here is similar to the proof of the Kahn Principle for I/O automata of [38]. We reproduce it here in outline, because it illustrates the connections between denotational and operational semantics and the essential properties of the KPN model. We first show that if the operational behaviors of individual processes of the KPN respect their functional specifications, then so does the KPN as a whole. For an execution σ with input i, we use the notation $h(\sigma, i)$ to denote the history identical to $h(\sigma)$ except for the input channels, which are mapped according to i, since some of the input of i offered to the network may not (yet) have been consumed. Thus $h(\sigma, i)$ is equal to the mapping $i \cup \sigma!(C \cup O)$. We can derive from an execution of the overall KPN how individual processes have contributed to that execution. $\sigma|p$ denotes execution σ projected on process p. If $\sigma = (\pi_0, \gamma_0) \xrightarrow{\alpha_0} (\pi_1, \gamma_1) \xrightarrow{\alpha_1} (\pi_2, \gamma_2) \xrightarrow{\alpha_2} \ldots$. Then $\sigma|p = \pi_{n_0}(p) \xrightarrow{\alpha_{n_0}} \pi_{n_1}(p)$ $\xrightarrow{\alpha_{n_1}} \pi_{n_2}(p) \xrightarrow{\alpha_{n_2}} \ldots$ where n_0, n_1 etcetera are such that $n_0 < n_1 < \ldots$ and α_{n_0}, $\alpha_{n_1}, \ldots$ are precisely the actions from process p.

Lemma 1. *If σ is a fair and maximal execution of a KPN with input i and p is a process of the KPN, then $\sigma|p$ is a fair and maximal execution of p with input $\sigma!I_p$.*

Proof. That $\sigma|p$ is an execution of p follows from the fact that if the KPN executes an action not from p, then the configuration does not change w.r.t. the state of p. If α does originate from p, then from $(\pi, \gamma) \xrightarrow{\alpha} (\pi', \gamma')$, it follows that $\pi(p) \xrightarrow{\alpha}_p \pi'(p)$. Fairness follows from the fact that an enabled read or write operation of the process induces an enabled action of the KPN. Fairness of the execution σ of the KPN prescribes that the action is executed at some point in σ and hence also in $\sigma|p$. Similarly, maximality is obtained from maximality of the execution of the network.

Lemma 2. *For every fair and maximal execution σ with input i, the history $h(\sigma, i)$ satisfies the network equations.*

Proof. Let σ be a fair and maximal execution with input i. It follows using Lemma 1 that $h(\sigma, i)|O_p = f_p(h(\sigma, i)|I_p)$. Thus, $h(\sigma, i)$ satisfies the network equations.

Lemma 3. *A history corresponding to a fair and maximal execution of the KPN with input i corresponds to the smallest solution to the network equations with input i.*

Proof. The history is unique (i.e., independent of the execution) by Proposition 1. In Lemma 2 we proved that it is a solution to the network equations. We have to prove that every solution to the network equations is an upper bound of the history of the execution. Let σ be a fair and maximal execution of the KPN with input i and let h be any history satisfying the network equations such that $h|I = i$. It suffices to prove that h is an upper bound of the history of every finite prefix σ' of the execution, $h(\sigma', i) \sqsubseteq h$, proved by induction on the length of the execution. This is trivial for the empty execution (π_0, γ_0); we proceed with the induction step. Let $\sigma' = \sigma'' \xrightarrow{\alpha} (\pi_n, \gamma_n)$,

- If α is internal to one of the processes or an input action of one of the processes, then $h(\sigma', i) = h(\sigma'', i)$ and the result follows by the induction hypothesis.
- If α is an output action of some process p, then by the induction hypothesis, $h(\sigma'', i)|I_p \sqsubseteq h|I_p$. By monotonicity of f_p and the fact that h satisfies the network equations, it follows that $f_p(h(\sigma'', i)|I_p) \sqsubseteq f_p(h|I_p) = h|O_p$. By monotonicity of f_p and $\sigma''?I_p \sqsubseteq h(\sigma'', i)|I_p$ we also have $f_p(\sigma''?I_p) \sqsubseteq f_p(h(\sigma'', i)|I_p)$. Combining everything, we then have that $h(\sigma', i)|O_p \sqsubseteq f_p(\sigma'?I_p) = f_p(\sigma''?I_p) \sqsubseteq f_p(h(\sigma'', i)|I_p) \sqsubseteq h|O_p$. Hence, it follows that $h(\sigma', i) \sqsubseteq h$.

Theorem 1. (KAHN PRINCIPLE) *The I/O relation of a KPN, derived from the operational semantics is a continuous function which corresponds to the denotational semantics of the KPN, i.e., to the least fixed point of the functional Φ defined in Sect. 2.*

Proof. It follows from Lemma 3 that for every input i, a fair and maximal execution of the KPN with input i yields the smallest channel history that satisfies the network equations. The least fixed point of Φ is the function that assigns to any input i precisely that smallest history.

5 Analyzability Results

Kahn Process Networks are a very expressive model of computation, despite the fact that it only allows specification of functional, determinate systems. In particular, compared to more restricted data-flow models such as Synchronous Data Flow, it allows that the rates at which a process communicates on its outputs are data dependent. Combined with unlimited storage capacity in its unbounded FIFO buffers, this makes KPN an expressive model. Expressiveness is usually in direct conflict with analyzability, the ability to (statically, off-line) analyze an application for its properties, such as deadlock-freedom (for buffers of a given, possibly unbounded, capacity), equivalence, minimum throughput (when adding timing information), static scheduling, and so on.

It is known for instance that the problem to decide for a given KPN (with unbounded buffer capacities) whether it is deadlock-free is undecidable. The proof of this fact [12, 43] relies on a reduction from the Halting Problem of Turing Machines. It is shown that a Boolean Dataflow (BDF) Graph (and therefore also a Kahn Process Network) can simulate a Universal Turing Machine; in other words KPN is *Turing Complete*. This allows the Halting Problem for Turing Machines to be reduced to deadlock-freedom of KPNs. Because the former is known to be an undecidable problem, so must the latter problem be undecidable. A sketch of the translation from Turing Machines to BDF is given in [12].

We can formalize the question for deadlock-free execution in the semantic framework of this chapter as follows.

Definition 5. (DEADLOCK) Given a KPN. Is there some input i, such that the KPN permits a finite, maximal execution with input i?

Note that in this case any maximal execution with input i is finite. Moreover, note that this does not always indicate a problem. Some KPNs may be designed to have only finite executions. The boundedness question can be formalized as follows.

Definition 6. (BOUNDEDNESS) Given a KPN with an input/output function f and channels C and input channels I. Do there exist finite channel capacities $B : C \to \mathbb{N}$, such that for every input history i, there is an execution σ_i such that $h(\sigma_i, i)|O = f(i)$ and at any state (π, γ) of σ_i, $|\gamma(c)| \leq B(c)$ for all $c \in C$?

A slightly weaker form of boundedness can also be defined in which for every input i there is a bounded execution, but there need not necessarily be a single bound for all input. A corresponding optimization problem, the buffer sizing problem, would be to try to find minimal such channel capacities. These capacities are not computable however, because of the undecidability of deadlock-freedom.

In the same way, *most non-trivial questions about KPNs are undecidable*. What is the minimum throughput or maximum latency of a given KPN extended with timing information? What are sufficient FIFO buffer capacities to execute with a guaranteed minimal throughput? Is a particular channel or process ever used or activated?

Some of these questions have to be answered to arrive at an implementation of a KPN. We discuss this further in the next section.

6   Implementing Kahn Process Networks

KPNs are a mathematical model of computation with conceptually unbounded FIFO buffers. KPNs are a very expressive model and we have seen that many of its properties are not statically decidable. Therefore, in some cases subclasses of KPN are used as a starting point for an (automated) synthesis trajectory. However, in many cases the expressiveness of the model is exploited and KPNs are used directly as the basis for synthesis, or for simulation, in which case conceptually the same problems need to be solved.

Besides the infinite buffer capacities, another issue to be addressed is that the semantics involves potentially infinite behaviors. The Fibonacci example if Fig. 1 represents a network that produces an infinite stream of numbers, the Fibonacci sequence. A real, physical implementation however will never be able to produce an infinite sequence in a finite amount of time. A judgement whether an implementation is correct should be based on its behavior in finite time to be of practical relevance. Sect. 9 discusses existing implementations of KPN.

6.1   Implementing Atomic Processes

The semantics of KPN assumes atomic processes which implement elementary continuous functions. Kahn and MacQueen suggest [30] that the atomic processes can be implemented with sequential program code which does not use any global variables shared with other processes and with explicit read and write operations added for reading symbols from input and writing symbols to output channels. An example is the code of the Inverse Zig Zag process in Fig. 2.

If multiple of such processes are required to run on a single processor, then typically a multi-threaded (light-weight) OS is used which spawns a single thread for every process. Determinacy of KPN guarantees that synchronization between these threads is only needed for communication on the FIFOs. When a read operation is executed on an empty FIFO, the reading process thread should stall until new symbols are written to the channel. (It is not allowed to do anything else than wait, because that would break determinacy, since scheduling order may have an impact on the outcome.) Similarly, if FIFO buffers of limited capacity are used in the implementation, a writing process thread may need to stall until sufficient space is available in the channel to complete the write operation. Threads can be, but need not be, scheduled in a preemptive manner.

6.2 Correctness Criteria

An implementation of a KPN should respect the formal (denotational and operational) semantics. It is not entirely trivial how to define correctness because the semantics talks about infinite executions and infinite streams as a convenient abstraction of streaming computation. However, in the real-world we will never be able to observe any actual infinitely long executions. We break down correctness of a KPN implementation into three aspects: *soundness*, *completeness* and *boundedness*. Soundness is the most basic requirement and it states that the KPN implementation should never produce any output that contains symbols *different* from the output predicted by the semantics, nor should it produce *more output* than predicted. For instance, if the semantics says that the Fibonacci KPN produces the outputs: 1, 1, 2, 3, etcetera, then the implementation should not produce: 1, 2, ..., and the JPEG decode should not produce any wrongly decoded blocks.

Definition 7. (SOUNDNESS) An implementation strategy for KPN is *sound* if for every KPN with input/output function f and every behavior the strategy may produce, if the strategy consumes input i and produces output o, then $o \sqsubseteq f(i)$.

Secondly, output should be complete. Intuitively this means that the implementation should produce *all* output predicted by the semantics. But of course we cannot expect an implementation to produce an infinite amount of output in a finite amount of time. However, every individual part (symbol) of the infinite output stream occurs after a finite amount of output produced before it. Hence, we do require from a good implementation that every bit of the predicted output is *eventually* (meaning after a *finite* amount of time) produced by the implementation. In other words, if we sample at finite time intervals the input consumed and output produced by the KPN implementation as a sequence of finite executions σ_k, then the limit of that sequence (its least upper bound) should be the *entire infinite* execution. If the JPEG decoder would be used ad infinitum, with an infinite stream of images to be decoded, it should not come to a complete halt after a finite amount of time.

Definition 8. (COMPLETENESS) An implementation strategy for KPN is *complete* if for every KPN and for every input i offered to this KPN and for every behavior the strategy may produce, if it is sampled at regular time intervals t_k having produced the finite executions σ_k, then $\sigma = \bigsqcup \{\sigma_k \mid k \geq 0\}$ is a maximal and fair execution with input i that is part of the operational semantics.

In particular, this implies that (a) any new amount of progress is made in a finite amount of time, and (b) no channels are excluded from making progress in finite time, there must be no starvation of parts of the network. We illustrate the importance of this constraint when we discuss run-time scheduling and buffer management below.

Thirdly, it is important that this is achieved within bounded memory *whenever possible* (we know that it is not always possible). This amounts to keeping the FIFOs bounded, also in (conceptually) infinite computations.

Fig. 4 Parks' scheduling method

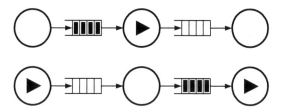

Definition 9. (BOUNDEDNESS) An implementation strategy for KPN is *bounded* if for every KPN and for every input offered to this KPN, if there exists a bounded execution according to the operational semantics, then this strategy will produce a bounded execution.

Not every KPN allows for bounded executions. In such a case, the strategy is allowed to produce an unbounded execution, but rather one should perhaps decide not to implement such a KPN, although one cannot, in general, automatically decide whether this will be the case!

6.3 Run-Time Scheduling and Buffer Management

Conceptually, the FIFO communication channels of a KPN have an unbounded capacity. A realization of a process network has to run within a finite amount of memory. An unbounded capacity can be mimicked using dynamic allocation of memory for a write operation, as suggested in [30], but rather than that, it is for reasons of efficiency better to allocate fixed amounts of memory to channels and change this amount only sporadically, if necessary. An added advantage of the fixed capacity of channels is that one can use an execution scheme introduced by Parks [43], where a write action on a full FIFO channel of limited capacity blocks until there is room freed up in the FIFO. Note that this behavior can be modeled within the KPN model using additional channels in opposite directions, similar to the well-known trick with Synchronous Data Flow graphs [43]. This importantly demonstrates that this does not impact determinacy of the model. This gives an efficient mixed form of data-driven and demand-driven scheduling [3,43], illustrated in Fig. 4, in which a process can produce output (in a data-driven way) until the channel it is writing to is full an/or the channel it is consuming from is empty. Then it blocks and other processes take over, effectively regulating the relative speeds of different processes. As discussed previously, it is undecidable in general, how much buffer capacity is needed in every channel [12]. If buffers are chosen too large, memory is wasted. If buffers are chosen too small, so-called *artificial deadlocks* may occur that are not present in the original KPN, when processes are being permanently blocked because of one or more full channels, an event which cannot occur in the original KPN. Therefore, a scheduler is needed that determines the order of execution of processes and that manages buffer sizes at run-time.

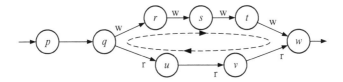

Fig. 5 An artificial deadlock

Figure 5 shows a process network in an artificial deadlock situation. Process w cannot continue because its input channel to process v is empty; it is blocked on a read action on the channel to v, denoted by the "r" in the figure. The required input should be provided by v, but this is in turn waiting for input from u. u is waiting for q. Process q is blocked, because it is trying to output a token on the full channel to r; the block on the write action is denoted by a "w". Similarly processes r, s and t are blocked by a full channel. Only w could start emptying these channels, but w is blocked. The processes are in artificial deadlock (q, r, s and t would not be blocked in the original KPN) and can only continue if the capacity of one of the full channels is increased. Note that a blocked process is dependent on a unique channel on which it is blocked either for reading or for writing and hence it depends on a unique other process that is also connected to that channel. This gives rise to a chain of dependencies and if this chain is cyclic, it is a deadlock. A *real* (non-artificial) deadlock, in contrast, is entirely due to the KPN specification and cannot be avoided by buffer capacity selection. Because the KPN with bounded FIFOs is also determinate we can conclude that the buffer management is independent from scheduling; an artificial deadlock, like a real deadlock, cannot be avoided by a different scheduling. (See [20] for more details and a proof.)

Thus, an important aspect of a run-time scheduler for KPNs is dealing with artificial deadlocks. We can discern different kinds of deadlocks. A process network is in *global deadlock* if none of the processes can make progress. In contrast, a *local deadlock* arises if some of the processes in the network cannot progress and their further progress cannot be initiated by the rest of the network.

Parks' strategy [43] for dealing with scheduling, artificial deadlocks and buffer management is the following procedure. First, select some arbitrary, fixed initial buffer capacities. Then, execute the KPN, with blocking read and blocking write operations. If and when a global deadlock occurs and it is an artificial deadlock, then increase the size of the smallest buffer and continue. (One can try to be more efficient in the selection of the buffer that needs to be enlarged [3, 20], but this does not fundamentally change the strategy.) Such a run-time scheduler thus needs to detect a global deadlock. In a single processor multi-threaded implementation, this is typically achieved by having a lowest priority thread of execution that becomes active only when all other threads are blocked, indicating a global deadlock. It then tests whether the deadlock is artificial and if so, increases the size of a selected

Fig. 6 A local deadlock

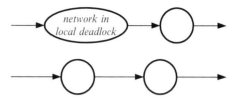

buffer, ultimately enabling one of the blocked processes again. In terms of the formulated correctness criteria, one can show this method to be sound and bounded, but not complete.

Proposition 2. *The scheduling strategy of [43] for KPNs is sound and bounded, but not complete.*

Proof. Soundness is straightforward. The appropriate code is executed and nothing else. The strategy of selecting the smallest buffer to increase guarantees that after a finite number of resolved artificial deadlocks, the buffer capacities must have grown beyond the bounds needed when a bounded execution exists. Incompleteness of the scheduling strategy follows from the counter example of Fig. 6, discussed below.

The strategy chosen in [3] is in this respect similar to the one of [43] and also leads to a bounded execution if one exists.

To guarantee the correct output of a network, a run-time scheduler must detect and respond to artificial deadlocks. Parks proposes to respond to global artificial deadlocks. In implementations of this strategy [1, 24, 31, 49, 55], global deadlock detection is realized by detecting that all processes are blocked, some of which by writing on a full FIFO. Although this guarantees that execution of the process network never terminates because of an artificial deadlock, it does not guarantee the production of *all* output required by the KPN semantics; output may not be complete. For example, in the network of Fig. 6, if the upper part reaches a local artificial deadlock, then the lower, independent part is not affected. Processes may not all come to a halt and the local deadlock is not detected and not resolved. The upper part may not produce the required output. Such situations exist in realistic networks. A particular example is the case when multiple KPNs are controlled by a single run-time scheduler, one entire process network may get stuck in a deadlock. Hence, to achieve completeness, a deadlock detection scheme has to detect *local* deadlocks as well.

It is shown in Sect. 4 that the Kahn Principle hinges on *fair* scheduling of processes [10, 29, 48]. Fairness means that all processes that can make progress should make progress at some point. This is often a tacit but valid assumption if the underlying realization is truly concurrent, or fairly scheduled. In the context however of bounded FIFO channels where processes appear to be inactive while they are blocked for writing, fairness of a schedule is no longer evident. This issue is neglected if one responds to global deadlocks only, leading to a discrepancy with the behavior of the conceptual KPN.

Fig. 7 Non effective network

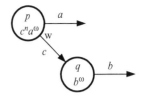

It is proved in [20] and illustrated below, that a perfect scheduler for KPN, satisfying all three correctness criteria for any KPN cannot exist.

Theorem 2. *([20]) A scheduling strategy for KPNs satisfying soundness, completeness and boundedness does not exist.*

The incompleteness of the scheduling method of [43] can have rather severe consequences, leading to starvation of parts of the network. The way to resolve this is to not wait until a global deadlock occurs before taking actions, but act (in finite time) when a local artificial deadlock occurs and resolve it. This is the adaptation that the strategy proposed by [20] makes to the original strategy of [43].

A problem for an implementation strategy for KPN is posed by the production of data that is never used. This is illustrated with Fig. 7. Process p writes n data elements (tokens) on channel c connecting p to process q; after that, it writes tokens to output channel a forever. q never reads tokens from c and outputs tokens to channel b forever. If the capacity of c is insufficient for n tokens, then output a will never be written to unless the capacity of c is increased. q doesn't halt and execution according to Parks' algorithm does not produce output on channel a, violating completeness, because in the KPN, infinite output is produced on both channels. The above suggests that a good scheduler should eventually increase the capacity of channel c so that it can contain all n tokens. However, such a scheduler fails to correctly schedule another KPN. Consider a process network with the same structure as the one of Fig. 7, but this time, p continuously writes tokens on output a, mixed with infinitely many tokens to channel c. q writes infinitely many tokens to b and reads infinitely many tokens from c, but at a different rate than p writes tokens to c. Note that a bounded execution exists; a capacity of one token suffices for channel c. If process p writes to c faster than q reads, channel c may fill up and the scheduler, not knowing if tokens on channel c will ever be read, decides to increase channel capacity. A process q exists that always postpones the read actions until after the scheduler decides to increase the capacity; the execution will be unbounded, although a bounded execution exists.

The way out of this dilemma taken by [20] is to assume that in a reasonable KPN, every token that is written to a channel, is eventually also read. Such KPNs are called *effective*.

The scheduling algorithm is based on the use of blocking write operations to full channels as in [43]. In order to define local deadlocks, it builds upon the notion of causal chains as introduced in [3]. Any blocked process depends for its further progress on a unique other process that must fill or empty the appropriate channel. These dependencies give rise to chains of dependencies. If such a chain of dependencies is cyclic, it indicates a local deadlock; no further progress can be made without external help. In Fig. 5, such a causal chain is indicated by the dashed ellipse indicating the cyclic mutual dependencies of the processes q, r, s, t, u, v and w, which form a local deadlock.

The scheduling strategy of [20] is identical to the scheduling strategy of [43] except that the search for an artificial deadlock does not occur only when the network is in global deadlock, but the scheduler needs to monitor the process network for the occurrence of local artificial deadlocks and resolve those in finite time. The behavior of this scheduling strategy has the following characteristic property, proved in [20].

Theorem 3. *The scheduling strategy of [20] for KPNs is sound, complete and bounded for effective KPNs.*

7   Extensions of KPN

KPN is an expressive model that allows the description of a wide class of data-flow applications. Its expressiveness leads to undecidability of various elementary questions about KPNs (Sect. 5). Yet still, there is sometimes the desire to extend the KPN model with additional features. The most prominent ones are *time* and *events* or *reactive behavior*. KPN describes the functional behavior of a process network, but makes no statements about the timing with which these outputs are produced, or the latency or throughput that can or should be attained. Another important element on the wish list is the ability to deal with sporadic messages or events. Often both are added at the same time to save determinacy of the model.

7.1   Events

The desire to deal with sporadic events is illustrated by the addition of the `select` statement in the KPN programming library YAPI [31]. This statement allows a process to see which of a number of input channels has data available and to read from that channel first. Similarly, [39] describes an extension of the data-flow model with the ability to probe a channel for the presence or absence of data. While both extensions clearly enhance the expressiveness of the model, they also destroy the property of determinacy, of independence of any concrete schedule or scheduling strategy, the denotational semantics of KPN and the Kahn Principle.

From a theoretical perspective, the essential ingredient to add to KPNs to deal with sporadic or reactive behavior is a *merge* process. A merge process has two inputs and one output and it copies data arriving on any of its inputs to the single output, in an order which is based on the arrival order of data, for instance in the

Fig. 8 Brock–Ackerman
counter example (from [9])

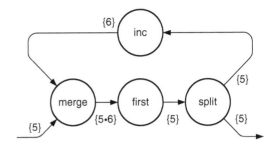

order in which they arrive on the inputs. In this case however, the input/output relation of the process is no longer a function; the same inputs can lead to different outputs, depending on the order in which they arrive. KPN denotational semantics captures behavior as continuous input/output functions and thus no longer works. It has been attempted to capture KPN with merge processes by input/output *relations*. It turns out however that this is not possible. This is commonly known as the Brock–Ackerman anomaly [9]. For non-determinate processes, their input/output relation does not sufficiently characterize the system's behavior to use it as a basis for a semantic framework. The counter example is reproduced in Fig. 8. It has a "merge" process as the only indeterminate (non-KPN) process in the network, a process "first" which passes on only the first symbol, a process "split" which splits the stream in two exact copies and a process "inc" which passes on all symbols after incrementing them by one. The network is provided with the input consisting of the single symbol 5. The merge process passes it on to its output and the processes "first" and "split" pass it on as well. "inc" turns it into 6 and passes it on to "merge". *Operationally speaking*, merge now passes the 6 to its output and because the token 6 is *causally dependent* on the token 5 passing to the output of the merge *first*, the only possible output of the merge process is $5 \cdot 6$. On the other hand, the denotational semantics of the merge process specifies that for the inputs $\{(5,6)\}$, the possible outputs are $\{6 \cdot 5, 5 \cdot 6\}$, but the $6 \cdot 5$ is not possible in any causal reality. The causality *information* which is necessary for providing a correct semantics to indeterminate processes is lost in the input/output relation.

Brock and Ackerman showed [9] that I/O relations do not provide a good semantics for the so-called *fair merge* [42] that merges streams in such a ways that all input tokens on either input are *eventually* produced at the output. Subtly different and weaker versions of merge processes (but with strictly different expressiveness) can be defined, *fair merge*, *infinity fair merge*, *angelic merge*, but it was later shown (see [46] for an overview), that they all suffer the same limitation, that input/output relations are insufficient. Only merge processes which interleave symbols from the inputs in a fixed, a priori determined order, for instance alternatingly, are determinate.

Alternatively to a denotational semantics, a trace-based semantics has been found to be able to serve as a fully abstract semantics [28], i.e., a semantic model which contains *enough* information, but *no redundant* information to represent behavior.

The use of a totally ordered trace indicates the loss of independence of scheduling of concurrent processes associated with the introduction of a merge construct, which is one of the strongest points of KPN. A theoretical analysis of relational models for indeterminate dataflow models is [54].

7.2   Time

A timed process network model is a model which not only describes the data transformations of the network, but also the timing of such a process [11, 56]. Such a model is typically made by labeling symbols on streams with time-stamps or tags from a particular time domain of choice (for instance non-negative integers or reals) or, more general, according to the tagged-signal model of [35], allowing for instance also partially ordered or super-dense time domains common in hardware description languages [37]. Alternatively, a stream can then be described as a mapping from the time-domain to the channel alphabet. If this function is total, it may also specify the absence of data at certain points in time, which can then be deterministically exploited by processes. Time-stamps may be interpreted as the exact timing of the production of tokens or as a specification of a deadline for the production.

From a semantics perspective, time-stamps can be exploited to "rescue" the merge process from indeterminacy and in general to make reactive behavior deterministic. Using time information, decisions can be taken in a deterministic way, based on the time-stamps of data, for instance to describe a merge process which merges symbols in the order in which they arrive (with some deterministic provision, for instance fixed priority, when tokens arrive at the same time) [11, 37, 56]. Alternative network equations can be formulated and different fixed-point theorems (such as Banach's) can be used to show them to have a unique solution. The (big) price one has to pay for salvaging determinacy of the model is that it requires a global notion of synchronized (logical) time, which puts constraints on implementation and requires additional synchronization.

By adopting such a notion of synchronized global logical time, we end up close to another end of the spectrum of data-flow languages, the domain of synchronous languages such as Esterel [6], Signal [5] or Lustre [26], where the execution of a network needs to be globally synchronized. This can be a strong disadvantage to a distributed implementation. Recent work [4, 13] in the synchronous language domain searches for conditions to relax the global synchrony constraints towards so-called GALS (globally asynchronous, locally synchronous) implementations, where particularly the most costly global synchronization may be eliminated.

8   Reactive Process Networks

The Reactive Process Networks (RPN) model of computation intends to provide a semantic framework and implementation model for data-flow networks with sporadic events, in the same way KPN is a reference for determinate data-flow models of computation. RPN integrates control and event processing with stream processing in a unifying model of computation with a compositional operational semantics. The model tries to find a balance in a trade-off between expressiveness, determinism and predictability, and implementability.

8.1   Introduction

We illustrate the Reactive Process Networks model by looking at the domain of multimedia applications, working with information streams such as audio, video or graphics. With modern applications, these streams and their encodings can be very dynamic. Smart compression, encoding and scalability features make these streams less regular than they used to be.

Streams are typically parts of larger applications. Other parts of these applications tend to be control-oriented and event-driven and interact with the streaming components. Modern (embedded) multimedia applications can often be seen as instances of the structure depicted in Fig. 9. At the heart of the application, computationally intensive data operations have to be performed in streams of for instance pixels, audio samples or video frames. Input and output of these processes are highly regular patterns of data. These data processing activities can often be statically analyzed and scheduled on efficient processing units. At a higher level, modern multimedia streams show a lot of dynamism. Object-based video (de)coders for instance work with dynamic numbers of objects that enter or leave a scene. Decoding of the individual objects themselves uses the static data processing functions, but they may need to be added, removed or adapted dynamically, for

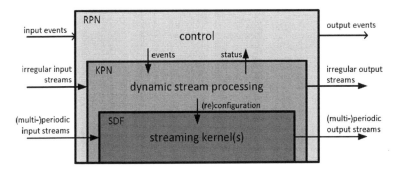

Fig. 9 Embedding of types of streams

Fig. 10 Classification of models of computation

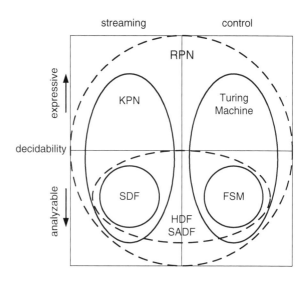

instance encoding modes or frame types in MPEG audio or video streams. These dynamic streams still compute functions and processing is determinate, i.e., the functional result is independent of the order in which operations are executed. In turn, the processing of these dynamic data streams is governed by control oriented components. This may for instance be used to convey user interactions to the streaming application or to respond to changing network conditions.

The three levels of an application require typical modeling and implementation techniques. A good candidate for instance to describe static computation kernels could be Synchronous Data Flow (SDF) [34], discussed in [25]. Dynamic stream processing can perhaps be best described using KPNs. They are capable of showing data dependent behavior and dynamic changes in their processing, but they are still determinate and can be executed fully asynchronously. To specify the control dominated parts of an application, there are many techniques, such as state machines and event-driven software models.

RPN defines a model and its formal operational semantics that allows for an integrated description and analysis of an application consisting of these three levels of computation. It is a unified model for streaming and control, that is also hierarchical and compositional. The goal is to be also able to incorporate more analyzable, less expressive models in the same way SDF fits within the KPN model, but still combining data-flow and control oriented behavior. For instance, a combination of SDF with finite state machines such as HDF or SADF [22, 23, 50, 51], which yields analyzable models, would fit within the framework. This is illustrated in Fig. 10. Vertically, it shows the trade-off between expressiveness and analyzability, with a border of decidability. Horizontally, it shows streaming vs. control oriented models.

Figure 11 shows the compositional integration of state machines with process networks. A process network is a component with stream input(s) (i in Fig. 11a), stream output(s) (o) and event input(s) (e). At any point in time, the network

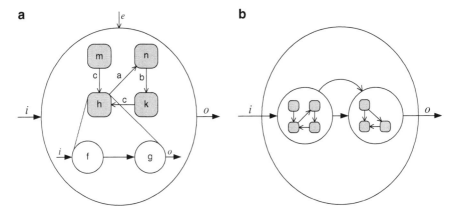

Fig. 11 Mixing state machines with process networks (**a**) states with different process networks. (**b**) processes may hierarchically have states of their own

operates in a mode that implements a particular streaming function, for instance mode h in Fig. 11a, implemented as the network drawn below it. At some later time, because of the occurrence of some event a, the function of the network needs to change to a mode n having a different streaming function. One could think for instance of a video system where a user changes settings or turns on or off special image processing features. One could view this as a (not necessarily finite) state machine where in every state, the network implements a particular function and external events force the state machine to move from one state to another. In every state, a particular process network performs operations on data streams. Similarly, such state machines can be embedded in components of a reactive process network as shown in Fig. 11b. Events that are communicated to such components can be generated from an output port of another component. Process networks can be hierarchical entities, we would like such a construct to be compositional and applicable at different levels of a network hierarchy. Note that this is the conceptual behavior of an RPN, but that does not mean that this is exactly how it is implemented. In many cases the events cause relatively small changes to the current streaming function, changing of parameters or activation or deactivation of individual functions.

8.1.1   A Reactive Process Network Example

An illustrative example of the type of application we are considering is shown in Fig. 12. It depicts an imaginary game, which includes modes of three-dimensional game play with streaming video based modes. The rendering pipeline, used in the 3D mode, is a dynamic streaming application. Characters or objects may enter or leave the scene because of player interaction, rendering parameters may be adapted to achieve the required frame rates based on a performance monitoring feedback

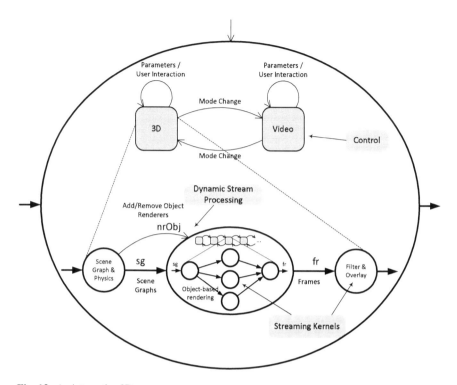

Fig. 12 An interactive 3D game

loop. Overlayed graphics (for instance text or scores) may change. This happens under control of the event-driven game control logic. At the processing core of the application, the streaming kernels, a lot of intensive pixel based operations are required to perform the various texture mapping or video filtering operations. Special hardware or processors may be available to execute these operations, which can be scheduled off-line, very efficiently.

The model is organized around the two main modes (3D graphics, video). In these modes, the game dynamics and mode changes are influenced or initiated by user interaction, game play and performance feedback. This is the control oriented part of the game depicted as the automaton at the top of Fig. 12. The self-loops on these states denote changes where the streaming network essentially stays the same, but its parameters may be changed.

In the 3D graphics mode (enlarged at the bottom of the figure), a scene graph, describing all entities and their positions in 3D space, is rendered to a two-dimensional view on the scene on some output device. The first process communicates the scene graphs with objects to the rendering component. The rendering component transforms the scene graphs into two-dimensional frames. The last process adds two-dimensional video processing such as filtering, overlays, and so forth. The output is shown by the display device.

Notice that the game as a whole has different modes of streaming (3D graphics, video), but similarly, components in the stream processing part have different modes or states of streaming execution. The object renderer for instance can be reconfigured to different modes depending on the number of objects that need to be rendered (using the event channel *nrObj* controlled by the scene graph process). This illustrates the need for a hierarchical, compositional approach to combining state/event based models with streaming and data-flow based models, as realized by the RPN model.

8.2 Design Considerations of RPN

We discuss the main concepts that have had an impact on the design of the model of Reactive Process Networks.

8.2.1 Streams, Events and Time

Streaming applications represent functions or data transformations. They absorb input and they produce the corresponding output. There is often no inherent notion of time, except for the ordering of tokens in the individual data streams. (We are aiming for an untimed model similar to KPN.) Tokens in different streams have no relation in time, except for causal relationships implicitly defined by the way processes operate on the tokens. These process networks are ideally determinate, i.e., the order and time in which processes execute is irrelevant for the functional result. The output of a process network is completely determined as soon as the input is known. The actual computation of this output introduces a certain latency in the reaction. This latency is not part of the functional specification, but merely a consequence of the computation process. It may be subject to constraints, such as a maximum latency. Time is sometimes implicitly present in the intention of steams. A stream may carry for instance, a sequence of samples of an audio signal that are 1/44100th of a second apart, or video frames of which there are 25 or 30 in every second. Such streams are called *periodic*. For final realizations, time-related notions such as throughput, latency and jitter are of course important.

Events, have a somewhat different relationship to time. An event is unpredictable and the moment when it arrives, in relation to the streams, is significant, but unknown in advance. In many event-based models, the *synchrony hypothesis* applies, which states that the response to an event can be completed before the following event arrives or is taken into account. This simplifies specifying how a system responds to events. A classical model for event-based systems are state machines, where events make it change from one state into another. Prominent characteristics are non-determinism and a total ordering of events. If events come from outside and are not predictable a priori, then the system evolves in a non-deterministic way under the influence of the events, even if the response to a particular event is deterministic.

Discrete changes due to events are alternated with stable periods of streaming. Conceptually, a discrete change occurs at a well defined point within the stream. Because of pipelining implementation of the stream processing however, there is not necessarily a point in time where the change can be applied instantaneously to the whole network. A video decoder for instance may be processing video frame by video frame in a pipelined fashion. A discrete change occurs between two frames such that the new frames are decoded according to new parameter settings. However, when the first new frame enters into the pipeline, there are still old frames in the pipeline ahead of it. The StreamIt language [52], which employs a fairly synchronized model of streaming, allows a mechanism of delivering asynchronous messages (events) to processes in accordance with the "information wavefront", i.e., in a pipelined fashion. In general, an important aspect of dealing with streaming computation and events is to coordinate their execution to implement a smooth transition.

There is a trade-off between predictability and synchronization overhead. Predictability of processing (non-deterministic) events is improved by added control over the moment when and the way how the event is processed relative to the streaming activities. Increased predictability requires more synchronization between processes and hence additional overhead. Such overhead is undesirable, especially if events occur only sporadically.

8.2.2 Semantic Model

Denotational semantics is often preferred to capture the intended functionality of a process network or to define the functional semantics of a system or programming language implementing process networks, without specifying unnecessary implementation details. The operational semantics on the other hand allows reasoning about implementation details, such as artificial deadlocks [20] or required buffer capacities [3, 8]. For RPN, a denotational semantics in terms of input/output relations is not possible (Sect. 7) and one based on sequential execution traces is possible, but hides the parallelism of the model. The detailed operational semantics of RPN can be found in [21]. In the next section, we discuss the main concepts.

8.2.3 Communicating Events

Processes or actors in data-flow graphs communicate via FIFO channels. We want to add communication of events and we have to decide what communication mechanism is used for events. It is often the case that what is perceived by a lower level process as an event, is considered to be part of streaming by higher level processes. For instance, a video decoder is decoding a stream of video frames and header information for every frame is an integral part of the data stream. For the lower level frame decoder processes, the frame header information is seen as an event that initializes the component to deal with the specific parameters of the following frame. For this reason, we use the same infrastructure for communicating streams also for events. For the sending process, there is no difference at all; at the receiving side, we distinguish *stream input* ports and *event input* ports. The former are used in ordinary streaming activity; tokens arriving on the latter will trigger discrete events.

8.3 Operational Semantics of RPN

The operational semantics of reactive process networks associates RPNs with a corresponding labeled transition system. An RPN, like a KPN, consists of processes, which may in turn be other RPNs, or primitive processes defined through other means, such as sequential code segments, and the LTS is constructed compositionally, similar to the operational semantics of KPN in Sect. 3. A detailed description of the operational semantics of RPN can be found in [21]; here we concentrate on the important concepts.

In general, two different types of things may happen to the network: data can be streaming through it, or it can encounter events that need to be processed. The events introduce non-determinism. The result should still be as predictable as possible. In particular, we want to guarantee that input consumed before the arrival of a new event will lead to the required output, also if the output has not been completed when the event arrives. In implementations, this is achieved at the expense of additional synchronization or coordination. In specific subclasses of the RPN model, this synchronization may be realized without much overhead. There are very regular subclasses of RPN, such as the SDF-based models of Scenario Aware Data Flow (SADF) [51] or Heterochronous Data Flow (HDF) [23]. SDF graphs execute periodic behavior, often called iterations. In-between such iterations is typically a good moment for reconfiguration. In [40] such moments are referred to as *quiescent states*, but as explained earlier, in a pipelined execution, such states may not naturally occur and may need to be enforced by stalling the pipeline.

The operational semantics of RPN models streaming as a sequence of individual read and write actions of the processes involved in the computation of the output, comparable to the semantics of KPN of Sect. 3. Since, conceptually, the reaction of a process network to incoming data is immediately determined, it may not be disturbed by the processing of events. To this end, in RPN semantics, these sequences of actions resulting from a data input are grouped together and represented as single, atomic transitions of the labeled transition system. Effectively, this gives internal actions (i.e., completing the reaction to already received input) priority over processing of events.

This leads to the concept of so-called *streaming transactions*, as illustrated in Fig. 13. The picture on the left shows the states and transitions of a process network performing individual input and output actions. Input actions are shown as horizontal arrows, output actions as vertical arrows. Because of pipelining, the network can perform several input actions before the corresponding output actions

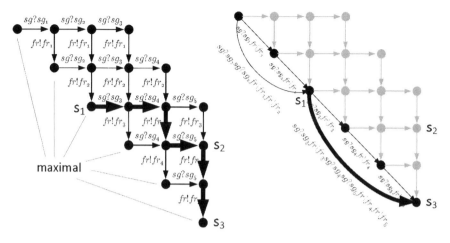

Fig. 13 Streaming transactions

are produced. Events should only be accepted in those states where no internal or output actions are available. In Fig. 13, these are all states on the lower-left boundary, such as states s_1 and s_3. In state s_2, some remaining output is still pending and an event cannot be processed yet. Sequences of actions starting from a state without enabled output or internal actions and going back to such a state represent complete reactions of the network to input stimuli. These sequences of actions are grouped together to form atomic transitions, the *streaming transactions*, that end in states where events can safely be processed. For example, the sequence of transitions with thick arrows in the picture from state s_1 to s_3 forms a streaming transaction; from the end state, only reading of new input is possible, all received input has been fully processed. (These states correspond to end-points of maximal executions for a particular finite input, and closely resemble the quiescent states of [40].) Some of the possible streaming transactions are shown in the picture on the right. The bold one corresponds to the bold path on the left. In this new LTS, the state space consist of only quiescent states.

The streaming transactions of an RPN are determined in two steps. Individual streaming transactions of the constituent processes are determined as well as the processing of events by these processes. Executions of these actions are then taken together to form maximal streaming transactions of the whole network. Such streaming transactions may not consume *input events*. However, they may include occurrences of internal events and hence they can be indeterminate.

Figure 14 shows a part of a possible execution of the 3D game example. The first transition is a streaming transaction, consisting of streaming actions of the processes, but also internal events ($nrObj(2)$). Note that we have abstracted from internal actions of for instance the rendering component, which might also be visible in these transactions. The second transition is an event; the player has reached the end of a level, and the game is reconfigured from 3D mode to the video mode.

Fig. 14 Execution of the 3D
game RPN

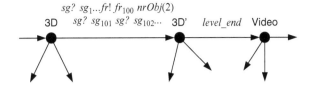

Such event transitions of an RPN are directly determined by the reception of input events, followed by the corresponding network transformation. The RPN semantics associates with every event an abstract transformation of the network structure. How such reconfigurations are precisely specified is left up to the concrete instances of models or languages based on this model. In *charts [23], which can be seen as an RPN instance, specification is done by direct refinement of FSM states by data-flow graphs. In SADF [51] actors or processes read control tokens sent to them by the controlling FSM and adapt or suspend their behavior accordingly.

The behavior of the network as a whole is formed by interleaving transitions of both kinds as in the example. An RPN thus has a labeled transition system with executions interleaving events and streaming transactions. In contrast with KPN, this LTS is indeterminate. With this LTS semantics, an RPN can be directly used as a process within a larger RPN and this ways supports hierarchical and compositional specification of RPNs.

8.4 Implementation Issues

The operational semantics of RPN defines the boundaries of the behavior that correct implementations of RPNs should adhere to. Within these boundaries there is still some room for making specific implementation decisions, depending on the application and the context. In this section, we briefly discuss some of these considerations.

8.4.1 Coordinating Streaming and Events

One of the most powerful aspects of KPNs is that execution can take place fully asynchronously. Processes need not synchronize and determinacy of the output is automatically guaranteed. The (deliberate) drawback of the generalization to RPN is that this advantage is (partially) lost. Before processing an input event, the streaming input to the network—conceptually—needs to be frozen and all data must be processed internally. Only when all data has been processed, the event can be applied and the data-flow can be continued. If implemented in this way, the pipelining of the data-flow may be disrupted and deadlines could potentially be missed because of this disruption if the nature of the application doesn't allow this.

In practice, one can do better for many classes of systems. Instead of processing an event for the whole process network at once, it may in some cases be possible to make the changes along with the "information flow". In particular, if the response of a network to an event is the forwarding of the event to one or more of its sub-processes, then this forwarding can be synchronized with the flow of data such that pipelining need not be interrupted. This is done for instance for the asynchronous messages following the stream wave-front in StreamIt [52].

The discussed operational semantics suggests that events must always be accepted by any process. In practice, it can be useful to allow a process some control w.r.t. the moment when events are accepted. For example, to allow it to accept events only at moments when the corresponding transformation is most easy to do, because the process is in a well-defined state, e.g. at frame boundaries.

Such an approach can for example be easily implemented if the underlying process network is an SDF graph that can be statically scheduled. It is then possible to define a cyclic schedule in such a way that one iteration of the cycle constitutes a single streaming transaction. Then, a test for newly arrived events can be inserted at the beginning of the cycle and event transitions can be safely executed.

8.4.2   Deadlock Detection and Resolution

The correct execution of KPNs using bounded FIFO implementations depends on the run-time environment to deal with artificial deadlock situations (see Sect. 6, [20, 43]). The same situation may arise in reactive process networks. The dependencies may now also include event channels. One can deal with these artificial deadlocks in a similar manner as for ordinary KPNs. Solving an artificial deadlock may be needed for completing the maximal transaction before processing an event.

8.5   *Analyzable Models Embedded in RPN*

KPN is a model of determinate stream processing, but due to its expressiveness has many undecidable aspects. In practice often Synchronous Data Flow Graphs (SDF) or Cyclo-Static Data Flow Graphs (CSDF) are used as restricted but analyzable subsets of KPN. Obviously, as an extension of KPN, RPN also has many undecidable aspects.

The SADF model of computation uses Synchronous Data Flow graphs as the streaming model, extended with time according to [47] to capture performance aspects. For the control aspects, it uses Markov chains which can be seen as finite state machines decorated with transition probabilities. These probabilities intend to capture the typical behavior of the application in a particular use case. As such it is comparable to the earlier Heterochronous Data Flow [23] model which also makes a combination of SDF and FSMs, but without time and probabilities.

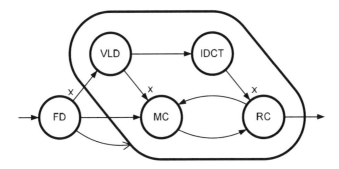

Fig. 15 SADF model of an MPEG-4 SP decoder

Figure 15 shows an example of an SADF graph of an MPEG-4 Simple Profile decoder. An input stream is analyzed by the FD (frame detector) process which can observe the frame type of the next incoming frame. Different frames may employ a different number of macro blocks. In the figure, this number is denoted by x. For every given number x, the sub-network of the other processes is an SDF graph with fixed rates, which allows a static schedule and the performance of which can be analyzed off-line. For every frame, the frame detector sends an event message to the rest of the network indicating the proper frame type and the sub-network makes the necessary changes and invokes the appropriate schedule.

The Markov model of the transitions between frame types can be used to study the expected performance of the given system [51]. Alternatively, a worst-case analysis can be done to establish a guaranteed performance using techniques such as introduced in [19, 22, 45].

Another example of an efficient combination of (restricted structures of) SDF with events is StreamIt [52] which employs a concept called *teleport messaging* [53] for sending sporadic messages along the information wavefront. Upstream actors or processes can send event messages to downstream actors to arrive at a specified iteration distance to the iteration at which the first data arrives which is dependent on the output currently being produced. In this case, the compiler and scheduler can automatically take care of the required synchronization.

9   Bibliography

Kahn Process Networks have been introduced by Kahn in [29]. Kahn and McQueen presented a programming/implementation paradigm for KPNs as the behavior of a model of (sequential) programs reading and writing tokens on channels. The Kahn Principle, stating that the operational behavior of such an implementation model conforms to the denotational semantics of [29], was introduced, but not proved, by Kahn in [29]. It was proved [17, 48] for an operational model of transition systems and in [38] for I/O automata. The operational semantics in terms of I/O automata is very much like the semantics in Sect. 3, except that in the I/O-automata semantics, FIFOs are not modeled explicitly; it is assumed that they are implicit in the transition system of the processes at the lowest level. This means that the processes will accept any input at any given time. Composition of networks is then achieved with synchronous communication. By making channels and their FIFOs more explicit one can reason about realizations including memory management (FIFO capacities).

KPN is often informally regarded as the upper bound of a hierarchy of data-flow models, although technically speaking it is not entirely obvious how to compare data-flow processes based on firing rules with either the denotational or operational semantics of KPN. The relationship between both types of models is elaborated in [36].

Scheduling and resource management, possibly in a distributed fashion are important subjects for realizing KPN implementations or simulators. Scheduling process networks using statically bounded channels is an important contribution towards this aim by Thomas Parks [43], introducing an algorithm that uses bounded memory if possible. Based on this scheduling policy, a number of tools and libraries have been developed for executing KPNs. YAPI [31] is a C++ library for designing stream-processing applications. Ptolemy II [33] is a framework for codesign using mixed models of computation. The process-network domain is described in [24]. The Distributed Process Networks of [55] form the computational back end of the Jade/PAGIS system for processing digital satellite images. [49] covers an implementation of process networks in Java. [1] is another implementation for digital signal processing. Common among all these implementations is a multi-threading environment in which processes of the KPN execute in their own thread of control and channels are allocated a fixed capacity. Semaphores control access to channels and block the thread when reading from an empty or writing to a full channel. This raises the possibility of a deadlock when one or more processes are permanently blocked on full channels. A special thread (preempted by the other threads) is used to detect a deadlock and initiate a deadlock resolution procedure when necessary. This essentially realizes the scheduling policy of [43]. The algorithm of Parks leaves some room for optimization of memory usage by careful selection of initial channel capacities (using profiling) and clever selection of channels when the capacity needs to be increased; see [3]. [3] also introduces causal chains, also used in this chapter to define deadlocks.

It is argued in [20] as discussed in Sect. 6 that a run-time scheduler for KPN should include local deadlock detection. Such deadlock detection has subsequently been implemented in a number of KPN implementations [2, 15, 27, 41]. To optimize the process it can be organized in a distributed fashion [2] or in a hierarchical way [27].

The desire to express non-deterministic behavior and event-based communication has led to additions to pure data-flow models. Examples are the probes of [31, 39]. Martin [39] describes an extension of the data-flow model with so-called probes, the possibility to test whether a channel has data available for reading. This can be a powerful construct, but it destroys the property of determinacy; the behavior is no longer independent of the concrete scheduling. This probe construct inspired the designers of YAPI [31] to introduce the select statement, having the same disadvantage.

In Ptolemy II [16, 24, 33], a framework is defined to connect multiple models of computation, including data-flow and event-based ones. The combination of the reactive and process network domains however, induce too much synchronization overhead to be directly used for implementation.

Many of the combinations of data-flow and reactive behavior are based on a combination of the Synchronous Data Flow model together with some form of reactive behavior [23, 32, 50–52]. The use of an analyzable model such as SDF is natural because it allows for a predictable and determinate combination. The reactive part is frequently specified using (hierarchical) state machines. Lee [32] describes a combination of hierarchical state machines and SDF models, where a complete iteration of the SDF is taken as an action of the state machine. A state machine inside an SDF is required to adhere to the SDF firing characteristics. FunState [50] defines a model, described as "functions driven by state machines", which deliberately tries to separate functional behavior from control. For FunState, the control part also extends to the coordination of the processes that form a data-flow graph and the functions are comparable to separate KPN processes. *charts [23] separates hierarchical finite state machine models from concurrency models. In the data-flow domain, a combination of Hierarchical Finite State Machines with Synchronous Data Flow is presented, called Heterochronous Data Flow (HDF). To combine the finite state transitions of an FSM with the typically infinite behavior associated with concurrent models of computations, the infinite behavior is split in finite pieces. For SDF, separate iterations are used. Similarly, FunState uses individual functions for that. For KPN inside RPN, we have used complete reactions in Sect. 8. Scenarios Aware Data Flow (SADF) [19, 51] models a system as a finite state machine (decorated with transition probabilities to a Markov Chain) of SDF graphs and is similar to HDF, although primarily with the intention to capture dynamic variation within a determinate execution rather than to capture indeterminate behavior. SADF focusses less on functional specification, but rather on performance modeling by including timing behavior and stochastic abstraction of the automaton behavior. Moreover, SADF deals explicitly with the effects of pipelining different iterations of the SDF graphs and the intermediate automaton transitions. StreamIt [52] employs an SDF-like model of computation. It allows sending sporadic event between processes in the network by a construct called teleport messaging [53]. The scheduling and compilation framework automatically takes care that all data in the pipeline between the sender and receiver is processed before the message is delivered to the receiving process, similar to the RPN model.

Bhattacharya and Bhattacharyya [7] describe parameterizable SDF models, allowing dynamic reconfiguration during run-time. Processes can change their data-flow behavior depending on parameter settings. In a given SDF configuration, actor executions are characterized by iterations, that fire sub-processes in a particular order that returns the internal buffer states in their original configuration. Such a process can (only) be reconfigured between such iterations and if the data-flow behavior has changed, a new schedule is determined (at run-time).

Similar to RPN in generality, is an approach of Neuendorffer and Lee [40]. It focusses on reconfiguration as a particular kind of event handling or change of model parameters. It defines *quiescent states* as the states where reconfigurations are allowed. These quiescent states are strongly related to the maximal streaming transactions. They also propose to use FIFO (First In First Out) channel communication also for events or parameters and to divide input ports in streaming input ports and parameter input ports. In contrast with [40], RPN considers more general event handling and reconfiguration than changing parameters. It also focusses on formalizing dependencies between parameters and quiescent points at different levels of a system hierarchy.

The operational semantics of RPN connects easily to implementations of the model. This operational semantics is however not fully abstract, i.e., it contains more details than strictly necessary. Russell [46] discusses fully abstract semantics of indeterminate data-flow models.

The traditionally separated branch of synchronous data-flow languages such as Lustre [26] or Signal [5] treat sporadic events in a completely different way. Because these models are based on a global synchrony hypothesis (all system components execute conceptually at the pace of a single global clock), absence of an event in a particular clock cycle is easily, deterministically detected. In this case, sporadic events do not necessarily lead to a non-deterministic model. The overhead of the globally synchronous clock may impact efficiency however.

Acknowledgements This work is supported in part by the EC through FP7 IST project 216224, MNEMEE and by the Netherlands Ministry of Economic Affairs under the Senter TS program in the Octopus project.

References

1. Allen, G., Evans, B., Schanbacher, D.: Real-time sonar beamforming on a UNIX workstation using process networks and POSIX threads. In: Proc. of the 32nd Asilomar Conference on Signals, Systems and Computers, pp. 1725–1729. IEEE Computer Society (1998)
2. Allen, G., Zucknick, P., Evans, B.: A distributed deadlock detection and resolution algorithm for process networks. In: Acoustics, Speech and Signal Processing, 2007. ICASSP 2007. IEEE International Conference on, vol. 2, pp. II–33–II–36 (2007). DOI 10.1109/ICASSP.2007. 366165
3. Basten, T., Hoogerbrugge, J.: Efficient execution of process networks. In: A. Chalmers, M. Mirmehdi, H. Muller (eds.) Proc. of Communicating Process Architectures 2001, Bristol, UK, September 2001, pp. 1–14. IOS Press (2001)
4. Benveniste, A., Caillaud, B., Carloni, L.P., Caspi, P., Sangiovanni-Vincentelli, A.L.: Composing heterogeneous reactive systems. ACM Trans. Embed. Comput. Syst. **7**(4), 1–36 (2008)
5. Benveniste, A., Guemic, P.L.: Hybrid dynamical systems theory and the signal language. IEEE Trans. Automat. Contr. **35**, 535–546 (1990)
6. Berry, G., Gonthier, G.: The Esterel synchronous programming language: Design, semantics, implementation. Sci. Comput. Program. **19**, 87–152 (1992)

7. Bhattacharya, B., Bhattacharyya, S.: Parameterized dataflow modeling for DSP systems. IEEE Transactions on Signal Processing **49**(10), 2408–2421 (2001)
8. Bhattacharyya, S., Murthy, P., Lee, E.: Synthesis of embedded software from synchronous dataflow specifications. J. VLSI Signal Process. Syst. **21**(2), 151–166 (1999)
9. Brock, J., Ackerman, W.: Scenarios: A model of non-determinate computation. In: J. Díaz, I. Ramos (eds.) Formalization of Programming Concepts, International Colloquium, Proceedings, vol. LNCS 107, pp. 252–259. Peniscola, Spain (1981)
10. Brookes, S.: On the Kahn principle and fair networks. Tech. Rep. CMU-CS-98-156, School of Computer Science, Carnegie Mellon University (1998)
11. Broy, M., Dendorfer, C.: Modelling operating system structures by timed stream processing functions. Journal of Functional Programming **2**(1), 1–21 (1992). URL citeseer.nj.nec.com/broy92modelling.html
12. Buck, J.: Scheduling dynamic dataflow graphs with bounded memory using the token flow model. Ph.D. thesis, University of California, EECS Dept., Berkeley, CA (1993)
13. Carloni, L.P., Sangiovanni-Vincentelli, A.L.: A framework for modeling the distributed deployment of synchronous designs. Form. Methods Syst. Des **28**, 93–110 (2006)
14. Davey, B.A., Priestley, H.A.: Introduction to Lattices and Order. Cambridge University Press, Cambridge, UK (1990)
15. Dulloo, J., Marquet, P.: Design of a real-time scheduler for Kahn Process Networks on multiprocessor systems. In: Proceedings of the International Conference on Parallel and Distributed Processing Techniques and Applications, PDPTA, pp. 271–277 (2004)
16. Eker, J., Janneck, J., Lee, E.A., Liu, J., Liu, X., Ludvig, J., Sachs, S., Xiong, Y.: Taming heterogeneity - the ptolemy approach. Proceedings of the IEEE **91**(1), 127–144 (2003). URL http://chess.eecs.berkeley.edu/pubs/488.html
17. Faustini, A.: An operational semantics for pure dataflow. In: M. Nielsen, E.M. Schmidt (eds.) Automata, Languages and Programming, 9th Colloquium, Aarhus, Denmark, July 12–16, 1982, Proceedings, LNCS Vol. 140, pp. 212–224. Springer Verlag, Berlin (1982)
18. Geilen, M.: An hierarchical compositional operational semantics of Kahn Process Networks and its Kahn Principle. Tech. rep., Electronic Systems Group, Dept. of Electrical Engineering, Eindhoven University of Technology (2009)
19. Geilen, M.: Synchronous data flow scenarios. Transactions on Embedded Computing Systems, Special issue on Model-driven Embedded-system Design, **10**(2), (2010)
20. Geilen, M., Basten, T.: Requirements on the execution of Kahn process networks. In: P. Degano (ed.) Proc. of the 12th European Symposium on Programming, ESOP 2003, vol. LNCS 2618. Warsaw, Poland (2003)
21. Geilen, M., Basten, T.: Reactive process networks. In: EMSOFT '04: Proceedings of the 4th ACM international conference on Embedded software, pp. 137–146. ACM, New York, NY, USA (2004). DOI http://doi.acm.org/10.1145/1017753.1017778
22. Geilen, M., Stuijk, S.: Worst-case performance analysis of synchronous dataflow scenarios. In: International Conference on Hardware-Software Codesign and System Synthesis, CODES+ISSS 10, Proc., Scottsdale, Az, USA, 24–29 October, 2010, pp. 125–134 (2010)
23. Girault, A., Lee, B., Lee, E.: Hierarchical finite state machines with multiple concurrency models. IEEE Transactions on Computer-aided Design of Integrated Circuits and Systems **18**(6), 742–760 (1999)
24. Goel, M.: Process networks in Ptolemy II. Technical Memorandum UCB/ERL No. M98/69, University of California, EECS Dept., Berkeley, CA (1998)
25. Ha, S., Oh, H.: Decidable dataflow models for signal processing: Synchronous dataflow and its extensions. In: S.S. Bhattacharyya, E.F. Deprettere, R. Leupers, J. Takala (eds.) Handbook of Signal Processing Systems, second edn. Springer (2013)
26. Halbwachs, N., Caspi, P., Raymond, P., Pilaud, D.: The synchronous programming language LUSTRE. Proceedings of the IEEE **79**, 1305–1319 (1991)
27. Jiang, B., Deprettere, E., Kienhuis, B.: Hierarchical run time deadlock detection in process networks. In: Signal Processing Systems, 2008. SiPS 2008. IEEE Workshop on, pp. 239–244 (2008). DOI 10.1109/SIPS.2008.4671769

28. Jonsson, B.: A fully abstract trace model for dataflow networks. In: POPL '89: Proceedings of the 16th ACM SIGPLAN-SIGACT symposium on Principles of programming languages, pp. 155–165. ACM, New York, NY, USA (1989)

29. Kahn, G.: The semantics of a simple language for parallel programming. In: J. Rosenfeld (ed.) Information Processing 74: Proceedings of the IFIP Congress 74, pp. 471–475. North-Holland, Amsterdam, Netherlands, Stockholm, Sweden (1974)

30. Kahn, G., MacQueen, D.: Coroutines and networks of parallel programming. In: B. Gilchrist (ed.) Information Processing 77: Proceedings of the IFIP Congress 77, pp. 993–998. North-Holland, Toronto, Canada (1977)

31. de Kock, E., et al.: YAPI: Application modeling for signal processing systems. In: Proc. of the 37th. Design Automation Conference, pp. 402–405. IEEE, Los Angeles, CA (2000)

32. Lee, B.: Specification and design of reactive systems. Ph.D. thesis, Electronics Research Laboratory, University of California, EECS Dept., Berkeley, CA (2000). Memorandum UCB/ERL M00/29

33. Lee, E.: Overview of the Ptolemy project. Technical Memorandum UCB/ERL No. M01/11, University of California, EECS Dept., Berkeley, CA (2001)

34. Lee, E., Messerschmitt, D.: Synchronous data flow. IEEE Proceedings **75**(9), 1235–1245 (1987)

35. Lee, E., Sangiovanni-Vincentelli, A.: A framework for comparing models of computation. Computer-Aided Design of Integrated Circuits and Systems, IEEE Transactions on **17**(12), 1217–1229 (Dec 1998). DOI 10.1109/43.736561

36. Lee, E.A., Matsikoudis, E.: The semantics of dataflow with firing. In: Y. Bertot, G. Huet, J.J. Lvy, G. Plotkin (eds.) From Semantics to Computer Science: Essays in Honour of Gilles Kahn, chap. 4. Cambridge University Press (2007). URL http://chess.eecs.berkeley.edu/pubs/428.html

37. Liu, X., Lee, E.A.: CPO semantics of timed interactive actor networks. Theor. Comput. Sci. **409**(1), 110–125 (2008). DOI http://dx.doi.org/10.1016/j.tcs.2008.08.044

38. Lynch, N., Stark, E.: A proof of the Kahn principle for Input/Output automata. Information and Computation **82**(1), 81–92 (1989). URL citeseer.nj.nec.com/lynch89proof.html

39. Martin, A.: The probe: An addition to communication primitives. Information Processing Letters **20**(3), 125–130 (1985)

40. Neuendorffer, S., Lee, E.A.: Hierarchical reconfiguration of dataflow models. In: Proc. Second ACM-IEEE International Conference on Formal Methods and Models for Codesign (MEMOCODE 2004) (2004)

41. Olson, A., Evans, B.: Deadlock detection for distributed process networks. In: Acoustics, Speech, and Signal Processing, 2005. Proceedings. (ICASSP '05). IEEE International Conference on, vol. 5, pp. v/73–v/76 Vol. 5 (2005). DOI 10.1109/ICASSP.2005.1416243

42. Park, D.: On the semantics of fair parallelism. In: Abstract Software Specifications, Volume 86 of Lecture Notes in Computer Science. Springer Verlag, Berlin (1979)

43. Parks, T.: Bounded Scheduling of Process Networks. Ph.D. thesis, University of California, EECS Dept., Berkeley, CA (1995)

44. Plotkin, G.: A structural approach to operational semantics. Tech. Rep. DAIMI FN-19, Århus University, Computer Science Department, Århus, Denmark (1981)

45. Poplavko, P., Basten, T., van Meerbergen, J.: Execution-time prediction for dynamic streaming applications with task-level parallelism. In: DSD '07: Proceedings of the 10th Euromicro Conference on Digital System Design Architectures, Methods and Tools, pp. 228–235. IEEE Computer Society, Washington, DC, USA (2007). DOI http://dx.doi.org/10.1109/DSD.2007.52

46. Russell, J.: Full abstraction for nondeterministic dataflow networks. In: Symposium on Foundations of Computer Science, pp. 170–175. Research Triangle Park, NC (1989). DOI http://doi.ieeecomputersociety.org/10.1109/SFCS.1989.63474

47. Sriram, S., Bhattacharyya, S.S.: Embedded Multiprocessors: Scheduling and Synchronization. Marcel Dekker, Inc., New York, NY, USA (2000)

48. Stark, E.: Concurrent transition system semantics of process networks. In: Proc. of the 1987 SIGACT-SIGPLAN Symposium on Principles of Programming Languages, Munich, Germany, January 1987, pp. 199–210. ACM Press (1987)

49. Stevens, R., Wan, M., Laramie, P., Parks, T., Lee, E.: Implementation of process networks in Java. Technical Memorandum UCB/ERL No. M97/84, University of California, EECS Dept., Berkeley, CA (1997)

50. Strehl, K., Thiele, L., Gries, M., Ziegenbein, D., Ernst, R., Teich, J.: FunState - an internal design representation for codesign. IEEE Transactions on Very Large Scale Integration (VLSI) Systems 9(4), 524–544 (2001). URL citeseer.nj.nec.com/strehl01funstate.html

51. Theelen, B.D., Geilen, M., Basten, T., Voeten, J., Gheorghita, S.V., Stuijk, S.: A scenario-aware data flow model for combined long-run average and worst-case performance analysis. In: MEMOCODE, pp. 185–194 (2006)

52. Thies, W., Karczmarek, M., Amarasinghe, S.: StreamIt: A language for streaming applications. In: R.N. Horspool (ed.) Proc. 11th International Conference Compiler Construction CC 2002, vol. LNCS 2306, pp. 179–196. Grenoble, France (2002)

53. Thies, W., Karczmarek, M., Sermulins, J., Rabbah, R., Amarasinghe, S.: Teleport messaging for distributed stream programs. In: PPoPP '05: Proceedings of the tenth ACM SIGPLAN symposium on Principles and practice of parallel programming, pp. 224–235. ACM, New York, NY, USA (2005). DOI http://doi.acm.org/10.1145/1065944.1065975

54. Thomas T. Hildebrandt Prakash Panangaden, G.W.: A relational model of non-deterministic dataflow. Mathematical Structures in Computer Science pp. 613–649 (2004)

55. Vayssière, J., Webb, D., Wendelborn, A.: Distributed process networks. Tech. Rep. TR 99-03, University of Adelaide, Department of Computer Science, South Australia 5005, Australia (1999)

56. Yates, R.K.: Networks of real-time processes. In: E. Best (ed.) CONCUR'93: Proc. of the 4th International Conference on Concurrency Theory, pp. 384–397. Springer Verlag, Berlin, Heidelberg (1993)

Decidable Dataflow Models for Signal Processing: Synchronous Dataflow and Its Extensions

Soonhoi Ha and Hyunok Oh

Abstract Digital signal processing algorithms can be naturally represented by a dataflow graph where nodes represent function blocks and arcs represent the data dependency between nodes. Among various dataflow models, decidable dataflow models have restricted semantics so that we can determine the execution order of nodes at compile-time and decide if the program has the possibility of buffer overflow or deadlock. In this chapter, we explain the synchronous dataflow (SDF) model as the pioneering and representative decidable dataflow model and its decidability focusing on how the static scheduling decision can be made. In addition the cyclo-static dataflow model and a few other extended models are briefly introduced to show how they overcome the limitations of the SDF model.

1 Introduction

Digital signal processing (DSP) algorithms are often informally, but intuitively, described by block diagrams in which a block represents a function block and an arc or edge represents a dependency between function blocks. While a block diagram is not a programming model, it resembles a formal dataflow graph in appearance. Figure 1 shows a block-diagram representation of a simple DSP algorithm, which can also be regarded as a dataflow graph of the algorithm.

A dataflow graph is a graphical representation of a dataflow model of computation in which a node, or an *actor*, represents a function block that can be executed,

S. Ha (✉)
Seoul National University, Seoul, Korea
e-mail: sha@snu.ac.kr

H. Oh
Hanyang University, Seoul, Korea
e-mail: hoh@hanyang.ac.kr

S.S. Bhattacharyya et al. (eds.), *Handbook of Signal Processing Systems*,
DOI 10.1007/978-1-4614-6859-2_33, © Springer Science+Business Media, LLC 2013

Fig. 1 Dataflow graph of a
simple DSP algorithm

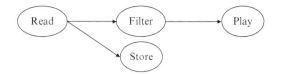

or *fired*, when enough input data are available. An arc is a FIFO channel that delivers
data samples, also called *tokens*, from an output port of the source node to an input
port of the destination node. If a node has no input port, the node becomes a source
node that is always executable. In DSP algorithms, a source node may represent
an interface block that receives triggering data from an outside source. The "Read"
block in Fig. 1 is a source block that reads audio data samples from an outside
source. A dataflow graph is usually assumed to be executed iteratively as long as the
source blocks produce samples on the output ports.

The dataflow model of computation was first introduced as a parallel programming model for the associated computer architecture called dataflow machines [6].
While the granularity of a node is assumed as fine as a machine instruction in
dataflow machine research, the node granularity can be as large as a well-defined
function block such as a filter or an FFT unit in a DSP algorithm representation. The
main advantage of the dataflow model as a programming model is that it specifies
only the true dependency between nodes, revealing the function-level parallelism
explicitly. There are many ways of executing a dataflow graph as long as data
dependencies between the nodes are preserved. For example, blocks "Filter" and
"Store" in Fig. 1 can be executed in any order after they receive data samples
from the "Read" block. They can be executed concurrently in a parallel processing
system.

To execute a dataflow graph on a target architecture, we have to determine where
and when to execute the nodes, which is called *scheduling*. Scheduling decision
can be made only at run-time for general dataflow graphs. A dynamic scheduler
monitors the input arcs of each node to check if it is executable, and schedules the
executable nodes on the appropriate processing elements. Thus dynamic scheduling
incurs run-time overhead of managing the ready nodes to schedule in terms of
both space and time. Another concern in executing a dataflow graph is resource
management. While a dataflow graph assumes an infinite FIFO queue on each arc, a
target architecture has a limited size of memory. Dynamic scheduling of nodes may
incur buffer overflow or a deadlock situation if buffers are not carefully managed. A
dataflow graph itself may have errors to induce deadlock or buffer overflow errors. It
is not decidable for a general dataflow program whether it can be executed without
buffer overflow or a deadlock problem.

On the other hand, some dataflow models have restricted semantics so that the
scheduling decision can be made at compile-time. If the execution order of nodes is
determined statically at compile-time, we can decide before running the program if
the program has the possibility of buffer overflow or deadlock. Such dataflow graphs
are called *decidable dataflow graphs*. More precisely, a dataflow is decidable if and
only if a schedule of which length is finite can be constructed statically. Hence,

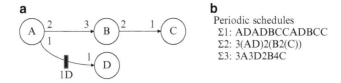

Fig. 2 (**a**) An SDF graph and (**b**) some periodic schedules for the SDF graph

in a decidable dataflow graph, the invocation number of each node is finite and computable at compile time. The *SDF (synchronous dataflow) model* proposed by Lee in [12], is a pioneering decidable model that has been widely used for DSP algorithm specification in many design environments including Ptolemy [5] and Grape II [11].

Subsequently, a number of generalizations of the SDF model have been proposed to extend the expression capability of the SDF model. The most popular extension is CSDF (cyclo-static dataflow) [4]. Three other extensions that aim to produce better software synthesis results will also be introduced in this chapter. In this chapter, we explain these decidable dataflow models and focus on their characteristics of decidability. For decidable dataflow graphs, the most important issue is to determine an optimal static schedule with respect to certain objectives since there are numerous ways to schedule the nodes.

2 Synchronous Dataflow

In a dataflow graph, the number of tokens produced (or consumed) per node firing is called the output (or the input) sample rate of the output (or the input) port. The simplest dataflow model is the single-rate dataflow (SRDF) in which all sample rates are unity. When a port may consume or produce multiple tokens, we call it multi-rate dataflow (MRDF). Among multi-rate dataflow graphs, synchronous dataflow (SDF) has a restriction that the sample rates of all ports are fixed integer values and do not vary at run time. Note that the SRDF is called sometimes the homogeneous SDF.

Figure 2a shows an SDF graph, which has the same topology as Fig. 1, where each arc is annotated with the number of samples produced and consumed by the incident nodes. There may be delay samples associated with an arc. The sample delay is represented as initial samples that are queued on the arc buffer from the beginning, and denoted as xD where x represents the number of initial samples, as shown on arc AD in Fig. 2a.

From the sample rate information on each arc, we can determine the relative execution rate of two end nodes of the arc. In order not to accumulate tokens unboundedly on an arc, the number of samples produced from the source node should be equal to the number of samples consumed by the destination node in the long run. In the example of Fig. 2a, the execution rate of node C should be twice as fast as the execution rate of node B on average. Based on this pair-wise information on the execution rates, we can determine the ratio of execution rates among all nodes. The resultant ratio of execution rates among nodes A, B, C and D in Fig. 2a becomes 3:2:4:3.

2.1 Static Analysis

The key analytical property of the SDF model is that the node execution schedule can be constructed at compile time. The number of executions of node A within a schedule is called the repetition count $x(A)$ of the node. A *valid* schedule is a finite schedule that does not reach deadlock and produces no net change in the number of samples accumulated on each arc. In a valid schedule, the ratio of repetition counts is equal to the ratio of execution rates among the nodes so that one iteration of a valid schedule does not increase the samples queued on all arcs. If there exists a valid schedule, the SDF graph is said *consistent* . We represent the repetition counts of nodes in a consistent SDF graph G by vector q_G. For the graph of Fig. 2a,

$$q_G = (x(A), x(B), x(C), x(D)) = (3, 2, 4, 3) \qquad (1)$$

Since an SDF graph imposes only partial ordering constraints between the nodes, the order of node invocations can be determined in various ways. Figure 2b shows three possible valid schedules of the Fig. 2a graph. In Fig. 2b, each parenthesized term $n(X_1 X_2 \ldots X_m)$ represents n successive executions of the sequence $X_1 X_2 \ldots X_m$, which is called a looped schedule. If every block appears exactly once in the schedule such as $\Sigma 2$ and $\Sigma 3$ in Fig. 2b, the schedule is called a *single appearance (SA) schedule*. An SA-schedule that has no nested loop is called a *flat SA-schedule*. $\Sigma 3$ of Fig. 2b is a flat SA-schedule, while $\Sigma 2$ is not. Consistency analysis of an SDF graph is performed by constructing a valid schedule; no valid schedule can be found for an erroneous SDF graph.

To construct a valid schedule, we first compute the repetition counts of all nodes. For a given arc e, we denote the source node as $src(e)$ and the destination node as $snk(e)$. The output sample rate of $src(e)$ onto the arc is denoted as $prod(e)$ and the input sample rate of $snk(e)$ as $cons(e)$. Then, the following equation, called a *balance* equation, should be held for a consistent SDF graph.

$$x(src(e))prod(e) = x(snk(e))cons(e) \ for \ each \ e. \qquad (2)$$

We can formulate the balance equations for all arcs compactly with the following matrix equation.

$$\Gamma q_G^T = 0 \qquad (3)$$

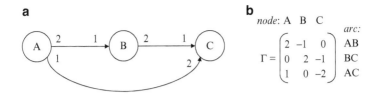

Fig. 3 (**a**) An SDF graph that is sample rate inconsistent and (**b**) the associated topology matrix

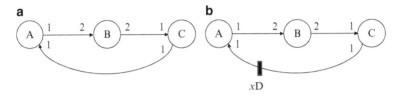

Fig. 4 (**a**) An SDF graph that is deadlocked and (**b**) the modified graph with initial samples on the feedback arc

where Γ, called the topology matrix of G, is a matrix of which rows are indexed by the arcs in G and columns are indexed by the nodes in G, An entry of the topology matrix is defined by

$$\Gamma(e,A) = \begin{cases} prod(e), & \text{if } A = src(e) \\ -cons(e), & \text{if } A = snk(e) \\ 0, & \text{otherwise} \end{cases} \tag{4}$$

A valid schedule exists only if Eq. (3) has a non-zero solution of repetition vector q_G. Mathematically, this condition is satisfied when the rank of the topology matrix Γ is $n-1$ [12], where n is the number of nodes in G. In case no non-zero solution exists, the SDF graph is called *sample rate inconsistent*. Figure 3a shows a simple SDF graph that is sample rate inconsistent, and its associated topology matrix. Note that the rank of the topology matrix is 3, not 2.

Sample rate consistency does not guarantee that a valid schedule exists. A sample rate consistent SDF graph can be deadlocked as illustrated in Fig. 4a if the SDF graph has a cycle with insufficient amount of initial samples. The repetition vector, $q_G = (x(A), x(B), x(C))$, is (2,1,2). However, there is no fireable node since all nodes wait for input samples from each other. So we modify the graph by adding initial samples on arc CA in Fig. 4b. Suppose that there is an initial sample on arc CA, or $x = 1$. Then node A is fireable initially. After node A is fired, one sample is produced and queued into the FIFO channel of arc AB. But the graph is deadlocked again since no node becomes fireable afterwards. In this example, the minimum number of initial samples is two in order to rescue the graph from the deadlock condition.

The simplest method to detect deadlock is to construct a static SDF schedule by simulating the SDF graph as follows:

1. At first, make an empty schedule list that will contain the execution sequence of nodes, and initialize the set of fireable nodes.
2. Select one of the fireable nodes and put it in the schedule list. If the set of fireable nodes is empty, exit the procedure.
3. Simulate the execution of the selected node by consuming the input samples from the input arcs and producing the output samples to the output arcs.
4. Examine each destination node of the output arcs, and add it to the set of fireable nodes only if it becomes fireable and its execution count during the simulation is smaller than its repetition count.
5. Go back to step 2 to repeat this procedure.

When we complete this procedure, we can determine if the graph is deadlocked by examining the schedule list. If there is any node that is scheduled fewer times than its repetition count in the schedule list, the graph is deadlocked. Otherwise, the graph is deadlock-free. In summary, an SDF graph is consistent if it is sample rate consistent and it is deadlock-free. Therefore, the consistency of an SDF graph can be statically verified by computing the repetition counts of all nodes (sample rate consistency) and by constructing a static schedule.

2.2   Software Synthesis from SDF Graph

An SDF graph can be used as a graphical representation of a DSP algorithm, from which target codes are automatically generated. Software synthesis from an SDF graph includes determination of an appropriate schedule and a coding style for each dataflow node, both of which affect the memory requirements of the generated software. One of the main scheduling objectives for software synthesis is to minimize the total (sum of code and data) memory requirements.

For software synthesis, the kernel code of each node (function block) is assumed already optimized and provided from a predefined block library. Then the target software is synthesized by putting the function blocks into the scheduled position once a schedule is determined. There are two coding styles, *inline* and *function*, depending on how to put a function block into the target code. The former is to generate an inline code for each node at the scheduled position, and the latter is to define a separate function that contains the kernel of each node. Figure 5 shows three programs based on the same schedule $\Sigma 2$ of Fig. 2b. The first two use the inline coding style, and the third the function coding style.

If we use function calls, we have to pay, at run-time, the function-call overhead which can be significant if there are many function blocks of small granularity. If inlining is used, however, there is a danger of large code size if a node is instantiated multiple times. For the example of Fig. 2, schedule $\Sigma 1$ is not adequate for inlining unlike SA-schedules, $\Sigma 2$ and $\Sigma 3$. Figure 5b shows an alternative code that uses

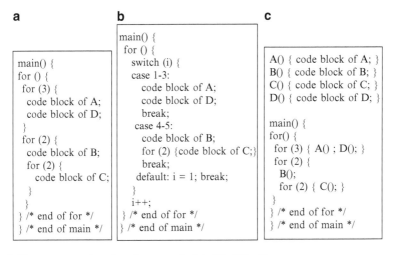

Fig. 5 Three programs based on the same schedule Σ2 of Fig. 2b

Table 1 Buffer requirements for three schedules of Fig. 2b

Schedule	Arc AB	Arc AD	Arc BC	Total
Σ1: ADADBCCADBCC	4	2	2	8
Σ2: 3(AD)2(B2(C))	6	2	2	10
Σ3: 3A3D2B4C	6	4	4	14

inlining without a proportional increase of code size to the number of instantiations of nodes. The basic idea is to make a simple run-time system that executes the nodes according to the schedule sequence. It pays the run-time overhead of switch-statements and code overhead for schedule sequence management. Hence an appropriate coding style should be selected considering the node granularity and the schedule.

For each schedule, the buffer size requirement can be computed. If we assume that a separate buffer is allocated on each arc, as is usually the case, the minimum buffer requirement of an arc becomes the maximum number of samples accumulated on the arc during an iteration of the schedule. For the example of Fig. 2, we can compare the buffer size requirements of three schedules as shown in Table 1.

From Table 1, we can observe that the SA schedules usually require larger buffers while they guarantee the minimum code size for inline code generation. In multimedia applications, frame-based algorithms are common where the size of a unit sample may be as large as a video frame or an audio frame. In these applications minimizing the buffer size is as important as minimizing the code size. In general, both code size and buffer size should be considered when we construct a memory-optimal schedule.

Buffering requirements can be reduced if we use buffer sharing. Arc buffers can be shared if their life-times are not overlapped with each other during an iteration of the schedule. The life-time of an arc buffer is defined by a set of durations from the source node invocation that starts producing a sample to the buffer to the completion of the destination node that empties the buffer. Consider schedule $\Sigma1$ of Fig. 2. The buffer life-time of arc BC consists of two durations, {BCC, BCC}, in the schedule. Since the buffer of arc AD is never empty, the buffer life-time of arc AD is the entire duration of the schedule. If we remove the initial sample on arc AD, the buffer life-time of arc AD consists of three durations, {AD, AD, AD}. Then we can share the two arc buffers of arc AD and arc BC since their life-times are not overlapped. A more aggressive buffer sharing technique has been developed by separating global sample buffers and local pointer buffers in case the sample size is large in frame-based applications [13]. The key idea is to allocate a global buffer whose size is large enough to store the maximum amount of live samples during an iteration of the schedule. Each arc is assigned a pointer buffer that stores pointers to the global buffer.

Code size can also be reduced by sharing the kernel of a function block when there are multiple instances of the same block [22] in a dataflow graph. Multiple instances of the same block are regarded as different blocks, and the same kernel, possibly with different local states, may appear several times in the generated code. A technique has been proposed to share the same kernel by defining a shared function. Separate state variables and buffers should be maintained for each instance, which define the *context* of each instance. The shared function is called with the context of an instance as an argument at the scheduled position of the instance. To decide whether sharing a code block is beneficial or not, the overhead and the gain of sharing should be compared. If Δ is an overhead that is incurred by function sharing, R is a code block size, and n is the number of instances of a block, the decision function for code sharing is summarized as the following inequality: $\frac{\Delta}{(n-1)R} < 1$.

For more detailed information on the code generation procedure and other issues related with software synthesis from SDF graphs, refer to [2].

2.3   Static Scheduling Techniques

Static scheduling of an SDF graph is the key technique of static analysis that checks the consistency of the graph and determines the memory requirement of the generated code. Since an SDF graph imposes only partial ordering constraints between the nodes, there exist many valid schedules and finding an optimal schedule has been actively researched.

2.3.1   Scheduling Techniques for Single Processor Implementations

The main objective for a single processor implementation is to minimize the memory requirement, considering both the code and the buffer size. Since the problem of finding a schedule with minimum buffer requirement for an acyclic graph is NP-complete, various heuristic approaches have been proposed. Since a single appearance schedule guarantees the minimum code size for inline code generation, a group of researchers have focused on finding a single appearance schedule that minimizes the buffer size. Bhattacharyya et al. developed two heuristics: APGAN and RPMC, to find an SA-schedule that minimizes the buffer requirements [3]. Ritz et al. used an ILP formulation to find a flat single appearance schedule that minimizes the buffer size [19] considering buffer sharing. Since a flat SA-schedule usually requires more data buffer than a nested SA-schedule, it is not evident which approach is better between these two approaches.

Another group of researches tries to minimize only the buffer size. Ade et al. presented an algorithm to determine the smallest possible buffer size for arbitrary SDF applications [1]. Though their work is mainly targeted for mapping an SDF application onto a Field Programmable Gate Array (FPGA) in the GRAPE environment, the computed lower bound on the buffer requirement is applicable to software synthesis. Govindarajan et al. [7] developed a rate optimal compile time schedule, which minimizes the buffer requirement by using linear programming formulation. Since the resultant schedule will not be an SA-schedule in general, a function coding style should be used to minimize the code size in the generated code.

No previous work exists that considers all design factors such as coding styles, buffer sharing, and code sharing. In spite of extensive prior research efforts, finding an optimal schedule that minimizes the total memory requirement still remains an open problem, even for single processor implementation.

2.3.2   Scheduling Techniques for Multiprocessor Implementations

A key scheduling objective for multiprocessor implementation is to reduce the execution length or to maximize the throughput of a given SDF graph. While there are numerous techniques developed for multiprocessor scheduling, they usually assume a single rate dataflow graph where each node is executed only once in a single iteration. And they primarily focus on exploiting the functional parallelism of an application to minimize the length of the schedule, called *makespan*. In stream-based applications, however, maximizing the throughput is more important than minimizing the schedule length. Pipelining is a popular way of improving the throughput of a dataflow graph. For example Hoang et al. have proposed a pipelined mapping/scheduling technique based on a list scheduling heuristic [8]. They maximize the throughput of a homogeneous dataflow graph on a homogeneous multi-processor architecture.

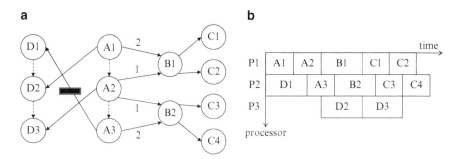

Fig. 6 (**a**) An APEG (acyclic precedence expanded graph) of the SDF graph in Fig. 2a, and (**b**) a parallel scheduling result displayed with a Gantt chart

To apply these techniques for an SDF graph directly, we need to translate an SDF graph to a single rate task graph, called an APEG (Acyclic Precedence Expanded Graph) or simply EG (Expanded Graph) [17]. A node of an SDF graph is expanded to as many nodes in the EG as the repetition counts of the node. The corresponding EG of the graph of Fig. 2a is shown in Fig. 6a where nodes A and D are expanded to three invocations respectively and node B to two and node C to four invocations. The number of samples communicated through each arc is unity unless specified otherwise; if an arc is annotated with a non-unity sample rate, such as an arc between nodes A1 and B2, the arc can be split into as many uni-rate arcs as the number to make a single-rate dataflow graph. If a node has any internal state, dependency arcs should be added between node invocations: in the figure, we assume that nodes A and D have internal states and the dependency between invocations is represented by dashed arcs. Note that the initial sample on arc AD is placed on the arc between A3 and D1. Since the initial sample breaks the execution dependency between A3 and D1 in the same iteration, D1 can be executed before A3.

One approach to schedule an SDF graph to a multiprocessor architecture is to generate an APEG and apply an existent multi-processor scheduling algorithm. Figure 6b shows a parallel scheduling result with a Gantt chart where the vertical axis represents the processing elements of the target system and the horizontal axis represents the elapsed time. There are some issues worth noting in this approach. First, loop-level data parallelism in an SDF graph is translated into functional parallelism in the EG. Nodes B and C in Fig. 2a express data-level parallelism since multiple invocations can be executed in parallel. But all invocations are translated into separate nodes that can be scheduled independently, ignoring the loop structure, in the EG. As a result, the same block can appear several times in the schedule, which may incur significant code size overhead if inline coding style is used. While it is a reasonable way to exploit the loop-level data parallelism, it may result in a very expensive solution.

Second, the total number of nodes in the EG is the sum of all repetition counts of the nodes. A simple SDF graph with non-trivial sample rate changes can be translated to a huge EG. Therefore the algorithm complexity of a multiprocessor scheduling technique should be low enough to scale well as the graph size increases.

Third, multiple invocations of the same node are likely to be mapped to different processors. If a node has internal states (for instance node D in Fig. 6b), the internal states should be transferred between invocations, which incurs significant run-time overhead. And additional code should be inserted to manage the internal states in the generated code.

Therefore several parallel scheduling techniques have been proposed that work with the SDF graph directly without APEG generation. A node with internal states is constrained to be mapped to a single processor. A recent approach considers functional parallelism, loop-level parallelism, and pipelining simultaneously to minimize the throughput of the graph [24]. In this work, a node can be mapped to multiple processors if the kernel code of a node has internal data parallelism.

Unlike single processor implementation, there is a trade-off between buffer size and throughput in multi-processor implementation of SDF graphs. Stuijik et al. have explored the trade-off and obtained the Pareto-optimal solutions in terms of the buffer size and the throughput, assuming that there is no constraint for the number of processors [21]. This work is extended to consider resource constraints in [20]. Other extensions can be found in a web-site (http://www.es.ele.tue.nl/sdf3/) that they manage to make the proposed technique open to public under the name of SDF3 (SDF for Free) [21].

While most work on multiprocessor scheduling assumes static scheduling, some recent researches consider static mapping and dynamic scheduling as a viable implementation technique of SDF task graphs. This approach maps nodes to processors at compile time and delegates the scheduling decision to a run-time scheduler on each processor. A main benefit of dynamic scheduling is that it may tolerate large variation of node execution times or communication delays. When we make a static schedule, we have to assume a fixed node execution time and a fixed communication delay; worst-case values are usually assumed for real-time systems. If the worst-case values are assumed, however, a processor may be idle while the current scheduled node is waiting for the arrival of input data from the other processors even though there are other executable nodes. Thus static scheduling may result in waste of resources while dynamic scheduling changes the execution sequence of nodes to increase resource utilization. A key issue for dynamic scheduling is how to assign priorities to the mapped nodes on each processor. There is a recent work to find an optimal static mapping and priority assignment to minimize the resource requirement under a throughput constraint [10].

If the execution time of each function block is fixed and known at compile-time, we can estimate the total execution length of the SDF graph. Hence the SDF graph is suitable for specifying an application that has real-time constraints on throughput or latency. When the execution time of a node varies at run-time, the worst case execution time (WCET) should be used in parallel scheduling in order to guarantee satisfaction of real-time constraints. Unfortunately, however, using WCETs is not enough to guarantee the schedulability if an application consists of multiple tasks running concurrently. If the execution time of a node becomes smaller than its WCET, the order of node firings may vary at run-time, which may lengthen the total execution time of the application. This behavior is known as "scheduling anomaly" of multiprocessor scheduling; If multiple tasks are running concurrently, resource contention may lengthen the execution of the graph when a node takes shorter time than its worst case execution time.

Since many DSP applications involve multiple tasks running concurrently, we need to consider multi-tasking in parallel scheduling. One solution is to statically schedule the multiple tasks up to their hyper-period that is defined as a least common multiple of all task periods. Since the hyper-period can be huge if the task periods are relatively prime, this approach may not be applicable. Multiprocessor scheduling of multi-tasking applications on heterogeneous multiprocessor systems still remains an open problem. With a given schedule, it is not yet possible to decide whether the schedule can satisfy the real-time constraints if some nodes have varying execution times and/or there are resource contentions during execution.

2.4   Hardware Synthesis from SDF Graph

While the target architecture is given as a constraint for software synthesis, the target hardware structure can be synthesized in hardware synthesis from an SDF graph. Therefore, we can achieve the *iteration bound* of an SDF graph in theory (see [15] to find the definition of the iteration bound of a graph) if there is no limitation on the hardware size. Since there is a trade-off between hardware cost and the throughput performance, however, architecture design and node scheduling should be considered simultaneously under given design constraints.

A key issue in hardware synthesis is to preserve the SDF semantics to maintain the correctness of the graph. In the SDF model, two samples that have the same value should be distinguished as separate samples while the same value is not identified as a new event in a hardware logic. So the arrival of an input sample should be notified somehow. And if a node has more than one input port, the node should wait until all input ports receive data samples before the node starts execution. It means that we need some control logic to perform scheduling of the nodes. There are two types of controllers: distributed controller and centralized controller. In the centralized control scheme, the execution timing of each node is controlled by a central scheduler. The execution timing can be determined at compile-time by static scheduling of the graph. In a distributed scheme, a node is associated with a control logic that monitors the input queues and triggers the node execution when all input queues have input samples to fire the node.

For hardware synthesis, a node should be specified by a hardware description language that will be synthesized by a CAD tool, or by a function block that is mapped to a pre-defined hardware IP. If an hardware IP is used, interface between the IP and the rest of the system should be designed carefully. Since the interface design is a laborious and error-prone task, extensive researches are being performed on the automatic interface synthesis.

In summary, hardware synthesis from a SDF graph involves the following problems: architecture and datapath synthesis, controller synthesis, and interface synthesis. The node granularity in a SDF graph also affects the hardware synthesis procedure. Various issues in hardware synthesis for a coarse grained graph is discussed in [9]. For FPGA synthesis from a fine-grained graph, see [23] for more detailed information.

3 Cyclo-Static Dataflow

The strict restriction of the SDF model, that all sample rates are constant, limits the expression capability of the model. Figure 7a shows a simple SDF graph that models a stereo audio processing application where the samples for the left and right channels are interleaved at the source. The interleaved samples are distributed by node D to the left (node L) and to the right (node R) channel. In this example, the distributor node (node D) waits until two samples are accumulated on its input arc to produce one sample at each output arc. A more natural implementation would be to make the distribution node route an input sample to two output ports alternatively at each arrival. A useful generalization of the SDF model, called the cyclo-static dataflow (CSDF), makes it possible [4].

In a cyclo-static dataflow graph, the sample rates of an input or output port may vary in a periodic fashion. Figure 7b shows how the CSDF model can specify the same application as Fig. 7a. To specify the periodic change of the sample rates, a tuple rather than an integer is annotated at the output ports of node D'. For instance, "$\{1,0\}$" on arc D'R denotes that the rate change pattern is repeated every other execution, where the rate is 1 at the first execution, and 0 at the second execution. Similarly, the periodic sample rate "$\{0,1\}$" means that the rate is 0 at every $(2n+1)$-th iteration and 1 at the other iterations.

Note that Fig. 7a, b represent the same application in functionality. One firing of node D in the SDF graph is broken down into two firings of node D' in the CSDF graph. Thus we have to split the behavior of node D' into phases. The number of phases is determined by the periods of the sample rate patterns of all input and output ports. In general, we can convert a CSDF graph to an *equivalent* SDF graph by merging as many firings of a CSDF node as the number of phases into a single firing of an equivalent SDF node. For instance, node D' in the CSDF graph repeats its behavior every two firings, and the number of phases becomes 2. So an equivalent

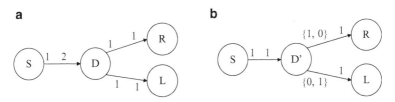

Fig. 7 (**a**) An SDF graph where node D is a distributor block and (**b**) a CSDF graph that shows the same behavior with a different version of a distributor block

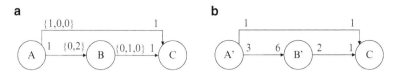

Fig. 8 (a) A cyclo-static dataflow graph and (b) its corresponding SDF graph

SDF node can be constructed by merging two firings of node D' into one, which is node D in the SDF model. The sample rates of input and output ports are adjusted accordingly by summing up the number of samples consumed and produced during one cycle of periodic behavior.

The CSDF model has a big advantage over the SDF model in that it can reduce the buffer requirement on the arcs. In the example shown in Fig. 7, the minimum size of input buffer for node D should be 2 in the SDF model while it is 1 in the CSDF model.

3.1   Static Analysis

Since we can construct an equivalent SDF graph, static analysis and scheduling algorithms developed for SDF are also applicable to CSDF. For formal treatment, we use vectors to represent the periodic sample rates in CSDF: for an arc e, the output sample rate vector of $src(e)$ and the input sample rate vector of $snk(e)$ are denoted by **prod**(e) and **cons**(e) in CSDF. Figure 8a shows another CSDF graph that has non-trivial periodic patterns of sample rates. The sample rate vectors for the graph become the following:

$$\mathbf{prod}(AC) = (1,0,0), \mathbf{cons}(AB) = (0,2), \mathbf{prod}(BC) = (0,1,0),$$
$$\mathbf{prod}(AB) = \mathbf{cons}(BC) = \mathbf{cons}(AC) = (1).$$

To make an equivalent SDF node for each CSDF node, we have to compute the repetition period for the phased operation of the CSDF node. First we obtain the period of the sample rate variation for each port, which is the size of the sample rate vector. Let $dim(\mathbf{v})$ be the dimension of the sample rate vector $\mathbf{v}$. Then the repetition period of a CSDF node A, denoted by $p(A)$ becomes the least common multiple (lcm) value of all $dim(\mathbf{v})$ values for the input and output ports of the node. For the example of Fig. 8a, the repetition periods become the following:

$$p(A) = lcm(dim(\mathbf{prod}(AB)), dim(\mathbf{prod}(AC))) = lcm(3,1) = 3.$$
$$p(B) = lcm(dim(\mathbf{cons}(AB)), dim(\mathbf{prod}(BC))) = lcm(2,3) = 6.$$
$$p(C) = lcm(dim(\mathbf{cons}(AC)), dim(\mathbf{prod}(BC))) = lcm(1,1) = 1.$$

If $p(A)$ firings of CSDF node A are merged into a single firing, an equivalent SDF actor A' is obtained. Hence the equivalent SDF graph is obtained as shown in Fig. 8b where node B' is obtained by merging six firings of node B in the CSDF graph. We denote this equivalence relation as $B' \approx 6B$. For an equivalent SDF node, the scalar sample rate of a port should be determined. Let $\sigma(\mathbf{v})$ be the sum of elements in vector $\mathbf{v}$. The total number of samples produced or consumed on arc e of CSDF node A per the corresponding SDF node execution is given by $p(A)\frac{\sigma(\mathbf{prod}(\mathbf{e}))}{dim(\mathbf{prod}(\mathbf{e}))}$ or $p(A)\frac{\sigma(\mathbf{cons}(\mathbf{e}))}{dim(\mathbf{cons}(\mathbf{e}))}$. So, we can construct a topology matrix for the equivalent SDF graph as follows:

$$\Gamma(e,A) = \begin{cases} p(A)\frac{\sigma(\mathbf{prod}(\mathbf{e}))}{dim(\mathbf{prod}(\mathbf{e}))}, & \text{if } A = src(e) \\ -p(A)\frac{\sigma(\mathbf{cons}(\mathbf{e}))}{dim(\mathbf{cons}(\mathbf{e}))}, & \text{if } A = snk(e) \\ 0, & \text{otherwise} \end{cases} \qquad (5)$$

We can check the sample rate consistency with this topology matrix. For the graph in Fig. 8b, the topology matrix and the repetition vector become:

$$\Gamma = \begin{pmatrix} 3 & -6 & 0 \\ 1 & 0 & -1 \\ 0 & 2 & -1 \end{pmatrix}$$

$$\mathbf{q_G} = (2,1,2)$$

Since rank of Γ is 2, the CSDF graph is sample rate consistent. Moreover, a valid schedule includes two invocations of nodes A' and C, and one invocation of node B'. This means that a valid CSDF schedule contains 6A, 6B and 2C since $A' \approx 3A$ and $B' \approx 6B$. The deadlock detection algorithm for an SDF graph in Sect. 2.1 is applicable for a CSDF graph, which is to construct a static schedule by simulating the graph.

3.2 Static Scheduling and Buffer Size Reduction

One strategy of scheduling a CSDF graph is to schedule the equivalent SDF graph and replace the execution of the equivalent node with the multiple invocations of the original CSDF node. We can obtain the following schedule for the graph in Fig. 8: $\Sigma 1 = 2A'B'2C = 6A6B2C$. Then the minimum buffer requirement on arc AB becomes 6. We can construct better schedules in terms of buffer requirements by utilizing the phased operation of a CSDF node. For the case of CSDF graph of Fig. 8a, we can construct a better schedule as follows.

1. Initially nodes A and B are fireable, so schedule nodes A and B: $\Sigma 2 = $ "AB".
2. Since node A is the only fireable node, we schedule node A again: $\Sigma 2 = $ "ABA"

3. Now two samples are accumulated on arc AB and the second phased of node B can start. So schedule node B: Σ2 = "ABAB".
4. Node C becomes fireable. Schedule node C for the first time: Σ2 = "ABABC".
5. We can schedule nodes A and B twice: Σ2 = "ABABCABAB"
6. At this moment, only one sample is stored on arc AC and we can fire nodes A and B. We choose to schedule the fifth invocation of node B to produce one sample on arc BC. Σ2 = "ABABCABABB"
7. Then, node C becomes fireable. Schedule node C: Σ2 = "ABABCABABBC"
8. Finally we schedule node A twice and node B once to complete one iteration of the schedule: Σ2 = "ABABCABABBCAAB"
9. Since schedule Σ2 contains 6A, 6B and 2C, scheduling is finished and no deadlock is detected.

Schedule Σ2 requires two buffers on arc AB, which is three times better than schedule Σ1. Generally, as sample rates vary more, the buffer size reduction becomes more significant. This gain is obtained by splitting the CSDF node into multiple phases. But we have to pay the overhead of code size since a single appearance schedule is given up. In general, there are more valid schedules for a CSDF graph than for the equivalent SDF model. Therefore, discussion on the SDF scheduling can be applied to the CSDF model, but with increased complexity of scheduling problems.

3.3 Hierarchical Composition

Another advantage of CSDF is that it offers more opportunities of clustering when constructing a hierarchical graph. It also allows a seemingly delay-free cycle of nodes, while no delay-free cycle is allowed in SDF. Figure 9a shows an SDF graph with four nodes A, B, C and D. All sample rates are unity since no sample rate is annotated on any arc. The graph can be scheduled without deadlock since there is an initial delay sample between nodes A and B. One unique valid schedule of this graph is "BCDA". Suppose we cluster nodes A and B into an hierarchical node W in CSDF and W' in SDF as illustrated in Fig. 9a and b respectively. Since CSDF node W fires node B at every $(2n+1)$-th cycle and node A at every $2n$-th cycle, the input and the output sample rate vectors of node W become "{0,1}" and "{1,0}" respectively. Therefore, the CSDF graph can be scheduled without deadlock and a valid static schedule is "WCDW" where node B is fired at the first invocation of node W and node A is fired at the second invocation. On the other hand, SDF node W' should execute both nodes A and B when it is fired. Therefore, the SDF graph as shown in Fig. 9b is deadlocked since nodes W', D, and C wait for each other.

Clustering of nodes may cause deadlock in SDF even though the original SDF graph is consistent. On the other hand, a CSDF graph that includes a cyclic loop without an initial delay can be scheduled without deadlock if the periodic rates are carefully determined. Therefore, the CSDF model allows more freedom of hierarchical composition of the graph.

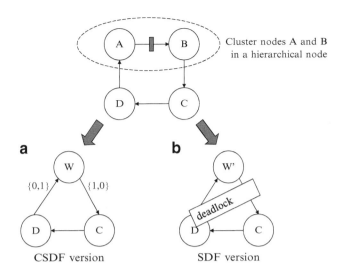

Fig. 9 Clustering of nodes A and B into an hierarchical node in (**a**) CSDF and (**b**) in SDF

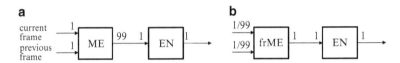

Fig. 10 A subgraph of an H.263 encoder graph (**a**) in SDF and (**b**) in FRDF

4 Other Decidable Dataflow Models

4.1 Fractional Rate Dataflow

The SDF model does not make any assumption on the data types of samples as long as the same data types are used between two communicating nodes. To specify multimedia applications or frame-based signal processing applications, it is natural to use composite data types such as video frames or network packets. If a composite data type is assumed, the buffer requirement for a single data sample can be huge, which amounts to 176×144 pixels for a QCIF video frame for instance. Then reducing the buffer requirement becomes more important than reducing the code size when we make a schedule.

Figure 10a shows an SDF subgraph of an H.263 encoder algorithm for QCIF video frames. A QCIF video frame consists of 11×9 macroblocks whose size is 16×16 pixels. Node ME that performs motion estimation consumes the current and the previous frames as inputs. Internally, the ME block divides the current frame into 99 macroblocks and computes the motion vectors and the pixel differences from the previous frame. And it produces 99 output samples at once where each output

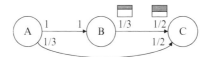

Fig. 11 An FRDF graph in which sample types on arc BC and AC are a composite and a primitive type respectively

sample is a macroblock-size data that represents a 16×16 array of pixel differences. Node EN performs macroblock encoding by consuming one macroblock at a time and produces one encoded macroblock as its output sample.

This SDF representation is not efficient in terms of buffer requirement and performance. Since node ME produces 99 macroblock-size samples at once after consuming a single frame size sample at each invocation, we need a 99-macroblock-size buffer or a frame-size buffer ($99 \times 16 \times 16 = 176 \times 144$) to store the samples on the arc between nodes ME and EN. Moreover node EN cannot start execution before node ME finishes motion estimation for the whole input frame. As this example demonstrates, the SDF model has inherent difficulty of efficiently expressing the mixture of a composite data type and its constituents: a video frame and macroblocks in this example. A video frame is regarded as a unit of data sample in integer rate dataflow graphs, and should be broken down into multiple macroblocks explicitly by consuming extra memory space.

To overcome this difficulty, the fractional rate dataflow (FRDF) model in which a fractional number of samples can be produced or consumed has been proposed [14]. In FRDF, a fraction number can be used as a sample rate as shown in Fig. 10b where the input sample rates of node frME is set to $\frac{1}{99}$. The fractional number means that the input data type of node frME is a macroblock and it corresponds to $\frac{1}{99}$ of a frame data.

Figure 11 shows an FRDF graph where the data type of arc BC is a composite type as illustrated in the figure and the data type of arc AC is a primitive type such as integer or float. A fractional sample rate has different meaning for a composite data type from a primitive type. For a composite data type, the fractional sample rate really indicates the partial production or consumption of the sample. In the example graph, one firing of node B fills $\frac{1}{3}$ of a sample on arc BC and node C reads the first half of the sample at every $(2n + 1)$-th firing and the second half at every $2n$-th firing. Hence, if we consider the execution order of nodes B and C, a schedule "BBCBC" is valid since $\frac{2}{3}$ of a sample is available after node B is fired twice and node C becomes fireable.

For primitive types, partial production or consumption is not feasible. Then statistical interpretation is applied for a fractional rate. In the example graph, the output sample rate of node A is $\frac{1}{3}$ on arc AC. This means that node A produces a single sample every three executions. Similarly node C consumes one sample every two executions. Note that a fractional rate does not imply at which firings samples are produced. So node C becomes fireable only after node A is executed three times. If we are concerned about the execution order of nodes A and C only,

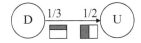

Fig. 12 If the consumer and the producer have the different access patterns for a composite type data then the type should be treated as atomic

schedule "3A2C" is valid while "2ACAC" is not. Consequently, a valid schedule for the FRDF graph of Fig. 11 is "3(AB)2C".

Regardless of the data type, a fractional sample rate $\frac{p}{q}$ guarantees that p samples are produced or consumed after q firings of the node. Similar to the CSDF graph, we can construct an equivalent SDF graph by merging q firings of an FRDF node into an equivalent SDF node that produces or consumes p samples per firing. Therefore, static analysis techniques for the SDF model can be applied to the FRDF model. For the analysis of sample rate consistency, however, we can use fractional sample rates directly in the topology matrix. For the FRDF graph of Fig. 11, the topology matrix and repetition vector are:

$$\Gamma = \begin{pmatrix} 1 & -1 & 0 \\ 0 & \frac{1}{3} & -\frac{1}{2} \\ \frac{1}{3} & 0 & -\frac{1}{2} \end{pmatrix}$$

$$\mathbf{q_G} = (3,3,2)$$

Since the rank of Γ is 2, the graph is sample rate consistent. Deadlock can be detected by constructing a static schedule similarly to the SDF case. If there exists a valid static schedule, the graph is free from deadlock. A static schedule is simply constructed by inserting a fireable node into the schedule list and simulating its firing. An FRDF node has different firing conditions depending on the data types of input ports. An FRDF node is fireable, or executable, if all input arcs satisfy the following condition depending on the data type:

1. If the data type is primitive, there must be at least as many stored samples as the numerator value of the fractional sample rate. An integer sample rate is a special fractional rate whose denominator is 1.
2. If the data type is composite, there must be at least as large a fraction of samples stored as the fractional input sample rate.

Special care should be taken for a composite type data. If the consumer and the producer have a different interpretation on the fraction, then a composite type should be regarded as atomic like a primitive type when the firing condition is examined. Suppose that for a composite data type of a two-dimensional array, the producer regards it as an array of row vectors while the consumer regards it as an array of column vectors as shown in Fig. 12. In this case, the two-dimensional array may not be regarded as a composite type data. Therefore, schedule "DDUDU" is not valid while "3D 2U" is.

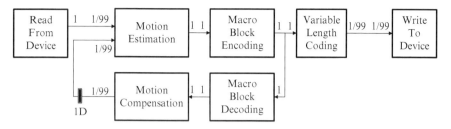

Fig. 13 H.263 encoder in FRDF

Fig. 14 An FRDF graph
corresponding to Fig. 8

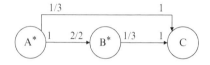

In general, the FRDF model results in an efficient implementation of a multi-media application in terms of buffer requirement and performance. Consider the example in Fig. 10b again. Since node frME uses a macroblock-size sample, the output arc requires only a single macroblock-size buffer. For each arrival of an input video frame, node frME is fired 99 times and consumes a macro-block size portion of the input frame per firing. Since node EN can be fired after each firing of node frME, shorter latency is experienced when compared with the SDF implementation. Figure 13 shows a whole H.263 encoding algorithm in FRDF where the sample types for "current frame" and "previous frame" are video frames. It is worth noting that the entire previous frame is needed to do motion estimation for each macroblock of the current frame while the sample rate of the bottom input port of node "Motion Estimation" is $\frac{1}{99}$. Hence, even though the previous frame is a composite data type, it should be regarded as atomic. Then node "Motion Estimation" is fireable only after the entire previous frame is available.

Figure 14 shows an FRDF graph that corresponds with Fig. 8a. Both have the same equivalent SDF graphs. Similar to the CSDF model, the FRDF model can reduce the buffer size when compared with the corresponding SDF model. Since node A^* produces a single sample and B^* consumes two samples every other execution, a valid schedule for the graph is "$2A^*$ $2B^*$ $2A^*$ B^* C B^* $2A^*$ $2B^*$ C". And the required buffer sizes on arcs A^*B^* and B^*C^* are equal to the sizes for the CSDF graph. The buffer size for arc A^*C is, however, larger than the CSDF graph since the FRDF model does not know when samples are produced and consumed, and the schedule for the FRDF model should consider the worst case behavior. For an output port, the worst case is when output samples are all produced at the last phase while it is when input samples are all consumed at the first phase for an input port. Therefore, in the FRDF model, rate $\frac{p}{q}$ for a primitive data type corresponds to "$\{(q-1) \times 0, p\}$" for an output sample rate and "$\{p, (q-1) \times 0\}$" for an input sample rate in the CSDF model, where "$(q-1) \times 0$" denotes "$\underbrace{0, 0, \cdots, 0}_{q-1}$". Hence, the CSDF model may generate better schedules than the associated FRDF model since we can split the node execution into finer granularity of phases at compile-time scheduling; this is not possible in the FRDF if the date type is primitive. The expression capability of two models is, however, different. The CSDF model allows only a periodic pattern to express sample rate variations while the FRDF model has no such restriction as long as the average value is constant during a period. So the FRDF model has more expression power than the CSDF model since it allows dynamic behavior of an FRDF node within a period and the periodic pattern can be regarded as a special case.

4.2   Synchronous Piggybacked Dataflow

The SDF model does not allow communication through a shared global variable since the access order to the global variable can vary depending on the execution order of nodes. Suppose a source block produces the frame header in a frame-based signal processing application that is to be used by several downstream blocks. A natural way of coding in C is to define a shared data structure that the downstream blocks access by pointers. But in a dataflow model, such sharing is not allowed. As a result, redundant copies of data samples are usually introduced in the automatically generated code from the dataflow model. Such overhead is usually not tolerable for embedded systems with tight resource and/or timing constraints. To overcome this limitation, an extended SDF model, called SPDF (Synchronous Piggybacked Dataflow) is proposed [16], by introducing the notion of "controlled global states" and by coupling a data sample with a pointer to the controlled global state.

The Synchronous Piggybacked Dataflow (SPDF) model was first proposed to support frame-based processing, or block-based processing, that frequently occurs in multimedia applications. In frame-based processing, the system receives input streams of frames that consist of a frame header and data samples. The frame header contains information on how to process data samples. So an intuitive implementation is to store the information in a global data structure, called *global states*, and the data processing blocks refer to the global states before processing the data samples.

Figure 15 shows a simple SPDF graph where node A reads a frame from a file and produces the header information and the data samples through different output ports. Suppose that a frame consists of 100 data samples in this example. The header information and the data samples are both connected to a special FRDF (Fractional Rate Dataflow) block, called *"Piggyback"* block, that has three parameters: *"statename"*, *"period"*, and *"offset"*. The Piggyback block updates the global state of *"statename"* with the received header information periodically with the specified *"period"* parameter. It piggybacks a pointer to the global state on each input data sample before sending it through the output port. Since it needs to receive the header information in order to update the global state only once per 100 executions, the sample rate of the input port associated with the header information

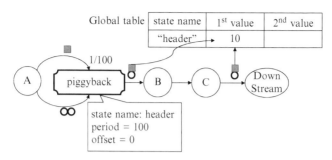

Fig. 15 An example SPDF graph that shows a typical frame-based processing: the Piggyback block writes the header information to the global state and the downstream blocks refer to the global state before processing the data samples

is set to the fractional value $\frac{1}{100}$, which means that it consumes one sample per 100 executions in the FRDF model. The input port of this fractional rate is called the *"state port"* of the Piggyback block. The sample rate of the data port, on the other hand, is unity.

The *"offset"* parameter indicates when to update the global state. The Piggyback block receives as many data samples as the *"offset"* value before updating the global state. In this example, the *"offset"* value is set to its default value, zero, which makes the Piggyback block consume the header information and update the global state before it piggybacks the data samples with a pointer to the global state.

Note that the Piggyback block with a fractional rate input port is the only extension to the SDF model. Since the sample rates of the SPDF graph are all static, the static analyzability of the SDF model is preserved even after addition of the Piggyback block. Also, piggybacking of data samples with pointers can be performed without any run-time overhead by utilizing the static schedule information of the graph. Suppose that the SDF graph in Fig. 15 has the following static schedule: "A 100(Piggyback, B, C, DownStream)". Then the pseudo code of the automatically generated code associated with the schedule is as follows:

```
code block of A
for (int i = 0; i < 100, i++) {
  if (i == offset_Piggyback)
    update the global state header
  code block of B
  code block of C
  if (i == offset_Piggyback)
    update the local state of DownSteam block
      from the global state information
  code block of DownStream
}
```

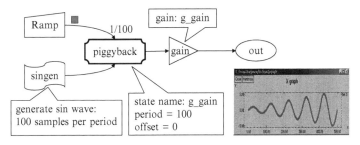

Fig. 16 An SPDF graph that produces a sinusoidal waveform with varying amplitude at run-time: the "gain" state of the "Gain" block is updated by the "Ramp" block through a global state, "global_gain"

Figure 16 shows another example that produces a sinusoidal waveform with varying amplitude at run-time. The "Singen" block generates a sinusoidal waveform (N samples per period) of unit amplitude and the "Gain" block amplifies the input samples by the "gain" parameter of the block. To control the amplitude, the graph uses a Piggyback block after the "Singen" block. Another source block, "Ramp", is connected to the state port of the Piggyback block. The "statename" of the Piggyback is named "global_gain" and the "gain" parameter of the "Gain" block is also set to "global_gain". Then, the "gain" parameter of the "Gain" block is updated with a global state named by "global_gain" whose value is determined by the "Ramp" block. In this example, the period of the Piggyback block is set to N so that the amplitude of the sinusoidal waveform is incremented by one every period as shown in Fig. 16. If we insert two initial samples on the input arc of the "Gain" block, the "offset" parameter of the Piggyback block should be 2.

Thus the SPDF model provides a safe mechanism to deliver state values through shared memory instead of message passing. Communication through shared memory is prohibited in conventional dataflow models since the access order to the shared memory may vary depending on the schedule. But the SPDF model gives another solution by observing that the side effect is caused by an implicit assumption that the global state is assigned a memory location before scheduling is performed. The SPDF model changes the order: allocate the memory for the global state after the schedule is made. Since the scheduling decision is made at compile-time, we know the access order to the variable and the life time of each global state variable. Suppose that the schedule of Fig. 16 becomes "2(100(Singen) Ramp Piggyback) 200(Gain Display)". From the static schedule, we know that we need to maintain two memory locations for the global state, "global_gain" since the Piggyback block writes the second value to the global state before the "Gain" reads the first global state.

Allowing shared memory communication without side effects gives a couple of significant benefits over conventional dataflow models. First, it can remove the unnecessary overhead of data copying of message passing communication since the global state can be shared by multiple blocks. Second, it greatly enhances the

a

```
void add()
{
   *addOut = *addIn1 + *addIn2;
}
```

b

```
void add()
{
   for(int i=0; i<Nb; i++)
      *addOut++ = *addIn1++ + *addIn2++;
}
```

Fig. 17 Code of an "Add" actor (**a**) in SDF and (**b**) in SSDF where N_b is the blocking factor

expression capability of the SDF model without breaking the static analyzability. It provides an explicit mechanism of affecting the internal behavior of a block from the outside through global states.

4.3   Scalable Synchronous Dataflow

DSP architectures have stringent on-chip memory limits and off-chip memory access is costly. They also allow vector processing of instructions and arithmetic pipelining like MAC in order to attain peak performance when the pipelining is fully utilized. Therefore, when an SDF block operates on primitive-type data and the granularity is small, it behaves inefficiently. For example, an adder block performs a single accumulation operation by reading two samples from memory and writing a sample into memory. It requires large run time overhead of off-chip memory access for two read and one write operations. In order to achieve efficient implementation, the scalable synchronous dataflow (SSDF) model is proposed [18]. The SSDF model has the same semantics as the SDF model except that a node may consume or produce any integer multiple of the fixed rates per invocation. The actual multiple, called *blocking factor*, is determined by a scheduler or an optimizer.

Figure 17a shows the code of an "Add" block in SDF. In SSDF, the code includes blocking factor N_b that is the number of iterations as shown in Fig. 17b. Since the function call overhead of "add()" is larger than the accumulation operation, the SSDF model amortizes the function call overhead by performing N_b accumulations per function call. When blocking factor N_b is 1, the SSDF graph degenerates to an SDF graph. Therefore, the SSDF model has the same sample rates as the SDF model and sample rate inconsistency can be checked using the topology matrix for the SDF model. Moreover, deadlock is detected by constructing a schedule by setting block factor $N_b = 1$. From the static analysis of the degenerated SDF graph, repetition vector q_G can be obtained, assuming N_b is 1 for all blocks. When $N_b > 1$, the repetition vector for SSDF becomes $N_b q_G$.

A straightforward scheduling technique for an SSDF graph is to increase the minimal scheduling period by an integer factor N_g where N_g is a global blocking factor. Each node A of the graph will be invoked $N_g x(A)$ times within one scheduling period, where $x(A)$ is the repetition count of node A. Increasing N_g reduces the function call overhead but requires larger buffer memory for graph execution. For

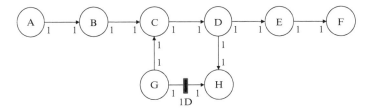

Fig. 18 A graph with feedback loop

instance, the "Add" node in Fig. 17 consumes N_b samples from each input port and produces N_b output samples, then all three buffers have size N_b while they have size 1 when the blocking factor is unity. Moreover, the increment of N_g delays the response time although it does not decrease the throughput.

Another major obstacle to increase the blocking factor is related with feedback loops. Vector processing is restricted to the number of initial delays on the feedback loop. If the number is smaller than N_g, the vector processing capability cannot be fully utilized. For example, the scheduling result for a graph shown in Fig. 18 is "A B G C D H E F" when the blocking factor is 1. If blocking factor N_g becomes 5 then the scheduling becomes "5A 5B 5(GCDH) 5E 5F" in which nodes G, C, D and H are repeated five times sequentially. Therefore, a scheduling algorithm for SSDF should consider the graph topology to minimize the program code size.

In case feedback loops exist, strongly-connected components are first clustered into a strong component. A *strong component* of graph G is defined as a subgraph $F \subset G$ if for all pairs of nodes $u, v \in F$ there exist paths p_{uv}(from u to v) and p_{vu}(from v to u). This clustering is performed in a hierarchical fashion until the top graph does not have any feedback loop. Then, a valid schedule for an SSDF graph can be constructed using the SDF scheduling algorithms. Each node is scheduled by applying the global blocking factor N_g. For the SSDF graph in Fig. 18, the top graph consists of five nodes "A B (CDGH) E F" where nodes C, D, G and H are merged into a clustered-node. When blocking factor N_g is set to 5, a schedule for the top graph becomes "5A5B5(clustered-node)5E5F".

Next, the strong components are scheduled. The blocking factor depends on the number of initial delay samples on a feedback loop. Let $N_l(L)$ denote the maximum bound of the blocking factor on feedback loop L. Since feedback loops can be nested, a feedback loop with the largest maximum bound $N_l(L)$ should be selected first. Subsequently, feedback loops are selected in a descending order of $N_l(L)$. Scheduling of the clustered subgraph starts with a node that has many initial delay samples on its input ports and allows a large blocking factor. When a strong component "(CDHG)" is scheduled in the SSDF graph, actor G should be fired since it has an initial delay sample.

For a selected strong component, we schedule the internal nodes as follows, depending on the number of delays on the feedback loop.

Case 1: N_g is an integer multiple of $N_l(L)$. The scheduling order needs to be repeated $N_g/N_l(L)$ times using $N_b = N_l(L)$ for the internal nodes. In the example of Fig. 18, since $N_l(L) = 1$, $N_g = 5$, and $N_g/N_l(L)$ is an integer, schedule of "GCDH" is repeated five times. Moreover, the blocking factor for each node N_b is 1. Hence, the final schedule is "5A 5B 5(GCDH) 5E 5F".

Case 2: $N_g \leq N_l(L)$. Blocking factor $N_b = N_g$ is applied for all actors in the strong component. For example, if the number of delay samples increases to 5 in Fig. 18, then blocking factor $N_l(L)$ is 5 which is equal to N_g, and the schedule becomes "5A 5B (5G 5C 5D 5H) 5E 5F". Therefore, the blocking factor can be fully utilized.

Case 3: If $N_g > N_l(L)$ but not an integer multiple. One of two scheduling strategies can be applied:

1. The schedule for the strong component is repeated N_g times using $N_b = 1$ internally, which produces the smallest code at the cost of throughput.
2. The schedule is repeated with blocking factor $N_b = N_l(L)$, and then once more for the remainder to N_g. This improves throughput but also enlarges the code size.

   When $N_l(L) = 2$ by increasing the number of delay samples to 2, a valid schedule is "5(GCDH)" if the first strategy is followed or "2(2G 2C 2D 2H) GCDH" if the second strategy is followed. Consequently, the final schedule is either "5A5B 5(GCDH) 5E 5F" or "5A 5B 2(2G 2C 2D 2H) GCDH 5E 5F".

Although the SSDF model is proposed to allow large blocking factors to utilize vector processing of simple operations in a node, the scheduling algorithm for SSDF is also applicable to an SDF graph in which every node has an inline style code specification. Without the modification of the SDF actor, the blocking factor can be applied to the SDF graph and the SDF schedule. For instance, when block factor $N_g = 3$ is applied to Fig. 2, a valid schedule is "9A 9D 6B 12C". For the given schedule with the blocking factor, programs can be synthesized as shown in Fig. 5 where each loop value in the codes will be multiplied by blocking factor N_g (=3).

References

1. Ade, M., Lauwereins, R., Peperstraete, J.A.: Implementing DSP applications on heterogeneous targets using minimal size data buffers. In: Proceedings of RSP'96, pp. 166–172 (1996)
2. Bhattacharyya, S.S., Murthy, P.K., Lee, E.A.: Software Synthesis from Dataflow Graphs. Kluwer Academic Publisher, Norwell MA (1996)
3. Bhattacharyya, S.S., Murthy, P.K., Lee, E.A.: APGAN and RPMC: Complementary heuristics for translating DSP block diagrams into efficient software implementations. Journal of Design Automation for Embedded Systems **2**, 33–60 (1997)
4. Bilsen, G., Engles, M., Lauwereins, R., Peperstraete, J.A.: Cyclo-static dataflow. IEEE Trans. Signal Processing **44**, 397–408 (1996)

5. Buck, J.T., Ha, S., Lee, E.A., Messerschimitt, D.G.: Ptolemy: A framework for simulating and prototyping heterogeneous systems. Int. Journal of Computer Simulation, Special issue on Simulation Software Development **4**, 155–182 (1994)
6. Dennis, J.B.: Dataflow supercomputers. IEEE Computer Magazine **13** (1980)
7. Govindarajan, R., Gao, G., Desai, P.: Minimizing memory requirements in rate-optimal schedules. In: Proceedings of the International Conference on Application Specific Array Processors, pp. 75–86 (1993)
8. Hoang, P.D., Rabaey, J.M.: Scheduling of DSP programs onto multiprocessors for maximum throughput. IEEE Transactions on Signal Processing pp. 2225–2235 (1993)
9. Jung, H., Yang, H., Ha, S.: Optimized RTL code generation from coarse-grain dataflow specification for fast HW/SW cosynthesis. Journal of Signal Processing Systems **52**, 13–34 (2008)
10. Kim, J., Shin, T., Ha, S., Oh, H.: Resource minimized static mapping and dynamic scheduling of SDF graphs. In: ESTIMedia (2011)
11. Lauwereins, R., Engels, M., Peperstraete, J.A., Steegmans, E., Ginderdeuren, J.V.: GRAPE: A CASE tool for digital signal parallel processing. IEEE ASSP Magazine **7**, 32–43 (1990)
12. Lee, E.A., Messerschmitt, D.G.: Static scheduling of synchronous dataflow programs for digital signal processing. IEEE Transactions on Computer **C-36**, 24–35 (1987)
13. Oh, H., Ha, S.: Memory-optimized software synthesis from dataflow program graphs with large size data samples. EURASIP Journal on Applied Signal Processing **2003**, 514–529 (2003)
14. Oh, H., Ha, S.: Fractional rate dataflow model for memory efficient synthesis. Journal of VLSI Signal Processing **37**, 41–51 (2004)
15. Parhi, K.K., Chen, Y.: Signal flow graphs and data flow graphs. In: S.S. Bhattacharyya, E.F. Deprettere, R. Leupers, J. Takala (eds.) Handbook of Signal Processing Systems, second edn. Springer (2013)
16. Park, C., Chung, J., Ha, S.: Extended synchronous dataflow for efficient DSP system prototyping. Design Automation for Embedded Systems **3**, 295–322 (2002)
17. Pino, J., Ha, S., Lee, E.A., Buck, J.T.: Software synthesis for DSP using Ptolemy. Journal of VLSI Signal Processing **9**, 7–21 (1995)
18. Ritz, S., Pankert, M., Meyr, H.: High level software synthesis for signal processing systems. In: Proceedings of the International Conference on Application Specific Array Processors (1992)
19. Ritz, S., Willems, M., Meyr, H.: Scheduling for optimum data memory compaction in block diagram oriented software synthesis. In: Proceedings of the ICASSP 95 (1995)
20. S. Stuijk, T.B., Geilen, M.C.W., Corporaal, H.: Multiprocessor resource allocation for throughput-constrained synchronous dataflow graphs. In: DAC, pp. 777–782 (2007)
21. Stuijk, S., Geilen, M.C.W., Basten, T.: Exploring trade-offs in buffer requirements and throughput constraints for synchronous dataflow graphs. In: DAC, pp. 899–904 (2006)
22. Sung, W., Ha, S.: Memory efficient software synthesis using mixed coding style from dataflow graph. IEEE Transactions on VLSI Systems **8**, 522–526 (2000)
23. Woods, R.: Mapping decidable signal processing graphs into FPGA implementations. In: S.S. Bhattacharyya, E.F. Deprettere, R. Leupers, J. Takala (eds.) Handbook of Signal Processing Systems, second edn. Springer (2013)
24. Yang, H., Ha, S.: Pipelined data parallel task mapping/scheduling technique for MPSoC. In: DATE (Design Automation and Test in Europe) (2009)

Systolic Arrays

Yu Hen Hu and Sun-Yuan Kung

Abstract This chapter reviews the basic ideas of systolic array, its design methodologies, and historical development of various hardware implementations. Two modern applications, namely, motion estimation of video coding and wireless communication baseband processing are also discussed.

1 Introduction

Systolic array [2, 13, 15] is an on-chip multi-processor architecture proposed by H. T. Kung in late 1970s. It is proposed as an architectural solution to the anticipated on-chip communication bottleneck of modern very large scale integration (VLSI) technology.

A systolic array features a mesh-connected array of identical, simple processing elements (PE). According to H. T. Kung [13], "*In a systolic system, data flows from the computer memory in a rhythmic fashion, passing through many processing elements before it returns to memory, much as blood circulates to and from the heart*". As depicted in Fig. 1, a systolic array is often configured into a linear array, a two-dimensional rectangular mesh array, or sometimes, a two-dimensional hexagonal mesh array. In a systolic array, every PE is connected only to its nearest neighboring PEs through dedicated, buffered local bus. This localized inter-connects, and regular array configuration allow a systolic array to grow in size without incurring excessive on-chip global inter-connect delays due to long wires.

Y.H. Hu (✉)
Department of Electrical and Computer Engineering, University of Wisconsin, Madison, WI 53706-1691, USA
e-mail: hu@engr.wisc.edu

S.-Y. Kung
Department of Electrical Engineer, Princeton University, Princeton, NJ 08544, USA
e-mail: kung@princeton.edu

S.S. Bhattacharyya et al. (eds.), *Handbook of Signal Processing Systems*,
DOI 10.1007/978-1-4614-6859-2_34, © Springer Science+Business Media, LLC 2013

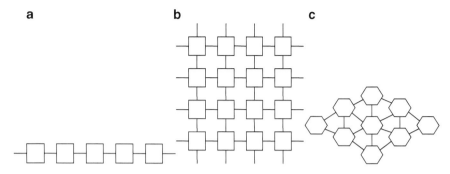

Fig. 1 Common configurations of systolic architecture: (**a**) linear array, (**b**) rectangular array, and (**c**) hexagonal array

Several key architectural concerns impacted on the development of systolic architecture [13]:

(a) *Simple and regular design*: In order to reduce design complexity, design cost, and to improve testability, fault-tolerance, it is argued that VLSI architecture should consist of simple modules (cores, PEs, etc.) organized in regular arrays.

(b) *Concurrency and communication*: Concurrent computing is essential to achieve high performance while conserving power. On-chip communication must be constrained to be local and regular to minimize excessive overhead due to long wire, long delay and high power consumption.

(c) *Balanced on-chip computation rate and on/off chip data input/output rate*: Moving data on/off chip remains to be a communication bottleneck of modern VLSI chips. A sensible architecture must balance the demand of on/off chip data I/O to maximize the utilization of the available computing resources.

Systolic array is proposed to implement *application specific* computing systems. Toward this goal, one must *map* the computing algorithm to a systolic array. This requirement stimulated two complementary research directions that have seen numerous significant and fruitful research results. The first research direction is to reformulate existing computing algorithms, or develop novel computing algorithms that can be mapped onto a systolic architecture to enjoy the benefit of systolic computing. The second research direction is to develop a systematic design methodology that would automate the process of algorithm mapping. In Sect. 2 of this chapter, we will provide a brief overview of these *systolic algorithms* that have been proposed. In Sect. 3, the formal design methodologies developed for automated systolic array mappings will be reviewed.

Systolic array computing was developed based on a globally synchronized, fine-grained, pipelined timing model. It requires a global clock distribution network free of clock skew to distribute the clock signal over the entire systolic array. Recognizing the technical challenge of developing large scale clock distribution network, Kung et al. [14–16] proposed a self-timed, data flow based *wavefront array processor architecture* that promises to alleviate the stringent timing constraint imposed by the global clock synchronization requirement. In Sect. 4, the wavefront array architecture and its related design methodology will be discussed.

These architectural features of systolic array have motivated numerous developments of research and commercial computing architectures. Notable examples include the WARP and iWARP project at CMU [1, 3, 7, 10]; Transputer™ of INMOS [8, 19, 22, 26]; and TMS 32040 DSP processor of Texas Instruments [23]. In Sect. 5 of this chapter, brief reviews of these systolic-array motivated computing architectures will be surveyed.

While the notion of systolic array was first proposed three decades ago, its impacts can be felt vividly today. Modern applications of the concept of systolic array can be found in field programmable gate array (FPGA) chip architectures, network-on-chip (NoC) mesh array multi-core architecture. Computation intensive special purpose architecture such as discrete cosine transform and block motion estimation algorithms in video coding standards, as well as the QR factorization for least square filtering in wireless communication standards have been incorporated in embedded chip designs. These latest real world applications of systolic array architecture will be discussed in Sect. 6.

2 Systolic Array Computing Algorithms

A systolic array exhibits characteristics of parallelism (pipelining), regularity, and local communication. A large number of signal processing algorithms, and numerical linear algebra algorithms can be implemented using systolic arrays.

2.1 Convolution Systolic Array

For example, consider a convolution of two sequences $\{x[n]\}$ and $\{h[n]\}$:

$$y[n] = \sum_{k=0}^{K-1} h[k] x[n-k] \ldots 0 \le n \le N-1. \tag{1}$$

A systolic array realization of this algorithm can be shown in Fig. 2 ($K = 4$). In Fig. 2a, the block diagram of the systolic array and the pattern of data movement are depicted. The block diagram of an individual processing element (PE) is illustrated in Fig. 2b where a shaded rectangle represents a buffer (delay element) that can be implemented with a register. The output $y[n]$ begins its evaluation at the upper left input with initial value 0. When it enters into each PE, the multiply-and-accumulate (MAC) operation

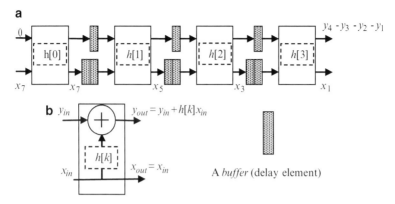

Fig. 2 (**a**) Fully pipelined systolic array for convolution and (**b**) internal architecture of a single processing element (PE)

$$y_{out} = y_{in} + h\,[k]\,x \qquad (2)$$

will be performed. The systolic array is of the same length as the sequence $\{h[k];\ 0 \leq k \leq K-1\}$ with each $h[k]$ resides in a register in each PE. The final result $\{y[n]\}$ appears at the upper right output port. Every other clock cycle, one output will be evaluated. The input $\{x[n]\}$ will be provided from the lower left input port. It will propagate toward the lower right output port without being modified ($x_{out} = x_{in}$). Along the way, it will be properly buffered to keep pace of the evaluation of the convolution.

2.2   Linear System Solver Systolic Array

Similar to above example, a systolic algorithm is often presented in the form of a high level block diagram of the systolic array configuration (e.g. Fig. 1) complemented with labels indicating data movement within the processor array, and a detailed block diagram explaining the operations performed within an individual PE.

Another example given in [13] is shown in Fig. 3. It consists of two systolic arrays for solving linear systems of equations. One triangular-configured systolic array is responsible for orthogonal triangulation of a matrix using QR factorization, and the other linear systolic array is responsible for solving a triangular linear system using back-substitution. A linear system of equations is represented as

$$\mathbf{Xb} = \mathbf{y}.$$

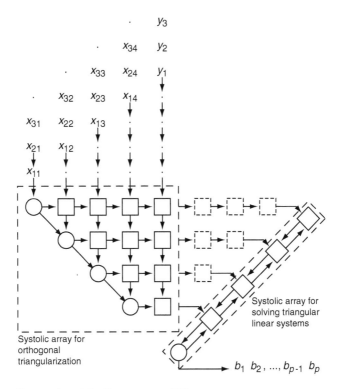

Fig. 3 Systolic array for solving linear systems [13]

Using a Jacobi's rotation method, the first column of the **X** matrix will enter the upper-left circular PE where an angle θ_k is evaluated such that

$$\begin{bmatrix} \cos \theta_k & -\sin \theta_k \\ \sin \theta_k & \cos \theta_k \end{bmatrix} \begin{bmatrix} x_{11}^{(k-1)} \\ x_{k,1} \end{bmatrix} = \begin{bmatrix} x_{11}^{(k)} \\ 0 \end{bmatrix}, k = 2, 3, \ldots \quad (3)$$

Clearly, in this circular PE, the operation to be performed will be

$$\theta_k = -\tan^{-1} \left(x_{k,1} \Big/ x_{11}^{(k-1)} \right). \quad (4)$$

This θ_k then will be propagated to the square PEs to the right of the upper left circular PE to perform rotation operations

$$\begin{bmatrix} \cos \theta_k & -\sin \theta_k \\ \sin \theta_k & \cos \theta_k \end{bmatrix} \begin{bmatrix} x_{12}^{(k-1)} & \cdots & x_{1N}^{(k-1)} & y_1^{(k-1)} \\ x_{k2}^{(k-1)} & \cdots & x_{k,N}^{(k-1)} & y_k^{(k-1)} \end{bmatrix} = \begin{bmatrix} x_{12}^{(k)} & \cdots & x_{1N}^{(k)} & y_1^{(k)} \\ x_{k2}^{(k)} & \cdots & x_{k,N}^{(k)} & y_k^{(k)} \end{bmatrix}. \quad (5)$$

The second row of the results of above equation will be propagated downward to the next row in the triangular systolic array repeating what has been performed on the first row of that array. After $N-1$ iterations, the results will be ready within the triangular array. Note that during this rotation process, the right hand size of the linear systems of equations $\mathbf{y}$ is also subject to the same rotation operation. Equivalently, these operations taken places at the triangular systolic array amount to pre-multiply the $\mathbf{X}$ matrix with a unitary matrix $\mathbf{Q}$ such that $\mathbf{QX} = \mathbf{U}$ is an upper triangular matrix, and $\mathbf{z} = \mathbf{Qy}$. This yields an upper triangular system $\mathbf{Ub} = \mathbf{z}$.

To solve this triangular system of equations, a *back-propagation* solver algorithm is used. Specifically, given

$$\mathbf{Ub} = \begin{bmatrix} u_{11} & u_{12} & \cdots & u_{1N} \\ 0 & u_{22} & & u_{2N} \\ \vdots & \ddots & \ddots & \vdots \\ 0 & \cdots & 0 & u_{NN} \end{bmatrix} \begin{bmatrix} b_1 \\ b_2 \\ \vdots \\ b_N \end{bmatrix} = \begin{bmatrix} z_1 \\ z_2 \\ \vdots \\ z_N \end{bmatrix} = \mathbf{z} \tag{6}$$

the algorithm begins by solving $u_{NN}b_N = z_N$ for $b_N = z_N/u_{NN}$. In the systolic array, u_{NN} are fed from the last (lower right corner) circular PE to the circular PE of the linear array to the right. The computed b_N then will be forwarded to the next square processor in the upper right direction to be substituted back into the next equation of

$$u_{N-1,N-1} \times b_{N-1} + u_{N-1,N} \times b_N = z_{N-1}. \tag{7}$$

In the rectangular PE, the operation performed will be $z_{N-1} - u_{N-1,N} \cdot b_N$. The result then will be fed back to the circular PE to compute $b_{N-1} = (z_{N-1} - u_{N-1,N} \cdot b_N)/u_{N-1,N-1}$.

2.3   Sorting Systolic Arrays

Given a sequence $\{x[n]; 0 \leq n \leq N-1\}$, the sorting algorithm will output a sequence $\{m[n]; 0 \leq n \leq N-1\}$ that is a permutation of the ordering of $\{x[n]\}$ such that $m[n] \geq m[n+1]$. There are many sorting algorithms available. A systolic array that implements the bubble sort algorithm is presented in Fig. 4.

Each PE in this systolic array will receive data a, b from both left and right sides. These two inputs will be compared and the maximum of the two will be output to the right side buffer while the minimum of the two output to the left side buffer. The input will be loaded into the upper buffer according to specific schedule. The left most input will be fixed at $-\infty$. It has been shown that systolic arrays of insertion sort and selection sort can also be derived using similar approach [15].

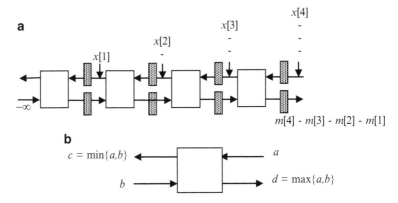

Fig. 4 (**a**) A bubble sort array and (**b**) operation performed within a PE

3 Formal Systolic Array Design Methodology

Due to the regular structure, localized interconnect, and pipelined operations, a systematic systolic array design methodology has been proposed that greatly simplified the systolic array design complexity and opened new avenue to seek optimized systolic architecture. In order to introduce the systematic systolic array design methodology in this section, a few important representations will be briefly surveyed.

3.1 Loop Representation, Regular Iterative Algorithm (RIA), and Index Space

Algorithms that are suitable for systolic array implementation must exhibit high degree of regularity, and require intensive computation. Such an algorithm often can be represented by a set of nested Do loops of the following general format:

```
L₁:  DO i₁ = p₁ , q₁
L₂:    DO i₂ = p₂ , q₂
 ⋮        ⋮
Lₘ:            DO iₘ = pₘ, qₘ
              H(i₁, i₂, ..., iₘ)
              Enddo
               ⋮
            Enddo
          Enddo
```

where $\{L_m\}$ specify the *level* of the loop nest, $\{i_m\}$ are *loop indices*, and $[i_1, i_2, \ldots, i_m]^T$ is a $m \times 1$ *index vector*, representing an index point in a m-dimensional

Fig. 5 Convolution

```
For n = 0 to N-1,
    y[n] = 0;
    For k = 0 to K-1,
        y[n] = y[n]+h[k]*x[n-k];
    end
end
```

lattice. $\{p_m, q_m\}$ are *loop bounds* of the mth loop nest. $H(i_1, i_2, \ldots, i_m)$ is the *loop body* and may have different *granularity*. That is, the loop body could represent bit-level operations, word-level program statements, or sub-routine level procedures. Whatever the granularity is, it is assumed that the loop body is to be executed in a single PE. For convenience, the execution time of a loop body in a PE will be assumed to be one clock cycle in this chapter. In other words, it is assumed that the execution of a loop body within a PE cannot be interrupted. All data needed to execute the loop body must be available before the execution of loop body can start; and none of the output will be available until the execution of the entire loop body is completed.

If the loop bounds are all constant, the set of indices corresponding to all iterations form a rectangular parallelepiped. In general, the loop bounds are linear (affine) function with integer coefficients of outer loop indices and can be represented with two inequalities:

$$\mathbf{p_0} \leq \mathbf{P\,i} \text{ and } \mathbf{P\,i} \leq \mathbf{q_0} \tag{8}$$

where $\mathbf{p_0}$ and $\mathbf{q_0}$ are constant integer-valued vectors, and $\mathbf{P}$, $\mathbf{Q}$ respectively, are integer-valued upper triangular coefficient matrices. If $\mathbf{P} = \mathbf{Q}$, then the corresponding loop nest can be *transformed* in the index space such that the transformed algorithm has constant iteration bounds. Such a nested loop is called a *regular* nested loop. If an algorithm is formulated to contain only regular nested loops, it is called a *regular iterative algorithm* (*RIA*).

Consider the convolution algorithm described in Eq. (1) of this chapter. The mathematical formula can be conveniently expressed with a 2-level loop nest as shown in Fig. 5.

In this formulation, n and k are *loop indices* having *loop bounds* $(0, N-1)$ and $(0, K-1)$ respectively. The *loop body* $H(\mathbf{i})$ consists of a single statement

$$y[n] = y[n] + h[k]x[n-k].$$

Note that

$$\mathbf{i} = \begin{bmatrix} n \\ k \end{bmatrix}; \quad \mathbf{p_0} = \begin{bmatrix} 0 \\ 0 \end{bmatrix} \leq \begin{bmatrix} 1 & 0 \\ 0 & 1 \end{bmatrix} \begin{bmatrix} n \\ k \end{bmatrix} = \mathbf{Pi} = \mathbf{Qi} \leq \begin{bmatrix} N-1 \\ K-1 \end{bmatrix} = \mathbf{q_0}. \tag{9}$$

Hence, this is a RIA.

3.2   *Localized and Single Assignment Algorithm Formulation*

As demonstrated in Sect. 2, a systolic array is a parallel, distributed computing platform where each PE will execute identical operations. By connecting PEs with specific configurations, and providing input data at right timing, a systolic array will be able to perform data intensive computations in a rhythmic, synchronous fashion. Therefore, to implement a given algorithm on a systolic array, its formulation may need to be adjusted. Specifically, since computation takes place at physically separated PEs, data movement in a systolic algorithm must be explicitly specified. Moreover, unnecessary algorithm formulation restrictions that may impede the exploitation of inherent parallelism must be removed.

A closer examination of Fig. 5 reveals two potential problems according to above arguments: (i) The variables $y[n]$, $h[k]$, $x[n-k]$ are one-dimensional arrays while each index vector $\mathbf{i} = (n, k)$ resides in a two-dimensional space. (ii) The memory address locations $y[n]$ will be repeatedly assigned with new values during each k loop K times before the final result is evaluated.

Having one-dimensional variable arrays in a two dimensional index space implies that the same input data will be needed when executing the loop body $H(\mathbf{i})$ at different iterations (index points). In a systolic array, it is likely that $H(\mathbf{i})$ and $H(\mathbf{j})$, $\mathbf{i} \neq \mathbf{j}$ may be executed at different PEs. As such, how these variables may be distributed to different index points where they are needed should be explicitly specified in the algorithm. Furthermore, a design philosophy that dominates the development of systolic array is to discourage on-chip global inter-connect due to many potential drawbacks. Hence, the data movement would be restricted to *local* communication. Namely passing the data from one PE to one or more of its nearest neighboring PEs in a systolic array. This restriction may be imposed by limiting the propagation of such a global variable from one index point to its nearest neighboring index points. For this purpose, we make the following modification of Fig. 5:

$$h[k] \rightarrow h1[n, k] \text{ such that } h1[0, k] = h[k], h1[n, k] = h1[n-1, k]$$
$$x[n] \rightarrow x1[n, k] \text{ such that } x1[n, 0] = x[n], x1[n, k] = x1[n-1, k-1].$$

Note that the equations for $h1$ and $x1$ are chosen based on the fact that $h[k]$ will be made available for the entire ranges of index n, and $x[n-k]$ will be made available to all (n', k') such that $n'-k' = n-k$. An algorithm with all its variables passing from one iteration (index point) to its neighboring index point is called a (variable) *localized* algorithm.

The repeated assignment of different intermediate results of $y[n]$ into the same memory location will cause an unwanted *output dependence* relation in the algorithm formulation. Output dependence is a type of false data dependence that would impede potential parallel execution of a given algorithm. The output dependence can be removed if the algorithm is formulated to obey a *single assignment* rule. That is, every memory address (variable name) will be assigned to a new value only once during the execution of an algorithm. To remedy, one would create new

Fig. 6 Convolution
(localized, single assignment
version)

```
h1(0,k)=h(k), k=0, ... ,K-1
x1(n,0)=x(n), n=0, ... ,N-1
y1(n,-1)=0, n=0,1, ... ,N-1
n=0,1,2,...,N-1 and k=0,...,K-1
      y1(n,k)=y1(n,k-1)+h1(n,k)*x1(n,k)
      h1(n,k)=h1(n-1,k)
      x1(n,k)=x1(n-1,k-1)
y(n)=y1(n,K), n=0,1,2,..., N-1
```

memory locations to be assigned to these intermediate results by expanding the one dimensional array $\{y[n]\}$ into a two-dimensional array $\{y1[n, k]\}$:

$$y[n] \rightarrow y1[n, k] \text{ such that } y1[n, -1] = 0, y1[n, k] = y1[n, k-1] + h1[n,k]x1[n,k]$$

where the previously localized variables $h1$ and $x1$ are used. With above modifications, Fig. 5 is reformulated as shown in Fig. 6.

3.3   Data Dependence and Dependence Graph

An iteration H(**j**) is *dependent* on iteration H(**i**) if H(**j**) will read from a memory location whose value is last written during execution of iteration H(**i**). The corresponding *dependence vector* **d** is defined as:

$$\mathbf{d} = \mathbf{j} - \mathbf{i}.$$

A matrix **D** consisting of all dependence vectors of an algorithm is called a *dependence matrix*. This *inter-iteration dependence relation* imposes a partial ordering on the execution of the iterative loop nest. From Fig. 6, three dependence vectors can be derived:

$$\mathbf{d}_1 = \begin{bmatrix} n \\ k \end{bmatrix} - \begin{bmatrix} n \\ k-1 \end{bmatrix} = \begin{bmatrix} 0 \\ 1 \end{bmatrix}; \quad \mathbf{d}_2 = \begin{bmatrix} n \\ k \end{bmatrix} - \begin{bmatrix} n-1 \\ k \end{bmatrix} = \begin{bmatrix} 1 \\ 0 \end{bmatrix};$$

$$\mathbf{d}_3 = \begin{bmatrix} n \\ k \end{bmatrix} - \begin{bmatrix} n-1 \\ k-1 \end{bmatrix} = \begin{bmatrix} 1 \\ 1 \end{bmatrix}. \tag{10}$$

In the index space, a lattice point whose coordinates fall within the range of the loop bounds represents the execution of the loop body of the particular loop index values. The dependence vectors may be represented by directed arcs starting from the iteration that produces the data to the iteration where the data is needed. Together with these index points and directed arcs, one has a *dependence graph* (DG) representing the computation tasks required of a localized RIA. The corresponding DG of the convolution algorithm is depicted in Fig. 7 for $K = 5$ and $N = 7$.

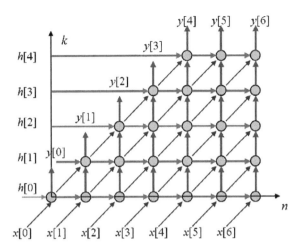

Fig. 7 Localized dependence graph of convolution algorithm

The dependence graph in Fig. 7 is *shift-invariant* in that the dependence vector structure is identical of each (circled) lattice point of the iteration space. This regularity and modularity is the key feature of a systolic computing algorithm that lends itself for efficient systolic array implementation.

Due to the shift invariant nature, a DG of a localized RIA algorithm can be conveniently represented by the set of indices $\{\mathbf{i}; \mathbf{p}_0 \leq \mathbf{Pi} \leq \mathbf{q}_0\}$ and the dependence vectors $\mathbf{D}$ at each index point.

3.4 Mapping an Algorithm to a Systolic Array

A *schedule* $S : \mathbf{i} \rightarrow t(\mathbf{i}) \in \mathbb{Z}^+$ is a mapping from each index point $\mathbf{i}$ in the index space R to a positive integer $t(\mathbf{i})$ which dictates *when* this iteration is to be executed. An *assignment* A: $\mathbf{i} \rightarrow p(\mathbf{i})$ is a mapping from each index point $\mathbf{i}$ onto a PE index $p(\mathbf{i})$ *where* the corresponding iteration will be executed. Given the dependence graph of a given algorithm, the development of a systolic array implementation amounts to find a mapping of each index point $\mathbf{i}$ in the DG onto $(p(\mathbf{i}), t(\mathbf{i}))$.

Toward this goal, two fundamental constraints will be discussed. First, it is assumed that each PE can only execute one task (loop body) at a time. As such, a *resource constraint* must be observed:

Resource Constraints

$$\text{if } t(\mathbf{i}) = t(\mathbf{j}), \mathbf{i} \neq \mathbf{j}, \text{then } p(\mathbf{i}) \neq p(\mathbf{j}); \text{ and if } p(\mathbf{i}) = p(\mathbf{j}), \mathbf{i} \neq \mathbf{j}, \text{then } t(\mathbf{i}) \neq t(\mathbf{j}).$$
(11)

In addition, the data dependence also imposes a partial ordering of schedule. This *data dependence constraint* can be summarized as follows:

Data Dependence Constraint

If index $\mathbf{j}$ can be reached from index $\mathbf{i}$ by following the path consisting of one or more dependence vectors, then H($\mathbf{j}$) should be scheduled after H($\mathbf{i}$). That is, if there exists a vector $\mathbf{m}$ consisting of non-negative integers such that

$$\text{if } \mathbf{j} = \mathbf{i} + \mathbf{Dm}, \text{ then } s(\mathbf{j}) > s(\mathbf{i}) \tag{12}$$

where $\mathbf{D}$ is the dependence matrix.

Since a systolic array often assumes a one or two dimensional regular configuration (cf. Fig. 1), the PE index p($\mathbf{i}$) can be associated with the lattice point in a *PE index space* just as each loop body in a loop nest is associated with an index point in the DG. To ensure the resulting systolic array features local interprocessor communication, the localized dependence vectors in the DG should not require global communication after the PE assignment $\mathbf{i} \rightarrow$ p($\mathbf{i}$). A somewhat restrictive constraint to enforce this requirement would be

Local Mapping Constraint

If $\mathbf{j} - \mathbf{i} = \mathbf{d}_k$ (a dependence vector), then

$$\|p(\mathbf{j}) - p(\mathbf{i})\|_1 \leq \|\mathbf{d}_k\|_1. \tag{13}$$

A number of performance metrics may be defined to compare the merits of different systolic array implementations. These include

Total Computing Time:

$$T_C = \max_{\mathbf{i}, \mathbf{j} \in DG} (t(\mathbf{i}) - t(\mathbf{j})). \tag{14}$$

PE Utilization:

$$U_{PE} = N_{DG} / (T_C \times N_{PE}) \tag{15}$$

where N_{DG} is the number of index points in the *DG*, and N_{PE} is the number of PEs in the systolic array.

Now we are ready to formally state the systolic array mapping and scheduling problem:

Systolic Array Mapping and Scheduling Problem

Given a localized, shift invariant DG, and a systolic array configuration, find a PE assignment mapping p($\mathbf{i}$), and a schedule t($\mathbf{i}$) such that the performance is optimized, namely, the total computing time T_C is minimized, and the PE utilization U_{PE} is maximized; subject to (i) the resource constraint, (ii) the data dependence constraint, and (iii) the local mapping constraints.

Thus the systolic array implementation is formulated as a discrete constrained optimization problem. By fully exploiting of the regular (shift invariant) structure of both the DG and the systolic array, this problem can be further simplified.

3.5   *Linear Schedule and Assignment*

A *linear schedule* is an integer-valued scheduling vector $\mathbf{s}$ in the index space such that

$$t(\mathbf{i}) = \mathbf{s}^T \mathbf{i} + t_0 \in \mathbb{Z}^+ \tag{16}$$

where t_0 is a constant integer. The data dependence constraint stipulates that

$$\mathbf{s}^T \mathbf{d} > 0 \text{ for any dependence vector } \mathbf{d}. \tag{17}$$

Clearly, all iterations that reside on a hyper-plane perpendicular to $\mathbf{s}$, called *equi-temporal hyperplane* must be executed in parallel at different PEs. The equi-temporal hyperplane is defined as $Q = \{\mathbf{i} \mid \mathbf{s}^T \mathbf{i} = t(\mathbf{i}) - t_0, \mathbf{i} \in DG\}$. According to the resource constraint, the maximum number of index points in Q determines the minimum size (number of PEs) of the systolic array.

Assume that the PE index space is a $m-1$ dimensional subspace in the iteration index space. Then the assignment of individual iterations $\mathbf{i}$ to a PE index $p(\mathbf{i})$ can be realized by projecting $\mathbf{i}$ onto the PE subspace along an integer-valued assignment vector $\mathbf{a}$. Define a $m \times (m-1)$ integer-valued *PE basis* matrix $\mathbf{P}$ such that $\mathbf{P}^T \mathbf{a} = \mathbf{0}$, then a *linear PE assignment* can be obtained via an affine transformation

$$p(\mathbf{i}) = \mathbf{P}^T \mathbf{i} + \mathbf{p}_0. \tag{18}$$

Combining Eqs. (16) and (18), one has a node mapping procedure:

$$Node\ mapping : \begin{bmatrix} \mathbf{s}^T \\ \mathbf{P}^T \end{bmatrix} \mathbf{i} = \begin{bmatrix} t(\mathbf{i}) \\ p(\mathbf{i}) \end{bmatrix}. \tag{19}$$

The node mapping procedure can also be extended to a subset of nodes where external data input and output take places. The same node mapping procedure will indicate where and when these external data I/O will take place in the systolic array. This special mapping procedure is also known as *I/O mapping*.

Different PEs in the systolic array are inter-connected by local buses. These buses are implemented based on the need of passing data from an index point (iteration) to another as specified by the dependence vectors. Hence, the orientation of these buses as well as buffers on them can be determined also using $\mathbf{P}$ and $\mathbf{s}$:

$$Arc\ mapping : \begin{bmatrix} \mathbf{s}^T \\ \mathbf{P}^T \end{bmatrix} \mathbf{D} = \begin{bmatrix} \tau \\ \mathbf{e} \end{bmatrix} \tag{20}$$

where τ is the number of first-in-first-out buffers required on each local bus, and $\mathbf{e}$ is the orientation of the local bus within the PE index space.

Consider two iterations $\mathbf{i}, \mathbf{j} \in DG$, $\mathbf{i} \neq \mathbf{j}$. If $p(\mathbf{i}) = p(\mathbf{j})$, it implies that

$$\mathbf{0} = p(\mathbf{i}) - p(\mathbf{j}) = \mathbf{P}^T(\mathbf{i} - \mathbf{j}) \Rightarrow \mathbf{i} - \mathbf{j} = k\mathbf{a}. \tag{21}$$

The resource constraint (cf. Eq. (11)) stipulates that if $p(\mathbf{i}) = p(\mathbf{j})$ $\mathbf{i} \neq \mathbf{j}$, then $t(\mathbf{i}) \neq t(\mathbf{j})$. Hence,

$$t(\mathbf{i}) - t(\mathbf{j}) = \mathbf{s}^T(\mathbf{i} - \mathbf{j}) = k \times \mathbf{s}^T\mathbf{a} \neq 0. \tag{22}$$

Example 1   Let us now use the convolution algorithm in Fig. 6 and its corresponding DG in Fig. 7 as an example and set $\mathbf{a}^T = [1\ 0]$, and $\mathbf{s}^T = [1\ 1]$. It is easy to derive the PE basis matrix $\mathbf{P}^T = [0\ 1]$. Hence, the node mapping becomes

$$\begin{bmatrix} \mathbf{s}^T \\ \mathbf{P}^T \end{bmatrix} \begin{bmatrix} n \\ k \end{bmatrix} = \begin{bmatrix} 1 & 1 \\ 0 & 1 \end{bmatrix} \begin{bmatrix} n \\ k \end{bmatrix} = \begin{bmatrix} n+k \\ k \end{bmatrix} = \begin{bmatrix} t(\mathbf{i}) \\ p(\mathbf{i}) \end{bmatrix}, \dots 0 \leq n \leq 6, 0 \leq k \leq \min(4, n).$$
$$\tag{23}$$

This implies every (n, k) iterations will be executed at PE #k of the systolic array and the scheduled execution time slot is $n + k$. Next, the arc mapping can be found as:

$$\begin{bmatrix} \mathbf{s}^T \\ \mathbf{P}^T \end{bmatrix} \mathbf{D} = \begin{bmatrix} 1 & 1 \\ 0 & 1 \end{bmatrix} \begin{bmatrix} 1 & 0 & 1 \\ 0 & 1 & 1 \end{bmatrix} = \begin{bmatrix} 1 & 1 & 2 \\ 0 & 1 & 1 \end{bmatrix}. \tag{24}$$

The second row of the right-hand-side (RHS) of Eq. (24) indicates that there are three local buses. The first one has an entry "0" implies that this is a bus that starts and ends at the same PE. The other two have an entry "1", indicating that they are local buses in the increasing k direction. The first row of the RHS gives the number of registers required on each local bus to ensure the proper execution ordering is obeyed. Thus, the first two buses have a single buffer, while the third bus has two buffers.

Note that the external data input $\{x[n]\}$ are fed into the DG at $\{(n, 0); 0 \leq n \leq 6\}$, and the final output $\{y[n]\}$ will be available at $\{(n, K); 0 \leq n \leq 6\}$ where $K = 4$. Thus, through I/O mapping, one has

$$\begin{bmatrix} \mathbf{s}^T \\ \mathbf{P}^T \end{bmatrix} \begin{bmatrix} n & n \\ 0 & K \end{bmatrix} = \begin{bmatrix} 1 & 1 \\ 0 & 1 \end{bmatrix} \begin{bmatrix} n & n \\ 0 & K \end{bmatrix} = \begin{bmatrix} n & n+K \\ 0 & K \end{bmatrix}. \tag{25}$$

This implies that the input $x[n]$ will be fed into the #0 PE of the systolic array at the nth clock cycle; and the output $y[n]$ will be available at the #K PE at the $(n + \mathrm{K})^{th}$ clock cycle. The node mapping, arc mapping and I/O mapping are summarized in Fig. 8.

At the left of Fig. 8, the original DG is overlaid with the equi-temporal hyperplane, which is depicted by parallel, dotted lines. To the right of Fig. 8 is the systolic

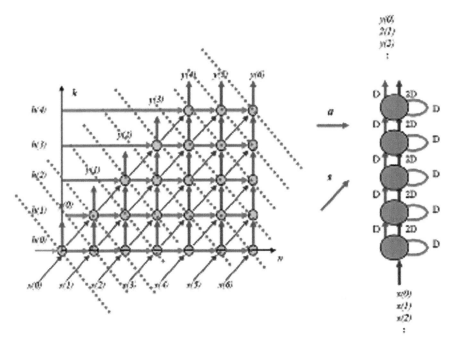

Fig. 8 Linear assignment and schedule of convolution algorithm

array, its local buses, and the number of buffers (delays) on each bus. This array is a more abstract version of what is presented in Fig. 2.

4 Wavefront Array Processors

4.1 Synchronous Versus Asynchronous Global On-Chip Communication

The original systolic array architecture adopted a globally synchronous communication model. It is assumed that a global clock signal is available to synchronize the state transition of every storage elements on chip. However, as predicted by the Moore's law, in modern integrated circuits, transistor sizes continue to shrink, and the number of transistors on a chip continues to increase. These trends make it more and more difficult to implement globally synchronized clocking scheme on chip. On the one hand, the wiring propagation delay does not scale down as transistor feature sizes reduce. As such, the signal propagation delay becomes very prominent compared to logic gate propagation delay. As on-chip clock frequency exceeds giga-hertz threshold, adverse impacts of clock skew become more difficult to compensate. On the other hand, as the number of on-chip transistors increases, so does the complexity and size of on-chip clock distribution network. The power consumption required to distribute giga-hertz clock signal synchronously over entire chip becomes too large to be practical.

In view of the potential difficulties in realizing a globally synchronous clocking scheme as required by the original systolic array design, a asynchronous array processor, known as *wavefront array processor* has been proposed.

4.2   Wavefront Array Processor Architecture

According to [14, 16], a wavefront array is a computing network with the following features:

- *Self-timed, data-driven computation*: No global clock is needed, as the computation is self-timed.
- *Regularity, modularity and local interconnection*: The array should consist of modular processing units with regular and (spatially) local interconnections.
- *Programmability in wavefront language or data flow graph (DFG)*: Computing algorithms implemented on a wavefront array processor may be represented with a data flow graph. Computation activities will propagate through the processor array as if a series of wavefronts propagating through the surface of water.
- *Pipelinability with linear-rate speed-up*: A wavefront array should exhibit a linear-rate speed-up. With M PEs, a wavefornt array promises to achieve an $O(M)$ speed-up in terms of processing rates.

The major distinction between the wavefront array the systolic array is that there is no global timing reference in the wavefront array. In the wavefront architecture, the information transfer is by mutual agreements between a PE and its immediate neighbors using, say, an asynchronous hand-shaking protocol [4, 16].

4.3   Mapping Algorithms to Wavefront Arrays

In general, there are three formal methodologies for the derivation of wavefront arrays [15]:

(1) Map a localized dependence graph directly to a data flow graph (DFG). Here a DFG is adopted as a formal abstract model for wavefront arrays. A systematical procedure can be used to map a dependence graph (DG) to a DFG.
(2) Convert an signal flow graph into a DFG (and hence a wavefront array), by properly imposing several key data flow hardware elements.
(3) Trace the computational wavefronts and pipeline the fronts through the processor array. This will be elaborated below.

The notion of computational wavefronts offers a very simple way to design wavefront computing, which consists of three steps:

(1) Decompose an algorithm into an orderly sequence of recursions;
(2) Map the recursions onto corresponding computational wavefronts in the array;
(3) Pipeline the wavefronts successively through the processor array.

4.4   Example: Wavefront Processing for Matrix Multiplication

The notion of computational wavefronts may be better illustrated by an example of the matrix multiplication algorithm where $\mathbf{A}$, $\mathbf{B}$, and $\mathbf{C}$, are assumed to be $N \times N$ matrices:

$$\mathbf{C} = \mathbf{A} \times \mathbf{B}. \tag{26}$$

The topology of the matrix multiplication algorithm can be mapped naturally onto the square, orthogonal $N \times N$ matrix array as depicted in Fig. 9. The computing network serves as a (data) wave-propagating medium. To be precise, let us examine the computational wavefront for the first recursion in matrix multiplication. Suppose that the registers of all the PEs are initially set to zero, that is, $C_{ij}(0) = 0$. The elements of $\mathbf{A}$ are stored in the memory modules to the left (in columns) and those of $\mathbf{B}$ in the memory modules on the top (in rows). The process starts with PE $(1, 1)$, which computes:

$$C_{11}(1) = C_{11}(0) + a_{11}b_{11}.$$

The computational activity then propagates to the neighboring PEs $(1, 2)$ and $(2, I)$, which execute:

$$C_{12}(1) = C_{12}(0) + a_{11}b_{12} \quad \text{and} \quad C_{21}(1) = C_{21}(0) + a_{21}b_{11}.$$

The next front of activity will be at PEs $(3,1)$, $(2,2)$, and $(1,3)$, thus creating a computation wavefront traveling down the processor array. This computational wavefront is similar to optical wavefronts (they both obey Huygens' principle), since each processor acts as a secondary source and is responsible for the propagation of the wavefront. It may be noted that wave propagation implies localized data flow. Once the wavefront sweeps through all the cells, the first recursion is over. As the first wave propagates, we can execute an identical second recursion in parallel by pipelining a second wavefront immediately after the first one. For example, the $(1, 1)$ processor executes

$$C_{11}(2) = C_{11}(1) + a_{12}b_{21} = a_{11}b_{11} + a_{12}b_{21}.$$

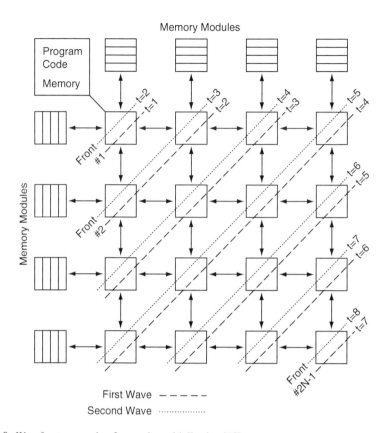

Fig. 9 Wavefront processing for matrix multiplication [15]

Likewise each processor (i, j) will execute (from $k = 1$ to N)

$$C_{ij}(k) = C_{ij}(k+1) + a_{ik}b_{kj} = a_{i1}b_{1j} + a_{i2}b_{2j} + \ldots + a_{ik}b_{kj}$$

and so on.

In the wavefront processing, the pipelining technique is feasible because the wavefronts of two successive recursions would never intersect. The processors executing the recursions at any given instant are different, thus any contention problems are avoided.

Note that the successive pipelining of the wavefronts furnishes additional dimension of concurrency. The separated roles of pipeline and parallel processing also become evident when we carefully inspect how parallel processing computational wavefronts are pipelined successively through the processor arrays. Generally speaking, parallel processing activities always occur at the PEs on the same front, whereas pipelining activities are perpendicular to the fronts. With reference to the wavefront processing example in Fig. 9, PEs on the anti-diagonals of the wavefront array execute in parallel, since each of the PEs process information independently. On the other hand, pipeline processing takes place along the diagonal direction, in which the computational wavefronts are piped.

In this example, the wavefront array consists of $N \times N$ processing elements with regular and local interconnections. Figure 9 shows the first 4×4 processing elements of the array. The computing network serves as a (data) wave propagating medium. Hence the hardware has to support pipelining the computational wavefronts as fast as resource and data availability allow. The (average) time interval T between two separate wavefronts is determined by the availability of the operands and operators.

4.5   Comparison of Wavefront Arrays against Systolic Arrays

The main difference between a wavefront array processor and a systolic array lies in hardware design, e.g., on clock and buffer arrangements, architectural expandability, pipeline efficiency, programmability in a high-level language, and capability to cope with time uncertainties in fault-tolerant designs.

As to the synchronization aspect, the clocking scheme is a critical factor for large-scale array systems, and global synchronization often incurs severe hardware design burdens in terms of clock skew. The synchronization time delay in systolic arrays is primarily due to the clock skew, which can vary drastically depending on the size of the array. On the other hand, in the data-driven wavefront array, a global timing reference is not required, and thus local synchronization suffices. The asynchronous data-driven model, however, incurs fixed time delay and hardware overhead due to hand-shaking.

From the perspective of pipelining rate, the data-driven computing in the wavefront array may improve the pipelinability. This becomes especially helpful in the case where variable processing times are used in individual PEs. A simulation study on a recursive least squares minimization computation also reports a speedup by a factor of almost two, in favor of the wavefront array over a globally clocked systolic array [4].

In general, a systolic array is useful when the PEs are simple primitive modules, since the handshaking hardware in a wavefront array would represent a non-negligible overhead for such applications. On the other hand, a wavefront array is more applicable when the modules of the PEs are more complex (such as floating-point multiply-and-add), when synchronization of a large array becomes impractical or when a reliable computing environment (such as fault tolerance) is essential.

5 Hardware Implementations of Systolic Array

5.1 Warp and iWARP

Warp [1] is a prototype linear systolic array processor developed at CMU in mid-1980s. As illustrated in Fig. 10, the Warp array contains 10 identical *Warp cells* inter-connected as a linear array. It is designed as an attach processor to a host processor through an interface unit. Each Warp cell has three inter-cell communication links: one address (Adr) link and two data links (X and Y). They are connected to nearest neighboring cells or the interface unit. Each cell contains two floating-point units (one for multiply and one for addition) with corresponding register files, two local memory banks (2K words each with 32 bits/word) for resident and temporary data, each communication link also has a 512 words buffer queue. All these function units are inter-connected via a cross-bar switch for intra-cell communication.

The Warp cell is micro-programmed with horizontal micro-code. Although all cells will execute the same cell program, broadcasting micro-code to all cells is not practical and would violate the basic principle of localized communication.

A noticeable feature of the WARP processor array is that its inter-cell communication is *asynchronous*. It is argued [1] that the synchronous, fine-grained inter-PE communication schemes of the original systolic array is too restrictive and is not suitable for practical implementations. Instead, a hard-ware assisted run-time flow control scheme together with a relatively large queue size would allow more efficient inter-cell communication without incurring excessive overheads.

The Warp array uses a specially designed programming language called "W2". It explicitly supports communication primitives such as "receive" and "send" to transfer data between adjacent cells. The program execution at a cell will stall if either the send or receive statement cannot be realized due to an empty receiving queue (nothing to receive from) or a full sent queue (nowhere to send to). Thus, the

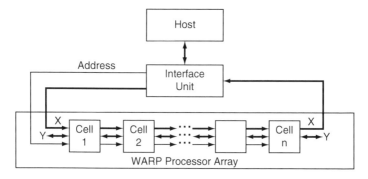

Fig. 10 Warp system overview [1]

Fig. 11 Photograph of iWarp chip [3]

programmer bears the responsibility of writing a deadlock free parallel program to run on the Warp processor array.

The performance of the Warp processor array is reported as hundreds of times faster than running the same type of algorithm in a VAX 11/780, a popular minicomputer at the time of Warp development.

The development of the Warp processor array is significant in that it is the first hardware systolic array implementation. Lessons learned from this project also motivated the development of iWarp.

The iWarp project [3, 7] was a follow-up project of WARP and started in 1988. The purpose of this project is to investigate issues involved in building and using high performance computer systems with powerful communication support. The project led to the construction of the iWarp machines, jointly developed by Carnegie Mellon University and Intel Corporation.

As shown in Fig. 11, the basic building block of the iWarp system is a full custom VSLI component integrating a LIW (long instruction word) microprocessor, a network interface and a switching node into one single chip of 1.2 × 1.2 cm silicon. The iWarp cell consists of a computation agent, a communication agent, and a local memory. The computation agent includes a 32-bit microprocessor with 96-bit wide instruction words, an integer/logic unit, a floating-point multiplier, and a floating-point adder. It runs at a clock speed of 20 MHz. The communication agent has four separate full duplex physical data links capable of transferring data at 40 MB/s.

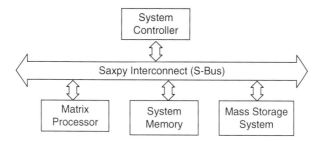

Fig. 12 Block diagram of a Matrix-1 system

These data links can be configured into 20 virtual channels. The clock speed of the communication agent is 40 MHz. Each cell is attached to a local memory sub-system including up to 4 MB static RAM (random access memory) or/and 16 MB DRAM. The iWarp system is designed to be configured as a $n \times m$ torus array. A typical system would have 64 cells configured as a 8×8 torus array and yields 1.2 GFlop/s peak performance.

The communication agent supports word-level flow control between connecting cells and transfers messages word by word to implement wormhole routing [18]. Exposing this mechanism to the computation agents allows programs to communicate systolically. Moreover, a communication agent can automatically route messages to the appropriate destination without the intervention of the computation agent.

5.2 SAXPY Matrix-1

Claimed to be "the first commercial, general-purpose, systolic computer", Matrix-1 [6] is a vector array processor developed by the SAXPY computer co. in 1987 for scientific, signal processing applications. It promises 1 GFLOP throughput by means of 32-fold parallelism, fast (64 ns) pipelined floating-point units, and fast and flexible local memories.

At system level, a Matrix-1 system (cf. Fig. 12) consists of a system controller, system memory, and mass storage in addition to the matrix processor. These system components are interconnected with a high-speed (320 MB/s) bus (S-bus). The system memory has a maximum capacity of 128 MB. It uses only physical addresses and hence allows faster access.

The Matrix Processor (Fig. 13) is a ring-connected linear array of 8, 16, 24, or 32 vector processors. Each processor is called a computational zone. All zones receive the same control and address instructions at each clock cycle. The Matrix Processor can function in a systolic mode (in which data are transferred from one zone to the next in a pipelined fashion) or in a block mode (in which all zones operate simultaneously to execute vector operations). Each zone has a pipelined, 32-bit floating-point multiplier; a pipelined, 32-bit floating-point adder with logic

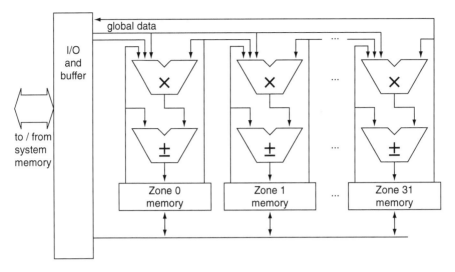

Fig. 13 The Matrix Processor Zone architecture of SAXPY Matrix-1 computer

capabilities, and a 4K-word local memory implemented as a two-way interleaved zone buffer. These components operate at a clock frequency of 16 MHz. With 32 zones, the maximum computing power would approach 960 MFLOP.

The Matrix-1 employs an application programming interface (API) approach to interface with the host processor. The user program will be written in C or Fortran and makes calls to the matrix processor subroutines. Experienced programmers may also write their own matrix processor subroutines or directly engage assembly level programming of the matrix processors.

5.3 Transputer

The Transputer (transistor computer) [8, 19, 22, 26] is a microprocessors developed by Inmos Ltd. in mid-1980s to support parallel processing. The name was selected to indicate the role the individual Transputers would play: numbers of them would be used as basic building blocks, just as transistors in integrated circuits.

A most distinct feature of a Transputer chip is that there are four serial links to communicate with up to four other Transputers simultaneously each at 5, 10 or 20 Mbit/s. The circuitry to drive the links is all on the Transputer chip and only two wires are needed to connect two Transputers together. The communication links between processors operate concurrently with the processing unit and can transfer data simultaneously on all links without the intervention of the CPU. Supporting the links was additional circuitry that handled scheduling of the traffic over them. Processes waiting on communications would automatically pause while

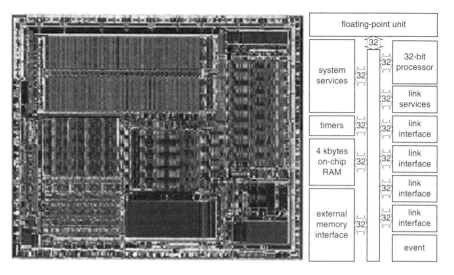

Fig. 14 INMOS T805 floating-point processor (http://www.classiccmp.org/transputer)

the networking circuitry finished its reads or writes. Other processes running on the transputer would then be given that processing time. These unique properties allow multiple Transputer chips to be configured easily into various topologies such as linear or mesh array, or trees to support parallel processing.

Depicted in Fig. 14 is a chip layout picture and a floor plan of Transputer T805. It has a 32-bit architecture running at 25 MHz clock frequency. It has an IEEE 754 64-bit on-chip floating-point unit, 4 Kbytes on-chip static RAM, and may connect to 4 GB directly addressable external memory (no virtual memory) at 33 Mbytes/s sustained data rate. It uses a 5 MHz clock input and runs on a single 5 V power supply.

Transputers were intended to be programmed using the Occam programming language, based on the CSP process calculus. Occam supported concurrency and channel-based inter-process or inter-processor communication as a fundamental part of the language. With the parallelism and communications built into the chip and the language interacting with it directly, writing code for things like device controllers became a triviality. Implementations of more mainstream programming languages, such as C, FORTRAN, Ada and Pascal, were also later released by both INMOS and third-party vendors.

5.4 TMS 32040

TMS 32040 [23] is Texas Instruments' floating-point digital signal processor developed in early 1990. The ***'320C40 has six on-chip communication ports for processor-to-processor communication with no external-glue logic. The communication ports remove input/output bottlenecks, and the independent smart DMA coprocessor is able to relieve the CPU input/output burden.

Each of the six serial communication ports is equipped with a 20 Mbytes/s bidirectional interface, and separate input and output eight-word-deep FIFO buffers. Direct processor-to-processor connection is supported by automatic arbitration and handshaking. The DMA coprocessor allows concurrent I/O and CPU processing for sustained CPU performance.

The processor features single-cycle 40-bit floating-point and 32-bit integer multipliers, 512-byte instruction cache, and 8 K bytes of single-cycle dual-access program or data RAM. It also contains separate internal program, data, and DMA coprocessor buses for support of massive concurrent input/output (I/O) program and data throughput.

The TMS 32040 is designed to support general-purpose parallel computation with different configurations. With six bidirectional serial link ports, it would directly support a hypercube configuration containing up to $2^6 = 64$ processing elements. It, of course, also can be easily configured to form a linear or two-dimensional mesh-connected processor array to support systolic computing.

6 Recent Developments and Real World Applications

6.1 Block Motion Estimation

Block motion estimation is a critical computation step in every international video coding standard, including MPEG-I, MPEG-II, MPEG-IV, H.261, H.263, and H.264. This algorithm consists of a very simple loop body (sum of absolute difference) embedded in a six-level nested loop. For real time, high definition video encoding applications, the motion estimation operation must rely on special purpose on-chip processor array structures that are heavily influenced by the systolic array concept.

The notion of block motion estimation is demonstrated in Fig. 15. To the left of this figure is the *current frame*, which is to be encoded and transmitted from the encoding end. To the right is the *reference frame*, which has already been transmitted and reconstructed at the receiver end. The encoder will compute a copy of this reconstructed reference frame for the purpose of motion estimation. Both the current frame and the reference frame are divided into *macro-blocks* as shown with dotted lines. Now focus on the *current block*, which is the shaded macro-block at the second row and the fourth column of the current frame. The goal of motion estimation is to find a matching macro-block in the reference frame, in the vicinity of the location of the current block such that it resembles the current block in the current frame. Usually, the current frame and the reference frame are separated by a couple of frames temporally, and are likely to contain very similar scene. Hence,

Fig. 15 Block motion
estimation

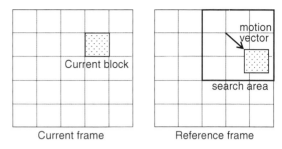

Current frame   Reference frame

there exists high degree of *temporal correlation* among them. As such, there is
high probability that the current block can find a very similar matching block in
the reference frame. The displacement between the location of the current block
and that of the matching macro-block is the *motion vector* of the current block. This
is shown to the right hand side of Fig. 15. By transmitting the motion vector alone
to the receiver, a predicted copy of the current block can be obtained by copying the
matching macro-block from the reference frame. That process is known as *motion
compensation*.

The similarity between the current block in the current frame and corresponding
matching block in the reference frame is measured using a *mean of absolute
difference (MAD) criterion*:

$$MAD(m,n) = \frac{1}{N^2} \sum_{i=0}^{N-1} \sum_{j=0}^{N-1} |x(i,j) - y(i+m,j+n)| \tag{27}$$

where the size of the macro-block is N pixels by N pixels. $x(i,j)$ is the value of the
$(i,j)^{th}$ pixel of the current frame and $y(i+m,j+n)$ is the value of the $(i+m,j+n)^{th}$
pixel of the reference frame. $MAD(m,n)$ is the mean absolute difference value
between the current block and the candidate matching block with a displacement of
(m,n), $-p \leq m, n \leq p$, where p is a bounded, pre-set *search range* which is usually
twice or thrice the size of a macro-block. The motion vector (MV) of the current
block is found as

$$MV = \arg \left\{ \min_{-p \leq m,n \leq p} MAD(m,n) \right\}. \tag{28}$$

We assume each video frame is partitioned into $N_h \times N_v$ macro-blocks. With Eqs.
(27) and (28), one may express the whole frame full-search block matching motion
estimation algorithm as a six-level nested loop as shown in Fig. 16.

The performance requirement for such a motion estimation operation is rather
stringent. Take MPEG-II for example, a typical video frame of 1080p format contains 1920×1080 pixels. With a macro-block size $N = 16$, one has $N_h = 1920/16 =
120$, $N_v = \lceil 1080/16 \rceil = 68$. Usually, $N = 16$, and $p = N/2$. Since there are 30
frames per second, the number of the sum of absolute difference operations that

```
Do h=0 to Nh-1
   Do v=0 to Nv-1
      MV(h,v)=(0,0)
      Dmin(h,v)= ∞
      Do m=-p to p
         Do n=-p to p
            MAD(m,n)=0
            Do i=h*N to (h+1)*N-1
               Do j=v*N to (v+1)*N-1
                  MAD(m,n)= MAD(m,n)+|x(i,j)-y(i+m,j+n)|
               End do j
            End do i
            If Dmin(h,v) > MAD(m,n)
               Dmin(h,v)=MAD(m,n)
               MV(h,v)=(m,n)
            End if
         End do n
      End do m
   End do v
End do h
```

Fig. 16 Full search block matching motion estimation

need to be performed would be around $30 \times N_h \times N_v \times (2p+1)^2 \times N^2 \approx 1.8 \times 10^{10}$ operation/s.

Since motion estimation is only part of video encoding operations, an application specific hardware module would be a desirable implementation option. In view of the regularity of the loop-nest formulation, and the simplicity of the loop-body operations (addition/subtraction), a systolic array solution is a natural choice. Toward this direction, numerous motion estimation processor array structures have been proposed, including 2D mesh array, 1D linear array, tree-structured array, and hybrid structures. Some of these realizations focused on the inner four-level nested loop formulation in Fig. 16 [12, 20], and some took the entire six-level loop nest into accounts [5, 11, 27]. An example is shown in Fig. 17. In this configuration, the search area pixel y is broadcast to each processing elements in the same column; and current frame pixel x is propagated along the spiral interconnection links. The constraint of $N = 2p$ is imposed to achieve low input/output pin count. A simple PE is composed of only two eight-bit adders and a comparator as shown in Fig. 18.

A number of video encoders micro-chips including motion estimation have been reported over the years. Earlier motion estimation architectures often use some variants of a pixel-based systolic array to evaluate the *MAD* operations. Often a fast search algorithm is used in lieu of the full search algorithm due to speed and power consumption concerns. One example is an MPEG-IV Standard profile encoder chip reported in [17]. Some chip characteristics are given in Table 1.

As shown in Fig. 19, the motion estimation is carried out with 16 adder tree (processing units, PU) for sum of absolute difference calculation and the motion vectors are selected based on these results. A chip micro-graph is depicted in Fig. 20.

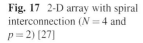

Fig. 17 2-D array with spiral interconnection ($N = 4$ and $p = 2$) [27]

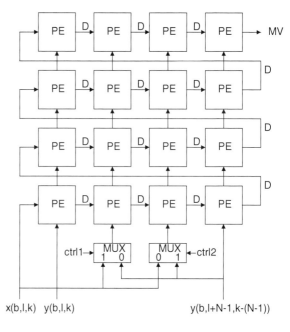

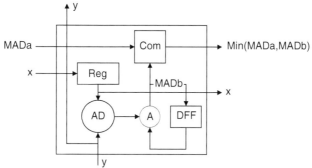

Fig. 18 Block diagram of an individual processing element [27]

Table 1 MPEG-IV motion estimation chip features [17]

Technology	TSMC 0.18 μm, 1P6M CMOS
Supply voltage	1.8 V (Core)/3.3 V (I/O)
Core area	1.78×1.77 mm^2
Logic gates	201 K (2-input NAND gate)
SRAMs	4.56 kB
Encoding feature	MPEG-4 SP
Search range	H[−16, +15.5] V[−16, +15.5]
Operating frequency	9.5 MHz CIF, 28.5 MHz VGA
Power consumption	5 mW (CIF, 9.5 MHz, 1.3 V), 18 mW (VGA, 28.5 MHz, 1.4 V)

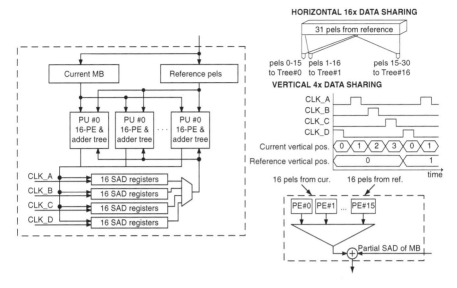

Fig. 19 Motion estimation architecture [17]

Fig. 20 MPEG-IV encoder
chip die micro-graph [17]

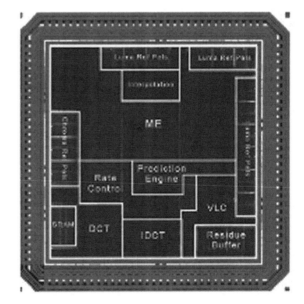

6.2   Wireless Communication

Systolic array has also found interesting applications in wireless communication
baseband signal processing applications. A typical block diagram of wireless
transceiver baseband processing algorithms is depicted in Fig. 21. It includes

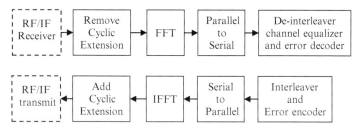

Fig. 21 A typical block diagram of wireless transceiver baseband processing

fast Fourier transform (FFT) and inverse fast Fourier transform (IFFT), channel estimator/equalizer, data inter-leaver, and channel encoder/decoder, etc.

In [21], a reconfigurable systolic array of CORDIC processing nodes (PN) is proposed to realize the computation intensive portion of the wireless baseband operations. CORDIC (Coordinate Rotion Digital Computer) [24, 25] is an arithmetic computing algorithm that has found many interesting signal processing applications [9]. Specifically, it is an efficient architecture to realize unitary rotation operations such as Jacobi rotation described in Eqs. (3), (4), and (5) in this chapter. With CORDIC, the rotation angle θ is represented with a weighted sum of a sequence of elementary angels $\{a(i); 0 \leq i \leq n-1\}$ where $a(i) = \tan^{-1} 2^{-i}$. That is,

$$\theta = \sum_{i=0}^{n-1} \mu_i a(i) = \sum_{i=0}^{n-1} \mu_i \tan^{-1} 2^{-i} \qquad \mu_i \in \{-1, +1\}.$$

As such, the rotation operation through each elementary angle may be easily realized with simple shift-and-add operations

$$\begin{bmatrix} x(i+1) \\ y(i+1) \end{bmatrix} = \begin{bmatrix} \cos a(i) & -\mu_i \sin a(i) \\ \mu_i \sin a(i) & \cos a(i) \end{bmatrix} \begin{bmatrix} x(i) \\ y(i) \end{bmatrix} = k(i) \begin{bmatrix} 1 & -m\mu_i 2^{-i} \\ \mu_i 2^{-i} & 1 \end{bmatrix} \begin{bmatrix} x(i) \\ y(i) \end{bmatrix}$$

where $k(i) = 1 / \sqrt{1 + 2^{-2i}}$.

A block diagram of a 4×4 CORDIC reconfigurable systolic array is shown in Fig. 22. The control unit is a general-purpose RISC (reduced instruction set computer) microprocessor. The PN array employs a data driven (data flow) paradigm so that globally synchronous clocking is not required. During execution phase, the address generator provides an address stream to the data memory bank. Accessed data is fed from the data memory bank to the PN array and back, via the memory interface, which adds a context pointer to the data. With the context pointer, dynamic reconfiguration of the PN array within a single clock cycle becomes possible.

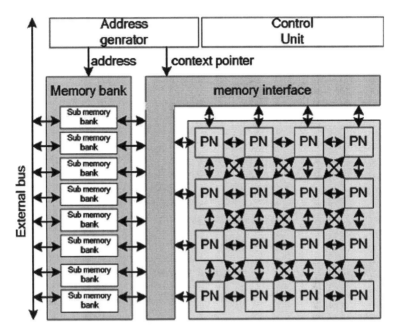

Fig. 22 CORDIC systolic array [21]

Fig. 23 Processing node
architecture [21]

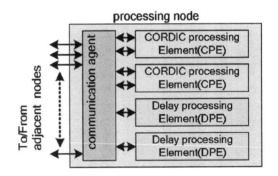

The PN architecture is depicted in Fig. 23 where two CORDIC processing elements, two delay processing elements are interconnected via the communication agent, which also handles external communications with other PNs.

Using this CORDIC reconfiguration systolic array, a minimum mean square error detector is implemented for an OFDM (orthogonal frequency division modulation) MIMO (multiple input, multiple output) wireless transceiver. A QR decomposition recursive least square (QRD–RLS) triangular systolic array is implemented on a FPGA prototype system and is shown in Fig. 24.

Fig. 24 QRD-RSL triangular systolic array [21]

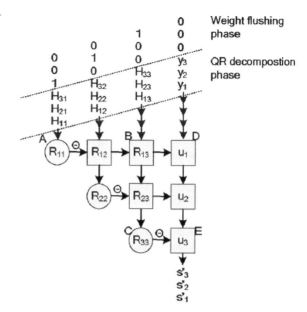

7   Conclusions

In this chapter, the historically important systolic array architecture is discussed. The basic systolic design methodology is reviewed, and the wavefront array processor architecture has been surveyed. Several existing implementations of systolic array like parallel computing platforms, including WARP, SAXPY Matrix-1, Transputer, and TMS320C40 have been briefly reviewed. Real world applications of systolic arrays to video coding motion estimation and wireless baseband processing have also been discussed.

References

1. Annaratone, M., et al.: The WARP computer: Architecture, implementation, and performance. IEEE Trans. Computers **36**, 1523–1538 (1987)
2. Arnould, E., Kung, H., et al.: A systolic array computer. In: Proc. IEEE International Conference on Acoustics, Speech, and Signal Processing, vol. 10, pp. 232 – 235 (1985)
3. Borkar, S., Cohn, R., Cox, G., Gross, T., Kung, H.T., Lam, M., Levine, M., Moore, B., Moore, W., Peterson, C., Susman, J., Sutton, J., Urbanski, J., Webb, J.: Supporting systolic and memory communication in iWarp. In: Proc. 17th Intl. Symposium on Computer Architecture, pp. 71–80 (1990)
4. Broomhead, D., Harp, J., McWhirter, J., Palmer, K., Roberts, J.: A practical comparison of the systolic and wavefront array processing architectures. In: Proc. Int'l Conf. Acoustics, Speech, and Signal Processing, vol. 10, pp. 296–299 (1985)

5. Chen, Y.K., Kung, S.Y.: A systolic methodology with applications to full-search block matching architectures. J. of VLSI Signal Processing **19**(1), 51–77 (1998)
6. Foulser, D.E.: The Saxpy Matrix-1: A general-purpose systolic computer. IEEE Computer **20**, 35–43 (1987)
7. Gross, T., O'Hallaron, D.R.: iWarp: Anatomy of a Parallel Computing System. MIT Press, Boston, MA (1998)
8. Homewood, M.,May, D., Shepherd, D., Shepherd, R.: The IMS T800 Transputer. IEEEMicro **7**(5), 10–26 (1987)
9. Hu, Y.H.: CORDIC-based VLSI architectures for digital signal processing. IEEE Signal Processing Magazine **9**, 16–35 (1992)
10. iWarp project. URL http://www.cs.cmu.edu/afs/cs/project/iwarp/archive/WWW-pages/iwarp.html
11. Kittitornkun, S., Hu, Y.: Systolic full-search block matching motion estimation array structure. IEEE Trans. Circuits Syst. Video Technology **11**, 248–251 (2001)
12. Komarek, T., Pirsch, P.: Array architectures for block matching algorithms. IEEE Trans. Circuits Syst. **26**(10), 1301–1308 (1989)
13. Kung, H.T.: Why systolic array. IEEE Computers **15**, 37–46 (1982)
14. Kung, S.Y.: On supercomputing with systolic/wavefront array processors. Proc. IEEE **72**, 1054–1066 (1984)
15. Kung, S.Y.: VLSI Array Processors. Prentice Hall, Englewood Cliffs, NJ (1988)
16. Kung, S.Y., Arun, K.S., Gal-Ezer, R.J., Bhaskar Rao, D.V.: Wavefront array processor: Language, architecture, and applications. IEEE Trans. Computer **31**(11), 1054–1066 (1982)
17. Lin, C.P., Tseng, P.C., Chiu, Y.T., Lin, S.S., Cheng, C.C., Fang, H.C., Chao, W.M., Chen, L.G.: A 5mW MPEG4 SP encoder with 2D bandwidth-sharing motion estimation for mobile applications. In: Proc. International Solid-State Circuits Conference, pp. 1626–1635. San Francisco, CA (2006)
18. Ni, L.M., McKinley, P.: A survey of wormhole routing techniques in direct networks. IEEE Computer **26**, 62–76 (1993)
19. Nicoud, J.D., Tyrrell, A.M.: The Transputer T414 instruction set. IEEE Micro **9**(3), 60–75 (1989)
20. Pan, S.B., Chae, S., , Park, R.: VLSI architectures for block matching algorithm. IEEE Tran. Circuits Syst. Video Technol. **6**(1), 67–73 (1996)
21. Seki, K., Kobori, T., Okello, J., Ikekawa, M.: A cordic-based reconfigurable systolic array processor for MIMO-OFDM wireless communications. In: Proc. IEEE Workshop on Signal Processing Systems, pp. 639–644. Shanghai, China (2007)
22. Taylor, R.: Signal processing with Occam and the Transputer. IEE Proceedings F: Communications, Radar and Signal Processing **131**(6), 610–614 (1984)
23. Texas Instruments: TMS320C40 Digital Signal Processors (1996). URL http://focus.ti.com/docs/prod/folders/print/tms320c40.html
24. Volder, J.E.: The CORDIC trigonometric computing technique. IRE Trans. on Electronic Computers **EC-8**(3), 330–334 (1959)
25. Walther, J.S.: A unified algorithm for elementary functions. In: Spring Joint Computer Conf. (1971)
26. Whitby-Strevens, C.: Transputers - past, present and future. IEEE Micro **10**(6), 16–19, 76–82 (1990)
27. Yeo, H., Hu, Y.: A novel modular systolic array architecture for full-search block matching motion estimation. IEEE Tran. Circuits Syst. Video Technol. **5**(5), 407–416 (1995)

Multidimensional Dataflow Graphs

Joachim Keinert and Ed F. Deprettere

1 Introduction

In many signal processing applications, the tokens in a stream of tokens have a dimension higher than one. For example, the tokens in a video stream represent images so that a video application is actually three- or four-dimensional: Two dimensions are required in order to describe the pixel coordinates, one dimension indexes the different color components, and the time finally corresponds to the last dimension. Static multidimensional (MD) streaming applications can be modeled using one-dimensional dataflow graphs [7], but these are at best cyclostatic dataflow graphs, often with many phases in the actor's vector valued token production and consumption patterns. These models incur a high control overhead. Furthermore such a notation hides many important algorithm properties such as inherent data parallelism, fine grained data dependencies and thus required memory sizes. Finally, the model is very implementation specific in that some of the degrees of freedom such as the processing order are already nailed down and cannot be changed easily without completely recreating the model.

All these problems can be solved by raising the level of abstraction and describing the application with a multidimensional model of computation. This offers the great benefit that the algorithms do not have to be translated into a one-dimensional representation, leading to more expressive and simpler models offering a greater design space. Different models of computation are available, ranging from a direct extension of the one-dimensional synchronous dataflow graph up to more complex and powerful description methods.

J. Keinert (✉)
Fraunhofer Institute for Integrated Circuits, Erlangen, Germany
e-mail: joachim.keinert@iis.fraunhofer.de

E.F. Deprettere
Leiden University, Leiden, the Netherlands
e-mail: edd@liacs.nl

S.S. Bhattacharyya et al. (eds.), *Handbook of Signal Processing Systems*,
DOI 10.1007/978-1-4614-6859-2_35, © Springer Science+Business Media, LLC 2013

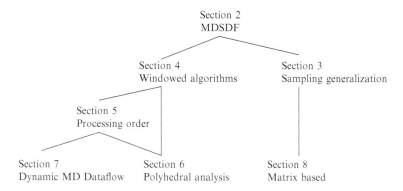

Fig. 1 Chapter structure

   The purpose of this chapter consists in highlighting the different aspects of multidimensional application design and analysis using the dataflow paradigm. Figure 1 depicts the corresponding chapter structure. Section 2 introduces the concept of multidimensional modeling by extending the concepts of one-dimensional synchronous dataflow, leading to *multidimensional synchronous dataflow graphs* (*MDSDF*). Next two generalizations are discussed. Section 3 describes techniques for more complex *sampling patterns* than rectangular grids, permitting the modeling of general up- and downsampling. Section 4 on the other hand, focus on the aspect of *local processing algorithms*, where an output array is generated by sampling one or more input arrays with possibly overlapping *filter windows*. This needs an extension of MDSDF, called *windowed synchronous dataflow* (*WSDF*). As illustrated in Fig. 1, those two sections are mostly independent of each other, meaning that Sect. 4 can be understood without consulting Sect. 3 and vice-versa.

   After introduction of these basics, Sect. 5 explains how such an approach offers a new degree of freedom during the design process in form of the processing order. It basically defines the temporal order by which tokens are generated and read. Being applicable to the WSDF model of computation, and given that WSDF is a true superset of MDSDF, the concepts can be directly applied to MDSDF as well.

   Once the pure dataflow model is annotated with the processing order, the system can be analyzed for properties such as required buffer sizes or parallel schedules. Compared to one-dimensional models of computation [7], this is complicated by the fact that for realistic applications an array contains a huge amount of tokens. Video applications with HD resolution for instance lead to $1,920 \times 1,080 = 2,073,600$ pixels, each associated with a corresponding token. Straightforward analysis on this level of granularity however overburdens most of the algorithms, since many problems in design analysis are NP-complete. Fortunately, multidimensional modeling permits usage of polyhedral methods that are very well adapted to such kinds of problems [16]. Since this is not possible with one-dimensional models of computation, it presents a true benefit of multidimensional dataflow and is hence discussed in Sect. 6.

While such a polyhedral analysis is only possible for static algorithms, multidimensional dataflow by itself does not suffer from this limitation. Instead it also permits the representation of dynamic decisions similar to one-dimensional dynamic dataflow [10]. Furthermore models can be created that contain both one- and multidimensional communication. Usage of finite state machines as explained in [4] helps to classify different actor behaviors. All these aspects are explained in Sect. 7. As illustrated in Fig. 1, Sects. 6 and 7 can be consulted independently of each other.

Section 8 finally concludes this chapter addressing the generalization of sampling patterns using a matrix notation. Consequently, it links the left and right parts of the tree depicted in Fig. 1.

2   Basics

MDSDF is—in principle—a straightforward extension of SDF. Figure 2 shows a simple example.

Actor A produces three tokens taken from three rows and a single column out of an array of tokens. We say that actor A produces $(3,1)$ tokens. Actor B consumes three tokens taken from three columns and a single row. We say that actor B consumes $(1,3)$ tokens. The SDF 1D repetitions vector [7] becomes a *repetitions matrix* in the MDSDF case:

$$R = \begin{pmatrix} r_{A,1} & r_{A,2} \\ r_{B,1} & r_{B,2} \end{pmatrix}.$$

$r_{A,1}$ defines the number of *executions* or *firings* of actor A in vertical direction (rows). $r_{A,2}$ equals the number of firings in horizontal dimension (columns). $r_{B,1}$ and $r_{B,2}$ do the same for actor B. Every firing of actor A yields in a $(3,1)$ token with 3×1 *data elements*, and every execution of actor B reads a $(1,3)$ token with 1×3 data elements. Demanding that the number of produced and read data elements shall be the same in all dimensions leads to the following balance equation:

$$r_{A,1} \times 3 \text{ (rows)} = r_{B,1} \times 1 \text{ (row)}$$

$$r_{A,2} \times 1 \text{ (column)} = r_{B,2} \times 3 \text{ (columns)}$$

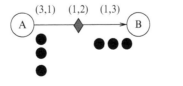   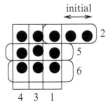

Fig. 2 A simple two-actor MDSDF graph

Thus actor A has to produce a number of times $(3,1)$ tokens, and actor B has to consume a number of times $(1,3)$ tokens. The equations can be solved to yield $(r_{A.1}, r_{A.2}) = (1,3)$, and $(r_{B.1}, r_{B.2}) = (3,1)$. In other words, actor A has to produce $(3,3)$ tokens, and actor B has to consume $(3,3)$ tokens.

These equations are independent of initial tokens (x, y) in the AB arc's buffer. Different interpretations of these initial tokens are possible. Figure 2 illustrates one of them. The buffer contains initial tokens in two columns of a single row, that is a *delay* of $(1,2)$. A possible schedule is shown at the right in the figure: A, B, 2(A), 2(B), labeled as $(1, 2, 3, 4,$ and $6)$. Note that the state returns to its initial value $(1,2)$.

An alternative interpretation of multidimensional initial tokens is discussed in [13, 14].

3   Arbitrary Sampling

Two canonical actors that play a fundamental role in MD signal processing are the *compressor* (or downsampler) and the *expander* (or upsampler) [13,14]. For the sake of pictorial representation, we will confine to the 2D case. Consider a 2D "signal" $\tilde{s}(\mathbf{k}), \mathbf{k} = (k_1, k_2) \in \mathbb{R}^2$, defined on a rectangular subspace $0 \le k_1 \le K_1 \wedge 0 \le k_2 \le K_2$. This bounded subspace is represented in Fig. 3 as a back-ground grid, with $K_1 = 7$, and $K_2 = 6$. Notice that although $\tilde{s}(\mathbf{k})$ is a continuous function, we do represent its supporting domain as a grid of k-points that can be arbitrary dense, see Fig. 3. The signal $\tilde{s}(\mathbf{k})$ can be sampled to yield the discrete signal $s(\mathbf{n}) = \tilde{s}(\mathbf{k} = V\mathbf{n})$, where V is a 2×2 non-singular *sampling matrix*, and $\mathbf{n} \in \mathbb{Z}^2$ is such that the sample points $\mathbf{k} = V\mathbf{n}$ fall within the rectangle $\mathcal{K}$ defined by the matrix $K = diag(K_1, K_2)$ ($diag(7,6)$ in the example). In fact, the set of n-points that satisfy this condition are the integral points in the parallelogram $\mathcal{Q}$ defined by the matrix $Q = V^{-1}K$. Thus, $VQ = K$ and $\mathbf{n} \in \mathcal{Q} \cap \mathbb{Z}^2 \leftrightarrow \mathbf{k} = V\mathbf{n} \in \mathcal{K} \cap \mathbb{Z}^2$. The matrix Q is called the *support matrix*.

A typical sampling matrix is

$$V_1 = \begin{pmatrix} v_{1.1} & 0 \\ 0 & v_{2.2} \end{pmatrix}$$

which yields a *rectangular* lattice of sample points $\mathbf{k} = V_1\mathbf{n}, \mathbf{n} \in \mathcal{Q}_1 \cap \mathbb{Z}^2$ (black dots). With $v_{1.1} = 2, v_{2.2} = 3$, the support matrix for this lattice is

$$Q_1 = \begin{pmatrix} 3.5 & 0 \\ 0 & 2 \end{pmatrix}.$$

A second typical sampling matrix is

$$V_2 = \begin{pmatrix} v_{1.1} = v_1 & v_{1.2} = v_2 \\ v_{2.1} = v_1 & v_{2.2} = -v_2 \end{pmatrix}$$

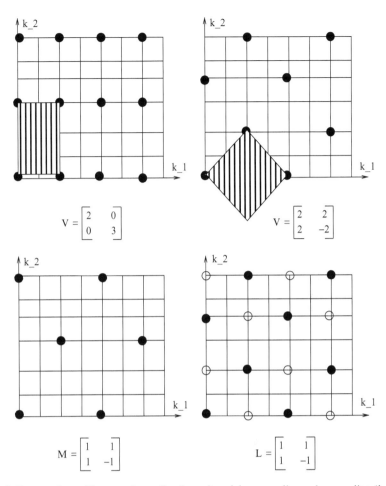

Fig. 3 Rectangular and hexagonal sampling (*upper*), and downsampling and upsampling (*lower*)

yielding a *hexagonal* lattice of sample points $\mathbf{k} = V_2\mathbf{n}$, $\mathbf{n} \in \mathscr{Q}_2 \cap \mathbb{Z}^2$ (black dots). Assuming $v_1 = 2, v_2 = 2$, the support matrix for this lattice is

$$Q_2 = \begin{pmatrix} 1.75 & 1.5 \\ 1.75 & -1.5 \end{pmatrix}$$

Both lattices are shown in the upper part of Fig. 3. Notice that for an integral sampling matrix V, the *volume* of the parallelogram $\mathscr{V}$ defined by the columns of the sampling matrix V is $|\det(V)|$. This is the number of integral points $V\mathbf{x}$, $\mathbf{x} = (x_1, x_2)$, $0 \le x_1, x_2 < 1$. $\mathscr{V}$ is commonly called the *basic lattice tile*. The two basic lattice tiles $\mathscr{V}$ are displayed in the upper part of Fig. 3 as a black and white striped rectangle and diamond, respectively. The number of integer points in $\mathscr{V}$ include 6 $\{(0,0), (1,0),$

(0,1), (1,1), (0,2) and (1,2) } and 8 {(0,0), (1,0), (2,0), (3,0), (1,1), (2, 1), (1,−1) and (2,−1) } integral points, respectively.

Notice that in both cases, only the origin is a sample point. The *sample density* is therefore $\frac{1}{|det(V)|}$ ($\frac{1}{6}$ and $\frac{1}{8}$, respectively). Sampling matrices can be arbitrary which makes MDSDF graphs more complicated than their uni-dimensional SDF counterparts.

The two-dimensional decimator is characterized by a non-singular integral matrix M. Downsampling converts the lattice built on the columns of the sampling matrix V to the lattice built on the columns of the sampling matrix VM. That is, it retains sample points $\mathbf{k} = V\mathbf{n}$, $\mathbf{n} \in \mathcal{Q} \cap \mathbb{Z}^2$ that also satisfy $\mathbf{k} = VM\mathbf{n}'$, $\mathbf{n}' \in \mathcal{Q}' \cap \mathbb{Z}^2$, where $\mathcal{Q}'$ is the parallelogram built on the support matrix $Q' = M^{-1}V^{-1}K$. The sampling matrix at the output of the decimator is $V_O = V_I M$, where $V_I = V$ is the sampling matrix at the input of the decimator.

Similarly, the two-dimensional expander is characterized by a non-singular integral matrix L. Upsampling converts the lattice built on the columns of the sampling matrix V to the lattice built on the columns of the sampling matrix VL^{-1}. That is, it retains all sample points $\mathbf{k} = V\mathbf{n}$, $\mathbf{n} \in \mathcal{Q} \cap \mathbb{Z}^2$, and creates additional sample points $\mathbf{k} = VL^{-1}\mathbf{n}' \wedge \mathbf{k} \neq V\mathbf{n}$, $\mathbf{n}' \in \mathcal{Q}' \cap \mathbb{Z}^2$, where $\mathcal{Q}'$ is the parallelogram built on the support matrix $Q' = LV^{-1}K$. The values of the samples at the additional lattice points are set to 0. The input lattice is a subset of the output lattice, and the number of added samples is equal to $|det(L)|$.

The sampling matrix at the output of the expander is $V_O = V_I L^{-1}$, where $V_I = V$ is the sampling matrix at the input of the expander.

Figure 3 illustrates how the upper left lattice is downsampled (lower left), and the upper right lattice is upsampled (lower right).

In summary, if Q_I is the support matrix at the input of a 2D decimator or expander, then the support matrix Q_O at the output is $Q_O = M^{-1}Q_I$ for the decimator, and $Q_O = LQ_I$ for the expander. In the example shown in Fig. 3, the output support matrix for the downsampling example is $M^{-1}V^{-1}K$, and for the upsampling example it is $LV^{-1}K$. They are

$$Q_O = \begin{pmatrix} 1.75 & 1.0 \\ 1.75 & -1.0 \end{pmatrix}$$

for the decimator (bottom left) in Fig. 3, and

$$Q_O = \begin{pmatrix} 3.5 & 0 \\ 0 & 3 \end{pmatrix}$$

for the expander (bottom right) in Fig. 3.

The characterization of input–output sampling matrices, and input–output support matrices is not enough to set up *balance equations* because it still remains to be determined how the points in the support parallelograms are ordered.

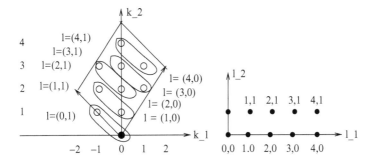

Fig. 4 Ordering of the upsampler's output tokens

For example, consider a source actor S producing $(3,3)$ samples. It thus has a support matrix $Q = diag(3,3)$. Assume that S is connected to an expander with input sampling matrix $V_I = I$, scanning its input in row scan order. Thus the upsampler has to output $|det(L)|$ samples in a parallelogram $\mathscr{L}$ defined by the columns of the matrix L for every input sample. Suppose

$$L = \begin{pmatrix} 2 & -2 \\ 3 & 2 \end{pmatrix}$$

$|det(L)| = 10$. These samples can be ordered as follows.

The first column of L can be interpreted as the upsampler's *horizontal direction*, and the second column as its *vertical direction* in a *generalized rectangle* $\mathscr{L}$. In $\mathscr{L}$, there are $L_1 = 5$ groups of $L_2 = 2$ samples (that is $L_1 \times L_2 = |det(L)| = 10$ samples): The sampling rate at the output of the upsampler can be chosen to be (L_2, L_1), that is L_2 rows and L_1 columns in *natural order*. The relation between the natural ordered samples and the samples in the generalized rectangle $\mathscr{L}$ is then

$$\begin{pmatrix} (l_1, l_2): (0,0) \ (1,0) \ (2,0) \ (3,0) \ (4,0) \ (0,1) \ (1,1) \ (2,1) \ (3,1) \ (4,1) \\ (k_1, k_2): (0,0) \ (0,1) \ (0,2) \ (1,2) \ (1,3) \ (-1,1) \ (-1,2) \ (-1,3) \ (0,3) \ (0,4) \end{pmatrix}$$

This is shown in Fig. 4.

In this figure, $\mathscr{L}$ is shown at the left, and the natural ordering of samples is shown at the right. The correspondence between samples in $\mathscr{L}$ and the natural ordering of these samples is displayed at the left as well.

3.1 A Complete Example

Consider the MDSDF graph shown in Fig. 5 [13, 14].

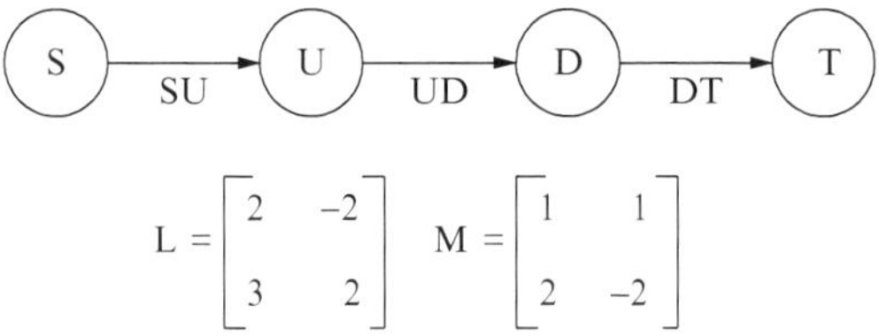

Fig. 5 A combined
upsampling and
downsampling example

$$L = \begin{bmatrix} 2 & -2 \\ 3 & 2 \end{bmatrix} \quad M = \begin{bmatrix} 1 & 1 \\ 2 & -2 \end{bmatrix}$$

It consists of a chain of four actors S, U, D, and T. S and T are source and sink actors, respectively. U is an expander

$$L = \begin{pmatrix} 2 & -2 \\ 3 & 2 \end{pmatrix}$$

and D is a compressor

$$M = \begin{pmatrix} 1 & 1 \\ 2 & -2 \end{pmatrix}$$

The arcs are labeled SU, UD, and DT, respectively. Let V_{SU} be the identity matrix, and $W_{SU} = diag(3,3)$ the support matrix. The expander consumes the samples from the source in row scan order, The output samples of the upsampler are ordered in the way discussed above, one at a time (that is as $(1,1)$), releasing $|det(L)|$ samples at the output on each invocation. The decimator can consume the upsampler's output samples in a rectangle of samples defined by a factorization $M_1 \times M_2$ of $|det(M)|$. This rectangle is deduced from the ordering given in Fig. 4. Thus, for a 2×2 factorization, the $(0,0)$ invocation of the downsampler consumes the (original) samples $(0,0)$, $(-1,1)$, $(0,1)$ and $(-1,2)$ or equivalently, the natural ordered samples $(0,0)$, $(0,1)$, $(1,0)$, and $(1,1)$.

Of course, other factorizations of $|det(M)|$ are possible, and the question is whether a factorization can be found for which the total number of samples output by the compressor equals the total number of samples consumed divided by $|det(M)|$ in a complete cycle as determined by the repetitions matrix which, in turn, depends on the factorizations of $|det(L)|$ and $|det(M)|$. Thus, the question is whether $N(W_{DT}) = \frac{N(W_{UD})}{|det(M)|}$. It can be shown [14] that this is not the case for the 2×2 factorization of $|det(M)|$. The condition is satisfied for a 1×4 factorization, though. Thus, denoting by $r_{X,1}$ and $r_{X,2}$ the repetitions of a node X in the *horizontal* direction and the *vertical* direction, respectively, the balance equations in this case become

$$\begin{pmatrix} 3 \times r_{S,1} = 1 \times r_{U,1} \\ 3 \times r_{S,2} = 1 \times r_{U,2} \\ 5 \times r_{U,1} = 1 \times r_{D,1} \\ 2 \times r_{U,2} = 4 \times r_{D,2} \\ r_{D,1} = r_{T,1} \\ r_{D,2} = r_{T,2} \end{pmatrix}$$

where it is assumed that the decimator produces (1,1) tokens at each firing, and the sink actor consumes (1,1) tokens at each firing.

The resulting elements in the repetitions matrix are

$$\begin{pmatrix} r_{S,1} = 1 & r_{S,2} = 2 \\ r_{U,1} = 3 & r_{U,2} = 6 \\ r_{D,1} = 15 & r_{D,2} = 3 \\ r_{T,1} = 15 & r_{T,2} = 3 \end{pmatrix}$$

Now, for a complete cycle we have,

$W_{SU} = diag(3,3) \times diag(r_{S,1}, r_{S,2})$, $W_{UD} = LW_{SU}$, and $W_{DT} = M^{-1}W_{UD}$.

Because W_{UD} is an integral matrix, $N(W_{UD}) = \mid det(W_{UD}) \mid = 180$. There is no simple way to determine $N(W_{DT})$ because this matrix is not an integral matrix. Nevertheless $N(W_{DT})$ can be found to be 45, which is, indeed, a quarter of $N(W_{UD}) = 180$. The reader may consult [14] for further details.

3.2   Initial Tokens on Arcs

In the example in the previous subsection, the arcs where void of initial tokens. In unidimensional SDF, initial tokens on an arc are a prefix to the stream of tokens output by the arc's source actor. Thus, if an arc's buffer contains two initial tokens, then the first token output by the arc's source node is actually the third token in the sequence of tokens to be consumed by the arc's destination actor. In multidimensional SDF this "shift" is a translation along the vectors of a support matrix W (in the generalized rectangle space) or along the vectors of a sampling matrix V.

Thus for a source node with output sampling matrix

$$\begin{pmatrix} 2 & 0 \\ 1 & 1 \end{pmatrix}$$

a multidimensional delay on its outgoing arcs of (1,2) means that the first token released on this arc is actually located at (2,2) in the sampling space.

Similarly, for a support matrix

$$\begin{pmatrix} 1 & 0 \\ -2 & 4 \end{pmatrix}$$

a delay of (1,2) means that the first token output is located at (1,2) in the generalized rectangle's natural order. See [13, 14] for further details.

4   Windowed Synchronous Dataflow (WSDF)

Processing multidimensional arrays is widely applied in digital signal processing. It occurs for instance extensively in image processing such as filtering, compression or medical analysis, as exemplified in Part I of this book. Arrays of pixels having two, three or more dimensions are analyzed and transformed into new result arrays. In this context, *point*, *local*, and *global* algorithms can be distinguished. For *point algorithms*, each output pixel depends on a single input pixel. Given a video as a sequence of images, thresholding is a simple example where each output pixel is computed according to the following equation:

$$o(x,y,t) = \begin{cases} 1 & \text{if } i(x,y,t) \geq 128 \\ 0 & \text{if } i(x,y,t) < 128 \end{cases}$$

$o(x,y,t)$ is the pixel at the coordinates (x,y) within the output image at time t and $i(x,y,t)$ is the pixel at the coordinates (x,y) within the input image at time t. Point algorithms can be well modeled with one-dimensional models of computation discussed in [2,7], because they basically transform a stream of pixels into another stream of pixels.

Global algorithms on the other hand, need in the worst case the complete input image before being able to compute a single pixel of the output. The rate control procedure applied in the JPEG 2000 standard [9] belongs to this category, since it needs to know all pixel values before being able to decide how many bits to spend for each of them. At least on a coarse granularity, this kind of algorithms is also well represented with one-dimensional models of computation when associating a token with a complete image.

For *local algorithms*, however, the situation is more difficult. In this case, an output pixel depends only on a small region of the input image. As a consequence, they show particular properties such as high degree of parallelism, low memory requirements or special data access patterns that can be exploited for efficient implementations. All these nice properties are hidden in the model, if a token represents a complete image. More fine-grained descriptions using 1D models of computation are often cumbersome. CSDF graphs for instance tend to end up with many phases in their vector valued token production and consumption cycle which incurs high control overhead [7].

The following subsections consider local algorithms in more detail and discuss how they can be modeled using an appropriate type of multidimensional dataflow. It extends MDSDF as presented in Sect. 2 by the ability to represent sampling with overlapping windows, to read token values multiple times, as well as by border processing.

Fig. 6 Subsampling of an image by a factor of two in horizontal and vertical direction

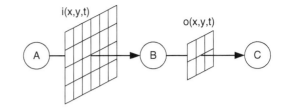

4.1 Motivation: Processing of Multidimensional Arrays by Sliding Window Algorithms

Figure 6 exemplarily depicts a local image processing algorithm in form of image subsampling. It aims to reduce the image size by a factor of two in both horizontal and vertical direction. The input image is supposed to have six columns and four rows. Consequently, the output image consists of three columns and two rows. Node A is the source of the system representing an image interface such as an input DVI connector. Node A thus forwards the unaltered image to the next node B. It performs the proper subsampling, producing an output image having half the width and height of the corresponding input image. Note C finally corresponds to a sink, representing for instance an output DVI connector.

In principle the subsampling can be implemented by simply discarding every other pixel in both horizontal and vertical direction. Section 3 introduced corresponding modeling techniques. However, such an approach typically leads to aliasing artifacts [15]. This can be avoided by first applying a low-pass filter that computes the weighted mean value over a number of neighbor pixels. The weights can be derived by corresponding filter design techniques. Mathematically, the output image can thus be expressed as

$$\begin{aligned}\forall 0 \leq x < w_o \\ \forall 0 \leq y < h_o\end{aligned} : o(x,y,t) = \sum_{a=-\alpha_1}^{\alpha_2} \sum_{b=-\beta_1}^{\beta_2} \left(\tilde{i}(2 \cdot x - a, 2 \cdot y - b, t) \cdot f(a,b)\right) \quad (1)$$

$$\tilde{i}(x,y,t) = \begin{cases} i(x,y,t) & \text{if } 0 \leq x < w_i, 0 \leq y < h_i \\ \dots & \text{otherwise} \end{cases} \quad (2)$$

where

- w_i is the input image width,
- $w_o = \left\lceil \frac{w_i}{2} \right\rceil$ is the output image width,
- h_i is the input image height,
- $h_o = \left\lceil \frac{h_i}{2} \right\rceil$ is the output image height,
- $\alpha_1 + \alpha_2 + 1$ is the filter width (typically $\alpha_1, \alpha_2 \geq 0$),
- $\beta_1 + \beta_2 + 1$ is the filter height (typically $\beta_1, \beta_2 \geq 0$),
- $f(a,b)$ are the filter weights or coefficients,

Fig. 7 Border processing by pixel mirroring. Each rectangle corresponds to a pixel. *Dashed arrows* represent the pixel mirroring

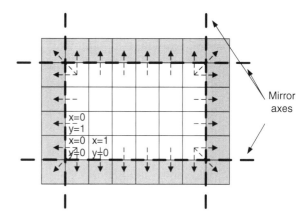

- $o(x,y,t)$ is the pixel at the coordinates (x,y) within the output image at time t,
- $i(x,y,t)$ is the pixel at the coordinates (x,y) within the input image at time t.

Equation (2) defines the *border processing* as explained later on. Consider first the case where $\tilde{i}(x,y,t) = i(x,y,t)$ and assume for a moment that $\beta_1 = \beta_2$ and $\alpha_1 = \alpha_2$, then obviously each output pixel $o(x,y,t)$ is computed from $(2 \cdot \alpha_1 + 1) \cdot (2 \cdot \beta_1 + 1)$ pixels centered at the input pixel $i(2 \cdot x, 2 \cdot y, t)$. Hence, only a local subset of the input image is required to compute an output pixel. In more detail, the filter represents a *sliding window* algorithm, where a window samples the input image $i(x,y,t)$, and for each window position, a corresponding output pixel $o(x,y,t)$ is generated. For the example given in Eq. (1), the factor of two causes that the sliding window moves by two pixels in horizontal and vertical direction.

While this operation is straightforward when the complete window is situated within the input image $i(x,y,t)$, computation of output pixel $o(x = 0, y = 0, t)$ for instance depends on the input pixels $i(x,y,t)$, $-\alpha \le x \le \alpha, -\beta \le y \le \beta$. Pixels with negative coordinates are situated outside of the input image and are thus not defined. Hence a border processing is required as expressed in Eq. (2). One might for instance set all pixels situated outside the image to a constant value. In this case, Eq. (2) gets

$$\tilde{i}(x,y,t) = \begin{cases} i(x,y,t) & \text{if } 0 \le x < w_i, 0 \le y < h_i \\ \iota = const & \text{otherwise} \end{cases}$$

Alternatively, often a mirroring of the pixels at the image border is performed as expressed by the following equation and graphically depicted in Fig. 2 (Fig. 7):

$$\tilde{i}(x,y,t) =$$

$$i\left(\begin{cases} x & \text{if } 0 \le x < w_i \\ -x-1 & \text{if } x < 0 \\ 2 \cdot w_i - 1 - x & \text{if } x \ge w_i \end{cases}, \begin{cases} y & \text{if } 0 \le y < y_i \\ -y-1 & \text{if } y < 0 \\ 2 \cdot h_i - 1 - y & \text{if } y \ge h_i \end{cases}, t\right)$$

A variant of this approach consists in not repeating the first respectively last image row and column:

$$\tilde{i}(x,y,t) =$$

$$i\left(\left\{\begin{array}{ll} x & \text{if } 0 \leq x < w_i \\ -x & \text{if } x < 0 \\ 2 \cdot w_i - 2 - x & \text{if } x \geq w_i \end{array}\right. , \left\{\begin{array}{ll} y & \text{if } 0 \leq y < y_i \\ -y & \text{if } y < 0 \\ 2 \cdot h_i - 2 - y & \text{if } y \geq h_i \end{array}\right. , t\right)$$

All these variants can be unified by assuming that the input image $i(x,y,t)$ is surrounded by a virtual extended border leading to the image $\tilde{i}(x,y,t)$.

Having summarized the principles of sliding window algorithms, the next sections will be dedicated to the question how those can be modeled via dataflow.

4.2 Semantics of the Windowed Synchronous Dataflow

As discussed in the previous section, local array processing shows a number of particularities that have to be taken into account when desiring a precise model showing important characteristics:

- Data elements—corresponding to pixels in the case of image processing—are possibly read multiple times.
- The processing can be represented by sampling a multidimensional array with possibly overlapping *sliding windows.*
- Since these sliding windows might exceed the image borders, special border processing is necessary.
- Often the algorithms show a large degree of possible data parallelism in that many data elements can be processed simultaneously.
- Because of the local data dependencies, sophisticated storage strategies can improve throughput by avoiding unnecessary memory accesses.

Unfortunately, one-dimensional models of computation either lead to cumbersome representations with many actors and phases, or they are too coarse grained to show important properties such as the available data parallelism. The latter is particularly important for efficient multi-core implementations. The multidimensional synchronous dataflow (MDSDF) as discussed in Sect. 2, on the other hand, is not suited to handle the sampling with overlapping windows, nor the extended border processing.

Solving these problems is the purpose of an extended dataflow model of computation called *windowed synchronous dataflow* (WSDF), which is a true superset of MDSDF. The following sections aim to explain its principles.

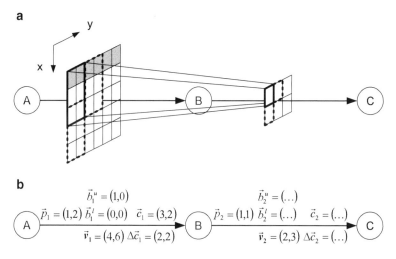

Fig. 8 Windowed synchronous dataflow (WSDF). (**a**) Token transport. *Gray shaded rectangles* belong to the extended border. (**b**) WSDF graph

4.2.1 WSDF Token Production and Consumption

Similar to other dataflow models of computation, *windowed synchronous dataflow* (WSDF) represents the application by a set of nodes, also called *actors*, that exchange data via edges. Each edge transports a multidimensional array that is generated by the source of the edge and sampled by the sink of the edge. For each individual sampling operation, the sink has access to a subset of data elements being part of a sliding window with well defined size. As a result of each sampling operation, an actor produces a multidimensional output token on each its associated output edges.

Figure 8 illustrates the corresponding concept for the application discussed in Sect. 4.1. Figure 8a corresponds to Fig. 6, except for the added extended border as discussed in Sect. 4.1. Furthermore, it depicts how actor B samples the input array with overlapping windows. For reasons of simplicity, a window size of three rows and two columns have been assumed. For each sliding window position, actor B generates one output token. For the considered example, this output token consists of one single data element. The individual output tokens from the different sampling operations are concatenated to an overall output array. In other words, for each firing of actor B, it reads the data elements belonging to one sliding window, performs the necessary computation, and appends one token with one data element to the output array. At this point it is important to state that horizontally adjacent input windows are related to horizontally adjacent output tokens. The same holds for the vertical dimension.[1] In the context of Fig. 8a, this means that the upper left input window

---

[1] This restriction will be released in Sect. 7.

leads to the upper left output token. The window right to the upper left input window also leads to the output token right to the upper left output token. In other words, there is a strict and static relation which input window contributes to which output token.

Because of the chosen window size, the windows do not overlap in horizontal direction, since the window moves by two data elements in the y-direction. In vertical direction, on the other hand, the windows overlap by one row, since the window moves by two data elements while having a size of three rows. In other words, while traditional schemes such as MDSDF assume that a data element will be read only once, this is not true any more for WSDF. In particular, data elements cannot always be discarded after the first reading operation.

The extended border handles the cases when the window exceeds the proper input array. Note that the size of the extended border has to be chosen in such a way, that sampling starts in the "upper left" corner of the extended array, and that all data elements of the extended arrays will be included in the sampling process. Note, furthermore that the streaming dimension is directed in x-direction, and that the model can handle more than two-dimensional arrays.

Figure 8b depicts a graphical notation and all necessary parameters in order to describe the above mentioned array transport. Obviously, it is necessary to define for each edge i the window size by means of the vector $\mathbf{c_i}$. Each of its components specifies the size in the corresponding coordinate direction. Vector $\Delta \mathbf{c_i}$ describes by how many data elements the window moves in the different directions, where the first coordinate corresponds to the x-dimension, the second to the y-dimension, and so on.

Vector $\mathbf{p_i}$ defines the produced output token size. Note that the produced token size does not need to correspond to the overall array size. Consider for instance actor B of Fig. 8a. For each firing, it only produces a $\mathbf{p_2} = (1,1)$ token, while the overall array consists of two rows and three columns. In other words, WSDF explicitly models that a huge array can be generated successively by smaller operations. By these means, it both reveals inherent data parallelism, and allows for memory efficient implementations, where the buffers associated to the edges can be smaller than one complete array.

4.2.2 Border Extension

As discussed in Sect. 4.1, border processing is a fundamental aspect of local algorithms. To this end, the vectors $\mathbf{b_i^l}$ and $\mathbf{b_i^u}$ in Fig. 8 define the size of the extended border. In case of two dimensional applications, both vectors $\mathbf{b_i^l} = \left(b_{i,1}^l, b_{i,2}^l \right)$ and $\mathbf{b_i^u} = \left(b_{i,1}^u, b_{i,2}^u \right)$ have two dimensions and $b_{i,1}^l$ defines the lower border, $b_{i,2}^l$

the left border, $b_{i,1}^u$ the upper border, and $b_{i,2}^u$ the right border.[2] Note that this extended border is not produced by the source of the dataflow edge. In other words, actor A in Fig. 8a generates only the data elements not marked by a gray shading. Instead, the extended border simply defines the number of pixels, by which the first sliding window exceeds the proper array produced by edge's source actor. These overlapping window positions will then be marked as border pixels, such that corresponding sink actor knows that they are not valid. Consequently it can apply the appropriate border processing algorithm, such as constant value insertion, or mirroring (see Sect. 4.1). In other words, the concrete border processing algorithm is not part of the WSDF model. Instead the user can freely decide which one to implement, with the restriction that an actor is only allowed to access data elements situated within the current sliding window. All other array elements have to be considered as unknown. This is important in order to allow proper data dependency analysis required for scheduling and buffer size determination (see also Sect. 6).

4.2.3 Specification of the Array Size

The WSDF semantics discussed in Sects. 4.2.1 and 4.2.2 allow to cover all parameters of Eq. (1), except for the array size. Recall to this end, that the parameter $\mathbf{p_i}$ does not define the extension of the processed arrays. Instead it specifies the number of data elements that are generated during one firing of the source actor. In general, the source actor needs to fire multiple times to produce an overall array. In other words, without specifying an array size the dataflow graph in Fig. 8 would for instance allow to downscale an 4×6 image into one having 2×3 pixels, or an 6×6 image into one having 3×6 pixels.

In order to compute the admissible array sizes, a balance equation can be computed according to the same idea described in Sect. 2. However, the WSDF variant would have different properties than the MDSDF one, because the solutions in general are not integer multiples anymore. Nevertheless such an approach would be completely feasible.

Unfortunately, this excludes applications where arrays need to be divided into sub-arrays that are to be processed independently of each other. The compression standard JPEG 2000, for instance, requires such a feature when using tiling [9]. Transferred to our simple example described in Sect. 4.1, the data elements produced by actor A need to be grouped into two independent images, a lower one and an upper one. These arrays shall then be processed as individual images. In other words, no sliding window shall contain pixels from the two different sub-images at the same time. This can be achieved by surrounding each sub-image with its own extended border, and then apply the local processing algorithm to the lower and upper image independently of each other.

---

[2]For more than two dimensions, see [12].

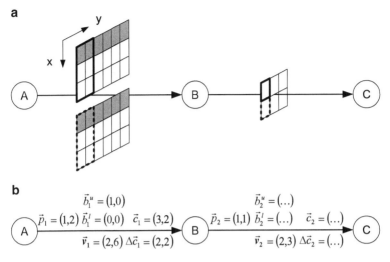

Fig. 9 Splitting of the input array into two parts. (**a**) Token transport. *Gray shaded rectangles* belong to the extended border. (**b**) WSDF graph

Figure 9 depicts the corresponding scenario. It exactly matches the scenario shown in Fig. 8, except that the input array to actor B is split into two parts. Note that from actor A's point of view, the situation has not changed. It still produces the same tokens in both scenarios. However, actor B groups these tokens into two subarrays, instead of one. Each of these subarrays is extended by its own extended border. In particular, the sampling with sliding windows restarts for each sub-array at the upper left corner. Note furthermore that the amount of tokens produced by actor B is the same for both scenarios, while their values will differ because of the different border processing. Note in addition that while the tokens generated by actor B will originate from two different input subarrays, actor C unifies the tokens again into one array. Such a behavior is again useful in applications like JPEG 2000, where the different tiles are combined into one output file.

In order to support such scenarios, WSDF edges are annotated by a vector v. It defines how many data elements are grouped to one subarray, the so called *virtual token*. Note that it does not impact the way how the source of an edge is behaving. Only the edge sink is concerned by performing the corresponding border processing and restarting the sampling at each upper left corner of each input subarray. This concept by the way leads to a balance equation with similar properties than for MDSDF. Further details on this aspect can be found in [12].

4.2.4 Graphs with Feedback Loops and Non-Rectangular Windows

Similar to cyclo-static dataflow graphs, feedback loops can be handled in WSDF either by placing so called initial tokens on the edges, or by carefully designing the border processing. The first solution is closely related to the initial tokens known for MDSDF. In the second case, the idea is to make the extended border big enough such that it can cover complete sliding windows. Then an actor can fire without needing any input tokens. The result of these firings can then be fed back to its own inputs. Further details on this aspect can be found in [11, 12]. The same holds for the modeling of windows that do not have a rectangular form.

5 Processing Order

One-dimensional dataflow models of computation describe streams of tokens that flow through a graph of actors. Their translation into a concrete implementation is thus relatively straightforward by mapping the edges of the graph to FIFOs. These FIFOs receive, store and forward the tokens without changing their order. The next step for translating a one-dimensional dataflow graph into an implementation consists in generation of an *actor schedule*. It defines the *temporal order* by which the dataflow actors are executed [7].

Compared to this approach, multidimensional dataflow graphs offer an additional degree of freedom that needs to be refined during system implementation as discussed in the following subsections.

5.1 Refinement of Multidimensional Dataflow Graphs

Multidimensional models of computation as discussed in Sects. 2 and 4 represent the transport of arrays, while these arrays are not atomic, but are produced and read in smaller parts. Compared to 1D models, this introduces a new degree of freedom, namely the *processing order*. The dataflow graph depicted in Fig. 8a, for instance, clearly defines which token produced by actor *B* depends on which input tokens generated by actor *A*. However, the order, by which the window samples the input array and hence the order by which the output tokens are generated, is still open.

Figure 10 shows some possible orders by which actor *B* of Fig. 8a might generate the output array. There are very well applications which assume a *horizontal raster scan order* that first moves in horizontal, and then in vertical direction (Fig. 10a). But the opposite is of course also viable (Fig. 10b). There are even applications

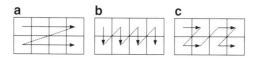

Fig. 10 Processing order of actor *B* belonging to Fig. 8a. (**a**) Horizontal raster scan. (**b**) Vertical raster scan. (**c**) Block-based

Fig. 11 Hierarchical
block-based processing order.
Each *rectangle* corresponds
to a firing producing one or
several data elements

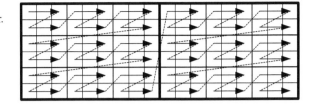

which divide the output image into smaller blocks, and scan these blocks in raster
scan order (Fig. 10c), because this reduces the amount of required internal chip
buffer [17].

Both WSDF and MDSDF do not impose any order. Hence, in addition to the
actor schedule, also the processing order has to be specified. For both models of
computation, specifying the processing order of one input or output array automatically defines the order for all other arrays related to the considered actor, because
there is a strict relation between produced output tokens and read input windows.[3]
Nevertheless, each actor can have its own processing order. Only feedback loops
within a dataflow graph might limit the admissible processing orders.[4]

5.2 Specification of the Processing Order

While scheduling multidimensional dataflow graphs will be subject to Sect. 6, this
section is dedicated to the processing order. Obviously, it is possible to define arbitrarily complex processing orders for each actor. However, this complicates further
analysis and implementation. Fortunately, many applications can be described by
relatively simple processing orders. They can be defined by clustering the actor
firings hierarchically into blocks.

Figure 11 depicts a corresponding example. In contrast to previous figures,
here each rectangle corresponds to one actor firing. Each of them generates a
multidimensional token on each output edge and reads a sliding window on each
input edge. Horizontally adjacent firings in Fig. 11 generate horizontally adjacent
output tokens and read horizontally adjacent input windows. The same holds for the
vertical dimension.

The processing order is now defined by clustering the actor firings into a
hierarchy of blocks as depicted in Fig. 11. The small rectangles with the thinnest
borders correspond to the individual firings (hierarchy level 0). The rectangles with
two rows and three columns having a medium sized border define the first level of

---

[3]Note that this relation will be broken in Sect. 7

[4]Section 3 introduced a processing order to enumerate sample points that are not part of a
rectangular grid. This section differs in that it considers rectangular instead of arbitrary sampling
patterns. The defined processing orders are hence purely to describe the application behavior, and
are not a prerequisite for the balance equation.

hierarchy and group 2×3 actor firings. The next level finally groups 3×3 blocks of the first level and is depicted with the thickest borders. Note that more levels of hierarchy are possible.

The processing order can then be specified by defining that within one block of hierarchy level i, the sub-blocks of level $i - 1$ are executed in horizontal raster scan order. This leads to the processing order as depicted in Fig. 11 by arrows. Note that this definition allows to represent all processing orders of Fig. 10.[5]

Using such a hierarchical block-based processing order allows to identify each actor firing with an m-dimensional *iteration vector*, where $m \in \mathbb{N}$ is a function of the array dimensions and the number of hierarchy levels. Such an iteration vector is similar to loop iterators and permits an efficient implementation as discussed in [12].

6   Translation into Polyhedral Model for Application Analysis

After having fixed the processing order, implementation of a multidimensional dataflow graph requires the creation of an actor schedule. It establishes a temporal order between the firings of different actors. Furthermore, is necessary to compute the buffer sizes associated to each edge. Since the arrays transported within a multidimensional dataflow graph might be very huge, an analysis on the level of individual data elements or actor firings can get computationally very expensive.

Polyhedral analysis is a tool that aims to alleviate these difficulties [16]. Its basic idea consists in putting different arrays into relation to each other, avoiding to consider each array point individually. While polyhedral analysis is well known in the domain of loop programs, the one-dimensional nature of most dataflow models prevents its direct application to the world of dataflow. The application of multidimensional models of computation, on the other hand, offers the possibility to employ this powerful tool also for dataflow based system design.

6.1   Principle

Polyhedral analysis can be applied to create schedules for multidimensional dataflow graphs, as well as computing for each edge the required buffer size. Therefore it is necessary to translate the dataflow graph into a polyhedral representation. This can be achieved by representing the firings of each actor in form of a point grid, and putting these point grids into relation to each other. This defines an actor schedule indicating the temporal order of the actor firings. The following paragraphs will give

---

[5]For Fig. 10b, a first level block with two rows and one column generates the desired processing order.

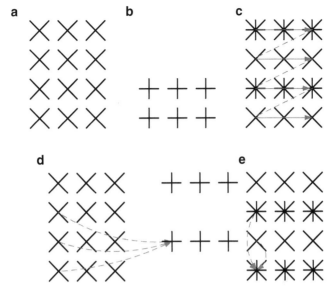

Fig. 12 Polyhedral analysis for Fig. 8. (**a**) Firing grid for actor A, (**b**) Firing grid for actor B, (**c**) Grids embedded into a common coordinate system, (**d**) Data dependencies, (**e**) Valid schedule

an overview about this process by means of examples. The main purpose consists in illustrating the major idea, while the rigorous mathematical formulation can be found in [12].

Figure 12 depicts the principles of polyhedral analysis for scheduling when applied to the dataflow graph given in Fig. 8. Figure 12a shows the *firing grid* for actor A. As defined in Fig. 8b, each firing of actor A causes the production of a $(1, 2)$ token. Hence, in order to generate a complete output array of size $(4, 6)$, three firings in horizontal and four firings in vertical directions are necessary. Each of these firings is represented by a cross in Fig. 12a. Horizontally adjacent firings read and write horizontally adjacent data elements. The same holds for the vertical direction.

A similar operation can be done for actor B (Fig. 12b), and all other actors of the dataflow graph. Next all these firing grids need to be put into relation to each other by embedding them into a common coordinate system as illustrated in Fig. 12c. In order to not overburden the illustrations, only actors A and B are displayed. Both use the same symbols than before such that the firings of actor A can be distinguished from those of actor B.

The embedded grid can now be used to derive an actor schedule. The idea is to traverse the embedded grid in a raster scan order and to fire the actors for each encountered cross. Depending on whether a concurrent or a sequential implementation is preferred, coinciding crosses are either executed concurrently, or sequentially. Applied to Fig. 12c and assuming a sequential implementation, this would mean to first fire actor A, followed by B. Next its again the turn of actor A, followed by B and so on until arriving at the last cross in the first row of Fig. 12c. By moving to the second row, the schedule requests to fire three times actor A without any execution of actor B.

At this point it is important to notice that the grid of actor B has been stretched in vertical direction compared to actor A, because A fires twice as often as actor B in this dimension. Otherwise actor B would show a period of huge activity with six firings, followed by a period with no activity at all. Since this leads to suboptimal implementations in terms of buffer size and required hardware resources, it is to avoid. Stretching the grid of actor B distributes the firings better over time.[6]

Obviously the schedule discussed above is not valid. As clearly shown in Fig. 8, the first execution of actor B requires the first two data elements (pixels) of the second array row. They are, however, only produced by the 4th execution of actor A. The schedule depicted in Fig. 12c, however, requests to immediately fire actor B after the first execution of actor A. In order to avoid this kind of problems, a data dependency analysis has to be performed. It establishes the dataflow between the different actor firings.

Figure 12d depicts a corresponding example for one firing of actor B. In order to simplify the illustration, the grid of actor B has been placed next to the grid of actor A. The depicted arrows show the flow of tokens from the firings of actor A to the firings of actor B. They can be derived from Fig. 8a by evaluating which firings of actor A produce data elements that are required by the considered firing of actor B. Obviously, the chosen firing of actor B requires the tokens produced by three different firings of actor A. The same analysis can be done for all other firings of actor B.

With this information it is then possible to compute a valid schedule by shifting the different actor grids relatively to each other. To this end it must be ensured that for each actor firing all required input data elements are already produced. Or in other words, a valid schedule must take care that no arrows point from source firings to sink firings that have already been executed in the past. Graphically, this is equivalent with all data dependency arrows pointing either to the right, or in bottom direction.[7] Figure 12e depicts the corresponding result. The schedule thus first requests to fire three times actor A, without any execution of actor B. Next actor A is to be fired, followed by actor B. Note that the latter disposes of all required input tokens after the previous execution of A.

---

[6]While the stretching depicted in Fig. 12c improves the situation, actor B still oscillates between a period of activity with three firings and a period with no activity. To solve this problem, more aggressive stretching techniques are discussed in [12].

[7]Mathematically, this can be expressed by the lexicographic order. See [16] as well as [12] for more details.

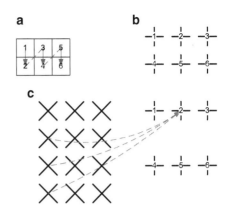

Fig. 13 Firing grid for processing orders not following the horizontal raster scan order. (**a**) Produced output tokens, (**b**) Firing grid, (**c**) Data dependency analysis

6.2 Different Processing Orders

The careful reader might have observed that the above reasoning is only valid as long as all actors follow horizontal raster scan order. As however discussed in Sect. 5, this is not necessarily the case. Solving this problem is possible by performing a grid point remapping. This means that the schedule is still constructed by traversing the common grid in raster scan order. However, when computing the data dependencies, it is taken into account that the actors themselves show a different processing order.

To illustrate the principles, assume that actor B of Fig. 8 now generates the output tokens in vertical raster order as shown in Fig. 13a. Every data element is produced in a single firing. Hence, an overall number 2×3 firings is necessary to generate the $(2,3)$ output array. Figure 13b depicts the corresponding firing grid of actor B, consisting of 2×3 crosses. It is defined that the firing grid is still traversed in raster scan order. Consequently, in order to take the correct processing order of actor B into account, a mapping between the firing grid points and the actually produced output token has to be established. The result is depicted by numbers in Fig. 13a, b. For instance, the second firing produces the first data element in the second row of the output array.

As discussed in Sect. 5, the processing order has no impact on the relation between input windows and output tokens. Hence the output token number 2 still depends on the lower left sliding window as illustrated in Fig. 8a. Since the processing order of actor A is assumed to be the same than in Sect. 6.1, this leads to the dependency analysis shown in Fig. 13c. It can be derived from the fact that the second firing of actor B reads the lower left sliding input window whose data elements are produced by the source firings being origins of the arrows in Fig. 13c. With this new dependency analysis, a valid schedule can be derived in the same manner as explained in Sect. 6.1.

Reference [12] shows how this can be mathematically expressed in form of integer programs solvable by cplex [8], or the PIP Library [1]. Furthermore it discusses the impact of non-overlapping grids. Since all these aspects exceed the scope of this book, the interested reader is refereed to [12].

Fig. 14 Buffer size
computation

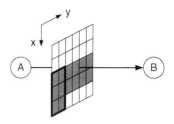

6.3   Buffer Size Computation

The implementation of local array algorithms as depicted in Fig. 8a requires buffer
memory for storing data elements produced by the individual source actors. To this
end, we assume that the data elements are stored in the order of production into
a memory with limited size. The question then arises how many data elements
must be buffered simultaneously, defining thus the necessary memory size. The
edge between actor A and B in Fig. 8a, for instance, requires at least a buffer
encompassing two rows and two pixels, when both are executed in horizontal raster
scan order. The reason can be seen by considering the firing of actor B reading the
lower left sliding window as depicted in Fig. 14. Obviously, the window covers
tokens from three different array rows. Since the tokens are generated in horizontal
raster scan order by actor A, all gray shaded data elements need to be kept in a buffer
for proper implementation.

While for this simple example the buffer computation seems to be easy, it
can quickly become challenging when considering complex graph topologies with
different processing orders. Fortunately, polyhedral analysis permits to compute the
buffer sizes associated to the different dataflow edges, once the actor schedule is
fixed. The principle idea is to find the data dependency vector that spans the largest
number of source actor firing grid points. In Fig. 12e for instance, the first firing of
actor A in the second row generates a token that will be read by the first sink firing
in the fourth grid row. Consequently, all data elements produced between these two
instances in time need to be kept into a memory buffer. In other words, the buffer
needs to encompass at least $2 \cdot 3 + 1$ tokens, because the arrow spans two rows, and
actor A needs to be executed before actor B in the fourth grid row. Since a token
consists of two data elements, an overall buffer size of 14 elements is necessary.
Note that this exactly corresponds to the results shown in Fig. 14.

Again this can be computed by integer programs that are explained in [12].

7   Dynamic Multidimensional Dataflow

The previous sections have explained how multidimensional dataflow eases the
description of applications processing arrays. In particular it introduces new degrees
of freedom during system design and enables powerful analysis methods that cannot

Fig. 15 Different read and
write orders

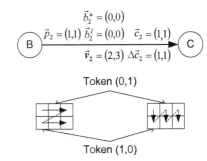

be directly applied to one-dimensional models of computation. In all these cases a fixed relation between produced output tokens and input windows has been assumed. In more detail, adjacent input windows have been related to adjacent output tokens. This strict relation reduces the decisions that can be performed during run-time of a dataflow graph.

The purpose of this section consists in alleviating this restriction. To this end it first discusses how multidimensional communication can be embedded into a FIFO-like structure. This permits to combine multidimensional dataflow with finite state machines which can express dynamic decisions as well as non-deterministic behavior. Furthermore, this allows to create dataflow models which use one-dimensional and multidimensional communication at the same time.

7.1   The Multidimensional FIFO

While one-dimensional dataflow models of computation can be refined into a concrete implementation using FIFO communication [7], the multidimensional models discussed in this chapter offer an additional degree of freedom in form of the processing order (see Sect. 5). This processing order defines the order by which the arrays are read and written. Since however every actor can have its own processing order, the array elements need not to be written and read in the same order. For instance, while the source actor of an edge might generate the array in horizontal raster scan order (Fig. 10a), the sink can read it in vertical raster scan order (Fig. 10b). Furthermore, array elements might be read multiple times because of the overlapping windows. As a consequence, data elements might be written and read in different orders.

Figure 15 depicts a corresponding example by means of a trivial multidimensional dataflow graph. The source actor B produces a $(2,3)$ array, where each firing leads to a $(1,1)$ token generated in horizontal raster scan order. This array is read by actor C using a $(1,1)$ window. In contrast to actor B, actor C uses a vertical raster scan order. Consider now two example tokens having the coordinates $(0,1)$ and $(1,0)$. Actor B writes token $(0,1)$ *before* token $(1,0)$. Actor C, however, reads

Fig. 16 Multidimensional
FIFO

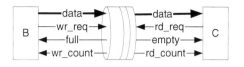

token $(0,1)$ *after* token $(1,0)$. In other words, the read and write order is not the same anymore, preventing the use of traditional FIFOs for communication synthesis.

However, as soon as the read and write orders are fixed, it is possible to embed this kind of communication into a FIFO-like primitive, called *multidimensional FIFO*. It provides the same interface than a classical FIFO as depicted in Fig. 16. The *data ports* transport the proper data in form of multidimensional arrays. In other words, the data port for the read interface provides access to all data elements belonging to the currently processed sliding window. The write interface accepts all data elements that need to be written for the currently generated output token. A *full query* reveals whether the source can write the next multidimensional output token. Similarly, and *empty query* indicates whether the next sliding window is already available or not. A *write counter* is able to predict how many multidimensional tokens can be written by the source without any firing of the sink. Similarly, a *read counter* indicates how many sliding windows can be accessed by the sink without any further firing of the source.

In order to clarify the internal operation of the multidimensional FIFO, consider again Fig. 15. The processing order and thus the read and write orders are considered to be fixed. Hence, after the first firing of actor B, the multidimensional FIFO can derive that the first sliding window is available for actor C. The FIFO is hence not empty anymore, and actor C can access all data elements belonging to this window. Once the read operation is terminated, actor C can request a new read. Because of the processing order of actor C, the next required data element is situated in the lower left corner of the input array. Since it is not already available, an empty-condition is signaled. This remains even true, when actor B fires the second time, because the data element produced during this invocation is not the one required next by actor C. Hence, the data element written by actor B needs to be buffered for later use. This situation only changes after actor B has finished the fourth firing, resulting in the data element required next by actor C. Note that this condition allows actor C to fire twice, because all required data elements are already available.

By tracking which data elements will still be read later on by the sink, the multidimensional FIFO can also perform the buffer management. This includes both the decision at which address to write a given data element as well as taking care that data elements will not be overwritten when still needed. The corresponding principles can be implemented both in hardware and software. The mathematical analysis necessary for this purpose can be found in [12]. Note that the same principles can also be applied to examples where both the source and the sink write or read tokens consisting of more than one data element. In particular, both overlapping windows as well border processing discussed in Sect. 4 are supported. It only requires to assume a fixed processing order, which might however vary for the different actors of the dataflow graph.

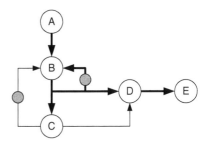

Fig. 17 WDF graph using one- and multidimensional token transport. *Bold arrays* correspond to multidimensional token transport, *thin arrays* to one-dimensional streams of data. *Dashed circles* define initial tokens respectively arrays

7.2 Windowed Dataflow (WDF)

The multidimensional FIFO described in the previous section permits to introduce a new class of multidimensional dataflow that allows for dynamic decisions, called *windowed dataflow (WDF)*. Such a dataflow graph consists of actors interconnected by two types of edges. These edges either represent one-dimensional streaming communication as employed for one-dimensional models of computation, or multidimensional communication represented by the multidimensional FIFO. In both cases actors can request one or more input tokens, and write one or more output tokens. The tokens might be one- or multidimensional. If only one of these tokens is not available, the actor blocks, leading to deterministic applications. In this case, we have a KPN-like behavior [6]. Alternatively, if non-determinism is allowed, an actor can also check whether enough input tokens are available on a given input edge. If not, an alternative processing path can be taken. This leads to the most general class of dynamic dataflow [2].

This general approach destroys the strict relation between input windows and output tokens, being fundamental for MDSDF and WSDF. Since an actor might decide to consume one input window, and then produce ten output tokens without reading any further input, it is not true anymore that adjacent input windows are related to adjacent output tokens. In particular, it is not possible anymore to define a processing order per actor. Instead each edge can have its own read and write order. In other words, an actor having two input ports might read the array on the first input edge in horizontal raster scan order, the array on the second input edge in vertical raster scan order. While this increases the expressiveness, it complicates the data dependency analysis and is hence not allowed in WSDF nor MDSDF.

Figure 17 illustrates a simple example graph demonstrating the interaction between one- and multidimensional dataflow. The purpose consists in iteratively filtering an input image until a certain exit condition is met. This corresponds to the behavior of a *while-loop*. Actor A is supposed to generate the multidimensional input array. Actor B has two multidimensional inputs. Depending on a one-dimensional control token, actor B either filters the array generated by actor A and discards the array on the second multidimensional input, or it filters its previous output and does not read anything from actor A. Actor C filters the output of actor B using a window of size $(1,1)$, in order to determine whether the exit condition is met.

Depending on the result, it generates a one-dimensional control token at the end of each array. This corresponds to a CSDF-like behavior of actor C. Actor D obtains the same control token in order to decide whether the result can be forwarded to the destination actor E, or whether the array represents an intermediate result and needs to be discarded. The graph contains hence a mixture of one- and multidimensional communication with dynamic decisions.

7.3 Combination of Multidimensional Dataflow with Finite State Machines

The model discussed in the previous section did not detail the inner structure of an actor. In fact they can be modeled by any sort of language or description technique. In particular, it is possible to define the dynamic behavior using finite state machines. This allows a general description technique for all sorts of actors, followed by a classification process whether an actor is static or dynamic. Depending on this classification, appropriate analysis techniques can be applied. The polyhedral analysis, for instance requires strictly static models of computation in order to permit the embedding into a common grid. Since the multidimensional FIFO uses the same interface than its one-dimensional counterpart, combination with finite state machines is very similar to the concepts described in [4].

8 Matrix-Based Multidimensional Dataflow

The WSDF model of computation as well as its subset MDSDF discussed in Sects. 4 and 2 produce and consume tokens along the Cartesian coordinate axes. While it is possible to define in which order these tokens are read and written (see Sect. 5), the ensemble of sliding windows and produced tokens still follow a rectangular grid for each edge. Furthermore non-rectangular windows need a decomposition into subwindows [12]. Finally, for both WSDF and MDSDF horizontally adjacent input windows are related to horizontally adjacent output tokens. The same holds for the other dimensions.

Section 3 discussed a generalization of the sampling patterns by introduction of corresponding sampling matrices. However overlapping sliding windows are not considered. Section 7, on the other hand, relaxed the strict relation between input and output tokens. However, this has to be paid by reduced analysis capabilities. Furthermore on each edge, the ensemble of all produced and consumed tokens still follow a rectangular grid.

For applications that need non-rectangular sampling patterns, use of a matrix based notation can be a corresponding remedy. Array-OL [5] follows such a

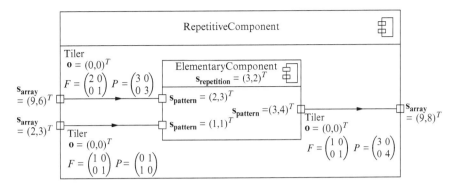

Fig. 18 Specification of repetitive tasks via tilers [5]. Each port is annotated with the produced and consumed array size. The number of sub-task invocations is defined by vector $s_{repetition}$. The tiler associated to each communication edge specifies the extraction of sub-arrays

methodology. Originally developed by Thomson Marconi Sonar,[8] *Array-OL* serves as model of computation for the *Gaspard2* design flow and has also been integrated into the Ptolemy framework [3]. Applications are modeled as a hierarchical task graph consisting of (i) *elementary*, (ii) *hierarchical*, and (iii) *repetitive* tasks. *Elementary* tasks are defined by their input and output ports as well as an associated function that transforms input arrays on the input ports to output arrays on the output ports. *Hierarchical* tasks connect *Array-OL* sub-tasks to acyclic graphs. Each connection between two tasks transports a multidimensional array.

Repetitive tasks as shown exemplarily in Fig. 18 are finally the key element for expressing data-parallelism. In principle they represent special hierarchical tasks in which a sub-task is repeatedly applied to the input arrays. Since the different sub-task invocations are considered as independent if not specified otherwise, they can be executed in parallel or sequentially in any order. So-called *input tilers* specify for each sub-task invocation the required part of the input arrays. These sub-arrays are also called *patterns* or *tiles*. The produced results are combined by *output tilers* to the resulting array.

Each tiler is defined by a paving and a fitting matrix as well as an origin vector **o** (see Fig. 18). The *paving matrix P* permits to calculate for each task invocation **i** the origin of the input or output pattern:

$$\forall \mathbf{i}, 0 \leq \mathbf{i} < s_{repetition} : \mathbf{o_i} = \mathbf{o} + P \cdot \mathbf{i} \bmod s_{array}$$

$s_{repetition}$ defines the number of repetitions for the sub-task. s_{array} is the size of the complete array, the modulo operation takes into account that a toroidal coordinate system is assumed. In other words, patterns that leave the array on the right border

---

[8]Now Thales.

re-enter it on the left side. Based on the origin o_i, the data elements belonging to a certain tile can be calculated by the *fitting matrix* F using the following equation:

$$\forall j, 0 \leq j < s_{pattern} : x_j = o_i + F \cdot j \bmod s_{array}$$

$s_{pattern}$ defines the pattern size and equals the required array size annotated at each port.

While such a matrix based approach generalizes the sampling pattern, it also adds new complexity. For instance, border processing as discussed in Sect. 4.2.2 cannot directly be covered. Furthermore, properties such as whether all tokens have been written exactly once by an actor are to be verified.

9   Conclusion

Applications processing multidimensional arrays are cumbersome to represent using one-dimensional models of computation, or lack enough precision for optimized analysis and synthesis. To solve this problem, a number of multidimensional models of computation are available that differ in their expressiveness and possibilities for analysis. By these means it is possible to profit from powerful analysis techniques that are incompatible with one-dimensional dataflow. At the same time systems with varying degree of dynamic decisions can be represented. Seamless interaction with one-dimensional models of computation enable the modeling of complete systems showing both data and task parallelism. This is an important benefit taking the unbowed trend towards parallel architectures and systems into account.

References

1. Piplib. http://www.piplib.org/
2. Bhattacharyya, S.S., Deprettere, E.F., Theelen, B.: Dynamic dataflow graphs. In: S.S. Bhattacharyya, E.F. Deprettere, R. Leupers, J. Takala (eds.) Handbook of Signal Processing Systems, second edn. Springer (2013)
3. Dumont, P., Boulet, P.: Another multidimensional synchronous dataflow: Simulating Array-OL in Ptolemy II. Tech. Rep. 5516, Institut National de Recherche en Informatique et en Automatique, Cité Scientifique, 59 655 Villeneuve d'Ascq Cedex (2005)
4. Falk, J., Haubelt, C., Zebelein, C., Teich, J.: Integrated modeling using finite state machines and dataflow graphs. In: S.S. Bhattacharyya, E.F. Deprettere, R. Leupers, J. Takala (eds.) Handbook of Signal Processing Systems, second edn. Springer (2013)
5. Gamatié, A., Rutten, E., Yu, H., Boulet, P., Dekeyser, J.L.: Synchronous modeling and analysis of data intensive applications. EURASIP Journal on Embedded Systems **2008**(561863), 1–22 (2008). DOI 10.1155/2008/561863
6. Geilen, M., Basten, T.: Kahn process networks and a reactive extension. In: S.S. Bhattacharyya, E.F. Deprettere, R. Leupers, J. Takala (eds.) Handbook of Signal Processing Systems, second edn. Springer (2013)

7. Ha, S., Oh, H.: Decidable dataflow models for signal processing: Synchronous dataflow and its extensions. In: S.S. Bhattacharyya, E.F. Deprettere, R. Leupers, J. Takala (eds.) Handbook of Signal Processing Systems, second edn. Springer (2013)

8. ILOG: cplex. http://www.ilog.com/products/cplex/

9. ISO/IEC JTC1/SC29/WG1: JPEG2000 Part I final committee draft version 1.0 (2002). N1646R

10. Keinert, J., Deprettere, E.F.: Multidimensional dataflow graphs. In: S.S. Bhattacharyya, E.F. Deprettere, R. Leupers, J. Takala (eds.) Handbook of Signal Processing Systems, second edn. Springer (2013)

11. Keinert, J., Falk, J., Haubelt, C., Teich, J.: Actor-oriented modeling and simulation of sliding window image processing algorithms. In: Proceedings of the 2007 IEEE/ACM/IFIP Workshop of Embedded Systems for Real-Time Multimedia (ESTIMEDIA 2007), pp. 113–118 (2007)

12. Keinert, J., Teich, J.: Design of Image Processing Embedded Systems Using Multidimensional Data Flow. Springer (2011)

13. Murthy, P.K.: Scheduling techniques for synchronous and multidimensional synchronous dataflow. Memorandum no. ucb/erl m96/79, Electronics Research Laboratory, College of Engineering, University of California, Berkeley (1996)

14. Murthy, P.K., Lee, E.A.: Multidimensional synchronous dataflow. IEEE Transactions on Signal Processing **50**(8), 2064–2079 (2002)

15. Vainio, O.: Mixed signal techniques. In: S.S. Bhattacharyya, E.F. Deprettere, R. Leupers, J. Takala (eds.) Handbook of Signal Processing Systems, second edn. Springer (2013)

16. Verdoolaege, S.: Polyhedral process networks. In: S.S. Bhattacharyya, E.F. Deprettere, R. Leupers, J. Takala (eds.) Handbook of Signal Processing Systems, second edn. Springer (2013)

17. Yu, H., Leeser, M.: Automatic sliding window operation optimization for FPGA-based computing boards. In: Proceedings of the 14th Annual IEEE Symposium on Field-Programmable Custom Computing Machines (FCCM '06), pp. 76–88. IEEE Computer Society, Washington, DC, USA (2006)

Compiling for VLIW DSPs

Christoph W. Kessler

Abstract This chapter describes fundamental compiler techniques for VLIW DSP processors. We begin with a review of VLIW DSP architecture concepts, as far as relevant for the compiler writer. As a case study, we consider the TI TMS320C62x™ clustered VLIW DSP processor family. We survey the main tasks of VLIW DSP code generation, discuss instruction selection, cluster assignment, instruction scheduling and register allocation in some greater detail, and present selected techniques for these, both heuristic and optimal ones. Some emphasis is put on phase ordering problems and on phase coupled and integrated code generation techniques.

1 VLIW DSP Architecture Concepts and Resource Modeling

In order to satisfy high performance demands, modern processor architectures exploit various kinds of parallelism in programs: *thread-level parallelism* (i.e., running multiple program threads in parallel on multi-core and/or hardware-multithreaded processors), *data-level parallelism* (i.e., executing the same instruction or operation on several parts of a long data word or on a vector of multiple data words together), *memory-level parallelism* (i.e., overlapping memory access latency with other, independent computation on the processor), and *instruction-level parallelism* (i.e., overlapping the execution of several instructions in time, using different resources of the processor in parallel at a time).

By pipelined execution of subsequent instructions, a certain amount of instruction level parallelism (ILP) can already be exploited in ordinary sequential RISC processors that issue a single instruction at a time. More ILP can often be leveraged by *multiple-issue* architectures, where execution of several independent instructions

---

C.W. Kessler (✉)
Department of Computer Science (IDA), Linköping University, 58183 Linköping, Sweden
e-mail: chrke@ida.liu.se

S.S. Bhattacharyya et al. (eds.), *Handbook of Signal Processing Systems*,
DOI 10.1007/978-1-4614-6859-2_36, © Springer Science+Business Media, LLC 2013

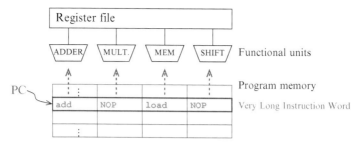

Fig. 1 A traditional VLIW processor with very long instruction words consisting of four issue slots, each one controlling one functional unit

can be started in parallel, resulting in a higher throughput (instructions per clock cycle, IPC). The maximum number of instructions that can be issued simultaneously is called the *issue width*, denoted by ω. In this chapter, we focus on multiple-issue instruction-level parallel DSP architectures, i.e., $\omega > 1$.

ILP in programs can either be given explicitly or implicitly. With *implicit ILP*, dependences between instructions are implicitly given in the form of register and memory addresses read and written by instructions in a sequential instruction stream, and it is the task of a run-time (usually, hardware) scheduler to identify instructions that are independent and do not compete for the same resource, and thus can be issued in parallel to different available functional units of the processor. Superscalar processors use such a hardware scheduler to analyze for data dependences and resource conflicts on-the-fly within a given fixed-size window over the next instructions in the instruction stream. While convenient for the programmer, superscalar processors require high energy and silicon overhead for analyzing dependences and dispatching instructions.

With *explicit ILP*, the assembler-level programmer or compiler is responsible to identify independent instructions that should execute in parallel, and group them together into *issue packets* (also known as *instruction groups* in the literature). The elementary instructions in an issue packet will be dispatched simultaneously to different functional units for parallel execution. In the following, we will consider explicit ILP architectures.

The issue packets do not necessarily correspond one-to-one to the units of instruction fetch. The processor's instruction fetch unit usually reads properly aligned, fixed-sized blocks of bytes from program memory, which contain a fixed number of elementary instructions, and decodes them together. We refer to these blocks as *fetch packets* (also known as *instruction bundles* in the literature). For instance, a fetch packet for the Itanium IA-64 processor family contains three instructions, and fetch packets for the TI 'C62x contain eight instructions.

In the traditional VLIW architectures (see Fig. 1), the issue packets coincide with the fetch packets; they have a fixed length of L bytes that are L-byte aligned in instruction (cache) memory, and are called *Very Long Instruction Words* (*VLIWs*). A VLIW contains $\omega > 1$ predefined slots for elementary instructions.

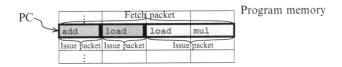

Fig. 2 Several issue packets may be accommodated within a single fetch packet. Here, the framed fetch packet contains three issue packets: the first two contain just one elementary instruction each, while the third one contains two parallel instructions

Each instruction slot may be dedicated to a certain kind of instructions or to controlling a specific functional unit of the processor. Not all instruction slots have to be used; unused slots are marked by NOP (no operation) instructions. While decoding is straightforward, code density can be low if there is not enough ILP to fill most of the slots; this wastes program memory space and instruction fetch bandwidth.

Instead, most explicit ILP architectures nowadays allow to pack and encode instructions more flexibly in program memory. An instruction of specific kind may be placed in several or all possible instruction slots of a fetch packet. Also, a fetch packet may accommodate several issue packets, as illustrated in Fig. 2; the boundaries between these may, for instance, be marked by special delimiter bits. The different issue packets in a fetch packet will be issued subsequently for execution. The processor hardware is responsible for extracting the issue packets from a stream of fetch packets.[1] In the DSP domain, the Texas Instruments TI TMS320C6x processor family uses such a flexible encoding schema, which we will present in Sect. 2.

The existence of multiple individual RISC-like elementary instructions as separate slots within an issue packet to express parallel issue is a key feature of VLIW and EPIC architectures. In contrast, consider dual-MAC (multiply-accumulate) instructions that are provided in some DSP processors but encoded as a single instruction (albeit a very powerful one) in a linear instruction stream. Such instructions are, by themselves, not based on VLIW but should rather be considered as a special case of SIMD (single instruction multiple data) instructions. Indeed, SIMD instructions can occur as elementary instructions in VLIW instruction sets. Generally, a *SIMD instruction* applies the same arithmetic or logical operation to multiple operand data items in parallel. These operand items usually need to reside in adjacent registers or memory locations to be treated and addressed as single long data words. Hence, SIMD instructions have only one opcode, while issue packets in VLIW/EPIC architectures have one opcode per elementary instruction.

The appropriate issue width and the number of parallel functional units for a VLIW processor design depends, beyond architectural constraints, on the characteristics of the intended application domain. While the average ILP degree achievable in general-purpose programs is usually low, it can be significantly higher in the

---

[1]Processors that decouple issue packets from fetch packets are commonly also referred to as *Explicitly Parallel Instruction set Computing* (EPIC) architectures.

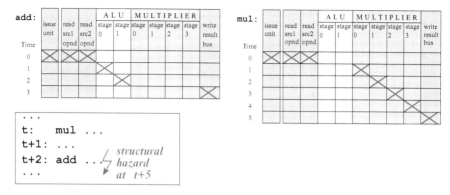

Fig. 3 *Top*: Example reservation tables for addition and multiplication on a pipelined processor with an ALU and multiplier unit. Resources such as register file access ports and pipeline stages on the functional units span the horizontal axis of the reservation tables while time flows downwards. Time slot 0 represents the issue time. *Bottom*: Scheduling an add instruction two clock cycles after a mul instruction would lead to conflicting subscriptions of the result resource (write back to register file). Here, the issue of add would have to be delayed to, say, time $t + 3$. If exposed to the programmer/compiler, a nop instruction could be added before the add to fill the issue slot at time $t + 2$. Otherwise, the processor will handle the delay automatically by stalling the pipeline for one cycle

computational kernels of typical DSP applications. For instance, Gangwar et al. [41] report for DSPstone and Mediabench benchmark kernels an achievable ILP degree of 20 on average for a (clustered) VLIW architecture with 16 ALUs and 8 load-store units. Moreover, program transformations can be applied to increase exploitable ILP; we will discuss some of these later in this chapter.

1.1 Resource Modeling

We model instruction issue and resource usage explicitly. An instruction i issued at time t occupies an *issue slot* (e.g., a slot in a VLIW) at time t and possibly[2] several resources (such as functional units or buses) at time t or later.

For each instruction type, its required resource reservations relative to the issue time t are specified in a *reservation table* [25], a boolean matrix where the entry in row j and column u indicates if the instruction uses resource u in clock cycle $t + j$.

If an instruction is issued at a time t, its reservations of resources are committed to a *global resource usage map* or *table*. Two instructions are in conflict with each other if their resource reservations overlap in the global resource usage map; this is also known as a *structural hazard*. See Fig. 3 for an example. In such a case, one of the two instructions has to be issued at a later time to avoid duplicate reservations of the same resource.

---

[2]NOP (no operation) instructions only occupy an issue slot but no further resources.

Non-pipelined resources that have to be reserved for more than one clock cycle in sequence can thus lead to delayed issuing of subsequent instructions that should use the same resource. The *occupation time* $o(i, j)$ denotes the minimum distance in issue time between two (data-independent) instructions i and j that are to be issued subsequently on the same issue unit and that subscribe to a common resource. Hence, the occupation time only depends on the instruction types. For instance, in Fig. 3, $o(\text{add}, \text{add}) = 1$. In fact, for most processors, occupation times are generally 1.

Sets of time slots on one or several physical resources (such as pipeline stages in functional units or buses) can often be modeled together as a single virtual resource if it is clear from analysis of the instruction set that, once an instruction is assigned the earliest one of the resource slots in this subset, no other instruction could possibly interfere with it in later slots or other resources in the same subset.

A processor is called *fully pipelined* if it can be modeled with virtual resources such that all occupation times are 1, there are no exposed structural hazards, and the reservation table for an instruction thus degenerates to a vector over the virtual resources. On regular VLIW architectures, these virtual resources often correspond one-to-one to functional units.

1.2 Latency and Register Write Models

Consider two instructions i_1 and i_2 where i_1 issued at time t_1 produces (writes) a value so that it is available at the beginning of time slot $t_1 + \delta_{w1}$ in some register r, which is to be consumed (read) by i_2 at time $t_2 + \delta_{r2}$. The time of writing the result relative to the issue time, δ_{w1} is called the *write latency*[3] of i_1, and δ_{r2} the *read latency* of i_2. For the earliest possible issue time t_2 of i_2 we have to preserve the constraint

$$t_2 \geq t_1 + \delta_{w1} - \delta_{r2}$$

to make sure that the operand value for i_2 is available in the register.

We refer to the minimum difference in issue times induced by data dependence, i.e.,

$$\ell(i_1, i_2) = \delta_{w1} - \delta_{r2}$$

as *latency* between i_1 and i_2. See Fig. 4 for an illustration. For memory data dependences between store and load instructions, latency is defined accordingly.

---

[3]For simplicity of presentation, we assume here that write latency and read latency are constants for each instruction. In general, they may in some cases depend on run-time conditions exposed by the hardware and vary in an interval between earliest and latest write resp. read latency. See also our remarks on the LE model further below. For a more detailed latency model, we refer to Rau et al. [87].

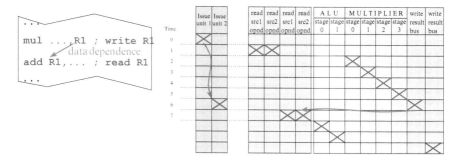

Fig. 4 A read-after-write (flow) data dependence forces the instruction scheduler to await the latency of six clock cycles between the producing and consuming instruction to make sure that the value written to register R1 is read

Latencies are normally positive, because operations usually read operands early in their execution and write results just before terminating. For the same reason, the occupation time usually does not exceed the latency. Only for uncommon combinations of an early-writing i_1 with a late-reading i_2, or in the case of write-after-read dependences, negative latencies could occur, which means that a successor instruction actually could be issued before its predecessor instruction in the data dependence graph and still preserve the data dependence. However, this only applies to the EQ model, which we now explain.

There exist two different latency models with respect to the result register write time: EQ (for "equal") and LE (for "less than or equal"). Both models are being used in VLIW DSP processors.

The *EQ model* specifies that the result register of an instruction i_1 issued at time t_1 will be written exactly at the end of time slot $t_1 + \delta_{w1} - 1$, not earlier and not later. Hence, the destination register r only needs to be reserved from time $t_1 + \delta_{w1}$ on.

In the *LE model*, $t_1 + \delta_{w1}$ is an upper bound of the write time, but the write could happen at any time between issue time t_1 and $t_1 + \delta_{w1}$, depending on hardware-related issues. In the LE model, the destination register r must hence be reserved already from the issue time on. The EQ model allows to better utilize the registers, but the possibility of having several in-flight result values to be written to the same destination register makes it more difficult to handle interrupts properly.

In some architectures, latency only depends on the instruction type of the source instruction. If the latency $\ell(i, j)$ is the same for all possible instructions j that may directly depend on i (e.g., that use the result value written by i) we set $\ell(i) = \ell(i, j)$. Otherwise, on LE architectures, we could instead set $\ell(i) = \max_j \ell(i, j)$, i.e., the maximum latency to any possible direct successor instruction consuming the output value of i. The assumption that latency only depends on the source instruction is then a conservative simplification and may lead in some cases to somewhat longer register live ranges than necessary.

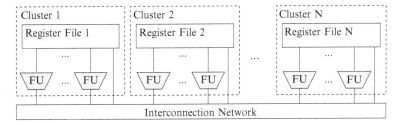

Fig. 5 Clustered VLIW processor with partitioned register set

The difference $\ell(i_1, i_2) - o(i_1, i_2)$ is usually referred to as the *delay*[4] of instruction i_1.

1.3 Clustered VLIW: Partitioned Register Sets

In VLIW architectures, possibly many instructions may execute in parallel, each accessing several operand registers and/or producing a result value to be written to some register. If each instruction should be able to access each register in a homogenous register set, the demands on the number of parallel read and write ports to the register set, i.e., on the access bandwidth to the register set, become extraordinarily high. Register files with many ports have very high silicon area and energy costs, and even access latency grows.

A solution is to constrain general accessibility and partition the set of functional units and likewise the register set to form *clusters*. A cluster consists of a set of functional units and a local register set, see Fig. 5. Within a cluster, each functional unit can access each local register. However, the number of accesses to a remote register is strictly limited, usually to one per clock cycle and cluster. A task for the programmer (or compiler) is thus to plan in which cluster data should reside at runtime and on which cluster each operation is to be performed to minimize the loss of ILP due to the clustering constraints.

1.4 Control Hazards

The pipelined execution of instructions on modern processors, including VLIW processors, achieves maximum throughput only in the absence of data hazards, structural hazards, and control hazards. In VLIW processors, these hazards are exposed to the assembler-level programmer or compiler. Data hazards and structural hazards have been discussed above.

---

[4]Note that in some papers and texts, the meanings of the terms *delay* and *latency* are reversed.

Control hazards denote the fact that branch instructions may disrupt the linear fetch-decode-execute pipeline flow. Branch instructions are detected only in the decoding phase and the branch target may, in the case of conditional branches, be known even later during execution. If subsequent instructions have been fetched, decoded and partly executed on the "wrong" control flow branch when the branch is detected or the branch target is known, the effect of these instructions must be rolled back and the pipeline must restart from the branch target. This implies a non-zero delay in execution that may differ depending on the type of branch instruction (nonconditional branch, conditional branch taken as expected, or conditional branch not taken as expected). There are basically two possibilities how processors manage branch delays:

1. *Delayed branch*: The branch instruction semantics is re-defined to take its effect on the program counter only after a certain number of delay time slots. It is a task for global instruction scheduling (see Sect. 7) to try filling these branch delay slots with useful instructions that need to be executed anyway but do not influence the branch condition. If no other instructions can be moved to a branch delay slot, it has to be filled with a NOP instruction as placeholder.

2. *Pipeline stall*: The entire processor pipeline is frozen until the first instruction word has been loaded from the branch target. The delay is not explicit in the program code and may vary depending on the branch instruction type.

In particular, conditional branches have a detrimental effect on processor through-put. For this reason, hardware features and code generation techniques that allow to reduce the need for (conditional) branches are important. The most prominent one is *predicated execution*: each instruction takes an additional operand, a boolean predicate, which may be a constant or a variable in a predicate register. If the predicate evaluates to true, the instruction executes as usual. If it evaluates to false, the effect of that instruction is rolled back such that it behaves like a NOP instruction.

1.5   Hardware Loops

Many innermost loops in digital signal processing applications have a fixed number of iterations and a fixed-length loop body consisting of straight-line code. Some DSP processors therefore support a hardware loop construct. A special hardware loop setup instruction at the loop entry initializes an iteration count register and also specifies the number of subsequent instructions that are supposed to form the loop body. The iteration count register is advanced automatically after every execution of the loop body; no separate add instruction is necessary for that purpose. A backward branch instruction from the end to the beginning of the loop body is now no longer necessary either, as the processor automatically resets its program counter to the first loop instruction, unless the iteration count has reached its final value. Hardware loops have thus no overhead for loop control per loop iteration and only a marginal constant loop setup cost. Also, they do not suffer from control hazards, as the processor hardware knows well ahead of time where and whether to execute the next backward branch.

1.6   Examples of VLIW DSP Processors

In the next section, we will consider the TI'C62x DSP processor as a case study. Other VLIW/EPIC DSP processors include, e.g., the HP Lx/STMicroelectronics ST200, Analog Devices TigerSHARC ADSP-TS20xS [5] and the NXP (formerly Philips Semiconductors) TriMedia [81].

2   Case Study: TI 'C62x DSP Processor Family

As a case study, we consider a representative of Texas Instrument's C62x™/C64x™/ C67x™ family of clustered VLIW fixed-point digital signal processors (DSPs) with the VelociTI™ instruction set.

The Texas Instruments TI TMS320C6201™[90] (shorthand: 'C6201) is a high-performance fixed-point digital signal processor (DSP) clocked at 200 MHz. It is a clustered VLIW architecture with issue width $\omega = 8$. A block diagram is given in Fig. 6.

The 'C6201 has eight functional units, including two 16-bit multipliers for 32-bit results (the .M-units) and six 32/40-bit arithmetic-logical units (ALUs), of which two (the .D-units) are connected to on-chip data cache memory. The 'C62x CPUs are load-store architectures, i.e., all operands of arithmetic and logical operations must be constants or reside in registers, but not in memory. The data addressing (.D) units are used to load data from (data) memory to registers and store register contents to (data) memory. The load and store instructions exist in variants for 32-bit, 16-bit and 8-bit data. The two .L units (logical units) mainly provide 32-bit and 40-bit arithmetic and compare operations and 32-bit logical operations like and,

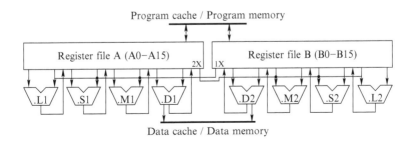

Fig. 6 The TI 'C6201 clustered VLIW DSP processor

| A | 1 | B | 1 | C | 0 | D | 1 | E | 1 | F | 0 | G | 1 | H | 0 |

Fig. 7 A fetch packet for the 'C62x can contain up to eight issue packets, as marked by the chaining bits. In this example, there are three issue packets: instructions A, B, C issued together, followed by D, E, F issued together and finally G and H issued together

or, xor. The two .S units (shift units) mainly provide 32-bit arithmetic and logical operations, 32-bit bit-level operations, 32-bit and 40-bit shifts, and branching.

The 'C62x architecture is fully pipelined. The reservation table of each instruction[5] is a 10×1 matrix, consisting of eight slots for the eight functional units and the two cross paths 1X and 2X (which will be described later) at issue time. In particular, each instruction execution occupies exactly one of the functional units at issue time. Separate slots for modeling instruction issue are thus not required, they coincide with the slots for the corresponding functional units. The occupation time is 1 for all instructions.

The 'C62x architecture offers the EQ latency model (there it is called "multiple assignment") for non-interruptible code, while the LE model (called "single assignment") should be used for interruptible code. Global enabling and disabling of interrupts is done by changing a flag in the processor's control status register.

The latency for load instructions is five clock cycles, for most arithmetic instructions it is one, and for multiply two clock cycles. Load and store instructions may optionally have an address autoincrement or -decrement side effect, which has latency one.

Some instructions are available on several units. For instance, additions can be done on the .L units, .S units and .D units.

Each of the two clusters A and B has 16 32-bit general purpose registers, which are connected to four units (including one multiplier and one load-store unit). The units of Cluster A are called .D1, .M1, .S1 and .L1, those of Cluster B are called .D2, .M2, .S2 and .L2. All units are fully pipelined (with occupation time 1), i.e., in principle, a new instruction could be issued to each unit in each clock cycle.

The 'C6201 has 128 KB on-chip static RAM, 64 KB for data and 64 KB for instructions.

An instruction fetch packet for the 'C62x family is 256 bit wide and is partitioned into eight instruction slots of 32 bit each. The least significant bit position of a 32-bit instruction slot is used as a *chaining bit* to indicate the limits of issue packets: if the chaining bit of slot i is 0, the instruction in the following slot $(i + 1)$ belongs to the next issue packet (see Fig. 7). Technically, issue packets cannot span several fetch

---

[5]Exception: For load and store instructions, two more resources are used to model load destination register resp. store source register access to the two register files, as only one loaded or stored register can be accessed per register file and clock cycle. Furthermore, load instructions can cause additional implicit delays (pipeline stalls) by unbalanced access to the internal memory banks (see later). This effect could likewise be modeled with additional resources representing the different memory banks. However, this will only be useful for predicting stalls where the alignment of the accessed memory addresses is statically known.

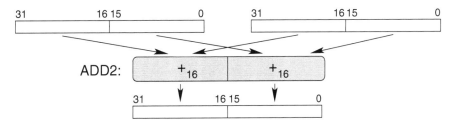

Fig. 8 The SIMD instruction ADD2 performs two 16-bit integer additions on the same functional unit in one clock cycle. The 32-bit registers shared by the operands and results are shown as *rectangles*

packets.[6] Hence, the maximum issue packet size of $\omega = 8$ occurs when all chaining bits (except perhaps the last one) are set in a fetch packet. As the other extreme, up to eight issue packets could occur in a fetch packet (if all chaining bits are cleared). The next fetch packet is not fetched before all issue packets of the previous one have been dispatched.

As each functional unit can do simple integer operations like addition, the 'C6201 can thus run up to eight integer operations per cycle, which amounts to 1600 MIPS (million instructions per second).

The ADD2 instruction, which executes on .S units, allows to perform a pair of 16-bit additions in a single clock cycle on the same functional unit, if the 16-bit operands (and results) each are packed into a common 32-bit register, see Fig. 8. One of these two 16-bit additions accesses the lower 16 bit (bits 0..15) of the registers, the other the higher 16 bit (bits 16..31). No carry is propagated from the lower to the higher 16-bit addition, which differs from the behavior of the 32-bit ADD instruction and therefore requires the separate opcode ADD2. The SUB2 instruction, also available on the .S units, works similarly for two 16-bit subtractions. Other instructions like bitwise AND, bitwise OR etc. work for 16-bit operand pairs in the same way as for 32-bit operands and thus do not need a separate opcode.

Within each cluster, each functional unit can access any register. At most one instruction per cluster and clock cycle can take one operand from the other cluster's register file, for which it needs to reserve the corresponding *cross path* (1X for accessing B registers from cluster A, and 2X for the other way), which is also modeled as a resource for this purpose. Assembler mnemonics encode the used resources as a suffix to the instruction's opcode: for instance, ADD.L1 is an ordinary addition on the .L1 unit using operands from cluster A only, while ADD.S2X

---

[6]Even though 'C62x assembly language allows an issue packet to start in a fetch packet and continue into the next one, the assembler will automatically create and insert a fresh fetch packet after the first one, move the pending issue packet there, and fill up the remainder of the first issue packet with NOP instructions.

Table 1 (a) Schedule generated by an early version of the TI-C compiler (12 cycles) [65] and (b) optimal schedule generated by OPTIMIST with dynamic programming (9 cycles) [58]

(a)	(b)
LDW.D1 *A4,B4	LDW.D1 *A0,A8 ǀǀ MV.L2X A1,B8
LDW.D1 *A1,A8	LDW.D2 *B8,B1 ǀǀ LDW.D1 *A2,A9 ǀǀ MV.L2X A3,B10
LDW.D1 *A3,A9	LDW.D2 *B10,B3 ǀǀ LDW.D1 *A4,A10 ǀǀ MV.L2X A5,B12
LDW.D1 *A0,B0	LDW.D2 *B12,B5 ǀǀ LDW.D1 *A6,A11 ǀǀ MV.L2X A7,B14
LDW.D1 *A2,B2	LDW.D2 *B14,B7 ǀǀ MV.L2X A8,B0
LDW.D1 *A5,B5	MV.L2X A9,B2
LDW.D1 *A7,A4	MV.L2X A10,B4
LDW.D1 *A6,B6	MV.L2X A11,B6
NOP 1	NOP 1 ; (last delay slot of LDW to B7)
MV.L2X A8,B1	
MV.L2X A9,B3	
MV.L2X A4,B7	

denotes an addition on the .S2 unit that accesses one A register via the cross path 2X. In total, there are 12 different instructions for addition (not counting the ADD2 option for 16-bit additions).

It becomes apparent that the problems of resource allocation (including cluster assignment) and instruction scheduling are not independent but should preferably be handled together to improve code quality. An example (adapted from Leupers [65]) is shown in Table 1: a basic block consisting of eight independent load (LDW) instructions is to be scheduled. The address operands are initially available (i.e., live on entry) in registers A0,...,A7 in register file A, the results are expected to be written to registers B0,...,B7 (i.e., live on exit) in register file B. Load instructions execute on the cluster containing the address operand register. The result can be written to either register file. However, only one load or store instruction can access a register file per clock cycle to write its destination register resp. read its source register; otherwise, the processor stalls for one clock cycle to serialize the competing accesses. Copying registers between clusters (which occupies the corresponding cross path) can be done by Move (MV), which is a shorthand for ADD with one zero operand, and has latency 1. As the processor has 2 load/store units and load has latency 5, a lower bound for the makespan (the time until all results are available) is eight clock cycles; it can be sharpened to nine clock cycles if we consider that at least one of the addresses has to be moved early to register file B to enable parallel computing, which takes one more clock cycle. A naive solution (a) sequentializes the computation by executing all load instructions on cluster A only. A more advanced schedule (b) utilizes both load/store units in parallel by transporting four of the addresses to cluster B as soon as possible, so the loads can run in parallel. Note also that no implicit pipeline stalls occur as the parallel load instructions always target different destination register files in their write-back phase, five clock cycles after issue time. Indeed, (b) is an optimal schedule; it was computed by the dynamic programming algorithm in OPTIMIST [58]. Generally,

Fig. 9 Interleaved internal data memory with four memory banks, each 16 bit (2 bytes) wide

Bank 0		Bank 1		Bank 2		Bank 3	
byte 0	byte 1	2	3	4	5	6	7
8	9	10	11	12	13	14	15
⋮	⋮	⋮	⋮	⋮	⋮	⋮	⋮

there can exist several optimal schedules. For instance, another one for this example is reported by Leupers [65], which was computed by a simulated annealing based heuristic.

Branch instructions on the 'C62x, which execute on the .S units, are delayed branches with a latency of six clock cycles, thus five delay slots are exposed. If two branches execute in the same issue packet (on .S1 and .S2 in parallel), control branches to the target for which the branch condition evaluates to true. This can be used to realize three-way branches. If both branch conditions evaluate to true, the behavior is undefined.

All 'C62x instructions can be predicated. The four most significant bits in the opcode form a *condition field*, where the first three bits specify the condition register tested, and the fourth bit specifies whether to test for equality or non-equality of that register with zero. Registers A1, A2, B0, B1 and B2 can serve as condition registers. The condition field code 0000 denotes unconditional execution.

Usually, branch targets will be at the beginning of an issue packet. However, branch targets can be any word address in instruction memory and thereby any instruction, which may also be in the middle of an issue packet. In that case, the instructions in that issue packet that appear in the program text before the branch target address will not take effect (are treated as NOPs).

Most 'C62x processor types use interleaved memory banks for the internal (on-chip) data memory. In most cases, data memory is organized in four 16-bit wide memory banks, and byte addresses are mapped cyclically across these (see Fig. 9). Each bank is single-ported memory, thus only one access is possible per clock cycle. If two load or store instructions try to access addresses in the same bank in the same clock cycle, the processor stalls for one cycle to serialize the accesses. For avoiding such delays, it is useful to know statically the alignment of addresses to be accessed in parallel, and make sure that these end up in different memory banks. Note also that load-word (LDW) and store-word (STW) instructions, which access 32-bit data, access two neighbored banks simultaneously. Word addresses must be aligned on word boundaries, i.e., the two least significant address bits are zero. Halfword addresses must be aligned on halfword boundaries.

TI 'C62x and 'C64x processors are fixed point DSP processors, where the 'C64x processors have architecture extensions that include, for instance, further support for SIMD processing (such as four-way 8 bit SIMD addition etc., four-way 16×16 bit multiply and eight-way 8×8 bit multiply), further instructions such as 32×32 multiply and complex multiply, compact (16-bit) instructions that can be mixed with 32-bit instructions [50], hardware support for software pipelining of loops, and more (2×32) registers. The TI 'C67x family also supports floatingpoint computations.

Beyond the 'C6x assembly language, TI provides three further *programming models* for 'C6x processors: (1) ANSI C, (2) C with calls to intrinsic functions that map one-to-one to 'C6x-specific instructions, such as _add2(), and (3) linear assembly code, which is RISC-like serial unscheduled code that uses 'C6x instructions but assumes no resource conflicts and only unit latencies. In general, the more processor-specific programming models allow to generate more efficient code. For instance, for an IIR filter example, TI reports that the software pipeline (see Sect. 7.2) generated from plain C code has a kernel length of five clock cycles, from C with intrinsics only four, while the linear assembly optimizer achieves three clock cycles and thus the best throughput [88].

3 VLIW DSP Code Generation Overview

In this section, we give a short overview of the main tasks in code generation that produce target-specific assembler code from a (mostly) target-independent intermediate representation of the program. We will consider these tasks and the main techniques used for them in some more detail in the following sections.

3.1 From Intermediate Representation to Target Code

Most modern compilers provide not just one but several *intermediate representations* (IR) of the program module being translated. These representations differ in their level of abstraction and degree of language independence and target independence. High-level representations such as abstract syntax trees follow the syntactic structure of the programs and represent e.g. loops and array accesses explicitly, while these constructs are, in low-level representations, lowered to branches and pointer arithmetics, respectively; such low-level IRs include control flow graphs, three-address code or quadruple sequences. A compiler supporting several different representations allows the different program analyses, optimizations and transformations to be implemented each on the level that is most appropriate for it. For instance, *common subexpression elimination* is best performed on a lower-level representation because more common subexpressions can be found after array accesses and other constructs have been lowered.

Code generation usually starts from a low-level intermediate representation (LIR) of the program. This LIR may be, to some degree, target dependent. For instance, IR operations for which no equivalent instruction exists on the target (e.g., there is no division instruction on the 'C62x) are lowered to equivalent sequences of LIR operations or to calls to corresponding routines in the compiler's run-time system.

For simple target architectures, the main tasks in code generation include instruction selection, instruction scheduling and register allocation:

- *Instruction selection* maps the abstract operations of the IR to equivalent instructions of the target processor. If we associate fixed resources (functional units, buses etc.) to be used with each instruction, this also includes the *resource allocation* problem. Details will be given in Sect. 4.
- *Instruction scheduling* arranges instructions in time, usually in order to minimize the overall execution time, subject to data dependence, control dependence and resource constraints. In particular, this includes the subproblems of *instruction sequencing*, i.e., determining a linear (usually, topological) order of instructions for scheduling, and *code compaction*, i.e. determining which independent instructions to execute in parallel and mapping these to slots in instruction issue packets and fetch packets. Local, loop-level and global instruction scheduling methods will be discussed in Sect. 7.
- *Register allocation* selects which values should, at what time during execution, reside in *some* target processor register. If there may not be enough registers available at a time, some values must be temporarily *spilled* to memory, which requires the generation and scheduling of additional spill code in the form of load and store instructions. The simpler problem of *register assignment* maps the run-time values that were allocated a register to a concrete register number. Details will be given in Sect. 6.

Advanced architectures such as clustered architectures may require additional tasks such as *cluster assignment* and data transfer generation (see Sect. 5), which may be considered separately or be combined with some of the above tasks. For instance, cluster assignment for instructions could be considered part of instruction selection, and cluster assignment for data may be modeled as an extension of register allocation.

Another task that is typical for DSP processors is that of *address code generation* for address generation units (AGUs). AGUs provide auto-increment and auto-decrement functionality as a parallel side effect to ordinary instructions that use special address registers for accessing data in memory. The AGUs may provide fixed offset values or offset registers to be used for in-/decrementing. A compiler could thus assign address registers and select offsets in an attempt to minimize the amount of residual addressing code that would still be computed with ordinary processor instructions on the main functional unit. See Sect. 3.1 in the chapter on *C Compilers and Code Optimizations for DSPs* [38] of this book for further details.

Further code generation problems frequently occuring with VLIW DSPs include exploiting available SIMD instructions, which can be regarded a subproblem of instruction selection, and optimizing memory data layout to avoid stalls caused by memory bank access conflicts. Also here we refer to the above-mentioned chapter [38], Sects. 3.3 and 3.4, for a discussion of SIMD code generation and optimization of memory bank assignment, respectively.

4 Instruction Selection and Resource Allocation

The instruction selection problem is often modeled as a pattern matching problem, where the processor-independent operations of the intermediate representation are to be covered by patterns that correspond to the semantics of target processor instructions. The description of the target processor instructions in terms of IR operations has to be provided by the compiler writer.

Each IR operation has to be covered by exactly one pattern node. Some examples are shown in Fig. 10. As intermediate results corresponding to inner edges of a multi-node pattern are no longer accessible in registers if that instruction is selected, such a pattern is only applicable as a cover if no other operations (such as SUB in Fig. 10b) access such an intermediate result. In other words, all outgoing edges from IR nodes covered by non-root pattern nodes must be covered by pattern edges.

Each pattern is associated with a *cost*, which is typically its occupation time or its latency as an a-priori estimation of the instruction's impact on overall execution time in the final code (the exact impact will only be known after the remaining tasks of code generation, in particular instruction scheduling, have been done). But also other cost metrics such as register space requirements are possible. The optimization goal is then to minimize the accumulated total cost of all covering patterns, subject to the condition that each IR operation is covered by exactly one pattern node.

The optimizing pattern matching problem can be solved in various ways. A common technique, *tree parsing*, is applicable if the patterns are tree-shaped and the data flow graph of the current basic block (for instruction selection, we usually consider one basic block of the input program at a time) is a *tree*. The patterns are modeled as tree rewrite rules in a tree grammar describing admissible derivations of coverings of the input tree. A heuristic solution could be determined by a LR-parser that selects, in each step of constructing bottom-up a derivation the input tree, in a greedy way whenever there are several applicable rules (patterns) to choose from [43]. An optimal solution (i.e., a minimum-cost covering of the tree with respect to the given costs) can be computed in polynomial time by a dynamic programming algorithm that keeps track of all possibly optimal coverings in a subtree [1, 39].

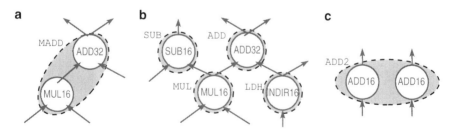

Fig. 10 Examples for covering IR nodes (*solid circles*) and edges (*arrows*) with patterns (*dashed ovals*) corresponding to instructions

Computing a minimum cost covering for a directed acyclic graph (DAG) is assumed to be NP-complete, but by splitting the DAG into trees processed separately and forcing the shared nodes' results to be stored in registers, dynamic-programming based bottom-up tree pattern matching techniques can still be used as heuristic methods. Ertl [35] gives an algorithm to check if a given processor instruction set (i.e., tree grammar) belongs to a special class of architectures (containing e.g. MIPS and SPARC processors) where the constrained tree pattern matching always produces optimal coverings for DAGs.

Another way to compute a minimum cost covering, albeit an expensive one, is to apply advanced optimization methods such as integer linear programming [11, 33, 67, 94] or partitioned boolean quadratic programming [28]. This may be an applicable choice if integer linear programming is also used for solving other subtasks such as register allocation or instruction scheduling, the basic block and the number of patterns are not too large, and a close integration between these tasks is desired to produce high-quality code. Furthermore, solving the problem by such general optimization techniques is by no means restricted to tree-shaped patterns or tree-shaped IR graphs. In particular, they work well with complex patterns such as forests (non-connected trees, as in Fig. 10c) and directed acyclic graph (DAG) patterns, as with autoincrement load and store instructions or memory read-modify-write instructions and even on cyclic IR structures, such as static single assignment (SSA) representation. Because covering several nodes with a complex pattern corresponds to merging these nodes, special care has to be taken with forest and DAG patterns to avoid the creation of artificial dependence cycles by the matching, which could lead to non-schedulable code [27].

Instruction selection can be combined with *resource allocation*, i.e., the binding of instructions to executing resources. For some instructions, there may be no choice: for instance, on 'C62x, a multiply instruction can only be executed on a .M unit. In case that the same instruction can execute on different functional units, each with its own cost, latency and resource reservations, one could model these variants as different instructions that just share the pattern defining the semantics. Like instruction selection, resource allocation is often done before instruction scheduling, but a tighter integration with scheduling would be helpful because resource allocation clearly constrains the scheduler.

A natural extension of this approach is to also model cluster assignment for instructions as a resource allocation problem and thus as extended instruction selection problem. However, cluster allocation has traditionally been treated as a separate problem; we will discuss it in Sect. 5.

Further target-level optimizations could be modeled as part of instruction selection. For instance, for special cases of IR patterns there could exist alternative code sequences that may be faster, shorter, or more appropriate for later phases in code generation. As an example, an ordinary integer multiplication by 2 can be implemented in at least three different ways (*mutations*): by a MUL instruction, maybe running on a separate multiplier unit, by an ADD instruction, and by a left-shift (SHL) by one, each having different latency and resource requirements. The ability to consider such mutations during instruction scheduling increases flexibility and can thus improve code quality [80].

Another extension of instruction selection concerns the identification of several independent IR operations with same operator and short operand data types that could be merged to utilize a SIMD instruction instead.

Also, the selection of short instruction formats to reduce code size can be considered a subproblem of instruction selection. While beneficial for instruction fetch bandwidth, short instruction formats constrain the number of operands, the size of immediate fields or the set of accessible registers, which can have negative effects on register allocation and instruction scheduling. For the 16-bit compact instructions of 'C64x, Hahn et al. [50] explore the trade-off between performance and code size.

5 Cluster Assignment for Clustered VLIW Architectures

Cluster assignment for clustered VLIW architectures can be done at IR level or at target level. It maps each IR operator or instruction, respectively, to one of the clusters. Also, variables and temporaries to be held in registers must be mapped to a register file. Indeed, a value could reside in several register files if appropriate data transfer instructions are added; this is also an issue of register allocation and instruction scheduling and typically solved later than instruction cluster assignment in most compilers, although there exist obvious interdependences.

There exist various techniques for cluster assignment for basic blocks. The goal is to minimize the number of transfer instructions. Usually, heuristic solutions are applied.

Ellis [31] gives a heuristic algorithm for cluster assignment called bottom-up greedy (BUG) for basic blocks and traces (see later) that is applied before instruction scheduling. Desoli [26] identifies sub-DAGs of the target-level dataflow graph that are mapped to a cluster as a whole. Gangwar et al. [41] first decompose the target-level dataflow graph into disjoint chains of nodes connected by dataflow edges. The nodes of a chain will always be mapped to the same cluster. Chains are grouped together by a greedy heuristic until there are as many chain groups left as clusters. Finally, chain groups are heuristically assigned to clusters so that the residual cross-chaingroup dataflow edges coincide with direct inter-cluster communication links wherever possible. For many-cluster architectures where no fully connected inter-cluster communication network is available, the algorithm tries to minimize the communication distance accordingly, such that communicating chain groups are preferably mapped to clusters that are close to each other.

Hierarchical partitioning heuristics are used e.g. by Aleta et al. [3] and Chu et al. [23]. Aleta et al. also consider replication of individual instructions in order to reduce the amount of communication.

Beg and van Beek [12] use constraint programming to solve the cluster assignment problem optimally for an idealized multi-cluster architecture with unlimited inter-cluster communication bandwidth.

Usually, cluster assignment precedes instruction scheduling in phase-decoupled compilers for clustered VLIW DSPs, because the resource allocation for instructions must be known for scheduling. On the other hand, cluster allocation could benefit from information about free communication slots in the schedule. The quality of the resulting code suffers from the separate handling of cluster assignment, instruction scheduling and register allocation. We will discuss phase-coupled and integrated code generation approaches for clustered architectures in Sect. 8.

6 Register Allocation and Generalized Spilling

In the low-level IR, all program variables that could be held in a register and all temporary variables are modeled as *symbolic registers*, which the register allocator then maps to the hardware registers, of which only a limited number is available.

A symbolic register s is *live* at a program point p if s is defined (written) on a control path from the procedure's entry point to p, and there exists a program point q where s is used (read) and s may not be (re-)written on the control path $p \ldots q$. Hence, s is live at p if it is used in a control flow successor q of p. The *live range* of s is the set of program points where s is live. The number of all symbolic registers live at a program point p is called the *register pressure* at p.

Live ranges could be defined on the low-level IR (if register allocation is to be done before instruction selection), but usually, they are defined at target instruction level, because instruction selection may introduce additional temporary variables to be kept in registers. If the schedule is given, the live ranges are fixed, which constrains the register allocator, and generated spill code has to be inserted into the schedule. If register allocation comes first, some pre-scheduling (sequencing) at LIR or target level is required to bring the operations/instructions of a basic block in a linear order that defines the live range boundaries. Early register allocation constrains the subsequent scheduler, but generated spill code will then be scheduled and compacted together with the remaining instructions.

Two live ranges *interfere* if they may overlap. Interfering live ranges cannot share the same register.

The *live range interference graph* is an undirected graph whose nodes represent the live ranges of a procedure and edges represent interferences. Register assignment now means coloring the live range interference graph by assigning a color (i.e., a specific physical register) to each node such that interfering nodes have different colors. Moreover, the coloring must not use more colors than the number K of machine registers available. Determining if a general graph is colorable with K colors is NP-complete for $K \geq 3$. If a coloring cannot be found, the register allocator must restructure the program to make the interference graph colorable.

Chaitin [19] proposed a heuristic for coloring the interference graph with K colors, where K denotes the number of physical registers available. The algorithm works iteratively. In each step, it tries to find a node with degree $< K$, because then there must be some color available for that node, and removes the node from the interference graph. If the algorithm cannot find such a node, the program must be transformed to change the interference graph into a form that allows the algorithm to continue. Such transformations include coalescing, live range splitting, and spilling with rematerialization of live ranges. After the algorithm has removed all nodes from the interference graph, the nodes are colored in reverse order of removal. The optimistic coloring algorithm by Briggs [16] improves Chaitin's algorithm by delaying spilling transformations.

Coalescing is a transformation applicable to copy instructions $s_2 \leftarrow s_1$ where the two live ranges s_1, s_2 do not overlap except at that copy instruction, which marks the end (last use) of s_1 and beginning (write) of s_2. Coalescing merges s_1 and s_2 together to a single live range by renaming all occurrences of s_1 to s_2, which forces the register allocator to store them in the same physical register, and the copy instruction can now be removed.

Long live ranges tend to interfere with many others and thus may make the interference graph harder to color. As coalescing yields longer live ranges, it should be applied with care. *Conservative coalescing* [17] merges live ranges only if the degree of the merged live range in the interference graph would still be smaller than K.

The reverse of coalescing is *live range splitting* i.e., insertion of register-to-register copy operations and renaming of accesses to split a long live range into several shorter ones. Splitting can make an interference graph colorable without having to apply spilling; this is often more favorable, as register-to-register copying is faster and less energy consuming than memory accesses. Live range splitting can be done considerably with a small number of sub-live-ranges, or aggressively, with one sub-live-range per access.

Spilling removes a live range as symbolic register, by storing its value in main memory (or other non-register location). For each writing access, a store instruction is inserted that saves the value to a memory location (e.g., in the variable's "home" location or in a temporary location on the stack), and for each reading access, a load instruction to a temporary register is inserted. This spill code leads to increased execution time, energy consumption and code size. In some cases, it could be more efficient to realize the *rematerialization* [17] of a spilled value not by an expensive load from memory but by recomputing it instead. The choice between several spill candidates could be made greedily by considering the *spill cost* for a live range s, which contains the number of required store and load (or other rematerialization) instructions (to model the code size penalty), often also weighted by predicted execution frequencies (to model the performance and energy penalty).

A live range may not have to be spilled everywhere in the program. For instance, even if a symbolic register has a long live range, it may not be accessed during major periods in its live range where register pressure is high, for instance during an inner loop. Such periods are good candidates for *partial spilling.*

Register allocation can be implemented as a two-step approach [6, 27], where a global *pre-spilling* phase is run first to limit the remaining register pressure at any program point to the available number of physical registers, which makes the subsequent coloring phase easier.

Coloring-based heuristics are used in many standard compilers. While just-in-time compilers and dynamic optimizations require fast register allocators such as linear scan allocators [84, 91], the VLIW DSP domain rather calls for static compilation with high code quality, which justifies the use of more advanced register allocation algorithms. The first register allocator based on integer linear programming was presented by Goodwin and Wilken [45].

Optimal spilling selects just those live ranges for spilling whose accumulated spill cost is minimal, while making the remaining interference graph colorable. Optimal selection of spill candidates (pre-spilling) and optimal a-posteriori insertion of spill code for a given fixed instruction sequence and a given number of available registers are NP-complete even for basic blocks and have been solved by dynamic programming or integer linear programming for various special cases of processor architecture and dependency structure [6, 53, 54, 75]. In most compilers, heuristics are used that try to estimate the performance penalty of inserted load and store instructions [13]. Recently, several practical method based on integer linear programming for general optimal pre-spilling and for optimal coalescing have been developed, e.g., by Appel and George [6].

A recent trend is towards register allocation on the SSA form: for SSA programs, the interference graph belongs to the class of chordal graphs, which can be K-colored in quadratic time [14, 18, 49]. The generation of optimal spill code and minimization of copy instructions by coalescing remain NP-complete problems also for SSA programs. For the problem of optimal coalescing in spill-free SSA programs, a good heuristic method was proposed by Brisk et al. [18], and an optimal method based on integer linear programming was given by Grund and Hack [48]. For optimal pre-spilling in SSA programs, Ebner [27] models the problem as a constrained min-cut problem and applies a transformation that yields a polynomial-time near-optimal algorithm that does not rely on an integer linear programming solver.

7   Instruction Scheduling

In this section, we review fundamental instruction scheduling methods for VLIW processors at basic block level, loop level, and global level. We also discuss the automatic generation of the most time consuming part of instruction schedulers from a formal description of the processor.

7.1   Local Instruction Scheduling

The *control flow graph* at the IR level or target level representation of a program is a graph whose nodes are the IR operations or target instructions, respectively, and its edges denote possible control flow transitions between nodes.

A *basic block* is any (maximum-length) sequence of textually consecutive operations (at IR level) or instructions (at target level) in the program that can be entered by control flow only via the first and left only via the last operation or instruction, respectively. Hence, branch targets are always at the entry of a basic block, and a basic block contains no branches except maybe its last operation or instruction.

Control flow executes all operations of a basic block from its entry to its exit. Hence, the data dependences in a basic block form a directed acyclic graph, the *data flow graph* of the basic block. This data flow graph defines the partial execution order that constrains instruction scheduling: the instruction/operation at the target of a data dependence must not be issued before the latency of the instruction/operation at the source has elapsed. Leaf nodes in the data flow graph do not depend on any other node and have therefore no predecessor (within the basic block), root nodes have no successor node (within the basic block).

A path from a leaf node with maximum accumulated latency over its edges towards a root node is called a *critical path* of the basic block; its length is a lower bound for the makespan of any schedule for the basic block.

Methods for instruction scheduling for basic blocks (i.e., local scheduling) are simpler than global scheduling methods because control flow in basic blocks is straightforward and can be ignored. Only data dependences and resource conflicts need to be taken into account. Interestingly, most basic blocks in real-world programs are quite small and consist of only a few instructions. However, program transformations such as function inlining, loop unrolling or predication can yield considerably larger basic blocks.

Traditionally, heuristic methods have been considered for local instruction scheduling, mostly because of fast optimization times. A simple and well-known heuristic technique is *list scheduling*.

List scheduling [47] is based on topological sorting of the operations or instructions in the basic block's data flow graph, taking the precedence constraints by data dependences into account. The algorithm schedules nodes iteratively and maintains a list of data-ready nodes, the *ready list*. Initially, it consists of the leaves of the data flow graph, i.e., those nodes that do not depend on any other and could be scheduled immediately. The nodes in the ready list are assigned priorities that could, for instance, be the estimated maximum accumulated latency on any path from that node to a root of the data flow graph. In each step, list scheduling picks, in a greedy way, as many highest-priority nodes as possible from the ready list that fit into the next issue packet and for which resource requirements can be satisfied. The resource reservations of these issued nodes are then committed to the global resource usage map, and the issued nodes are removed from the data flow graph. Some further nodes may now become data ready in the next steps after the latency after all their predecessors has elapsed. The ready list is accordingly updated, and the process repeats until all nodes have been scheduled. The above description is for *forward scheduling*. *Backward scheduling* starts with the roots of the data flow graph and works in an analogous way in reversed topological order.

Another technique is *critical path scheduling*. First, a critical path in the basic block is detected; the nodes of that path are removed from the data flow graph and scheduled in topological order, each in its own issue packet. For the residual data flow graph, a critical path is determined etc., and this process is repeated until all nodes in the data flow graph have been scheduled. If there is no appropriate free slot in an issue packet to accommodate a node to be scheduled, a new issue packet is inserted.

Time-optimal instruction scheduling for basic blocks is NP-complete for almost any nontrivial target architecture, including most VLIW architectures. For special combinations of simple target architectures and restricted data flow graph topologies such as trees, polynomial-time optimal scheduling algorithms are known.

In the last decade, more expensive optimal methods for local instruction scheduling have become more and more popular, driven by (1) the need to generate high-quality code for embedded applications, (2) the fact that modern computers offer the compiler many more CPU cycles that can be spent on advanced optimizations, and (3) advances in general optimization problem solver technology, especially for integer linear programming. For local instruction scheduling on general acyclic data flow graphs, optimal algorithms based on integer linear programming [11, 34, 57, 68, 93], branch-and-bound [22, 51], constraint logic programming [10] and dynamic programming [59, 61] have been developed. Also, more expensive heuristic optimization techniques such as genetic programming [34, 72, 99] have been used successfully.

Even if the scope limitation to a single basic block is too restrictive in practice, local instruction scheduling techniques are nevertheless significant because they are also used in several global scheduling algorithms for larger acyclic code regions and even in certain cyclic scheduling algorithms, which we will discuss in Sect. 7.2.

7.2 Modulo Scheduling for Loops

Most DSP programs spend most of their execution time in some (inner) loops. Efficient loop scheduling techniques are therefore key to high code quality.

Loop unrolling is a simple transformation that can increase the scope of a local scheduler (and also other code optimizations) beyond the iteration boundaries, such that independent instructions from different iterations could be scheduled in parallel. However, loop unrolling increases code size considerably, which is often undesirable in embedded applications.

Software pipelining is a technique to overlap the execution of subsequent loop iterations such that independent instructions from different iterations can be scheduled in parallel on an instruction-level parallel architecture, without having to replicate the loop body code as in unrolling. As most scheduling problems with resource and dependence constraints, (rate-)optimal software pipelining is NP-complete.

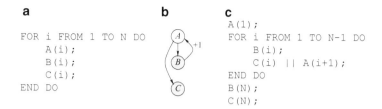

Fig. 11 Simple example: (**a**) Original loop, where $A(i)$, $B(i)$, $C(i)$ denote operations in the loop body that may compete for common resources, in this example $B(i)$ and $C(i)$, and that may involve both loop-independent data dependences, here $A(i) \rightarrow B(i)$ and $A(i) \rightarrow C(i)$, and loop-carried data dependences, here $B(i) \rightarrow A(i+1)$, see the dependence graph in (**b**). (**c**) After software pipelining

Software pipelining has been researched intensively, both as a high-level loop transformation (performed in the middle end of a compiler or even as source-to-source program transformation) and as low-level optimization late in the code generation process (performed in the back end of a compiler), after instruction selection with resource allocation has been performed. The former approaches are independent of particular instructions and functional units to be selected for all operations in the loop, and thus have to rely on inaccurate cost estimations for execution time, energy, or register pressure when comparing various alternatives, while the actual cost will also depend on decisions made late in the code generation process. The latter approaches are bound to fixed instructions and functional units, and hence the flexibility of implementing the same abstract operation by a variety of different target machine instructions, with different resource requirements and latency behavior, is lost. In either case, optimization opportunities are missed because interdependent problems are solved separately in different compiler phases. Approaches to integrate software pipelining with other code generation tasks will be discussed in Sect. 8.

Software pipelining, also called *cyclic scheduling*, transforms a loop such as that in Fig. 11a into an equivalent loop whose body contains operations from different iterations of the original loop, such as that in Fig. 11c, which may result in faster code on an instruction-level parallel architecture as e.g. $C(i)$ and $A(i+1)$ could be executed in parallel (| |) because they are statically known to be independent of each other and not to subscribe to the same hardware resources. This parallel execution was not possible in the original version of the loop because the code generator usually treats the loop body (a basic block) as a unit for scheduling and resource allocation, and furthermore separates the code for $C(i)$ and $A(i+1)$ by a backward branch to the loop entry. The body of the transformed loop is called the *kernel*, the operations before the kernel that "fill" the pipeline (here $A(1)$) are called the *prologue*, and the operations after the kernel that "drain" the pipeline (here $B(N)$ and $C(N)$), are called the *epilogue* of the software-pipelined loop. Software pipelining thus overlaps in the new kernel the execution of operations originating from different iterations of the original loop, as far as permitted by given dependence and resource constraints, in order to solicit more opportunities for parallel execution on instruction-level parallel architectures, such as superscalar, VLIW or EPIC processors. Software pipelining can be combined with loop unrolling.

In their survey of software pipelining methods, Allan et al. [4] divide existing approaches into two general classes. Based on a lower bound determined by analyzing dependence distances, latencies, and resource requirements, the *modulo scheduling* methods, as introduced by Rau and Glaeser [86] and refined in several approaches [63, 71], first guess the kernel size (in terms of clock cycles), called the *initiation interval (II)*, and then fill the instructions of the original loop body into a modulo reservation table of size *II*, which produces the new kernel. If no such modulo schedule could be found for the assumed *II*, the kernel is enlarged by incrementing *II*, and the procedure is repeated. The *kernel-detection methods*, such as those by Aiken and Nicolau [2] (no resource constraints) and Vegdahl [92], continuously peel off iterations from the loop and schedule their operations until a pattern for a steady state emerges, from which the kernel is constructed.

Modulo scheduling starts with an initial initiation interval given by the lower bound *MinII* (minimum initiation interval), which is the maximum of the recurrence-based minimum initiation interval (*RecMinII*) and the resource-based minimum initiation interval (*ResMinII*).

RecMinII is the maximum accumulated sum of latencies along any dependence cycle in the dependence graph, divided by the number of iterations spanned by the dependence cycle. If there is no such cycle, *RecMinII* is 0.

ResMinII is the maximum accumulated number of reserved slots on any resource in the loop body.

Modulo scheduling attempts to find a valid modulo schedule by filling all instructions in the modulo reservation table for the current *II* value. Priority is usually given to dependence cycles in decreasing order of accumulated latency per accumulated distance.

If the first attempt fails, most heuristic methods allow backtracking for a limited number of further attempts. An exhaustive search is usually not feasible because of the high problem complexity. Instead, if no attempt was successful, the *II* is incremented and the procedure is repeated with a one larger modulo reservation table. As there exists a trivial upper bound for the *II* (namely, the accumulated sum of all latencies in the loop body), this iterative method will eventually find a modulo schedule.

The main goal of software pipelining is to maximize the throughput by minimizing *II*, i.e., *rate-optimal* software pipelining. Moreover, minimizing the makespan (the elapsed time between the first instruction issue and last instruction termination) of a single loop iteration in the modulo scheduled loop is often a secondary optimization goal, because it directly implies the length of prologue and epilogue and thereby has an impact on code size (unless special hardware support for rotating predicate registers allows to represent prologue and epilogue code implicitly with the predicated kernel code).

Register allocation for software pipelined loops is another challenge. Software pipelining tends to increase register pressure. If a live range is longer than *II* cycles, it will interfere with itself (e.g., with its instance starting in the next iteration of the kernel) and thus a single register will not be sufficient; special care has to be taken to access the "right" one at any time. There are two kinds of techniques for such self-overlapping live ranges: hardware based techniques, such as rotating register sets and register queues, and software techniques such as modulo variable expansion [63] and live range splitting [89]. With *modulo variable expansion*, the kernel is unrolled and symbolic registers renamed until no live range self-overlaps any more: if μ denotes the maximum length of a self-overlapping live range, the required unroll factor is $\rho = \lceil \mu/II \rceil$, and the new initiation interval of the expanded kernel is $II' = \rho\, II$. The drawback of modulo variable expansion is increased code size and increased register need. An alternative approach is to avoid self-overlapping live ranges *a priori* by splitting long live ranges on dependence cycles into shorter ones, by inserting copy instructions.

Optimal methods for software pipelining based on integer linear programming have been proposed e.g. by Badia et al. [8], Govindarajan et al. [46] and Yang et al. [97]. Combinations of modulo scheduling with other code generation tasks will be discussed in Sect. 8.

Software pipelining is often combined with loop unrolling. Especially if the lower bound *MinII* is a non-integer value, loop unrolling before software pipelining can improve throughput. Moreover, loop unrolling reduces loop overhead (at least on processors that do not have hardware support for zero-overhead loops). The downside is larger code size.

7.3   Global Instruction Scheduling

Basic blocks are the units of (procedure-)global control flow analysis. The *basic block graph* of a program is a directed graph where the nodes correspond to the basic blocks and edges show control flow transitions between basic blocks, such as branches or fall-through transitions to branch targets.

Global instruction scheduling methods consider several basic blocks at a time and allow to move instructions between basic blocks. The (current) scope of a global scheduling method is referred to as a *region*. Regions used for global scheduling include traces, superblocks, hyperblocks and treegions. Local scheduling methods are extended to address entire regions. Because the scope is larger, global scheduling has more flexibility and may generate better code than local scheduling. Program transformations such as function inlining, loop unrolling or predication can be applied to additionally increase the size of basic blocks and regions.

The idea of *trace scheduling* [36] is to make the most frequently executed control flow paths fast while accepting possible performance degradations along less frequently used paths. Execution frequencies are assigned to the outgoing edges at branch instructions based on static predictions or on profile data. A *trace* is a linear path (i.e., free of backwards edges and thereby of loops) through the basic block graph, where, at each basic block B_i in the trace except for the last one, its successor

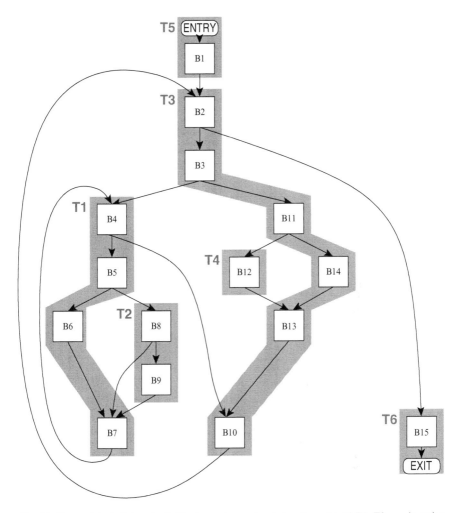

Fig. 12 Traces (*shaded*) in a basic block graph, constructed and numbered T1, T2, ... in order of decreasing predicted execution frequency. A trace ends at a backwards branch or at a join point with another trace of higher execution frequency (which thus was constructed earlier). Trace T1 represents the more frequent control path in the inner loop starting at basic block B4

B_j in the trace is the target of the more frequently executed control flow edge leaving B_i. Traces may have side entrances and side exits of control flow from outside the trace. Forward edges are possible, and likewise backwards edges to the first block in the trace. See Fig. 12 for an example.

Trace scheduling repeatedly identifies a maximum-length trace in the basic block graph, removes its basic blocks from the graph and schedules the instructions of the trace with a local scheduling method as if it were a single basic block. As instructions are moved across a control flow transition, either upwards or downwards,

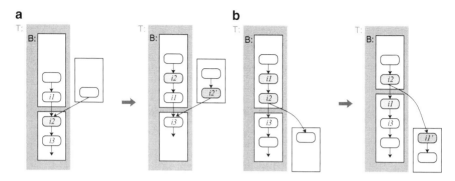

Fig. 13 Two of the main cases in trace scheduling where compensation code must be inserted. The trace T is being compacted. Case (**a**) Hoisting instruction $i2$ into the predecessor basic block B requires inserting a copy $i2'$ of $i2$ into the other predecessor block(s). Case (**b**) Moving instruction $i1$ forward across the branch instruction $i2$ requires inserting a copy $i1'$ of $i1$ into the other branch target basic block

correctness of the program must be re-established by inserting compensation code into the other predecessor or successor block of the original basic block of that instruction, respectively. Two of the possible cases are shown in Fig. 13. The insertion of compensation code may lead to considerable code size expansion on the less frequently executed branches. After the trace has been scheduled, its basic blocks are removed from the basic block graph, the next trace is determined, and this process is repeated until all basic blocks of the program have been scheduled.

Superblocks [55] are a restricted form of traces that do not contain any branches into it (except possibly for backwards edges to the first block). This restriction simplifies the generation of compensation code in trace scheduling. A trace can be converted into a superblock by replicating its tail parts that are targets of branches from outside the trace. Tail duplication is a form of generating compensation code ahead of scheduling, and can likewise increase code size considerably.

While traces and superblocks are linear chains of basic blocks, *hyperblocks* [73] are regions in the basic block graph with a common entry block and possibly several exit blocks, with acyclic internal control flow. Using predication, the different control flow paths in a hyperblock could be merged to a single superblock.

A *treegion* [52], also known as *extended basic block* [76], is an out-tree region in the basic block graph. There are no side entrances of control flow into a treegion, except to its root basic block.

Recently, optimal methods for global instruction scheduling on instruction-level parallel processors have become popular. Winkel used integer linear programming for optimal global scheduling for Intel IA-64 (Itanium) processors [95] and showed that it can be used in a production compiler [96]. Malik et al. [74] proposed a constraint programming approach for optimal scheduling of superblocks.

7.4   Generated Instruction Schedulers

Whenever a forward scheduling algorithm such as list scheduling inserts another data-ready instruction at the end of an already computed partial schedule, it needs to fit the required resource reservations of the new instruction against the already committed resource reservations, and likewise obey pending latencies of predecessor instructions where necessary, in order to derive the earliest possible issue time for the new instruction relative to the issue time of the last instruction in the partial schedule. While the impact of dependence predecessors can be checked quickly, determining that issue time offset is more involved with respect to the resource reservations. The latter calculation could be done, for instance, by searching through the partial schedule's resource usage map, resource by resource. For advanced scheduling methods that try lots of alternatives, faster methods for detecting of resource conflicts resp. computing the issue time offset are desirable.

Note that the new instruction's issue time relative to the currently last instruction of the partial schedule only depends on the most recent resource reservations, not the entire partial schedule. Each possible contents of this still relevant, recent part of the pipeline can be interpreted as a pipeline *state*, and appending an instruction will result in a new state and an issue time offset for the new instruction, such that scheduling can be described as a finite state automaton. The initial state is an empty pipeline. The set of possible states and the set of possible transitions depend only on the processor, not on the input program. Hence, once all possible states have been determined and encoded and all possible transitions with their effects on successor state and issue time have been precomputed once and for all, scheduling can be done very fast by looking up the issue time offset and the new state in the precomputed transition table for the current state and inserted instruction.

This automaton-based approach was introduced by Müller [77] and was improved and extended in several works [9, 29, 85]. The automaton can be generated automatically from a formal description of the processor's set of instructions with their reservation tables. A drawback is that the number of possible states and transitions can be tremendous, but there exist techniques such as standard finite state machine minimization, automata factoring, and replacement of several physical resources with equivalent contention behavior by a single virtual resource, to reduce the size of the scheduling automaton.

8   Integrated Code Generation for VLIW and Clustered VLIW

In most compilers, the subproblems of code generation are treated separately in subsequent *phases* of the compiler back-end. This is easier from a software engineering point of view, but often leads to suboptimal results because decisions made in earlier phases constrain the later phases.

For instance, early instruction scheduling determines the live ranges for a subsequent register allocator; when the number of physical registers is not sufficient, spill code must be inserted a-posteriori into the existing schedule, which may compromise the schedule's quality. Conversely, early register allocation introduces additional ("false") data dependences, which are an artifact caused by the reuse of registers but constrain the subsequent instruction scheduling phase.

Interdependences exist also between instruction scheduling and instruction selection: in order to formulate instruction selection as a separate minimum-cost covering problem, phase-decoupled code generation assigns a fixed, context-independent cost to each instruction, such as its expected or worst-case execution time, while the actual cost also depends on interference with resource occupation and latency constraints of other instructions, which depends on the schedule. For instance, a potentially concurrent execution of two independent IR operations may be prohibited if instructions are selected that require the same resource.

Even the subdivision of instruction scheduling into separate phases for sequencing and compaction can have negative effects on schedule quality if instructions with non-block reservation tables occur [60].

Furthermore, on clustered VLIW processors, concurrent execution may be possible only if the operands reside in the right register sets at the right time, as discussed earlier. While instruction scheduling and register allocation need to know about the cluster assignment of instructions and data, cluster assignment could profit from knowing about free slots where transfer instructions could be placed, or free registers where transferred copies could be held. Any phase decoupled approach may result in bad code quality because the later phases are constrained by decisions made in the early ones.

Hence, the integration of these subproblems to solve them as a single optimization problem, as visualized in Fig. 14, is highly desirable, but unfortunately this increases the overall complexity of code generation considerably. Despite the recent improvements in general optimization problem solver technology, this ambitious approach is limited in scope to basic blocks and loops. Other methods take a more conservative approach based on a phase-decoupled code generator and make, heuristically, an early phase aware of possibly different goals of later phases. For instance, register pressure aware scheduling methods, which trade less instruction level parallelism for shorter live ranges where register pressure is predicted to be high, can lead to better register allocations with less spill code later.

8.1 Integrated Code Generation at Basic Block Level

There exist several heuristic approaches that aim at a better integration of instruction scheduling and register allocation [15, 40, 44, 62]. For the case of clustered VLIW processors, the heuristic algorithm proposed by Kailas et al. [56] integrates cluster assignment, register allocation, and instruction scheduling.

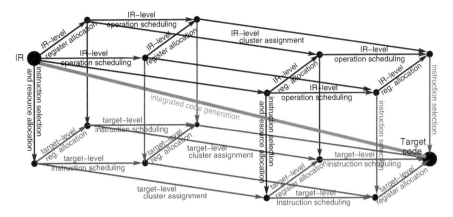

Fig. 14 Phase-decoupled vs. fully integrated code generation for clustered VLIW processors. Only the four main tasks of code generation (cluster assignment, instruction selection, instruction scheduling, and register allocation) are shown. While often performed in just this order, many phase orderings are possible in phase-decoupled code generation, visualized by the paths along the edges of the four-dimensional hypercube from the processor-independent low-level IR to final target code. Fully integrated code generation solves all tasks simultaneously as a monolithic combined optimization problem, thus following the main diagonal

Heuristic methods that couple or integrate instruction scheduling and cluster assignment were proposed by Özer et al. [83], Leupers [66], Chu et al. [23], and by Nagpal and Srikant [79]. For example, Leupers [66] uses a simulated-annealing based approach where cluster allocation and instruction scheduling are applied alternatingly in an iterative optimization loop.

For the computationally intensive kernels of DSP application programs to be used in an embedded product throughout its lifetime, the manufacturer is often willing to afford spending a significant amount of time in optimizing the code during the final compilation. However, there are only a few approaches that have the potential—given sufficient time and space resources—to compute an optimal solution to an integrated problem formulation, mostly combining local scheduling and register allocation [10, 57, 64].

Some of these approaches are also able to partially integrate instruction selection problems, even though for rather restricted machine models. For instance, Wilson et al. [94] consider architectures with a single, non-pipelined ALU, two non-pipelined parallel load/store/move units, and a homogeneous set of general-purpose registers. Araujo and Malik [7] consider integrated code generation for expression trees with a machine model where the capacity of each sort of memory resource (register classes or memory blocks) is either one or infinity, a class that includes, for instance, the TI C25.

The integrated method adopted in the retargetable framework AVIV [51] for clustered VLIW architectures builds an extended data flow graph representation of the basic block that explicitly represents all alternatives for implementation; then, a branch-and-bound heuristic selects an alternative among all representations that is optimized for code size.

Chang et al. [20] use integer linear programming for combined instruction scheduling and register allocation with spill code generation for non-pipelined, non-clustered multi-issue architectures.

Kessler and Bednarski [59] propose a dynamic programming algorithm for fully integrated code generation for clustered and non-clustered VLIW architectures at the basic block level, which was implemented in the retargetable integrated code generator OPTIMIST. Bednarski and Kessler [11] and Eriksson et al. [32, 34] solve the problem with integer linear programming; the latter work also gives a heuristic approach based on a genetic algorithm.

8.2   Loop-Level Integrated Code Generation

There exist several heuristic algorithms for modulo scheduling that attempt to reduce register pressure, such as Hypernode Resource Modulo Scheduling [71] and Swing Modulo Scheduling [70]. Nyström and Eichenberger [82] couple cluster assignment and modulo scheduling for clustered VLIW architectures. Codina et al. [24] give a heuristic method for modulo scheduling integrated with register allocation and spill code generation for clustered VLIW processors. Zalamea et al. [98] consider the integration of register pressure aware modulo scheduling with register allocation, cluster assignment and spilling for clustered VLIW processors and present an iterative heuristic algorithm with backtracking. Aleta et al. [3] use a phase-coupled heuristic approach to cluster-assignment and modulo scheduling that also considers replication of instructions for reduced inter-cluster communication. Stotzer and Leiss [89] propose a preprocessing transformation for modulo scheduling for the 'C6x clustered VLIW DSP architecture that attempts to reduce self-overlapping cyclic live ranges in a preprocessing phase and thereby eliminate the need for modulo variable expansion or rotating register files.

Eisenbeis and Sawaya [30] propose an integer linear programming method for modulo scheduling integrated with register allocation, which gives optimal results if the number of schedule slots is fixed. Nagarakatte and Govindarajan [78] provide an optimal method for integrating register allocation and spill code generation with modulo scheduling for non-clustered architectures. Eriksson and Kessler [32, 33] give an integer linear programming method for optimal, fully integrated code generation for loops, combining modulo scheduling with instruction selection, cluster assignment, register allocation and spill code generation, for clustered VLIW architectures.

9 Concluding Remarks

Compilers for VLIW DSP processors need to apply a considerable amount of advanced optimizations to achieve code quality comparable to hand-written code. Current advances in general optimization problem solver technology are encouraging, and heuristic techniques developed for standard compilers are being complemented by more aggressive optimizations. For small and medium sized program parts, even optimal solutions are within reach. Also, most problems in code generation are strongly interdependent and should be considered together in an integrated or at least phase-coupled way to avoid poor code quality due to phase ordering effects. We expect further improvements in optimized and integrated code generation techniques for VLIW DSPs in the near future.

Acknowledgements The author thanks Mattias Eriksson and Dake Liu for discussions and commenting on a draft of this chapter. The author also thanks Eric Stotzer from Texas Instruments for interesting discussions about code generation for the TI 'C6x DSP processor family. This work was funded by Vetenskapsrådet (project *Integrated Software Pipelining*) and SSF (project *DSP platform for emerging telecommunication and multimedia*).

References

1. Alfred V. Aho, Mahadevan Ganapathi, and Steven W.K. Tjiang. Code Generation Using Tree Matching and Dynamic Programming. *ACM Transactions on Programming Languages and Systems*, 11(4):491–516, October 1989.
2. Alexander Aiken and Alexandru Nicolau. Optimal loop parallelization. *SIGPLAN Notices*, 23(7):308–317, July 1988.
3. Alex Aleta, Josep M. Codina, Jesus Sanchez, Antonio Gonzalez, and David Kaeli. AGAMOS: A graph-based approach to modulo scheduling for clustered microarchitectures. *IEEE Transactions on Computers*, 58(6):770–783, June 2009.
4. Vicki H. Allan, Reese B. Jones, Randall M. Lee, and Stephen J. Allan. Software pipelining. *ACM Computing Surveys*, 27(3), September 1995.
5. Analog Devices. TigerSHARC embedded processor ADSP-TS201S. Data sheet, www.analog.com/en/embedded-processing-dsp/tigersharc, 2006.
6. Andrew W. Appel and Lal George. Optimal Spilling for CISC Machines with Few Registers. In *Proc. ACM conf. on Programming language design and implementation*, pages 243–253. ACM Press, 2001.
7. Guido Araujo and Sharad Malik. Optimal code generation for embedded memory non-homogeneous register architectures. In *Proc. 7th Int. Symposium on System Synthesis*, pages 36–41, September 1995.
8. Rosa M. Badia, Fermin Sanchez, and Jordi Cortadella. OSP: Optimal Software Pipelining with Minimum Register Pressure. Technical Report UPC-DAC-1996-25, DAC Dept. d'arquitectura de Computadors, Univ. Polytecnica de Catalunya, Barcelona, Campus Nord. Modul D6, E-08071 Barcelona, Spain, June 1996.
9. Vasanth Bala and Norman Rubin. Efficient instruction scheduling using finite state automata. In *Proc. 28th int. symp. on miocroarchitecture (MICRO-28)*, pages 46–56. IEEE, 1995.
10. Steven Bashford and Rainer Leupers. Phase-coupled mapping of data flow graphs to irregular data paths. *Design Automation for Embedded Systems (DAES)*, 4(2/3):119–165, 1999.

11. Andrzej Bednarski and Christoph Kessler. Optimal integrated VLIW code generation with integer linear programming. In *Proc. Int. Euro-Par 2006 Conference*. Springer LNCS, August 2006.

12. Mirza Beg and Peter van Beek. A constraint programming approach for instruction assignment. In *Proc. Int. Workshop on Interaction between Compilers and Computer Architectures (INTERACT-15)*, pp. 25–34, February 2011.

13. D. Bernstein, M.C. Golumbic, Y. Mansour, R.Y. Pinter, D.Q. Goldin, H. Krawczyk, and I. Nahshon. Spill code minimization techniques for optimizing compilers. In *Proc. Int. Conf. on Progr. Lang. Design and Implem.*, pages 258–263, 1989.

14. F. Bouchez, A. Darte, C. Guillon, and F. Rastello. Register allocation: what does the NP-completeness proof of Chaitin et al. really prove? [...]. In *Proc. 19th int. workshop on languages and compilers for parallel computing, New Orleans*, November 2006.

15. Thomas S. Brasier, Philip H. Sweany, Steven J. Beaty, and Steve Carr. Craig: A practical framework for combining instruction scheduling and register assignment. In *Proc. Int. Conf. on Parallel Architectures and Compilation Techniques (PACT'95)*, 1995.

16. Preston Briggs, Keith Cooper, Ken Kennedy, and Linda Torczon. Coloring heuristics for register allocation. In *Proc. Int. Conf. on Progr. Lang. Design and Implem.*, pages 275–284, 1989.

17. Preston Briggs, Keith Cooper, and Linda Torczon. Rematerialization. In *Proc. Int. Conf. on Progr. Lang. Design and Implem.*, pages 311–321, 1992.

18. Philip Brisk, Ajay K. Verma, and Paolo Ienne. Optimistic chordal coloring: a coalescing heuristic for SSA form programs. *Des. Autom. Embed. Syst.*, 13:115–137, 2009.

19. G.J. Chaitin, M.A. Auslander, A.K. Chandra, J. Cocke, M.E. Hopkins, and P.W. Markstein. Register allocation via coloring. *Computer Languages*, 6:47–57, 1981.

20. Chia-Ming Chang, Chien-Ming Chen, and Chung-Ta King. Using integer linear programming for instruction scheduling and register allocation in multi-issue processors. *Computers Mathematics and Applications*, 34(9):1–14, 1997.

21. Chung-Kai Chen, Ling-Hua Tseng, Shih-Chang Chen, Young-Jia Lin, Yi-Ping You, Chia-Han Lu, and Jenq-Kuen Lee. Enabling compiler flow for embedded VLIW DSP processors with distributed register files. In *Proc. LCTES'07*, pages 146–148. ACM, 2007.

22. Hong-Chich Chou and Chung-Ping Chung. An Optimal Instruction Scheduler for Superscalar Processors. *IEEE Trans. on Parallel and Distr. Syst.*, 6(3):303–313, 1995.

23. Michael Chu, Kevin Fan, and Scott Mahlke. Region-based hierarchical operation partitioning for multicluster processors. In *Proc. Int. Conf. on Progr. Lang. Design and Implem. (PLDI'03)*, pp. 300–311, ACM, June 2003.

24. Josep M. Codina, Jesus Sánchez, and Antonio González. A unified modulo scheduling and register allocation technique for clustered processors. In *Proc. PACT-2001*, September 2001.

25. Edward S. Davidson, Leonard E. Shar, A. Thampy Thomas, and Janak H. Patel. Effective control for pipelined computers. In *Proc. Spring COMPCON75 Digest of Papers*, pages 181–184. IEEE Computer Society, February 1975.

26. Giuseppe Desoli. Instruction assignment for clustered VLIW DSP compilers: a new approach. Technical Report HPL-98-13, HP Laboratories Cambridge, February 1998.

27. Dietmar Ebner. *SSA-based code generation techniques for embedded architectures*. PhD thesis, Technische Universität Wien, Vienna, Austria, June 2009.

28. Erik Eckstein, Oliver König, and Bernhard Scholz. Code instruction selection based on SSA-graphs. In A. Krall, editor, *Proc. SCOPES-2003, Springer LNCS 2826*, pages 49–65, 2003.

29. Alexandre E. Eichenberger and Edward S. Davidson. A reduced multipipeline machine description that preserves scheduling constraints. In *Proc. Int. Conf. on Progr. Lang. Design and Implem. (PLDI'96)*, pages 12–22, New York, NY, USA, 1996. ACM Press.

30. Christine Eisenbeis and Antoine Sawaya. Optimal loop parallelization under register constraints. In *Proc. 6th Workshop on Compilers for Parallel Computers (CPC'96)*, pages 245–259, December 1996.

31. John Ellis. Bulldog: A Compiler for VLIW Architectures. MIT Press, Cambridge, MA, 1986.

32. Mattias Eriksson. Integrated Code Generation. PhD thesis, Linköping Studies in Science and Technology Dissertation No. 1375, Linköping University, Sweden, 2011.

33. Mattias Eriksson and Christoph Kessler. Integrated modulo scheduling for clustered VLIW architectures. In *Proc. HiPEAC-2009 High-Performance and Embedded Architecture and Compilers, Paphos, Cyprus*, pages 65–79. Springer LNCS 5409, January 2009.

34. Mattias Eriksson, Oskar Skoog, and Christoph Kessler. Optimal vs. heuristic integrated code generation for clustered VLIW architectures. In *Proc. 11th int. workshop on software and compilers for embedded systems (SCOPES'08)*. ACM, 2008.

35. M. Anton Ertl. Optimal Code Selection in DAGs. In *Proc. Int. Symposium on Principles of Programming Languages (POPL'99)*. ACM, 1999.

36. Joseph A. Fisher. Trace scheduling: A technique for global microcode compaction. *IEEE Trans. Comput.*, C–30(7):478–490, July 1981.

37. Joseph A. Fisher, Paolo Faraboschi, and Cliff Young. *Embedded computing: a VLIW approach to architecture, compilers and tools*. Elsevier/Morgan Kaufmann, 2005.

38. Björn Franke. C Compilers and Code Optimization for DSPs. In S. S. Bhattacharyya, E. F. Deprettere, R. Leupers, and J. Takala, eds., Handbook of Signal Processing Systems, Second Edition, Springer 2013.

39. Christopher W. Fraser, David R. Hanson, and Todd A. Proebsting. Engineering a Simple, Efficient Code Generator Generator. *Letters of Programming Languages and Systems*, 1(3):213–226, September 1992.

40. Stefan M. Freudenberger and John C. Ruttenberg. Phase ordering of register allocation and instruction scheduling. In *Code Generation: Concepts, Tools, Techniques [42]*, pages 146–170, 1992.

41. Anup Gangwar, M. Balakrishnan, and Anshul Kumar. Impact of intercluster communication mechanisms on ILP in clustered VLIW architectures. *ACM Trans. Des. Autom. Electron. Syst.*, 12(1):1, 2007.

42. Robert Giegerich and Susan L. Graham, editors. *Code Generation - Concepts, Tools, Techniques*. Springer Workshops in Computing, 1992.

43. R.S. Glanville and S.L. Graham. A New Method for Compiler Code Generation. In *Proc. Int. Symposium on Principles of Programming Languages*, pages 231–240, January 1978.

44. James R. Goodman and Wei-Chung Hsu. Code scheduling and register allocation in large basic blocks. In *Proc. ACM Int. Conf. on Supercomputing*, pages 442–452. ACM press, July 1988.

45. David W. Goodwin and Kent D. Wilken. Optimal and near-optimal global register allocations using 0–1 integer programming. *Softw. Pract. Exper.*, 26(8):929–965, 1996.

46. R. Govindarajan, Erik Altman, and Guang Gao. A framework for resource-constrained rate-optimal software pipelining. *IEEE Trans. Parallel and Distr. Syst.*, 7(11):1133–1149, November 1996.

47. R. L. Graham. Bounds for certain multiprocessing anomalies. *Bell System Technical Journal*, 45(9):1563–1581, November 1966.

48. Daniel Grund and Sebastian Hack. A fast cutting-plane algorithm for optimal coalescing. In *Proc. 16th int. conf. on compiler construction*, pages 111–125, March 2007.

49. Sebastian Hack and Gerhard Goos. Optimal register allocation for SSA-form programs in polynomial time. *Information Processing Letters*, 98:150–155, 2006.

50. Todd Hahn, Eric Stotzer, Dineel Sule, and Mike Asal. Compilation strategies for reducing code size on a VLIW processor with variable length instructions. In *Proc. HiPEAC'08 conference*, pages 147–160. Springer LNCS 4917, 2008.

51. Silvina Hanono and Srinivas Devadas. Instruction scheduling, resource allocation, and scheduling in the AVIV retargetable code generator. In *Proc. Design Automation Conf.* ACM, 1998.

52. W. A. Havanki. Treegion scheduling for VLIW processors. M.S. thesis, Dept. Electrical and Computer Engineering, North Carolina State Univ., Raleigh, NC, USA, 1997.

53. L.P. Horwitz, R. M. Karp, R. E. Miller, and S. Winograd. Index register allocation. *Journal of the ACM*, 13(1):43–61, January 1966.

54. Wei-Chung Hsu, Charles N. Fischer, and James R. Goodman. On the minimization of loads/stores in local register allocation. *IEEE Trans. Softw. Eng.*, 15(10):1252–1262, October 1989.

55. Wen-Mei Hwu, Scott A. Mahlke, William Y. Chen, Pohua P. Chang, Nancy J. Warter, Roger A. Bringmann, Roland G. Ouellette, Richard E. Hank, Tokuzo Kiyohara, Grant E. Haab, John G. Holm, and Daniel M. Lavery. The superblock: an effective technique for VLIW and superscalar compilation. *J. Supercomput.*, 7(1–2):229–248, 1993.

56. Krishnan Kailas, Kemal Ebcioglu, and Ashok Agrawala. CARS: A new code generation framework for clustered ILP processors. In *Proc. 7th Int. Symp. on High-Performance Computer Architecture (HPCA'01)*, pages 133–143. IEEE Computer Society, June 2001.

57. Daniel Kästner. *Retargetable Postpass Optimisations by Integer Linear Programming*. PhD thesis, Universität des Saarlandes, Saarbrücken, Germany, 2000.

58. Christoph Kessler and Andrzej Bednarski. Optimal integrated code generation for clustered VLIW architectures. In *Proc. ACM SIGPLAN Conf. on Languages, Compilers and Tools for Embedded Systems/Software and Compilers for Embedded Systems, LCTES-SCOPES'2002*. ACM, June 2002.

59. Christoph Kessler and Andrzej Bednarski. Optimal integrated code generation for VLIW architectures. *Concurrency and Computation: Practice and Experience*, 18:1353–1390, 2006.

60. Christoph Kessler, Andrzej Bednarski, and Mattias Eriksson. Classification and generation of schedules for VLIW processors. *Concurrency and Computation: Practice and Experience*, 19:2369–2389, 2007.

61. Christoph W. Keßler. Scheduling Expression DAGs for Minimal Register Need. *Computer Languages*, 24(1):33–53, September 1998.

62. Tokuzo Kiyohara and John C. Gyllenhaal. Code scheduling for VLIW/superscalar processors with limited register files. In *Proc. 25th int. symp. on miocroarchitecture (MICRO-25)*. IEEE CS Press, 1992.

63. Monica Lam. Software pipelining: An effective scheduling technique for VLIW machines. In *Proc. CC'88*, pages 318–328, July 1988.

64. Rainer Leupers. *Retargetable Code Generation for Digital Signal Processors*. Kluwer, 1997.

65. Rainer Leupers. *Code Optimization Techniques for Embedded Processors*. Kluwer, 2000.

66. Rainer Leupers. Instruction scheduling for clustered VLIW DSPs. In *Proc. PACT'00 int. conference on parallel architectures and compilation*. IEEE Computer Society, 2000.

67. Rainer Leupers and Steven Bashford. Graph-based code selection techniques for embedded processors. *ACM TODAES*, 5(4):794–814, October 2000.

68. Rainer Leupers and Peter Marwedel. Time-constrained code compaction for DSPs. *IEEE Transactions on VLSI Systems*, 5(1):112–122, 1997.

69. Dake Liu. *Embedded DSP processor design*. Morgan Kaufmann, 2008.

70. Josep Llosa, Antonio Gonzalez, Mateo Valero, and Eduard Ayguade. Swing Modulo Scheduling: A Lifetime-Sensitive Approach. In *Proc. PACT'96 conference*, pages 80–86. IEEE, 1996.

71. Josep Llosa, Mateo Valero, Eduard Ayguade, and Antonio Gonzalez. Hypernode reduction modulo scheduling. In *Proc. 28th int. symp. on miocroarchitecture (MICRO-28)*, 1995.

72. M. Lorenz and P. Marwedel. Phase coupled code generation for DSPs using a genetic algorithm. In *Proc. conf. on design automation and test in Europe (DATE'04)*, pages 1270–1275, 2004.

73. Scott A. Mahlke, David C. Lin, William Y. Chen, Richard E. Hank, and Roger A. Bringmann. Effective compiler support for predicated execution using the hyperblock. In *Proc. 25th int. symp. on microarchitecture (MICRO-25)*, pages 45–54, December 1992.

74. Abid M. Malik, Michael Chase, Tyrel Russell, and Peter van Beek. An application of constraint programming to superblock instruction scheduling. In *Proc. 14th Int. Conf. on Principles and Practice of Constraint Programming*, pages 97–111, September 2008.

75. Waleed M. Meleis and Edward D. Davidson. Dual-issue scheduling with spills for binary trees. In *Proc. 10th ACM-SIAM Symposium on Discrete Algorithms*, pages 678–686. Society for Industrial and Applied Mathematics, Philadelphia, PA, USA, 1999.

76. Steven S. Muchnick. *Advanced Compiler Design and Implementation*. Morgan Kaufmann, 1997.

77. Thomas Müller. Employing finite automata for resource scheduling. In *Proc. 26th int. symp. on microarchitecture (MICRO-26)*, pages 12–20. IEEE, December 1993.

78. S. G. Nagarakatte and R. Govindarajan. Register allocation and optimal spill code scheduling in software pipelined loops using 0–1 integer linear programming formulation. In *Proc. int. conf. on compiler construction (CC-2007)*, pages 126–140. Springer LNCS 4420, 2007.

79. Rahul Nagpal and Y. N. Srikant. Integrated temporal and spatial scheduling for extended operand clustered VLIW processors. In *Proc. 1st conf. on Computing Frontiers*, pages 457–470. ACM Press, 2004.

80. Steven Novack and Alexandru Nicolau. Mutation scheduling: A unified approach to compiling for fine-grained parallelism. In *Proc. Workshop on compilers and languages for parallel computers (LCPC'94)*, pages 16–30. Springer LNCS 892, 1994.

81. NXP. Trimedia TM-1000. Data sheet, www.nxp.com, 1998.

82. Erik Nyström and Alexandre E. Eichenberger. Effective cluster assignment for modulo scheduling. In *Proc. 31st annual ACM/IEEE Int. symposium on microarchitecture (MICRO-31)*, IEEE CS Press, 1998.

83. Emre Özer, Sanjeev Banerjia, and Thomas M. Conte. Unified assign and schedule: a new approach to scheduling for clustered register file microarchitectures. In *Proc. 31st annual ACM/IEEE Int. Symposium on Microarchitecture*, pages 308–315. IEEE CS Press, 1998.

84. Massimiliano Poletto and Vivek Sarkar. Linear scan register allocation. *ACM Transactions on Programming Languages and Systems*, 21(5), September 1999.

85. Todd A. Proebsting and Christopher W. Fraser. Detecting pipeline structural hazards quickly. In *Proc. 21st symp. on principles of programming languages (POPL'94)*, pages 280–286. ACM Press, 1994.

86. B. Rau and C. Glaeser. Some scheduling techniques and an easily schedulable horizontal architecture for high performance scientific computing. In *Proc. 14th Annual Workshop on Microprogramming*, pages 183–198, 1981.

87. B. Ramakrishna Rau, Vinod Kathail, and Shail Aditya. Machine-description driven compilers for EPIC and VLIW processors. *Design Automation for Embedded Systems*, 4:71–118, 1999. Appeared also as technical report HPL-98-40 of HP labs, Sep. 1998.

88. Richard Scales. Software development techniques for the TMS320C6201 DSP. Texas Instruments Application Report SPRA481, www.ti.com, December 1998.

89. Eric J. Stotzer and Ernst L. Leiss. Modulo scheduling without overlapped lifetimes. In *Proc. LCTES-2009*, pages 1–10. ACM, June 2009.

90. Texas Instruments, Inc. TMS320C62x DSP CPU and instruction set reference guide. Document SPRU731, focus.ti.com/lit/ug/spru731, 2006.

91. Omri Traub, Glenn Holloway, and Michael D. Smith. Quality and Speed in Linear-scan Register Allocation. In *Proc. ACM SIGPLAN Conf. on Progr. Lang. Design and Implem. (PLDI'98)*, pages 142–151, 1998.

92. Steven R. Vegdahl. A Dynamic-Programming Technique for Compacting Loops. In *Proc. 25th annual ACM/IEEE Int. symposium on microarchitecture (MICRO-25)*, pages 180–188. IEEE CS Press, 1992.

93. Kent Wilken, Jack Liu, and Mark Heffernan. Optimal instruction scheduling using integer programming. In *Proc. Int. Conf. on Progr. Lang. Design and Implem. (PLDI'00)*, pages 121–133, 2000.

94. Tom Wilson, Gary Grewal, Ben Halley, and Dilip Banerji. An integrated approach to retargetable code generation. In *Proc. Int. Symposium on High-Level Synthesis*, pages 70–75, May 1994.

95. Sebastian Winkel. *Optimal global instruction scheduling for the Itanium processor architecture*. PhD thesis, Universität des Saarlandes, Saarbrücken, Germany, September 2004.

96. Sebastian Winkel. Optimal versus heuristic global code scheduling. In *Proc. 40th annual ACM/IEEE Int. symposium on microarchitecture (MICRO-40)*, pages 43–55, 2007.

97. Hongbo Yang, Ramaswamy Govindarajan, Guang R. Gao, George Cai, and Ziang Hu. Exploiting schedule slacks for rate-optimal power-minimum software pipelining. In *Proc. Workshop on Compilers and Operating Systems for Low Power (COLP-2002)*, September 2002.

98. Javier Zalamea, Josep Llosa, Eduard Ayguade, and Mateo Valero. Modulo scheduling with integrated register spilling for clustered VLIW architectures. In *Proc. ACM/IEEE Int. symp. on microarchitecture (MICRO-34)*, pages 160–169, 2001.

99. Thomas Zeitlhofer and Bernhard Wess. Operation scheduling for parallel functional units using genetic algorithms. In *Proc. Int. Conf. on ICASSP '99: Proceedings of the Acoustics, Speech, and Signal Processing (ICASSP'99)*, pages 1997–2000. IEEE Computer Society, 1999.

Software Compilation Techniques for MPSoCs

Rainer Leupers, Weihua Sheng, and Jeronimo Castrillon

Abstract The increasing demands such as high-performance and energy-efficiency for future embedded systems result in the emerging of heterogeneous Multiprocessor System-on-Chip (MPSoC) architectures. To fully enable the power of those architectures, new tools are needed to take care of the increasing complexity of the software to achieve high productivity. An *MPSoC compiler* is the tool-chain to tackle the problems of expressing parallelism in applications' modeling/programming, mapping/scheduling and generating the software to distribute on an MPSoC platform for efficient usage, for a given (pre-)verified MPSoC platform. This chapter talks about the various aspects of MPSoC compilers for heterogeneous MPSoC architectures, using a comparison to the well-established uni-processor C compiler technology. After a brief introduction to MPSoC and MPSoC compilers, the important ingredients of the compilation process, such as programming models, granularity and partitioning, platform description, mapping/scheduling and code-generation, are explained in detail. As the topic is relatively young, a number of case studies from academia and industry are selected to illustrate the concepts at the end of this chapter.

1 Introduction

1.1 MPSoCs and MPSoC Compilers

Lately it has become clear that Multiprocessor System-on-Chip (MPSoC) is the most promising way to keep on exploiting the high level of integration provided

R. Leupers (✉) • W. Sheng • J. Castrillon
Institute for Software for Systems on Silicon, RWTH Aachen University, SSS-611910,
Templergraben 55, 52056 Aachen, Germany
e-mail: leupers@ice.rwth-aachen.de; sheng@ice.rwth-aachen.de; castrill@ice.rwth-aachen.de

S.S. Bhattacharyya et al. (eds.), *Handbook of Signal Processing Systems*,
DOI 10.1007/978-1-4614-6859-2_37, © Springer Science+Business Media, LLC 2013

by the semiconductor technology and, at the same time, matching the constraints imposed by the embedded system market in terms of performance and power consumption. Take a look at today's standard cell phone—it is integrated with a great number of functions such as camera, personal digital assistant applications, voice/data communications and multi-band wireless standards. Moreover, like many other consumer electronic products, many non-functional parameters are evenly critical for their successes in the market, e.g. energy consumption and form factor. All those requirements necessitate the emerging of the heterogeneous MPSoC architectures, which usually consist of programmable processors, special hardware accelerators and efficient Networks on Chips (NoCs) and run a large amount of the software, in order to catch up with the next wave of integration.

Compared with high-performance computing systems in supercomputers and computer clusters, embedded computing systems require a different set of the constraints that need to be taken into consideration in the design process:

- *Real-time constraints*: Real-time performance is key to the embedded devices especially in the signal processing domain such as wireless and multimedia. Meeting real-time constraints requires not only the hardware being capable of meeting the demands of high-performance computations but also the predictable behavior of the running applications.
- *Energy-efficiency*: Most handheld equipments are driven by batteries and energy-efficiency is one of the most important factors during the system design.
- *Area-efficiency*: How to efficiently use the limited chip area becomes critical, especially for the consumer electronics where portability is a must-to-have.
- *Application/Domain-specific*: Unlike general-purpose computing products, the embedded products usually target at specific market segments, which in turn asks for the specialization of the system design to tailor for specific applications.

With those design criteria mentioned, heterogeneous MPSoC architectures are believed to outperform the previous uni-processor or homogeneous solutions. For a detailed discussion on the architectures, the readers are referred to Chap. [15]. MPSoC design methodologies, also referred as ESL (electronic system-level) tools, are growing in importance to tackle the challenge of exploring the exploding design space brought by the heterogeneity [59]. Many different tools are required for completing a successful MPSoC design (or a series of MPSoC product generations, which is more often seen now in the industry e.g. Texas Instruments (TI) OMAP [79]). The *MPSoC compiler* is one important tool among those, which will be discussed in detail in this chapter.

Firstly, what is an *MPSoC Compiler*? Today's compilers are targeted to uni-processors and the design and implementation of special compilers optimized for the various processor types (RISC, DSP, VLIW, etc.) has been well understood and practiced. Now, the trend moving to MPSoCs raises the level of complexity of the compilers for utilizing MPSoCs. The problems of expressing parallelism in applications' modeling/programming, and then mapping, scheduling and generating the software to distribute on an MPSoC platform for efficient usage, remain largely unaddressed [26]. We define *MPSoC Compiler* as the tool-chain to tackle those problems for a given (pre-)verified MPSoC platform.

It is worth mentioning that our definition of MPSoC compiler is subtly different from the term *software synthesis* as it appears in the hardware–software co-design community [30]. In this context, *software synthesis* emphasizes that starting from a single high-level system specification, the tools perform hardware/software partitioning and automatically synthesize the software part so as to meet the system performance requirements of the specifications. The flow is also called an application-driven "top-down" design flow. The MPSoC compiler is, however, used mostly in the platform-based design, where the system and semiconductor suppliers evolve the MPSoC designs in generations targeting a specific application domain for the programmers to develop software. The function of an MPSoC compiler is very close to that of a uni-processor compiler, where the compiler translates the high-level programming language (e.g. C) into the machine binary code. The difference is that an MPSoC compiler needs to perform additional (and much more complex) jobs over the uni-processor one such as partitioning/mapping, in that the underlying MPSoC machine is magnitudes more complex. Though software synthesis and MPSoC compilers share some similarities (both generate code), the major difference is that they exist in different methodologies' contexts, thus focusing on different objectives. Please refer to [18] for detailed discussions on this topic.

The rest of the chapter is organized as follows. Section 1.2 briefly introduces the challenges of building MPSoC compilers, using a comparison of an MPSoC compiler to a uni-processor compiler, followed by Sect. 2 where detailed discussions are carried out. Section 3 looks into how the challenges are tackled by demonstrating a number of case studies of MPSoC compiler construction in academia and industry.

1.2 Challenges of Building MPSoC Compilers

Before the multi-processor or multi-core era, the uni-processor-based hardware system has been very successful in creating a comfortable and convenient programming environment for software developers. The success is largely due to the fact that the sequential programming model is very close to the natural way humans think and that it has been taught for decades in the basic engineering courses. Also, the compilers of high-level programming languages like C for uni-processors are well studied, which hide nearly all the hardware details from the programmers as a holistic tool [38]. The user-friendly graphical integrated development environments (IDEs) like Eclipse [1] and mature debugging tools like gdb [2] also contribute to the good ecosystem of hardware and software in the uni-processor age.

The complexity of programming and compiling for MPSoC architectures has increased a great deal, compared to the uni-processor. The reasons are many-fold and the most important two are as follows. On one hand, MPSoCs inherently ask for applications being written in parallel programming models so as to efficiently utilize the hardware resources. Parallel programming (or thinking) has been proven to be difficult for programmers, despite many years' efforts in high-performance computing. On the other hand, the heterogeneity of MPSoC architectures requires the compilation process to be ad-hoc. The programming models for different processing elements (PEs) can be different. The granularity of the parallelism might vary. The compiler tool-chains can come from different vendors for PEs and Networks on Chips (NoCs). All those make the MPSoC compilation an extremely sophisticated process, which is most likely not "the holistic compiler" anymore for the end users. The software tool-chains are not yet fully prepared to well handle MPSoC systems, plus the lack of productive multi-core debugging solutions.

An MPSoC compiler, as the key tool to enable the power of MPSoCs, is known to be difficult to build. Then what are the fundamental challenges to construct an MPSoC compiler with respect to the uni-processor's compiler construction? A brief list is provided below, with an in-depth discussion in the following Sect. 2.

1. *Programming Models*: Evidently the transition to the parallel programming models impacts the MPSoC compiler fundamentally.
2. *Granularity of Parallelism*: While instruction-level parallelism (ILP) is exploited by the uni-processor compiler, the MPSoC compiler focuses on more coarse-grained levels of parallelism.
3. *Platform Description*: The traditional uni-processor compiler requires the architecture information such as the instruction-set and latency table in the backend to generate code. The MPSoC compiler needs another set of the platform description to work with, and the utilization phases will go beyond just the backend.
4. *Mapping/Scheduling*: An MPSoC compiler needs to map and schedule the tasks (or called code blocks) while the uni-processor compiler performs those at instruction-level.
5. *Code generation*: It is yet another leaping complexity for the MPSoC compiler— to be able to generate the final binaries for heterogeneous PEs plus the NoC vs. to generate the binary for the one-ISA architecture.

2 Foundation Elements of MPSoC Compilers

In this section we delve into the details of the problems mentioned in the introduction of this chapter. We base our discussion on the general structure of a compiler, shown in Fig. 1. We highlight the issues that make the tasks of an MPSoC compiler particularly challenging, taking the case of a uni-processor compiler as a reference.

A compiler is typically divided into three phases: The *front end*, the *middle end* and the *back end*. The front end checks for the lexical, syntactic and semantic correctness of the application. Its output is an abstract *intermediate representation* (IR) of the application which is suitable for optimizations and for code generation in the following phases of the compiler. The middle end, sometimes conceptually included within the front end, performs different analyses on the IR. These analyses enable several target-independent optimizations that mainly aim at improving the performance of the posterior generated code. The back end is in charge of the

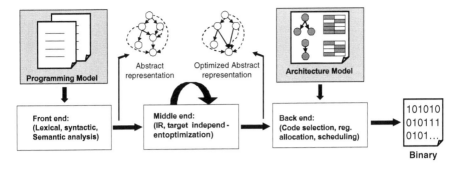

Fig. 1 Coarse view of uni-processor compiler phases

actual code generation and is divided into phases as well. Typical back end phases include *code selection*, *register allocation* and *instruction scheduling*. These phases are machine dependent and therefore require a model of the target architecture.

As in the uni-processor case, MPSoC compilers are also divided into phases in order to manage complexity. Throughout this section more details about these phases will be given, as they help to understand the differences between uni-processor and multi-processor compilers. The overall structure of the compiler (in Fig. 1) will suffer some changes though. On the one hand, if the programming model makes some architecture features visible to the programmer, the front end will need information of the target platform. On the other hand, the back end should be aware of formal properties of the programming model that may ease the code generation.

2.1 Programming Models

The main entry for any compiler is a representation of an application using a given programming model (see Fig. 1). A programming model is a bridge that provides humans access to the resources of the underlying hardware platform. Designing such a model is a delicate art, in which hardware details are hidden for the sake of productivity and usually at the cost of performance. In general, the more details remain hidden, the harder the job of the compiler is to close the performance gap. In this sense, a given programming model may reduce the work of the compiler but will never circumvent using one. Figure 2 shows an implementation of an FIR filter with different programming languages representing different programming models. The figure shows an example of the productivity-performance trade-off. On one extreme, the Matlab implementation (Fig. 2a) features high simplicity and no information of the underlying platform. The C implementation (Fig. 2b) provides more information, having types and the memory model visible to the programmer. On the other extreme, the DSP-C implementation (Fig. 2c) has explicit memory bank allocation (through the *memory qualifiers* X and Y) and dedicated data

```
a  function S = fir(coeff,X)
       S = coeff * X';
```

```
b  float coeff[N] = { 0.7, ... };

   float fir(float *ptr) {
     int i;
     float sum = 0;
     for( i = 0 ; i < N ; i++ )
       sum += coeff[i] * (*ptr++)
     return sum;
   }
```

```
c  X fract coeff[N] = { 0.7, ... };

   fract fir(Y fract * Y ptr) {
     int i;
     accum sum = 0;
     for( i = 0 ; i < N ; i++ )
       sum += coeff[i] *
               (accum)(*ptr++);
     return sum;
   }
```

Fig. 2 FIR implementation on different programming languages. (**a**) Matlab, (**b**) C, (**c**) DSP-C (adapted from [7])

types (`accum`, `fract`). Programming at this level requires more knowledge and careful thinking, but will probably lead to better performance. Without this explicit information, a traditional C compiler would need to perform complex memory disambiguation analysis in order to place the arrays in separate memory banks.

In [10], the authors classify programming models as being either hardware-centric, application-centric or formalism-centric. Hardware-centric models strive for efficiency and usually require a very experienced programmer (e.g. Intel IXP-C [57]). Application-centric models strive for productivity allowing fast application development cycles (e.g. Matlab [72], LabView [62]), and formalism-centric models strive for safeness due to the fact of being verifiable (e.g. Actors [34]). Practical programming models for embedded MPSoCs cannot pay the performance overhead brought by a pure application-centric approach and will seldom restrict programmability for the sake of *verifiability*. As a consequence of this, programming models used in industry are typically hardware-centric and provide some means to ease programmability as will be discussed later in this section.

Orthogonal to the previous classification, programming models can be broadly classified into sequential and parallel ones. The latter being of particular interest for MPSoC programming and this chapter's readers, though having its users outnumbered by the sequential programming community. As a matter of fact, nearly 90 % of the software developers nowadays use C and C++ [23], which are languages with underlying sequential semantics. Programmers have been educated for decades to program sequentially. They find it difficult to describe an application in a parallel manner, and when doing so, they introduce a myriad of (from their point of view) unexpected errors. Apart from that, there are millions of lines of sequential legacy code that will not be easily rewritten within a short period of time to make use of the new parallel architectures.

Compiling a sequential application, for example written in C, for a simple processor is a very mature field. Few people would program an application in assembly language for a single issue embedded RISC processor such as the ARM7 or the MIPS32. In general, compiler technology has advanced a lot in the uni-processor arena. Several optimizations have been proposed for superscalar processors [44], for DSPs [55], for VLIW processors [25] and for exploiting *Single*

Instruction Multiple Data (SIMD) engines [56]. Nonetheless, high performance routines for complex processor architectures with complex memory hierarchies are still hand-optimized and are usually provided by processor vendors as library functions. In the MPSoC era, the optimization space is too big to allow hand-crafted solutions across different cores. The MPSoC compiler has to help the programmer to optimize the application, probably taking into account the presence of optimized routines for some of the processing elements.

In spite of the efforts invested in classical compiler technology, plain C programming is not likely to be able to leverage the processing power of future MPSoCs. When coding a parallel application in C, the parallelism is hidden due to the inherent sequential semantics of the language and its centralized flow of control. Retrieving this parallelism requires complex dataflow and dependence analyses which are usually NP-complete and sometimes even undecidable (see Sect. 2.2). For this reason MPSoC compilers need also to cope with parallel programming models, some of which will be introduced in the following.

2.1.1 Parallel Programming Models

There are manifold parallel programming models. In this section only a couple of them will be introduced in order to give the reader an idea of the challenges that these models pose to the compiler. Part 4 of this book, Design Methods, provides in depth discussion of several models of computation (MoCs) relevant to programming models for signal processing MPSoCs.

Most current parallel programming models are built on top of traditional sequential languages like C or Fortran by providing libraries or by adding language extensions. These models are usually classified by the underlying memory architecture that they support; either shared or distributed. Prominent examples used in the high performance computing (HPC) community are:

- *POSIX Threads* (pThreads) [77]: A library-based shared memory parallel programming model.
- *Message Passing Interface* (MPI) [74]: A library-based distributed memory parallel programming model.
- *Open Multi-Processing* (OpenMP) [66]: A shared memory parallel programming model based on language extensions.[1]

These programming models are widely used in the HPC community, but their acceptance in the embedded community is quite limited. Apart from the well-known problems of programming with threads [54], the overhead of embedded implementations of these standards and the complex and hybrid (shared-distributed) nature of embedded architectures have prevented the adoption of such programming models. Embedded applications do not usually feature the regularity present in

---

[1]OpenMP has also distributed memory extensions.

scientific applications, and even if they do, their granularity is too fine to afford the overhead introduced by complex software stacks. Even in the HPC community and as a reaction to the problems of threaded programming, some new programming models are being proposed. The two most prominent ones, inspired from transactions in data bases, are: *transactional memory* (TM) [8] and *thread level speculation* (TLS) [46]. How successfully these models can be deployed in embedded MPSoCs is still unknown. Both models feature a high performance penalty if not supported by hardware, and the power overhead of such hardware may be a showstopper when adopting these techniques.

2.1.2   Embedded Parallel Programming Models

In the embedded domain, parallel programming models based on concurrent MoCs are gaining acceptance and appear to be one promising choice for describing signal processing applications. The theory of MoC originated from theoretical computer science in order to formally describe a computing system and was initially used to compute bounds on complexity (the "big O" notation). MoCs were thereafter used in the early 1980s to model VLSI circuits and only in the 1990s started to be utilized for modeling parallel applications. Dataflow programming models based on concurrent MoCs like *Synchronous Dataflow* (SDF) [52] and some extensions (like *Boolean Dataflow* (BDF) [53]) have been deeply studied in [76]. More general dataflow programming models based on *Dynamic Dataflow* (DDF) and *Kahn Process Networks* (KPN) [39] MoC have also been proposed [50, 65] (see also Chaps. [11, 28]).

- *KPN Programming Model*: In a KPN programming model, an application is represented as a graph $G = (V, E)$ like the one in Fig. 3a. In such a graph, a node $n \in V$ is called *process* and represents certain data processing. The edges represent unbounded FIFO channels through which processes communicate by sending data items or *tokens*. Processes can only be in one of two states: ready or blocked. The blocked state can only be reached by reading from *only one* empty input channel—*blocking reads* semantics. A KPN is said to be determinate: the history of tokens produced on the communication channels does not depend on the scheduling [39].
- *DDF Programming Model*: In this programming model, an application is also represented as a graph $G = (V, E, R)$ with R a family of sets, one set for every node in V. Edges have the same semantics as in the KPN model. Nodes are called *actors* and do not feature the blocking read semantics of KPN. Instead, every actor $a \in V$ has a set of *firing rules* $R_a \in R, R_a = \{R_{a.1}, R_{a.2}, \dots\}$. A firing rule for an actor $a \in V$ with p inputs is a p-tuple $R_{a.i} = (c_1, \dots, c_p)$ of conditions. A condition describes a sequence of tokens that has to be available at the given input FIFO. Parks introduced a notation for such conditions in [70]. The condition $[X_1, X_2, \dots, X_n]$ requires n tokens with values $X_1, X_2, \dots, X_n$ to be available at the top of the input FIFO. The conditions $[*], [*, *], [*_{(1)}, \dots, *_{(m)}]$

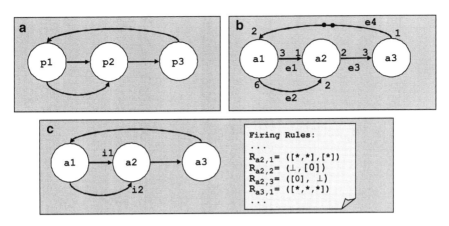

Fig. 3 Example of concurrent MoCs. (**a**) KPN, (**b**) SDF, (**c**) DDF

require at least 1, 2 and m tokens respectively with arbitrary values to be available at the input. The symbol $\perp$ represents any input sequence, including an empty FIFO. For an actor a to be in the ready state at least one of its firing rules need to be satisfied. An example of a DDF graph is shown in Fig. 3c. In this example, the actor a_2 has three different firing rules. This actor is ready if there are at least two tokens in input i1 and at least 1 token in input i2, or if the next token on input i2 or i1 has value 0. Notice that more than one firing rule can be activated, in this case the dataflow graph is said to be non-determinate.

- *SDF Programming Model*: An SDF can be seen as a simplification of DDF model,[2] in which an actor with p inputs has only one firing rule of the form $R_{a,1} = (n1,\dots,n_p)$ with $n \in \mathbb{N}$. Additionally, the amount of tokens produced by one execution of an actor on every output is also fixed. An SDF can be defined as a graph $G = (V,E,W)$ where $W = \{w_1,\dots,w_{|E|}\} \subset \mathbb{N}^3$ associates three integer constants $w_e = (p_e,c_e,d_e)$ to every channel $e = (a_1,a_2) \in E$. p_e represents the number of tokens produced by every execution of actor a_1, c_e represents the number of tokens consumed in every execution of actor a_2 and d_e represents the number of tokens (called delays) initially present on edge e. An example of an SDF is shown in Fig. 3b with delays represented as dots on the edges. For the SDF in the example, $W = \{(3,1,0),(6,2,0),(2,3,0),(1,2,2)\}$.

Different dataflow models differ in their expressiveness, some being more general, some being more restrictive. By restricting the expressiveness, models possess stronger formal properties (e.g. determinism) which make them more amenable to analysis. For example, the questions of *termination* and *boundedness*, i.e. whether an application runs forever with bounded memory consumption, are decidable for SDFs but undecidable for BDFs, DDFs and KPNs [70]. Also, since

---

[2]Being more closely related to the so-called *Computation Graphs* [41]

a

```
 1: int b[N];
 2: int A[4][N];
 3: int c[4];
 4:
 5: void foo()
 6: {
 7:    int i,j,s;
 8:
 9:    b[0] = f1();
10:    for (i = 0; i < N; i++)
11:       f2(&b[0], &b[i], &A[i%4][0]);
12:
13:    s = b[0]; b[0] = 0;
14:    for (i = 0; i < 4*16; i++) {
15:       j = i % 4;
16:       f3(s, &A[j][0]);
17:       c[j] = f4(&A[j][0]);
18:       if (j == 3)
19:          s += sum(&c[0]);
20:    }
21:}
```

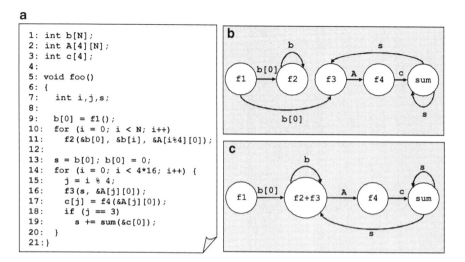

Fig. 4 Sample program. (a) C implementation, (b) a "good" KPN representation. (c) A "bad" KPN representation

the token consumption and production of an SDF actor are known beforehand, it is possible for a compiler to compute a plausible static schedule for an SDF. For a KPN instead, due to control dependent access to channels, it is impossible to compute a pure static schedule.

Apart from explicitly exposing parallelism, dataflow programming models became attractive mainly for two reasons. On the one hand, they are well suited for graphical programming, similar to the block diagrams used to describe signal processing algorithms. On the other hand, some of the underlying MoC's properties enable tools to perform analysis and compile/synthesize the specification into both software and hardware. Channels explicitly expose data dependencies among computing nodes (processes or actors), and these nodes display a distributed control flow which is easily mapped to different PEs.

To understand how dataflow models can potentially reduce the compilation effort, an example of an application written in a sequential and in two parallel forms is shown in Fig. 4. Let us assume that the KPN parallel specification in Fig. 4b represents the desired output of a parallelizing compiler. In order to derive this KPN from the sequential specification in Fig. 4a, complex analyses have to be performed. For example, the compiler needs to identify that there is no dependency on array A among lines 11 and 16 (i.e. between processes f2 and f3), which is a typical example of dataflow analysis (see Sect. 2.2.3). Only for a restricted subset of C programs, namely *Static Affine Nested Loop Programs* (SANLP), similar transformations to that shown in Fig. 4 have been successfully implemented in [83]. For more complex programs, starting from a parallel specification would greatly simplify the work of the compiler.

However, we argue that even with a parallel specification at hand, an MPSoC compiler has to be able to look inside the nodes in order to attain higher performance. With the applications becoming more and more complex, a compiler cannot completely rely on the programmer's knowledge when decomposing the application into blocks. A block diagram can hide lots of parallelism in the interior of the blocks and thus, computing nodes cannot always be considered as *black boxes* but rather as *gray/white boxes* [58]. As an example of this, consider the KPN shown in Fig. 4c. Assume that this parallel specification was written by a programmer to represent the same application logic in Fig. 4a. This KPN might seem appropriate to a programmer, because the communication is reduced (five instead of six edges). However, notice that if functions f2 and f3 are time consuming, running them in parallel could be greatly advantageous. In this representation, the parallelism remains hidden inside block f2+f3.

Summary

Currently, MPSoC compilers need to support sequential programming models, both because of the great amount of sequential legacy code and because of a couple of generations of programmers that were taught to program sequentially. At the same time, dataflow models need to be also taken into account and the compiler needs to be aware of their formal properties. An MPSoC compiler should profit from the techniques for *Instruction Level Parallelism* (ILP) present in each of the uniprocessor compilers. Therefore, the granularity of the analysis need to be re-thought in order to extract coarser parallelism as will be discussed in the next section.

2.2   *Granularity, Parallelism and Intermediate Representation*

In a classical compiler, the front end translates application code into an abstract representation—the IR. Complex constructs of the original high level programming languages are lowered into the IR while keeping machine independence. The IR serves as basis for performing analysis (e.g. control and data flow), upon which many compiler optimizations can be performed. Although there is no *de facto* standard for IRs, most compiler IRs use graph data structures to represent the application. The fundamental analysis units used in traditional compilers are the so-called *basic blocks* (BB), where a BB is defined as *a maximal sequence of consecutive statements in which flow of control enters at the beginning and leaves at the end without halt or possibility of branching except at the end* [9]. A procedure or function is represented as a *control flow graph* (CFG) whose nodes are BBs and edges represent the control transitions in the program. Data flow is analyzed inside a BB and as a result a *data flow graph* (DFG) is produced, where nodes represent

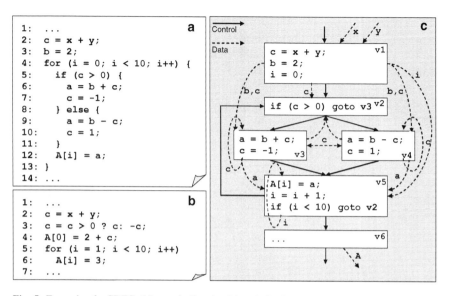

Fig. 5 Example of a CDFG. (**a**) sample C code, (**b**) optimized code, (**c**) CDFG for (**a**)

statements (or instructions) and edges represent data dependencies (or precedence constraints). With intra-procedural analysis, data dependencies that span across BB borders can be identified. As a result both control and data flow information can be summarized in a *Control Data Flow Graph* (CDFG). A sample CDFG for the code in Fig. 5a is shown in Fig. 5c. BBs are identified with the literals v1,v2,...,v6. For this code it is easy to identify the data dependencies by simple inspection. Notice however, that the self-cycles because of variable c in v3 and v4 will never be executed, i.e. the definition of c in line 7 will never *reach* line 6. Moreover, notice that the code in Fig. 5b is equivalent to that in Fig. 5a even if not evident at the first sight. Even for such a small program, a compiler needs to be equipped with powerful analysis to derive such an optimization.

For simple processors, the analysis at the BB *granularity* has worked perfectly for the last 20–30 years. The instructions inside a BB will always be executed one after another in an in-order processor, and for that reason BBs are very well suited for exploiting ILP. Already for more complex processors, like VLIW, BBs fall short to leverage the available ILP. *Predicated execution*, *software pipelining* and *trace scheduling* [25] are just some examples of optimizations that cross the BB borders seeking for more parallelism. This quest for parallelism is even more challenging in the case of MPSoC compilers.

An MPSoC compiler must go beyond ILP, which will be exploited by the individual compilers on each one of the processors. The question of granularity and its implication on parallelism for MPSoCs is still open. The *ideal* granularity for the analysis depends in part on the characteristics of the target platform. The interconnect will define the communication characteristics, and thus will dictate which granularity is reasonable and which not. The following sections will give more insights on the granularity issue, will present different kinds of parallelism and will briefly discuss the problem of dataflow analysis.

2.2.1 Granularity and Partitioning

Granularity is a major issue for MPSoC compilers and has a direct impact on the kind and degree of parallelism that can be achieved. We define partitioning as the process of analyzing an application and fragmenting it into blocks with a given granularity suitable for parallelism extraction. In this sense, the process of constructing CFGs out of an application as discussed before can be seen as a partitioning. The most intuitive granularities and their drawbacks for MPSoC compilers are:

- *Statement*: A statement is the smallest standalone entity of a programming language. An application can be broken to the level of statements and the relations among each them. This granularity provides the highest degree of freedom to the analysis but could prevent ILP from being exploited at the uni-processor level. Besides, the complexity could be prohibitively big for real life applications.
- *Basic Block*: As already discussed, traditional compilers work on the level of BBs, for they are well suited for ILP. However, in practice it is easy to find examples where BBs are either too big or too small for MPSoC parallelism extraction. A BB composed of a sequence of function calls inside a loop would be seen as a single node by the MPSoC compiler, and potential parallelism will be therefore hidden. On the other extreme, a couple of small basic blocks divided by simple control constructs could be better handled by a uni-processor compiler with support for predicated execution.
- *Function*: A function is simply defined as a subroutine with it own stack. At this level, only function calls are analyzed and the rest of the code is considered as irrelevant for the analysis. As with BBs, this granularity can be too coarse or too fine-grained depending on the application and on the programmer. It is possible to force a coding style where parallelism is explicitly coded in the way the behavior is factorized into functions. However, an MPSoC compiler should not make any assumption on the coding style.
- *Task/process*: For parallel programming models, the granularity is determined by the tasks (processes or actors) defined in the specification itself. As mentioned in Sect. 2.1, a powerful MPSoC compiler might inspect the inside of the tasks in order to look for parallelism hidden in the specification.

As an example, different partitions at different granularity levels for the program introduced in Fig. 4a are shown in Fig. 6. The BBs of function foo are shown in Fig. 6a. This partition is a good example of what discussed above. The BB on line 9 is too light weight in comparison to the other BBs, whereas the BB in lines 16-17 may be too coarse. The partition at function level is shown in Fig. 6b. This partition

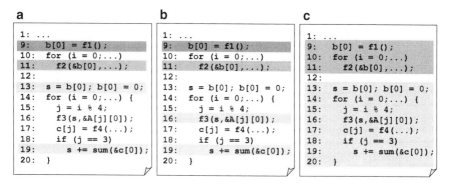

Fig. 6 Granularity example. (**a**) Basic block level, (**b**) function level, (**c**) task level

happens to match the KPN derivation introduced in Fig. 4b. Whether this granularity is appropriate or not, depends on the amount of data flowing between the functions and the timing behavior of each one of the functions as will be discussed later in this chapter. Finally, Fig. 6c shows a coarser task level partitioning, where the first task processes mostly array b and the second task processes array A. Again, this task level granularity might turn out to be inefficient if the time spent processing array A is too large. If this is the case, exploiting the parallelism inside the second task would be required.

As illustrated with the example, it is not clear what will be the ideal granularity for an MPSoC compiler to work on. Most likely, there will be no exact definition as in the case of BBs, and the final granularity will depend on the MPSoC, the programming model and the application domain. Some research groups have performed steps towards the identification of a suitable granularity for particular platforms. The approach is usually based on partitioning the application into *code blocks* by means of heuristics or clustering algorithms [20]. Irrespective of the granularity chosen, in the remaining of this chapter we refer to the outcome of the partitioning as *code blocks*. A code block may be a BB, a function, a task, a process, an actor or the result of intricate clustering algorithms.

2.2.2 Parallelism

While a traditional compiler tries to exploit ILP, the goal of an MPSoC compiler is to extract coarser parallelism. The most prominent types of coarse or macroscopic parallelism are illustrated in Fig. 7 and are described in the following.

- Data Level Parallelism *(DLP)*: In DLP the same computational task is carried out on several disjoint *Data Sets* as illustrated in Fig. 7a. This kind of parallelism can be seen as a generalization of SIMD. DLP is typically present in multimedia applications, where a decoding task performs the same operations on different portions of an image or video (e.g. *Macroblocks* in H.264). Several programming

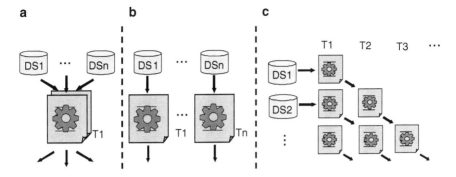

Fig. 7 Kinds of parallelism: (**a**) DLP, (**b**) TLP, (**c**) PLP

models provide support for DLP. For instance, an *OpenMP for* directive tells the compiler or the runtime system that the iterations of the loop are independent from each other and can therefore be parallelized.

- Task (or functional) Level Parallelism *(TLP)*: In TLP different tasks can compute in parallel on different data sets as shown in Fig. 7b. This kind of parallelism is inherent to programming models based on concurrent MoCs (see Sect. 2.1.1). Tasks may have dependencies to each other, but once a task has its data ready, it can execute in parallel with the already running tasks in the system.

- Pipeline Level Parallelism *(PLP)*: In PLP a computation is broken into a sequence of tasks and these tasks are repeated for different data sets. In this sense, PLP can be seen as a mixture of DLP and TLP. Following the principle of pipelining, a task with DLP can be segmented in order to achieve a higher throughput. At a given time, different tasks of the original functionality are executed concurrently on different data sets. This kind of parallelism is e.g. exploited in the programming models targeting network processors.

Exploiting these kinds of parallelism is a must for an MPSoC compiler, which has to be therefore equipped with powerful flow and dependence analysis capabilities.

2.2.3   Flow and Dependence Analysis

Flow analysis includes both control and data flow. The result of these analyses can be summarized in a CDFG, as discussed at the beginning of this section. Data flow analysis serves to gather information at different program points, e.g. about available defined variables (*reaching definitions*) or about variables that will be used later in the control flow (*liveness analysis*). As an example, consider the CDFG in Fig. 5c in which a reaching definitions analysis is carried out. The analysis tells, for example, that the value of variable c in line 5 can come from three different definitions in lines 2, 7 and 10.

Data flow analysis deals mostly with scalar variables, like in the previous example, but falls short when analyzing the flow of data when explicit memory accesses are included in the program. In practice, memory accesses are very common through the use of pointers, structures or arrays. Additionally, in the case of loops, data flow analysis only says if a definition reaches a point but does not specify exactly in which iteration the definition is made. The analyses that answer these questions are known as *array analysis*, *loop dependence analysis* or simply *dependence analysis*.

Given two statements $S1$ and $S2$, dependence analysis determines if $S2$ depends on $S1$, i.e. if $S2$ cannot execute before $S1$. If there is no dependency, $S1$ and $S2$ can execute in any order or in parallel. Dependencies are classified into control and data:

- Control Dependence: A statement $S2$ is control dependent on $S1$ ($S1 \ \delta^c \ S2$) if whether or not $S2$ is executed depends on $S1$'s execution. In the following example, $S1 \ \delta^c \ S2$:

  ```
  S1: if (a > 0) goto L1;        S2: a = b + c;
  S3: L1: ...
  ```

- Data Dependencies:

  - *Read After Write* (RAW, also *true/flow dependence*): There is RAW dependence between statements $S1$ and $S2$ ($S1 \ \delta^f \ S2$) if $S1$ modifies a resource that $S2$ reads thereafter. In the following example, $S1 \ \delta^f \ S2$:

    ```
    S1: a = b + c;                 S2: d = a + 1;
    ```

  - *Write After Write* (WAW, also *output dependence*): There is WAW dependence between statements $S1$ and $S2$ ($S1 \ \delta^o \ S2$) if $S2$ modifies a resource that was previously modified by $S1$. In the following example, $S1 \ \delta^o \ S2$:

    ```
    S1: a = b + c;                 S2: a = d + 1;
    ```

  - *Write After Read* (WAR, also *anti-dependence*): There is WAR dependence between statements $S1$ and $S2$ ($S1 \ \delta^a \ S2$) if $S2$ modifies a resource that was previously read by $S1$. In the following example, $S1 \ \delta^a \ S2$:

    ```
    S1: d = a + 1;                 S2: a = b + c;
    ```

Obviously, two statements can exhibit different kinds of dependencies simultaneously. Computing these dependencies is one of the most complex tasks inside a compiler, either for uni-processor or for multi-processors systems. For a language like C, the problem of finding all dependencies *statically* is NP complete and in some cases undecidable. The main reason for this being the use of pointers [35] and indexes to data structures that can only be resolved at runtime. Figure 8 shows three sample programs to illustrate the complexity of dependence analysis. In Fig. 8a, in order to determine if there is a RAW dependence between $S3$ and $S4$ ($S3 \delta^f S4$) across iterations, one has to solve a constrained integer linear system of equations, which is NP complete. For the example, the system of equations is:

```
S1: for (i = N; i > X; i--) {                                    a
S2:     for (j = M; j > Y ; j--) {
S3:         A[3*i + 2*j + 2][i + j + 1] = ...;
            ...
S4:         ... = A[4 * i + j + 1][2*i + j + 1] ...;
S5: }}
```

```
S1: while (i) {              b
S2:     f1(&A[i]);
S3:     ... = A[f2(i)];
        ...
S4: }
```

```
S1: scanf("%d",&i);          c
S2: scanf("%d",&j);
S3: A[i] = ...;
S4: ...   = A[j];
    ...
```

Fig. 8 Examples of dependence analysis. (**a**) NP complete, (**b**) requires inter-procedural analysis, (**c**) undecidable

$$3x_1 + 2x_2 + 2 = 4y_1 + y_2 + 1$$

$$x_1 + x_2 + 1 = 2y_1 + y_2 + 1$$

subject to $X < x_1, y_1 < N$ and $Y < x_2, y_2 < M$. Notice for example that there is a RAW dependence between iterations $(1,1)$ and $(2,-2)$ on A[7][3]. In order to analyze the sample code in Fig. 8b, a compiler has to perform inter-procedural analysis to identify if f1 potentially modifies the contents of A[i] and to sort out the potential return values of f2(i). Depending on the characteristics of f1 and f2 this problem can be undecidable. Finally, the code in Fig. 8c is an extreme case of the previous one, in which it is impossible to know the values of the indexes at compile time. The complexity of dependence analysis motivated the introduction of means for memory disambiguation at the programming language level, such as the restrict keyword in C99 standard [85].

For an MPSoC compiler, the situation is not different. The same kind of analysis has to be performed at the granularity produced by the partitioning step. Array analysis could still be handled by a vectorizing compiler for one of the processors in the platform. The MPSoC compiler has to perform the analysis at a coarser granularity level in which function calls will not be an exception. This is for example the case for the code in Fig. 4a. In order to derive KPN representations like those presented in Fig. 4b,c, the compiler needs to be aware of the side effects of all functions. For example, it has to make sure that function f2 does not modify the array A, otherwise there would be a dependency (an additional channel in the KPN) between processes f2 and f3 in Fig. 4b. The dependence analysis should also provide additional information, for example, that the sum function is only executed every four iterations of the loop. This means that every four instances of f3 followed by f4 can be executed in parallel. This is illustrated in Fig. 9. A summarized version of the CDFG with the dependence information is shown in Fig. 9a. In this graph, data edges are annotated with the variable that generates the dependency and, in the case of *loop-carried* dependencies, with the *distance* of

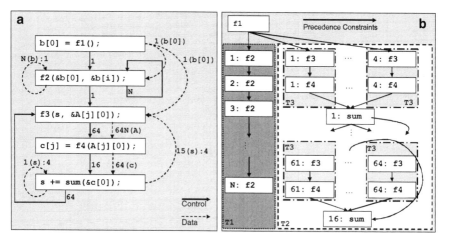

Fig. 9 Dependence analysis on example in Fig. 4a. (**a**) Summarized CDFG, (**b**) unrolled dependencies

the dependence [61]. The distance of a dependence tells after how many iterations a defined value will be actually used. With the dependence information, it is possible to represent the precedence constraints along the execution of the whole program as shown in Fig. 9b. In the figure, n: f represents the n-th execution of function f. With this partitioning, it is possible to identify two different kinds of parallelism: T1 and T2 are a good example of TLP, whereas T3 displays DLP. This is a good example where flow and dependence analysis help determining a partitioning that exposes coarse grained parallelism.

Due to the complexity of static analyses, research groups have recently started to take decisions based on *dynamic data flow analysis* DDFA [20, 81]. Unlike static analyses, where dependencies are determined at compile time, DDFA uses traces obtained from profiling runs. This analysis is of course not sound and cannot be directly used to generate code. Instead, it is used to obtain a coarse measure of the data flowing among different portions of the application in order to derive plausible partitions and in this way identify DLP, TLP and/or PLP. Being a profile-based technique, the quality of DDFA depends on a careful selection of the input stimuli. In interactive programming environments, DDFA can provide hints to the programmer about where to perform code modifications to expose more parallelism.

Summary

Traditional compilers work at the basic block granularity which is well suited for ILP. MPSoC compilers in turn need to be equipped with powerful flow analysis techniques, that allow to partition the application into a suitable granularity.

This granularity may not match any mainstream granularity and may depend on the target platform. The partitioning step must break the application into code blocks from which coarse level parallelism such as DLP, TLP and PLP can be extracted.

2.3 Platform Description for MPSoC Compilers

After performing optimizations in the middle end, a compiler back end generates code for the target platform based on a model of it. Such a platform model is also required by an MPSoC compiler, but, in contrast to a uni-processor compiler flow, the architecture model or a subset of it may also be used by other phases of the compiler. For example, if the programming model exposes some hardware details to the user, the front end needs to be able to cope with that and probably perform consistency checks. Besides, some MPSoC optimizations in the middle end may need some information about the target platform as discussed in Sect. 2.2.

Traditionally an architecture model describes:

(i) Available *operations*: In form of an abstract description of the *Instruction Set Architecture* (ISA). This information is mainly used by the code selection phase.
(ii) Available resources: A list of hardware resources such as registers and functional units (in case of a superscalar or a VLIW). This information is used, for example, by the register allocator and the scheduler.
(iii) Communication links: Describe how data can be moved among functional units and register files (e.g. cross paths in a cluster VLIW processor).
(iv) Timing behavior: In form of *latency* and *reservation tables*. For each available operation, an entry in the latency table tells the compiler how long it takes to generate a result whereas an entry in the reservation table tells the compiler which resources are blocked and for how long. This information is mainly used to compute the schedule.

In the case of an MPSoC, a platform description has to provide similar information but at a different level. Instead of a representation of an ISA, the available operations describe which kinds of processors and hardware accelerators are in the platform. Instead of a list of functional units, the model provides a list of PEs and a description of the memory subsystem. The communication links represent no longer interconnections among functional units and register files, but possibly a complex Network-On-Chip (NoC) that interconnects the PEs among them an with the memory elements. Finally, the timing behavior has to be provided not for individual operations (instructions) but for code blocks.

Usually, the physical description (items i–iii) is a graph representation provided in a given format (usually an XML file, see Sect. 3 for practical examples). The timing behavior cannot be represented as application independent static tables due to the presence of programmable processing elements that can, in principle, execute arbitrary pieces of code. Getting the timing behavior of the code blocks obtained after partitioning is a major research topic and a requisite for an MPSoC compiler. Several techniques, under the name of *Performance Estimation*, are applied in order to get such execution times: *Worst/Best/Average Case Execution Time* (W/B/ACET) [84]. These techniques can be roughly categorized as:

- *Analytical*: Analytical or static performance estimation tries to find theoretical bounds to the WCET, without actually executing the code. Using compiler techniques, all possible control paths are analyzed and bounds are computed by using an abstract model of the architecture. This task is particularly difficult in the presence of caches and other non-deterministic architectural features. For such architectures, the WCET might be too pessimistic and thus induce bad decisions (e.g. wrong schedules). There are already some commercial tools available, aiT [6] is a good example.
- *Simulation-based*: In this case the execution times are measured on a simulator. Usually cycle accurate virtual platforms are used for this purpose [24,78]. Virtual platforms allow full system simulation, including complex chip interconnects and memory subsystems. Simulation-based models suffer from the *context-subset* problem, i.e. the measurements depend on the selection of the inputs.
- *Emulation-based*: The simulation time of cycle accurate models can be prohibitively high. Typical simulation speeds range from 1 to 100 KIPS (Kilo Instructions per Second). Therefore, some techniques emulate the timing behavior of the target platform in the host machine without modeling every detail of the processor by means of instrumentation. Source level timing estimation has proven to be useful for simple architectures [37, 42], the accuracy for VLIW or DSP processors is however not satisfactory. The authors in [27] use so-called *virtual back ends* to perform timing estimation by emulating the effects of the compiler back end and thus improving the accuracy of source level timing estimation considerably. With these techniques, simulation speeds of up to 1 GIPS are achievable.

Performance estimation plays an important role when compiling for MPSoCs. Granularity, discussed in Sect. 2.2, is directly related to the timing of the code blocks. Is the execution time of a code block too short compared to the other blocks, then it can be merged with into other blocks or simply ignored during the timing analysis. Additionally, the timing information is key for driving the scheduling and mapping decisions. As an example consider a platform with two types of PEs, RISCs and DSPs, for the application in Fig. 4a. Assume that the performance estimation determines an average time $\hat{t}_f^p$ for the execution of function f on processing type p. Figure 10 shows possible scheduling traces for the sample application with different partitioning and mapping. The following timing variables are used in the figure to simplify the visualization:

$$\Delta t_1 = \hat{t}_{f3}^{risc} + \hat{t}_{f4}^{risc} + 3 \cdot max(\hat{t}_{f3}^{risc}, \hat{t}_{f4}^{risc})$$
$$\Delta t_2 = \hat{t}_{f2}^{risc}$$

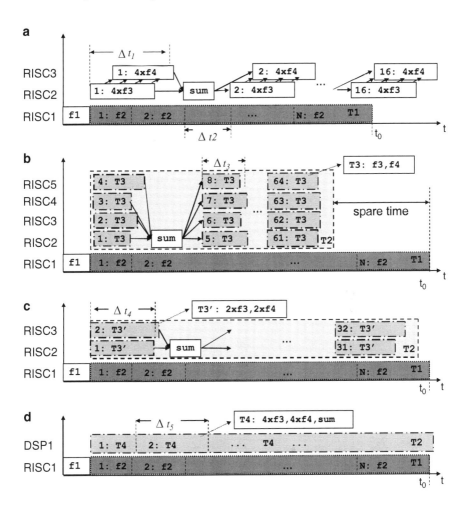

Fig. 10 Example of impact of timing in the mapping and scheduling decisions on application code in Fig. 4a. (**a**) Partition with PLP, direct implementation of Fig. 4b, (**b**) full parallelism exposed in Fig. 9b, (**c**) improved processor utilization, (**d**) improved mapping

$$\Delta t_3 = \hat{t}_{f3}^{risc} + \hat{t}_{f4}^{risc}$$

$$\Delta t_4 = 2 \cdot \hat{t}_{f3}^{risc} + 2 \cdot \hat{t}_{f4}^{risc}$$

$$\Delta t_5 = 4 \cdot \left(\hat{t}_{f3}^{dsp} + \hat{t}_{f4}^{dsp} \right) + \hat{t}_{sum}^{dsp}$$

The trace in Fig. 10a corresponds to the KPN representation introduced in Figs. 4b and 6b. This partition exhibits PLP, but the fact that some instances of f3 and f4 can be executed in parallel, as suggested by the dependence graph in

Fig. 9a, is not exploited. Besides, depending on the architecture, this partition may require the array A to be sent through the interconnect, which could be too costly. The schedule in Fig. 10b uses the information gathered by the dependence analysis in Fig. 9 and exploits fully the TLP and the DLP available in the application. This behavior can be obtained by unrolling the loop at the granularity shown in Fig. 6b. In this example, task T1 dominates the execution of the complete application and therefore processors RISC2-RISC5 feature a low utilization. To improve the utilization, two tasks T3 are merged into T3' as shown in Fig. 10c. Finally, if the DSP is used to execute task T2, a similar execution time of the application can be obtained by using only two processing elements as illustrated in Fig. 10d. Note that the last partition corresponds to the task-level granularity in Fig. 6c.

The previous example is of course too simplistic and serves only as an illustration of the importance of timing information. In this example no interconnect nor access to share resources such as peripherals or memory were modeled. For real applications it is much more difficult to characterize the timing behavior and the resource usage. It might be impractical to characterize the timing behavior of a code block with a simple average value. For some applications, a discrete set of values for different *scenarios* might be used [29,58]. For other applications a more detailed statistical description in form of a probability density function could be required [43]. Access to shared resources can be modeled in a similar way to reservation tables. Several code blocks might access the same resources in the MPSoC (e.g. analog-to-digital converters, serial communication links, graphical display) and these *structural hazards* need to be taken into account by the MPSoC compiler.

Summary

Platform models for MPSoC compilers describe similar features to those of traditional compilers but at a higher level. Processing elements and NoCs take the place of functional units, register files and their interconnections. On an MPSoC compiler, the platform model is no longer restricted to be used on the back end but a subset of it may be used by the front end and the middle end. Out of the information needed to describe the platform, the timing behavior is the most challenging. This timing information is needed for performing successfully mapping and scheduling, as will be described in the next section.

2.4 Mapping and Scheduling

Mapping and scheduling in a traditional compiler is done in the back end provided a description of the architecture. Mapping refers to the process of assigning operations to instructions and functional units (code selection) and variables to registers (register allocation). Scheduling refers to the process of organizing the instructions in a timed sequence. The schedule can be computed statically (for RISC, DSPs and VLIWs) or dynamically at runtime (for Superscalars) whereas the mapping of operations to instructions is always computed statically. The main purpose of mapping and scheduling in uni-processor compilers had been always to improve performance. Code size is also an important objective for embedded (specially VLIW) processors. Only recently, power consumption became an issue. However, the reduction in power consumption with back end techniques does not have a big impact on the overall system power consumption.

In an MPSoC compiler similar operations have to be performed. Mapping, in this context, refers to the process of assigning code blocks to PEs and logical communication links to physical ones. In contrast to the uni-processor case, mapping can be also dynamic. A code block could be mapped at runtime to different PEs, depending on availability of resources. Scheduling for MPSoCs has a similar meaning as for uni-processors, but instead of scheduling instructions, the compiler has to schedule code blocks. The presence of different application classes, e.g. real time, add complexity to the optimizations in the compiler. Particularly, there is much more room for improving power consumption in an MPSoC; after all, power consumption is one of the MPSoC drivers in the first place.

The result of scheduling and mapping is typically represented in form of a *Gantt Chart*, similar to the ones presented in Fig. 10. The PEs are represented in the vertical axis and the time in the horizontal axis. Code blocks are located in the plane, according to the mapping and the scheduling information. In Fig. 10a functions f1 and f2 are mapped to processor RISC1, the functions f3 and sum are mapped to processor RISC2 and function f4 to processor RISC3.

Given that code blocks have a higher time variability than instructions, scheduling can be rarely performed statically. Pure static scheduling requires full knowledge of the timing behavior and is only possible for very predictable architectures and regular computations, like in the case of *systolic arrays* [48]. If it is not possible to obtain a pure static schedule, some kind of synchronization is needed. Different scheduling approaches require different synchronization schemes with different associated performance overhead. In the example of the previous section, the timing information of task T3 is not known precisely as shown in Fig. 10b. Therefore the exact starting time of function sum cannot be determined and a synchronization primitive has to be inserted to ensure correctness of the result. In this example, a simple *barrier* is enough in order to ensure that the execution of T3 in processors RISC3, RISC4 and RISC5 has finished before executing function sum.

2.4.1 Scheduling Approaches

Which scheduling approach to utilize depends on the characteristics of the application and the properties of the underlying MoC used to describe it. Apart from pure static schedules, one can distinguish among the following scheduling approaches:

- *Self-timed Scheduling*: Typical for applications modeled with dataflow MoCs. A self-timed schedule is close to a static one. Once a static schedule is computed, the code blocks are ordered on the corresponding PEs and synchronization primitives are inserted that ensure the presence of data for the computation. This kind of scheduling is used for SDF applications. For a more detailed discussion the reader is referred to [76].
- *Quasi-static Scheduling*: Used in the case where control paths introduce a predictable time variation. In this approach, unbalanced control paths are balanced and a self-timed schedule is computed. Quasi-static scheduling for dynamically parameterized SDF graphs is explored in [13] (see also Chap. [14]).
- *Dynamic Scheduling*: Used when the timing behavior of the application is difficult to predict and/or when the number of applications is not known in advance (like in the case of general purpose computing). The scheduling overhead is usually higher, but so is also the average utilization of the processors in the platform. There are many dynamic scheduling policies. *Fair queue scheduling* is common in general purpose operating systems (OSs), whereas different flavors of priority based scheduling are typically used in embedded systems with real time constraints, e.g. *rate monotonic* (RM) and *earliest deadline first* (EDF).
- *Hybrid Scheduling*: Term used to refer to scheduling approaches in which several static or self-timed schedules are computed for a given application at compile time, and are switched dynamically at run-time depending on the *scenario* [29]. This approach is applied to streaming multimedia applications, and allows to adapt at runtime making it possible to save energy [58].

Virtually every MPSoC platform provides support for implementing mapping and scheduling. The support can be provided in software or in hardware and might restrict the available policies that can be implemented. This has to be taken into account by the compiler, which needs to generate/synthesize appropriate code (see Sect. 2.5). Software support can be provided by a full fledged OS or by light micro kernels. Several commercial OS vendors provide solutions for MPSoC platforms, but full efficient support for highly heterogeneous MPSoCs remains a major challenge. In order to reduce the overhead introduced by software stacks, hardware support for mapping and scheduling has been proposed both in the HPC [32,47] and in the embedded communities [19,69].

2.4.2 Computing a Schedule

Independent of which scheduling approach and how this is supported, the MPSoC compiler has to compute a schedule (or several of them). This problem is known to be NP-complete even for simple *directed acyclic graphs* (DAGs). Uni-processor compilers therefore employ heuristics, most of them being derived from the classical *List Scheduling* algorithm [36]. Computing a schedule for multiprocessor platforms is by no means simpler. The requirements and characteristics of the schedule depend on the underlying MoC with which the application was modeled. In this chapter we distinguish between application modeled with centralized and distributed control flow.

Centralized Control Flow

Uni-processor compilers deal with centralized control flow, i.e. instructions are placed in memory and a central entity dictates which instructions to execute next, e.g. the *program counter* generator. The scheduler in a traditional uni-processor compiler leaves the control decisions out of the analysis and focus on scheduling instructions inside a BB. Since the control flow inside a BB is linear, there are no circular data dependencies and the data dependence graph is therefore acyclic. The resulting DAG is typically scheduled with a variant of the list scheduling algorithm.

Given a DAG $G = (V, E)$, the list scheduling algorithm determines an starting time t_s^n and a resource r^n to every node $n \in V$. This is done by iterating the following steps until a valid schedule is obtained:

(i) Compute the *ready set*: Set of nodes for which all the predecessors in G are already scheduled.
(ii) Select node: Picks the node from the ready set with the highest priority. There are several heuristics to compute the priority of a node. This can be done, for example, by associating a higher priority to nodes in the critical path.
(iii) Allocate node: The selected node n is mapped to a resource r^n at time t_s^n. There are also several heuristics for choosing the resource. This can be done, for example, by selecting a resource that minimizes the finishing time of the node.

In order to achieve a higher level of parallelism, uni-processor compilers apply different techniques that goes beyond BBs. Typical examples of this techniques include *loop unrolling* and *software pipelining* [51]. An extreme example of loop unrolling was introduced in the previous section, where the dependence graph in Fig. 9a was completely unrolled in Fig. 9b. Note that the graph in Fig. 9b is acyclic and could be scheduled with the list scheduling algorithm. The results of list scheduling with 5 and 3 resources would look similar to the scheduling traces in Fig. 10b,c respectively.

In principle, the same scheduling approach can be used for MPSoCs. However, since every processor in the MPSoC has its own control flow, a mechanism has to be implemented to transfer control. In the example in Fig. 10b, some processor has the control code for the loop in line 14 of Fig. 4 and activates the four parallel tasks T3. There are several ways of handling this distribution of control. Parallel programming models like pthreads offer source level primitives to implement *forks* and *joins*. Some academic research platforms offer dedicated instructions to send so-called *control tokens* among processors [86].

Distributed Control Flow

Parallel programming models based on concurrent MoCs like the ones discussed in Sect. 2.1.2 feature distributed control flow. For applications represented in this way, the issue of synchronization is greatly simplified and can be added to the logic of the channel implementation. Simple applications represented as acyclic task precedence graphs with predictable timing behavior can be scheduled with a list scheduling algorithm or with one of many other available algorithms for DAGs. For a survey on DAG scheduling algorithms the reader is referred to [49]. Applications, where precedence constraints are not explicit in the programming model and where communication can be control dependent, e.g. KPNs, are usually scheduled dynamically. Finally, for applications represented as SDF, a self-timed schedule can be easily computed.

- *KPN scheduling*: KPNs are usually scheduled dynamically. There are two major ways of scheduling a KPN: *data* and *demand* driven. In data driven scheduling, every process in the KPN with available data at its input is in the ready state. A dynamic scheduler then decides which process gets executed on which processor at runtime. A demand driven scheduler first schedules processes with no output channels. This processes execute until a read blocks in one of the input channels. The scheduler triggers then only the processes from which data has been requested (*demanded*). This process continues recursively. For further details de reader is referred to [70].
- *SDF scheduling*: As mentioned before, SDFs are usually scheduled using a self-timed schedule, which requires a static schedule to be computed in the first place. There are two major kinds of schedules: *blocked* and *non-blocked schedules*. In the former, a schedule for one cycle is computed and is repeated without overlapping, whereas in the latter, the execution of different iterations of the graph are allowed to overlap. For computing a blocked schedule, a *complete cycle* in the SDF has to be determined. A complete cycle is a sequence of actor firings that brings the SDF to its initial state. Finding a complete cycles requires that (*i*) enough initial tokens are provided in the edges and (*ii*) there is a non trivial solution for the system of equations $\Gamma \cdot \mathbf{r} = 0$, where $[\Gamma_{ij}] = p_{ij} - c_{ij}$, and $p_{ij}\, c_{ij}$ are the number of tokens that actor i produces to and consumes from channel j respectively. In the literature, $\mathbf{r}$ is called *repetition vector* and Γ *topology matrix*. As an example, consider the SDF in Fig. 3b. This SDF has a topology matrix:

$$\Gamma = \begin{pmatrix} 3 & -1 & 0 \\ 6 & -2 & 0 \\ 0 & 2 & -3 \\ -2 & 0 & 1 \end{pmatrix}$$

and a repetition vector is $\mathbf{r} = [1\ 3\ 2]^T$. By *unfolding* the SDF according to its repetition vector and removing the feedback edges (those with delay tokens) one obtains the DAG shown in Fig. 11a with a possible schedule on two processors sketched in Fig. 11b. Using this procedure, the problem of scheduling an SDF is turned into DAG scheduling, and once again, one of the many heuristics for DAGs can be used. See Chap. [31] for further details.

For general application models and for obtaining better results than with human-designed heuristics, several optimization methods are used. *Integer Linear*

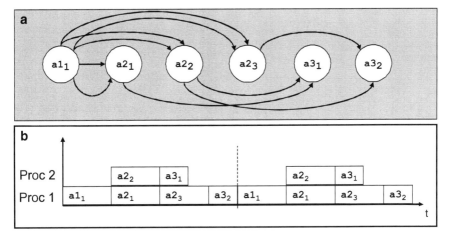

Fig. 11 Example of SDF scheduling, for SDF in Fig. 3b, (**a**) derived DAG with $\mathbf{r} = [1\ 3\ 2]^T$, (**b**) possible schedule on two PEs

Programming is used in [67] and a combination of Integer Linear Programming and *Constraint Programming* (CP) is employed in [12]. *Genetic Algorithms* have also been used for this purpose, see Chap. [11]. Apart from scheduling and mapping code blocks and communication, a compiler also needs to map data. Data locality is already an issue for single processor systems with complex memory architectures: caches and *scratch pad memories* (SPM). In MPSoCs, maximizing data locality and minimizing *false sharing* is an even bigger challenge [40].

Summary

Mapping and scheduling is one of the major challenges of MPSoC compilers. Different application constraints lead to new optimization objectives. Besides, different programming models with their underlying MoC allow different scheduling approaches. Most of these techniques work under the premise of accurate performance estimation (Sect. 2.3) which is by itself a hard problem. Due to the high heterogeneity of signal processing MPSoCs, mapping of data represents a bigger challenge than in uni-processor systems.

2.5 Code Generation

The code generation phase of an MPSoC compiler is ad-hoc due to the heterogeneity of MPSoCs. To name a few examples: the PEs are heterogeneous where the programming models may differ, the communications networks (thus the APIs) are heterogeneous, and the OS-service libraries implementations can vary from one to another. After the partitioning, mapping and scheduling, the code generation of an MPSoC compiler acts like a *meta-compiler*, distributing the code blocks to the PE compilers, generating the code for communications and schedulers, linking with the low-level libraries and so on.

The distinct feature of heterogeneous MPSoCs is that different types of processors, which are usually programmable, are deployed. They can be developed in-house or acquired through third-party IP vendors. They are usually shipped with software tool-chains such as C compilers, linkers and assemblers. After partitioning and mapping phases, the MPSoC compiler needs to generate the mapped-to-processor code, in the form of either a high-level programming language like C or PE-specific assembly, to be further compiled by the PE tool-chain. The PE compiler can turn on its own optimization features to further improve the code quality. In this sense, an MPSoC compiler is more like a meta-compiler on top of many off-the-shelf compilers to coordinate the compilation process. One example of the many initial efforts made in this area is the Real Time Software Components (RTSC) project initially driven by TI to help standardizing the embedded components [5]. A RTSC package is a named collection of files that form a unit of versioning, update, and delivery from a producer to a consumer. Important information such as codec library, building/linker parameters and instruction-set architecture for this particular codec are specified and packaged. Therefore, different codecs from different vendors are possible to be packaged seamlessly together, giving further benefits such as helping the automation of building/linking.

The PEs will communicate with each other using the NoC, which requires communication/synchronization primitives (e.g. semaphores, message-passing) correctly set in place of the code blocks that the MPSoC compiler distributes to the PEs. Again, due to the heterogeneous nature of the underlying architecture, the same communication link may look very different in the implementation, e.g. when the sending/receiving points are in the different PEs. The embedded applications often need to be implemented in a portable fashion for the sake of software re-use. *Abstraction* of the communication functions to a higher level into the programming model is widely practiced, though it is still very ad-hoc and platform-specific. Recently the Multicore Association has published the first draft of Multicore Communications API (MCAPI), which is a message-passing API to capture the basic elements of communication and synchronization that are required for closely distributed embedded systems [3]. This might have been a good first step in this area.

As discussed in Sect. 2.4, the scheduling decision is a key factor in the MPSoC compiler, especially in embedded computing where real-time constraints have to be met. No matter which scheduling policy is determined for the final design, the functionality has to be implemented, in hardware, or software, or in a hybrid manner. A common approach is to use a ready OS, often a RTOS, as a scheduler. There are many commercial solutions available such as QNX and WindRiver. The scheduler implementation in hardware is not uncommon for embedded devices, as software solutions may lead to larger overhead, which is not acceptable for

RT-constrained embedded systems. Industrial companies as well as academia have delivered some results in this area which are promising though more successful stories are still needed to justify. A hybrid solution is a mixture, where some acceleration for the scheduler is implemented in hardware while flexibility is provided by software programmability, therefore customizing a trade-off between efficiency and flexibility. If the scheduling is not helped by e.g. an OS or a hardware scheduler, the code generation phase needs to generate or synthesize the scheduler e.g. [22, 50].

Summary

Code generation is a complicated process, where many efforts are made to hide the compilation complexity via layered SW stacks and APIs. Heterogeneity will cause ad-hoc tool-chains to exist for a long time, further aggravating the situation.

3 Case Studies

As discussed in Sect. 2, MPSoC compilers' complexity grows rapidly compared to the compilers of uni-processors. Considering MPSoC compilers in the infancy, nowadays the MPSoC compiler constructions of different industrial platforms and academic prototypes are very much ad-hoc. This section surveys some typical case studies for the readers to see how real-world implementations address the various MPSoC compiler challenges.

- *Academic Research*: In academia, many research groups have looked into the topic of MPSoC compilers recently. Because the topic in nature is very heterogeneous, it has called for interests from different research communities, such as real-time computing, compiler optimization, parallelization and fast simulation. A considerable amount of efforts have been put in the areas such as MoCs, automatic parallelization, retargetable compilers and virtual simulation platforms. Compared to their counterparts in industry, the academic researchers focus mostly on the *upstream* of the MPSoC compilation flow, e.g., using MoCs to model applications, automatic task-to-processor mapping, early system performance estimation and holistic construction of the MPSoC compilers.
- *Industrial Cases*: Several large semiconductor houses have already a few mature product lines aiming at different segments of the market due to the application-specific nature of the embedded devices. The stringent time-to-market window calls for the necessity to adopt platform-based MPSoC design methodology. That is, a new generation of an MPSoC architecture is based on a previous successful model with some evolutionary improvements. Compared to their counterparts in academia, the MPSoC SW architects in industry focus more on the SW tools

re-use (considering the huge amount of certified code), providing abstractions and conveniences to the programmers for SW development and efficient code-generation.

Two case studies are selected here to be explained in detail, followed by a number of other research projects described in short. From academia, a detailed look on the MPSoC software compilation flow of the SHAPES project [68] is given. From industry, the compilation tool-chain of TI OMAP [79] is demonstrated.

SHAPES

SHAPES [68] is an European Union FP6 Integrated Project whose objective is to develop a prototype of a *tiled* scalable HW and SW architecture for embedded applications featuring inherent parallelism. The major SHAPES building block, the RISC-DSP tile (RDT), is composed of an Atmel Magic VLIW floating-point DSP, an ARM9 RISC, on-chip memory, and a network interface for on- and off-chip communication. On the basis of RDTs and interconnect components, the architecture can be easily scaled to meet the computational requirements of the application at minimum cost.

The SHAPES software development environment is illustrated in Fig. 12. The starting point is the Model-driven compiler/Functional simulator, which takes an application specification in process networks as input. High-level mapping exploration involves the trace information from the Virtual Shapes Platform (VSP) and the analytical performance results from the Analytical Estimator, based on multi-objective optimization considering the various conflicting criteria such as throughput, delay, predictability and efficiency. With the mapping information, the Hardware dependent Software (HdS) phase then generates the necessary dedicated communication and synchronization primitives, together with OS services.

The central part of the SHAPES software environment is the distributed operation layer (DOL) framework [75]. The DOL structure and interactions with other tools and elements are shown in the Fig. 13. DOL mainly provides the MPSoC software developers two main services: system level performance analysis and process-to-processor mapping exploration.

- *DOL Programming Model*: DOL uses process networks as the programming model—the structure of the application is specified in XML consisting of processes, software channels and connections, while the application functionality is specified in C/C++ language and process communications are performed by the DOL APIs, e.g. DOL_read() and DOL_write(). DOL uses a special *iterator* element to allow the user to instantiate several processes of the same type. For the process functionality in C/C++, a set of coding rules needs to be followed. In each process there must be a *init* and a *fire* procedure. The *init* procedure allocates and initializes data, which is called once during the initialization of the application. The *fire* procedure is called repeatedly afterwards.

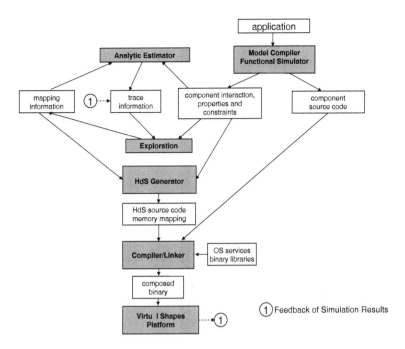

Fig. 12 The SHAPES software development environment

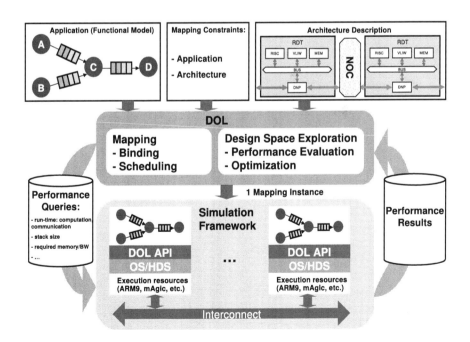

Fig. 13 The DOL framework

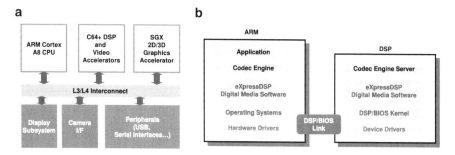

Fig. 14 (**a**) A detailed block diagram of the OMAP3530 processor, (**b**) TI OMAP software stack

- *Architecture Description*: DOL aims at mapping, therefore its architecture description abstracts away several details of the underlying architecture. The architecture description in XML format contains three kinds of information: *structural elements* such as processors/memories, *performance data* such as bus throughputs, and *parameters* such as memory sizes.
- *Mapping Exploration*: DOL mapping includes two phases: performance evaluation and optimization. Performance evaluation collects the data from both analytical performance evaluation and the detailed simulation. The designer a-priori fixes the optimization objectives and the DOL uses evolutionary algorithms and the PISA interface [4] to generate the mapping in XML format.

With the mapping descriptor the HdS layer generates hardware dependent implementation codes and makefiles. Thereafter the application can be compiled and linked against communication libraries and OS services. The final binary can be run on the VSP or on the SHAPES hardware prototype.

TI OMAP

TI OMAP (Open Multimedia Application Platform) family consists of multiple SoC products targeting at portable and mobile multimedia applications. The recent OMAP35x series [79] offers a good combination of general purpose, multimedia and graphics processing in a single-chip combination and delivers good performance for advanced user interfaces, web browsing, enhanced graphics and multimedia.

Figure 14a shows the detailed block diagram of the OMAP3530 processor. OMAP3530 has an ARM Cortex A8 CPU, a TI C64x DSP and a 2D/3D graphics accelerator. Generally speaking, the application programming strategy [45] for OMAP is to allocate computation-intensive tasks (e.g. many video/audio signal-processing algorithms) to the DSP processor and to keep the ARM loaded with high-level OS tasks and I/O. Like many other embedded SoC solutions, TI OMAP solutions carry more and more hardware accelerators for special processing along the product line. For instance, the SGX graphics accelerator is available on OMAP3530, which enables the programmers to migrate OpenGL code. TI provides a rich set of tools for the programmer to develop the applications using different PEs and provide many abstractions to ease the programmer's job by not going deep into many platform details. Nevertheless, how to partition and allocate different tasks/applications onto the platform so as to efficiently leverage the underlying processor power remains an art, which highly relies on the knowledge and experience of the programmer.

Figure 14b illustrates the software development tools and layers that TI provides. TI promotes the philosophy of *abstractions* among the software layers to hide just enough details for the developers at different roles/layers.

- OS Level (low-level developer): At OS level, the choices of OSes on the ARM are diverse including Linux and WinCE. The so-called DSP/BIOS developed by TI is a real-time multi-tasking kernel which is especially optimized for real-time DSP task scheduling. DSP/BIOS has a very small foot-print in size, fast speed optimized for DSP tasks, and a low latency which is also deterministic. DSP/BIOS Link is a special layer developed by TI to handle inter-processor communication.
- Algorithm/Codec Level (algorithm developer): Algorithms/codecs are usually allocated onto the DSP due to its computation power. TI has standardized the interface of the algorithms/codecs to the upper levels like XDAIS (eXpressDSP Algorithm Interoperability Standard [80]) as the *abstractions*. Those standards provide the coding rules and guidelines for the algorithm developers. There are also helper tools which can check the compliance of the given algorithms/codes to the standards and provide the standard *packaging* for delivery.
- Codec Engine Server Level (system integrator): Codec Engine Server needs to be developed by the system integrator by combining the core codecs along with the other infrastructure pieces (DSP/BIOS, DSP/BIOS Link, and so on) and ultimately produces an executable that is callable from another core (usually ARM) where the Codec Engine is running. It is critical that the system resource requirements (MIPS, memory, DMA and so on) are evaluated to ensure that the codec combination can co-exist and run as required.
- Custom Application Development Level (application developer): The application developer uses Codec Engine APIs to leverage the codec functionality, on top of the layers introduced earlier. Third-party tools/frameworks that provide valuable addons like GUI or streaming frameworks can be ported here. TI also provides the DMAI, the DaVinci (and OMAP) Multimedia Applications Interface solution, to ease the programmer's job by providing a collection of useful modules such as abstraction of common peripherals.

The *abstractions* among the layers are realized by the standardized interfaces. Therefore, different teams can work in different domains at the same time thus boosting the productivity. Moreover, this also enables the possibility of third-parties participating in the TI software stack to provide valuable/commercial solutions, e.g. special codec packages, application-level GUI frameworks.

HOPES

HOPES is a parallel programming framework based on a novel programming model called Common Intermediate Code (CIC) [50]. The proposed work-flow of MPSoC software development is as follows. The software development process starts with specifying the application tasks in CIC, by either manually writing or deriving from an initial model-based specification. The CIC has separate sections to describe the application's functionality (in terms of both functional parallelism and data parallelism) as well as the architecture information. The next step is mapping the tasks to PEs. Afterwards the CIC translator translates the CIC program into the executable C codes for each processor core, which includes the phases such as generic API translation and scheduler code generation. In the end, the target dependent executables can be generated and further run in the simulator or the hardware.

The core of the HOPES is the CIC programming model which is designed as the central IR enabling the retargetable parallel compilation. Similar to most of the existing uni-processor retargetable tool-sets, the CIC separates the target-independent part and target-dependent part as follows. The *Task Code* is a target-independent unit which will be mapped to a PE. The default inter-task communication model is a streaming-based channel-like structure, which uses ports to send/receive data between tasks. OpenMP pragmas can be used to specify DLP or loop parallelism inside task code. Generic APIs, e.g. channel access and file I/O, are used in the task code to keep code-reusability. A special pragma to indicate the possibility of using a hardware accelerator for certain functions is also available. All those constructs will later be translated or implemented by target dependent translators, depending on the mapping. The *Architecture Information* has three subsections: the *hardware* defines e.g processor lists, memory map and hardware accelerators information, the *constraints* defines global constraints like power consumption as well as task-level constraints like period and deadline, and the *structure* section describes the application structure in terms of interconnecting all the tasks and the task-to-PE mapping.

The CIC translator *marries* the two sections in the CIC by taking care of the implementing the target-independent constructs in the Task Code section by the target dependent libraries/code in the Architecture Information section. All the interface code and PE-executable C code will be generated and further compiled by the PE C compilers. To support the retargetable compilation flow for a new target architecture, a new CIC translator needs to be written to realize the translation of the target-independent constructs in the CIC to the new target, which could be non-trivial.

Daedalus

Daedalus framework [65] is a tool-flow developed at Leiden University for automated design, programming and implementation of MPSoCs starting at a high level of abstraction. Daedalus design-flow consists of three key tools, KPNgen tool [83], Sesame (Simulation of Embedded System Architectures for Multilevel Exploration) [71] and ESPAM (Embedded System-level Platform synthesis and Application Modeling) [64], which work together to offer the designers a single environment for rapid system-level architectural exploration and automated programming and prototyping of multimedia MPSoC architectures.

The ESPAM tool-set is the core synthesis tool in the Daedalus framework, which accepts the following inputs:

- *Application Specification.* In Daedalus, the application is specified as a KPN. For applications specified as parameterized static affine loop programs in C, KPNgen tool can automatically derive KPN descriptions from the C programs. Otherwise, i.e. for the applications not in this class, application specifications need to be created by hand.
- *Mapping Specification.* The relation between the application and the platform, i.e. how all the processes in the KPN description of the application are mapped to the components in the MPSoC platform, are specified here.
- *Platform Description.* The platform description represents the topology of the MPSoC platform—the components are taken from a library of IP components, which are pre-defined and pre-verified including a variety of programmable and dedicated processors, memories and interconnects. The Sesame tool is used here for design space exploration to automatically generate the mapping and the platform description. It allows fast performance evaluation for different application-to-architecture mappings, hardware/software partitioning and target platforms. The good candidates, as a result, are given as input to the ESPAM tool.

The ESPAM tool uses the inputs mentioned above together with the low-level component models (in RTL) from the IP library to automatically generate synthesizable VHDL code that implements the hardware architecture. It also generates, from the XML specification of the application, the C code for those processes that are mapped on to programmable cores, together with the code for synchronization of the communication between the processors. Furthermore, the well developed commercial synthesis tools and the component compilers can be used to process the outputs for fast hardware/software prototyping.

TCT

The TCT model [86] provides a simple programming model based on C language that allows designers to partition the application directly in the C code. The TCT compiler analyzes the control and data dependencies, automatically inserts the inter-thread communications and generates the concurrent execution model. Therefore, the productivity in the iterative process of application parallelization is greatly improved.

The TCT programming model is very simple. The construct THREAD (name) {...} is used to group statements to form a *TCT thread*, which is to be executed in a (separate) concurrent process. With few exceptions such as goto statements and dynamic memory allocations, any compound statement in C is allowed inside the thread scope. With minimum code modifications, the sequential C code is quickly turned into a parallel specification of the application. The TCT execution model is a program-driven MIMD (Multiple Instruction Multiple Data) model where each thread is statically allocated to one processor. The execution model has multiple flows of control simultaneously, which enables modeling different parallelism schemes such as functional pipelining and task parallelism. Three thread synchronization instructions implement the decentralized thread interaction in the generated parallel code. The TCT compiler takes the input of annotated C source code, performs the dependency analysis among the threads and generates the parallel code with inserted thread communications to guarantee the correct concurrent execution. The TCT MPSoC, which is based on an existing AMBA SoC platform with a homogeneous distributed-memory multiprocessor-array, is supported as backend for code generation: a hardware prototype and a software tool called TCT trace scheduler are available for the designer to simulate and get the performance results given the partitioning.

MPA

The MPSoC Parallelization Assist (MPA) tool [60] is developed at IMEC to help designers map the applications onto an embedded multicore platform efficiently, and explore the solution space fast. The exploration flow's starting point is the sequential C source code of an application. The initial parallelization specification is assisted by profiling the application using the instruction set simulator (ISS) or source code instrumentation and execution on a target platform. The key input to the tool is the Parallelization Specification (ParSpec), which is a separate file from the application source code and orchestrates the parallelization. The ParSpec is composed of one or more parallel sections to specify the computational partitioning of the application: outside a parallel section the code is sequentially executed, whereas inside a parallel section all code must be assigned to at least one thread. Both functional and data-level splits can be expressed in the ParSpec. Keeping the ParSpec separate from the application code allows the designer to try out different mapping strategies easily.

The parallel C code is then generated by the MPA tool, where scalar dataflow analysis is done to insert communication primitives into the generated code. The *shared* variables have to be explicitly specified by the designer and the necessary synchronization has to be added using e.g. *LoopSync*. The MPA tool targets two execution environments for the generated parallel code—hardware or virtual platform in case the platform is available, and a High-Level Simulator in case the platform is not ready. The execution trace is generated, which the designer analyzes to decide whether changes are needed for the ParSpec to improve the mapping result.

CoMPSoC/C-HEAP

The CoMPSoC (Composable and Predictable Multi-Processor System on Chip) platform template is proposed in [33] which aims at providing a scalable hardware and software MPSoC template that removes all the interferences between applications through resource reservations. The hardware design features admission control and budget enforcement employed in the building blocks in the shared architecture in order to achieve *composability* and *predictability*. We will focus on the software platform of the CoMPSoC here in this chapter. The interested reader is referred to [33] for the details on the hardware CoMPSoC template.

The software stack of the CoMPSoC is mainly composed of two parts: the *application middleware* and the *host software*. Like other works introduced in this chapter, in order to accomplish the synchronization and communication between the tasks running simultaneously on the processors, an implementation of the C-HEAP [63] is bundled as a part of the CoMPSoC template offering (C-HEAP itself is a complete API which performs not only the communication protocol, such as synchronization and data transportation, but also the reconfiguration protocol such as task management and configuration of the channels.) The communication API is implemented on the shared memory, with synchronization done using pointers in a memory-mapped FIFO-administration structure [33]. The *host software*, i.e. all the administrative tasks for the processor tiles and the control registers of the communication infrastructure, is carried out on a central host tile, implemented in portable C libraries.

MAPS

The MAPS (MPSoC Application Programming Studio) project is part of RWTH Aachen University's Ultra high speed Mobile Information and Communication (UMIC) research cluster [82]. It is a research prototype tool-set for semi-automatic code parallelization and task-to-processor mapping, inspired by a typical problem setting of SW development for wireless multimedia terminals, where multiple applications and radio standards can be activated simultaneously and partially compete for the same resources. Applications can be specified either as sequential C code or in the form of pre-parallelized processes [73], where real-time properties such as latency and period as well as preferred PE types can be optionally annotated. MAPS uses advanced dataflow analysis to extract the available parallelism from the sequential codes (in [20] a more detailed description of code partitioning is presented) and to form a set of fine-grained task graphs based on a coarse model of the target architecture. Mapping and scheduling is critical in MPSoC compilation as they determine systems performance largely. Besides manually specifying spatial and temporal task-to-processor mapping, MAPS provides means to compute scheduling and mapping automatically to meet given constraints [16]. This applies not only to single applications but also to multiple application scenarios. Besides directly running the platform simulator, the partitioning/mapping can also be exercised and refined with a fast, high-level SystemC based simulation environment MVP (MAPS Virtual Platform [21]), which has been put in practice to evaluate different software settings in a multi-application scenario. After further refinement, a code generation phase translates the tasks into C codes for compilation onto the respective PEs with their native compilers and OS primitives. The SW can then be executed, depending on availability, either on the real HW or on a cycle-approximate virtual platform incorporating instruction-set simulators. MAPS also develops a dedicated task dispatching ASIP (OSIP, operating system ASIP [19]) in order to enable higher PE utilization via more fine-grained tasks and lower context switching overhead. Early evaluation case studies exhibited great potential of the OSIP approach in lowering the task-switching overhead, compared to an additional RISC performing scheduling, in a typical MPSoC environment. MAPS has been used and extended to support programming the OSIP-based platform and provide debugging support. The cast study [17] has proved that MAPS is flexible and extensible enough to support complicated heterogeneous MPSoCs with specialized APIs and configurations. It also has shown the productivity increase provided by MAPS. The MAPS compiler allows to test different configurations faster than if coding each of them by hand using the OSIP-APIs. The debugging facilities greatly simplifies application development cycles as well.

Summary

In this chapter we presented an overview of the challenges for building MPSoC compilers and described some of the techniques, both established and emerging, that are being used to leverage the computing power of current and yet to come MPSoC platforms. The chapter concluded with selected academic and industrial examples that show how the concepts are applied to real systems.

We have seen how new programming models are being proposed that change the requirements of the MPSoC compiler. We discussed that, independent of the programming model, an MPSoC compiler has to find a suitable granularity to expose parallelism beyond the instruction level (ILP), demanding advanced analysis of the data and control flows. Computing a schedule and a mapping of an application is one of the most complex tasks of the MPSoC compiler and can only be achieved successfully with accurate performance estimation or simulation. Most of these analyses are target specific, hence the MPSoC itself needs to be abstracted and fed to the compiler. With this information, the compiler can tune the different optimizations to the target MPSoC and finally generate executable code.

The whole flow shares similarities with that of a traditional uni-processor compiler, but is much more complex in the case of an MPSoC. We have presented some foundations and described approaches to deal with these problems. However, there is still a lot of research to be done to make the step from a high level specification to executable code as transparent as it is in the uni-processor case.

Acknowledgements The authors would like to thank Jianjiang Ceng, Anastasia Stulova, Stefan Schürmans, and all ISS research members for their valuable comments.

References

1. Eclipse. http://www.eclipse.org/. Visited on Jan. 2010
2. GDB: The GNU Project Debugger. http://www.gnu.org/software/gdb/. Visited on Jan. 2010
3. MCAPI - Multicore Communications API. http://www.multicore-association.org/workgroup/comapi.php. Visited on Nov. 2009
4. PISA - A Platform and Programming Language Independent Interface for Search Algorithms. http://www.tik.ee.ethz.ch/sop/pisa/. Visited on Nov. 2009
5. Real Time Software Components. http://www.eclipse.org/dsdp/rtsc. Visited on Jan. 2010
6. AbsInt: aiT worst-case execution time analyzers. http://www.absint.com/ait/. Visited on Nov. 2009
7. ACE: Embedded C for high performance DSP programming with the CoSy compiler development system. a-qual.com/topics/2005/EmbeddedCv2.pdf. Visited on Jan. 2010
8. Adl-Tabatabai, A.R., Kozyrakis, C., Saha, B.: Unlocking concurrency. Queue **4**(10), 24–33 (2007)
9. Aho, A.V., Sethi, R., Ullman, J.D.: Compilers: Principles, Techniques, and Tools. Addison-Wesley Longman Publishing Co., Inc., Boston, MA, USA (1986)
10. Asanovic, K., Bodik, R., Catanzaro, B.C., Gebis, J.J., Husbands, P., Keutzer, K., Patterson, D.A., Plishker, W.L., Shalf, J., Williams, S.W., Yelick, K.A.: The landscape of parallel computing research: A view from Berkeley. Tech. rep., EECS Department, University of California, Berkeley (2006)
11. Bacivarov, I., Haid, W., Huang, K., Thiele, L.: Methods and tools for mapping process networks onto multi-processor systems-on-chip. In: S.S. Bhattacharyya, E.F. Deprettere, R. Leupers, J. Takala (eds.) Handbook of Signal Processing Systems, second edn. Springer (2013)
12. Benini, L., Bertozzi, D., Guerri, A., Milano, M.: Allocation and scheduling for MPSoCs via decomposition and no-good generation. Principles and Practices of Constrained Programming - CP 2005 (DEIS-LIA-05-001), 107–121 (2005)
13. Bhattacharya, B., Bhattacharyya, S.S.: Parameterized dataflow modeling for DSP systems. IEEE Transactions on Signal Processing **49**(10), 2408–2421 (2001)
14. Bhattacharyya, S.S., Deprettere, E.F., Theelen, B.: Dynamic dataflow graphs. In: S.S. Bhattacharyya, E.F. Deprettere, R. Leupers, J. Takala (eds.) Handbook of Signal Processing Systems, second edn. Springer (2013)
15. Carro, L., Rutzig, M.B.: Multicore systems on chip. In: S.S. Bhattacharyya, E.F. Deprettere, R. Leupers, J. Takala (eds.) Handbook of Signal Processing Systems, second edn. Springer (2013)
16. Castrillon, J., Leupers, R., Ascheid, G.: MAPS: Mapping concurrent dataflow applications to heterogeneous MPSoCs. IEEE Transactions on Industrial Informatics p. 19 (2011). DOI 10.1109/TII.2011.2173941
17. Castrillon, J., Shah, A., Murillo, L., Leupers, R., Ascheid, G.: Backend for virtual platforms with hardware scheduler in the MAPS framework. In: Circuits and Systems (LASCAS), 2011 IEEE Second Latin American Symposium on, pp. 1–4 (2011). DOI 10.1109/LASCAS.2011.5750280

18. Castrillon, J., Sheng, W., Leupers, R.: Trends in embedded software synthesis. In: SAMOS, pp. 347–354 (2011)
19. Castrillon, J., Zhang, D., Kempf, T., Vanthournout, B., Leupers, R., Ascheid, G.: Task management in MPSoCs: An ASIP approach. In: ICCAD 2009 (2009)
20. Ceng, J., Castrillon, J., Sheng, W., Scharwächter, H., Leupers, R., Ascheid, G., Meyr, H., Isshiki, T., Kunieda, H.: MAPS: an integrated framework for MPSoC application parallelization. In: DAC '08: Proceedings of the 45th annual conference on Design automation, pp. 754–759. ACM, New York, NY, USA (2008)
21. Ceng, J., Sheng, W., Castrillon, J., Stulova, A., Leupers, R., Ascheid, G., Meyr, H.: A high-level virtual platform for early MPSoC software development. In: CODES+ISSS '09: Proceedings of the 7th IEEE/ACM international conference on Hardware/software codesign and system synthesis, pp. 11–20. ACM, New York, NY, USA (2009)
22. Cesario, W., Jerraya, A.: Multiprocessor Systems-on-Chips, chap. Chapter 9. Component-Based Design for Multiprocessor Systems-on-Chip, pp. 357–394. Morgan Kaufmann (2005)
23. Collette, T.: Key technologies for many core architectures. In: 8th International Forum on Application-Specific Multi-Processor SoC (2008)
24. CoWare: CoWare Virtual Platforms. http://www.coware.com/products/virtualplatform.php. Visited on Apr. 2009
25. Fisher, J., P., F., Young, C.: Embedded Computing: A VLIW Approach to Architecture, Compilers and Tools. Morgan-Kaufmann (Elsevier) (2005)
26. Flake, P., Davidmann, S., Schirrmeister, F.: System-level exploration tools for MPSoC designs. In: Design Automation Conference, 2006 43rd ACM/IEEE, pp. 286–287 (2006)
27. Gao, L., Huang, J., Ceng, J., Leupers, R., Ascheid, G., Meyr, H.: TotalProf: a fast and accurate retargetable source code profiler. In: CODES+ISSS '09: Proceedings of the 7th IEEE/ACM international conference on Hardware/software codesign and system synthesis, pp. 305–314. ACM, New York, NY, USA (2009)
28. Geilen, M., Basten, T.: Kahn process networks and a reactive extension. In: S.S. Bhattacharyya, E.F. Deprettere, R. Leupers, J. Takala (eds.) Handbook of Signal Processing Systems, second edn. Springer (2013)
29. Gheorghita, S., T. Basten, H.C.: An overview of application scenario usage in streaming-oriented embedded system design
30. Gupta, R., Micheli, G.D.: Hardware-software co-synthesis for digital systems. In: IEEE Design & Test of Computers, pp. 29–41 (1993)
31. Ha, S., Oh, H.: Decidable dataflow models for signal processing: Synchronous dataflow and its extensions. In: S.S. Bhattacharyya, E.F. Deprettere, R. Leupers, J. Takala (eds.) Handbook of Signal Processing Systems, second edn. Springer (2013)
32. Hankins, R.A., et al.: Multiple instruction stream processor. SIGARCH Comp. Arch. News **34**(2) (2006)
33. Hansson, A., Goossens, K., Bekooij, M., Huisken, J.: CoMPSoC: A template for composable and predictable multi-processor system on chips. ACM Trans. Des. Autom. Electron. Syst. **14**(1), 1–24 (2009)
34. Hewitt, C., Bishop, P., Greif, I., Smith, B., Matson, T., Steiger, R.: Actor induction and meta-evaluation. In: POPL '73: Proceedings of the 1st annual ACM SIGACT-SIGPLAN symposium on Principles of programming languages, pp. 153–168. ACM, New York, NY, USA (1973)
35. Hind, M.: Pointer analysis: Haven't we solved this problem yet? In: PASTE '01, pp. 54–61. ACM Press (2001)
36. Hu, T.C.: Parallel sequencing and assembly line problems. Operations Research **9**(6), 841–848 (1961). URL http://www.jstor.org/stable/167050
37. Hwang, Y., Abdi, S., Gajski, D.: Cycle-approximate retargetable performance estimation at the transaction level. In: DATE '08: Proceedings of the conference on Design, automation and test in Europe, pp. 3–8. ACM, New York, NY, USA (2008)
38. Hwu, W.M., Ryoo, S., Ueng, S.Z., Kelm, J.H., Gelado, I., Stone, S.S., Kidd, R.E., Baghsorkhi, S.S., Mahesri, A.A., Tsao, S.C., Navarro, N., Lumetta, S.S., Frank, M.I., Patel, S.J.: Implicitly

parallel programming models for thousand-core microprocessors. In: DAC '07: Proceedings of the 44th annual conference on Design automation, pp. 754–759. ACM, New York, NY, USA (2007)

39. Kahn, G.: The semantics of a simple language for parallel programming. In: J.L. Rosenfeld (ed.) Information Processing '74: Proceedings of the IFIP Congress, pp. 471–475. North-Holland, New York, NY (1974)

40. Kandemir, M., Dutt, N.: Multiprocessor Systems-on-Chips, chap. Chapter 9. Memory Systems and Compiler Support for MPSoC Architectures, pp. 251–281. Morgan Kaufmann (2005)

41. Karp, R.M., Miller, R.E.: Properties of a model for parallel computations: Determinacy, termination, queuing. SIAM Journal of Applied Math 14(6) (1966)

42. Karuri, K., Al Faruque, M.A., Kraemer, S., Leupers, R., Ascheid, G., Meyr, H.: Fine-grained application source code profiling for ASIP design. In: DAC '05: Proceedings of the 42nd annual conference on Design automation, pp. 329–334. ACM, New York, NY, USA (2005)

43. Kempf, T., Wallentowitz, S., Ascheid, G., Leupers, R., Meyr, H.: A workbench for analytical and simulation based design space exploration of software defined radios. In: Proc. Int. Conf. VLSI Design, pp. 281–286. New Delhi, India (2009)

44. Kennedy, K., Allen, J.R.: Optimizing compilers for modern architectures: A dependence-based approach. Morgan Kaufmann Publishers Inc., San Francisco, CA, USA (2002)

45. Kloss, N.: Application Programming Strategies for TI's OMAP Solutions. Embedded Edge (2003)

46. Krishnan, V., Torrellas, J.: A chip-multiprocessor architecture with speculative multithreading. IEEE Trans. Comput. 48(9), 866–880 (1999)

47. Kumar, S., Hughes, C.J., Nguyen, A.: Carbon: Architectural support for fine-grained parallelism on chip multiprocessors. SIGARCH Comp. Arch. News 35(2) (2007)

48. Kung, H.T.: Why systolic architectures? Computer 15(1), 37–46 (1982)

49. Kwok, Y.K., Ahmad, I.: Static scheduling algorithms for allocating directed task graphs to multiprocessors. ACM Comput. Surv. 31(4), 406–471 (1999)

50. Kwon, S., Kim, Y., Jeun, W.C., Ha, S., Paek, Y.: A retargetable parallel-programming framework for MPSoC. ACM Trans. Des. Autom. Electron. Syst. 13(3), 1–18 (2008)

51. Lam, M.: Software pipelining: An effective scheduling technique for VLIW machines. SIGPLAN Not. 23(7), 318–328 (1988)

52. Lee, E., Messerschmitt, D.: Synchronous data flow. Proceedings of the IEEE 75(9), 1235–1245 (1987)

53. Lee, E.A.: Consistency in dataflow graphs. IEEE Trans. Parallel Distrib. Syst. 2(2), 223–235 (1991)

54. Lee, E.A.: The problem with threads. Computer 39(5), 33–42 (2006). URL http://portal.acm.org/citation.cfm?id=1137232.1137289

55. Leupers, R.: Retargetable Code Generation for Digital Signal Processors. Kluwer Academic Publishers, Norwell, MA, USA (1997)

56. Leupers, R.: Code selection for media processors with SIMD instructions. In: DATE '00, pp. 4–8. ACM (2000)

57. Li, L., Huang, B., Dai, J., Harrison, L.: Automatic multithreading and multiprocessing of C programs for IXP. In: PPoPP '05: Proceedings of the tenth ACM SIGPLAN symposium on Principles and practice of parallel programming, pp. 132–141. ACM, New York, NY, USA (2005)

58. Ma, Z., Marchal, P., Scarpazza, D.P., Yang, P., Wong, C., Gmez, J.I., Himpe, S., Ykman-Couvreur, C., Catthoor, F.: Systematic Methodology for Real-Time Cost-Effective Mapping of Dynamic Concurrent Task-Based Systems on Heterogenous Platforms. Springer Publishing Company, Incorporated (2007)

59. Martin, G.: ESL requirements for configurable processor-based embedded system design. Design and Reuse

60. Mignolet, J.Y., Baert, R., Ashby, T.J., Avasare, P., Jang, H.O., Son, J.C.: MPA: Parallelizing an application onto a multicore platform made easy. IEEE Micro 29(3), 31–39 (2009)

61. Muchnick, S.S.: Advanced Compiler Design and Implementation. Morgan Kaufmann Publishers Inc., San Francisco, CA, USA (1997)
62. National Instruments: LabView. http://www.ni.com/labview/. Visited on Mar. 2009
63. Nieuwland, A., Kang, J., Gangwal, O.P., Sethuraman, R., Busa, R.S.C.N., Goossens, K., Llopis, R.P.: C-HEAP: A heterogeneous multi-processor architecture template and scalable and flexible protocol for the design of embedded signal processing systems. Design Automation for Embedded Systems (7), 233–270 (2002)
64. Nikolov, H., Stefanov, T., Deprettere, E.: Systematic and automated multiprocessor system design, programming, and implementation. Computer-Aided Design of Integrated Circuits and Systems, IEEE Transactions on 27(3), 542–555 (2008)
65. Nikolov, H., Thompson, M., Stefanov, T., Pimentel, A., Polstra, S., Bose, R., Zissulescu, C., Deprettere, E.: Daedalus: Toward composable multimedia MP-SoC design. In: DAC '08: Proceedings of the 45th annual conference on Design automation, pp. 574–579. ACM, New York, NY, USA (2008)
66. The OpenMP specification for parallel programming: www.openmp.org. Visited on Nov. 2009
67. Palsberg, J., Naik, M.: Multiprocessor Systems-on-Chips, chap. Chapter 12. ILP-based Resource-aware Compilation, pp. 337–354. Morgan Kaufmann (2005)
68. Paolucci, P.S., Jerraya, A.A., Leupers, R., Thiele, L., Vicini, P.: SHAPES:: a tiled scalable software hardware architecture platform for embedded systems. In: CODES+ISSS '06: Proceedings of the 4th international conference on Hardware/software codesign and system synthesis, pp. 167–172. ACM, New York, NY, USA (2006)
69. Park, S., sun Hong, D., Chae, S.I.: A hardware operating system kernel for multi-processor systems. IEICE 5(9) (2008)
70. Parks, T.M.: Bounded scheduling of process networks. Ph.D. thesis, Berkeley, CA, USA (1995)
71. Pimentel, A.D., Erbas, C., Polstra, S.: A systematic approach to exploring embedded system architectures at multiple abstraction levels. IEEE Transactions on Computers 55(2), 99–112 (2006)
72. Sharma, G., Martin, J.: MATLAB (R): A language for parallel computing. International Journal of Parallel Programming 37(1) (2009)
73. Sheng, W., Schürmans, S., Odendahl, M., Leupers, R., Ascheid, G.: Automatic calibration of streaming applications for software mapping exploration. In: Proceedings of the International Symposium on System-on-Chip (SoC) (2011)
74. Snir, M., Otto, S.: MPI-The Complete Reference: The MPI Core. MIT Press (1998)
75. Sporer, T., Franck, A., Bacivarov, I., Beckinger, M., Haid, W., Huang, K., Thiele, L., Paolucci, P., Bazzana, P., Vicini, P., Ceng, J., Kraemer, S., Leupers, R.: SHAPES - a scalable parallel HW/SW architecture applied to wave field synthesis. In: Proc. 32nd Intl Audio Engineering Society (AES) Conference, pp. 175–187. Audio Engineering Society, Hillerod, Denmark (2007)
76. Sriram, S., Bhattacharyya, S.S.: Embedded Multiprocessors: Scheduling and Synchronization. Marcel Dekker, Inc., New York, NY, USA (2000)
77. Standard for information technology - portable operating system interface (POSIX). Shell and utilities. IEEE Std 1003.1–2004, The Open Group Base Specifications Issue 6, section 2.9: IEEE and The Open Group
78. Synopsys: Synopsys Virtual Platforms. http://www.synopsys.com/Tools/SLD/VirtualPlatforms/Pages/default.aspx. Visited on May 2009
79. TI: OMAP35x Product Bulletin. http://www.ti.com/lit/sprt457. Visited on Mar. 2009
80. TI: TI eXpressDSP Software and Development Tools. http://focus.ti.com/general/docs/gencontent.tsp?contentId=46891. Visited on Jan. 2010
81. Tournavitis, G., Wang, Z., Franke, B., O'Boyle, M.: Towards a holistic approach to auto-parallelization – integrating profile-driven parallelism detection and machine-learning based mapping. In: PLDI 0–9: Proceedings of the Programming Language Design and Implementation Conference. Dublin, Ireland (2009)
82. UMIC: Ultra high speed Mobile Information and Communication. http://www.umic.rwth-aachen.de. Visited on Nov. 2009

83. Verdoolaege, S., Nikolov, H., Stefanov, T.: pn: A tool for improved derivation of process networks. EURASIP J. Embedded Syst. **2007**(1), 19–19 (2007)
84. Wilhelm, R., Engblom, J., Ermedahl, A., Holsti, N., Thesing, S., Whalley, D., Bernat, G., Ferdinand, C., Heckmann, R., Mitra, T., Mueller, F., Puaut, I., Puschner, P., Staschulat, J., Stenström, P.: The worst-case execution-time problem - overview of methods and survey of tools. ACM Trans. Embed. Comput. Syst. **7**(3), 1–53 (2008)
85. Working Group ISO/IEC JTC1/SC22/WG14: C99, Programming Language C ISO/IEC 9899:1999
86. Zalfany Urfianto, M., Isshiki, T., Ullah Khan, A., Li, D., Kunieda, H.: Decomposition of task-level concurrency on C programs applied to the design of multiprocessor SoC. IEICE Trans. Fundam. Electron. Commun. Comput. Sci. **E91-A**(7), 1748–1756 (2008)

Embedded C for Digital Signal Processing

Bryan E. Olivier

Abstract The majority of micro processors in the world do not sit inside a desktop personal computer or laptop as general purpose processor, but have a dedicated purpose inside some kind of apparatus, like a mobile telephone, modem, washing machine, cruise missile, hard disk, DVD player, etc. Such processors are called embedded processors. They are designed with their application in mind and therefore carry special features. With the high volume and strict real time requirements of mobile communication the digital signal processor (DSP) emerged. These embedded processors featured special hardware and instructions to support efficient processing of the communication signal. Traditionally these special features were programmed through some assembly language, but with the growing volume of devices and software a desire arose to access these features from a standardized programming language. A work group of the International Organization for Standardization (ISO) has recognized this desire and came up with an extension of their C standard to support those features. This chapter intends to explain this extension and illustrate how to use them to efficiently use a DSP.

1 Introduction

Traditionally the core algorithms in digital signal processing are programmed in assembly or by using intrinsics because this gives access to the special features offered by dedicated digital signal processors (DSP). These algorithms are becoming increasingly complex, because of error correction and encryption requirements. This increasing complexity makes the tradition of programming in assembly or

B.E. Olivier (✉)
ACE Associated Compiler Experts bv., The Netherlands
e-mail: bryan@ace.nl

S.S. Bhattacharyya et al. (eds.), *Handbook of Signal Processing Systems*,
DOI 10.1007/978-1-4614-6859-2_38, © Springer Science+Business Media, LLC 2013

by using intrinsics more and more of a burden on maintenance and development. Therefore the desire to support such DSP features in the C-language, as defined in [7], is growing.

With the Embedded C extension as described in the ISO technical report "Extensions for the programming language C to support embedded processors" [8] it has become possible to write such algorithms in portable and standard C, without compromising the required performance. This chapter exhibits the support offered by this Embedded C extension and shows some examples of how to use it.

The Embedded C extensions add four features to the C language [7].

- Fixed point types
- Memory spaces
- Named Registers
- Hardware I/O addressing

2   Typical DSP Architecture

Figure 1 shows the essence of architectures found in digital signal processors. The arithmetical unit can perform a multiply accumulate in one cycle. The operands can be obtained from separate memories, which means both can be loaded from memory in parallel. This is called a Harvard architecture. Conventional processors have one memory and both operands would have to be loaded from memory one after another,

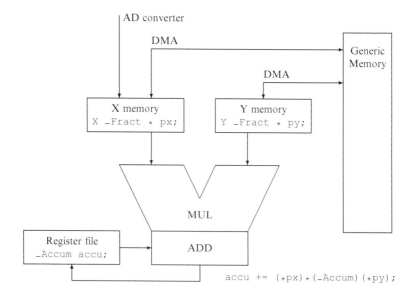

Fig. 1 Typical DSP multiply/accumulate architecture

referred to as a von Neumann architecture. The operands are of fixed point type, resulting in less chip surface for the arithmetical operations in comparison with conventional floating point.

This architecture is motivated by the often used *finite impulse response* (FIR) filter. The core of this algorithm consists of the inner product of two vectors. For performance reasons the elements of these vectors are of fixed point type and both vectors are stored in two separate memories so they can be accessed simultaneously on a Harvard architecture. One of these memories can be filled with samples taken by an analog-digital (AD) converter. Such samples are also easily represented as fixed point values.

Figure 1 uses the Embedded C syntax to show the elements in the architecture. Two pointers px and py are defined to point to two separate memory spaces X and Y. These two memories X and Y contain, for this example, the two vectors and the multiply/accumulate (MAC) unit multiplies two values and adds the product to an accumulator register. Often these X and Y memories can only be filled with data through some Direct Memory Access (DMA) hardware. This means that these memories are not part of the global generic memory of the processor. If the architecture also supports zero overhead loops, then the inner product can be computed with one cycle for each pair of elements in the vectors. Zero overhead loops are supported by special hardware on the processor. With one instruction this hardware is set up to iterate a predetermined number of times. This way costly branches at the end of a loop body are avoided.

The key elements in this architecture are the following:

• Zero overhead loops
• Multiply/accumulate unit
• Separate memories
• Fixed point types

Zero overhead loops can be recognized relatively easy by a compiler, as can the operations for the MAC unit. For these features of the architecture no language extensions are required. It is a lot harder to distribute data appropriately over separate memories and virtually impossible to recognize fixed point operations written in standard ISO-C. Embedded C is an extension of ISO-C allowing the programmer to specify these last two elements in the application program in a portable way.

3 Fixed Point Types

As described in Sect. 2, most DSP architectures support fixed point types as an efficient alternative for floating point. Also the samples taken by an analog-digital converter are easily represented as fixed point values. In the Embedded C report therefore two new and related arithmetic types are introduced, _Fract and _Accum. The names of the types start with an underscore and capital to prevent pollution of the name-space of existing programs. Such names are reserved by the

Fig. 2 Value ranges of fixed
point types. *n* and *m* are some
powers of 2

Type	Value range
signed _Fract	$[-1.0.+1.0)$
unsigned _Fract	$[0.0.+1.0)$
signed _Accum	$[-n.+n)$
unsigned _Accum	$[0.0.+m)$

ISO C standard and should not be used in an application. Embedded C requires however that every compiler environment should define the more natural spelling fract, accum and sat (see below) in the include file stdfix.h.

Different sizes of these types can be specified through short and long. These types can also be specified as signed and unsigned, where the unspecified default type is signed. These types can also be qualified as _Sat for saturated arithmetic. Saturation means that if the result of an arithmetical operation overflows (underflows), it will stick to the highest (lowest) possible value instead of wrapping around. This is often used in digital signal processing, because the induced error from saturation is smaller than that following from wrap around.

Figure 2 describes the ranges of values that different fixed point types may have. These ranges include the lower bound, but exclude the upper bound.

The _Accum types are introduced to safely accumulate a set of _Fract without the risk of overflow. Only when this accumulator type is converted back to a regular _Fract type overflow has to be accounted for. It should be noted that even though the _Accum types also come in a saturating flavor, this is hardly ever supported by the hardware. Application programmers are therefore advised to carefully use the saturating accumulator types, because arithmetic will often be diverted to expensive emulation routines. Saturation on _Fract types however is often very well supported by the hardware.

3.1   Fixed Point Sizes

The _Fract types only have a number of fractional bits and the _Accum types additionally have a number of integral bits. Figure 3 gives the constraints set by Embedded C on these number of bits for various types.

Roughly, these constraints state that _Fract is at least as big as the size of short _Fract and long _Fract is at least as big as the size of _Fract. Similar constraints are stated for the different flavors of _Accum. It also states that a _Fract has the same number of fractional bits as a _Accum, again for all flavors. Finally, it sets some minimum requirements on the number of fractional and integral bits. Such minimum sizes can be used for highly portable applications only relying on such minimums.

The signed and unsigned versions are required to have the same size. It is however allowed to use the sign bit as an additional fractional or integral bit in the unsigned version.

Fig. 3 Constraints on the various sizes of the fixed point types

$f(\text{short \_Fract}) \leq f(\text{\_Fract}) \leq f(\text{long \_Fract})$
$i(\text{short \_Fract}) = i(\text{\_Fract}) = i(\text{long \_Fract}) = 0$
$f(\text{short \_Accum}) = f(\text{short \_Fract})$ (recommended)
$f(\text{\_Accum}) = f(\text{\_Fract})$ (recommended)
$f(\text{long \_Accum}) = f(\text{long \_Fract})$ (recommended)
$i(\text{short \_Accum}) \leq i(\text{Accum}) \leq i(\text{longAccum})$
$f(\text{short \_Fract}) \geq 7$
$f(\text{\_Fract}) \geq 15$
$f(\text{long \_Fract}) \geq 23$ (31 recommended)
$i(\text{short \_Accum}) \geq 4$ (8 recommended)
$i(\text{\_Accum}) \geq 4$ (8 recommended)
$i(\text{long \_Accum}) \geq 4$ (8 recommended)

$f(\text{type})$ is the number of fractional bits,
$i(\text{type})$ the number integral bits of a type.

The recommended values and constraints do not have to be supported by a compiler, however if supported it would improve portability of applications across different architectures.

The actual sizes of the implementation will be highly dependent on the architecture targeted by the compiler. It can be expected that different architectures for the same application domain will have converging sizes of their fixed point hardware, because the application domain will dictate a certain precision.

3.2 Fixed Point Constants

Fixed point constants look just like floating point constants, except for a suffix that specifies the exact type of the constant. The r is used to specify a _Fract and the k for the _Accum. The h is used for short and the l is used for long. The u is used for unsigned. So 0.5uhr specifies an unsigned short _Fract constant, whereas 0.51k specifies a signed long _Accum. All constants must be within the range of their types, with the exception of 1.0 that can be used to specify the largest possible value of a _Fract type, that is $1.0 - 1/2^{f(\text{\_Fract})}$, where $f(\text{\_Fract})$ is the number of fractional bits as described in Sect. 3.1.

Figure 4 lists all the suffices for fixed point constants and their implied type.

3.3 Fixed Point Conversions and Arithmetical Operations

Conversions involving fixed point follow similar rules in C as apply to integer. This means that the types of the operands determine the type of the operation through a ranking order and the rounding and overflow are dictated by this destination type. Conversion from a fixed point type to an integral type uses rounding towards zero.

Meaning of a suffix letter	
r	_Fract
k	_Accum
h	short
l	long
u	unsigned

Suffix	Fixed-point type	Suffix	Fixed-point accumulator type
hr	short _Fract	hk	short _Accum
uhr	unsigned short _Fract	uhk	unsigned short _Accum
r	_Fract	k	_Accum
ur	unsigned _Fract	uk	unsigned _Accum
lr	long _Fract	lk	long _Accum
ulr	unsigned long _Fract	ulk	unsigned long _Accum

Fig. 4 The constant suffices with their implied types

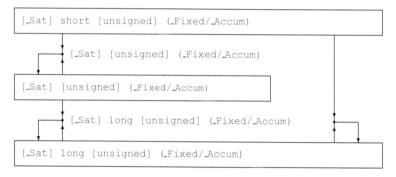

Fig. 5 Conversion in expressions with mixed sized fixed point types. Convert the smaller to the bigger type

All sizes of _Fract are ranked as smaller than _Accum. This means that the resulting type of adding a (long _Fixed) with a (short _Accum) will result in a short _Accum, thereby loosing precision of the long _Fixed. See Fig. 5 for a visualization.

Figures 5–9 use brackets around elements in the types to denote that these are optional. However per figure such an optional element should be left out in all occurrences or none at all. Boxes represent objects and types without a box represent the type in which an operation is performed. Arrows going out of a box denote the types of the two operands. The arrow pointing to a box denotes the type of the result.

Conversions from and to fixed point types in mixed expressions follow similar rules as for other C types, with a few exceptions. If a fixed point type is combined with a floating point type in an operation, then the fixed point type is first converted to a floating point. See Fig. 6 for a visualization.

If two _Fract types are combined, the one with the lower precision is converted to the type with the higher precision before the operation is performed. The exception is made for combining integral types or _Accum types with _Fract

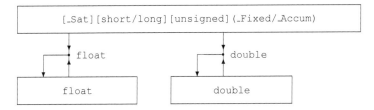

Fig. 6 Conversion in expressions with mixed floating point and fixed point types. Convert the fixed point type to the floating point type

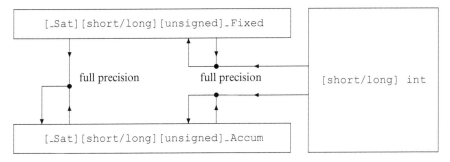

Fig. 7 Conversion in expressions with mixed integral and fixed point types. Compute the result in full precision and convert to the resulting type

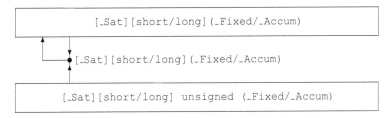

Fig. 8 Conversion in expressions with mixed signedness of fixed point types. Convert the unsigned fixed point type to its corresponding signed type

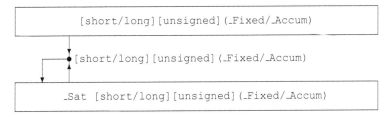

Fig. 9 Conversion in expressions with mixed saturation of fixed point types. The operation and the result are saturating

types. In this case *the result of the operation is calculated using the values of the two operands, with their full precision.* The result is rounded back to the required fixed point type dictated by the ranking of the fixed point types. The same holds for relational operators (`<`, `<=` etc.). See Fig. 7 for a visualization.

The reason for this exception is that the normal conversion rules in C would result in pointless expressions. For example `3*0.1r` would either result in `0.1r`, if 3 was converted to a `_Fract`, or it would result in `0`, if `0.1r` was first converted to an integral type. With this exception the result is `0.3r`, which was deemed to make more sense.

If an `unsigned` fixed point type is combined with `signed` fixed point type, then the `unsigned` value is first converted to its corresponding `signed` value and the result will also have this type. See Fig. 8 for a visualization. If any of the operands was of a saturated type, then the result of the operation is also saturated.

If one of the operands in an expression is a saturating fixed point type, then the result is also a saturating fixed point type. See Fig. 9 for a visualization.

The efficiency of the machine code generated for such combinations highly depends on the support in the architecture. Most processors with fixed point hardware support do not support such mixed operations, and the generated code will emulate the calculation using regular integral arithmetic.

With the exception of bitwise and `&`, or `|` and exclusive or `^` all operations on fixed point types are supported.

3.4 Fixed Point Support Functions

As described in the previous section, all operations on combinations of integral and fixed point types will yield a fixed point type. Other combinations are supported through special support functions. The value returned by these functions are defined similarly as for the regular operators: the result value is first calculated using the exact value of the operands and then rounded according the result type. If the result type is an integral type, then overflow results in undefined behavior.

Embedded C also specifies some useful support functions for which there is no operator.

Figure 10 describes the various support functions, where the replacement for *fx* refers to a qualified fixed point type using the suffices as also used for fixed point constants. The integral types involved in these functions use the same qualifiers as the involved fixed point type, unless stated otherwise.

Figure 11 shows the four new conversion specifiers used for the new fixed point types and the standard length modifiers that also apply to the conversions specifiers for fixed point types. The specification of precision and length is similar to that of floating point formatter `f`.

muli*fx*	Multiply 2 fixed point types resulting in an integral type.
	Example: `int mulik(_Accum, _Accum)`
divi*fx*	Divide an integral by a fixed point type resulting in an integral type.
	Example: `int divir(int, _Fract)`
*fx*divi	Divide 2 integral types resulting in a fixed point type.
	Example: `long _Accum lkdivi(long int, long int)`
idiv*fx*	Divide 2 fixed point types resulting in an integral type.
	Example: `int idivr(_Fract, _Fract)`
for *fx* in r, lr, k, lk, ur, ulr, uk, ulk	
abs*fx*	Compute the absolute value of a fixed point type.
	Example: `long _Fract abslr(long _Fract)`
for *fx* in hr, r, lr, hk, k, lk	
round*fx*	Round a fixed point value to a number of fractional bits given by an `int` parameter.
	Example: `long _Fract roundlr(long _Fract, int)`
countls*fx*	Count the number of bits a fixed point value can be shifted without overflow. Always returns an `int`.
	Example: `int countlsr(_Fract)`
bits*fx*	Compute an integral value with the same bit representation as a fixed point value.
	Example: `int bitsr(_Fract)`, where `int` is big enough
*fx*bits	Compute a fixed point value with the same bit representation as an integral value.
	Example: `_Fract rbits(int)`, the inverse of `bitsr`
strtof*fx*fx	Convert a string to a fixed point value.
	Example: `short _Fract strtofxhr(const char * restrict nptr, char ** restrict endptr)`
for *fx* in hr, r, lr, hk, k, lk, uhr, ur, ulr, uhk, uk, ulk	

Fig. 10 Various support functions as defined by Embedded C

Fig. 11 Additional formatting letters for fixed point types

Conversion specifiers	Length modifiers
`%r` (signed) _Fract	h short
`%R` unsigned _Fract	l long
`%k` (signed) _Accum	
`%K` unsigned _Accum	

4 Memory Spaces

The C language only knows two memory spaces, one for data and one for code. Embedded-C facilitates multiple memory spaces for data. As mentioned in Sect. 2 most DSPs can access two memories simultaneously. Embedded C supports such architectures by allowing the programmer to locate objects in special memory spaces, called *named memory spaces*. Although Embedded C has provisions to declare new named memory spaces in the program code, few compilers will actually support that and therefore a detailed description is omitted here.

Memory spaces must comply with one important rule: for any two memory spaces, either both memory spaces are disjoint or one is a subset of the other. The unqualified (generic) memory space should also always be supported. If this were not the case regular portable C-code suddenly would be illegal C in the context of

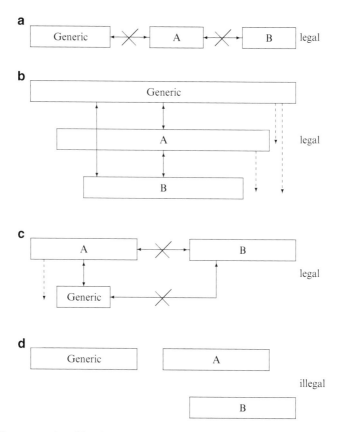

Fig. 12 Some examples of legal and illegal memory space configurations

a compiler that only supports named memory spaces. The named memory spaces however do not need to be a subset or superset of this generic memory space.

Let A and B be two memory spaces. If A is a subset of B, then every pointer to A is compatible with a pointer to B and can be automatically converted. Conversion of a pointer to B to a pointer to A will also be automatic but the result is undefined if the actual memory pointed to is not part of memory space A. If A and B are disjoint, then such conversions are not legal in Embedded C.

Figure 12 shows some legal and one illegal memory space configuration. Arrows are included in the picture representing pointer conversions from one space to another. Regular arrows represent legal conversions. Arrows with a cross represent conversions that a compiler will flag as an error. Dashed arrows represent conversions of which the result is undefined, but will pass through a compiler.

Heap memory management routines, such as `malloc`, for named memory spaces can be supplied through dedicated routines but Embedded C does not require such support to exist. Of course heap management in the generic memory space must be supported. Embedded C also stipulates that the standard C-library shall only use generic pointers.

It should be noted that applications using such named memory spaces are not portable to other architectures. The spaces only effect the data layout but not their values. So it is possible to emulate such an application on some general purpose platform by using an appropriate library and set of include files. Such include files basically will use a #define to remove the memory qualifiers and supply emulation routines for the special support functions required to handle such spaces. This is very useful to test an application on a native workstation. Often the hardware is not available early in the application development or is hard to operate in a testing environment.

Following are a couple of examples of pointer declaration using named memory spaces X and Y. The memory space qualifier are put in the same place as other type qualifiers, for example const and volatile. As a rule of thumb: the closer the qualifier to the variable identifier, the more it relates to the object itself. There is no difference between placing a qualifier directly before or after the base type, so const int c means the same as int const c.

Following are a couple of examples of a pointer p pointing to an int.

```
int * X p;      p in memory X pointing to an int in generic memory
int X * p;      p in generic memory pointing to an int in memory X
int X * Y p;    p in memory Y pointing to an int in memory X
```

5   Applications

Any floating point algorithm can be converted to use fixed point for the sake of performance. However this comes at a price in precision and range. It also requires careful thought about the range of the values involved in the computation, because if too much overflow/saturation takes place the quality of the results will deteriorate. This section exhibits two examples of such algorithms and explains the analytic aspects involved.

5.1   FIR Filter

The most common DSP algorithms are the finite impulse response (FIR) filter and the fast Fourier transform (FFT). The FIR filter uses the inner product at its core (see below) and the FFT algorithm also has a variation of the inner product (now on complex numbers) at its core.

This algorithm is the main motivation for the typical DSP architecture as described in Sect. 2. It uses two memories X and Y to stream data through a multiply accumulate operation.

```
#define N 16
X _Fract coeff[N] = { 0.7r, ... };

_Fract inner_product(Y _Fract *inp)
{
    int     i;
    _Accum  sum = 0.0k;
    for (i = 0; i < N; i++)
    {
        sum += coeff[i] * (_Accum)*(inp++);
    }
    return (_Sat _Fract)sum;
}
```

The above example sums the products in an _Accum typed variable sum. It is assumed that this summation will not overflow, because N is smaller or equal than the integral part of the _Accum. Only the final result is saturated back to a _Fract.

The cast to _Accum inside the loop makes sure that the multiplication is done in the _Accum type. Without this cast the multiplication would have been done in _Fract type and if both the coefficient and the input would have the exact value −1.0, then the result would be 1.0 which is out of range for a regular _Fract, resulting in an overflow. In a normal FIR filter there will not however be a −1.0 in the coefficients. So in that case the _Accum cast could be safely removed and the multiplication would be performed within the _Fract type.

The example also uses the memory spaces X and Y from where respectively the inp and the coeff constants are retrieved. Assuming the target architecture looks like the one described in Sect. 2, every iteration of this loop could thus be performed in one cycle.

5.2 Sine Function in Fixed Point

Another example of converting a floating point algorithm into fixed point is the computation of the sine function.

```
#define F 5
static _Accum sinfract(_Accum x)
{
    int     f;
    _Accum result = 1.0k;

    for (f = 2*F+1; f > 1; f -= 2)
    {
        result = 1.0k - result*x*x/(f*(f-1));
    }
```

```
        result *= x;

        return result;
}
```

The above algorithm is based on the Taylor expansion of the sine function evaluated using a Horner scheme.

$$\sin(x) = x - \frac{x^3}{3!} + \frac{x^5}{5!} - \frac{x^7}{7!} + \frac{x^9}{9!} - \frac{x^{11}}{11!} \cdots$$

It can be shown that r will always stay in range of the _Accum if x is in the range of $[-\pi, \pi]$. The precision of this approximation however deteriorates dramatically as x approaches $-\pi$ and π. With a 31 bits precision fixed point and 11 iterations, the error at π is around $3.5 \cdot 10^{-4}$. To improve the results one can use the symmetry of the sine function and only use the approximation for $[-\pi/2, \pi/2]$. Still the approximation becomes worse as x approaches $[-\pi/2]$ and $[\pi/2]$. For a 31 bits precision fixed point and 11 iterations, the error at $\pi/2$ is around $4.5 \cdot 10^{-8}$. The results can be further improved by using the formula

$$\sin(2x) = 2\sin(x)\cos(x)$$

The approximation for $\sin(x)$ and $\cos(x)$ will now only be used for $x \in [-\pi/4, \pi/4]$. For this range only a few iterations are required to obtain high precision. With the 31 bits precision fixed point and 11 iterations, the error is erratic over the whole range and is maximally $9 \cdot 10^{-10}$, which is around the precision of the fixed point.

Note that if $x \in [-\pi/4, \pi/4]$ all values involved in the computation will stay smaller than 1.0, except for the constant 1.0k itself. This suggests that using a _Fixed type instead of _Accum would suffice. Experimental results show that replacing the 1.0k constant by the slightly smaller 1.0r hurts the precision only within the precision of the _Fixed itself.

An optimizing compiler would take x*x out of the loop and compute it only once. It could also avoid the integer multiplication and complicated mixed type division (integer and fixed point) by putting the values for 1.0k/(f*(f-1)) in some table. Alternatively the compiler could simply fully unroll the whole loop. In that case f*(f-1) will be computed at compile time. Standard compiler optimizations next can turn the division by a constant into a multiplication by a constant. The resulting loop then only contains one multiplication and a multiply subtract operation.

If one takes a careful look at the accuracy of the above algorithm, then one will see that there is a trend in the error that is the result of the used Taylor expansion. Variations of the above Horner scheme exist, where the constants $1/(f*(f-1))$ are slightly changed to obtain a uniform distribution of the error over the whole range of x. As a final note it must be said that the CORDIC (coordinate rotation digital computer) algorithm for computing the trigonometric functions is very often used as it requires only addition and shift operations.

6 Named Registers

The C syntax is extended by Embedded C to include a storage-class specifier for
named registers. It can be used to identify a variable with a machine register.
A variable declared this way has external linkage and static storage. The names
of the registers must be reserved identifiers, so they are outside the name space
of existing programs. A reserved identifier either starts with two underscores or
with one underscore and a capital. For each supported register only one variable
can be associated with this register in every scope. The size of the type of the
variable should be big enough to contain the full machine register. Otherwise the
part contained in the variable is implementation defined.

The following is an example of a named register declaration, where _SP is a
intrinsic name of the compiler referring to the stack pointer.

```
register _SP volatile void * stack_pointer;
```

stack_pointer is an object stored in register _SP and is a pointer to volatile
void.

7 Hardware I/O Addressing

The Embedded-C report also contains a section on supporting I/O hardware by
a compiler. Apart from standardizing an interface to such I/O hardware, it is an
elaborate description on how such support could be engineered. The goal of this
part of the Embedded-C report is to facilitate the separate development of support
in a compiler and the device driver itself.

Specialized I/O hardware is often connected to a processor through memory
mapped I/O, that is they are accessed through a reserved part of the memory
space. Other processors however have dedicated instructions to communicate with
specialized I/O hardware. The programmer of a device driver would prefer to write
his driver so that it is portable across all platforms, independent of the way this
hardware is accessed by the processor. This part of the Embedded-C report aims to
facilitate this through a standardized set of functions to access such devices. It also
describes the way such an interface can be supported by a compiler.

Figure 13 lists all functions that should be supported by a compiler. Central in
this interface are the *ioreg*-designators. Basically these are identifiers, one for each
register in the I/O device. These identifiers can correspond to an object in the device
driver, but they are more often used for token concatenation inside macros. Below
some examples are given.

The engineering of I/O hardware support consists of intrinsic compiler support
to access I/O registers from the processor through for example memory mapped I/O
or dedicated instructions. Next to that the compiler offers an include file iohw.h
mapping the standardized functions to this intrinsic compiler support. For each
combination of I/O device and processor architecture an include file is written

```
void iogroup_acquire(iogroup);
void iogroup_release(iogroup);
void iogroup_map(iogroup, iogroup);

unsigned int iord(ioreg);
unsigned long iordl(ioreg);
void iowr(ioreg, unsigned int a);
void iowrl(ioreg, unsigned long a);

unsigned int iordbuf(ioreg, ioindex_t ix);
unsigned long iordbufl(ioreg, ioindex_t ix);
void iowrbuf(ioreg, ioindex_t ix, unsigned int a);
void iowrbufl(ioreg, ioindex_t ix, unsigned long a);

void ioand(ioreg, unsigned int a);
void ioor(ioreg, unsigned int a);
void ioxor(ioreg, unsigned int a);
void ioandl(ioreg, unsigned long a);
void ioorl(ioreg, unsigned long a);
void ioxorl(ioreg, unsigned long a);

void ioandbuf(ioreg, ioindex_t ix, unsigned int a);
void ioorbuf(ioreg, ioindex_t ix, unsigned int a);
void ioxorbuf(ioreg, ioindex_t ix, unsigned int a);
void ioandbufl(ioreg, ioindex_t ix, unsigned long a);
void ioorbufl(ioreg, ioindex_t ix, unsigned long a);
void ioxorbufl(ioreg, ioindex_t ix, unsigned long a);
```

Fig. 13 Functions to access I/O hardware

specifying how each *ioreg* maps onto the intrinsic compiler support. As a result a device driver can be written in a portable way, since it will only use the functions from the Embedded-C report.

In the following two simple examples are given. One for regular memory mapped I/O that does actually not require any additional intrinsic compiler support and another example for the PowerPC that uses dedicated instructions to access I/O registers. For the latter the CoSy style inline assembly is used to implement the intrinsic compiler support.

```
/* iohw.h */
#define MM_ACCESS(TYPE,ADR)                \
    (*((TYPE volatile *)(ADR)))

#define MM_DIR_8_DEV8_RD(ADR)              \
    (MM_ACCESS(uint8_t,ADR))
#define MM_DIR_8_DEV8_WR(ADR,VAL)    \
    (MM_ACCESS(uint8_t,ADR) = (uint8_t)VAL)

#define iord(NAME)                         \
    NAME##_TYPE_RD(NAME##_ADR)
#define iowr(NAME,VAL)                     \
    NAME##_TYPE_WR(NAME##_ADR,(VAL))
```

```
/* iohw_device.h */
#define MYPORT1_TYPE_RD              MM_DIR_8_DEV8_RD
#define MYPORT1_TYPE_WR              MM_DIR_8_DEV8_WR
#define MYPORT1_ADR                  0xBEAF
```

The compiler support is written in the `iohw.h` part. The macro `MM_ACCESS` dereferences a memory mapped I/O address. The `MM_DIR..` macros map a read and write operation onto this dereference. The prototypes as specified in Embedded C are mapped onto names in the device driver part by using token concatenation. The device driver part maps them back onto the `MM_DIR..` macros. The actual memory mapped address is also specified through token concatenation.

```
/* iohw.h */

/* Write to a SPR (Special Purpose Register) */
asm void __mtspr(int spr, unsigned int v)
{
@[ .spr-serialize; .constant spr]
    mtspr    @{spr},@{v}
}

/* Read from a SPR (Special Purpose Register) */
asm unsigned int __mfspr(int spr)
{
@[ .spr-serialize; .constant spr]
    mfspr    @{}, @{spr}
}

#define IO_DIR_32_DEV32_RD(ADR)           \
    (__mfspr(ADR))
#define IO_DIR_32_DEV32_WR(ADR,VAL)       \
    (__mtspr(ADR,(unsigned int)(VAL)))

#define iord(NAME)                        \
    NAME##_TYPE_RD(NAME##_ADR)
#define iowr(NAME,VAL)                    \
    NAME##_TYPE_WR(NAME##_ADR,VAL)

/* iohw_device.h */
#define MYPORT2_TYPE_RD IO_DIR_32_DEV32_RD
#define MYPORT2_TYPE_WR IO_DIR_32_DEV32_WR
#define MYPORT2_ADR      311
```

Above is an example of I/O registers accessed through special instructions on the PowerPC architecture. The main difference with the memory mapped example is that an address in this case is the number of a special purpose register (SPR). The inline assembly routines above contain a clause to instruct the compiler to serialize the accesses to the SPRs and they specify that the number of the SPR should be a constant, as the processor does not support a dynamically computed SPR number.

Embedded C also provides methods of dynamically (at runtime) initialization of the I/O device and grouping of a collection of I/O registers to represent a complete device.

8   History and Future of Embedded C

Embedded C [8] is rooted in DSP-C [4]. DSP-C was and is still used as industry standard to support features of DSP processors. DSP-C supports fixed point types and memory spaces. The most important difference between DSP-C and Embedded C is that the latter supports the special combinations of integral and fixed point type operations. The DSP-C extension also supports circular buffers, which was not included in the Embedded C report, because there were too many different implementations of such buffers. Hardware supporting circular buffers automatically sets a pointer to this buffer to the start of the buffer when it runs of the end of the buffer. This is also very useful in the implementation of FIR filters and fast Fourier transforms. ACE Associated Compiler Experts bv developed DSP-C in collaboration with Philips Semiconductors and submitted it for standardization by the ISO working group on C. This group adopted the rationale of DSP-C and this resulted in Embedded C in February 2004 [8]. The standardized interface to I/O hardware doesn't seem to be widely accepted by the industry.

Embedded C is supported by the CoSy compiler development system [2], the Byte Craft Limited compilers [9], the EDG front-end [6], the Dinkumware libraries [10] the Nullstone compiler performance analysis tools [5] and test suites are supplied in Perennial [11] and SuperTest [3]. With these tools a wide range of compiler support is covered. For more information on Embedded C see [1].

The report on the Embedded C extension mentions two features that may be supported in future versions. One is that if an approach for implementing circular buffers in hardware emerges in the industry, then this may be abstracted into an extension of Embedded C. Another is complex fixed point data types as a future extension, because these complex types are used very often in the industry and several DSP algorithms could be more efficiently described using complex fixed point data types.

As fixed point types are used in the industry to trade between precision and speed every application domain sets its own standard for the number of integral and fractional bits. From a hardware point of view there is very little difference in the complexity of implementing division for integer and fixed point arithmetic. From a programming point of view however, there is a big difference. Conversions have to be made explicit using shifts and the programmer must be aware what division between bits is actually used at every point in the program. This means that the industry would benefit if Embedded C would support every conceivable partitioning between integral and fractional bits.

9 Conclusions

Embedded C removes the need to use intrinsics or assembly to target fixed point arithmetic found on typical DSP architectures. It thereby facilitates more readable and portable programs. It facilitates a natural specification of memory spaces, where one otherwise would have to resort to non-portable pragmas.

There will still be features of the architecture that may have to be targeted through intrinsics, assembly, pragmas or other compiler-supported extensions. Good examples are complex fixed point types or special Viterbi instructions. A Viterbi instruction takes two arguments and produces two values and a flag. Such operations are not easily captured in an expression.

References

1. www.embedded-c.org.
2. ACE Associated Compiler Experts bv. Cosy compiler development system.
3. ACE Associated Compiler Experts bv. Supertest.
4. ACE Associated Compiler Experts bv. DSP-C, An extension to ISO/IEC IS 9899:1990, www.dsp-c.org. 2005.
5. Nullstone Corporation. www.nullstone.com.
6. Edison Design Group, Inc. www.edg.com.
7. ISO/IEC. International Standard ISO/IEC 9899:1999, Programming languages – C. 1999.
8. ISO/IEC. ISO/IEC TR 18037, Programming languages – C – Extensions to support embedded processors. 2008.
9. Byte Craft Limited. www.bytecraft.com.
10. Dinkumware Ltd. www.dinkumware.com.
11. Perennial. www.peren.com.

Signal Flow Graphs and Data Flow Graphs

Keshab K. Parhi and Yanni Chen

Abstract This chapter first introduces two types of graphical representations of digital signal processing algorithms including signal flow graph (SFG) and data flow graph (DFG). Since SFG and DFG are in general used for analyzing structural properties and exploring architectural alternatives using high-level transformations, such transformations including retiming, pipelining, unfolding and folding will then be addressed. Finally, their real-world applications to both hardware and software design will be presented.

1 Introduction

Signal processing programs differ from the traditional computing programs in the sense that these programs are referred to as non-terminating programs. In other words, input samples are processed periodically (typically with a certain iteration period or sampling period) and the tasks are repeated infinite number of times. A traditional dependence graph representation of such a program would require infinite number of nodes. Signal flow graphs and data flow graphs are powerful representations of signal processing algorithms and signal processing systems because these can represent the operations using a finite number of nodes.

K.K. Parhi (✉)
University of Minnesota, Department of Electrical and Computer Engineering, 200 Union St.
S.E., Minneapolis, MN 55455, USA
e-mail: parhi@umn.edu

Y. Chen
Marvell Semiconductor Inc., 5488 Marvell Lane, Santa Clara, CA 95054, USA
e-mail: yannic@marvell.com

S.S. Bhattacharyya et al. (eds.), *Handbook of Signal Processing Systems*,
DOI 10.1007/978-1-4614-6859-2_39, © Springer Science+Business Media, LLC 2013

Signal flow graphs have been used for a long time to analyze transfer functions of linear systems. Data flow graphs are more general and are used to represent both linear and non-linear systems. These can also be used to represent multi-rate systems that contain rate-altering components such as interpolators and decimators. The z^{-1} elements in signal flow graphs and delay elements in data flow graphs describe *inter-iteration* precedence constraints, i.e., constraints between two tasks of different iterations. A simple edge without any z^{-1} or delay element represents a precedence constraint within the same iteration. These are referred as *intra-iteration* precedence constraints. Both types of precedence constraints describe the causality constraints among different operations. These causality constraints are important both in hardware designs [1] and software implementations. In a hardware implementation, the critical path time is the time needed for satisfying the causality constraints among all tasks within the same iteration, and is a lower bound on the clock period. In a software implementation, the scheduling algorithm must satisfy these causality constraints by satisfying both intra-iteration and inter-iteration constraints.

Another important property of these flow graphs is that these can be transformed into equivalent forms by high-level transformations such as pipelining, retiming, unfolding and folding. These equivalent forms have same input-output characteristics but have different sets of constraints. Pipelining and retiming can be used to reduce clock period in a circuit. These transformations can also be used as a preprocessing step for folding where the objective is to design time-multiplexed or folded architectures. Unfolding transformation can lead to lower iteration periods in software implementations as it can unravel some of the concurrency hidden in the original data flow graph. In hardware implementations, unfolding leads to parallel implementations that can increase the effective sample speed while the clock speed remains unaltered. Both pipelined and parallel implementations can be used to reduce power consumption if achieving higher speed is less important. Alternatively, in deep submicron technologies with low supply voltage, pipelining and parallel processing can be used to meet the critical path requirements.

This chapter provides a brief overview of signal flow graphs and data flow graphs, and is organized as follows. Section 2 introduces signal flow graphs, and illustrates how signal flow graphs are used for deriving transfer functions, either by using Mason's gain formula or by using a set of equations. This section also illustrates retiming and pipelining of signal flow graphs. Section 3 addresses single-rate and multi-rate data flow graphs, and illustrates how an equivalent single-rate data flow graph can be obtained from a multi-rate data flow graph. Section 4 provides an overview of unfolding and folding transformations, and their applications to hardware design. Section 5 illustrates how the causality constraints imposed by the intra-iteration and inter-iteration constraints are exploited for scheduling in software implementations. Sections 3–5 are adapted from the text book [2].

2 Signal Flow Graphs

In this section, the notation of signal flow graph (SFG) is first overviewed. Then two useful approaches, i.e., the Mason's gain formula and the equation-solving, are explained in detail to derive the corresponding transfer function for a given SFG.

2.1 Notation

Signal flow graphs have been used for the analysis, representation, and evaluation of linear digital networks, especially digital filter structures. An SFG is a collection of nodes and directed edges [3], where the nodes represent computations or tasks and a directed edge (j, k) denotes a branch originating from node j and terminating at node k. With input signal at node j and output signal at node k, the edge (j, k) denotes a linear transformation from the signal at node j to the signal at node k. An example of SFG is shown in Fig. 1, where both nodes j and k represent summing operations and the edge (j, k) denotes a unit gain transformation.

Two types of nodes exist in SFG, source nodes and sink nodes. A source node is a node with no entering edges, and is used to represent the injection of external inputs into a graph. A sink node is a node with only entering edges, and is used to extract outputs from a graph.

2.2 Transfer Function Derivation of SFG

For a given SFG, there are mainly two approaches to derive its corresponding transfer function. One is the Mason's gain formula, which provides a step-by-step method to obtain the transfer function. The other is the equation-solving approach by labeling each intermediate signal, writing down the equation for that signal with dependency on other signals, and then solving the multiple equations to

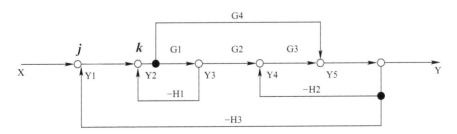

Fig. 1 An example of signal flow graph

represent the output signal *only* in terms of the input signal. Note that the variables used in the signal flow graphs and the equations correspond to frequency-domain representations of signals.

2.2.1 Mason's Gain Formula

First of all, a few useful terminologies in Mason's gain formula have to be defined related to an SFG.

- Forward path: a path that connects a source node to a sink node in which no node is traversed more than once.
- Loop: a closed path without crossing the same point more than once.
- Loop gain: the product of all the transfer functions in the loop.
- Non-touching or non-interacting loops: two loops are nontouching or noninteracting if they have no nodes in common.

In general, Mason's gain formula [4] is presented as below:

$$M = \frac{Y}{X} = \frac{\sum_{j=1}^{N} M_j \Delta_j}{\Delta} \tag{1}$$

where

- M = transfer function or gain of the system
- Y = output node
- X = input node
- N = total number of forward paths between X and Y
- Δ = determinant of the graph = $1 - \sum$ loop gains + $\sum$ non-touching loop gains taken two at a time $- \sum$ non-touching loop gains taken three at a time + ...
- M_j = gain of the jth forward path between X and Y
- Δ_j = 1-loops remaining after eliminating the jth forward path, i.e., eliminate the loops touching the jth forward path from the graph. If none of the loops remains, $\Delta_j = 1$.

To illustrate the actual usage of Mason's gain formula, the transfer function for the example SFG shown in Fig. 1 is derived by following the steps below:

1. Find the forward paths and their corresponding gains
   Two forward paths exist in this SFG:
   $M_1 = G_1 G_2 G_3$ and $M_2 = G_4$
2. Find the loops and their corresponding gains
   There are four loops in this example:
   $Loop_1 = -G_1 H_1,$
   $Loop_2 = -G_3 H_2,$

$Loop_3 = -G_1G_2G_3H_3,$
$Loop_4 = -G_4H_3$

3. Find the Δ_j
   If we eliminate the path $M_1 = G_1G_2G_3$ from the SFG, no complete loops remain, so $\Delta_1 = 1$. Similarly, if the path $M_2 = G_4$ is eliminated from the SFG, no complete loops remain neither, so $\Delta_2 = 1$ as well.

4. Find the determinant Δ
   Only one pair of non-touching loops is in this SFG, i.e., $Loop_1$ and $Loop_2$, thus $\sum$ non-touching loop gains taken two at a time $= (-G_1H_1)(-G_3H_2)$.
   Therefore,
   $\Delta = 1 - \sum$ loop gains $+ \sum$ non-touching loop gains taken two at a time
   $= 1 - (-G_1H_1 - G_3H_2 - G_1G_2G_3H_3 - G_4H_3) + (-G_1H_1)(-G_3H_2)$
   $= 1 + G_1H_1 + G_3H_2 + G_1G_2G_3H_3 + G_4H_3 + G_1G_3H_1H_2$

5. The final step is to apply the Mason's gain formula using the terms found above

$$M = \frac{Y}{X} = \frac{\sum_{j=1}^{N} M_j \Delta_j}{\Delta} = \frac{G_1G_2G_3 + G_4}{1 + G_1H_1 + G_3H_2 + G_1G_2G_3H_3 + G_4H_3 + G_1G_3H_1H_2} \quad (2)$$

2.2.2   Equations-Solving Based Transfer Function Derivation

As an alternative approach, the equations-solving based method follows different set of steps. Note that the intermediate signals have already been appropriately labeled in Fig. 1.

1. Write down the equations for each labeled signal with dependency on other signals:
   $Y_1 = X - YH_3$
   $Y_2 = Y_1 - Y_3H_1$
   $Y_3 = G_1Y_2$
   $Y_4 = G_2Y_3 - YH_2$
   $Y_5 = G_3Y_4 + G_4Y_2$
   $Y = Y_5$

2. Solve all the equations above and derive the relationship between output node Y and input node X:
   $Y = Y_5 = G_3Y_4 + G_4Y_2 = G_3(G_2Y_3 - YH_2) + G_4Y_2$
   $= G_3(G_2G_1Y_2 - YH_2) + G_4Y_2 = -G_3H_2Y + (G_1G_2G_3 + G_4)Y_2$

   Therefore,

$$Y = \frac{G_1G_2G_3 + G_4}{1 + G_3H_2}Y_2 \quad (3)$$

   Note that

$$Y_2 = Y_1 - Y_3H_1 \quad (4)$$

By substituting both Y_1 and Y_3 into (4), we obtain

$$Y_2 = X - YH_3 - Y_3H_1 = X - YH_3 - G_1H_1Y_2 \tag{5}$$

Consequently,

$$Y_2 = \frac{X - YH_3}{1 + G_1H_1} \tag{6}$$

Then by substituting (6) into (3),

$$Y = \frac{G_1G_2G_3 + G_4}{1 + G_3H_2} \cdot \frac{X - YH_3}{1 + G_1H_1} = \frac{(G_1G_2G_3 + G_4)(X - YH_3)}{(1 + G_3H_2)(1 + G_1H_1)}$$

$$= \frac{G1G2G3 + G4}{(1 + G_3H_2)(1 + G_1H_1)}X - \frac{(G_1G_2G_3 + G_4)H_3}{(1 + G_3H_2)(1 + G_1H_1)}Y$$

As a result,

$$1 + \frac{(G_1G_2G_3 + G_4)H_3}{(1 + G_3H_2)(1 + G_1H_1)}Y = \frac{G_1G_2G_3 + G_4}{(1 + G_3H_2)(1 + G_1H_1)}X$$

$$\frac{Y}{X} = \frac{G_1G_2G_3 + G_4}{(1 + G_3H_2)(1 + G_1H_1) + (G_1G_2G_3 + G_4)H_3}$$

Finally, the transfer function between the output node Y and input node X is

$$M = \frac{G_1G_2G_3 + G_4}{1 + G_1H_1 + G_3H_2 + G_1G_3H_1H_2 + G_4H_3 + G_1G_2G_3H_3} \tag{7}$$

which is exactly the same as the derived transfer function using Mason's gain formula in (2).

3   Data Flow Graphs

In this section, the notation of data flow graph (DFG) is introduced and is followed by an overview of the single-rate DFG and the multi-rate DFG. How to construct an equivalent single-rate DFG from the multi-rate DFG is then explained in detail. After that, the concepts of retiming and pipelining are briefly introduced to derive equivalent DFGs.

3.1   Notation

In data flow graph representations, the nodes represent computations or functions or subtasks and the directed edges represent data paths (communications between nodes) and each edge has a nonnegative number of delays associated with it.

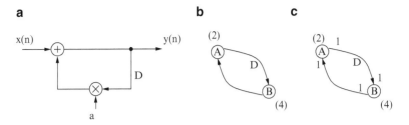

Fig. 2 (**a**) Block diagram description of the computation $y(n) = ay(n-1) + x(n)$. (**b**) Conventional DFG representation. (**c**) Synchronous DFG representation

For example, Fig. 2b is the DFG corresponding to the block diagram in Fig. 2a. Compared to the SFG in Sect. 2.1, DFG can be seen as a more generalized form of SFG in that it can effectively describe both linear single-rate and nonlinear multi-rate DSP systems.

DFG describes the data flow among subtasks or elementary computations modeled as nodes in a signal processing algorithm. Similar to SFG, various DFGs derived for one algorithm can be obtained from each other through high-level transformations. DFGs are generally used for high-level synthesis to derive concurrent implementations of DSP applications onto parallel hardware, where subtask scheduling and resource allocation are of major concerns.

3.2   Synchronous Data Flow Graph

A synchronous data flow graph (SDFG) is a special case of data-flow graph where the number of data samples produced or consumed by each node in each execution is specified a priori [5]. For example, Fig. 2c is an SDFG for the computation $y(n) = ay(n-1) + x(n)$, which explicitly specifies that one execution of both nodes A and B consumes one data sample and produces one output sample, which is a single-rate system.

In addition, the SDFG can describe multi-rate systems in a simple way. For example, Fig. 3a shows an SDFG representation of a multi-rate system, where nodes A, B and C are operated at different frequencies f_A, f_B and f_C, respectively. Note that A processes f_A input samples and produces $3f_A$ output samples per time unit. Node B consumes five input samples during each execution, hence consumes $5f_B$ input samples per time unit. Using the equality $3f_A = 5f_B$, we have $f_B = 3f_A/5$. Similarly, the operating rate of node C can be computed as $f_C = 2f_B/3 = 2f_A/5$. For a specified input sampling rate, the operating frequencies for nodes A, B and C can be computed. An equivalent single-rate DFG for the multi-rate DFG in Fig. 3a is shown in Fig. 3b. In contrast, this single-rate DFG contains 10 nodes and 30 edges, as compared to 3 nodes and 4 edges in the SDFG representation.

a

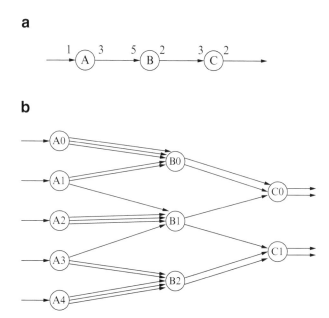

b

Fig. 3 Multi-rate DFG in (**a**) can be converted into single-rate DFG in (**b**), which can then be represented using linear SFG

Fig. 4 An edge $U \rightarrow V$ in a multi-rate DFG

3.3 Construct an Equivalent Single-Rate DFG from the Multi-Rate DFG

By definition, each node in a single-rate DFG (SRDFG) is executed exactly once per iteration. In contrast, multi-rate DFG (MRDFG) allows each node to be executed more than once per iteration, and 2 nodes are not required to execute the same number of times in an iteration. However, SRDFG can still be used to represent multi-rate systems by first unfolding the multi-rate systems to single-rate.

An edge from the node U to the node V in an MRDFG is shown in Fig. 4, where the value O_{UV} is the number of samples produced on the edge by an invocation of the node U, the value I_{UV} is the number of samples consumed from the edge by an invocation of the node V and the value i_{UV} is the number of delays on the edge.

If the nodes U and V are invoked k_U times and k_V times in one iteration, respectively, then the number of samples produced on the edge from the node U to the node V in this iteration is $O_{UV}k_U$, and the number of samples consumed from the edge by the node V in this same iteration is $I_{UV}k_V$. Intuitively, to avoid a buildup or deficiency of samples on the edge, the number of samples produced

Fig. 5 A multi-rate DFG

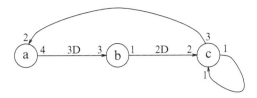

in one iteration must equal the number of samples consumed in one iteration. This relationship can be described mathematically as

$$O_{UV}k_U = I_{UV}k_V \qquad (8)$$

An algorithm for constructing an equivalent SRDFG from an MRDFG is described as follows:

1. For each node U in the MRDFG
2. For $k = 0$ to $k_U - 1$
3. Draw a node U^k in the SRDFG with the same computation time as U in the MRDFG
4. For each edge $U \overset{i_{UV}}{\to} V$ in the MRDFG
5. For $j = 0$ to $O_{UV}k_U$-1
6. Draw an edge $U^{j/O_{UV}} \to V^{((j+i_{UV})/I_{UV})\%k_V}$ in the SRDFG with $(j+i_{UV})/(I_{UV}k_V)$ delays

To determine how many times each node must be executed in an iteration, the set of equations found by writing (8) for each edge in the MRDFG must be solved so the number of invocations for the nodes are coprime. For example, the set of equations for the MRDFG in Fig. 5 is

$$4k_a = 3k_b$$

$$k_b = 2k_c$$

$$k_c = k_c$$

$$3k_c = 2k_a$$

which has a solution of $k_a = 3$, $k_b = 4$, $k_c = 2$. Once the number of invocations of the nodes has been determined, an equivalent SRDFG can be constructed for the MRDFG. For the MRDFG in Fig. 5, the equivalent SRDFG is shown in Fig. 6.

3.4 Equivalent Data Flow Graphs

Data flow graphs can be transformed into different yet equivalent forms. Two common techniques to derive the equivalent DFGs are introduced here, one is the retiming and the other is pipelining. The transfer functions in these equivalent

Fig. 6 An equivalent
SRDFG for the MRDFG in
Fig. 5

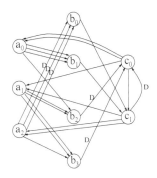

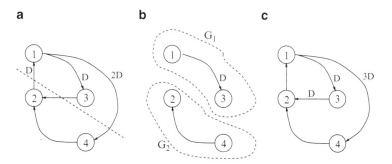

Fig. 7 (**a**) The unretimed DFG with a cutset shown as a *dashed line*. (**b**) The 2 graphs $G1$ and $G2$ formed by removing the edges in the cutset. (**c**) The retimed graph found using cutset retiming with $k = 1$

forms are either unaltered or differ only by a factor of z^{-i}, i.e., they may contain i additional delay elements. Note that both retiming and pipelining transformations can be applied to DFGs as well as SFGs in an identical manner.

3.4.1 Retiming

Retiming [6] is a transformation technique that changes the locations of delay elements in a circuit without affecting its input/output characteristics. For example, although the DFGs in Fig. 7a, c have different number of delays at different locations, they share the same input/output characteristics. Furthermore, these two DFGs can be derived from one another using retiming.

Retiming has many applications in synchronous circuit design including reducing the clock period of the circuit by reducing the computation time of the critical path, decreasing the number of registers in the circuit, reducing the dynamic power consumption of the circuit by placing the registers at the inputs of nodes with large capacitances to reduce the switching activity, and logic synthesis.

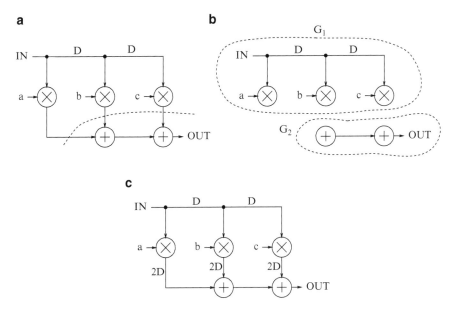

Fig. 8 (**a**) The unretimed DFG with a cutset shown as a *dashed line*. (**b**) The 2 graphs G1 and G2 formed by removing the edges in the cutset. (**c**) The graph obtained by cutset retiming with $k = 2$

Cutset retiming is a special case of retiming and it only affects the weights of the edges in the cutset, which is a set of edges that can be removed from the graph to create two disconnected subgraphs. If these two disconnected subgraphs are labeled G_1 and G_2 as depicted in Fig. 7b, then cutset retiming consists of adding k delays to each edge from G_1 and G_2, and removing k delays from each edge from G_2 to G_1. In the case of $k = 1$, the DFG in Fig. 7a can then be transformed to the DFG in Fig. 7c by using cutset retiming technique.

3.4.2 Pipelining

Pipelining is a special case of cutset retiming where there are no edges in the cutset from the subgraph G_2 to the subgraph G_1 as shown in Fig. 8b, i.e., pipelining applies to graphs without loops. These cutsets are referred to as feed-forward cutsets, where the data move in the forward direction on all the edges of the cutset. Consequently, registers can be arbitrarily placed on a feed-forward cutset without affecting the functionality of the algorithm. If two registers are inserted to each edge, the DFG in Fig. 8a can then be transformed into the DFG in Fig. 8c by applying pipelining technique. Therefore, complex retiming operations can be described by multiple simple cutset retiming or pipelining operations applied in a step-by-step manner.

Fig. 9 (**a**) A datapath. (**b**)
The 2-level pipelined
structure of (**a**)

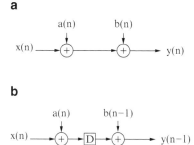

Pipelining transformation leads to a reduction in the effective critical path by introducing pipelining registers along the datapath, which can be exploited to either increase the clock speed or sample speed or to reduce power consumption at same speed. Consider the simple structure in Fig. 9a, where the computation time of the critical path is $2T_A$. Figure 9b shows the 2-level pipelined structure, where one register is placed between two adders and hence the critical path is reduced by half.

Obviously, in an M-level pipelined system the number of delay elements in any path from input to output is $(M-1)$ greater than that in the same path in the original sequential circuit. While pipelining offers the benefit of critical path reduction, its two drawbacks lie in the increase in the number of registers and the system latency, which is the time difference in the availability of the first output data in the pipelined system and the sequential system.

4 Applications to Hardware Design

In this section, two DFG-based high-level transformations applicable to practical hardware design such as field programmable gate array (FPGA) or application specific integrated circuit (ASIC) implementations are introduced, one is the unfolding transformation, and the other is the folding transformation. Examples will be given to demonstrate their usage in hardware design. For more details on how to map the decidable signal processing graphs to FPGA implementation, the reader is referred to [1].

4.1 Unfolding

Unfolding is a transformation technique that can be applied to a DSP program to create a new program describing more than one iteration of the original program. More specifically, unfolding a DSP program by the unfolding factor J creates a new program that describes J consecutive iterations of the original program. As a result, in unfolded system each delay is J-slow.

Unfolding has applications in designing high-speed and low-power VLSI architectures. One application is to unfold the program to reveal hidden concurrencies so that the program can be scheduled to a smaller iteration period, thus increasing the throughput of the implementation. Another application is to design parallel architectures at the word level and bit level from serial counterpart to increase the throughput or decrease the power consumption of the implementation.

4.1.1 The DFG Based Unfolding

Two approaches can be used to derive the J-unfolded DFG. One is to write equations for the original and the J-unfolded programs and then draw the corresponding unfolded DFG. This method could be tedious for large value of J. The other approach is to use a graph-based technique which directly unfolds the original DFG to create the DFG of the J-unfolded program without explicitly writing the equations describing the original unfolded system.

For each node U in the original DFG, there are J nodes with the same function as U in the J-unfolded DFG. Additionally, for each edge in the original DFG, there are J edges in the J-unfolded DFG. Consequently, the DFG of the J-unfolded program contains J times as many nodes and edges as the DFG of the original DFG. The following two-step algorithm could be used to construct a J-unfolded DFG:

1. For each node U in the original DFG, draw the J nodes $U_0, U_1, \ldots, U_J - 1$.
2. For each edge $U \to V$ with w delays in the original DFG, draw the J edges $U_i \to V_{(i+w)\%J}$ with $\lfloor \frac{i+w}{J} \rfloor$ delays for $i = 0, 1, \ldots, J - 1$. Apparently, if an edge has $w < J$ delays in the original DFG, unfolding produces $J - w$ edges with no delays and w edges with 1 delay in the J-unfolded DFG.

To demonstrate the unfolding algorithm, the DFG in Fig. 10b that corresponds to the DSP algorithm in Fig. 10a will serve as an example, where the nodes A and B represent input and output, respectively, and the nodes C and D represent addition and multiplication by a, respectively. To unfold this DFG in Fig. 10b by unfolding factor of 2 to obtain the 2-unfolded DFG as shown in Fig. 10c, the two steps of the unfolding algorithm are performed:

1. The 8 nodes A_i, B_i, C_i and D_i for $i = 0, 1$ are first drawn according to the 1st step of the unfolding algorithm.
2. After these nodes have been drawn, for an edge $U \to V$ such as $D \to C$ with no delays, this step reduces to drawing the J edges $U_i \to V_i$ with no delays. Additionally, for the edges $C \to D$ with $w = 9$ delay, there are the edges $C_0 \to D_{(0+9)\%2} = D_1$ with $\lfloor (\frac{0+9}{2}) \rfloor = 4$ delays and $C_1 \to D_{(1+9)\%2} = D_0$ with $\lfloor (\frac{1+9}{2}) \rfloor = 5$ delays.

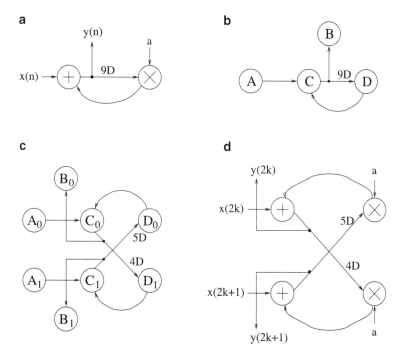

Fig. 10 (**a**) The original DSP program describing $y(n) = ay(n-9) + x(n)$ for $n = 0$ to ∞. (**b**) the DFG corresponding to DSP program in (**a**). (**c**) The 2-unfolded DFG. (**d**) The 2-unfolded DSP program describing $y(2k) = ay(2k-9) + x(2k)$ and $y(2k+1) = ay(2k-8) + x(2k+1)$ for $n = 0$ to ∞

Referring to Fig. 10c, the nodes C_0 and C_1 in the 2-unfolded DFG represent addition as the node C in the original DFG. Similarly, the nodes D_0 and D_1 in the 2-unfolded DFG represent multiplications as the node D in the original DFG. The node A in the original DFG represents the input $x(n)$. The k-th iteration of the node A_i in the unfolded DFG executes the $Jk + i$-th iteration of the node A in the original DFG for $i = 0, 1, \ldots, J - 1$ and $k = 0$ to ∞. Similarly, the node B_0 corresponds to the output samples $y(2k + 0)$ and the node B_1 corresponds to the output sample $y(2k + 1)$. Therefore, the 2-unfolded DFG in Fig. 10c corresponds to the 2-unfolded DSP program in Fig. 10d.

4.1.2 Applications to Parallel Processing

A direct application of the general unfolding transformation is to design parallel processing architectures from serial processing architectures. At the word level, this means that word-parallel architectures can be designed from word-serial architectures. At the bit level, it means that bit-parallel and digit-serial architecture can be designed from bit-serial architectures.

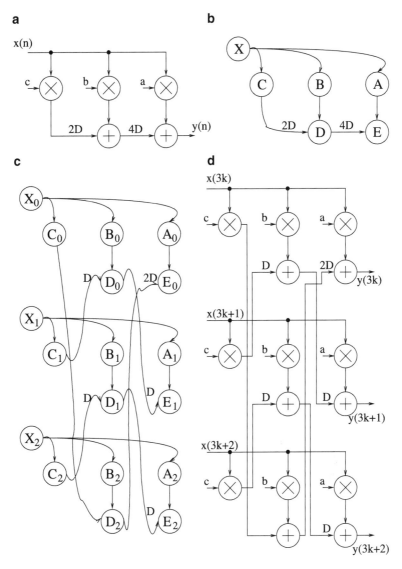

Fig. 11 (**a**) The original DSP program. (**b**) The DFG for (**a**). (**c**) The 3-unfolded DFG. (**d**) The 3-parallel DSP program

Word-Level Parallel Processing

In general, unfolding a word-serial architecture by J creates a word-parallel architecture that processes J words per clock cycle. As an example, consider the DSP program $y(n) = ax(n) + bx(n-4) + cx(n-6)$ shown in Fig. 11a. To create an architecture that can process more than 1 word per clock cycle, the first step is to

Fig. 12 A switch

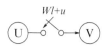

draw a corresponding DFG as in Fig. 11b. The next step is to unfold the DFG to the
3-unfolded DFG as in Fig. 11c by following the steps described in 4.1.1. The final
step is to draw the corresponding 3-unfolded DSP program as in Fig. 11d. The exact
details are omitted here and left to the reader as an exercise.

Bit-Level Parallel Processing

Assume the wordlength of the data is W bits, the hardware implementation could
have following possible architectures:

- Bit-serial processing: One bit is processed per clock cycle and hence a complete
  word is processed in W clock cycles.
- Bit-parallel processing: one word of W bits is processed every clock cycle.
- Digital serial processing: N bits are processed per clock cycle and a word is
  processed in W/N clock cycles, which is referred to as the digit size.

Most bit-serial architecture contains an edge with a switch, which corresponds to
a multiplexer in hardware. Consider the edge $U \to V$ in Fig. 12. To unfold this edge
with unfolding factor J, two basic assumptions are made:

- The wordlength W is a multiple of the unfolding factor J, i.e., $W = W'J$
- All edges into and out of the switch have no delays.

With these two assumptions in mind, the edge in Fig. 12 can be unfolded using
the following two steps:

1. Write the switching instance as
   $$Wl + u = J(W'l + \lfloor \tfrac{u}{J} \rfloor) + (u\%J)$$
2. Draw an edge with no delays in the unfolded graph from the node $U_{u\%J}$ to the
   node $V_{u\%J}$, which is switched at time instance $(W'l + \lfloor \tfrac{u}{J} \rfloor)$.

If the switch has multiple instances, then each switching instance is treated
separately. In addition, if an edge contains a switch and a positive number of delays,
a dummy node can be used to reduce this problem to the case where the edge
contains no delay elements.

Using these techniques for unfolding switches, bit-parallel and digit-serial
architectures can be systematically designed from bit-serial architectures. This is
demonstrated using the bit-serial adder in Fig. 13a. A DFG corresponding to this
adder is shown in Fig. 13b. The 2-unfolded version of this DFG is shown in Fig. 13c
and the architectures corresponding to this unfolded DFG is in Fig. 13d, where the
2-unfolded architecture is a digit-serial adder with digit size equal to 2.

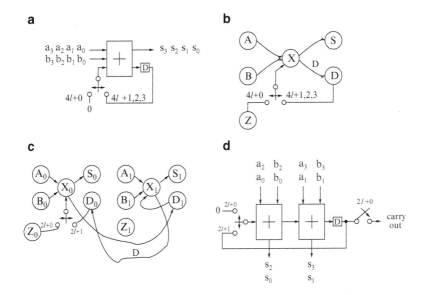

Fig. 13 (**a**) Bit-serial addition $s = a + b$ for wordlength $W = 4$. (**b**) The DFG corresponding to the bit-serial adder. (**c**) The DFG resulting from unfolding the DFG using unfolding factor of $J = 2$. (**d**) The digit-serial adder designed by unfolding the bit-serial adder using $J = 2$

For more details on how to unfold the DFG when the unfolding factor J is not the divisor of the wordlength W, the reader is referred to Sect. 5.5.2.2 in [2].

4.1.3 Infinite Unfolding of DFG

Any DFG can be unfolded by a factor of ∞. This infinitely unfolded DFG explicitly represents all intra-iteration and inter-iteration constraints. These DFGs correspond to dependence graphs (DGs) of traditional terminating programs. The DG or the infinitely unfolded DFG cannot contain any delay elements. Figure 14a shows a DFG and Fig. 14b shows the corresponding infinitely unfolded DFG.

4.2 Folding

The folding transformation is used to systematically determine the control circuits in DSP architectures where multiple algorithm operations such as additions are time-multiplexed to a single functional unit such as pipelined adder. By executing multiple algorithm operations on a single functional unit, the number of functional units in the implementation is reduced, resulting in an integrated circuit with low

a

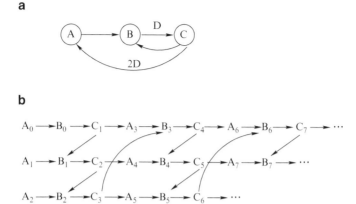

b

Fig. 14 (**a**) The original DFG. (**b**) The infinitely unfolded DFG

a   **b**

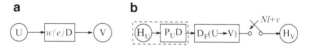

Fig. 15 (**a**) An edge $U \xrightarrow{e} V$ with $w(e)$ delays. (**b**) The corresponding folded datapath. The data begin at the functional unit H_U which has P_U pipelining stage, pass through $D_F(U \xrightarrow{e} V)$ delays, and are switched into the functional unit H_V at the time instances $Nl + v$

silicon area. In general, folding can be used to reduce the number of hardware functional units by a factor of N at the expense of increasing the computation time by a factor of N.

Consider the edge e connecting the nodes U and V with $w(e)$ delays, as shown in Fig. 15a. Let the executions of the l-th iteration of the nodes U and V be scheduled at the time units $Nl + u$ and $Nl + v$, respectively, where u and v are folding orders of the nodes U and V that satisfy $0 \leq u, v \leq N - 1$ and N is the folding factor. The folding order of a node is the time partition to which the node is scheduled to execute in hardware. The functional units that execute the nodes U and V are denoted as H_U and H_V, respectively. If H_U is pipelined by P_U stages, then the result of the l-th iteration of the node U is available at the time unit $Nl + u + P_U$. Since the edge $U \xrightarrow{e} V$ has $w(e)$ delays, the result of the l-th iteration of the node U is used by the $(l + w(e))$-th iteration of the node V, which is executed at $N(l + w(e)) + v$. Therefore, the result must be stored for

$$D_F(U \xrightarrow{e} V) = [N(l + w(e)) + v] - [Nl + P_U + u] = Nw(e) - P_U + v - u \quad (9)$$

time units, which is independent of the iteration number l. The edge $U \xrightarrow{e} V$ is implemented as a path from H_U to H_V in the architecture with $D_F(U \xrightarrow{e} V)$ delays, and data on this path are input to H_V at $Nl + v$, as in Fig. 15b.

Fig. 16 The retimed biquad filter with valid folding sets assigned

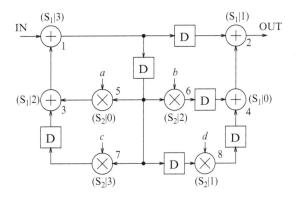

Fig. 17 The folded biquad filter using the folding sets given in Fig. 16

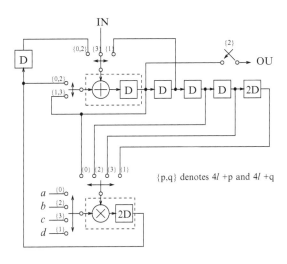

A folding set is an ordered set of operations executed by the same functional unit. Each folding set contains N entries, some of which may be null operations. The operation in the j-th position within the folding set, where j goes from 0 to $N-1$, is executed by the functional unit during the time partition j. The biquad filter example shown in Fig. 16 is folded with folding factor of $N=4$ and the folding sets shown in the figure can be written as $S_1 = \{4, 2, 3, 1\}$ and $S_2 = \{5, 8, 6, 7\}$, where the folding set S_1 and S_2 contains only addition operations using the same hardware adder and multiplication operation using the same hardware multiplier, respectively. To obtain the folded architecture as shown in Fig. 17 corresponding to the DFG in Fig. 16, the folding equations for each of the 11 edges are written as below:

$$D_F(1 \rightarrow 2) = 4(1) - 1 + 1 - 3 = 1$$
$$D_F(1 \rightarrow 5) = 4(1) - 1 + 0 - 3 = 0$$
$$D_F(1 \rightarrow 6) = 4(1) - 1 + 2 - 3 = 2$$

$$D_F(1 \rightarrow 7) = 4(1) - 1 + 3 - 3 = 3$$
$$D_F(1 \rightarrow 8) = 4(2) - 1 + 1 - 3 = 5$$
$$D_F(3 \rightarrow 1) = 4(0) - 1 + 3 - 2 = 0$$
$$D_F(4 \rightarrow 2) = 4(0) - 1 + 1 - 0 = 0$$
$$D_F(5 \rightarrow 3) = 4(0) - 2 + 2 - 0 = 0$$
$$D_F(6 \rightarrow 4) = 4(1) - 2 + 0 - 2 = 0$$
$$D_F(7 \rightarrow 3) = 4(1) - 2 + 2 - 3 = 1$$
$$D_F(8 \rightarrow 4) = 4(1) - 2 + 0 - 1 = 1$$

For a folded system to be realizable, $D_F(U \xrightarrow{e} V) \geq 0$ must hold for all of the edges in the DFG. Once valid folding sets have been assigned, retiming can be used to either satisfy this property or determine that the folding sets are infeasible.

In general, the original DFG and the N-unfolded version of the folded DFG are retimed and/or pipelined versions of each other. Furthermore, an arbitrary DFG can be unfolded by factor N and the unfolded DFG can be folded with many possible folding sets to generate a family of architectures. In order to obtain the original DFG from the unfolded DFG via folding transformation, an appropriate folding set has to be chosen.

5 Applications to Software Design

In this section, the precedence constraints in DFG will first be introduced followed by the definition of critical path and the iteration bound. A DFG based scheduling algorithm is then explained in detail.

5.1 Intra-iteration and Inter-iteration Precedence Constraints

The DFG captures the data-driven property of DSP algorithms where any node can fire (perform its computation) whenever all the input data are available. This implies that a node with no input edges can fire at any time. Thus many nodes can be fired simultaneously, leading to concurrency. Conversely, a node with multiple input edges can only fire after all its precedent nodes have fired. The latter case imposes the precedence constraints between two nodes described by each edge. This precedence constraint is an *intra-iteration* precedence constraint if the edge has no delay elements or an *inter-iteration* precedence constraint if the edge has one or more delays. Together, the intra-iteration and inter-iteration precedence constraints specify the order in which the nodes in the DFG can be executed.

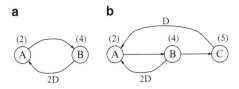

Fig. 18 (**a**) A DFG with one loop that has a loop bound of $6/2 = 3$ u.t. The iteration bound for this DFG is 3 u.t. (**b**) A DFG with iteration bound $T_\infty = max\{6/2, 11/1\} = 11$ u.t.

For example, the edge from node A to node B in Fig. 2b enforces the inter-iteration precedence constraint, which states that the execution of the k-th iteration of A must be completed before the $(k+1)$-th iteration of B. On the other hand, the edge from B to A enforces the intra-iteration precedence constraint, which states that the k-th iteration of B must be executed before the k-th iteration of A.

5.2 Definition of Critical Path and Iteration Bound

The critical path of a DFG is defined to be the path with the longest computation time among all paths that contain no delay elements. The critical path in the DFG in Fig. 18a is the path $A \rightarrow B$, which requires 6 u.t. Since the critical path is the longest path for combinational rippling in the DFG, the computation time of the critical path is the minimum computation time for one iteration of the DFG, which is the execution of each node in the DFG exactly once.

A loop is a directed path that begins and ends at the same node and the amount of time required to execute a loop can be determined from the precedence relation described by the edges of the DFG. According to these precedence constraints, iteration k of the loop consists of the sequential execution of A_k and B_k. Given that the execution times of nodes A and B are 2 and 4 u.t., respectively, one iteration of the loop requires 6 u.t. This is the loop bound, which represents the lower bound on the loop computation time. Formally, the loop bound of the l-th loop is defined as $\frac{t_l}{w_l}$, where t_l is the loop computation time and w_l is the number of the delays in the loop. As a result, the loop bound for the loop in Fig. 18a is $6/2 = 3$ u.t.

As another example, the DFG in Fig. 18b contains two loops, namely, the loops $l_1 = A \rightarrow B \rightarrow A$ and $l_2 = A \rightarrow B \rightarrow C \rightarrow A$. Therefore, the loop bounds for l_1 and l_2 are $6/2 = 3$ u.t. and $11/1 = 11$ u.t., respectively. The loop with the maximum loop bound is called the critical loop and its corresponding loop bound is the iteration bound of the DSP program, which is the lower bound on the iteration or sample period of the DSP program regardless of the amount of the computing resources available. Formally, the iteration bound is defined as

$$T_\infty = max_{l \in L}\{\frac{t_l}{w_l}\} \tag{10}$$

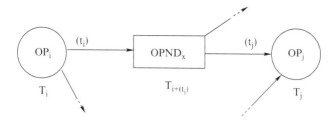

Fig. 19 A pair of operators showing the timing information

where L is the set of the loops in the DFG, t_l and w_l are the computation time and the number of delays of the loop l, respectively.

To compute the iteration bound of a DFG by locating all the loops and directly compute T_∞ by using (10) is rather straightforward. However, the number of loops in a DFG grows exponentially with the number of nodes, and therefore polynomial-time algorithms are desired for computing the iteration bound. Examples of polynomial-time algorithms include longest path matrix algorithm [7] and the minimum cycle mean algorithm [8].

5.3 Scheduling

DFGs are generally used for high-level synthesis to derive concurrent implementations of DSP applications onto parallel hardware, where a major concern is subtask scheduling which determines when and in which hardware units nodes can be executed.

Scheduling algorithm consists of assigning a scheduling time to every operator in an architecture, where the time represents when the operation will normally take place. Each operator must be scheduled at a time after all of its inputs become available. Consequently, the scheduling problem can be formulated as a linear programming problem. Furthermore, because all scheduled time of the operators must be integers, the scheduling algorithm must find the optimal integer solution to this problem.

5.3.1 The Scheduling Algorithm

Consider a pair of operators joined by an edge shown in Fig. 19. One of the outputs produced by the operator OP_i is the operand $OPND_x$, which in turn is the input to the operator OP_j. The scheduled times of operators OP_i and OP_j are denoted by T_i and T_j, respectively. In addition, the timing specifications of the relevant output and input ports of OP_i and OP_j are denoted by t_i and t_j, respectively. Since OP_i is scheduled at time T_i, the output $OPND_x$ will become available at time $T_i + t_i$.

Further, the same operand will be required as an input to the operator OP_j at time $T_i + t_j$. By the requirement that the operand can not be used by operator OP_j before it is produced by operator OP_i, the following inequality can be derived:

$$T_j + t_j \geq T_i + t_i \qquad (11)$$

Such an inequality holds for each pair of operands joined by an edge in the DFG. In these inequalities, T_i and T_j are the unknowns, where the value $t_i - t_j$ is a known constant. A solution to the set of inequalities can be determined by using common techniques for solving linear programming problems. Once a solution is found to this set of inequalities, the circuit may be correctly synchronized by inserting a delay equal to $T_j - T_i - (t_i - t_j)$ clock cycles between the operators OP_i and OP_j.

In general, many solutions exist to satisfy the set of inequalities for a given architecture and hence there are many ways to synchronize the circuit. The goal of optimal scheduling is to generate a solution that provides the minimal cost, where cost is defined to be the total number of shift-register delays required to properly synchronize the circuit. If a linear cost function can be defined, the minimum cost problem can be easily formulated as a linear programming problem.

5.3.2   Minimum Cost Solution

Minimizing the total number of synchronization delays required for reach edge between functional units of a circuit is not sufficient. Note that there exists the possibility of multiple fanout from any functional unit. Therefore, the delays from a multiple fanout output should be allocated sequentially instead of in parallel. Consider a simple case where an output of some operator OP_o is used as an input to three other operators, OP_A, OP_B and OP_C as shown in Fig. 20a with delays of 10, 12 and 25 clock cycles. respectively. The total number of delays is equal to $10 + 12 + 25 = 47$. An alternative sequential arrangement of the delays is shown in Fig. 20b, where the total number of delays is 25, which is the length of the longest delay. Therefore, the total number of delays that need to be allocated to any node in the circuit is equal to the maximum delay that must be allocated.

Let D_x represents the maximum delay that must be allocated to an operand $OPND_x$ of width w_x. A total cost function to be minimized can now be defined:

$$Cost = \sum_x D_x w_x \qquad (12)$$

where the sum is over each operand node in the circuit. For each node as shown in Fig. 20b, there exists a constraint equivalent to (11),

$$T_j - T_i \geq t_i - t_j \qquad (13)$$

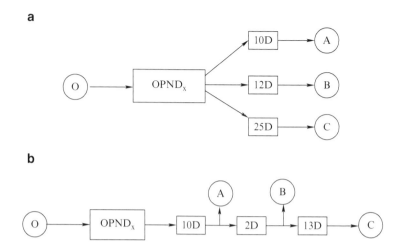

Fig. 20 (**a**) Operator O with a fanout of three and no delay sharing. (**b**) Operator O with a fanout of three and delay sharing

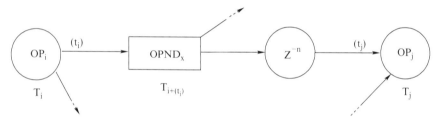

Fig. 21 General code to be scheduled including z^{-1} *operators*

In addition, the maximum delay on operand OPNDx from OP_i to OP_j is less than the maximum delay D_x, which is described by the constraint:

$$T_j - T_i - (t_i - t_j) \leq D_x \qquad (14)$$

These two constraints along with the cost function that will be minimized as in (12), describe a linear programming problem capable of providing the minimum cost scheduling.

5.3.3 Scheduling of Edges with Delays

The scheduling algorithm will generate optimal solutions for DFGs consisting of delay-free edges. To handle edges that contain delays, a preferable method is to incorporate the word delays right into the linear programming solutions, which is achieved by slightly modifying equations to take into account the presence of z^{-1} operators. Specifically, Fig. 21 shows a general situation in which the output of some operator OP_i undergoes a z^{-n} transformation before being used as an input to the operator OP_j. The modified equations describing these scheduling constraints thus become:

$$T_j - T_i \geq t_i - t_j - nW \tag{15}$$

and

$$T_j - T_i - (t_i - t_j) + nW \leq D_x \tag{16}$$

where W is the number of clock cycles in a word or wordlength. In this case, $T_j - T_i - (t_i - t_j) + nW$ is the delay applied to the connection shown in the diagram, and D_x is the maximum delay applied to this variable.

6 Conclusions

This chapter has introduced the signal flow graphs and data flow graphs. Several transformations such as pipelining, retiming, unfolding and folding have been reviewed, and applications of these transformations on signal flow graphs and data flow graphs have been demonstrated for both hardware and software implementations. It is important to note that any DFG that can be pipelined can be operated in a parallel manner, and *vice versa*. Any transformation that can improve performance in a hardware system can also improve the performance of a software system. Thus, preprocessing of the DFGs and SFGs by these high-level transformations can play a major role in the system performance.

Acknowledgements Many parts of the text and figures in this chapter are taken from the text book in [2]. These have been reprinted with permission of John Wiley & Sons, Inc. The authors are grateful to John Wiley & Sons, Inc., for permitting the authors to use these figures and parts of the text from [2]. They are also grateful to George Telecki, associate publisher at Wiley for his help in this regard.

References

1. Woods, R.: Mapping decidable signal processing graphs into FPGA implementations. In: S.S. Bhattacharyya, E.F. Deprettere, R. Leupers, J. Takala (eds.) Handbook of Signal Processing Systems, second edn. Springer (2013)
2. Parhi, K.: VLSI Digital Signal Processing Systems, Design and Implementation. John Wiley & Sons, New York (1999)
3. Crochiere, R., Oppenheim, A.: Analysis of linear digital networks. Proc. IEEE **64**(4), 581–595 (1975)
4. Bolton, W.: Newnes Control Engineering Pocketbook. Newnes, Oxford, UK (1998)
5. Lee, E., Messerschmitt, D.: Synchronous data flow. Proc. IEEE, special issue on hardware and software for digital signal processing **75**(9), 1235–1245 (1987)

6. Leiserson, C., Rose, F., Saxe, J.: Optimizing synchronous circuitry by retiming. In: Third Caltech Conference on VLSI, pp. 87–116 (1983)
7. Gerez, S., Heemstra de Groot, S., Herrmann, O.: A polynomial-time algorithm for the computation of the iteration-period bound in recursive data flow graphs. IEEE Trans. on Circuits and Systems-I: Fundamental Theory and Applications **39**(1), 49–52 (1992)
8. Ito, K., Parhi, K.: Determining the minimum iteration period of an algorithm. Journal of VLSI Signal Processing **11**(3), 229–244 (1995)

Optimization of Number Representations

Wonyong Sung

Abstract In this section, automatic scaling and word-length optimization procedures for efficient implementation of signal processing systems are explained. For this purpose, a fixed-point data format that contains both integer and fractional parts is introduced, and used for systematic and incremental conversion of floating-point algorithms into fixed-point or integer versions. A simulation based range estimation method is explained, and applied to automatic scaling of C language based digital signal processing programs. A fixed-point optimization method is also discussed, and optimization examples including a recursive filter and an adaptive filter are shown.

1 Introduction

Many embedded processors do not equip floating-point units, thus it is needed to develop integer versions of code for real-time execution of signal processing applications. Integer programs run much faster than the floating-point versions, but integer versions can suffer from overflows and quantization effects. Converting a floating-point program to an integer version requires scaling of data, which is known to be difficult and time-consuming. VLSI implementation of digital signal processing algorithms demands fixed-point arithmetic for reducing the chip area, circuit delay, and power consumption. With fixed-point arithmetic, it is possible to use the fewest number of bits possible for each signal and save the chip area. However, if the number of bits is too small, quantization noise will degrade the

W. Sung (✉)
Department of Electrical and Computer Engineering, Seoul National University,
599 Gwanangno, Gwanak-gu, Seoul, Republic of Korea
e-mail: wysung@snu.ac.kr

S.S. Bhattacharyya et al. (eds.), *Handbook of Signal Processing Systems*,
DOI 10.1007/978-1-4614-6859-2_40, © Springer Science+Business Media, LLC 2013

system performance to an unacceptable level. Thus, fixed-point optimization that minimizes the hardware cost while meeting the fixed-point performance is very needed.

In Sect. 2, the data format for representing a fixed-point data is presented. This format contains both integer and fractional parts for representing a data. Thus, this format is very convenient for data conversion between floating-point and fixed-point data types. Section 3 contains the range estimation methods that are necessary for integer word-length determination and scaling. A simulation based range estimation method is explained, which can be applied to not only linear but also non-linear and time-varying systems. In Sect. 4, a floating-point to integer C program conversion process is shown. This code conversion process is especially useful for C program language based implementation of signal processing systems. Section 5 presents the word-length optimization flow for signal processing programs, which should be important for VLSI or 16-bit programmable digital signal processor (DSP) based implementations. In Sect. 6, the summary and related works are described.

2 Fixed-Point Data Type and Arithmetic Rules

A fixed-point data type does not contain an exponent term, which makes the hardware for fixed-point arithmetic much simpler than that for floating-point arithmetic. However, fixed-point data representation only allows a limited dynamic range, hence scaling is needed when converting a floating-point algorithm into a fixed-point version. In this section, fixed-point data formats, fixed-point arithmetic rules, and a simple floating-point to fixed-point conversion example will be shown. The two's complement format is used when representing negative numbers.

2.1 Fixed-Point Data Type

A widely used fixed-point data format is the integer format. In this format, the least significant bit (LSB) has the weight of 1, thus the maximum quantization error can be as large as 0.5 even if the rounding scheme is used. As a result, small numbers cannot be faithfully represented with this format. Of course, there can be overflows even with the integer format because an N-bit signed integer has a value that is between -2^{N-1} and $2^{N-1} - 1$. Another widely used format is the fractional format, in which the magnitude of a data cannot exceed 1. This format seems convenient for representing a signal whose magnitude is bounded by 1, but it suffers from overflow or saturation problems when the magnitude exceeds the bound. Figure 1 shows two different interpretations for a binary data "01001000."

With either the integer or the fractional format, an expert can design an optimized digital signal processing system by incorporating proper scaling operations, which is, however, very complex and difficult to manage. This conversion flow is not easy

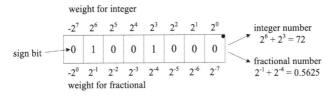

Fig. 1 Integer and fractional data formats

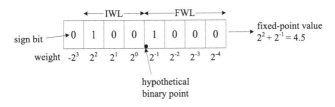

Fig. 2 Generalized fixed-point data format

because all of the intermediate variables or constants should be represented with only integers or fractions whose represented values are usually much different from those of the corresponding floating-point data. The difference of the representation format also hinders incremental conversion from a floating-point design to a fixed-point one. For seamless floating-point to integer or fixed-point conversion, the semantic gap between the floating-point and fixed-point data formats needs to be eliminated.

To solve these problems, a generalized fixed-point data-type that contains both integer and fractional parts can be used [17]. This fixed-point format contains the attributes specified as follows.

$$< wordlength, integer\_wordlength, sign\_overflow\_quantization\_mode > \qquad (1)$$

The word-length (WL) is the total number of bits for representing a fixed-point data. The integer word-length (IWL) is the number of bits to the left of the (hypothetical) binary-point. The fractional word-length (FWL) is the number of bits to the right of the (hypothetical) binary point. The sign can be either unsigned("u") or two's complement("t"). Thus, the word-length (WL) corresponds to "IWL+FWL+1" for signed data, and is "IWL+FWL" for unsigned data. If the fractional word-length is 0, the data with this format can be represented with integers. At the same way, it becomes the fractional format when the IWL is 0. Note that the IWL or FWL can be even larger than the WL; in this case, the other part has a minus word-length. Figure 2 shows an interpretation of an 8-bit binary data employing the fixed-point format with the IWL of 3.

The overflow and quantization modes are needed for arithmetic or quantization operations. The overflow mode specifies whether no treatment ("o") or saturation ("s") scheme is used when overflow occurs, and the quantization mode denotes whether rounding ("r") or truncation ("t") is employed when least significant bits are quantized.

Most of all, this fixed-point data representation is very convenient for translating a floating-point algorithm into a fixed-point version because the data values is not limited to integers or fractions. The range (R) and the quantization step (Q) are dependent on the IWL and FWL, respectively: $-2^{IWL} \le R < 2^{IWL}$ and $Q = 2^{-FWL} = 2^{-(WI-1-IWL)}$ for the signed format. Assigning a large IWL to a variable can prevent overflows, but it increases the quantization noise. Thus, the optimum IWL for a variable should be determined according to its range or the possible maximum absolute value. The minimum IWL for a variable x, $IWL_{min}(x)$, can be determined according to its range, $R(x)$, as follows.

$$IWL_{min}(x) = \lceil \log_2 R(x) \rceil, \tag{2}$$

where $\lceil x \rceil$ denotes the smallest integer which is equal to or greater than x. Note that preventing overflow and saturation is very critical in fixed-point arithmetic because the magnitude of the error caused by them is usually much larger than that produced by quantization.

2.2   Fixed-Point Arithmetic Rules

Since the generalized fixed-point data format allows a different integer word-length, two variables or constants that do not have the same integer word-length cannot be added or subtracted directly. Let us assume that $x1$ is "01001000" with the IWL of 3 and $x2$ is "00010000" with the IWL of 2. Since the interpreted value of $x1$ is 4.5 and that of $x2$ is 0.5, the result should be a number that corresponds to 5.0 or a close one. However, direct addition of 01001000 ($x1$) and 00010000 ($x2$) does not yield the expected result. This is because the two data have different integer word-lengths. The two fixed-point data should be added after aligning their hypothetical binary points. The binary point can be moved, or the integer word-length can be changed, by using arithmetic shift operations. Arithmetic right shift by one bit increases the integer word-length by one, while arithmetic left shift decreases the integer word-length. The number of shifts required for addition or subtraction can easily be obtained by comparing the integer word-lengths of the two input data format. In the above example, $x2$, with the IWL of 2, should be shifted right by 1 bit before performing the integer addition to align the binary-point locations. As illustrated in Fig. 3, this results in a correct value of 5.0 when the output is interpreted with the IWL of 3. Note that the result of addition or subtraction sometimes needs an increased IWL. If the IWL of the added result is greater than those of two input operands, the inputs should be scaled down to prevent overflows. Subtraction can be treated the same way as addition. The scaling rules for addition and subtraction

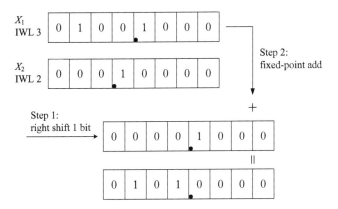

Fig. 3 Fixed-point addition with different IWL's

Table 1 Fixed-point arithmetic rules

	Floating-point	Fixed-point $I_x > I_y, I_z$	$I_y > I_x, I_z$	$I_z > I_x, I_y$	Result IWL
Assignment	$x = y$	$x = y \gg$ $(I_x - I_y)$	$x = y \ll$ $(I_y - I_x)$	–	I_x
Addition/ Subtraction	$z = x + y$	$z = (x + (y \gg$ $(I_x - I_y)))$ $\ll (I_x - I_z)$	$z = ((x \gg (I_y - I_x)) + y) \ll$ $(I_y - I_z)$	$z = (x \gg$ $(I_z - I_x)) +$ $(y \gg$ $(I_z - I_y))$	$max(I_x, I_y, I_z)$
Multiplication	$x * y$	$mulh(x, y)$			$I_x + I_y + 1$ or $I_x + I_y$

z: a variable storing the result

are shown in Table 1, where Ix and Iy are the IWL's of two input operands x and y, respectively, and Iz is that of the result, z.

In fixed-point multiplication, the word-length of the product is equal to the sum of two input word-lengths. In two's complement multiplication, two identical sign bits are generated except for the case that both input data correspond to the negative minimum, "$100\cdots0$." Ignoring this case, the IWL of the two's complement multiplied result becomes $Ix + Iy + 1$. Figure 4 shows the multiplied result of two 8-bit fixed-point numbers. By assuming the IWL of 5, we can obtain the interpreted value of 2.25.

2.3 Fixed-Point Conversion Examples

To illustrate the fixed-point conversion process, a floating-point version of the recursive filter shown in Fig. 5a is transformed to a fixed-point hardware system. Assume that the input signal has the range of 1, which implies that it is between -1

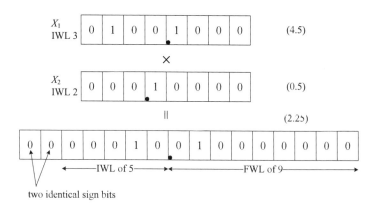

Fig. 4 Fixed-point multiplication

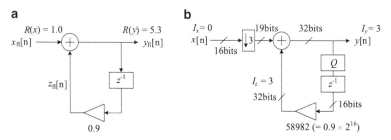

Fig. 5 Floating-point to fixed-point conversion of a recursive filter. (**a**) Floating-point filter, (**b**) Fixed-point filter

and 1. The output signal is also known to be between -5.3 and 5.3. The output signal range is obtained from floating-point simulation results. The coefficient is 0.9, and is unsigned. Hence, the range of the multiplied signal, $z[n]$, will be $4.77\ (= 0.9 * 5.3)$. From the given range information, we can assign the IWL's of 0 for $x[n]$, 3 for $y[n]$, 3 for $z[n]$ and 0 for the coefficient. The coefficient a is coded as "58982," which corresponds to the unsigned number 0.9×2^{16}. Since the multiplication of $y[n]$ and a is conducted between signed and unsigned numbers, the IWL of $z[n]$ is 3, which is $I_y + I_a$. If the coefficient a is coded with the two's complement format, the IWL of $z[n]$ would be 4 due to the extra sign generated in the multiplication process. Since the precision of the hardware multiplier is 16-bit, only the upper 16 bits, including the sign bit, of $y[n]$ is used for the multiplication. The quantizer (Q) in this figure takes the upper 16 bits among 32 bits of $y[n]$. Since the difference between I_x and I_z is 3, $x[n]$ is scaled down or arithmetic shift-righted by 3 bits, as the hardware in Fig. 5b shows.

There are a few different fixed-point implementations. One example is a fixed-point implementation without needing shift operations. Note that no shift operation is needed when adding or subtracting two fixed-point data with the same integer word-length. In this case, the IWL of 3 is assigned to the input $x[n]$, even though

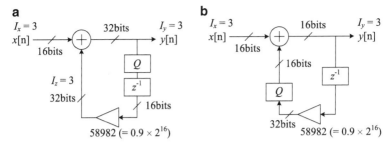

Fig. 6 Fixed-point filters with reduced complexity. (**a**) Fixed-point filter without shift, (**b**) Fixed-point filter with a 16 bit adder

the range of $x[n]$ is 1.0. This means that the input $x[n]$ is de-normalized and the upper 3 bits to the sign bit are unused. Since the IWL's of $x[n]$ and $z[n]$ are the same, there is no need of inserting a shifter. Figure 6a shows the resulting hardware. The SQNR (Signal to Quantization Noise Ratio) of the input is obviously lowered by employing the de-normalized scheme. Another fixed-point implementation in Fig. 6b shows the hardware using a 16-bit adder. In this case, the quantizer (Q) is moved to the output of the multiplier. Note that the SQNR of this scheme is even lower than that of Fig. 6a.

In the above example, the range of 1 is assumed to the input $x[n]$, which is from the floating-point design. However, assuming the range of 2 as for the input $x[n]$ does not change the resultant hardware because the output range should be doubled in this case.

3 Range Estimation for Integer Word-Length Determination

The floating-point to fixed-point conversion examples in the previous section shows that estimating the ranges of all of variables is most crucial for this conversion process. There are two different approaches for range estimation. One is to calculate the L1-norm of the system and the other is using the simulation results of floating-point systems [12, 17].

3.1 L1-Norm Based Range Estimation

The L1-norm of a linear shift-invariant system is the maximum value of the output when the absolute value of the input is bounded to 1. If the unit-pulse response of a system is $h[n]$, where $n = 0, 1, 2, 3, \cdots \infty$, the L1-norm of this system is defined as:

$$L1\text{-}norm(h[n]) = \sum_{n=0}^{\infty} |h[n]| \tag{3}$$

Obviously, the L1-norm can easily be estimated for an FIR system. There are also several analytical methods that compute the L1-norm of an IIR (infinite impulse response) system [12]. Since the unit-pulse response of an IIR system usually converges to zero when thousands of time-steps elapse, it is practically possible to estimate the L1-norm of an IIR system with a simple C code or a Matlab program that sums up the absolute value of the unit-pulse response, instead of conducting a contour integration [11, 12]. Since L1-norm cannot easily be defined to time-varying or non-linear systems, the L1-norm based range estimation method is hardly applicable to systems containing non-linear and time-varying blocks. Another characteristic of the L1-norm is that it is a very conservative estimate, which means that the range obtained with the L1-norm is the largest one for any set of the given input, and hence the result can be an over-estimate. For example, the L1-norm of the first order recursive system shown in Fig. 5a is 10, which corresponds to the case that the input is a DC signal with the maximum value of 1. For example, if we design a speech processing system, the input with this characteristic is not likely to exist. With an over-estimated range, the data should be shift-down by more bits, which will increase the quantization noise level. For a large scale system, the L1-norm based scaling can be impractical because accumulation of extra-bits at each stage may seriously lower the accuracy of the output. However, if a very reliable system that should not experience any overflow is needed, the L1-norm based scaling can be considered. The L1-norm based scaling is limited in use for real applications because most practical systems contain time-varying or non-linear blocks.

3.2 Simulation Based Range Estimation

The simulation based method estimates the ranges by simulation of floating-point design while applying realistic input signal-samples [17]. This method is especially useful when there is a floating-point simulation model, which can be a C program or a CAD system based design. This method can be applied to general, including non-linear and time-varying, systems. Thus, provided that there is a floating-point version of a designed system and various input files for simulation, a CAD tool can convert a floating-point design to a fixed-point version automatically. One drawback of this method is that it needs extensive simulations with different environmental parameters and various input signal files. The scaling with this approach is not conservative, thus there can be overflows if the statistics of the real input signal differ much from the ones used for the range estimation. Therefore, it is needed to employ various input files for simulation or give some additional integer bits, called the guard-bits, to secure overflow-free design. This simulation based method can also be applied to word-length optimization.

For unimodal and symmetric distributions, the range of a variable can be effectively estimated by using the mean and the standard deviation, which are obtained from the floating-point simulation results, as follows.

$$R = |\mu| + n \times \sigma, \quad n \propto k \tag{4}$$

Specifically, we can use n as $k + 4$, where k is the kurtosis [17]. However, the above rule is not satisfactory for other distributions, which may be multimodal or non symmetric. As an alternate rule, we can consider

$$R = \hat{R}_{99.9\%} + g \tag{5}$$

where g is a guard for the range. A partial maximum, $\hat{R}_{P\%}$, indicates a sub-maximum value, which covers $P\%$ of the entire samples. Note that various sub-maxima are collected during the simulation. The more different $\hat{R}_{100\%}$ and $\hat{R}_{99.9\%}$ are, the larger guard value is needed.

3.3 C++ Class Based Range Estimation Utility

A range estimation utility for C language based digital signal processing programs is explained, which is freely available [3]. This range estimation utility is not only essential for automatic integer C code generation, but also useful for determining the number of shifts in assembly programming of fixed-point DSPs [16]. With this utility, users develop a C application program with floating-point arithmetic. The range estimator then finds the statistics of internal signals throughout floating-point simulation using real inputs, and determines the integer word-lengths of variables. Although we can develop a separate program that traces the range information during simulation, this approach may demand too much program modification. The developed range estimation class uses the operator overloading characteristics of C++ language, thus a programmer does not need to change the floating-point code significantly for range estimation.

To record the statistics during simulation, a new data class for tracing the possible maximum value of a signal, i.e., the range, has been developed and named as fSig. In order to prepare a range estimation model of a C or C++ digital signal processing program, it is only necessary to change the type of variables, from float to fSig. The fSig class not only computes the current value, but also keeps the records of a variable using private members for it. When the simulation is completed, the ranges of the variables declared as fSig class are readily available from the records stored in the class.

The fSig class has several private members including Data, Sum, Sum2, Sum3, Sum4, AMax, and SumC. Data keeps the current value, while Sum and Sum2 record the summation and the square summation of past values, respec-

a

```
void
iir1(short argc, char *argv[])
{
    float Xin;
    float Yout;      // fSig()
    float Ydly;      // fSig()
    float Coeff;

    Coeff = 0.9;
    Ydly = 0.;
    for( i=0  ;i<    1000; i++){
        infile >> Xin ;
        Yout = Coeff * Ydly + Xin;
        Ydly = Yout ;
        outfile << Yout << '\n';
    }
}
```

b

```
void
iir1(short argc, char *argv[])
{
    float Xin;
    static fSig   Yout("iir1/Yout");
    static fSig   Ydly("iir1/Ydly");
    float Coeff;

    Coeff = 0.9;
    Ydly = 0.;
    for( i=0  ;i<    1000; i++){
        infile >> Xin ;
        Yout = Coeff * Ydly + Xin ;
        Ydly = Yout ;
        outfile << Yout << '\n';
    }
}
```

Fig. 7 C++ programs for a first order IIR filter. (**a**) The original C++ program, (**b**) A version for the range estimator

tively. Sum3 and Sum4 store the third and fourth moments, respectively. They are needed to calculate the statistics of a variable, such as mean, standard deviation, skewness, and kurtosis. AMax stores the absolute maximum value of a variable during the simulation. The class also keeps the number of modifications during the simulation in SumC field.

The fSig class overloads arithmetic and relational operators. Hence, basic arithmetic operations, such as addition, subtraction, multiplication, and division, are conducted automatically for fSig variables. This property is also applicable to relational operators, such as "==," "!=," ">," "<," ">=," and "<=." Therefore, any fSig instance can be compared with floating-point variables and constants. The contents, or private members, of a variable declared by the fSig class is updated when the variable is assigned by one of the assignment operators, such as "=," "+=," "−=," "*=," and "/=." The *range estimator* is executed after preparing the simulation model that only modifies the variable declaration. After the simulation is completed, the mean (μ), standard deviation (σ), skewness (s), and kurtosis (k) can be calculated using Sum, Sum2, Sum3, Sum4 and SumC information. Then, the statistical range of fSig variable x can be estimated. The integer word-lengths of all signals are then obtained from their ranges.

As an example, let us consider a first order digital IIR filter. The original C++ program for the filter and a translated version for range estimation are given in Fig. 7.

As shown in Fig. 7, it is only necessary for developing a range estimation program to modify the declaration part of the original floating-point version. In this example, a white noise sequence uniformly distributed between −1 and +1 is used for the input data, and four times of standard deviation is used for estimating the ranges. Note that the integer word-length for X_{in} is known, 0. The range estimation result is shown in Fig. 8.

```
Statistics:
   VarName     Mean   StdDev  Skewness  Kurtosis    R99.9%     R100%  Update
iir1/Ydly  -0.1133  +1.3076  +0.0220   -0.0258   +4.2638   +4.4214    3001
iir1/Yout  -0.1134  +1.3078  +0.0220   -0.0268   +4.2638   +4.4214    3000

Integer  word-lengths:
      VarName            Range      IWL
     iir1/Ydly       +5.309891       +3
     iir1/Yout       +5.309515       +3
```

Fig. 8 The result of the range estimator for the IIR filter

The elements of an array variable are assumed to have the same IWL for simple code generation. If it is not, the scaled integer codes need to check the array index, which can slow down program execution significantly. For a pointer variable, the IWL is defined as that of the pointed variables. For example, when the pointer variables p and q have the IWL of 2 and 3, respectively, the expression $*q = *p$ can be converted to $*q = *p \gg 1$. Since the IWL of a pointer variable is not changed at runtime, a pointer cannot support two variables having different IWL's. In this case, the IWL's of these pointers are equalized automatically at the integer C code generation step that will be described in the next section.

4   Floating-Point to Integer C Code Conversion

C language is most frequently used for developing digital signal processing programs. Although C language is very flexible for describing algorithms with complex control flows, it does not support fixed-point data formats. In this section, a floating-point to integer C program conversion procedure is explained [13, 24]. As shown in Fig. 9, the conversion flow utilizes the simulation based range estimation results for determining the number of shift operations for scaling. In addition, the number of shift operations is minimized by equalizing the IWL's of corresponding variables or constants for the purpose of reducing the execution time.

4.1   Fixed-Point Arithmetic Rules in C Programs

As summarized in Table 1, the addition or subtraction of two input data with different IWL's needs arithmetic shift before conducting the operation. Fixed-point multiplication in C language needs careful treatment because integer multiplication in ANSI C only stores the lower-half of the multiplied result, while fixed-point multiplication needs the upper-half part. Integer multiplication is intended to prevent any loss of accuracy in multiplication of small numbers, and hence it can generate an overflow when large input data are applied. However, for signal processing purpose, the upper part of the result is needed to prevent overflows and keep accuracy. Integer and fixed-point multiplication operations are compared in Fig. 10a,b [14, 15].

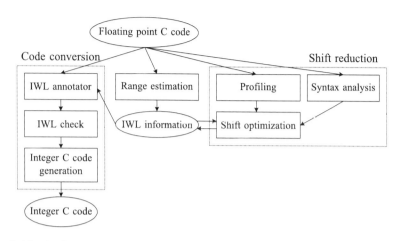

Fig. 9 Fixed-point addition with different IWL's

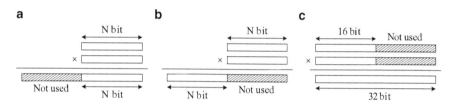

Fig. 10 Integer and fixed-point multiplications. (**a**) ANSI C integer multiplication. (**b**) Fixed-point multiplication. (**c**) MPYH instruction of TMS320C60

In traditional C compilers, a double precision multiplication operation followed by a double to single conversion is needed to obtain the upper part, which is obviously very inefficient [28]. However, in recent C compilers for some DSPs such as Texas Instruments' TMS320C55 ('C55), the upper part of the multiplied result can be obtained by combining multiply and shift operations [6]. In the case of TMS320C60 ('C60), which has 16 by 16-bit multipliers as well as 32-bit registers and ALU's, the multiplication of the upper 16-bit parts of two 32-bit operands is efficiently supported by C intrinsics as depicted in Fig. 10c [7]. If there is no support for obtaining the upper part of the multiplied result in the C compiler level, an assembly level implementation of fixed-point multiplication is useful. For the Motorola 56000 processor, fixed-point multiplication is implemented with a single instruction using inline assembly coding [5]. Note that, in Motorola 56000, the IWL of the multiplication result is $I_x + I_y$, because the output of the multiplier is one bit left shifted in hardware. The implementation of the macro or inline function for fixed-point multiplication, *mulh( )*, is dependent on the compiler of a target processor as illustrated in Table 2.

Table 2 Implementation of fixed-point multiplication

Target processor	Implementation
TMS320C50	`#define mulh(x,y)  ((x)*(y)>>16)`
TMS320C60	`#define mulh(x,y) _mpyh(x,y)`
Motorola 56000	`__inline int mulh(int x, int y) {`
	`    int z;`
	`    __asm("mpy %1,%2,%0":"=D"(z):"R"(x),"R"(y));`
	`    return z;`
	`}`

Fig. 11 An example of expression conversion

y=(a+b)*c; ⟹ tmp = a+b;
y=tmp*c;

4.2 Expression Conversion Using Shift Operations

The most frequently used expression in digital signal processing is the accumulation of product terms, which can be generally modeled as follows.

$$x_i = \sum_{j,k} x_j \times x_k + \sum_l x_l \tag{6}$$

Complex expressions in C programs are converted to several expressions having this form. Figure 11 shows one example.

Assuming that there is no shifter at the output of the adder, the IWL of the added result is determined by the maximum value of two input operands and the result, as shown in Table 1. From this, the IWL of the right hand side expression, I_{rhs}, is represented by the maximum IWL of the terms as shown in Eq. (7).

$$I_{rhs} = \max_{j,k,l}(I_{x_j} + I_{x_k} + 1, I_{x_l}, I_{x_i}), \tag{7}$$

where $I_x + I_y + 1$ represents the IWL of the multiplied results. The number of scaling shifts for the product, addition, or assignment, which is represented as, $s_{j,k}$, s_l or s_i, respectively, is determined as follows.

$$s_{j,k} = I_{rhs} - (I_{x_j} + I_{x_k} + 1) \tag{8}$$

$$s_l = I_{rhs} - I_{x_l} \tag{9}$$

$$s_i = I_{rhs} - I_{x_i} \tag{10}$$

Equation (6) is now converted to the scaled expression as follows.

$$x_i = \{\sum_{j,k}((x_j \times x_k) \gg s_{j,k}) + \sum_l (x_l) \gg s_l)\} \ll s_i \tag{11}$$

4.3   Integer Code Generation

The IWL information file generated in the range estimation step includes the scope of a variable that indicates whether it is global or local, the variable name and its IWL. In the automatic scaling integer program converter for C, this information is attached to the symbol table of the floating-point program. The conversion of types and expression trees is conducted in the integer C code generation stage. The symbol tables are modified to replace floating-point types with integer types. Not only the float type but also the float-based types such as pointers to float, float arrays, and float functions are converted to corresponding integer-based types. The expression tree conversion that inserts scaling shifts uses the fixed-point arithmetic rules shown in Table 1. It is performed from the bottom to the top of a parse tree, and the IWL information of each tree node is also propagated in the same way. In this step, the pointer operations that involve different IWL's are also checked.

4.3.1   Shift Optimization

In many programmable DSP's, an implementation that needs no or less scaling shift operations is not only faster but also requires a smaller code size. Since no shift operation is needed for addition or assignment of operands having the same IWL, the number of scaling shifts can be reduced by equalizing the IWL's of relevant variables. An example implementation with shift reduction is illustrated in Fig. 6a. Note that it is only allowed to increase the initial IWL's that are determined according to Eq. (2), thus the equalization can increase the quantization noise level. Scaling shift reduction requires global optimization because IWL modification of a variable in an expression can incur additional scaling shifts in other expressions. Shift optimization also depends on the architecture of a DSP. For example, if a DSP has a barrel shifter, the number of bits for one scaling shift, unless it is zero, does not affect the number of execution cycles. However, if it has no barrel shifter and should conduct the scaling by employing one-bit shift operations, the shift cost is also affected by the number of bits for one scaling operation. It is also needed for minimizing the execution time to reduce the number of scaling operations that are inside of a long loop. Thus, this optimization requires program-profiling results.

The IWL modification that minimizes the overhead for scaling is conducted as follows. First, the number of shifts for each expression is formulated with the IWL's of the relevant variables and constants. Second, the cost function that corresponds to the total overhead of scaling shifts is made based on the results of the first step, the target DSP architecture, and the program-profiling information. Finally, the cost function is minimized by modifying the IWL's using the integer linear programming or the simulated annealing algorithms. Note that shift reduction using a DSP architecture without a barrel shifter can be modeled as an integer linear programming problem. The simulated annealing algorithm is a general optimization method, but the optimization can take much time. Detailed methods for shift reduction can be found in [13].

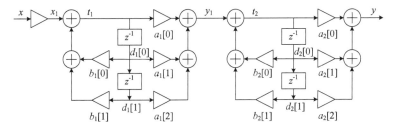

Fig. 12 A fourth order IIR filter

4.4 Implementation Examples

A fourth order IIR filter is implemented for TI's 'C50, 'C60 and Motorola 56000 digital signal processors using the developed scaling method. Floating-point C code for a fourth order IIR filter shown in Fig. 12 is given in Fig. 13a.

After the range estimation, the original IWL's are determined as shown in Table 3. With these initially determined IWL's, the integer C code shown in Fig. 13b is generated using the IWL model of $I_x + I_y + 1$ for the multiplied results. At first, the coefficients a1, a2, b1, and b2 are all converted with the two's complement format having the IWL of 1. Although some coefficients, such as 0.35 or -0.75504, can be encoded with the IWL of 0, the same IWL is given to all the coefficients to reduce the number of shifts for scaling. As a result, the coefficient value of 1 is converted to 2^{30} ($= 1073741824$), the value of 0.355407 is translated to 0.355407×2^{30} ($= 381615360$), and so on. The expression x1 = 0.01 * *x is converted as follows. The constant "0.01" is changed to a two's complement integer "1374389534" ($= 0.01 \times 2^{37}$) because the decimal number 0.01 is represented in binary as "0.0 0000 0101 0001 1110 1011 $\cdots$," where there are six zeroes below the binary point. Thus, the IWL of -6 is given to this constant, and translates 0.01 to $0.01 \times 2^{(31+6)}$. The input x is read from the file and has the IWL of 17 with the two's complement format. The multiplied result of a and x has the IWL of 12 ($= -6 + 17 + 1$), but x1 has the IWL of 10. Thus, there needs two bit left shift before assigning the multiplied result to x1 and the expression becomes x1=sll(mulh(1374389534, *x), 2). Next, since the IWL of t1 or d1 is 12, the IWL of b1[0]*d1[0] and b1[1]*d1[1] is 14 ($= 1 + 12 + 1$). At the same way, since the IWL of x1 is 10, and that of mulh(*b1, *d1) or mulh(b1[1], d1[1]) is 14, x1 needs to be right shifted by 4 bits for addition. The added result is again assigned to t1, which has the IWL of 12, thus it needs two bit left shift before assigning to t1, and forms the equation of t1 = sll((x1≫ 4)+ mulh(*b1, *d1) + mulh(b1[1], d1[1]), 2). Because the IWL of t1 and d1 is 12, the multiplication with the coefficients a1 produces the result having the IWL of 14 ($= 12 + 1 + 1$), and the multiplied result can be assigned to y1 without any shift. The rest of the code can be interpreted at the same way.

a

```
float a1[3] = { 1, 0.355407, 1.0 };
float a2[3] = { 1, -1.091855, 1.0 };
float b1[2] = { 1.66664, -0.75504 };
float b2[2] = { 1.58007, -0.92288 };
float d1[2], d2[2];
void iir4(float *x, float *y)
{
  float x1, y1, t1, t2;
  x1 = 0.01* *x;
  t1 = x1 + b1[0]*d1[0] + b1[1]*d1[1];
  y1 = a1[0]*t1 + a1[1]*d1[0] + a1[2]*d1[1];
  d1[1] = d1[0];
  d1[0] = t1;
  t2 = y1 + b2[0]*d2[0] + b2[1]*d2[1];
  *y = a2[0]*t2 + a2[1]*d2[0] + a2[2]*d2[1];
  d2[1] = d2[0];
  d2[0] = t2;
}
```

b

```
#define sll(x,y) ((x)<<(y))
int a1[3] = { 1073741824, 381615360, 1073741824   };
int a2[3] = { 1073741824, -1172370379, 1073741824   };
int b1[2] = { 1789541073, -810718027   };
int b2[2] = { 1696587243, -990934855   };
int d1[2];
int d2[2];
extern void iir4(int *x, int *y)
{
  int x1;
  int y1;
  int t1;
  int t2;
  x1 = sll(mulh(1374389534, *x), 2);
  t1 = sll((x1 >> 4) + mulh(*b1, *d1) + mulh(b1[1], d1[1]), 2);
  y1 = mulh(*a1, t1) + mulh(a1[1], *d1) + mulh(a1[2], d1[1]);
  d1[1] = *d1;
  *d1 = t1;
  t2 = sll((y1 >> 4) + mulh(*b2, *d2) + mulh(b2[1], d2[1]), 2);
  *y = sll(mulh(*a2, t2) + mulh(a2[1], *d2)
       + mulh(a2[2], d2[1]), 3);
  d2[1] = *d2;
  *d2 = t2;
}
```

Fig. 13 The C codes for the fourth order IIR filter. (**a**) The floating-point C code, (**b**) The integer C code before shift reduction, (**c**) The integer C code after shift reduction

c

```
#define sll(x,y) ((x)<<(y))
int a1[3] = { 134217728, 47701920, 134217728 };
int a2[3] = { 1073741824, -1172370379, 1073741824 };
int b1[2] = { 1789541073, -810718027 };
int b2[2] = { 1696587243, -990934855 };
int d1[2];
int d2[2];
extern void iir4(int *x, int *y)
{
  int x1;
  int y1;
  int t1;
  int t2;
  x1 = mulh(1374389534, *x);
  t1 = sll(x1 + mulh(*b1, *d1) + mulh(b1[1], d1[1]), 2);
  y1 = mulh(*a1, t1) + mulh(a1[1], *d1) + mulh(a1[2], d1[1]);
  d1[1] = *d1;
  *d1 = t1;
  t2 = sll(y1 + mulh(*b2, *d2) + mulh(b2[1], d2[1]), 2);
  *y = mulh(*a2, t2) + mulh(a2[1], *d2) + mulh(a2[2], d2[1]);
  d2[1] = *d2;
  *d2 = t2;
}
```

Fig. 13 (continued)

Table 3 IWL determined by the range estimation for the fourth order filter

Variable	Original IWL	Optimized IWL	IWL increment
x	17	20	3
y	15	18	3
x1	10	15	5
y1	14	18	4
t1, d1	12	13	1
t2, d2	16	16	0
b1, b2, a2	1	1	0
a1	1	4	3

Shift reduction is performed with a DSP architecture that employs a barrel shifter, and the integer C code shown in Fig. 13c is generated by controlling the IWL's. When considering the expression of $x1 = sll(mulh(1374389545, *x), 2)$, we can eliminate the shift left operation by increasing the IWL of $x1$ by 2 bits. Note that the IWL reduction process requires simultaneous IWL modification of several variables or constants. Table 3 gives the optimized IWL's and Fig. 13c shows the optimized integer C code. Although there can be a loss of bit resolution, we can find that several redundant shift operations are successfully eliminated. In the optimized code, left shift operations are eliminated when calculating $x1$ and y. We can also find that 4 bit right shift operations are removed when using $x1$ and $y1$.

Table 4 Performance comparison for the fourth order IIR filter

	# of cycles			SQNR	
	Floating-p.	Integer	Speed-up	Floating-p.	Integer (dB)
'C50	2,980	100	29.8	–	49.3
'C60	3,659	9	406.6	–	57.9
56000	26,282	921	28.5	–	78.5

Table 5 Shift reduction results of the fourth order IIR filter

IWL increment upper bound		0 (No shift reduction)	3	Infinite
# of shifts in C codes		7	4	2
'C50	# of cycles	100	96	94
	Speedup	–	4%	6%
	SQNR (dB)	49.3	51.2	54.1
C60	# of cycles	9	6	8
	Speedup	–	33%	11%
	SQNR (dB)	57.9	57.1	54.2
# of shifts in C codes		5	3	2
56000	# of cycles	921	675	577
	Speedup	–	27%	37%
	SQNR (dB)	78.5	78.5	78.5

In the fourth order IIR filter, the speed-up, which is the ratio in the execution time of the integer to the floating-point versions, was 29.8, 406, and 28.5 for 'C50, 'C60, and Motorola 56000, respectively, as shown in Table 4. The remarkable speed-up of 'C60 is mainly due to the deeply pipelined VLIW architecture having a large register file and an efficient C compiler. This machine can execute up to 8 integer operations in one cycle and store all the variables of a small loop kernel in the registers, but needs a large number of no-operation cycles for floating-point function calls to flush pipeline registers. The compiler for 'C60 is very efficient because it has several compiler friendly components, such as large general purpose register files, an orthogonal instruction set and a VLIW scheduler [7]. The developed shift reduction technique is applied to this example. The number of shift operations in the converted C code is reduced from 7 to 2 without imposing an IWL upper bound for TI's 'C50 and 'C60. The number of shifts in the C code for Motorola 56000 is different from that of TI's DSP's, because the IWL of multiplication results is different as described in the previous section. The cycle counts of the shift reduced codes are shown in Table 5. As shown in this Table, 'C60 achieves 33% of speed-up increase using the shift optimization. 'C50 shows a relatively low speed-up because the shifts can be performed by load-store instructions with no additional cycle in 'C50. For the Motorola 56000, a high speed-up can be achieved because its shift cost is much higher than that of the other DSP's employing barrel shifters. The SQNR of the fixed-point implementations was measured as 49.3 dB, 57.9 dB and

78.5 dB for 'C50, 'C60 and Motorola 56000, respectively. Note that 'C50 uses a 16-bit word-length for internal memory, while 'C60 supports a native 32-bit word-length, although both machines have only 16-bit multipliers. In 'C60, the upper 16 bits of the 32-bit data are multiplied to produce a 32-bit result. Thus, the 'C50 based implementation generates more quantization noise due to the truncation of the lower 16 bits of the 32-bit multiplied result. Motorola 56000 uses a 24-bit data type for both addition and multiplication. The upper 24 bits of the multiplier output are used for the multiplication results. When the IWL's are increased for the shift reduction, the fixed-point performance is slightly degraded in the 'C60 examples because the FWL's are decreased, but the 'C50 example results do not agree with our expectation. It is because the quantization noises due to the scale down shifts are eliminated. In this example, the input is scaled instead of the internal signals to reduce the scaling shifts, and it results in less quantization noise at the output signal. Since the results of optimization as a function of the IWL increase are not simple, we need to try a few different upper bounds to find the best one.

5 Word-Length Optimization

VLSI implementation of digital signal processing algorithms requires fixed-point arithmetic for the sake of circuit area minimization, speed, and low-power consumption. Word-length optimization is also used for 16-bit programmable DSP and SIMD processor based implementations because some SIMD arithmetic instructions for embedded processors employ shortened word-length, 8-bit or 16-bit, data for increasing the data-level parallelism. In the word-length optimization, it is necessary to reduce the quantization effects to an acceptable level without increasing the hardware cost too much. If the number of bits assigned is too small, quantization noise will degrade the system performance too much; on the other hand, if the number of bits is too large, the hardware cost will become too high. Word-length optimization for linear time-invariant systems may be conducted by analytical methods [9]. However, the simulation based approach is preferred because these methods allow not only linear but also non-linear and time-varying blocks [23]. In this sub-section, the finite word-length effects will be described, a C++ class library based fixed-point simulation tool will be presented, and the simulation based word-length optimization method will be explained.

5.1 Finite Word-Length Effects

Fixed-point arithmetic introduces quantization noise, and by which the final system performance is inevitably degraded in most cases. Thus, the needed performance of a system with fixed-point arithmetic should be defined first for word-length optimization. Finite word-length effects for implementing a digital filter can be

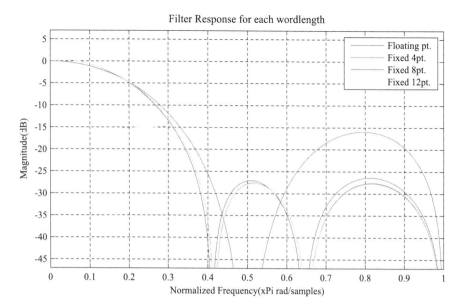

Fig. 14 Coefficient quantization effects of an FIR filter

classified into coefficient and signal quantization [8]. The coefficient quantization changes the system transfer function, thus its effect can be observed by displaying the frequency response of the transfer function. Frequency responses of an FIR filter with floating-point, 12-bit, 8-bit, and 4-bit coefficients are shown in Fig. 14. The filter structure also affects the quantization effects. When implementing a high order recursive filters, the second order cascaded forms usually show better results when compared to the direct forms.

The signal quantization can be considered as adding noise, instead of distorting the system transfer function. One of the most widely used fixed-point performances is the SQNR that is defined as Eq. (12), where $y_{fl}[n]$ is the floating-point result and $y[n]$ is the fixed-point result.

$$SQNR = \frac{P_{signal}}{P_{noise}} = \frac{E[y_{fl}{}^2]}{E[(y_{fl} - y)^2]} \tag{12}$$

The SQNR is a convenient measure, but it can usually be applied to linear time-invariant systems where the output quantization noise can be modeled as additive terms.

Figure 15 shows a simple additive noise model for the first order filter shown in Fig. 5a. Note that the quantization noise can be roughly modeled as uniformly distributed between $-\delta/2$ and $\delta/2$ when the rounding scheme is employed, where δ corresponds to 2^{-FWL}. Thus the maximum noise amplitude is halved when increasing the FWL by 1 bit. The quantization scheme that simply eliminates the low significant bits in the word-length reduction creates DC biased noise, which

Fig. 15 An additive noise model

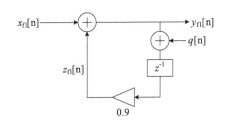

can result in a serious problem when an accumulation circuit that has a very high DC gain is followed. Thus, as the FWL is decreased by 1 bit, the output noise power becomes quadrupled, and the SQNR is decreased by 6 dB. At the same way, the increase of the FWL by 1 bit increases the SQNR by 6 dB. For more complex systems containing multiple quantization noise sources, the additive quantization noise model can also be built in a similar way, and the output noise is the sum of all the quantization noise sources [12]. Therefore, the increase of all the fractional word-lengths by 1 bit also raises the output SQNR by 6 dB in this case, too. The SQNR as the fixed-point performance measure can also be used for waveform coders, such as the adaptive delta modulators(ADMs) or CELP vocoders; however the increase of all the fractional word-lengths by 1 bit does not reduce the output noise power by 6 dB because these systems are not linear with respect to the quantization noise sources. For the case of an adaptive filter, word-length reduction causes slow convergence and higher steady state noise power. Thus, the fixed-point performance for optimizing an adaptive filter can be modeled as the noise power after some time-off period [22].

5.2   Fixed-Point Simulation Using C++ gFix Library

Although several analytical methods for evaluating the fixed-point performance of a digital signal processing algorithm have been developed by using the statistical model of quantization noise, they are not easily applicable to practical systems containing non-linear and time-varying blocks[26, 27]. The analysis is more complicated when a specific kind of input signal, such as speech, is required for the evaluation. In order to relieve these problems, simulation tools can be used for evaluating the fixed-point characteristics of a digital signal processing algorithm. There are a few commercially available fixed-point simulation tools for signal processing. The SPD (Signal Processing Designer) of CoWare and the Simulink of MathWorks provide fixed-point simulation libraries [1, 4]. Mixed simulation of floating-point and fixed-point blocks is allowed with these libraries. The fixed-point block can be a simple adder or a quite complex one, such as FFT or digital filtering. In order to assign a fixed-point format for each block, it is just needed to open a block by mouse clicking and edit the fixed-point attributes for the block, such as the word-length, integer or fractional word-length, overflow or saturation mode, rounding or quantization mode, and so on.

a

```
gFix a(12,0,"tsr");
gFix b(12,0,"tsr");
gFix c(12,0,"tsr");
gFix d(10,1,"tsr");

d = a + b + c;
```

b

```
gFix a(12,0,"tsr");
gFix b(12,0,"tsr");
gFix tmp(10,1,"tsr");
gFix c(12,0,"tsr");
gFix d(10,1,"tsr");

tmp = a + b;
d = tmp + c;
```

Fig. 16 Three operand addition using different architectures

However, the widely used C programming language does not support fixed-point arithmetic. Fixed-point simulation of a C program for signal processing can be conducted with a fixed-point class that also uses the operator overloading characteristic of the C++ language. A new fixed-point data class, gFix, and its operators are developed to prepare a fixed-point version of a floating-point program, and to know its finite word-length and scaling effects by simulation [17]. The gFix class follows the generalized fixed-point format. For example, gFix (10, 2, "tsr") represents a format with the WL of 10, IWL of 2, the two's complement, saturation for overflow handling, and rounding for quantization. The gFix class supports all of the assignment and arithmetic operations supported in C or C++ languages. There is no loss of accuracy during the fixed-point add or multiply operations. However, arithmetic right shift or arithmetic left shift may cause loss of accuracy or overflows. The assignment operator, "=," converts the input data according to the fixed-point format of the left side variable, and assigns the format converted data to this variable. If the given format of the left side variable does not support enough precision for representing the input data, the data is modified according to the attributes of the left side variable, such as saturation, rounding, or truncation. Since the format conversion is occurred only at the assignment operator, two programs shown in Fig. 16a,b can have different fixed-point results. In Fig. 16b, the result of $a + b$ is format converted to 10 bit data, and then added to the operand c, and then format converted again. The fixed-point performance of different implementations with the same algorithm can be compared by utilizing the above characteristics.

A fixed-point simulation model of a simple IIR filter converted from the floating-point C program is shown in Fig. 17. Note that only the type of variables is converted to gFix, but the other parts of the program are not changed.

5.3 Word-Length Optimization Method

Word-length optimization usually tries to find the word-length vector that minimizes the hardware cost while meeting the system performance with fixed-point arithmetic. Since the optimum word-length is dependent on the desired fixed-point performance of a system, a performance measurement block has to be included as

Fig. 17 A fixed-point
C++ program for a first
order IIR filter

```
void
iir1(short argc, char *argv[])
{
    gFix   Xin(12,0);
    gFix   Yout(16,3);
    gFix   Ydly(16,3);
    gFix   Coeff(10,0);

    Coeff = 0.9;
    Ydly = 0.;
    for( i = 0; i < 1000; i++ ) {
      infile >> Xin ;
      Yout = Coeff * Ydly + Xin ;
      Ydly = Yout ;
      outfile << Yout << '\n';
    }
}
```

a part of a system set-up. The performance measurement block must generate a positive result (or pass) when the quantization effects are acceptable. A hardware cost library is also needed to estimate the total complexity when implementing the system with the given word-length vector. Note that not only the algorithm but also the system architecture affects the hardware cost.

In the simplest case, only one word-length is used for all arithmetic operations, which is called the uniform word-length optimization. In the uniform word-length optimization, fixed-point simulation with a shorter (or longer) word-length than the optimum one should yield a fixed-point performance which is lower (or higher) than the needed performance. Thus, it is possible to arrive at the optimum word-length by increasing (or decreasing) the word-length when the obtained performance is lower (or higher) than the needed fixed-point performance. In the case of linear time-invariant systems, it is possible to reduce the number of simulations by considering that the SQNR becomes higher by 6 dB with the word-length increase of one bit.

Usually, there are multiple word-lengths to optimize in implementing a fixed-point system. As the number of word-lengths to optimize increases, optimization of them should take a longer time. In other words, minimizing the number of variables is very important for reducing the optimization time. In this optimization method, the number of different word-lengths is reduced by signal grouping that assigns the same word-length to signals, for example, connected with a delay or a multiplexer block. The word-length sensitivity of a signal needs to be considered for optimization. Some signals are very sensitive to quantization, thus they need a long word-length. The minimum bound of the word-length for each signal group is in inverse relation with the sensitivity, and can be used to reduce the search space. Finally, the optimization of different word-lengths requires a hardware cost model. The word-length optimization method in this section consists of four steps: signal grouping, sign and integer word-length determination, minimum word-length determination, and cost optimum word-length search.

As an example, Fig. 18 shows the setup for fixed-point optimization of a first order filter. The SQNR is used as the measure of the fixed-point performance.

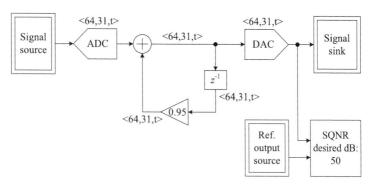

Fig. 18 Fixed-point setup for a first order filter

5.3.1 Signal Grouping

The optimization method preprocesses the netlist of a signal flow block diagram to group the signals that can have the same fixed-point attributes and, as a result, to minimize the number of variables for optimization. Both automatic and manual grouping functions are employed. The automatic grouping rules are as follows.

1. Signals connected by a delay, a multiplexer, or a switch are grouped.
2. Input and output signals of an adder or a subtracter are grouped.
3. Signals connected by a multiplier, a quantizer, or a format converter can have different fixed-point data formats unless these signals are grouped together via the other path in a signal flow block diagram.

In [18], more grouping rules were implemented to reduce the number of groups as much as possible. But, for the Fixed Point Optimizer of SPW, a manual grouping mechanism is added to give more flexibility. By manually adding the same prefix to each fixed-point attribute parameter, we can combine signals into one group that would otherwise be assigned to different groups. For example, the ADC and the DAC in Fig. 7 can have the same fixed-point data format if they are manually grouped.

In high-level synthesis, it is necessary to bind operations in order to reduce the number of hardware components [20]. In this case, we can use the binding results for grouping.

5.3.2 Determination of Sign and Integer Word-Length

If the minimum value for a signal is not negative, the sign can be "u" (unsigned). Otherwise, it should be "t" (two's complement). The integer word-length for a signal can be determined from the range of a signal, $R(x)$, using Eq. (2).

Although each signal can have a separate integer word-length, one common integer word-length is assigned to all the signals in the same group in order to lower the use of shifters for implementation. All the signs and the integer word-lengths are determined in just one simulation.

5.3.3 Determination of the Minimum Word-Length for Each Group

Assume a word-length vector w, whose component is the word-length in each group.

$$w = (w_1, w_2, \cdots, w_N), \tag{13}$$

where N is the number of groups.

The performance of a fixed-point system, such as SQNR or negative of mean squared error, is represented by $p(w)$, and the hardware cost, usually the number of gates, is $c(w)$. Then, the optimum word-length vector, w_{opt}, should have the minimum value of $c(w)$ while $p(w)$ is larger than $p_{desired}$. We assume the following relation between the word-length vector and the fixed-point performance.

$$p((w_1, w_2, \cdots, w_i, \cdots, w_N)) \geq p((w_1, w_2, \cdots, w_i - 1, \cdots, w_N)), \tag{14}$$

where $1 \leq i \leq N$. The above equation represents that reducing a word-length of a group decreases, or at least does not increase, the fixed-point performance of a system. Then, the number of simulations required for the search can be reduced greatly by decomposing the procedure into two steps: the minimum word-length and the cost optimum word-length determination.

The minimum word-length for a group, $w_{i,min}$, is the smallest word-length that satisfies the fixed-point performance of a system when the word-lengths of all other groups are very large, typically a 64 bit fixed-point or the floating-point type. By the assumption shown in Eq. (14), this minimum word-length is not larger than the optimum word-length for the group. The minimum word-length determination procedure for the first order digital filter is illustrated in Table 6. The word-length vector for this filter consists of three components, which are "ADC," "DAC" and filter word-lengths. The uniform word-length that satisfies the fixed-point performance is first determined. The uniform word-length is not only useful for some architecture, such as bit-serial implementations, but also the upper limit of the minimum word-length according to the assumption in Eq. (14). Thus, the search for finding the minimum word-length for each group goes downward from the determined uniform word-length. The number of simulations required for the minimum word-length determination procedure is typically two to four times the number of groups. The minimum word-length determination does not require the hardware cost model.

Table 6 Search sequence for the first order recursive filter

Word-length					
ADC	Filter	DAC	Simulation result	Hardware cost	Comment
16	16	16	Pass		Uniform word-length determination
14	14	14	Fail		
15	15	15	Fail		
14	64	64	Pass		Group ADC minimum word-length determination
12	64	64	Pass		
10	64	64	Fail		
11	64	64	Pass		
64	14	64	Fail		Group filter minimum word-length determination
64	15	64	Fail		
64	64	14	Pass		Group DAC minimum word-length determination
64	64	12	Pass		
64	64	10	Fail		
64	64	11	Pass		
11	16	11	Fail	8054	Minimum word-length
12	16	11	Fail	8254	Exhaustive search
11	16	12	Fail	8256	
11	17	11	Fail	8281	
13	16	11	Fail	8454	
12	16	12	Fail	8456	
11	16	13	Fail	8458	
12	17	11	Pass	8481	Determined word-length

5.3.4 Determination of the Minimum Hardware Cost Word-Length Vector

The word-length vector that satisfies the system performance while requiring the minimum hardware cost is determined at the next step. If the simulation using the minimum word-length vector, w_{min}, satisfies the desired performance, the minimum word-length vector w_{min} becomes the optimum word-length vector, w_{opt}. If not, the word-length is increased, and the fixed-point system performance is measured by simulation. Three search strategies, an exhaustive and two heuristic methods, are developed for this step. The search sequences for the first order recursive filter will be illustrated as an example. Usually, the optimum word-length for each group is, at most, two bits larger than the minimum word-length.

The hardware cost of the components for implementing a DSP system, such as ADC, adder, multiplier, and delay, has to be known for the minimum cost word-length optimization of a system. For example, when the price of an ADC is high, it may be justified to minimize the word-length of the ADC by increasing the word-lengths for multipliers and adders. Adders, constant coefficient multipliers, and delays are modeled to consume the hardware resources in proportion to their word-lengths. The number of gates for a multiplier is dependent on the product of two input word-lengths. Note that the hardware cost model is affected not only by the process technology but also by the implementation architecture. When a signal processing algorithm is implemented in a time-domain multiplexed mode, the per-bit cost of arithmetic blocks that are shared can be lowered. Since the hardware cost model is stored at an external file, a designer can assign an appropriate value to each component according to the implementation architecture and the ASIC library. In this example, the hardware cost model using VLSI Technologies' cell library is used, and the hardware cost of 200 and 202 is assumed for each bit of ADC and DAC, respectively. According to these models, the hardware cost increase for each bit is 200 for group "ADC," 227 for group "filter," and 202 for group "DAC."

In the exhaustive search algorithm, the word-length vector that requires the least cost is selected in a priority for the fixed-point performance measurement starting from the minimum word-length vector. For example, the word-length for group "ADC" will be increased first as shown in Table 6. If the test is failed, the word-length for group "DAC" will be increased instead of group "ADC." The search sequence using the exhaustive algorithm is shown in Table 6. The minimum cost word-length vector determined is (12, 17, 11), and the hardware cost required is 8481. Although the result is the minimum cost solution, this method demands many fixed-point performance measurements. The search procedure is a combinatorial algorithm as a function of the number of groups and is practical only when the number of groups is small, usually less than 6.

In the first heuristic search algorithm, all the word-lengths are increased by one, e.g. the word-length vector is increased to (12, 17, 12) from the minimum word-length vector (11, 16, 11), and the fixed-point performance is measured. If the result is not satisfactory, the above procedure is repeated. This step requires typically one to two simulations. After then, the word-length for a group whose cost saving is the greatest is decreased. If it does not satisfy the performance, the word-length is restored. This procedure is conducted for all the group, and, as a result, requires N simulations. The maximum number of simulations required for the heuristic search based word-length optimization, including the minimum word-length determination, is only linearly proportional to the number of groups, N. The word-length vector determined for the first order digital filter shown in Fig. 7 is (12, 17, 11), which is the same as the minimum cost word-length. According to the experiments using eight examples, the additional hardware cost is usually less than 5% of that required for the exhaustive search algorithm.

In the second heuristic search algorithm, the word-length vector that shows the best ratio of performance increment to cost increase is selected. The word-length of each group is increased by one bit, and the fixed-point performance is measured using simulations for each case, and the best result is selected for the next iteration until the desired fixed-point performance is satisfied. Since the fixed-point performance increment is used in this method, this heuristic can be applied when the fixed-point performance can be quantified such as SQNR. When the fixed-point

Table 7 Search sequences of performance/cost directed heuristic method

Word-length					
ADC	Filter	DAC	Performance SQNR(dB)	Hardware cost	Comment
11	16	11	34.27	8054	Minimum word-length
12	16	11	38.56	8254	Selected
11	17	11	34.83	8281	
11	16	12	34.59	8256	
13	16	11	39.65	8454	
12	17	11	40.36	8481	Determined word-length
12	16	12	37.43	8456	

Fig. 19 An LMS channel identification system

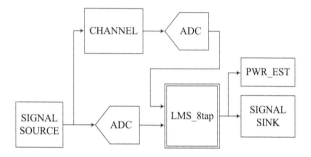

performance is tested using binary decision, e.g. pass or fail, this method cannot be used. The search sequences of this heuristic method for the first order recursive filter is shown in Table 7.

5.4   Optimization Example

The impulse response of a channel can be identified by using an adaptive digital filter which adjusts the coefficients to minimize the channel modeling error. When the word-length of the filter coefficients or the input data is not sufficient, the system not only converges slowly but also the magnitude of error becomes large. A power estimation block that measures the average power of the error signal after some time-off period is employed as for the fixed-point performance measure [22] (Fig. 19).

When we apply the automatic grouping rules, the coefficient in each tap becomes a separate group. In order to reduce the number of groups, the coefficients are manually grouped by assigning a fixed-point attribute of "coef." The grouping result for each tap is shown in Fig. 20.

The "tap_in" and "tap_out" signals are automatically grouped because they are connected by a delay, and assigned to "u005." The "sum_in" and "sum_out" signals are also grouped, and assigned to the "u001" group. The integer word-length, minimum word-length, and optimum word-length for each group of signals

Fig. 20 Grouping result for one tap of LMS adaptive filter

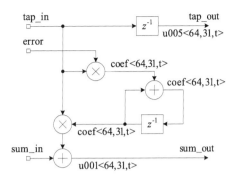

Table 8 Determined fixed-point attributes for the LMS adaptive filter

Group	Signal	Integer word-length	Minimum word-length	Optimum word-length
u001	Sum	3	12	12
u005	Tap	2	7	7
u002	Error	−4	5	5
coef	Coefficient	1	13	13
u004	ADC (channel)	3	8	10
u006	ADC (source)	2	7	13

are shown in Table 8. The optimization results are compared for three search algorithms: uniform word-length, exhaustive search, heuristic search based on uniformly increasing the minimum word-length vector. The uniform word-length optimization requires 13 bits for all the signals and the hardware cost required is 96,265 gates. The exhaustive and heuristic search algorithms, both are based on the minimum cost criterion, need the cost of 46,278 and 49,520 gates, respectively. The hardware cost required using the minimum cost optimization is just 48% of that needed for the uniform word-length determination in this example. The number of simulations for searching the optimum word-length vector from the minimum word-length vector is 27 and 5 for the exhaustive and heuristic search algorithms, respectively.

6 Summary and Related Works

Fixed-point hardware or integer arithmetic based implementation of digital signal processing algorithms is important for not only hardware cost minimization but also power consumption reduction. The conversion of a floating-point algorithm into a fixed-point or an integer version has been considered very time-consuming; it often takes more than 50% of the algorithm to hardware or software implementation procedure [25]. The fixed-point format discussed in Sect. 2 bridges the gap between the floating-point and fixed-point data types, and allows seamless and incremental conversion. Note that incremental conversion can help a designer very much because the effects of each small conversion can readily be verified. There are other fixed-point formats used for similar purposes, and one of them is the Q format [2]. In the Q format, the integer and fractional bits are given, while the sign is usually implied as the two's complement. For example, "Q1,30" describes a fixed-point data with 1 integer bit and 30 fractional bits stored as a 32-bit two's complement number. The Q format has been used widely for assembly programming of Texas Instruments digital signal processors.

The simulation based range estimation and automatic conversion process becomes more and more popular as the computing power for simulation becomes cheaper. The FRIDGE, a fixed-point design and simulation environment, supports interpolative approach to reduce the range estimation overhead [25]. This tool also supports integer and VHDL code generation path from a floating-point C program. In these days, application oriented language, such as Matlab from MathWorks, is frequently used for signal processing application development. The Simulink package for Matlab and SPD (Signal Processing Designer) of CoWare support easy to use, GUI (Graphic User Interface) supported, fixed-point arithmetic [1, 4]. With these tools, it is possible to convert a floating-point design into a fixed-point version in an incremental and convenient way.

For the word-length optimization of VLSI and FPGA based design, several different search methods have been studied recently [10, 21]. For linear systems, the word-length optimization is converted to an integer programming problem by exploiting the additive quantization noise model [9]. A fixed-point optimization flow that conducts both high-level synthesis and fixed-point optimization simultaneously has been developed [19]. As one of the future works, integer code generation for SIMD (Single Instruction Multiple Data) architecture is needed; this work requires combined processing of scaling, word-length optimization, and automatic vectorization. Also, fast word-length optimization of very large systems is an interesting problem. The word-length optimization or scaling software needs to be integrated into widely used CAD tools. The capability of fixed-point optimization tools currently supported by commercial CAD software can hardly meet users' expectation.

References

1. URL http://www.coware.com/products/signalprocessing.php
2. URL http://en.wikipedia.org/wiki/Q_(number_format)
3. Fixed-Point C++ class. URL http://msl.snu.ac.kr/fixim/
4. Simulink. URL http://www.mathworks.com/products/simulink/
5. DSP56KCC User's Mannual. Motorola Inc. (1992).
6. TMS320C2x/C2xx/C5x Optimizing C Compiler (Version 6.60). Texas Instruments Inc., TX (1995).
7. TMS320C6x Optimizing C Compiler. Texas Instruments Inc., TX (1997).

8. Catthoor, F., Vandewalle, J., Man, H.D.: Simulated Annealing based Optimization of Coefficient and Data Word-Lengths in Digital Filters. Int. J. Circuit Theory and Applications **16**, 371–390 (1988).

9. Constantinides, G., Cheung, P., Luk, W.: Wordlength optimization for linear digital signal processing. Computer-Aided Design of Integrated Circuits and Systems, IEEE Transactions on **22**(10), 1432–1442 (2003).

10. Han, K., Evans, B.L.: Optimum wordlength search using sensitivity information. EURASIP J. Appl. Signal Process. **2006**, 76–76 (2006).

11. Han, K., Olson, A., Evans, L.: Automatic floating-point to fixed-point transformations. In: Signals, Systems and Computers, 2006. ACSSC '06. Fortieth Asilomar Conference on, pp. 79–83 (2006).

12. Jackson, L.B.: On the Interaction of Roundoff Noise and Dynamic Range in Digital Filters. The Bell System Technical Journal pp. 159–183 (1970).

13. Kum, K.I., Kang J., Sung, W.: AUTOSCALER for C: an optimizing floating-point to integer C program converter for fixed-point digital signal processors. IEEE Trans. Circuits and Systems-II: Analog and Digital Signal Processing **47**(9), 840–848 (2000).

14. Kang, J., Sung, W.: Fixed-point C language for digital signal processing. In: Proc. of the 29th Annual Asilomar Conference on Signals, Systems and Computers, vol. 2, pp. 816–820 (1995).

15. Kang, J., Sung, W.: Fixed-point C compiler for TMS320C50 digital signal processors. In: Proc. of 1997 IEEE International Conference on Acoustics, Speech, and Signal Processing, pp. 707–710 (1997).

16. Kim, S., Sung, W.: A Floating-point to Fixed-point Assembly Program Translator for the TMS 320C25. IEEE Trans. on Circuits and Systems **41**(11), 730–739 (1994).

17. Kim, S., Sung, W.: Fixed-point optimization utility for C and C++ based digital signal processing programs. IEEE Trans. on Circuits and Systems **45**(11), 1455–1464 (1998).

18. Kum, K.I., Sung, W.: VHDL based Fixed-point Digital Signal Processing Algorithm Development Software. In: Proceeding of International Conference on VLSI and CAD '93, pp. 257–260. Korea (1993).

19. Kum, K.I., Sung, W.: Combined word-length optimization and high-level synthesis of digital signal processing systems. Computer-Aided Design of Integrated Circuits and Systems, IEEE Transactions on **20**(8), 921–930 (2001).

20. Micheli, G.D.: Synthesis and Optimization of Digital Circuits. McGraw-Hill, Inc., NJ (1994).

21. Shi, C., Brodersen, R.: Automated fixed-point data-type optimization tool for signal processing and communication systems. In: Design Automation Conference, 2004. Proceedings. 41st, pp. 478–483 (2004).

22. Sung, W., Kum, K.I.: Word-Length Determination and Scaling Software for a Signal Flow Block Diagram. In: Proceeding of the International Conference on Acoustics, Speech, and Signal Processing '94, vol. 2, pp. 457–460. Adelaide, Australia (1994).

23. Sung, W., Kum, K.I.: Simulation-Based Word-Length Optimization Method for Fixed-Point Digital Signal Processing Systems. IEEE Trans. on Signal Processing **43**(12), 3087–3090 (1995).

24. Willems, M., Bürsgens, V., Grötker, T., Meyr, H.: FRIDGE: An interactive code generation environment for HW/SW codesign. In: Proc. of 1997 IEEE International Conference on Acoustics, Speech, and Signal Processing, pp. 287–290 (1997).

25. Willems, M., Bürsgens, V., Meyr, H.: FRIDGE: Floating-point programming of fixed-point digtal signal processors. In: Proc. of the International Conference on Signal Processing Applications and Technology (1997).

26. Wong, P.W.: Quantization and roundoff noises in fixed-point FIR digital filters. IEEE Trans. Signal Processing **39**, 1552–1563 (1991).

27. Yun, I.D., Lee, S.U.: On the fixed-point error analysis of several fast DCT algorithms. IEEE Trans. Circuits and Systems for Video Technology **3**(1), 27–41 (1993).

28. Zivŏjnovic, V.: Compilers for Digital Signal Processors. DSP & Multimedia Technology **4**(5), 27–45 (1995).

Polyhedral Process Networks

Sven Verdoolaege

Abstract Reference implementations of signal processing applications are often written in a sequential language that does not reveal the available parallelism in the application. However, if an application satisfies some constraints then a parallel specification can be derived automatically. In particular, if the application can be represented in the polyhedral model, then a *polyhedral process network* can be constructed from the application. After introducing the required polyhedral tools, this chapter details the construction of the processes and the communication channels in such a network. Special attention is given to various properties of the communication channels including their buffer sizes.

1 Introduction

Signal processing applications are prime candidates for a parallel implementation. As we have seen in previous chapters, there are several parallel models of computation that can be used for specifying such applications. However, many programmers are unfamiliar with these parallel models of computation. Writing parallel specifications can therefore be a difficult, time consuming and error prone process. For this reason, many application developers still prefer to specify an application as a sequential program, even though such a specification may not be suitable for a direct mapping onto a parallel multiprocessor platform.

In this chapter, we present a technique for automatically extracting a parallel specification from a sequential program, provided the program fragments that are to be converted satisfy some conditions. In particular, the control flow of these program fragments needs to be static, meaning that the control flow should not

S. Verdoolaege (✉)
Department of Computer Science, Katholieke Universiteit Leuven,
Celestijnenlaan 200A, B-3001 Leuven, Belgium
e-mail: sven@cs.kuleuven.be

S.S. Bhattacharyya et al. (eds.), *Handbook of Signal Processing Systems*,
DOI 10.1007/978-1-4614-6859-2_41, © Springer Science+Business Media, LLC 2013

depend on the signals being processed. Furthermore, all loop bounds, conditions and array index expressions need to be such that they can be expressed using affine constraints. All functions called from within the program fragment should be pure. In particular, they should not change the values of loop iterators or arrays other than through their output arguments. These requirements ensure that we can represent all relevant information about the program using mathematical objects that we will call polyhedral sets and relations. The name derives from related objects called polyhedra [26]. The resulting parallel model, a variation on Kahn process networks [9, 11], can then also be described using such polyhedral sets and relations, whence the name *polyhedral process networks*.

The concept of a polyhedral process network was developed in the context of the Compaan project [14, 20]. The name was coined in [15]. The exposition in this chapter closely follows that of [24].

2 Overview

This section presents a high-level overview of the process of extracting a process network from a sequential program. The extracted process network represents the task-level parallelism that is available in the program. The input program is assumed to consist of a sequence of nested loops performing various "operations". These operations may be calls to functions that can be arbitrarily complicated. The operations are performed on data that has been computed in some iteration of another operation or in a previous iteration of the same operation. The output process network consists of a set of processes, each encapsulating all iterations of a given operation, and communication channels connecting the processes and representing the dataflow. The processes in the network can be executed independently of each other, as long as data is available from the channels from which the process reads and as long as buffer space is available in the channels to which the process writes. That is, the communication primitives implement blocking reads and blocking writes.

Example 1. As a simple example, consider the code for performing Sobel edge detection in Fig. 1. The first loop of this program reads the input image, while the

```
1    for (i = 0; i < K; i++)
2      for (j = 0; j < N; j++)
3    R:   a[i][j] = ReadImage();
4      for (i = 1; i < K-1; i++)
5        for (j = 1; j < N-1; j++) {
6    S:   Sbl[i][j] = Sobel(a[i-1][j-1], a[i][j-1], a[i+1][j-1],
7                           a[i-1][ j], a[i][ j], a[i+1][ j],
8                           a[i-1][j+1], a[i][j+1], a[i+1][j+1]);
9    W:   WriteImage(Sbl[i][j]);
10     }
```

Fig. 1 Source code of a Sobel edge detection program

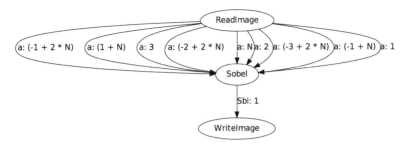

Fig. 2 A process network corresponding to the sequential program in Fig. 1

second loop performs the actual edge detection and writes out the output image. A process network that corresponds to this program is shown in Fig. 2. There are three processes in the network, each corresponding to one of the three "operations" performed by the program, i.e., reading, edge detection and writing. Data flows from the reading process to the edge detection process and from the edge detection process to the writing process, resulting in the communication channels shown in the figure. The annotations on the edges will be explained later.

The extraction of a polyhedral process network consists of several steps summarized below and explained in more detail in the following sections:

- In a first step, a model is extracted from the input program on which all further analysis will be performed. In particular, the program is represented by a *polyhedral model*. This model allows for an efficient analysis, but imposes some restrictions on the input programs. The polyhedral model is explained in Sect. 3, while some basic components for the analysis of polyhedral models are explained in Sect. 4.
- Dataflow analysis is performed to determine which processes communicate with which other processes and how, i.e., to determine the communication channels. For example, the results of the call to ReadImage in Line 3 of Fig. 1 are stored in the a array, which is read by the call to Sobel in Line 6. Dataflow analysis therefore results in one or more communication channels from the reading process to the edge detection process. Dataflow analysis is explained in Sect. 5.
- In the next step, the type of each communication channel is determined. For example, the channel may be a FIFO, in which case the processes connected to the channel simply need to write to and read from the channel, or it may not be a FIFO, in which case additional processing will be required. The classification of channels is discussed in Sect. 6.
- The communication channels may need to buffer some data to ensure a deadlock-free execution of the network. Especially for a hardware implementation of process networks, it is important to know how large these buffers may need to grow. The computation of buffer sizes is the subject of Sect. 8.

- The number of processes in the network may exceed the number of processing elements available. Some processes may therefore need to be merged. This merging requires the construction of a combined schedule, which is the subject of Sect. 7. Depending on the kind of dataflow analysis that was performed in constructing the network, some of this analysis may need to be updated or redone based on the merging decisions.
- Finally, code needs to be written out for each of the processes in the network. This code needs to execute all iterations of the single or multiple (in case of merging) operations and needs to read from and write to the appropriate communication channels at the appropriate times. The main difficulty in this step is writing out code to scan overlapping polyhedral domains, which is discussed in Sect. 4.5.

The buffer size computation itself (Sect. 8) consists of several substeps. First a global schedule is computed (Sect. 7), assigning a global time point to each iteration of each process. This global schedule is only used during the buffer size computation and not during the actual execution. For each channel and each time point, the number of tokens in the channel at that time point is then computed (Sect. 4.3). Finally, for each channel, an upper bound is computed for the maximal number of tokens in the channel throughout the whole execution (Sect. 4.4).

3 Polyhedral Concepts

The key to an efficient transformation from sequential code to a process network is the polyhedral model used to represent the sequential program and the resulting process network. This section defines both models and related concepts.

3.1 Polyhedral Sets and Relations

Each statement inside a loop nest is executed many times when the sequential program is run. Each of these executions can be represented by the values of the iterators in the loops enclosing the statement. This sequence of iterator values is called the *iteration vector* associated to a given execution of the statement. The set of all such iteration vectors is called the *iteration domain* of the statement. Assuming that each iterator is an integer that is incremented by one in each iteration of the loop, this iteration domain can be represented very succinctly by simply collecting the lower and upper bounds of each of the enclosing loops.

Example 2. The iteration domain associated to the `ReadImage` statement in Line 3 of Fig. 1 is

$$D_1 = \{\, R(i,j) \mid 0 \le i < K \wedge 0 \le j < N \,\}, \tag{1}$$

where R refers to the label on the statement and i and j are integer variables.

Fig. 3 Source code of a loop nest with iteration domains requiring existentially quantified variables to represent

```
1   for (i = 0; i < N; i++)
2     if (i % 3 == 0)
3   S1: a[i] = f1();
4     else
5   S2: a[i] = f2();
```

The extraction of a process network requires several manipulations of iteration domains and related sets. To ensure that these manipulations can be performed efficiently or even performed at all, we need to impose some restrictions on how these sets are represented. In particular, we require that the sets are described by integer *affine* inequalities and equalities over integer variables. An affine inequality is an inequality of the form $a_1x_1 + a_2x_2 + \ldots + a_dx_d + c \geq 0$, i.e., it expresses that some degree-1 polynomial over the variables is greater than or equal to zero. When dealing with several such inequalities in general, it will be convenient to represent them using a matrix notation $A\mathbf{x} + \mathbf{c} \geq \mathbf{0}$.

Besides iterators and parameters, we may also need additional variables to accurately describe an iteration domain. These variables are not used to identify a given iteration, but rather to restrict the possible values of the iterators. This means that we do not care about the values of these variables, but rather that *some* integer value exists for these variables that satisfies the constraints. These variables may therefore be existentially quantified. The need for such variables arises especially when loop bounds or guards contain modulos or integer divisions. Expressions that contain such constructs but that can still be expressed using affine constraints are called *quasi-affine*.

Example 3. Consider the program shown in Fig. 3. While the modulo constraint "i % 3 == 0" is not in itself an affine constraint, it can be represented as an affine constraint $i = 3\alpha$ by introducing an extra integer variable $\alpha \in \mathbb{Z}$, with $\mathbb{Z}$ the set of integers. The iteration domain of the statement in Line 3 can therefore be represented as

$$D_1 = \{\, \text{S1}(i) \mid \exists \alpha \in \mathbb{Z} : 0 \leq i < N \wedge i = 3\alpha \,\}. \tag{2}$$

Similarly, the iteration domain of the statement in Line 5 can be represented as

$$D_2 = \{\, \text{S2}(i) \mid \exists \alpha \in \mathbb{Z} : 0 \leq i < N \wedge 1 \leq i - 3\alpha \leq 2 \,\}.$$

Let us now also consider how we could represent such a set in a practical polyhedral tool. In particular, we will show how to represent the set D_1 in the iscc tool [22], which is distributed along with the barvinok distribution.[1] The same tool will also be used in many of the later examples in this chapter. The notation is very similar to that used above. The main difference is that square brackets are used

---

[1] http://freecode.com/projects/barvinok/

to enclose i. We also need to explicitly list the parameters used in the description of the set and we can use the available syntactic sugar for representing modulo constraints. In particular D_1 can be represented as

```
[N] -> { S1[i] : 0 <= i < N and i % 3 = 0 };
```

Whenever the result of an expression is not assigned to a variable, iscc will print the result of the expression. In this case, the expression is just the given set and iscc will print the set, albeit using a slightly different representation:

```
[N] -> { S1[i] : exists (e0 = [(i)/3]: 3e0 = i and
         i >= 0 and i <= -1 + N) }
```

Note that iscc has made the use of an existentially quantified variable explicit and that it has figured out that this existentially quantified variable is always equal to $\lfloor i/3 \rfloor$, i.e., the greatest integer part of $i/3$.

In general then, the *"polyhedral sets"* such as D_1 and D_2 that are used in the polyhedral representation of both the input program and the resulting process network, are defined as follows.

Definition 1 (Polyhedral Set). A *polyhedral set* S is a finite union of basic sets $S = \bigcup_i S_i$, each of which can be represented using affine constraints

$$S_i : \mathbb{Z}^n \to \left(\Sigma \times 2^{\mathbb{Z}^{d_i}} \right) : \mathbf{s} \mapsto S_i(\mathbf{s}) = \{ \ell_i(\mathbf{x}) \mid \mathbf{x} \in \mathbb{Z}^{d_i} \wedge \exists \mathbf{z} \in \mathbb{Z}^e : A\mathbf{x} + B\mathbf{s} + D\mathbf{z} \geq \mathbf{c} \},$$

with $A \in \mathbb{Z}^{m \times d_i}$, $B \in \mathbb{Z}^{m \times n}$, $D \in \mathbb{Z}^{m \times e}$, $\mathbf{c} \in \mathbb{Z}^m$ and $2^{\mathbb{Z}^{d_i}}$ the power set of $\mathbb{Z}^{d_i}$, i.e., the set of all subsets of $\mathbb{Z}^{d_i}$. Σ is a finite set of labels and ℓ_i is a specific label in that set uniquely determined by S_i. Note that S_i is a parametric description of a family of sets, one for each value of the parameters $\mathbf{s}$. The set of all parameter values $\mathbf{s}$ for which $S(\mathbf{s})$ is non-empty, i.e.,

$$\text{pdom}\, S = \{ \mathbf{s} \in \mathbb{Z}^n \mid S(\mathbf{s}) \neq \emptyset \}.$$

is called the *parameter domain*.

For ease of notation, we will sometimes consider a set as a boolean function so that we can write $\mathbb{Z}^n \to (\Sigma \times \mathbb{Z}^d) \to \mathbb{B}$, with $\mathbb{B}$ the set of boolean values, instead of $\mathbb{Z}^n \to \left(\Sigma \times 2^{\mathbb{Z}^d} \right)$. Note that any polyhedral set can be represented in infinitely many ways. The operations we apply to polyhedral sets in this chapter do not depend on the chosen representation.

In a similar fashion, we can define *"polyhedral relations"* over pairs of sets.

Definition 2 (Polyhedral Relation). A *polyhedral relation* R is a finite union of basic relations $R = \bigcup_i R_i$ of type $\mathbb{Z}^n \to \left(\left(\Sigma \times 2^{\mathbb{Z}^{d_1}} \right) \times \left(\Sigma \times 2^{\mathbb{Z}^{d_2}} \right) \right)$, each of which can be represented using affine constraints $R_i = \mathbf{s} \mapsto R_i(\mathbf{s})$, with $R_i(\mathbf{s})$ equal to

$$\{ \ell_1(\mathbf{x}_1) \to \ell_2(\mathbf{x}_2) \mid \mathbf{x}_1 \in \mathbb{Z}^{d_1} \wedge \mathbf{x}_2 \in \mathbb{Z}^{d_2} \wedge \exists \mathbf{z} \in \mathbb{Z}^e : A_1\mathbf{x}_1 + A_2\mathbf{x}_2 + B\mathbf{s} + D\mathbf{z} \geq \mathbf{c} \},$$

with $A_i \in \mathbb{Z}^{m \times d_i}$, $B \in \mathbb{Z}^{m \times n}$, $D \in \mathbb{Z}^{m \times e}$ and $\mathbf{c} \in \mathbb{Z}^m$. The *parameter domain* of R is the (non-parametric) polyhedral set $\mathrm{pdom}\, R = \{\, \mathbf{s} \in \mathbb{Z}^n \mid R(\mathbf{s}) \neq \emptyset \,\} = \{\, \mathbf{s} \in \mathbb{Z}^n \mid \exists \mathbf{x}_1 \in \mathbb{Z}^{d_1}, \mathbf{x}_2 \in \mathbb{Z}^{d_2} : (\mathbf{x}_1, \mathbf{x}_2) \in R(\mathbf{s}) \,\}$.

We will apply the following operations to polyhedral relations.

- The *domain* of a relation R is the polyhedral set $\mathrm{dom}\, R = \mathbf{s} \mapsto \{\, \ell_1(\mathbf{x}_1) \mid \exists \ell_2 \in \Sigma, \mathbf{x}_2 \in \mathbb{Z}^{d_2} : (\ell_1(\mathbf{x}_1) \to \ell_2(\mathbf{x}_2)) \in R(\mathbf{s}) \,\}$.
- The *range* of a relation R is the polyhedral set $\mathrm{ran}\, R = \mathbf{s} \mapsto \{\, \ell_2(\mathbf{x}_2) \mid \exists \ell_1 \in \Sigma, \mathbf{x}_1 \in \mathbb{Z}^{d_1} : (\ell_1(\mathbf{x}_1) \to \ell_2(\mathbf{x}_2)) \in R(\mathbf{s}) \,\}$.
- The *image* of an element $\ell_1(\mathbf{t}) \in \mathrm{dom}\, R(\mathbf{s})$ is the set

$$R(\mathbf{s}, \ell_1(\mathbf{t})) = \{\, \ell_2(\mathbf{x}_2) \mid (\ell_1(\mathbf{t}) \to \ell_2(\mathbf{x}_2)) \in R(\mathbf{s}) \,\}.$$

Given two polyhedral sets, we also define an operator $\to$ that constructs a universal relation over those two sets, i.e.,

$$S_1 \to S_2 = \{\, \ell_1(\mathbf{x}) \to \ell_2(\mathbf{y}) \mid \ell_1(\mathbf{x}) \in S_1 \wedge \ell_2(\mathbf{y}) \in S_2 \,\}$$

Polyhedral sets and polyhedral relations have essentially the same type if we set $d = d_1 + d_2$. The difference is mainly a matter of interpretation. In polyhedral relations, we make a distinction between two sets of variables, whereas in polyhedral sets, there is no such distinction. Any polyhedral set can also be treated as a polyhedral relation with a zero-dimensional domain, i.e., by setting $d_1 = 0$, $d_2 = d$ and $A_2 = A$. Any statement about polyhedral relations will therefore also hold for polyhedral sets. We will usually only treat the general case of polyhedral relations.

3.2 Lexicographic Order

For a proper analysis of a sequential program, we not only need to know for which iterator values a given statement is executed, but also in which order these instances are executed. A given instance is executed before another instance if they are executed in the same iteration of zero or more outermost loops and if the second instance is executed in a later iteration of the next outermost loop. In other words, the execution order corresponds to the *lexicographic order* on iteration vectors.

Definition 3 (Lexicographic order). A vector $\mathbf{a} \in \mathbb{Z}^n$ is said to be *lexicographically smaller* than $\mathbf{b} \in \mathbb{Z}^n$ if for the first position i in which $\mathbf{a}$ and $\mathbf{b}$ differ, we have $a_i < b_i$, or, equivalently,

$$\mathbf{a} \prec \mathbf{b} \equiv \bigvee_{i=1}^{n} \left(a_i < b_i \wedge \bigwedge_{j=1}^{i-1} a_j = b_j \right). \tag{3}$$

Note that the lexicographic order can be represented as a polyhedral relation

$$L_n = \{\mathbf{a} \to \mathbf{b} \mid \mathbf{a}, \mathbf{b} \in \mathbb{Z}^n \wedge \mathbf{a} \prec \mathbf{b}\} \tag{4}$$

$$= \bigcup_{i=1}^{n} \{\mathbf{a} \to \mathbf{b} \mid \mathbf{a}, \mathbf{b} \in \mathbb{Z}^n \wedge a_i < b_i \wedge \bigwedge_{j=1}^{i-1} a_j = b_j\}. \tag{5}$$

Example 4. In `iscc`, we can use the `<<` operator on a pair of sets to construct a relation between pairs of elements in the two sets such that the element of the first set is lexicographically smaller than the element of the second set. Applying this operation to universe sets, we obtain the pure lexicographic order. In particular,

```
{ [i,j] } << { [i,j] };
```

results in

```
{ [i, j] -> [i', j'] : i' >= 1 + i;
  [i, j] -> [i, j'] : j' >= 1 + j }
```

3.3   Polyhedral Models

The polyhedral model of a sequential program mainly consists of the iteration domains of the statements (as explained in Sect. 3.1), and *access relations*. An access relation is a polyhedral relation $R \subseteq (\Sigma \times \mathbb{Z}^d) \to (\Sigma \times \mathbb{Z}^a)$ that maps the iteration vector of the corresponding statement to an array element. Here, d is the dimension of the iteration domain and a is the dimension of the array. A scalar can be considered as a zero-dimensional array. The access relation of an access to a scalar is therefore simply the Cartesian product of the iteration domain and the entire zero-dimensional space.

Example 5. The access relation of the first argument to the call to `Sobel` in Line 6 of Fig. 1 is

$$\{S(i, j) \to a(a, b) \mid 1 \le i < K - 1 \wedge 1 \le j < N - 1 \wedge a = i - 1 \wedge b = j - 1\}. \tag{6}$$

Finally, the model should contain information about the relative execution order of any pair of statements in order to determine whether an iteration of one statement precedes or follows an iteration of another statement. A common way of representing this relative position is through a *schedule*, which assigns a timestamp to each iteration of each statement. It is usually convenient to allow these timestamps to be multi-dimensional, with the order determined by the lexicographic order. In other words, a schedule is a relation that maps each of the iteration domains to some common domain.

The schedule should reflect the execution order of the input program. Since only the relative order implicit in the schedule is of importance, such a schedule can be constructed in many different ways. Here, we will describe just one of those. Let d_i be the dimension of the iteration domain D_i and let d be the maximal such dimension, i.e., $d = \max_i d_i$. We define a relation between on one side each of the iteration domains and on the other side a $(2d + 1)$-dimensional space. The odd dimensions $2k + 1$ (1 to $2d - 1$) in this space are equated to iterators k. The even dimensions $2k$ (0 to $2d$) represent the "position" of the kth loop surrounding the given statement (for $k < d$), or the statement itself (for $k = d$). The position could be a line number or a sequence number. For iteration domains with $d_i < d$, only dimensions up $2d_i$ are defined as above. The remaining dimensions can be assigned an arbitrary value, say zero.

To see that the lexicographic order on the target space corresponds to the execution order of the input program, note that if the first dimension in which two schedule iteration vectors differ is $2n$, then the two corresponding statements share n loops, the iterators of these n loops have the same value in both vectors and the statements have a different location inside the nth loop. If the first dimension in which two schedule iteration vectors differ is $2n + 1$, then the two corresponding statements share at least $n + 1$ loops, the iterators of first n loops have the same value in both vectors, but the iterator of the next loop has a different value in the two vectors.

Example 6. If we take line numbers as positions, then the schedule corresponding to the original execution order of the program in Fig. 1, is

$$\{\text{R}(i,j) \rightarrow (1,i,2,j,3)\} \cup \{\text{S}(i,j) \rightarrow (4,i,5,j,6)\} \cup \{\text{W}(i,j) \rightarrow (4,i,5,j,9)\}. \quad (7)$$

Combining all this information, we have the following definition.

Definition 4 (Polyhedral Model). The *polyhedral model* of a sequential program consists of a list of statements, where each statement is in turn represented by

- An identifier ℓ_i,
- A dimension d_i,
- An iteration domain (a polyhedral set) $D_i \subseteq \{\ell_i\} \times \mathbb{Z}^{d_i}$,
- A list of accesses and
- A schedule.

Finally, each array access is represented by an identifier, an access relation (a polyhedral relation) and a type (read or write).

The use of a polyhedral model imposes the following restrictions on the input program:

- Static control flow,
- Pure functions and
- Loop bounds, conditions and index expressions are quasi-affine expressions in the parameters and the iterators of enclosing loops.

The extraction of a polyhedral model from a sequential program is available in several modern industrial and research compilers, e.g., [1, 10, 17, 19].

Example 7. The parse_file operation of iscc parses a C file and constructs a polyhedral model that is similar to that of Definition 4. In particular, it returns a list with four elements. The first contains the union of all iteration domains. The second contains the union of all write access relation. The third contains the union of all read access relation. The final element contains the schedule. Applying this operation to the program in Fig. 8 (with the required declarations), results in the following output. The program will be discussed in more detail in Sect. 6.1.

```
([N]  -> { S1[i]  : i >= 0 and i <= N;
           S2[i]  : i >= 0 and i <= N },
 [N]  -> { S1[i]  -> a[i] : i >= 0 and i <= N;
           S2[i]  -> b[i] : i >= 0 and i <= N },
 [N]  -> { S2[i]  -> a[N - i] : i >= 0 and i <= N },
 [N]  -> { S2[i]  -> [4, i]; S1[i]  -> [3, i] })
```

3.4 Piecewise Quasi-Polynomials

As explained in Sect. 2, each channel in the process network has a buffer of a bounded size. If the network is parametric, then this bound will in general not simply be a number, but rather an expression in the parameters. Some of the intermediate steps in computing these bounds may also result in expressions involving both the parameters and the iterators, or just the iterators if the network is non-parametric. In both cases the expressions will be *piecewise quasi-polynomials*. Quasi-polynomials are polynomial expressions that may involve integer divisions of affine expressions in the variables. Piecewise quasi-polynomials are quasi-polynomials defined over polyhedral pieces of the domain. More formally, they can be defined as follows.

Definition 5 (Quasi-Polynomial). A *quasi-polynomial* $q(\mathbf{x})$ in the integer variables $\mathbf{x}$ is a polynomial expression in greatest integer parts of affine expressions in the variables, i.e., $q(\mathbf{x}) \in \mathbb{Q}[\lfloor \mathbb{Q}[\mathbf{x}]_{\leq 1} \rfloor]$.

A note on the notation used in this definition. $\mathbb{Q}$ is the set of rational numbers. $\mathbb{Q}[\mathbf{z}]$ is the set of polynomial expressions in $\mathbf{z}$. A subscript on this notation indicates a bound on the degree. That is, $\mathbb{Q}[\mathbf{x}]_{\leq 1}$ represents the degree-1 polynomials in $\mathbf{x}$, i.e., the affine expressions in $\mathbf{x}$. Here, we take the greatest integer part of each of these affine expressions and then consider the polynomial expressions in those greatest integer parts

Definition 6 (Piecewise Quasi-Polynomial). A *piecewise quasi-polynomial* $q(\mathbf{x})$, with $\mathbf{x} \in \mathbb{Z}^d$ consists of a finite set of pairwise disjoint polyhedral sets $K_i \subseteq \mathbb{Q}^d$, each with an associated quasi-polynomial $q_i(\mathbf{x})$. The value of the piecewise quasi-polynomial at $\mathbf{x}$ is the value of $q_i(\mathbf{x})$ with K_i the polyhedral set containing $\mathbf{x}$, i.e.,

$$q(\mathbf{x}) = \begin{cases} q_i(\mathbf{x}) & \text{if } \mathbf{x} \in K_i \\ 0 & \text{otherwise.} \end{cases}$$

Note that the usual polynomials are special cases of quasi-polynomials as $\lfloor x_j \rfloor = x_j$ for x_j integer.

Example 8. Consider the statement in Line 3 of Fig. 3. The number of times this statement is executed can be represented by the piecewise quasi-polynomial

$$\left\{ \left\lfloor \tfrac{N+2}{3} \right\rfloor \quad \text{if } N \geq 0. \right.$$

Note that there is only one polyhedral "piece" in this example.

3.5 Polyhedral Process Networks

Now we can finally define the structure of a polyhedral process network. For simplicity we assume that no merging of processes has been performed, i.e., that each process corresponds to a statement in the polyhedral model of the input.

Definition 7 (Polyhedral Process Network). A *polyhedral process network* is a directed graph with as vertices a set of *processes* $\mathscr{P}$ and as edges *communication channels* $\mathscr{C}$. Each process $P_i \in \mathscr{P}$ has the following characteristics

- A statement identifier s_i,
- A dimension d_i,
- An iteration domain $D_i \subseteq \{ s_i \} \times \mathbb{Z}^{d_i}$.

Each channel $C_i \in \mathscr{C}$ has the following characteristics

- A source process $S_i \in \mathscr{P}$,
- A target process $T_i \in \mathscr{P}$,
- A source access identifier corresponding to one of the accesses in the statement s_{S_i},
- A target access identifier corresponding to one of the accesses in the statement s_{T_i},
- A polyhedral relation $M_i \subseteq D_{S_i} \to D_{T_i}$ mapping iterations from the source domain to the target domain,
- A type (e.g., FIFO),
- A piecewise quasi-polynomial buffer size.

The identifiers in the process network can be used to obtain more information about the statements and the accesses when constructing a hardware or software realization of the network. The mapping M_i identifies which iterations of the source process write to the channel, which iterations of the target process read from the channel and how these iterations are related. The buffer sizes are such that the network can be executed without deadlocks.

Example 9. Consider the network in Fig. 2, derived from the code in Fig. 1. All processes have dimension $d_i = 2$. The iteration domain D_1 of the ReadImage process was given in Example 2. The iteration domains of the other two processes are

$$D_2 = \{ \, \text{S}(i, j) \mid 1 \leq i < K - 1 \wedge 1 \leq j < N - 1 \, \} \tag{8}$$

and

$$D_3 = \{ \, \text{W}(i, j) \mid 1 \leq i < K - 1 \wedge 1 \leq j < N - 1 \, \}.$$

There are nine communication channels between the first and the second process and one communication channel between the second and the third. All communication channels in this network are FIFOs. How these communication channels are constructed is explained in Sect. 5.1 and how their types are determined is explained in Sect. 6. The arrows in Fig. 2 representing the communication channels are annotated with the name of the array that has given rise to the channel and the buffer size. These buffer sizes are computed in Sect. 8.

4 Polyhedral Analysis Tools

The construction of a polyhedral process network relies on a number of fundamental polyhedral operations. This section provides an overview of these operations. In each case, one or more tools are mentioned with which the operation can be performed. This list of tools is not exhaustive.

4.1 *Parametric Integer Programming*

By far the most fundamental step in the construction of a process network is figuring out which processes communicate data with each other and in what way. At the level of the input source program, this means figuring out for each value read in the program where the value was written. That is, for each read access to an array element, we need to know what was the last write access to the same array element that occurred before the given read access. In particular, if many iterations of the same statement write to the same array element, we need the (lexicographically) last iteration of that statement. Computing such a lexicographically maximal (or minimal) element of a polyhedral set can be performed using *parametric integer programming*.

Let us first define a lexmax operator on polyhedral relations.

Definition 8 (Lexicographic Maximum). Given a polyhedral relation R, the *lexicographic maximum* of R is a polyhedral relation with the same parameter domain and the same domain as R that maps each element $\mathbf{i}$ of the domain to the unique lexicographically maximal element that corresponds to $\mathbf{i}$ in R, i.e.,

$$\operatorname{lexmax} R = \{\, \ell_1(\mathbf{i}) \to \ell_2(\mathbf{j}) \in R \mid \neg(\exists \mathbf{j}' : \ell_1(\mathbf{i}) \to \ell_2(\mathbf{j}') \in R \wedge \mathbf{j} \prec \mathbf{j}') \,\}. \qquad (9)$$

The lexicographic maximum of set S is similarly defined as

$$\operatorname{lexmax} S = \{\, \ell(\mathbf{i}) \in S \mid \neg(\exists \mathbf{i}' : \ell(\mathbf{i}') \in S \wedge \mathbf{i} \prec \mathbf{i}') \,\}.$$

Now we can define an operation for computing $\operatorname{lexmax} R$ on a given domain as a polyhedral relation.

Operation 1 (Partial Lexicographic Maximum).
Input: • *a basic polyhedral relation*
$\quad\quad R : \mathbb{Z}^n \to (\Sigma \times \mathbb{Z}^{d_1}) \to (\Sigma \times \mathbb{Z}^{d_2}) \to \mathbb{B}$,
$\quad\quad$• *a basic polyhedral set* $S : \mathbb{Z}^n \to (\Sigma \times \mathbb{Z}^{d_1}) \to \mathbb{B}$
Output: • *a polyhedral relation* $M = \operatorname{lexmax}(R \cap (S \to (\Sigma \times \mathbb{Z}^{d_2})))$
$\quad\quad$• *a polyhedral set* $E = S \setminus \operatorname{dom} R$
The polyhedral relation M satisfies the following additional conditions:

- *Every existentially quantified variable in M is explicitly represented as the greatest integer part of an affine expression in the parameters and the domain variables,*
- *Every variable in the range of M is explicitly represented as an affine expressions in the parameters, the domain variables and the existentially quantified variables.*

By essentially removing "$\Sigma \times \mathbb{Z}^{d_1}$" from the above operation, we obtain a similar operation on polyhedral sets. Operation 1 can be performed using `isl` [21],[2] or, with some additional transformations, using `piplib` [7].[3] The use of the output set E will become clear in Sect. 5.

Example 10. Let us compute the lexicographically maximal element of the polyhedral set D_1 (2) from Example 3. S may be taken as the universal parameter domain, i.e., $S = N \mapsto \{\mid\}$. The output is the lexicographically maximal element of the set D_1:

$$M = \operatorname{lexmax} D_1(N) = N \mapsto \left\{\, \mathtt{S1}(i) \mid \exists \alpha = \left\lfloor \frac{N+2}{3} \right\rfloor : i = 3\alpha - 3 \wedge N \geq 1 \,\right\}.$$

The set E describes the (parameter) values for which there is no element in the set D_1, i.e., $E = N \mapsto \{\mid N \leq 0\}$.

Let us now use the `isl_pip` application to compute this result. We call `isl_pip` as follows.

---

[2] http://freecode.com/projects/isl/
[3] http://www.piplib.org/

```
$ isl_pip --format=affine
[N] -> { : }
-1
[N] -> { S1[i] : exists a : 0 <= i < N and i = 3 a }
Maximize
```

The first line invokes isl_pip. The remaining lines represent the input. The first line of the input is the parameter domain. The second line is always -1 and is only present for compatibility with the original pip application. The third line is the actual input set and the last line specifies that we want to compute the lexicographic maximum. (By default, isl_pip will compute the lexicographic minimum.) The output is as follows.

```
[N] -> { S1[(-3 + 3*[(2 + N)/3])] : N >= 1 }
no solution: [N] -> { : N <= 0 }
```

Note that the $\lfloor \cdot \rfloor$ operation is represented using square brackets.

Note that Operation 1 can be used to compute an explicit representation of the existentially quantified variables. As an alternative to Operation 1, the lexicographic maximum can be computed using Omega [12],[4] by expressing the lexicographic order in (9) using linear constraints, as in (5). However, the result will not necessarily satisfy the two conditions of Operation 1.

4.2 Emptiness Check

A very basic and frequently used operation is that of checking whether a given polyhedral set or relation contains any elements for any value of the parameters.

Operation 2 (Emptiness Check).
Input: *a polyhedral relation* $R : \mathbb{Z}^n \to (\Sigma \times \mathbb{Z}^{d_1}) \to (\Sigma \times \mathbb{Z}^{d_2}) \to \mathbb{B}$
Output: true *if* $\forall \mathbf{s} \in \mathbb{Z}^n : R(\mathbf{s}) = \emptyset$ *and* false *otherwise*

Operation 2 can be performed by applying Operation 1 (Partial Lexicographic Maximum) on each of the basic polyhedral sets in a polyhedral set $S \in \mathbb{Z}^{n+d_1+d_2}$ with the same description as R, but where all parameters and input variables are treated as set variables. The relation R is empty iff S is empty iff in turn none of the basic polyhedral sets has a lexicographically maximal element. Since S is (a union of) integer linear programming (ILP) problem(s), any other algorithm for testing the feasibility of an ILP problem will work as well. For example, isl uses generalized basis reduction [4].

---

[4]http://www.chunchen.info/omega/

4.3   Parametric Counting

An important step in the buffer size computation is the computation of the number of elements in the buffer before a given read. We will be able to reduce this computation to that of counting of the number of elements in the image of a polyhedral relation R, denoted #R, for which we will use the following operation.

Operation 3 (Number of Image Elements).
Input: *a polyhedral relation* $R : \mathbb{Z}^n \to \left(\Sigma \times \mathbb{Z}^{d_1}\right) \to \left(\Sigma \times \mathbb{Z}^{d_2}\right) \to \mathbb{B}$
Output: *a piecewise quasi-polynomial*
$q : \mathbb{Z}^n \to (\Sigma \times \mathbb{Z}^{d_1}) \to \mathbb{Q} : (\mathbf{s}, \ell(\mathbf{t})) \mapsto q(\mathbf{s}, \ell(\mathbf{t})) = \#R(\mathbf{s}, \ell(\mathbf{t}))$

Example 11. Consider the set D_1 from (2) in Example 3. Applying Operation 3 to this polyhedral set results in the piecewise quasi-polynomial of Example 8. In iscc, we can perform

```
card [N] -> { S1[i] : 0 <= i < N and i % 3 = 0 };
```

to obtain

```
[N] -> { ([(2 + N)/3]) : N >= 1 }
```

Operation 3 can be performed using barvinok [23, 25].

4.4   Computing Parametric Upper Bounds

As explained in the previous section, Operation 3 can be used to compute the number of elements in a buffer at any given time. When allocating memory for this buffer, we need to know the maximal number of elements that will ever have to reside in the buffer. As usual, we want to perform this computation parametrically. In general, computing the actual maximum may be too difficult, however. We therefore settle for computing an upper bound that is hopefully reasonably close to the maximum.

Operation 4 (Upper Bound on a Quasi-Polynomial).
Input: • *A piecewise quasi-polynomial* $q : \mathbb{Z}^n \to (\Sigma \times \mathbb{Z}^d) \to \mathbb{Q}$
• *A bounded polyhedral set* $S : \mathbb{Z}^n \to (\Sigma \times \mathbb{Z}^d) \to \mathbb{B}$, *the domain over which to compute the upper bound*
Output: *A piecewise quasi-polynomial* $u : \mathbb{Z}^n \to \mathbb{Q}$ *such that*

$$u(\mathbf{s}) \geq \max_{\ell(\mathbf{t}) \in S(\mathbf{s})} q(\mathbf{s}, \ell(\mathbf{t})) \qquad \forall \mathbf{s} \in \text{pdom}\, S$$

Fig. 4 A simple program
reading the same array
elements several times

```
1   for (i = 0; i < N; ++i)
2     for (j = 0; j < i; ++j)
3   S:  b[i][j] = f(a[i + j]);
```

Operation 4 can be performed using `bernstein`[3][5] or `isl`. While `bernstein` only implements the basic technique which only applies to polynomials, `isl` also implements an extension to quasi-polynomials [6]. Alternatively, the quasi-polynomials, which are usually the result of a counting problem such as Operation 3, can be approximated by a polynomial during the counting process [16].

Combining Operations 3 and 4 results in the following operation, by taking $S = \text{dom} R$.

Operation 5 (Upper Bound on the Number of Image Elements).
Input: *A polyhedral relation* $R : \mathbb{Z}^n \to (\Sigma \times \mathbb{Z}^{d_1}) \to (\Sigma \times \mathbb{Z}^{d_2}) \to \mathbb{B}$
Output: *A piecewise quasi-polynomial* $u : \mathbb{Z}^n \to \mathbb{Q}$ *such that*

$$u(\mathbf{s}) \geq \max_{\ell(\mathbf{t}) \in \text{dom} R(\mathbf{s})} \#R(\mathbf{s}, \ell(\mathbf{t})) \qquad \forall \mathbf{s} \in \text{pdom} R$$

Example 12. Consider the program in Fig. 4 and assume we want to know the maximal number of times an unspecified element of array a is read. This number is the maximal number of domain elements in the access relation that map to the same array element. In terms of the operations we have defined above, it is the maximal number of image elements in the inverse of the access relation. The access relation is

$$A = \{ \mathsf{S}(i, j) \to \mathsf{a}(a) \mid 0 \leq i < N \wedge 0 \leq j < i \wedge a = i + j \}.$$

Its inverse is

$$A^{-1} = \{ \mathsf{a}(a) \to \mathsf{S}(i, j) \mid 0 \leq i < N \wedge 0 \leq j < i \wedge a = i + j \}.$$

Applying Operation 3 yields the number of times a given array element is read:

$$\#A^{-1}(N, \mathsf{a}(a)) = \begin{cases} a - \left\lfloor \frac{a}{2} \right\rfloor & \text{if } 0 \leq a < N \\ N - \left\lfloor \frac{a}{2} \right\rfloor - 1 & \text{if } N \leq a \leq 2N - 3. \end{cases}$$

An upper bound on this number can then be computed using Operation 4, yielding,

$$\max_{\mathsf{a}(a) \in \text{dom} A^{-1}} \#A^{-1}(N, \mathsf{a}(a)) \leq u(N) = \left\{ \frac{N}{2} \quad \text{if } N \geq 2. \right.$$

---

[5]http://icps.u-strasbg.fr/pco/bernstein.htm

Using `iscc` to perform these calculations, we have as input

```
A := [N] -> { S[i, j] -> a[i + j] : 0 <= i < N and
                                     0 <= j < i };
card (A^-1);
ub (card (A^-1));
```

and as output

```
[N] -> { a[i0] -> ((-1 + N) - [(i0)/2]) :
                  i0 <= -3 + 2N and i0 >= N;
         a[i0] -> (i0 - [(i0)/2]) :
                  i0 <= -1 + N and i0 >= 1 }
([N] -> { max(1/2 * N) : N >= 3;
          max(1/2 * N) : N = 2 }, False)
```

Note that the ub operation produces two results, one is the actual upper bound and the other is a boolean that indicates whether the upper bound is definitely reached, i.e., whether the upper bound is known to be equal to the maximum. In this example, the upper bound may not be equal to the maximum.

4.5 Polyhedral Scanning

When writing out code for a process in a process network, we not only need to make sure that an operation is performed for each element of its iteration domain, but we also need to insert the appropriate reading and writing primitives in the appropriate places. In particular, if C_i is a communication channel with the given process as its source, then a write to the communication channel needs to be inserted in each element of the domain of its mapping relation M_i. Similarly, if C_j is a communication channel with the given process as its target, then a read from the communication channel needs to be inserted in each element of the range of its mapping relation M_j. Each of these domains and ranges is a polyhedral set and we see that we need to generate code for visiting each element of these sets. That is, we need the following operation.

Operation 6 (Code Generation).
Input: *A polyhedral relation R with* $\operatorname{ran} R \subseteq \{\ell\} \times \mathbb{Z}^d$
Output: *Code for visiting each element in* $\operatorname{dom} R$ *in lexicographic order of their image*

The condition that the range of R needs to live in a single space, i.e., with a fixed label and a fixed dimension, ensures that we can evaluate the lexicographic order.

Fig. 5 A process with a one-dimensional iteration domain

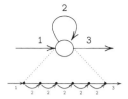

Operation 6 can be performed using CLooG [2],[6] CodeGen [13] (part of the Omega library) or isl.

Essentially, the input of Operation 6 is a schedule. When generating code for a single process in a process network, we can essentially use an identity schedule. However, we not only need to generate code for applying the main operation of the process, but also for reading and writing to the communication channels. Furthermore, we need to ensure that within a given iteration of the process, the reads happen *before* the actual operation and that the writes happen *after* the operation. To enforce this extra ordering constraint, we introduce an extra innermost dimension in the range of the schedule, assigning it the value 0 for reads, 1 for the iteration domain and 2 for writes. The domains for the read and write operations are obtained as ranges and domains of the associated communication channel mappings.

Example 13. Consider a process with a one-dimensional iteration domain

$$D = \{ S(i) \mid 0 \leq i < N \}$$

that reads a value from some other process in its first iteration

$$M_1 = \{ P() \rightarrow S(i) \mid i = 0 \wedge N > 0 \},$$

propagates a value from one iteration to the next

$$M_2 = \{ S(i) \rightarrow S(i') \mid 0 \leq i < N - 1 \wedge i' = i + 1 \}$$

and then sends a value to some other process in its last iteration

$$M_3 = \{ S(i) \rightarrow C() \mid i = N - 1 \wedge N > 0 \}.$$

The process is shown in Fig. 5. The code that results from scanning D at 1, ran M_1 at 0, dom M_2 at 2, ran M_2 at 0 and dom M_3 at 2 is shown schematically in Fig. 6.

In iscc, we could generate such code using the following script.

```
D  := [N] -> { S[i]  : 0 <= i < N };
M1 := [N] -> { P[] -> S[0]  : N > 0 };
M2 := [N] -> { S[i] -> S[i+1]  : 0 <= i < N - 1 };
M3 := [N] -> { S[N-1] -> C[]  : N > 0 };
```

---

[6]http://www.cloog.org/

Fig. 6 Code generated for
the process in Fig. 5

```
 1  for (i = 0; i < N; i++) {
 2      if (i == 0)
 3          Read(1);
 4      if (i >= 1)
 5          Read(2);
 6      f();
 7      if (i < N - 1)
 8          Write(2);
 9      if (i == N - 1)
10          Write(3);
11  }
```

```
R1 := { S[i] -> R1[i] } (ran M1);
R2 := { S[i] -> R2[i] } (ran M2);
W2 := { S[i] -> W2[i] } (dom M2);
W3 := { S[i] -> W3[i] } (dom M3);
S  := { D[i] -> [i,1]; R1[i] -> [i,0]; R2[i] -> [i,0];
         W2[i] -> [i,2]; W3[i] -> [i,3] }
      * (D + R1 + R2 + W2 + W3);
codegen S;
```

The first four lines define the iteration domain and the communication channel
mappings. The next four lines compute the domains of the reads and writes by
computing domains and ranges of the mappings and performing a renaming. The
next line defines the schedule. Note that $*$ is here applied to a relation and a set, in
which case the domain of the relation is intersected with the set. The final line prints
the code corresponding to S.

5 Dataflow Analysis

This section describes how the communication channels in the process network are
constructed using exact dataflow analysis [8]. We first discuss the standard dataflow
analysis and then explain how some inter-process communication can be avoided
by considering reuse.

5.1 Standard Dataflow Analysis

Standard exact dataflow analysis [8] is concerned with finding for each read of
a value from an array element in the program, the write operation that wrote
that value to that array element. By replacing the write to the array by a write
to one or more communication channels (as many as there are accesses in the
program where the value is read) and the read from the array to a read from

the appropriate communication channel, we will have essentially constructed the communication channels. The effect of dataflow analysis can be described as the following operation.

Operation 7 (Dataflow Analysis).
Input: • A read access relation $R \subseteq \mathbb{Z}^n \to (\Sigma \times \mathbb{Z}^d) \to (\Sigma \times \mathbb{Z}^a)$
  • A list of [write] access relations $W_i \subseteq \mathbb{Z}^n \to (\Sigma \times \mathbb{Z}^{d_i}) \times (\Sigma \times \mathbb{Z}^a)$
  • A schedule $S \subseteq \mathbb{Z}^n \to (\Sigma \times \bigcup_i \mathbb{Z}^{d_i}) \to (\Sigma \times \mathbb{Z}^k)$ such that $\mathrm{dom}(R \cup \bigcup_i W_i) \subseteq \mathrm{dom}\, S$
Output: • A list of polyhedral relations $M_i \subseteq \mathbb{Z}^n \to (\Sigma \times \mathbb{Z}^{d_i}) \to (\Sigma \times \mathbb{Z}^d)$
  • A polyhedral set $U \subseteq \mathbb{Z}^n \to (\Sigma \times \mathbb{Z}^d)$
The output satisfies the following constraints

- Each element in the domain of R is an element either of U or of the range of exactly one of the mappings M_i, i.e., $\{\,\mathrm{ran}\, M_i\,\}_i \cup \{\,U\,\}$ partitions $\mathrm{dom}\, R$,
- If a particular element in the domain of R is in the range of one of the mappings, i.e., $\ell(\mathbf{j}) \in \mathrm{dom}\, R \cap \mathrm{ran}\, M_k$, then the corresponding [write] iteration of W_k, i.e., $M_k^{-1}(\ell(\mathbf{j}))$, was the last iteration (according to the schedule S) in any of the domains of the input access relations W_i that accessed the same array element as $\ell(\mathbf{j})$, [i.e., it wrote the value read by $\ell(\mathbf{j})$] and was executed before $\ell(\mathbf{j})$,
- If a read iteration belongs to U, i.e., $\ell(\mathbf{j}) \in \mathrm{dom}\, R \cap U$, then this iteration accesses an array element that was not accessed by any of the W_i, [i.e., this iteration reads an uninitialized value].

Operation 7 (available in `isl`) is applied for each read access in the program. The list of access relations required in the input is constructed from all write accesses in the program that access the same array. For each $M_i \neq \emptyset$, a communication channel is created from the process corresponding to the writing statement to the process corresponding to the reading statement, with the given M_i as mapping. The type and buffer size are defined in the following sections. The set U in the output is assumed to be empty.

We will now briefly sketch how Operation 7 can be implemented. For this purpose, we will assume that the access relations have been scheduled, i.e., that the schedule S has been applied to the domains of the access relations. We will further assume that the input contains a single (scheduled) write access relation W. The composition of the read access relation with the inverse of the write access relation yields a mapping from read iterations to write iterations that wrote to the array element accessed by the read. The one that actually wrote the value that is read, is the one that was the (lexicographically) last to write to the element before the read. That is, we want to compute

$$\mathrm{lexmax}\left(\left(W^{-1} \circ R\right) \cap L_{2d+1}^{-1}\right),$$

with L_{2d+1} the lexicographically smaller relation from (5). The result of this computation is the inverse of the relation M in the output.

Unfortunately, we cannot directly apply Operation 1 to compute this lexico-graphic maximum because L_{2d+1} (5) is a union of basic relations as soon as $d \geq 1$. (If W or R are unions then the basic relations in these unions can be treated as separate writes or reads, so we need not worry about unions in this part of the equation.) However, a scheduled write iteration that shares i_1 iterator values with the scheduled read iteration will always be executed after a scheduled write iteration that only shares $i_2 < i_1$ iterator values with the scheduled read iteration. We can therefore first apply Operation 1 to the writes that share $2d$ iterator values and with S set to the domain of the access relation. If the resulting set E is not empty then we can continue with the writes that share $2d - 1$ iterator values, with S set to each of the basic sets in E. This process continues until all the resulting sets E are empty or until we have finished the case of 0 common iterator values.

Example 14. Consider once more the access relation (6) of the first argument to the call to Sobel in Line 6 of Fig. 1 from Example 5 and apply the schedule (7) from Example 6. The result is

$$R = \{ (4,i,5,j,6) \to a(a,b) \mid 1 \leq i < K-1 \land 1 \leq j < N-1 \land a=i-1 \land b=j-1 \}.$$

The only write to the a array occurs in Line 3, with scheduled access relation

$$W = \{ (1,i,2,j,3) \to a(a,b) \mid 0 \leq i < K \land 0 \leq j < N \land a=i \land b=j \}.$$

Composition of these two relations yields

$$W^{-1} \circ R = \{ (4,i,5,j,6) \to (1,i',2,j',3) \mid 1 \leq i < K-1 \land 1 \leq j < N-1 \land$$
$$i' = i-1 \land j' = j-1 \}.$$

We see that the range iterators are uniquely defined in terms of the domain iterators, so in this case there is no need to compute the lexicographic maximum as $\text{lexmax}\, W^{-1} \circ R$ would be identical to $W^{-1} \circ R$. However, let us consider what would happen if we were to apply the above algorithm anyway. Since the first iterators in domain and range are distinct constants, the two iteration vectors never share any initial iterator values. The first $2d = 4$ applications of Operation 1 therefore operate on an empty basic polyhedral relation R' and simply return the input basic polyhedral set $S = \text{dom}\, R$ as E. The final application returns $M_1^{-1} = R' = W^{-1} \circ R$ and $E = \emptyset$. Applying the inverse of the schedule, we obtain

$$M_1 = \{ R(i',j') \to S(i,j) \mid 1 \leq i < K-1 \land 1 \leq j < N-1 \land i' = i-1 \land j' = j-1 \}. \tag{10}$$

iscc provides an operation that calls isl to perform these computations for you (including the computations mentioned after the example).

```
D := [K,N] -> { R[i,j] : 0 <= i < K and 0 <= j < N;
                S[i,j] : 1 <= i < K - 1 and
                         1 <= j < N - 1 };
```

```
W := { R[i,j] -> a[i,j] } * D;
R := { S[i,j] -> a[i-1,j-1] } * D;
S := { R[i,j] -> [1,i,2,j,3]; S[i,j] -> [4,i,5,j,6] };
last W before R under S;
```

produces

```
([K, N] -> { R[i, j] -> S[1 + i, 1 + j] : i >= 0 and
                i <= -3 + K and j >= 0 and j <= -3 + N },
 [K, N] -> {  })
```

The output contains both the union of the M_i and U. Here, U is the empty set because all of the read accesses read a value that was previously written.

If there is more than one write access relation in the input of Operation 7, then the computation is a little bit more complicated. After computing the lexicographically maximal element of a write that shares i iterator values with the read, we still need to check that there is no other write in between. That is, we need to check if there is a write from a different write access that also shares i iterator values with the read, also occurs before the read, but occurs after the already found write. Again we have to consider all possible cases of shared iterator values between the two writes, now from i to $2d + 1$, as a write that occurs *after* a given iteration and that has fewer iterator values in common with the given iteration will be executed after an iteration that has more iterator values in common.

The above sketch of an algorithm can still be significantly improved by checking the write access relations in the appropriate order and by taking into account the fact that the dimensions that correspond to the statement locations have fixed values. For a more detailed description, we refer to [8], where a variant of the above algorithm is applied directly on "quasts", the output structure of piplib which we will not explain here.

5.2 Reuse

A process network constructed using the standard dataflow analysis of the previous section will always send data from the process that corresponds to the statement that wrote a given value in the sequential program to the process that corresponds to the statement that read the given value, even if the same value is read multiple times from the same process. Now, in practically any form of implementation of process networks, sending data from a process to itself (or simply keeping it) is much cheaper than sending data to another process. It is therefore better to only transmit data from one process to another the first time it is needed.

Obtaining a network that only transmits data to a given process once is actually fairly simple. When applying Operation 7 from the previous section, instead of only supplying a list of write accesses to the array read by the read access R, we extend the list with all read accesses to the given array (including R itself) from

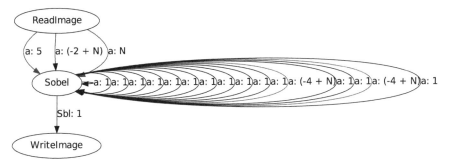

Fig. 7 A process network corresponding to the sequential program in Fig. 1 exploiting reuse

the same statement. Any value that was already read by the given process will then be obtained from the last of those reads from the same process instead of from the process that actually wrote the value.

Example 15. In Example 1, we have shown a process network (Fig. 2) derived from the code in Fig. 1 without exploiting reuse inside the Sobel process. Figure 7 shows a process network derived from the same code, but in this case with reuse. The first network was created by only considering the write to array a in Line 13 of Fig. 1 as a possible source for the reads in Line 6, while the second was created by also considering all the reads in Line 6 as possible sources. It is clear from the figures that in the first network the Sobel process does not communicate with itself, while in the second network it does. What may not be apparent, however, is that in both networks there are nine channels from the ReadImage process to the Sobel process. The second figure only shows three channels because the software used to generate these process networks automatically combines some communication channels if this combination does not affect the type of the channels, as explained at the end of Sect. 6.1. The number of channels from the first process to the second has not changed because each read in Line 6 reads at least one value that is not read by any of the other reads. However, even though the number of channels may not have changed, the number of values transmitted over these channels has been drastically reduced. In particular, the image read by the ReadImage process is now transmitted only once, instead of nearly nine times.

Performing

```
last (W+R) before R under S;
```

with W, R and S as in Example 14 produces the same results as before because R only includes a single access relation and there is no reuse within a single access. If we change it to include all reads from the given statement, then this operation would exploit the reuse in the combined accesses.

6   Channel Types

In the previous section, we showed how to determine the communication channels between the processes in the network. The practical implementation of these channels in hardware depends on certain properties that may or may not hold for the channels. In particular, it will be important to know if a communication channel can be implemented as a FIFO. Other properties include multiplicity and special kinds of internal channels. In this section, we describe how to check these properties.

6.1   FIFOs and Reordering Channels

The communication channels derived in Sect. 5 are characterized by a mapping M (see Definition 7) relating write iterations in one process to the corresponding read iterations in the same or another process. The channel needs to ensure that when the second process reads from the channel it is given the correct value and therefore needs to keep track of which iteration in the first process wrote a given value. There is, however, no need to keep track of this extra information if it can be shown that the values will always be read in the same order as that in which they were written. In such cases, the communication channel can simply be implemented as a FIFO.

To check if the writes and reads are performed in the same order, we consider what happens when the order is not the same, i.e., when reordering occurs on the channel. In this case, there is a pair of writes $(\mathbf{w}_1, \mathbf{w}_2)$ such that the corresponding reads $(\mathbf{r}_1, \mathbf{r}_2)$ are executed in the opposite order, i.e., $\mathbf{w}_1 \succ \mathbf{w}_2$ while $\mathbf{r}_1 \prec \mathbf{r}_2$. That is, reordering occurs if the relation

$$T = \left\{ \mathbf{w}_1 \to \mathbf{w}_2 \mid \exists \mathbf{r}_1, \mathbf{r}_2 : (\mathbf{w}_1 \to \mathbf{r}_1), (\mathbf{w}_2 \to \mathbf{r}_2) \in M \wedge \mathbf{w}_1 \succ \mathbf{w}_2 \wedge \mathbf{r}_1 \prec \mathbf{r}_2 \right\} \quad (11)$$

is non-empty, which can be verified by applying Operation 2 (Emptiness Check). Note that T will only be considered empty if there is no reordering for any value of the parameters. Also note that we only need the lexicographic order within iteration domains and not across iteration domains, as we only compare reads to other reads and writes to other writes. That is, we do not need a global schedule. If we identify two (or more) communication channels C_1 and C_2 as FIFOs, we can check if we can combine them into a single FIFO by performing two emptiness checks on a relation T as in (11), except that in one relation $(\mathbf{w}_1 \to \mathbf{r}_1)$ is taken from M_1 and $(\mathbf{w}_2 \to \mathbf{r}_2)$ is taken from M_2 and vice versa in the other relation.

Example 16. Consider the program in Fig. 8. The corresponding process network has two processes and one communication channel connecting them. The mapping on the channel is

Fig. 8 A program resulting in a network with a reordering channel

```
1   for (i = 0; i <= N; i++)
2   S1:  a[i] = g(i);
3   for (i = 0; i <= N; i++)
4   S2:  b[i] = f(a[N-i]);
```

$$M = \{ \texttt{S1}(w) \rightarrow \texttt{S2}(r) \mid 0 \le w \le N \wedge r = N - w \}$$

and we have

$$T = \{ \texttt{S1}(w_1) \rightarrow \texttt{S1}(w_2) \mid \exists r_1, r_2 : (\texttt{S1}(w_1) \rightarrow \texttt{S2}(r_1)), (\texttt{S1}(w_2) \rightarrow \texttt{S2}(r_2)) \in M \wedge$$

$$w_1 > w_2 \wedge r_1 < r_2 \}$$

$$= \{ \texttt{S1}(w_1) \rightarrow \texttt{S1}(w_2) \mid 0 \le w_1, w_2 \le N \wedge w_1 > w_2 \wedge N - w_1 < N - w_2 \}$$

$$= \{ \texttt{S1}(w_1) \rightarrow \texttt{S1}(w_2) \mid 0 \le w_2 < w_1 \le N \}.$$

For $N > 0$, this relation is clearly non-empty. We conclude that we are dealing with a reordering channel. See Example 21 for a variation on this example where we find a FIFO.

In iscc, this test can be performed as follows.

```
M := [N] -> { S1[w] -> S2[N - w] : 0 <= w <= N };
T := (M . (ran M << ran M) . M^-1) * (dom M >> dom M);
T;
sample T;
```

That is, we construct a relation that maps reads (ran M) to later reads, map both of these reads to the corresponding writes and intersect the result with a relation that maps writes to earlier writes. At the end, we perform the sample operation to obtain an element of the set T, thereby verifying that it is non-empty. The result is

```
[N] -> { S1[w] -> S1[w'] : w >= 0 and w <= N and
              w' <= -1 + w and w' <= N and w' >= 0 }
[N] -> { S1[1] -> S1[0] : N = 1 }
```

6.2 Multiplicity

The standard dataflow analysis from Sect. 5.1 may result in communication channels that are read more often than they are written to when the same value is used in multiple iterations of the reading process. Such channels require special treatment and this multiplicity condition should therefore be detected by checking whether there is any write iteration $\mathbf{w}$ that is mapped to multiple read iterations through the mapping M. In particular, let T be the relation

$$T = \{ \mathbf{w}_1 \rightarrow \mathbf{w}_2 \mid \exists \mathbf{r}_1, \mathbf{r}_2 : (\mathbf{w}_1 \rightarrow \mathbf{r}_1), (\mathbf{w}_2 \rightarrow \mathbf{r}_2) \in M \wedge \mathbf{w}_1 = \mathbf{w}_2 \wedge \mathbf{r}_1 \prec \mathbf{r}_2 \}.$$

This relation contains all pairs of identical write iterations that correspond to a pair of distinct read iterations. If this relation is non-empty, then multiplicity occurs. As in the previous section, we can check the emptiness using Operation 2 (Emptiness Check).

Fig. 9 Outer product source code

```
1  for (i = 0; i < N; ++i)
2  a:   a[i] = A(i);
3  for (j = 0; j < N; ++j)
4  b:   b[j] = B(j);
5  for (i = 0; i < N; ++i)
6      for (j = 0; j < N; ++j)
7  c:       c[i][j] = a[i] * b[j];
```

Fig. 10 Outer product network with multiplicity

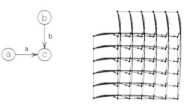

Fig. 11 Outer product network without multiplicity

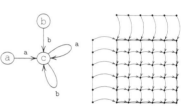

It should be noted that if we use the dataflow analysis from Sect. 5.2, i.e., taking into account possible reuse, then the resulting communication channels will never exhibit multiplicity. Instead, two channels would be constructed, one for the first time a value is read and one for propagating the value to later iterations. In general, it is preferable to detect reuse rather than multiplicity, because the analysis results in fewer types of channels and because taking advantage of reuse may split channels that are both reordering and with multiplicity into a pair of FIFOs.

Example 17. Consider the code for computing the outer product of two vectors shown in Fig. 9. Figures 10 and 11 show the results of applying standard dataflow analysis (Sect. 5.1) and dataflow analysis with reuse (Sect. 5.2) respectively. Each figure shows the network on the left and the dataflow between the individual iterations of the processes on the right. The iterations are executed top-down, left-to-right. In the first network, the channel between b and c has mapping

$$M_{b \to c} = \{ b(w) \to c(r_1, r_2) \mid 0 \leq r_1, r_2 < N \land w = r_2 \}.$$

For testing multiplicity, we have the relation

$$T = \{ r_{1,2} \to r_{2,2} \mid \exists r_{1,1}, r_{2,1} \in \mathbb{Z} : 0 \leq r_{1,1}, r_{1,2}, r_{2,1}, r_{2,2} < N \land r_{1,2} = r_{2,2} \land r_{1,1} < r_{2,1} \},$$

which is clearly non-empty, while for testing reordering we use the relation

$$T' = \{ r_{1,2} \to r_{2,2} \mid \exists r_{1,1}, r_{2,1} \in \mathbb{Z} : 0 \leq r_{1,1}, r_{1,2}, r_{2,1}, r_{2,2} < N \land r_{1,2} > r_{2,2} \land r_{1,1} < r_{2,1} \},$$

which is also non-empty. In the second network, the channel between b and c has mapping

$$M_{\text{b}\rightarrow\text{c}} = \{\, \text{b}(w) \rightarrow \text{c}(r_1, r_2) \mid 0 \leq r_2 < N \wedge r_1 = 0 \wedge w = r_2 \,\}. \qquad (12)$$

There is no multiplicity since each write corresponds to exactly one read. There is also no reordering, since the writes and reads occur for increasing values of $w = r_2$ in both processes.

6.3 Internal Channels

Internal channels, i.e., channels from a process to itself, can typically be implemented more efficiently than external channels. In particular, since processes are scheduled independently there is no guarantee that when a value is read, it is already available, or that when a value is written, there is still enough room in the channel to hold the data. The external channels therefore need to implement both blocking on read and blocking on write. For internal channels, there is no need for blocking. In fact, blocking would only lead to deadlocks. Besides the blocking issue, there are some special cases that allow for a more efficient implementation than a general FIFO buffer.

6.3.1 Registers

The first special case is that where the FIFO buffer holds at most one value. The FIFO buffer can then be replaced by a register. In Sect. 8 we will see how to compute a bound on a FIFO buffer in general, but the case where this bound is one can be detected in a simpler way. If each write $\mathbf{w}_1$ to a channel is followed by the corresponding read $\mathbf{r}_1$, without any other read occurring in between, then we indeed only need room for one value. On the other hand, if there *is* an intervening read, $\mathbf{w}_1 \prec \mathbf{r}_2 \prec \mathbf{r}_1$, then we will need more storage space. As usual, we can detect this situation by performing an emptiness check. Consider the set

$$T = \{\, \mathbf{w}_1 \mid \exists \mathbf{r}_1, \mathbf{r}_2 : (\mathbf{w}_1 \rightarrow \mathbf{r}_1) \in M \wedge \mathbf{r}_2 \in \text{ran}\, M \wedge \mathbf{w}_1 \prec \mathbf{r}_2 \prec \mathbf{r}_1 \,\}.$$

This set contains all writes $\mathbf{w}_1$ such that there exists a read $\mathbf{r}_2$ that occurs before the read $\mathbf{r}_1$ that corresponds to the write $\mathbf{w}_1$. Note that it is legitimate to compare read and write iterations because we are dealing with internal channels and so these iterations belong to the same iteration domain.

6.3.2 Shift Registers

Another special case occurs when the number of iterations between a write to an internal channel and the corresponding read from the channel is constant. If so, the FIFO can be replaced by a shift register, shifting the data in the channel by one in every iteration, independently of whether any data was read or written in that iteration. Such shift registers can be implemented more efficiently than FIFOs in hardware.

Checking whether we can use a shift register is as easy as writing down the relation

$$R = \{\, \mathbf{w} \to \mathbf{i} \mid \exists \mathbf{r} \in \mathbb{Z}^d : (\mathbf{w} \to \mathbf{r}) \in M \wedge \mathbf{i} \in D \wedge \mathbf{w} \prec \mathbf{i} \prec \mathbf{r} \,\},$$

where M is the mapping on the channel and D is the iteration domain of the process, and applying Operation 3 (Number of Image Elements). The result is a piecewise quasi-polynomial q of type $\mathbb{Z}^n \times \mathbb{Z}^d \to \mathbb{Q}$, where n is the number of parameters. If the expression is independent of the last d variables, i.e., if q is defined purely in terms of the parameters and not in terms of the write iterators, then we can use a shift register. Note that we also count iterations in which no value is read from or written to the channel. This means that the size of the shift register may be larger than the buffer size of the FIFO, as computed in Sect. 8.

7 Scheduling

Although the processes in a process network are scheduled independently during their execution, there are two occasions where we may want to compute a common schedule for two or more processes. The first such occasion is when there are more processes in the network than there are processing elements to run them on. The second is when we want to compute safe buffer sizes as will be explained in Sect. 8. In principle, we could use the original schedule from Sect. 3.3 as the common schedule, resulting in the same execution order as that of the input sequential program. This schedule may lead to very high overestimates of the buffer sizes, however, and we therefore prefer constructing a schedule that is more suited for the buffer size computation. There are many ways of obtaining such a schedule. Below we discuss a simple incremental technique.

7.1 Two Processes

Let us first consider the case where there are only two processes, P_1 and P_2, that moreover have the same dimension $d = d_1 = d_2$. We will relax these conditions in Sects. 7.2 and 7.4. For ease of notation, we will not work with a separate schedule, but instead apply the schedule to the iteration domains. Since the two iteration domains have the same dimension, we can consider the two iteration domains as being part of the same iteration space and start off with an identity schedule. However, if there are any communication channels between the two processes, then we need to make sure that that the second does not read a value from any channel before it has been written. Let $M_i \subseteq D_{P_1} \times D_{P_2}$ be the mapping of one of these channels and consider the delays between a write to the channel and the corresponding read

$$\Delta_i = \{ \mathbf{t} \in \mathbb{Z}^d \mid \exists (\mathbf{w} \to \mathbf{r}) \in M_i : \mathbf{t} = \mathbf{r} - \mathbf{w} \}.$$

Note that the concept of a delay, in particular the subtraction $\mathbf{r} - \mathbf{w}$, is only valid because we now consider both iteration domains as being part of the same iteration space. If any of these delays is lexicographically negative, i.e., if the smallest delay

$$\delta_i = \text{lexmin} \, \Delta_i \tag{13}$$

is lexicographically negative, then some reads do occur before the corresponding writes and the schedule would be invalid. The solution is to delay the whole second process by just enough to make all the delays over the communication channels lexicographically non-negative. In particular, let

$$\delta^{1 \to 2} = \text{lexmin} \{ \delta_i \}_i, \tag{14}$$

with i ranging over all communication channels between processes P_1 and P_2. Then, replacing the iteration domain D_2 of P_2 by $D_2 - \delta^{1 \to 2}$ (and adapting all mappings M_i accordingly) will ensure that all delays over channels are lexicographically non-negative.

In principle, we could choose any delay on the second iteration domain that is lexicographically larger than $-\delta^{1 \to 2}$. However, delaying the reads more than strictly necessary would only increase the time the values stay inside the communication channel and could therefore only increase the buffer size needed on the channel. Furthermore, if there are also communication channels from P_2 to P_1, then we need to make sure the delays on these channels are also non-negative. Fortunately, if $\delta^{2 \to 1}$ is the minimal delay over any such channel, then

$$\delta^{1 \to 2} + \delta^{2 \to 1} \succ \mathbf{0}. \tag{15}$$

Otherwise there would be a read operation in process P_1 that indirectly depends on a write operation from the same process that occurs later or at the same time, which is clearly impossible if the process network was derived from a sequential program. By choosing a delay on process P_2 of $-\delta^{1 \to 2}$, the minimal delay on any channel from P_2 to P_1 becomes exactly $\delta^{1 \to 2} + \delta^{2 \to 1}$ and is therefore safe.

There are still some details we glanced over in the discussion above. First, it should be clear that δ_i (13) can be computed using a variation of Operation 1 (Partial

Lexicographic Maximum) or even directly as $-\operatorname{lexmax}(-\Delta_i)$. The minimal delay over all communication channels $\delta^{1\to 2}$ (14) can be computed as

$$\operatorname{lexmin}\{\delta_i\}_i = \bigcup_i \{\delta_i \mid \delta_i \prec \delta_j \text{ for all } j < i \text{ and } \delta_i \preccurlyeq \delta_j \text{ for all } j > i\},$$

where both i and j range over all communication channels between the two processes. That is, we select δ_i for the values of the parameters where it is strictly smaller than all previous δ_j and smaller than or equal to all subsequent δ_j. The result is a polyhedral set that contains a single vector for each value of the parameters.

Finally, we need to deal with the fact that the above procedure can set the delay over a channel to zero, which means that the write and corresponding read would happen simultaneously. The solution is to assign an order of execution to the statements *within* a given iteration by introducing an extra innermost dimension as we did in Sect. 4.5. The values for the extra dimension can be set based on a topological sort of a direct graph with as only edges those communication channels with a zero delay. The absence of cycles in this graph is guaranteed by (15).

Example 18. Consider a network composed of the first two processes in Fig. 2. There are nine channels between the two processes. For one of these channels, we computed the mapping M_1 (10) in Example 14. The delay between writes and reads on this channel is constant and we obtain $\{\delta_1\} = \Delta_1 = \{(1,1)\}$. The other channels also have constant delays and the smallest of these yields $\delta^{1\to 2} = (-1,-1)$. We therefore replace the second iteration domain D_2 (8) by $D_2 + (1,1)$, resulting in a new $\delta^{1\to 2}$ of $(0,0)$. To make this delay lexicographically positive (rather than just non-negative), we introduce an extra dimension and assign it the value 0 in P_1 and 1 in P_2. The new (scheduled) iteration domains are therefore

$$D_1 = \{(i,j,0) \in \mathbb{Z}^3 \mid 0 \le i < K \wedge 0 \le j < N\},$$
$$D_2 = \{(i,j,1) \in \mathbb{Z}^3 \mid 2 \le i < K \wedge 2 \le j < N\}.$$

The corresponding schedule is

$$\{\mathsf{R}(i,j) \to (i,j,0) \mid 0 \le i < K \wedge 0 \le j < N\} \cup$$
$$\{\mathsf{S}(i,j) \to (i+1,j+1,1) \mid 1 \le i < K-1 \wedge 1 \le j < N-1\}.$$

7.2   *More than Two Processes*

If there are more than two processes in the network, then we can still apply essentially the same technique as that of the previous section, but we need to be careful about how we define the minimal delay $\delta^{1\to 2}$ (14) between two processes

P_1 and P_2. It will not be sufficient to consider only the communication channels between P_1 and P_2 themselves. Instead, we need to consider all paths in the process network from P_1 to P_2, compute the sum of the minimal delays on the communication channels in each path and then take the minimum over all paths. In effect, this is just the shortest path between P_1 and P_2 in the process network with as weights the minimal delays on the communication channels. As in the case of a two process network (15), the minimal delay over a cycle is always (lexicographically) positive. We can therefore apply the Bellman–Ford algorithm (see, e.g., [18, Sect. 8.3]) to compute this shortest path.

If the network not only contains more than two processes, but we also want to combine more that two processes, then we can apply the above procedure incrementally, in each step combining two processes into one process until we have performed all the required combinations. If many or all processes need to be combined, it may be more efficient to compute the all-pairs shortest paths using the Floyd–Warshall algorithm [18, Sect. 8.4] and then to update the minimal delays in each combination step.

Example 19. Consider once more the network in Fig. 2. There are three processes, so we will need two combination steps. Let us for illustration purposes combine the first and the last process first, even though there is no direct edge between these two processes. (Usually, we would only combine processes that are directly connected.) The delay between P_2 and P_3 is constant and equal to $\delta^{2\to3} = (0,0)$, while the minimal delay between P_1 and P_2, as computed in Example 18, is $\delta^{1\to2} = (-1,-1)$. The minimal delay between P_1 and P_3 is therefore $\delta^{1\to3} = \delta^{1\to2} + \delta^{2\to3} = (-1,-1)$. We therefore replace the third (scheduled) iteration domain D_3 by $D_3 + (1,1)$, resulting in a new $\delta^{1\to3}$ of $(0,0)$. $\delta^{1\to2}$ remains unchanged, but $\delta^{2\to3}$ changes to $(1,1)$. In the next combination step, D_2 is again replaced by $D_2 + (1,1)$ and $\delta^{1\to2}$ and $\delta^{2\to3}$ both become $(0,0)$. There are now two edges with a zero minimal delay, so we assign the processes an increasing value in a new innermost dimension. The final scheduled iteration domains of the first two processes are as in Example 18, while that of the third process is

$$D_3 = \{\, (i,j,2) \in \mathbb{Z}^3 \mid 2 \leq i < K \wedge 2 \leq j < N \,\}.$$

The corresponding schedule is

$$\{\, \mathtt{W}(i,j) \to (i+1, j+1, 2) \mid 1 \leq i < K-1 \wedge 1 \leq j < N-1 \,\}.$$

Figure 12 shows the code for the completely merged process. For simplicity, we have written out this code in terms of the original statements and their original accesses rather than using read and writes to communication channels. If we were to compute buffer sizes based on the original schedule (Fig. 1), then each channel between the first and the second process would get assigned a buffer size of $(K-2)(N-2)$ because in this original schedule the ReadImage process runs to completion before the Sobel process starts running. The significantly smaller buffer sizes that we obtain using the procedure of Sect. 8 based on the schedule computed here are shown in Fig. 2.

```
1    for (i = 0; i < K; i++)
2      for (j = 0; j < N; j++) {
3        a[i][j] = ReadImage();
4        if (i >= 2 && j >= 2) {
5          Sbl[i-1][j-1]=Sobel(a[i-2][j-2],a[i-1][j-2],a[i][j-2],
6                              a[i-2][j-1],a[i-1][j-1],a[i][j-1],
7                              a[i-2][ j],a[i-1][ j],a[i][ j]);
8          WriteImage(Sbl[i-1][j-1]);
9        }
10   }
```

Fig. 12 Merged code for the process network of Fig. 2

7.3 Blocking Writes

In deriving combined schedules, we have so far only been interested in avoiding deadlocks. When merging several processes into a single process this is indeed all we need to worry about. However, when using a global schedule during the computation of valid buffer sizes, we may end up with buffer sizes that, while not introducing any deadlocks, may impose blocking writes, impacting negatively on the throughput of the network. In particular, iterations of different processes that are mapped onto the same iteration in the global schedule (ignoring the extra innermost dimension), can usually be executed in a pipelined fashion, except when non-adjacent processes in this pipeline communicate with each other. In this case, the computed buffer sizes may be so small that the first of these non-adjacent processes needs to wait until the second reads from the communication channel before proceeding onto the next iteration.

The solution is to enforce a delay between a writing process and the corresponding reading process that is equal to one iteration of the writing process. This delay ensures that the delay between non-adjacent processes in a pipeline will be large enough to avoid blocking writes that hamper the throughput. The required delay is the smallest variation in the iteration domain D of a process. Let P be the relation that maps a given iteration of D to the previous iteration of D, i.e.,

$$P(D) = \text{lexmax}\{\mathbf{i} \to \mathbf{i}' \mid \mathbf{i}, \mathbf{i}' \in D \wedge \mathbf{i}' \prec \mathbf{i}\}. \tag{16}$$

The relation $P(D)$ can be computed using Operation 1 (Partial Lexicographic Maximum). The required delay η is then the smallest difference between two subsequent iterations, i.e.,

$$\eta = \text{lexmin}\{\mathbf{t} \mid \exists(\mathbf{i} \to \mathbf{i}') \in P(D) \wedge \mathbf{t} = \mathbf{i} - \mathbf{i}'\}.$$

The delay is enforced by subtracting the η corresponding to the writing process from each channel delay δ_i (13). Note, however, that in the presences of cycles, we

Fig. 13 A program
illustrating blocking writes

```
1   for (i = 0; i < N; ++i) {
2       a[i] = f(i);
3       b[i] = g(a[i]);
4       c[i] = h(a[i], b[i]);
5   }
```

Fig. 14 Time diagram with
blocking

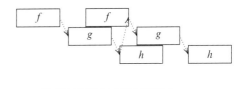

Fig. 15 Time diagram
without blocking

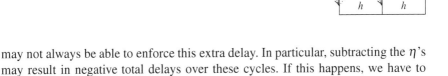

may not always be able to enforce this extra delay. In particular, subtracting the η's may result in negative total delays over these cycles. If this happens, we have to resort to the scheduling of Sect. 7.2, without the additional delays.

Example 20. Consider the code in Fig. 13. There are three statements, f, g, and h with three FIFO communication channels between them: one from f to g, one from g to h and one from f to h. The delays on these channels are all zero. The scheduling of Sect. 7.2 therefore leaves everything in place and the resulting code is identical to the original. It is clear that in this schedule the size of each of the FIFOs can be set to 1. However, if we try to run the resulting process network, then we see that the second iteration of process f needs to wait for the first iteration of h to empty the FIFO before it can write to the FIFO of size 1. Because of this blocking write, the three processes cannot be fully pipelined, as illustrated in Fig. 14. Since all domains are one-dimensional, the internal delay in each process is 1. Subtracting this internal delay from each channel delay, we see that the minimal delay between process f and h is $(-1) + (-1) = -2$. Process h is therefore scheduled at position 2 relative to process f. This means that the channel between these two process needs to be of size at least 2. This size in turn then allows a fully pipelined execution of the three processes, as illustrated in Fig. 15.

7.4 *Linear Transformations*

The schedules that we have seen so far are all of the form $\theta(\mathbf{j}) = I_{d+1,d}\mathbf{j} + \mathbf{b}$, where $I_{d+1,d}$ is $(d+1) \times d$ matrix with $i_{k,k} = 1$ and $i_{k,l} = 0$ for $k \neq l$, and $\mathbf{b} \in \mathbb{Z}^{d+1}$. That is, we add an extra dimension $(I_{d+1,d}\mathbf{j})$ and apply a shift ($\mathbf{b}$). Note that we have also assumed so far that all iteration domains have the same dimension. In principle, we could apply more general affine schedules $\theta(\mathbf{j}) = A\mathbf{j} + \mathbf{b}$, i.e., including an arbitrary linear transformation $A\mathbf{j}$. Not only would this allow us to compute potentially tighter bounds on the buffer sizes, but it may even change the types of the communication channels.

Example 21. Consider once more the program in Fig. 8. In Example 16 we have shown that the only channel in the corresponding network has reordering. If we reverse the second loop, however, i.e., if we apply the transformation $\mathbf{j} \rightarrow -\mathbf{j} + N$, then the mapping on the channel becomes

$$M = \{ w \rightarrow r \mid 0 \leq w \leq N \wedge r = w \}$$

and we have

$$T = \{ w_1 \rightarrow w_2 \mid 0 \leq w_1, w_2 \leq N \wedge w_1 > w_2 \wedge w_1 < w_2 \}.$$

This set is clearly empty and so the communication channel has become a FIFO.

The example shows that if we were to apply general linear transformations, then we would need to do this before the detection of channel types of Sect. 6. In fact, if we want to detect reuse, as we did in Sect. 5.2, then we would need to do this reuse detection *after* any linear transformation. The reason is that a linear transformation may change the internal order inside an iteration domain, meaning that what used to be a previous iteration, from which we could reuse data, may have become a later iteration. On the other hand, the dataflow in the sequential program imposes some restrictions on which linear transformations are valid, so we need to perform dataflow analysis (Sect. 5) *before* any linear transformation. The solution is to first apply standard dataflow analysis (Sect. 5.1), then to perform linear transformations and finally to detect reuse inside the communication channels that were constructed in the first step. A full discussion of general linear transformations is beyond the scope of this chapter.

There is however one case where we are forced to apply a linear transformation, and that is the case where not all iteration domains have the same dimension. Before we can merge two iteration domains, they have to reside in the same space. In particular, they need to have the same dimension. Let d be the maximal dimension of any iteration domain, then we could could expand a d_i-dimensional iteration domain with $d_i \leq d$ using the transformation I_{d,d_i}. This transformation effectively pads the iteration vectors with zeros at the end. This may, however, not be the best place to insert the zeros. A simple heuristic to arrive at better positions for the zeros is to look at the communication channels between a lower-dimensional domain and a full-dimensional domain and to insert the zeros such that iterators that are related to each other through the communication channels are aligned. Note that inserting zeros does not change the internal order inside the iteration domain. The same holds for any scaling that we may want to apply based on essentially the same heuristic.

Example 22. Consider the communication channel between processes b and c in the network of Fig. 11 (see Example 17). The first process has a one-dimensional iteration domain, while the second process has a two-dimensional iteration domain. The mapping on the channel (12) shows that there is a relation $w = r_2$ between the single iterator of the writing process and the second iterator of the reading process. We therefore insert a zero before the iteration vectors of the writing process such that the original iterator ends up in the second position. For the channel between a and c the relation is $w = r_1$, so in this case we insert a zero *after* the original iterator. These choices are reflected by the orientations of these two one-dimensional iteration domains in Fig. 11. With these orientations, the writes may be moved on top of the corresponding reads by simply shifting the iteration domains, meaning that a buffer size of at most 1 is needed. Had we chosen a different orientation for the two one-dimensional iteration domains, then this would not have been possible.

8 Buffer Size Computation

This section describes how to compute valid buffer sizes for all communication channels. The buffer sizes are valid in the sense that imposing them does not cause deadlocks. The first step is to compute a global schedule for all processes in the network, e.g., using the techniques of Sect. 7. This scheduling step effectively makes all communication channels internal (to the single combined process). We then compute buffer sizes for this particular schedule. The schedule itself is not used during the actual execution of the network. However, we know that there is at least one schedule for which the buffer sizes are valid. The blocking reads and writes on the communication channels will therefore not introduce any deadlocks.

8.1 FIFOs

We are given an internal communication channel that has been identified as a FIFO using the technique of Sect. 6.1 and we want to know how much buffer space is needed on the FIFO. Recall that any channel may be considered internal after a global scheduling step that maps all iteration domains to a common iteration space. The buffer should be large enough to hold the number of tokens that are in transit between a write and a read at any point during the execution. We therefore first count this number of tokens in terms of an arbitrary iteration and then compute an upper bound over all iterations.

Let us look at these steps in a bit more detail. The communication channel is described by a mapping M from the write iteration domain D_w to the read iteration domain D_r. The maximal number of tokens in the buffer will occur after some write to the buffer and before the first subsequent read from the buffer. It is therefore sufficient to investigate what happens right before a token is read from the buffer. Let W be the relation mapping any read iteration $\mathbf{r}$ to all write iterations that occur before this read and let R be the relation mapping the same read iteration to all previous read iterations, i.e.,

$$W = \{ \mathbf{r} \to \mathbf{w}' \mid \mathbf{r} \in \operatorname{ran} M \wedge \mathbf{w}' \in \operatorname{dom} M \wedge S(\mathbf{w}') \prec S(\mathbf{r}) \} \qquad (17)$$

$$R = \{ \mathbf{r} \to \mathbf{r}' \mid \mathbf{r}, \mathbf{r}' \in \operatorname{ran} M \wedge \mathbf{r}' \prec \mathbf{r} \}. \qquad (18)$$

Then the number $n(\mathbf{s}, \mathbf{r})$ of elements in the buffer right before the execution of read $\mathbf{r}$ is the number of writes to the buffer before the read, i.e., $\#W(\mathbf{s}, \mathbf{r})$, minus the number of reads from the buffer before the given read, i.e., $\#R(\mathbf{s}, \mathbf{r})$, where as usual, $\mathbf{s}$ are the parameters. Both of these computation can be performed using Operation 3 (Number of Image Elements). Finally, we apply Operation 4 (Upper Bound on a Quasi-Polynomial) to the piecewise quasi-polynomial $n(\mathbf{s}, \mathbf{r}) = \#W(\mathbf{s}, \mathbf{r}) - \#R(\mathbf{s}, \mathbf{r})$ and the polyhedral set $\operatorname{ran} M$, resulting in a piecewise quasi-polynomial $u(\mathbf{s})$ that is an upper bound on the number of elements in the FIFO channel during the whole execution.

Example 23. Consider once more the network in Fig. 2. A global schedule for this network was derived in Examples 18 and 19, and code corresponding to this schedule is shown in Fig. 12. Writing the mapping (10) on the channel constructed from the first argument to the call to Sobel in Line 6 of Fig. 1 in Example 14 in terms of the scheduled iterators, we obtain

$$M = \{ (i', j', 0) \to (i, j, 1) \mid 2 \leq i < K \wedge 2 \leq j < N \wedge\ i' = i - 2 \wedge j' = j - 2 \}.$$

From this relation we derive,

$$W = \{ (i, j, 1) \to (i', j', 0) \mid 2 \leq i < K \wedge 2 \leq j < N \wedge$$
$$0 \leq i' < K - 2 \wedge 0 \leq j' < N - 2 \wedge$$
$$((i' < i) \vee (i' = i \wedge j' \leq j)) \}.$$

The number of image elements of this relation can be computed as

$$\#W(K, N, i, j) = \begin{cases} i(N-2) + j + 1 & \text{if } (i, j, 1) \in \operatorname{ran} M \wedge i < K - 2 \wedge j < N - 2 \\ (i+1)(N-2) & \text{if } (i, j, 1) \in \operatorname{ran} M \wedge i < K - 2 \wedge j \geq N - 2 \\ (K-2)(N-2) & \text{if } (i, j, 1) \in \operatorname{ran} M \wedge i \geq K - 2. \end{cases}$$

For the number of reads before a given read $(i, j, 1)$, we similarly find

$$\#R(K, N, i, j) = \left\{ (i-2)(N-2) + j - 2 \quad \text{if } (i, j, 1) \in \operatorname{ran} M. \right.$$

Taking the difference yields

$$n(K,N,i,j) = \begin{cases} 2(N-2)+3 & \text{if } (i,j,1) \in \operatorname{ran} M \wedge i < K-2 \wedge j < N-2 \\ 3(N-2)-j+2 & \text{if } (i,j,1) \in \operatorname{ran} M \wedge i < K-2 \wedge j \geq N-2 \\ (K-i)(N-2)-j+2 & \text{if } (i,j,1) \in \operatorname{ran} M \wedge i \geq K-2. \end{cases}$$

The maximum over all reads in $\operatorname{ran} M$ occurs in the first domain and is equal to $2(N-2)+3 = 2N-1$, which is the value shown on the first edge in Fig. 2.

In iscc, the computation can be performed a follows.

```
M  := [K, N]  -> { R[i, j]  -> S[1+i, 1+j] : i >= 0 and
                  i <= -3 + K and j >= 0 and j <= -3 + N };
S  := { R[i,j]  -> [i,j,0]; S[i,j]  -> [i+1,j+1,1] };
W  := (S >> S) * (ran M -> dom M);
R  := (ran M) >> (ran M);
ub (((card W) - (card R)) * [K,N] -> { : K,N >= 10 });
```

The first two lines specify the input communication channel mapping from Example 14 and the schedule from Example 18. The next two lines construct the relations W and R. When the >> operator is applied to a pair of relations, it returns a relation between the domains of the input relations based on the lexicographic order in the corresponding images. Note that we only need the schedule during the construction of W since R is a relation between elements from the same iteration domain. In the final line, we intersect the result of the counting problem with constraints on the parameters. This intersection is only applied in order to produce nicer output. Without the intersection, the output would contain several special cases involving small values of the parameters. The final output is as follows.

```
([K, N]  -> { max((-1 + 2 * N)) : N >= 10 and K >= 10 },
 True)
```

Note that in this case the upper bound is determined to be equal to the maximum.

8.2 Reordering Channels

For channels that have been identified as exhibiting reordering using the technique of Sect. 6.1, we need to choose where we want to perform the reordering of the tokens. One option is to perform the reordering inside the channel itself. In this case we can apply essentially the same technique as that of the previous section to compute an upper bound on the minimal number of locations needed to store all elements in the buffer at any given time. However, since the tokens now have to be reordered, we need to be able to address them somehow. An efficient mapping from read or write iterators to the internal buffer of the channel may require more space than strictly necessary. We refer to [5] for an overview and a mathematical framework for finding good memory mappings.

Another option is to perform the reordering inside the reading process. In this case, a process wanting to read a value from the reordering channel first looks in its internal buffer associated with the channel. If the value is not in the buffer, it reads values from the channel in the order in which they were written, storing all values it does not need yet in the local buffer until it has read the value it is actually looking for. In other words, the original reordering channel is split into a FIFO channel and a local reordering buffer. There are therefore two buffer sizes to compute in this case.

Before embarking upon the computation of these buffer sizes, it should be noted that nothing really interesting happens to the buffers during iterations that do not read from the FIFO. In particular, the maximal number of elements in the FIFO buffer will be reached right before a read from the FIFO and the maximal number of elements in the reordering buffer will be reached right before the read of the value from the FIFO that the process is actually interested in, i.e., after it has copied all intermediate data to the reordering buffer. The uninteresting iterations U are those reads $\mathbf{r}$ for which there is an earlier read $\mathbf{r}'$ that reads something that was written *after* the value read by $\mathbf{r}$. This latter value will have been put in the reordering buffer at or perhaps even before read $\mathbf{r}'$. That is,

$$U = \{\, \mathbf{r} \mid \exists \mathbf{w}, \mathbf{r}', \mathbf{w}' : (\mathbf{w} \to \mathbf{r}) \in M \wedge (\mathbf{w}' \to \mathbf{r}') \in M \wedge \mathbf{w} \prec \mathbf{w}' \wedge \mathbf{r}' \prec \mathbf{r} \,\}$$

and

$$S = \operatorname{ran} M \setminus U$$

is the set of "interesting" iterations. Note that in order not to overload the notation, we assume here that domain and range of M refer to the scheduled domains.

For the first of these interesting iteration, i.e., the first read $\mathbf{r}^*$ of a value from the FIFO, the number of tokens in the FIFO is equal to the number of writes that have occurred before that read, i.e., $\#W(\mathbf{s}, \mathbf{r}^*)$, where W is as defined in (17). For any other read $\mathbf{r} \in S$, we will also have read some values from the FIFO. In particular, we will have read all values that were written up to and including the value that we needed in the previous read $\mathbf{r}'$. Let W' map a given read $\mathbf{r}$ to all these previously read values. Then the number of tokens in the FIFO right before $\mathbf{r}$ is $\#W(\mathbf{s}, \mathbf{r}) - \#W'(\mathbf{s}, \mathbf{r})$. The relation W' can be computed as

$$W' = \{\, \mathbf{r} \to \mathbf{w}'' \mid \exists \mathbf{r}' : \mathbf{r} \in S \wedge (\mathbf{r} \to \mathbf{r}') \in P \wedge (\mathbf{r}' \to \mathbf{w}') \in M \wedge \mathbf{w}'' \preccurlyeq \mathbf{w}' \,\},$$

where $P = P(S)$ is the relation that maps a read in S to the previous read in S, as defined in (16). As before, the relation P can be computed using Operation 1 (Partial Lexicographic Maximum). As a side effect, we obtain a polyhedral set E containing the first read $\mathbf{r}^*$. The counts can be computed using Operation 3 (Number of Image Elements) and finally Operation 4 (Upper Bound on a Quasi-Polynomial) needs to be applied to obtain a bound on the FIFO buffer size that is valid for all reads.

As for the internal buffer, the number of tokens in this buffer after a read $\mathbf{r}$ from the FIFO is equal to the number of subsequent reads that read something (from the internal buffer) that was written (to the FIFO) before the token that was actually needed by $\mathbf{r}$, i.e.,

$$\#\{\,\mathbf{r} \rightarrow \mathbf{r}' \mid \exists \mathbf{w}, \mathbf{w}' : (\mathbf{w}, \mathbf{r}), (\mathbf{w}', \mathbf{r}') \in M \wedge \mathbf{w}' \prec \mathbf{w} \wedge \mathbf{r} \prec \mathbf{r}'\,\}.$$

Again, this number can be computed using Operation 3 and a bound on the buffer size can then be computed using Operation 4.

8.3   Accuracy of the Buffer Sizes

Although the buffer sizes computed using the methods described above are usually fairly close to the actual minimal buffer sizes, there are some possible sources of inaccuracies that we want to highlight in this section. These inaccuracies will always lead to over-approximations. That is, the computed bounds will always be safe. For internal channels, the only operation that can lead to over-approximations is Operation 4 (Upper Bound on a Quasi-Polynomial). There are three causes for inaccuracies in this operation: the underlying technique [3] is defined over the rationals rather than the integers; in its basic form it only handles polynomials and not quasi-polynomials; and the technique itself will only return the actual maximum (rather than just an upper bound) if the maximum occurs at one of the extremal points of the domain. For external channels, we rely on a scheduling step to effectively make them internal. Since we only consider a limited set of possible schedulings, the derived buffer sizes may in principle be much larger than the absolute minimal buffer sizes that still allow for a deadlock-free schedule.

9   Summary

In this chapter we have seen how to automatically construct a polyhedral process network from a sequential program that can be represented in the polyhedral model. The basic polyhedral tools used in this construction are parametric integer programming, emptiness check, parametric counting, computing parametric upper bounds and polyhedral scanning. The processes in the network correspond to the statements in the program, while the communication channels are computed using dependence analysis. Several types of channels can be identified by solving a number of emptiness checks and/or counting problems. Safe buffer sizes for the channels can be obtained by first computing a global schedule and then computing an upper bound on the number of elements in each channel at each iteration of this schedule.

Acknowledgements This work was supported by FWO-Vlaanderen, project G.0232.06N. The author would like to thank Maurice Bruynooghe and Sjoerd Meijer for their feedback on earlier versions of this chapter.

References

1. Amarasinghe, S.P., Anderson, J.M., Lam, M.S., Tseng, C.W.: The SUIF compiler for scalable parallel machines. In: Proceedings of the Seventh SIAM Conference on Parallel Processing for Scientific Computing (1995)
2. Bastoul, C.: Code generation in the polyhedral model is easier than you think. In: PACT '04: Proceedings of the 13th International Conference on Parallel Architectures and Compilation Techniques, pp. 7–16. IEEE Computer Society, Washington, DC, USA (2004). DOI 10.1109/ PACT.2004.11
3. Clauss, P., Fernández, F.J., Gabervetsky, D., Verdoolaege, S.: Symbolic polynomial maximization over convex sets and its application to memory requirement estimation. IEEE Transactions on Very Large Scale Integration (VLSI) Systems **17**, 983–996 (2009)
4. Cook, W., Rutherford, T., Scarf, H.E., Shallcross, D.F.: An implementation of the generalized basis reduction algorithm for integer programming. ORSA Journal on Computing **5**(2) (1993)
5. Darte, A., Schreiber, R., Villard, G.: Lattice-based memory allocation. IEEE Trans. Comput. **54**(10), 1242–1257 (2005). DOI 10.1109/TC.2005.167
6. Devos, H., Van Campenhout, J., Stroobandt, D.: Finding bounds on ehrhart quasi-polynomials. In: Architecture and Compilers for Embedded Systems (ACES 2007), Edegem. 2007 (2007). DOI 1854/11101
7. Feautrier, P.: Parametric integer programming. Operationnelle/Operations Research **22**(3), 243–268 (1988)
8. Feautrier, P.: Dataflow analysis of array and scalar references. International Journal of Parallel Programming **20**(1), 23–53 (1991)
9. Geilen, M., Basten, T.: Kahn process networks and a reactive extension. In: S.S. Bhattacharyya, E.F. Deprettere, R. Leupers, J. Takala (eds.) Handbook of Signal Processing Systems, second edn. Springer (2013)
10. Grosser, T., Zheng, H., A, R., Simbürger, A., Grösslinger, A., Pouchet, L.N.: Polly - polyhedral optimization in LLVM. In: First International Workshop on Polyhedral Compilation Techniques (IMPACT'11). Chamonix, France (2011)
11. Kahn, G.: The semantics of a simple language for parallel programming. In: Proc. of the IFIP Congress 74, pp. 471–475. North-Holland Publishing Co. (1974)
12. Kelly, W., Maslov, V., Pugh, W., Rosser, E., Shpeisman, T., Wonnacott, D.: The Omega library. Tech. rep., University of Maryland (1996)
13. Kelly, W., Pugh, W., Rosser, E.: Code generation for multiple mappings. In: Frontiers'95 Symposium on the Frontiers of Massively Parallel Computation. McLean (1995)
14. Kienhuis, B., Rijpkema, E., Deprettere, E.: Compaan: Deriving process networks from Matlab for embedded signal processing architectures. In: CODES '00: Proceedings of the eighth international workshop on Hardware/software codesign, pp. 13–17. ACM Press, New York, NY, USA (2000). DOI 10.1145/334012.334015
15. Meijer, S., Nikolov, H., Stefanov, T.: Throughput modeling to evaluate process merging transformations in polyhedral process networks. In: Proceedings of the Conference on Design, Automation and Test in Europe, DATE '10, pp. 747–752. European Design and Automation Association, 3001 Leuven, Belgium (2010)
16. Meister, B., Verdoolaege, S.: Polynomial approximations in the polytope model: Bringing the power of quasi-polynomials to the masses. In: J. Sankaran, T. Vander Aa (eds.) Digest of the 6th Workshop on Optimization for DSP and Embedded Systems, ODES-6 (2008)

17. Pop, S., Cohen, A., Bastoul, C., Girbal, S., Jouvelot, P., Silber, G.A., Vasilache, N.: Graphite: Loop optimizations based on the polyhedral model for GCC. In: 4th GCC Developer's Summit. Ottawa, Canada (2006)
18. Schrijver, A.: Combinatorial Optimization - Polyhedra and Efficiency. Springer (2003)
19. Schweitz, E., Lethin, R., Leung, A., Meister, B.: R-stream: A parametric high level compiler. In: J. Kepner (ed.) Proceedings of HPEC 2006, 10th Annual Workshop on High Performance Embedded Computing. Lincoln Labs, Lexington, MA (2006)
20. Turjan, A.: Compaan - A Process Network Parallelizing Compiler. VDM Verlag (2008)
21. Verdoolaege, S.: isl: An integer set library for the polyhedral model. In: K. Fukuda, J. Hoeven, M. Joswig, N. Takayama (eds.) Mathematical Software - ICMS 2010, *Lecture Notes in Computer Science*, vol. 6327, pp. 299–302. Springer (2010)
22. Verdoolaege, S.: Counting affine calculator and applications. In: First International Workshop on Polyhedral Compilation Techniques (IMPACT'11). Chamonix, France (2011)
23. Verdoolaege, S., Beyls, K., Bruynooghe, M., Catthoor, F.: Experiences with enumeration of integer projections of parametric polytopes. In: R. Bodik (ed.) Proceedings of 14th International Conference on Compiler Construction, Edinburgh, Scotland, *Lecture Notes in Computer Science*, vol. 3443, pp. 91–105. Springer-Verlag, Berlin (2005). DOI 10.1007/b107108
24. Verdoolaege, S., Nikolov, H., Stefanov, T.: pn: A tool for improved derivation of process networks. EURASIP Journal on Embedded Systems, special issue on Embedded Digital Signal Processing Systems **2007** (2007). DOI 10.1155/2007/75947
25. Verdoolaege, S., Seghir, R., Beyls, K., Loechner, V., Bruynooghe, M.: Counting integer points in parametric polytopes using Barvinok's rational functions. Algorithmica **48**(1), 37–66 (2007). DOI 10.1007/s00453-006-1231-0
26. Ziegler, G.M.: Lectures on Polytopes. Springer-Verlag, Berlin (1995)

Mapping Decidable Signal Processing Graphs into FPGA Implementations

Roger Woods

Abstract Field programmable gate arrays (FPGAs) are examples of complex programmable system-on-chip (PSoC) platforms and comprise dedicated DSP hardware resources and distributed memory. They are ideal platforms for implementing computationally complex DSP systems for image processing and radar, sonar and signal processing. The chapter describes how decidable signal processing graphs are mapped onto such platforms and shows how parallelism and pipelining can be controlled to achieve the required speed using minimal hardware resource. The work shows how the techniques outlined there are used to build efficient FPGA implementations. The process is demonstrated for a number of DSP circuits including a finite impulse response (FIR) filter, lattice filter and a more complex adaptive signal processing design, namely a least means squares (LMS) filter.

1 Introduction

Typically, DSP systems are implemented on either application specific integrated circuits (ASICs) or dedicated DSP microprocessor platforms. With the emergence of coarse grained processing elements such as dedicated multipliers, adders and DSP blocks in field programmable gate arrays (FPGAs), they have now emerged as a highly attractive platform for implementing high performance, complex DSP systems. In addition to the processing elements, distributed memory both in the form of small distributed memory e.g. registers and look up tables (LUTs), as well as embedded RAMs allows high performance to be achieved. This is done by distributing the system memory requirements and permitting high parallel, pipelined implementations to be realized on the technology.

R. Woods (✉)
ECIT Institute, Queen's University of Belfast, Queen's Road, Queen's Island, Belfast, UK
e-mail: r.woods@qub.ac.uk

S.S. Bhattacharyya et al. (eds.), *Handbook of Signal Processing Systems*,
DOI 10.1007/978-1-4614-6859-2_42, © Springer Science+Business Media, LLC 2013

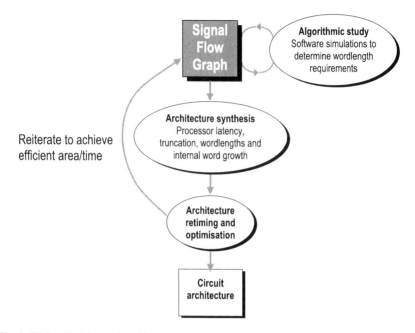

Fig. 1 SFG to FPGA circuit architecture design flow

With non-recurrent engineering (NRE) costs and increasing design challenges for sub-65 nm CMOS technology, programmability is becoming increasingly important. The existence of programmable platforms such as DSP processors allow users to develop software descriptions and then use commercially available compilers and assemblers to create the resulting designs. This approach has numerous advantages including software portability, ease of modification and product migration, and reduced design time compared to ASIC or FPGA solutions. FPGAs also offer programmability but unlike processors, involve the creation of the architecture which is typically a complex design process. However, this offers the key advantage to achieve very high performance but with the use of prefabricated technology; this avoids the non-recurrent engineering (NRE) costs associated with implementing designs using ASIC technology.

The main issue therefore, is to map the DSP functionality efficiently onto the hardware effectively creating a *circuit architecture* which best matches the system requirements. This is ideally achieved by making modifications to the signal flow graph (SFG) representation of the algorithm such as those highlighted in chapters [3, 12], so that an efficient circuit architecture can be produced. The basic process for realizing a circuit architecture from a SFG description is given in Fig. 1. The initial stage involves investigating performance using suitable design software e.g. Simulink® or LabVIEW, to determine issues such as suitable wordlengths, etc.; the next stage is to determine performance in terms of the sampling rate required, latency, etc. and then use these parameters to generate the circuit architecture from the SFG description; a key step is to iterate around this loop to optimize the performance typically minimizing area requirements for the throughput rate required. As will be demonstrated in the chapter, the generation of the circuit architecture can be complicated when the SFG representation involves feedback loops.

Given the make-up of FPGA architectures with either separate multiplier and adder circuits or in the case of more recent FPGA architectures, dedicated DSP blocks with fixed wordlength representations and also distributed memory, the design focus used in this chapter, is to optimize the circuit architecture in terms of these building blocks. It should be noted that some synthesis tools now offer some level of pipelining and retiming as an option but these features have limited impact as shown by the synthesis results given.

1.1   Chapter Breakdown

The chapter is organized as follows. A brief review of FPGA hardware platforms is given in Sect. 2 concentrating on the features most relevant to DSP in recent and powerful FPGA families from Xilinx and Altera namely the Virtex® and Stratix® families. The focus of the section is to highlight the key features of the families and not provide a detailed description of the technologies. Section 3 highlights the key steps in create FPGA circuit architectures for various speed and area requirements, covering *retiming* and *delay scaling*. Optimizing the circuit architecture to meet certain performance criteria is achieved by applying *folding* and *unfolding* either to increase speed or if the speed requirement has been met, to achieve better overall performance by reducing area requirements; finally, the LMS filter is used to demonstrate the various optimization and covers various FPGA design trade-offs; this is covered in Sect. 4. The chapter concludes by looking at how some of core features are exposed to the system designer and also how some of the techniques can be used to reduce dynamic power consumption.

2   FPGA Hardware Platforms

Historically, FPGAs were viewed as a glue logic component where the design focus was on maximizing LUT utilization and minimizing programmable routing to provide interconnectivity at the lowest possible delay. With regard to structure, FPGAs are categorized by the following programmable components although the granularity of some of blocks may have changed over the years:

- Programmable logic units that can be programmed to realize different digital functions.
- Programmable interconnect that allow different blocks to be connected together.
- Programmable I/O pins.

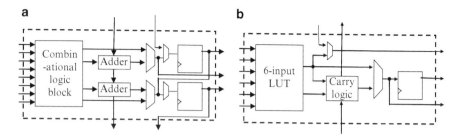

Fig. 2 FPGA programmable logic blocks. (**a**) Stratix® ALM block diagram. (**b**) Xilinx Virtex®-5 CLB

The evolution of programmable adders in the 1990s meant that effort was concentrated on mapping particular DSP algorithms such as fixed FIR filters and transforms using distributed arithmetic techniques [8, 16]. In this era, the focus was to exploit the nature of some fixed DSP operations and transforms to allow fixed coefficient multipliers to be derived. These *fixed coefficient* structures could be implemented efficiently using a small number of adders and evolutions such as the reduced coefficient multipliers (RCMs) [15]; very efficient multipliers which could operate on a number of multiplicands rather than a single one, could then be created.

More recent evolutions such as dedicated multipliers and more recently DSP blocks, have now changed the design focus once again [18]; the challenge is now to map the signal flow graph (SFG) not to LUTs and adders as previous but to multiplier and adder blocks, meaning that pipelining is applied at the processor rather than the adder and LUT level; this is the focus taken here. Moreover, overall performance has improved. For example, the Xilinx Virtex®-5 family can now be clocked at 550 MHz and has a number of on-board IP blocks and DSP48E slices which combine to give a maximum of 352 GMACS performance. With regard to connectivity, RocketIO GTP transceivers deliver 100 Mbps to 3.2 Gbps. The Altera Stratix® III E family offers 48,000–338,000 equivalent LEs, up to 20 Mbits of memory and a number of high-speed DSP blocks that provide implementation of multipliers, multiply accumulate blocks and FIR filters. It is worthwhile looking at this aspect of FPGA functionality in a little more detail.

2.1   FPGA Logic Blocks

The programmable logic block in FPGAs has largely remained unchanged over the years and comprises a LUT followed by dedicated circuitry for implementing fast addition, and a programmable storage unit i.e. flip-flop (see Fig. 2). The only major development has been growth in LUT size and now 7-input or 6-input LUTs are commonplace. This can be seen in the Altera Stratix® III adaptive logic module

(ALM) in Fig. 2a and in the Xilinx Virtex®-5 configurable logic block (CLB) in Fig. 2b. These structures are similar to those used in the latest Virtex-7 and Stratix V FPGA families. The adder configuration in FPGAs is typically based on a ripple carry adder structure as this can be easily scaled in terms of input/output wordlength even though it is conventionally the slowest of the adder structures [10]. In the case of the ALM, the adders can perform two 2-bit addition or a single ternary addition and with the Xilinx Virtex®, it is possible to perform a fast addition using the fast carry logic circuitry. This gives a range of speeds from 1 ns for an 8-bit addition to 2.5 ns for a 64-bit addition.

2.2   FPGA DSP Functionality

In addition to the programmable adder structures, both FPGA families have developed dedicated on-board DSP hardware as shown in Fig. 3. A simplified version of the Xilinx DSP48 hardware block (Fig. 3a) comprises multiplicative and accumulation hardware which can be configured to implement one tap of an FIR filter. In addition, the programmable connectivity provided by the multiplexers (some of which are not shown), allows a variety of modes of operation including single stage multiplication, single stage addition, multiply accumulation and increased arithmetic computations (achieved by chaining DSP blocks together using the connectivity shown at the bottom of the figure). A pattern detector is also provided on the output which gives support for a number of numerical features including convergent rounding, overflow/underflow, block floating-point, and accumulator terminal count (counter auto reset) with pattern detector outputs.

The Altera Stratix® equivalent circuit is shown in Fig. 3b. Altera have opted for a more complex structure with four multipliers and adders/accumulators per stage. It has been clearly developed to support a number of specific DSP functions such as a 4-tap FIR filter, an FFT butterfly and a complex multiplication namely $(a + ib)$x$(c + jd)$. A number of wordlengths are supported as indicated in Table 1. As with the Xilinx DSP hardware, functionality is also provided to support a number of modes of operation including looping back from the output register (useful for recursion), connectivity of DSP block from above (as DSP blocks are connected in columns) and dedicated rounding, underflow and overflow circuitry.

2.3   FPGA Memory Organization

A key aspect in FPGA is memory distribution which is important for DSP applications as pipelining is commonly used to speed up computations. The availability of registers in each ALM and CLB as shown in Fig. 2, allows for direct implementation

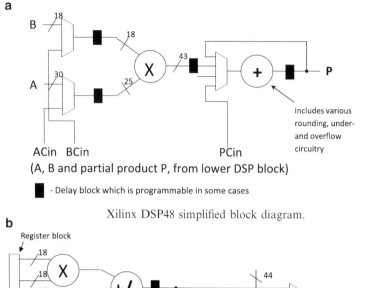

Fig. 3 DSP dedicated FPGA hardware. (**a**) Xilinx DSP48 simplified block diagram. (**b**) Altera Stratix® Half-DSP block

Table 1 DSP block operation modes for Altera Stratix®III FPGA technology [1]

Mode	Multiplier width (bits)	No. of multipliers	Per block
Independent multiplier	9/12/18/36/double	1	8/6/4/2
Two-multiplier adder	18	2	4
Four-multiplier adder	18	4	2
Multiply-accumulate	18	4	2
Shift	36	1	2

of pipeline registers. However, there is also other forms of distributed memory in terms of distributed RAM blocks which can be used to store local data and also performing hardware sharing where the FPGA may have to store the internal values for numerous independent calculations.

Table 2 Stratix® memory types and usage [1]

Type	Bits/block	No. of blocks	Suggested usage
MLAB	640	6,750	Shift registers, small FIFO buffers, filter delay lines
M9K	9,216	1,144	General-purpose applications, cell buffers
M144K	147,456	48	Processor code storage, video frames

Table 3 Virtex®-5 memory types and usage

Type	Unit size (Kb)	Total size (Kb)	Access time (ns)	Location
DDR/DDR2	4,096,000	4,096,000	10	Off-chip
Block RAM	18	1,728	2	On-chip
Distributed RAM	0.016	288	2	On-chip
Registers	0.001	36	2s	On-chip

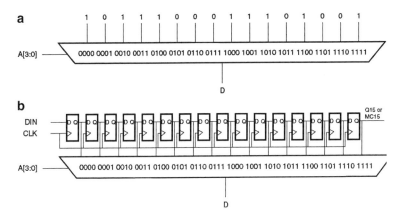

Fig. 4 Configurations for Xilinx CLB LUT [19]. (**a**) 16:1 multiplexer. (**b**) Shift register

The various forms of memory for the Altera Stratix® is shown in Table 2. This shows a number of different size memories that can be used in FPGAs. A higher level approach for use of this memory is given in Table 3 for the Xilinx Virtex®-4 family and shows the speed and area size treatment much more clearly. This gives a range of memories for storing coefficient data or internal data either in block processing applications or in cases where cores may be time-shared in relatively low speed applications.

In addition to the implementation of combinational logic circuits, the LUT can be used to implement a single bit RAM and also an efficient shift register. The principle by which the LUT can be used as a shift register is covered in detail in the Xilinx application note [19] and given in Fig. 4; it shows how a 16:1 multiplexer is realized where the 4-bit address input is used to address the specific input stored in the RAM and Fig. 4b outlines how this can be extended to an addressable shift register. This provides an efficient form for implementing shift registers.

2.4   FPGA Design Strategy

The availability of dedicated multipliers, adders, multiply-accumulate blocks and distributed memory either in the form of distributed RAM blocks or registers, represents an ideal platform for implementing DSP systems. A number of points emerge from the process of implementing DSP systems on FPGA.

- The availability of dedicated multipliers, adders and memory elements suggest a direct mapping from the processing graphs to FPGAs where each function can be implemented by a separate processing element thereby allowing high levels of parallelism. Historically, a lot of effort had been dedicated to implementing multiplicative functionality using the LUT-based programming element (Fig. 2) [8,15,16] but this has now been superseded by the recent developments in FPGA architectures, namely the DSP blocks.
- The plethora of small, distributed memories suggest a highly pipelined approach is applicable in FPGA. Pipelining not only has the benefits of improving the throughput rate of many systems but can also act to reduce dynamic power consumption [17,18] as it acts to reduce average net length and switching activity.
- Given that typical FPGA implementations can be clocked at between 200 and 500 MHz depending on application requirements, a key aspect is to apply hardware sharing through folding to allow better utilization of the FPGA hardware. This acts to reduce the computational hardware but acts to increase memory requirements (to allow storage of current state for the many functions being implemented). Efficient implementation using the available memory resource is therefore a requirement.

3   Circuit Architecture Derivation

The availability of processing elements and distributed memory on FPGA provides a clear focus to investigate mapping of DSP functions into highly pipelined, parallel circuit architectures. This is demonstrated using the simple examples of FIR and lattice filters which highlight non-recursive and recursive computations.

3.1   Basic Mapping of DSP Functionality to FPGA

As highlighted in chapter [12], DSP systems can be represented using a SFG. This is illustrated for the FIR filter description given in Fig. 5a and represented by Eq. (1). The directed edge (j,k) denotes a linear transform usually given as *multiplications*, *additions* or *delay* elements, from the signal at node j to the signal at node k. The data flow graph (DFG) representation is shown in Fig. 5b and it shown here, as it is a

Fig. 5 Representations of
N-tap FIR filter. (**a**) SFG
representation. (**b**) DFG
representation

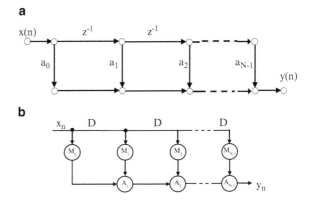

Table 4 64-Tap FIR filter
implementation In Virtex®-5
XC5VLX30

Wordlength	DSP48E	No. of registers	No. of LUTs
8	32	834	5,384
16	32	1,338	8,367
24	32	1,842	11,683

Table 5 64-Tap FIR filter
implementation In Stratix®
III EP3SE50 FPGA

Wordlength	DSP blocks	No. of registers	No. of LUTs
8	8	1,126	220
12	8	1,622	284
16	64	2,118	1,061

more useful representation for applying much of the retiming optimizations applied
later in the chapter.

$$y_n = \sum_{i=0}^{N-1} a_i x_{n-i} \qquad (1)$$

A direct mapping of the algorithm into hardware results in the multiplications
and additions being mapped into the dedicated DSP blocks and the delay blocks
into registers in the CLB/ALM functional blocks. Multiplication of an m-bit
multiplicand by an n-bit multiplier results in a $m + n$ output if the output is not
truncated. The output wordlength of n successive additions of two x-bit input is
given as $x + log_2 n$. Typically, truncation or more probably rounding is applied
throughout the structure and modeled effectively in Simulink® or LabVIEW using
suitable fixed-point libraries to effectively model the wordlength effects.

Tables 4 and 5 give Xilinx Virtex®-5 and Altera Stratix® III area and speed
figures respectively, for the 64-tap FIR filter of the form given in Fig. 5; they
have been synthesized using Synplicity® Synplify Pro 9.4 synthesis tools from
Synopsys®. In the comparison, a 12-bit coefficient wordlength has been chosen
and data wordlengths of 8-bit, 16-bit and 24-bit input wordlengths respectively
to illustrate how the mapping changes for each technology. The input wordlength

exceeds the multiplier size in the Altera DSP blocks when the wordlength grows above 18-bits as clearly seen in Fig. 3b, so the number of blocks goes up as well as the ALMs in the form of registers and LUTs. With the Xilinx device, the number of DSP48 blocks remains constant as they can support one input word size of 25-bits as can be seen in Fig. 3a. The register and LUT counts then increase accordingly with word size. This highlights the importance of relating the wordlength chosen during the algorithmic development stages to the implementation on specific FPGA platform.

3.2 Retiming

The throughput rates of the FIR filter implementations is not comparable to a single DSP block performance. This is because the critical path of the FIR filter structure is given as one multiplier and 63 additions. Of course, a key advantage is that this speed is not the clock rate but the sampling rate as 64 multiplications and 63 additions are performed on each clock cycle, unlike a sequential processor where the clock rate has to be divided by the number of operations.

Pipelining can be applied to achieve improved throughput, by performing retiming [5] as highlighted in chapter [3]. Retiming has also been applied in synchronous designs for clock period reduction [5], power consumption reduction [9], and logical synthesis in general.

For a circuit with two edges U and V and ω delays between them, a retimed circuit can be derived where ω_r delays are now present on the retimed edge. ω_r value is computed using Eq. (2) where $r(U)$ and $r(V)$ are the retimed values for nodes U and V respectively.

$$\omega_r(e) = \omega(e) + r(U) - r(V) \tag{2}$$

Retiming has a number of properties [11].

1. Weight of any retimed path is given by Eq. (2).
2. Retiming does not change the number of delays in a cycle.
3. Retiming does not alter the iteration bound in a DFG as the number of delays in a cycle does not change.
4. Adding the constant value, j, to the retiming value of each node does not alter the number of delays in the edges of the retimed graph.

Of course, the key trick is to be able to work out the retiming values $r(U)$ and $r(V)$ which results in a better retimed graph. A number of retiming routines have been developed but the most intuitive one is the *cut-set retiming* technique defined in [4].

Fig. 6 Cut-set theorem
application

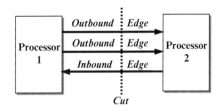

3.3   Cut-Set Theorem

A cut-set in an SFG (or DFG) is a minimal set of edges, which partitions the SFG
into two parts. The procedure is based upon two simple rules.

> Rule 1: *Delay Scaling.* All delays D presented on the edges of an original SFG
> may be scaled by a single positive integer, α, known as the pipelining period of
> the SFG meaning that $D' \longrightarrow \alpha D$. Correspondingly, the input and output rates
> also have to be scaled by a factor of α (with respect to the new time unit D').
> Time scaling does not alter the overall timing of the SFG.
>
> Rule 2: *Delay Transfer* [5]. Given any cut-set of the SFG, which partitions the
> graph into two components, we can group the edges of the cut-set into inbound
> and outbound, as shown in Fig. 6, depending upon the direction assigned to the
> edges. The delay transfer rule states that a number of delay registers, say k, may
> be transferred from outbound to inbound edges, or vice versa, without affecting
> the global system timing.

Consider the application of this to the DFG of the FIR filter given earlier. Aspects
of this were considered in chapter [12] but are given here again for completeness.
Its application is better understood by redrawing the FIR filter structure to show the
connectivity to the input source, x_n, and the individual delay connections to each of
the multipliers (see Fig. 7a). Application of a horizontal cut separates the multipliers
and adders into two separate graphs and pipelining after the multipliers is achieved.
The application of several vertical cut-sets partitions the DFG into $N - 2$ separate
sub-graphs which allows the adders to be pipelined. A better approach is to apply the
same cut-sets to the transposed version of the FIR filter where the data is fed from
right to left resulting in a transfer of delays from the x_n input to the adder chain rather
than adding additional delays. This results in the various figures given in Table 6 for
a Virtex®-5 target technology. The figures included in brackets represent the values
when the pipelining and retiming synthesis options are applied when synthesizing
the designs using Synplicity® Synplify Pro 9.4 synthesis tools from Synopsys®.

Fig. 7 SFG representation of
3-tap FIR filter [11].
(**a**) Original FIR DFG.
(**b**) Horizontal cut-set.
(**c**) Vertical cut-sets.
(**d**) Pipelined transposed FIR

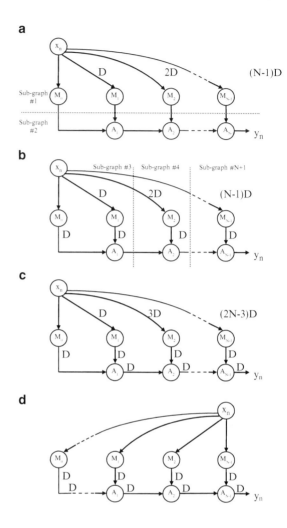

Table 6 Various 64-tap FIR filter implementations with 8-bit input, 10-bit coefficient in a Xilinx Virtex®-5 XC5VSX50T FPGA

Circuit	DSP blocks	Registers	LUTs	Throughput (MHz)
Fig. 7a	64 (64)	1,147 (1,341)	599 (516)	93 (107)
Fig. 7b	64 (64)	2,119 (1,580)	1,105 (829)	122 (151)
Fig. 7d	64 (64)	1,651 (1,640)	68 (68)	271 (271)

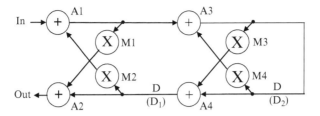

Fig. 8 Lattice filter structure

3.4 Application to Recursive Structures

The previous structure was non-recursive and simply involved application of delay transfer. Many DSP functions involve feedback loops like the lattice filter structure given in Fig. 8 which has a number of feedback loops.

Examining the DSP units in Fig. 3, it is clear that a good strategy for FPGA is to apply pipelining at the processor level. The first step is to determine the pipelining period of the structure which is done by working out the possible delays between D_1 and D_1 and then D_1 and D_2, etc. The possible paths are given below.

D1 to D1 $\Rightarrow$ D1 $\to M4 \to A3 \to$ D1
D1 to D2 $\Rightarrow$ D1 $\to A4 \to$ D2 and D1 $\to M4 \to A3 \to M3 \to A4 \to$ D2
D2 to D1 $\Rightarrow$ D2 $\to M2 \to A1 \to A3 \to$ D1
D2 to D2 $\Rightarrow$ D2 $\to M2 \to A1 \to A3 \to M3 \to A4 \to$ D2

The next stage is to determine the pipelining period by applying the *longest path matrix* algorithm [11]. This involves constructing a series of matrices to determine the pipelining period. If d is assumed to be the number of delays in the DFG, then a series of matrices, $L(m), m = 1, 2, \ldots, d$ is created such that the element $l_{(i,j,)}^{(m)}$ represents the longest path from delay element D_i to D_j which passes through exactly $m - 1$ delays (not including D_i and D_j); if no path exists, then $l_{(i,j,)}^{(m)}$ is -1. The iteration bound is found by examining the diagonal elements. The underlying assumption is pipelining at the processor level and therefore a delay is required at processing unit.

Examining the paths shown above, we generate matrices, L(1) as shown below and the compute higher level matrices. The higher order matrices do not need to derived from the DFG and can be recursively computed using the computation $l_{(i,j,)}^{(m+1)} = \overset{max}{k \in K} (-1, l_{(i,k,)}^{(1)} + l_{(j,k,)}^{(m)})$. This is given below for the lattice structure of Fig. 3 as:

$$\begin{bmatrix} 2 & 4 \\ 3 & 5 \end{bmatrix} \begin{bmatrix} 2 & 4 \\ 3 & 5 \end{bmatrix} \Rightarrow \begin{bmatrix} 7 & 9 \\ 8 & 10 \end{bmatrix}$$
$$L^{(1)} \qquad L^{(1)} \qquad L^{(2)}$$

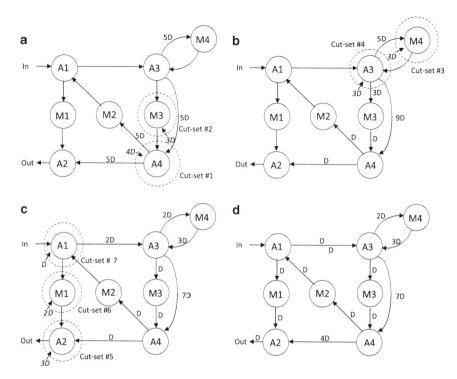

Fig. 9 Retiming applied to lattice filter. (**a**) Redrawn lattice filter. (**b**) Application of first cut-set. (**c**) Application of second cut-set. (**d**) Final pipelined structure

The pipelining period is then calculated by determining the following equation:

$$T_\infty = \overset{max}{i, m \varepsilon 1, 2, \ldots D} \left\{ \frac{l^m_{1,k}}{m} \right\} \tag{3}$$

For this example, this gives the pipelining period of 5, as shown below:

$$T_\infty = \left\{ \frac{2}{1}, \frac{5}{1}, \frac{7}{2}, \frac{10}{2} \right\} = 5$$

Using this value, the circuit is retimed using the cut-theorem as illustrated in Fig. 9. The process is more easily demonstrated by redrawing the lattice filter structure as shown in Fig. 9a. Firstly, we transfer 4D from the output to the input edges resulting in 4D on edge M3-A4 and 9D on edge A3-A4; the outputs are then reduced to D which can be embedded into the processor to allow pipelining. The 3D delays are then transferred in cut-set#2 giving the revised form in Fig. 9b. Two additional cut-sets are employed in Fig. 9b to ensure that a single D is left available on the output of the processors as shown in Fig. 9c. In cut-set#7, applying

Table 7 Synthesis results for the lattice filter using a Xilinx Virtex®-5 XC5VSX50T FPGA

Circuit	DSP blocks	LUTs	Clock (MHz)	Throughput (MHz)
Original (Fig. 8)	4	109	88	88
Scaled and retimed (Fig. 9d)	4	–	266	53.2
Hardware shared (Fig. 10)	1	50	240	48

the cut will result in a delay of $-D$ on edge $A1$-$M1$, so to preempt this we transfer $3D$ from the output in cut-set#5 and transfer the delays as shown, giving Fig. 9d. The presence of a single delay on each output edge means that the circuit can now be pipelined at the processor level.

Synthesis results of the original function and the pipelined version using the Xilinx Virtex®-5 FPGA technology is given in Table 7. The addition of the pipeline allows a neat mapping to the DSP48E processors, allowing the blocks to be fully used and reducing the amount of LUTs needed. Of course, the throughput is reduced as inputs are only required once very four clock cycles although the clock rate has increased considerably; this will be addressed later.

4   Circuit Architecture Optimization

The techniques so far have involved the use of pipelining to create the required sampling rates, but some applications operate at lower sampling rates than those indicated in Tables 6 and 7. In addition, redundancy may have occurred as a result of pipelining as illustrated for the lattice filter example which runs at a high clock rate but at comparable lower sampling rates when compared to the original design. In these scenarios, the aim would be to share the hardware by folding.

4.1   Folding

In folding, the aim is to perform hardware sharing on the DFG graph thereby allowing trade-offs to be made at a reasonably high level as it is relatively straightforward to work out the clock rate for the lattice filter after pipelining (by computing the delay for a pipelined multiplication and accumulation), without the need to perform FPGA synthesis. Given that a scaling factor of 5 has been applied to the lattice filter, it is clear that we can schedule each multiplication and addition operation to take place at different cycles; so it is possible to share hardware between these operations using one multiplier and adder without losing performance. This is achieved by *folding* [11] which was described in detail in

Fig. 10 Hardware sharing
applied to lattice filter

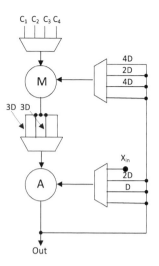

chapter [12]. By systematically applying folding to the block diagram of Fig. 7, the circuit of Fig. 10 is derived.

The synthesis results for the hardware shared version show a considerable reduction in hardware where one 48E DSP block is required rather than four. The clock rate is reduced when compared to the retimed version in Fig. 9d due to the existence of the multiplexers but given the resources used, it could used to give the best performance of all three solutions.

4.2 Unfolding

In addition to reducing the hardware, it is also possible to increase the amount of hardware by unfolding the algorithm to allow a greater throughput rate. Given that the speed is limited by the loops in this circuit, this does not make sense for this example unless a lot of independent computations are to be performed. For this reason, it is not appropriate to apply the transform here but the reader is referred to other texts which highlight relevant examples of its application [11, 18].

4.3 Adaptive LMS Filter Example

To date, the examples have been quite simple and so a delayed LMS filter is now used to show how algorithmic design and architecture development are closely linked. An 8-tap transposed form of the delayed LMS (TF-DLMS) adaptive algorithm [14] is given in Fig. 11a. It shows that the DLMS filter is constructed from an adder and eight instances of a tap cell highlighted in dashed lines. One of

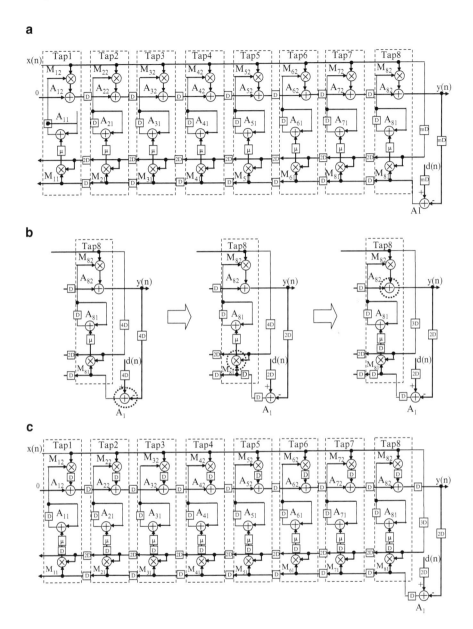

Fig. 11 LMS filter. (**a**) Detailed LMS filter SFG. (**b**) Application of retiming. (**c**) Retimed LMS filter

the reasons that a TF-LMS filter has been chosen is that the delay is increased to allow pipelining to be applied; however this compromises performance and must be judged carefully. Thus, a key aspect of the design process is to determine the minimum value of m needed to create the pipelining.

There are in effect, 16 delay loops in the circuit. Eight of the loops are trivial and refer to the adders $A_{11}, A_{21}, \ldots, A_{81}$ that have one processing unit and one delay, thus giving a pipelining period of 1. The other eight loops are given below and the pipelining period is calculated as before. For loop 1, the path is given by $y_n \rightarrow mD \rightarrow A_1 \rightarrow M_{81} \rightarrow A_{81} \rightarrow M_{82} \rightarrow A_{82}$ which is equivalent to 5D if pipelining is to be applied at the processor level. Given the programmable delay is $m + 1$, then this gives a pipelining period m of 4D. By considering all of the seven other loops (only some of which are shown), it is seen that this represents the worse case delay.

Loop 1: $y_n \rightarrow mD \rightarrow A_1 \rightarrow M_{81} \rightarrow A_{81} \rightarrow M_{82} \rightarrow A_{81}$

$$\alpha_1 = \left\lceil \frac{5}{m+1} \right\rceil \Rightarrow m_1 = 4$$

Loop 2: $y_n \rightarrow mD \rightarrow A_1 \rightarrow M_{71} \rightarrow A_{71} \rightarrow M_{72} \rightarrow A_{71} \rightarrow A_{81}$

$$\alpha_2 = \left\lceil \frac{6}{m+3} \right\rceil \Rightarrow m_2 = 3$$

Loop 3: $y_n \rightarrow mD \rightarrow A_1 \rightarrow M_{61} \rightarrow A_{61} \rightarrow M_{62} \rightarrow A_{61} \rightarrow A_{71} \rightarrow A_{81}$

$$\alpha_3 = \left\lceil \frac{7}{m+5} \right\rceil \Rightarrow m_3 = 2$$

$$\vdots$$

Loop 8: $y_n \rightarrow mD \rightarrow A_1 \rightarrow M_{11} \rightarrow A_{11} \rightarrow M_{12} \rightarrow A_{11} \rightarrow A_{21} \rightarrow A_{31} \rightarrow A_{41} \rightarrow A_{51} \rightarrow A_{61} \rightarrow A_{71} \rightarrow A_{81}$

$$\alpha_8 = \left\lceil \frac{12}{m+15} \right\rceil \Rightarrow m_8 = -3$$

$$m = max \ all \ m_c) = 4D$$

The delay of 4D is then applied to the mD delay block and retiming is performed as illustrated in Fig. 11b which shows the right most tap of the filter; the process is repeated throughout the structure in the same way to give the final retimed circuit in Fig. 11c. The first cut is made around processor A_1 around which two delays are transferred from input to output giving the second figure in Fig. 11b. This leaves a delay at the output of A_1 which symbolizes that it can be pipelined. A second cut is applied around multiplier M_{81} and a delay transferred from the output. Due to the fork in output from A_1, a delay appears as shown in the third figure in Fig. 11b

Table 8 64-Tap LMS filter
results in Virtex®-5 FPGA

	Area DSP blocks	Throughput registers	LUTs	(MHz)
Fig. 11a	128	4,029	2,440	123
Fig. 11c	129	20,985	2,952	203

which is in effect, at the input to multiplier M_{71} (not shown on the diagram). A delay already exists around adder A_{81}, so it does not require transfer. The final cut is made around A_{82}, allowing transfer from the output of A_{82} to introduce delays to the two inputs, one of which is the output of multiplier M_{82}. The same procedure is then applied to each tap in the circuit to produce a fully pipelined version as shown in Fig. 11c.

The circuits of Fig. 11a and 11c have been coded using VHDL and synthesized using the Xilinx Virtex®-5 FPGA family. The results are given in Table 8. The throughput rate clearly shows the speed improvement when applying pipelining with the only effective area increase in the number of registers employed. This can be mitigated by employing the shift registers shown earlier and is covered in detail in [18].

4.4 Towards FPGA-Based IP Core Generation

A key concern in modern electronic design is the design time, a feature which is most critical in SoC design as the complexity grows. One way to reduce design time is to explore *design reuse*, allowing designs or partial designs to be exploited in the creation of future systems. This issue has been recognized by the International Technology Roadmap for Semiconductors 2005 [13] which indicates that levels of reused logic will increase from 36% quoted in 2008, to an expected value of 48% by 2015.

One of the ways to increase design reuse is to use of silicon *intellectual property* (IP) cores. *IP cores* can range from dedicated circuit layout known as *hard IP* through to efficient code targeted to programmable DSP or RISC processors (*firm IP*) right through to hardware description language (HDL) descriptions of architectures such as those derived in this chapter, known as *soft IP*. A key aspect of soft IP is that scalability and parameterization can be built into the cores; this feature makes the cores particularly powerful as FPGA-based synthesis tools such as Synplify DSP high-level synthesis from Synopsys® allow the detailed performance to be quickly evaluated. This latter form of cores is know as *parameterizable soft IP cores* and can be designed so that they may be synthesized in hardware for a range of specifications and processes such as filter tap size, transform point size, or wordlength in DSP applications by setting generic parameters in the HDL descriptions [6].

Whilst parameters such as input/output and coefficient wordlength can be easily identified as a parameter, the implementation of the *level of pipelining* is much more difficult to achieve but is important, as typically this is a key parameter in determining the throughput rate. Therefore a sound approach is to allow a level of pipelining to be set by the user. From a SFG perspective, this requires the determination of the delay to be computed in terms of α for each edge of the SFG, as retiming usually results in a delay in every SFG edge following a processor and possibly, in edges between delays. In this way, *level of pipelining* can be introduced as a IP core parameter to the design, thereby giving the designer the control to adjust the speed of the design.

5 Conclusions

The chapter has shown how decidable signal processing graphs are mapped into FPGA hardware by adjusting the pipelining period and subsequently exploring folding and unfolding. The availability of hierarchical levels of memory but specifically flip-flops and distributed RAM, makes pipelining an attractive optimization to explore in FPGA implementations, particularly with programmable interconnect having such an impact on the critical path. The chapter has shown this for a number of applications including a simple FIR filter and a more complex adaptive LMS filter. In most cases, real synthesis figures from Xilinx Virtex®-5 FPGAs have been used to back up the theoretical analysis.

As much of the material in this book illustrates, implementation of pipelining within such structures increasingly only represents a component in the design of complex DSP systems; a key feature therefore, is the efficient incorporation of cores created for these system components in a higher level design flow which is considered in Sect. 5.1. Whilst the approaches in the chapter have been explored for mostly optimization of speed against area, the implications for power are given in Sect. 5.2.

5.1 Incorporation of FPGA Cores into Data Flow Based Systems

One of the knock-on effects of applying pipelining for system level DSP implementation, is the change in latency in the DSP component. For example, in the FIR filter of Fig. 7a, there is no delay in generating the output; however for the transposed pipelined version (Fig. 7d), the impact of retiming has been to improve the throughput rate but also to increase the latency to N delays. This latency needs to be reflected at a higher system level. In many cases, this will have no impact as the FIR filter most probably will be used in the creation of a more complex system and this latency could have an impact on performance, if it is in the critical path. Therefore, there may be a requirement to adjust the latency of the FIR filter at a later stage in the design flow.

For the lattice filter example, the obvious choice would be to select the hardware shared version of the filter (Fig. 10) as it uses less hardware, but only if it gives the required speed. However, if the system requirements need several lattice filtering operations, the implementation of Fig. 9d may be preferable as it allows several operations to be interlaced and has a higher clock rate than Fig. 10.

An initial viewpoint is to treat the synthesized circuit as a black box where all the latencies are coded as fixed values or even programmable values if the delay is dependent on a programmable value such as the number of taps in the filter. The aim of the system level designer would then be to set this value from system level requirements. However, if the aim is to use the core to process several streams of data as in the case of the lattice filter highlighted above, then state information would also need to be incorporated into the core. In this case, the aim would be to develop a strategy based around the "white box reuse" strategy [2], where previously synthesized systems allow some possibility of changing internal architecture for pipelining or retiming reasons. Increasingly this will become an issue as high level design tools emerge which increasingly have to interface with cores that have been developed effectively in a "bottom-up" fashion.

5.2 Application of Techniques for Low Power FPGA

Power consumption is becoming a critical issue in current systems and is one of the reasons why FPGAs are recommended over processors for DSP systems. Generally speaking, power consumption is separated into *static* power consumption which is the power consumed when the circuit is in static mode, switched on but necessarily processing data, and *dynamic* power consumption which is the power used when the systems is processing data. Static power consumption is important for battery life in standby mode, as it represents the power consumed when the device is powered up and dynamic power is important for battery life when operating, as it represents the power consumed when processing data.

One of the interesting aspects of the techniques outlined in this chapter is that they generally result in solutions that have a reduced power reduction when compared to the original circuits. At first, this is surprising as pipelining would appear to increase hardware and therefore capacitance due to the additional registers added. However, the dynamic power consumption does not just depend on capacitance but on *switched capacitance* i.e. the product of the switching on the node and the capacitance of the node. The impact of pipelining as a system level technique is to provide a mechanism by which the FPGA place and route tools can reduce the interconnect length and therefore capacitance.

This can have a major impact on performance as demonstrated in Table 9 which gives power figures for a 32-tap version of the FIR filter implementations given

Table 9 Internal signal/logic
power consumption of
various FIR filters [7]

Circuit	Power consumption (mW)
Fig. 7a	964.2
Fig. 7b	391.7 (−59.3%)
Fig. 7c	16.4 (−98%)

earlier and have been taken from real hardware [7]. The figures are taken from implementations on a Virtex®-2 FPGA and show substantial power gain due to reduction in the length of interconnect. Obviously, this reduction would be reduced in more modern FPGA structures, where the DSP blocks have dedicated routing.

Acknowledgements The author would like to acknowledge the effort and help of colleagues Dr Ying Yi and Dr Stephen McKeown in generating the results for the adaptive LMS and FIR filter examples respectively. Acknowledgement is also given to the numerous researchers who worked on this topic over the years, Dr Tim Courtney, Dr Richard Turner, Dr John McAllister and Dr Lok Kee Ting.

References

1. Altera Corp.: Stratix III device handbook. Web publication downloadable from http://www.altera.com (2007)
2. Bringmann, O., Rosenstiel, W.: Resource sharing in hierarchical synthesis. In: International Conference on Computer Aided Design, pp. 318–325 (1997)
3. Hu, Y.H., Kung, S.Y.: Systolic arrays. In: S.S. Bhattacharyya, E.F. Deprettere, R. Leupers, J. Takala (eds.) Handbook of Signal Processing Systems, second edn. Springer (2013)
4. Kung, S.Y.: VLSI Array Processors. Prentice Hall Int., Englewood Cliffs, NJ (1988)
5. Leiserson, C., Rose, F., Saxe, J.: Optimizing synchronous circuitry by retiming. In: Proceedings of the 3rd Caltech Conference on VLSI, pp. 87–116 (1983)
6. McCanny, J., Hu, Y., Ding, T., Trainor, D., Ridge, D.: Rapid design of DSP ASIC cores using hierarchical VHDL libraries. In: Thirtieth Asilomar Conference on Signals, Systems and Computers, pp. 1344–1348 (1996)
7. McKeown, S., Fischaber, S., Woods, R., McAllister, J., Malins, E.: Low power optimisation of DSP core networks on FPGA for high end signal processing systems. In: Proceedings on International Conference on Military and Aerospace Programmable Logic Devices (2006)
8. Meyer-Baese, U.: Digital Signal Processing with Field Programmable Gate Arrays. Springer, Germany (2001)
9. Monteiro, J., Devadas, S., Ghosh, A.: Retiming sequential circuits for low power. In: Proceedings of IEEE Int'l Conf. on Computer Aided Design, pp. 398–402 (1993)
10. Omondi, A.R.: Computer Arithmetic Systems. Prentice Hall Int., New York (1994)
11. Parhi, K.K.: VLSI digital signal processing systems: design and implementation. John Wiley and Sons, Inc., New York (1999)
12. Parhi, K.K., Chen, Y.: Signal flow graphs and data flow graphs. In: S.S. Bhattacharyya, E.F. Deprettere, R. Leupers, J. Takala (eds.) Handbook of Signal Processing Systems, second edn. Springer (2013)
13. Semiconductor Industry Association: International technology roadmap for semiconductors: Design. Web publication downloadable from http://www.itrs.net/Links/2005ITRS/Design2005.pdf (2005)

14. Ting, L., Woods, R., Cowan, C., Cork, P., Sprigings, C.: High-performance fine-grained pipelined LMS algorithm in Virtex FPGA. In: Advanced Signal Processing Algorithms, Architectures, and Implementations X: SPIE San Diego, pp. 288–299 (2000)
15. Turner, R.H., Woods, R.: Highly efficient, limited range multipliers for LUT-based FPGA architectures. IEEE Trans. on VLSI Systems **12**, 1113–1118 (2004)
16. White, S.A.: Applications of distributed arithmetic to digital signal processing. IEEE ASSP Magazine pp. 4–19 (1989)
17. Wilton, S.J.E., Luk, W., Ang, S.S.: The impact of pipelining on energy per operation in field-programmable gate arrays. In: Proceedings of Int'l conf. on Field Programmable Logic and Application, pp. 719–728 (2004)
18. Woods, R., McAllister, J., Lightbody, G., Yi, Y.: FPGA-based Implementation of Signal Processing Systems. Wiley, UK (2008)
19. Xilinx Inc.: Using look-up tables as shift registers (SRL-16) in Spartan-3 generation FPGAs. Web publication downloadable from http://www.xilinx.com (2005)